Television Production Handbook + Workbook

FROM THE WADSWORTH SERIES IN BROADCAST AND PRODUCTION

Albarran, *Management of Electronic Media*, Third Edition

Alten, *Audio in Media*, Ninth Edition

Craft/Leigh/Godfrey, *Electronic Media*

Eastman/Ferguson, *Media Programming: Strategies and Practices*, Eighth Edition

Gross/Ward, *Digital Moviemaking*, Seventh Edition

Hausman/Benoit/Messere/O'Donnell, *Announcing: Broadcast Communicating Today*, Fifth Edition

Hausman/Benoit/Messere/O'Donnell, *Modern Radio Production*, Eighth Edition

Hilliard, *Writing for Television, Radio, and New Media*, Ninth Edition

Hilmes, *Only Connect: A Cultural History of Broadcasting in the United States*, Third Edition

Mamer, *Film Production Technique: Creating the Accomplished Image*, Fifth Edition

Meeske, *Copywriting for the Electronic Media: A Practical Guide*, Sixth Edition

Zettl, *Sight Sound Motion: Applied Media Aesthetics*, Sixth Edition

Zettl, *Television Production Handbook*, Eleventh Edition

Zettl, *Television Production Workbook*, Eleventh Edition

Zettl, *Video Basics*, Sixth Edition

Zettl, *Video Basics Workbook*, Sixth Edition

Zettl, *Zettl's VideoLab 4.0*

Television Production Handbook + Workbook

ELEVENTH EDITION

Herbert Zettl

San Francisco State University

Australia • Brazil • Japan • Korea • Mexico • Singapore • Spain • United Kingdom • United States

Television Production Handbook + Workbook
Eleventh Edition, International Edition
Herbert Zettl

Senior Publisher: Lyn Uhl

Publisher: Michael Rosenberg

Associate Development Editor: Megan Garvey

Assistant Editor: Erin Pass

Editorial Assistant: Rebecca Donahue

Media Editor: Jessica Badiner

Senior Marketing Manager: Amy Whitaker

Marketing Coordinator: Gurpreet Saran

Marketing Communications Manager: Caitlin Green

Senior Content Project Manager: Michael Lepera

Art Director: Linda Helcher

Senior Print Buyer: Justin Palmeiro

Rights Acquisition Specialist: Mandy Groszko

Production Service/Compositor: Ideas to Images

Copy Editor: Elizabeth von Radics

Cover and Text Designer: Gary Palmatier, Ideas to Images

Cover Image: © iStockphoto.com/craftvision

International Edition:
ISBN-13: 978-1-111-34788-8
ISBN-10: 1-111-34788-3

Cengage Learning International Offices

Asia
www.cengageasia.com
tel: (65) 6410 1200

Australia/New Zealand
www.cengage.com.au
tel: (61) 3 9685 4111

Brazil
www.cengage.com.br
tel:(55) 11 3665 9900

India
www.cengage.co.in
tel: (91) 11 4364 1111

Latin America
www.cengage.com.mx
tel: (52) 55 1500 6000

UK/Europe/Middle East/Africa
www.cengage.co.uk
tel: (44) 0 1264 332 424

Represented in Canada by
Nelson Education, Ltd.
www.nelson.com
tel: (416) 752 9100 / (800) 668 0671

Cengage Learning is a leading provider of customized learning solutions with office locations around the globe, including Singapore, the United Kingdom, Australia, Mexico, Brazil, and Japan. Locate your local office at:
www.cengage.com/global

For product information and free companion resources:
www.cengage.com/international
Visit your local office: **www.cengage.com/global**
Visit our corporate website: **www. cengage.com**

Printed in Canada
1 2 3 4 5 6 7 13 12 11 10

For Erika

Brief Contents of Handbook

Brief Contents of Workbook

Contents of Handbook

Contents of Workbook

About the Author

PHOTO BY EDWARD AIONA

Herbert Zettl is a professor emeritus of the Broadcast and Electronic Communication Arts Department at San Francisco State University (SFSU). He taught there for many years in the fields of video production and media aesthetics. While at SFSU he headed the Institute of International Media Communication. For his academic contributions, he received the California State Legislature Distinguished Teaching Award and, from the Broadcast Education Association, the Distinguished Education Service Award.

Prior to joining the SFSU faculty, Zettl worked at KOVR (Stockton-Sacramento) and as a producer-director at KPIX, the CBS affiliate in San Francisco. While at KPIX he participated in a variety of CBS and NBC network television productions. Because of his outstanding contributions to the television profession, he was elected to the prestigious Silver Circle of the National Academy of Television Arts and Sciences (NATAS), Northern California Chapter. He is also a member of the Broadcast Legends of the NATAS Northern California Chapter.

In addition to this book, Zettl has authored *Television Production Workbook, Sight Sound Motion,* and *Video Basics.* All of his books have been translated into several languages and published overseas. His numerous articles on television production and media aesthetics have appeared in major media journals worldwide. He has lectured extensively on television production and media aesthetics at universities and professional broadcast institutions in the United States and abroad and has presented key papers at a variety of national and international communication conventions.

Zettl developed an interactive DVD-ROM, *Zettl's VideoLab 4.0,* published by Wadsworth Cengage. His previous CD-ROM version won several prestigious awards, among them the *Macromedia* People's Choice Award, the *New Media* Invision Gold Medal for Higher Education, and Invision Silver Medals in the categories of Continuing Education and Use of Video.

Preface to Handbook

Today you can video-record an event in high-definition with a camcorder that easily fits into your shirt pocket, use your laptop to stabilize the shaky video, edit your story, and e-mail it to your friend—all while having your morning latte at your favorite coffee shop. Why, then, bother learning about television production systems, high-end equipment, and various studio and field production techniques? Because I don't want you to remain a gifted amateur who occasionally sparkles with a home video. Rather, I would like to see you become a true media professional. But what makes a true professional? One who has acquired these five fundamental qualities:

- **Production knowledge.** You need to know how to move from basic idea to the finished production—and how to do it effectively and efficiently. This requires a thorough knowledge of preproduction, production, and postproduction activities, the necessary equipment, and the people involved.

- **Consistency.** Television requires your full engagement day in and day out. Your value as a media professional is measured not by your occasional spectacular brain thrusts but by your dependability for delivering the goods on time and within budget.

- **Creativity.** When working in television, the clock dictates when you must be creative. You do not have the luxury of waiting for divine inspiration.

- **Efficiency.** You must come to every assignment fully prepared. If the production is live or live-recorded, you have no second chance; your production must be error-free the first time around. Even if the event is recorded for postproduction editing, the usual time and budget constraints will not tolerate indecisiveness in your decision-making or, worse, your "making it up as you go along."

- **Responsibility.** Even if you assume that the majority of your audience might not recognize or care about the finer points of production, you have a responsibility to help the audience achieve a higher level of aesthetic and emotional literacy. This means your shots must be correctly framed, your lighting and audio in sync with the feel of the scene, and your editing tight and rhythmically structured, even if you could get your point across without such artistry. With such a professional approach, your audience cannot help but learn, however subconsciously, what production quality is all about.

This Eleventh Edition of the *Television Production Handbook* is designed to help you become such a media professional.

PURPOSE OF A TELEVISION PRODUCTION TEXT

No textbook—even this one—can replace practical experience. But by first reading about the complex process of television production, you will be a hundred steps ahead of everybody else when you get involved in actual productions. You will have the knowledge of how to move from initial idea to the various production stages and of what you can and should do with a specific type of equipment; this will give you the confidence to make the right choices. Getting to this point by trial and error would take much too long and therefore be counterproductive to your goals of becoming a video professional.

The following sections are intended primarily for the instructor.

ORGANIZATION OF THE ELEVENTH EDITION

Much like the previous edition, this text is presented in four major parts:

- Part I Introduction: Process and System
- Part II Preproduction
- Part III Production
- Part IV Postproduction

Part I Introduction: Process and System

The problem of learning and teaching television production is that to really understand a single production element or piece of equipment, students should ideally know the function of all the others. To prevent students from getting lost in the myriad equipment and production details, and to help them understand television production as a system, part I gives an overview of the preproduction, production, and postproduction phases, the technical and nontechnical production personnel, and the major television equipment.

Part II Preproduction

Part II stresses the importance of preproduction. Most beginning production students, armed with a digital camcorder and a laptop computer loaded with special effects, are itching to run outside and shoot an Emmy award–caliber documentary after the first class. They are, understandably, reluctant to spend much time on thinking about exactly what it is they want to say and how best to say it. But even the most eager students will soon find out that the success or failure of a production is largely determined in the preproduction phase. Part II therefore emphasizes how to move through the entire preproduction process, starting with generating worthwhile ideas, developing the program proposal, and, finally, writing the script. The discussion of preproduction includes major information resources and introduces students to unions, legal matters, and audience ratings. To make this text reflect as closely as possible an actual production sequence, on the recommendation of two astute reviewers the director's preproduction activities have been moved from Part III to Part II.

Part III Production

This part includes a detailed explanation of what digital is all about, the various scanning systems, the aspect ratios, the major production tools and their effective use, and what the talent, director, and floor manager do during the actual production. Part III also includes tips and suggestions for directing single- and multicamera productions for television and digital cinema both in the studio and in the field. The field production section shows detailed camera and mic setups for major sporting events and covers site surveys and signal distributions for big remotes.

Part IV Postproduction

The laptop computer and relatively inexpensive editing software have made postproduction editing simply another phase of even the most modest video productions. The aesthetic principles of how to make a sequence of images look seamless or especially dramatic, however, have not changed with the new technology. Production students still have to learn how and why various shots cut together well regardless of what nonlinear software they use. The last two chapters explain nonlinear editing and the principles of continuity and complexity editing. The section on linear editing has been retained because some of its principles are also applicable to nonlinear editing and because there are still linear editing systems in use.

PEDAGOGICAL FEATURES

To ensure optimal student learning, this edition of the *Handbook* again incorporates important pedagogical principles.

Short sections Each chapter is broken down into relatively short sections marked by separate headings. The layout is intended to combat short attention spans and keep the reader's attention without fracturing chapter content.

Two-tier approach The *Handbook* is designed to serve beginning students as well as those who are more adept at television production. To prevent the less advanced students from getting bogged down by the multitude of technical details, each chapter is divided into two sections. Section 1 contains the basic information about a specific topic; section 2 presents more detailed material. The two sections can be assigned and read together or independently. Some chapters, however, use section 2 for important material that would otherwise have made section 1 too long and cumbersome.

Chapter order The chapter order in this edition is guided by the three major steps of how we ordinarily move from the initial idea to the finished production: preproduction, production, and postproduction. I realize, however, that this organization my not suit the specific way an instructor wants or has to teach this subject. Although in the television production system each part is necessary for the proper functioning of all the others, the individual chapters are sufficiently self-contained to be taught in any desired order. In fact, the first chapter is intended as a reference guide to the whole production process, irrespective of the order in which the parts of the system (the following chapters) are taught.

Reinforcement As we all know from advertising, a certain amount of repetition is essential for making a product name stick. The same redundancy principle holds true when learning the language and the production concepts of television. The key terms of each chapter are first listed after the chapter introduction; they appear in bold italic in the text and again in the extensive glossary at the back of the book. To benefit from this learning aid, students should read the key terms before committing to the chapter and use the chapter summaries as a checklist of what they are expected to know.

Illustrations The numerous full-color pictures and diagrams are included to bridge the gap between description and the real thing. All appropriate illustrations that simulate TV images are in the 16 × 9 HDTV aspect ratio. This is done to help students visualize shots in the wide-screen aspect ratio. In most cases the pictures of equipment represent a generic type rather than a specific preferred model.

ACCOMPANYING RESOURCES

As with previous editions, the Eleventh Edition of the *Television Production Handbook* offers a variety of support materials for both students and instructors.

Television Production Workbook

The *Television Production Workbook* is intended primarily to test students' comprehension of various equipment and production concepts. Each chapter contains a review of the key terms and the basic content covered in the text as well as a true/false quiz. The problem-solving applications are intended mainly for in-class discussion. I have often used the *Workbook* as a diagnostic tool by having students answer some of the questions in the first or second class meeting without grading them. In this way I could quickly determine who knows what about television production. Not too surprisingly, the diagnostic tests frequently produced results that contradicted the students' self-evaluations. Because of the immediate feedback, such results served as a positive wakeup call without having to wave a finger in admonition.

All *Workbook* problems except the problem-solving applications are based on a binary system, which means the questions can be answered by filling in the appropriate bubbles. This approach makes checking the answers relatively easy, even without the aid of a computer, and provides a basis for objective evaluation.

Instructor's Manual with Answer Key to Workbook

The *Instructor's Manual* is intended primarily for the instructor who may be quite experienced in television production but relatively new to teaching. Even experienced instructors, however, may find information that makes the difficult job of teaching television production just a little easier. The *Instructor's Manual* comprises four parts:

- Part I, General Approach to Teaching Television Production, presents information on teaching approaches and ideas about how to teach the subject most effectively.
- Part II, Key Concepts, Activities, and Tests, contains expanded definitions of the key concepts introduced in each chapter, appropriate activities for reinforcing them, and multiple-choice problems to test student retention of the material.
- Part III, Additional Resources, is a compact reference that recommends additional teaching and learning resources.
- Part IV consists of the answer key to all the *Television Production Workbook* problems.

Zettl's VideoLab DVD-ROM

Both releases of this Windows- and Mac-compatible DVD-ROM—3.0 and the new 4.0—contain an interactive program of camera operation, studio and field lighting, audio capture and control, switching, linear and nonlinear editing, and the major parts of the production process. *Zettl's VideoLab* is intended to give students virtual hands-on practice and a proven shortcut from reading about production techniques to actually applying them in the studio and in the field. The in-text ZVL cues in the *Handbook* work with both *Zettl's VideoLab 3.0* and *4.0.* The new 4.0 release incorporates additional advanced exercises.

Additional Online Resources

To access additional course materials and companion resources, please visit *www.cengagebrain.com*. At the CengageBrain.com home page, search for the ISBN of this book (from the back cover), using the search box at the top of the page. This will take you to the product page, where free companion resources can be found.

Wadsworth's ExamView® Computerized Testing

ExamView offers both a *Quick Test Wizard* and an *Online Test Wizard* that guide you step-by-step through the process of creating tests, while its "what you see is what you get" interface allows you to see the test you are creating onscreen exactly as it will print or display online. For additional information please contact your local Wadsworth Cengage Learning representative or the Wadsworth Academic Resource Center at (800) 423-0563.

ACKNOWLEDGMENTS

As with previous editions, I had a great number of experienced people help me with this Eleventh Edition of the *Television Production Handbook.* My sincere thanks go to these key people at Wadsworth Cengage: Michael Rosenberg, publisher; Megan Garvey, development editor; Erin Pass, assistant editor; Michael Lepera, senior content project manager; Linda Helcher, art director; Carly Bergey, photo researcher; Mandy Groszko, rights acquisition specialist; and Justin Palmeiro, senior print buyer.

I was indeed lucky to have, once again, my truly superb production team for this newest edition of the *Handbook:* Gary Palmatier of Ideas to Images, art director and project manager; Elizabeth von Radics, copy editor; Ed Aiona, photographer; and Mike Mollett, proofreader. All of them gave their best to make this edition even better than the previous one. I am lucky because Gary, Elizabeth, and Ed are not only recognized experts in their fields but also knowledgeable about video production. Best of all, they are fun to work with.

In this edition I benefited once again from colleagues who were kind enough to review the Tenth Edition of the *Television Production Handbook* for the Eleventh Edition: Jim Alchediak, North Carolina State University; Steven Enfield, Brigham Young University; Billy M. Oliver, Miami Dade College; Phillip Powell, Valparaiso University; and Brian L. Shleton, Rock Valley College. Their expertise and teaching experience helped me fine-tune some areas and brought about a change in chapter order.

I have also had generous help for this edition from television experts at various universities, television stations, and independent production companies. They know how much their contributions mean to me but may not always realize how much they help the many students and the campus community at large in learning about this demanding discipline: Marty Gonzales, Hamid Khani, Steve Lahey, Val Sakovich, Vinay Shrivastava, Douglas Smith, and Winston Tharp of San Francisco State University.

In addition to my San Francisco State University support team, the following colleagues and friends were always ready to provide me with specific and reliable information: Stanley Alten, Syracuse University; Manfred Wolfram, University of Cincinnati; Rudolf Benzler of T.E.A.M., Lucerne, Switzerland; Manfred Muckenhaupt, Media Studies, University of Tuebingen, Germany; Elan Frank of E'lan Productions, Los Angeles; John Beritzhoff and Greg Goddard of Snader and Associates, San Rafael, California; Ed Cosci, Jeff Green, Jim Haman of KTVU, Oakland; and the many people at the NAB/BEA conventions in Las Vegas who kept me informed through panel discussions and equipment demonstrations at their display areas. A very big thank-you to all of you.

Every time I look at the people who appear in the many pictures, I wish I could walk up to them and express my gratitude again for lending their talent during the many photo sessions: Talia Aiona, Karen Austin, Ken Baird, Jerome Bakum, Rudolf Benzler, Tiemo Biemueller, Monica Caizada, William Carpenter, Andrew Child, Laura Child, Rebecca Child, Renee Child, Skye Christensen, Ed Cosci, Carla Currie, Sabrina Dorsey, Tammy Feng, Jedediah Gildersleeve, Cassandra Hein, Poleng Hong, Sangyong Hong, Akiko Kajiwara, Hamid Khani, Philip Kipper, Christine Lojo, Orcun Malkoclar, Michael Mona, Johnny Moreno, Anita Morgan, Jacqueline Murray, Tuan Nguyen, Richard Piscitello, Matthew Prisk, Marlin Quintero, Kerstin Riediger, Suzanne Saputo, Alisa Shahonian, Steve Shlisky, Talisha Teague, Takako Thorstadt, and Yanlan Wu.

Finally, my wife, Erika, gets a big hug not only for tolerating my writing but also for supporting me when I encountered the inevitable bumps along the road.

Herbert Zettl

Television Production Handbook

PART I

Introduction: Process and System

CHAPTER

1

The Television Production Process

When watching television, somebody's vacation videos, a blogger's video podcast, or even a movie, you probably feel that you could do just as well or even better than what's on the screen. This may be true, but it is more likely that you will be surprised to find how difficult it is to match the high production values of the average television show, even if the content begs for improvement. The automatic features of most equipment may even fool you into believing that television production is relatively easy—until your luck runs out. Even if your short vacation video looks pretty good to you, it may need much more effort and production skills to make it look good to somebody else. A seemingly simple 55-second chat between a news anchor in Portland and a soccer star in Madrid presents a formidable challenge even for experienced production personnel. This book will help you meet such a challenge.

The digital era has brought a general convergence of digital video and the necessary production processes, regardless of whether you are working in broadcast television, in digital cinema, or independently on small video projects. Fortunately, this convergence has a common base: multicamera and single-camera television production. Learning the ins and the outs of television production allows you to readily adapt to other forms of digital video production.

The major problem with learning television production is that to understand the function of one piece of equipment or production phase, you should already know all the others. Chapter 1 is designed to help you with this chicken-and-egg problem. It provides you with an overview of the initial production process, the people involved in small and large productions, and the necessary tools to generate the screen images and sound—the standard television equipment. Later chapters provide more detailed descriptions and explanations of equipment and production processes.

Section 1.1, What Television Production Is All About, walks you through the three phases of production, demonstrates a useful production model, and introduces you to the major nontechnical and technical production personnel.

Section 1.2, Technical Production Systems, introduces you to the basic, expanded, and field production television systems and their major production equipment.

KEY TERMS

effect-to-cause model Moving from idea to desired effect on the viewer, then backing up to the specific medium requirements to produce such an effect.

EFP Stands for *electronic field production.* Television production outside of the studio that is normally shot for postproduction (not live). Part of field production.

ENG Stands for *electronic news gathering.* The use of portable camcorders or cameras with separate portable video recorders, lights, and sound equipment for the production of daily news stories. ENG is usually not planned in advance and is often transmitted live or immediately after postproduction editing.

linear editing Analog or digital editing that uses tape-based systems. The selection of shots is nonrandom.

medium requirements All content elements, production elements, and people needed to generate the defined process message.

news production personnel People assigned exclusively to the production of news, documentaries, and special events.

nonlinear editing (NLE) Allows instant random access to shots and sequences and easy rearrangement. The video and audio information is stored in digital form on computer hard drives or other digital recording media.

nontechnical production personnel People concerned primarily with nontechnical production matters that lead from the basic idea to the final screen image.

postproduction Any production activity that occurs after the production. Usually refers to either video editing or audio sweetening (a variety of quality adjustments of recorded sound).

preproduction The preparation of all production details.

process message The message actually perceived by the viewer in the process of watching a television program. The program objective is the defined process message.

production The actual activities in which an event is recorded and/or televised.

technical personnel People who operate and maintain the technical equipment.

television system Equipment and people who operate the equipment for the production of specific programs. The basic television system consists of a television camera and a microphone, which convert pictures and sound into electrical signals, and a television set and a loudspeaker, which convert the signals back into pictures and sound.

SECTION

1.1

What Television Production Is All About

As a painter it is relatively easy to get your idea onto the canvas. All you need is something to paint on, some paints, a brush, and, of course, a little technique. You are the only one involved in the translation process from idea to image. Such a translation process for even a simple television production, however, is considerably more complex. You are seldom alone in the production process, face strict deadlines, and are always forced to work with a variety of complex equipment. This section gives a brief overview of the three phases of production—preproduction, production, and postproduction—suggests a production model that will streamline the use of people and equipment, and charts the major nontechnical and technical personnel. ZVL1 PROCESS→ Process introduction

- **THREE PRODUCTION PHASES**
  Preproduction, production, and postproduction
- **PRODUCTION MODEL**
  Effect-to-cause model, medium requirements, and process message shaping medium requirements
- **PRODUCTION PEOPLE**
  Nontechnical production personnel, technical production personnel, and news production personnel

THREE PRODUCTION PHASES

Regardless of whether you are part of the nontechnical or technical personnel, or whether you work with a big production team or all by yourself, you will inevitably be involved in one or all of the three production phases: preproduction, production, and postproduction.

Preproduction

Preproduction includes all the preparations and activities before you actually move into the studio or the field on the first day of production. It usually happens in two stages. Stage 1 consists of all the activities necessary to transform the basic idea into a workable concept or script. In stage 2 all the necessary production details, such as location, crews, and equipment for a single-camera or multicamera production, are worked out.

Production

As soon as you open the studio doors for rehearsal or a video-recording session, or load a camcorder into the van for a field shoot, you are in ***production***. Except for rehearsals, production involves equipment and normally a crew—people who operate the equipment. It includes all activities in which an event is video-recorded or televised.

Postproduction

The major activity of ***postproduction*** consists of video and audio editing. It may also include color correction of video clips (so that the red shirt of an actor looks the same from one shot to the next), the selection of appropriate background music, and the creation of special audio effects. When using a single camera film-style, which means that a scene is built by recording one shot after another with only one camera, the postproduction activities may take longer than the actual production. ZVL2 PROCESS→ Phases→ preproduction | production | postproduction

PRODUCTION MODEL

Like any other model, a production model is meant to help you move from the original idea to the finished production as efficiently as possible. It is designed to help you decide on the most effective approach the first time around, evaluate each major production step, and finish on time. Its function is similar to that of a road map: you don't have to follow it to get from here to there, but it makes finding your way much easier. If you feel that it is restrictive and cramps your creativity or style, don't use it.

Effect-to-cause Model

Like most other production models, the ***effect-to-cause model*** starts with a basic idea, but instead of moving from the basic idea directly to the production process, it jumps to the desired communication effect on the target audience—the general program objective. This program objective can be reached through a specific message that, ideally, the viewer will actually receive, internalize, or act on. Because this all-important message is generated by the process of the viewer's watching the video and audio content of your television program and attaching meaning to it, we call it the ***process message***. This process requires that you as a producer have a fairly clear idea of what you want the target audience to learn, do, and feel before you think about the necessary technical requirements. The model suggests that you move from the general idea directly to the desired effect and then back up and think about how to bring about—cause—this effect. ZVL3 PROCESS→ Effect-to-cause→ basic idea | desired effect | cause | actual effect

The more the actual process message (viewer effect) matches the defined one, the more successful the communication. **SEE 1.1**

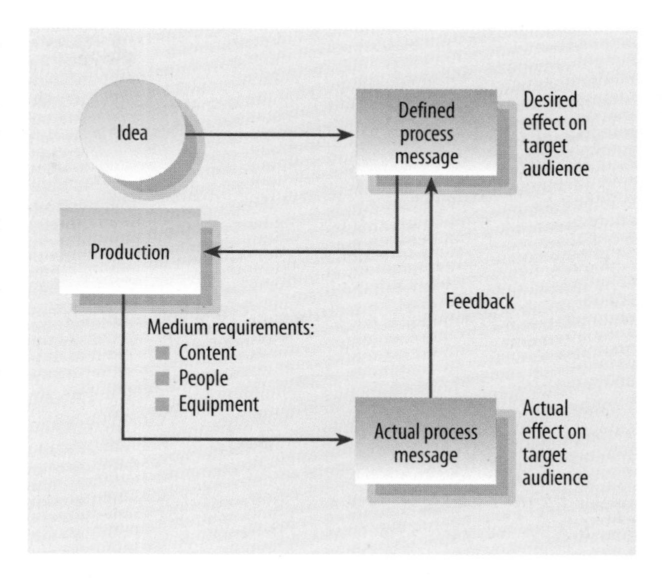

1.1 EFFECT-TO-CAUSE PRODUCTION MODEL
The effect-to-cause production model jumps from the initial idea and story angle directly to the desired effect—the process message. It then backs up to the medium requirements that suggest the production elements and processes necessary to produce the defined process message.

Defined process message Rather than being driven by the initial idea, the production process is now driven by the defined process message—the desired effect on the target audience. At this point you could proceed to the ***medium requirements***—the people, facilities, and equipment necessary for the preproduction, production, and postproduction phases. To further streamline the production process, you should find a useful angle.

Angle As you undoubtedly know, the angle is a specific story focus, a point of view from which to look at and describe an event. It can lead to an obvious bias of who tells the story, or it can be subtler and make a story more interesting to watch.

If a dog bites the letter carrier, the dog owner's story angle might be the rising crime in the neighborhood and the dog's attempt to protect his master. The letter carrier, on the other hand, may have quite a different view of the same event. He may well focus on the viciousness of the neighborhood dogs and the need for stricter leash laws. Both angles contain a strong and unacceptable bias.

You can also use an angle that gives the story a specific approach without introducing a strong bias. For example, you could document a popular singing star by watching her give a concert for a large enthusiastic audience or by observing her during a studio recording session. The first version would be a more public "looking-at" point of view, the second a more private "looking-into" point of view. This would change not only what equipment you need (a multicamera setup with live switching or extensive postproduction for the first version) but also your shooting style (many more close-ups for version 2 than for version 1).

Medium Requirements

The advantage of this model is that the precise definition of the process message and a specific angle will help the content and production people work as a team and facilitate selecting the necessary production personnel and equipment. By first carefully defining the desired effect on the audience, you can then decide quite easily on the specific people you need to do the job (content expert, writer, director, and crew), on where to do the production most effectively (studio or field), and on the necessary equipment (studio or field cameras, types of mics, and so forth).

Process Message Shaping Medium Requirements

Let's assume that you are to produce a 15-minute segment of a live morning show. You are told by the show's executive producer to get a lawyer who is willing to talk about an ongoing high-profile murder trial.

The usual and intuitive way to approach this assignment would be to contact a well-known criminal lawyer and have the art director design a set that looks like a well-to-do lawyer's office, with an elegant desk, leather chairs, and lots of law books in the background. You would then have to arrange for the recording date, studio time, transportation for the guest, talent fees, and other such details.

When using the effect-to-cause model, on the other hand, you might come up with two different angles: one that shows the intellectual brilliance of a defense lawyer and her skill to engender reasonable doubt in the jurors, and another that reveals the emotional makeup and the inner conflict of a lawyer defending a suspect despite the overwhelming evidence that he is guilty.

Here's how the two different angles might influence the resulting process messages, and, in turn, dictate different production approaches:

Process message 1: *The viewer should gain insight into some of the major defense strategies used by the guest.*

In this case, the questions would revolve around some of the lawyer's former cases and the reasons for their success or failure. Would you need an interviewer who understands the law? Yes. The interviewer could interpret the legal language for the audience or immediately challenge the lawyer's ethics within the framework of the law. The elaborate studio set resembling the lawyer's office would also be appropriate. You may even consider conducting this interview on-location in the lawyer's actual office.

Process message 2: *The viewer should gain deeper insight into the conscience and the feelings of the lawyer when handling an especially difficult case as well as how she deals with personal ethics when applying specific defense strategies.*

Do you now need a host who is a legal expert? Not at all. In fact, a psychologist would probably be better suited to conduct this interview. You might opt for close-ups of the lawyer throughout most of the show. You may even stay on a close-up of the guest when the host asks questions. Reaction shots (the lawyer listening to questions) are often more telling than action shots (the lawyer answering). Does this interview require an elaborate set? No. Because the interview deals primarily with the lawyer as a person rather than the person as a lawyer, you can conduct it in any environment. Two comfortable chairs on an interview set are all you would need.

Some unsolicited advice: There has been a great reluctance in television production to show "talking heads"—people talking on close-ups without any supporting visual material, special effects, or a constant dribble of background music. Do not blindly adopt this prejudice. So long as the heads talk well, there is no need for additional visual or aural clutter. ZVL4 PROCESS→ Effect-to-cause→ basic idea | desired effect | cause

As you can see, in this case the angle was not stated separately but rather embedded in the defined process message. But would you need a process message if you were to write a play? Of course not. Even a nicely formulated process message would not help you write a more effective drama. Any dramatic presentation has its own internal structure that does not benefit from stating its desired effect on the audience. It is more important to think about character development and conflict than defining whether you want the audience to cry or laugh. More goal-directed program forms, however, such as instructional shows, interviews, documentaries, and certainly advertising, can benefit greatly from a precisely stated process message.

PRODUCTION PEOPLE

Even the most sophisticated television production equipment and computer interfaces will not replace *you* in the television production process; you and those working with you still reign supreme—at least so far. The equipment cannot make ethical and aesthetic judgments for you; it cannot tell you exactly which part of the event to select and how to present it for optimal communication. You make such decisions within the context of the general communication intent and through interaction with other members of the production team—the production staff, technical crews, engineers, and administrative personnel. You may soon discover that the major task of television production is working not so much with equipment as with people. In general, we can divide the production personnel into nontechnical personnel and technical personnel. Because news departments work independently of the regular production personnel, we list them separately.

Nontechnical Production Personnel

The ***nontechnical production personnel*** are generally involved in translating a script or an event into effective television images. They are also called above-the-line personnel because they fall under a different budget category from the technical crew, who are called below-the-line personnel. The above- and below-the-line distinction is anything but absolute or even uniform, however, and it changes depending on the crewmembers' union affiliations and the budgetary practices of the production company. We therefore use here the more self-evident division of nontechnical and technical personnel. Figure 1.2 shows the principal functions of the major nontechnical production personnel. SEE 1.2

You should realize, however, that in smaller television operations one person might carry out several different functions. For example, the producer may also write and direct the show, and the floor manager may take on the responsibilities of the line producer. You may find an AD (associate director) in the production of soap operas or a digital movie but rarely during most routine television shows. The art director may also function as a graphic artist, and most medium-sized or smaller production companies have little use for a permanent costume designer, wardrobe person, property manager, or sound designer.

1.2 NONTECHNICAL PRODUCTION PERSONNEL

PERSONNEL	FUNCTION
NONTECHNICAL PRODUCTION PERSONNEL	
Executive producer	In charge of one or several large productions or program series. Manages budget and coordinates with client, station management, advertising agencies, financial supporters, and talent and writers' agents.
Producer	In charge of an individual production. Responsible for all personnel working on the production and for coordinating technical and nontechnical production elements. Often serves as writer and occasionally as director.
Associate producer (AP)	Assists producer in all production matters. Often does the actual coordinating jobs, such as telephoning talent and confirming schedules.
Line producer	Supervises daily production activities on the set.
Field producer	Assists producer by taking charge of remote operations (away from the studio). At small stations function may be part of producer's responsibilities.
Production manager	Schedules equipment and personnel for all studio and field productions. Also called *director of broadcast operations.*
Production assistant (PA)	Assists producer and director during actual production. During rehearsal takes notes of producer's and/or director's suggestions for show improvement.
Director	In charge of directing talent and technical operations. Is ultimately responsible for transforming a script into effective video and audio messages. At small stations may often be the producer as well.
Associate director (AD)	Assists director during the actual production. In studio productions does timing for director. In complicated productions helps "ready" various operations (such as presetting specific camera shots or calling for a video recorder to start). Also called *assistant director.*
Floor manager	In charge of all activities on the studio floor. Coordinates talent, relays director's cues to talent, and supervises floor personnel. Except for large operations, is responsible for setting up scenery and dressing the set. Also called *floor director* and *stage manager.*
Floor persons	Set up and dress sets. Operate cue cards and other prompting devices, easel cards, and on-camera graphics. Sometimes help set up and work portable field lighting instruments and microphone booms. Assist camera operators in moving camera dollies and pulling camera cables. At small stations also act as wardrobe and makeup people. Also called *grips, stagehands,* and *utilities personnel.*

1.2 NONTECHNICAL PRODUCTION PERSONNEL *(continued)*

PERSONNEL	FUNCTION
ADDITIONAL PRODUCTION PERSONNEL	
In small operations these production people are not always part of the permanent staff or their functions are fulfilled by other personnel.	
Writer	At smaller stations and in corporate television, the scripts are often written by the director or producer. Usually hired on a freelance basis.
Art director	In charge of the creative design aspects of show (set design, location, and/or graphics).
Graphic artist	Prepares computer graphics, titles, charts, and electronic backgrounds.
Makeup artist	Does the makeup for all talent. Usually hired on a freelance basis.
Costume designer	Designs and sometimes even constructs various costumes for dramas, dance numbers, and children's shows. Usually hired on a freelance basis.
Wardrobe person	Handles all wardrobe matters during production.
Property manager	Maintains and manages use of various set and hand properties. Found in large operations only. Otherwise, props are managed by the floor manager.
Sound designer	Constructs the complete sound track (dialogue and sound effects) in postproduction. Usually hired on a freelance basis for large productions.

Television talent—the performers and actors who work in front of the camera—are usually considered part of the nontechnical production personnel (discussed in chapter 16). ZVL5 PROCESS→ People→ nontechnical

Technical Personnel and Crew

The ***technical personnel*** consist of people who are primarily concerned with operating equipment. They are usually part of the crew. The technical personnel include camera operators, audio and lighting people, video recorder (VR) operators, video editors, C.G. (character generator) operators, and people who set up communication and signal transmission equipment. The term *technical* does not refer to electronic expertise but rather to operating the equipment with skill and confidence. The true engineers, who understand electronics and know where to look when something goes wrong with a piece of equipment, usually do not operate equipment; rather they ensure that the whole system runs smoothly, supervise its installation, and maintain it. You may find that in larger professional operations, however, the technical production people are still called engineers, mainly to satisfy the traditional job classification established by the labor unions.

The DP (director of photography) is sometimes listed as part of the nontechnical personnel and sometimes as part of the technical team. The term, borrowed from film production, has found its way into television. In standard theatrical film production, the DP is mainly responsible for lighting and the proper exposure of the film rather than for running the camera. In smaller digital film productions and EFP (electronic field production), the DP operates the camera as well as does the lighting. So when you hear that an independent television producer/director is looking for a reliable and creative DP, he or she is primarily referring to an experienced EFP camera operator. SEE 1.3 ZVL6 PROCESS→ People→ technical

As mentioned, many of the functions of technical and nontechnical production people overlap and even change, depending on the size, location, and relative complexity of the production. For example, you may initially have acted as a producer when setting up the video recording of the semiannual address of a corporation president;

1.3 TECHNICAL PERSONNEL

PERSONNEL	FUNCTION
ENGINEERING STAFF	
These people are actual engineers who are responsible for the purchase, installation, proper functioning, and maintenance of all technical equipment.	
Chief engineer	In charge of all technical personnel, budgets, and equipment. Designs system, including transmission facilities, and oversees installations and day-to-day operations.
Assistant chief engineer	Assists chief engineer in all technical matters and operations. Also called *engineering supervisor.*
Studio or remote engineer-in-charge	Oversees all technical operations. Usually called *EIC.*
Maintenance engineer	Maintains all technical equipment and troubleshoots during productions.
NONENGINEERING TECHNICAL PERSONNEL	
Although skilled in technical aspects, the following technical personnel do not have to be engineers but usually consist of technically trained production people.	
Technical director (TD)	Does the switching and usually acts as technical crew chief.
Camera operators	Operate the cameras; often do the lighting for simple shows. When working primarily in field productions (ENG/EFP), they are sometimes called *videographers* and *shooters.*
Director of photography (DP)	In film productions, in charge of lighting. In EFP, operates EFP camera.
Lighting director (LD)	In charge of lighting; normally found mostly in large productions.
Video operator (VO)	Adjusts camera controls for optimal camera pictures (shading). Sometimes takes on additional technical duties, especially during field productions and remotes. Also called *shader.*
Audio technician	In charge of all audio operations. Works audio console during the show. Also called *audio engineer.*
Video-record operator	Runs the video recorder.
Character generator (C.G.) operator	Types and/or recalls from the computer the names and other graphic material to be integrated with the video image.
Video editor	Operates postproduction editing equipment. Often makes or assists in creative editing decisions.
Digital graphic artist	Renders digital graphics for on-air use. Can be nontechnical personnel.

then, on the day of the shoot, you may find yourself busy with such technical matters as lighting and running the camera. In larger productions, such as soap operas, your job responsibility is much more limited. When acting as a producer, you have nothing to do with lighting or equipment operation. When working the camera, you may have to wait patiently for the lighting crew to finish, even if the production is behind schedule and you have nothing else to do at the time.

NEWS PRODUCTION PERSONNEL

Almost all television broadcast stations produce at least one daily newscast; in fact, the newscasts are often the major production activity at these stations. Because news departments must be able to respond quickly to a variety of production tasks, such as covering a downtown fire or a protest at city hall, there is generally little time to prepare for such events. News departments therefore have their own ***news production personnel***. These people are dedicated exclusively to the production of news, documentaries, and special events and perform highly specific functions. SEE 1.4

Don't be puzzled if you hear the assignment editor of a news department sending several VJs to cover breaking stories. *VJ* stands for *video journalist*—an individual who must combine the functions of reporter, videographer, writer, and editor. This rather demanding job was obviously not instituted to improve news coverage but to save money. Nevertheless it's apparent that you can no longer afford a narrowly focused training but must be fluent in all aspects of television production.

As in any other organization, television and corporate video involve many more people than what you see listed in this section, such as clerical personnel and the people who answer phones, schedule events, sell commercial time, negotiate contracts, build and paint the sets, and clean the building. Because these support personnel operate outside of the basic production system, their functions aren't discussed here.

1.4 NEWS PRODUCTION PERSONNEL

PERSONNEL	FUNCTION
News director	In charge of all news operations. Bears ultimate responsibility for all newscasts.
Producer	Directly responsible for the selection and the placement of the stories in a newscast so that they form a unified, balanced whole.
Assignment editor	Assigns reporters and videographers to specific events to be covered.
Reporter	Gathers the stories. Often reports on-camera from the field.
Video journalist	Reporter who shoots and edits his or her own footage.
Videographer	Camcorder operator. In the absence of a reporter, decides on what part of the event to cover. Also called *news photographer* and *shooter.*
Writer	Writes on-the-air copy for the anchors. The copy is based on the reporter's notes and the available video.
Video editor	Edits video according to reporter's notes, writer's script, or producer's instructions.
Anchor	Principal presenter of newscast, normally from a studio set.
Weathercaster	On-camera talent, reporting the weather.
Traffic reporter	On-camera talent, reporting local traffic conditions.
Sportscaster	On-camera talent, giving sports news and commentary.

MAIN POINTS

- The three production phases are preproduction, production, and postproduction.
- Preproduction includes the preparation of a show before the actual production activities take place. It usually happens in two stages: the first is the move from the basic idea to the script; the second is the designation of the necessary equipment (cameras, microphones, and so forth), facilities (studio or field production), and people to transform the script into a television show.
- Production includes all the activities in which equipment and the crew operating it create the actual program or program segments. The program can be video-recorded or put on the air; the segments are usually video-recorded for postproduction.
- Postproduction involves mostly video and audio editing. The various program sections that were recorded in the production phase are put into the proper sequence. It can also include the enhancement of the pictures and the sound.
- The effect-to-cause model facilitates the production approach. It moves from the basic idea to the process message (the desired effect on the viewer) and from there to the medium requirements (content, people, and equipment) necessary to actually cause the process message. The closer the defined and actual process messages match, the more successful the program.
- The nontechnical production personnel include a variety of people who design the program (writers, art director, sound designer, and so forth) and execute the program (producers, director, floor manager, and assistants).
- The technical personnel include the engineers, who install and maintain the equipment, and the nonengineering technical personnel, who operate the equipment.
- The news department has its own production personnel, who consist of a variety of producers, writers, assignment editors, graphic artists, reporters, and videographers as well as video journalists, who report, operate the camera, and write and edit the story.

SECTION

1.2

Technical Production Systems

To make sense of the various pieces of television equipment and how they interact in a multicamera or single-camera production, you should consider them as part of a system. This way you can relate how they function together, even though they are presented here individually. This section gives an overview of the studio and field production systems and an introduction to the major equipment.

- **BASIC TELEVISION SYSTEM**
  How a program host appears on the television receiver
- **EXPANDED TELEVISION SYSTEM**
  Multicamera studio system
- **FIELD PRODUCTION SYSTEMS**
  ENG (electronic news gathering) and EFP (electronic field production) systems
- **MAJOR EQUIPMENT**
  Camera, audio, lighting, switcher, video recorder, and postproduction editing

BASIC TELEVISION SYSTEM

A system is a collection of elements that work together to achieve a specific purpose. Each element depends on the proper functioning of the others, and none of the individual elements can do the job alone. The ***television system*** consists of equipment and people who operate that equipment for the production of specific programs.

How a Program Host Appears on the Television Receiver

Whether the production is simple or elaborate or originates in the studio or in the field—that is, on-location—the television system works on the same basic principle: the television camera converts whatever it "sees" (optical images) into electrical signals that can be temporarily stored or directly reconverted by the television set into visible screen images. The microphone converts whatever it "hears" (actual sounds) into electrical signals that can be temporarily stored or directly reconverted into sounds by the loudspeaker. In general, the basic television system transduces (converts) one state of energy (optical image, actual sound) into another (electric energy). **SEE 1.5** The picture signals are called video signals, and the sound signals are called audio signals. Any small consumer camcorder represents such a system.

EXPANDED TELEVISION SYSTEM

The expanded system includes more equipment in a variety of configurations. Productions such as news, interviews, game shows, and soap operas use the multicamera studio system.

Multicamera Studio System

The multicamera studio system in its most elementary stage includes two or more cameras, camera control units (CCUs), preview monitors, a switcher, a line monitor, one or more video recorders, and a line-out that transports the video signal to the video recorder and/or the transmission device. **SEE 1.6**

Usually integrated into the expanded system are computer servers or videotape machines for playback, character or graphic generators that produce various forms of lettering or graphic art, and an editing system.

The audio portion of the expanded system consists of one or more microphones, an audio mixer or console, an audio monitor (speaker), and a line-out that transports the sound signal to the video recorder and/or the transmitter (see figure 1.6).

FIELD PRODUCTION SYSTEMS

Except for big-remote telecasts that are used for the transmission of live sports or special events, the field production systems are much less complex than even a simple studio show. These field productions usually consist of ***ENG*** (electronic news gathering) or the more elaborate ***EFP*** (electronic field production).

ENG System

Electronic news gathering is usually done with a camcorder, which houses an entire video system in an amazingly

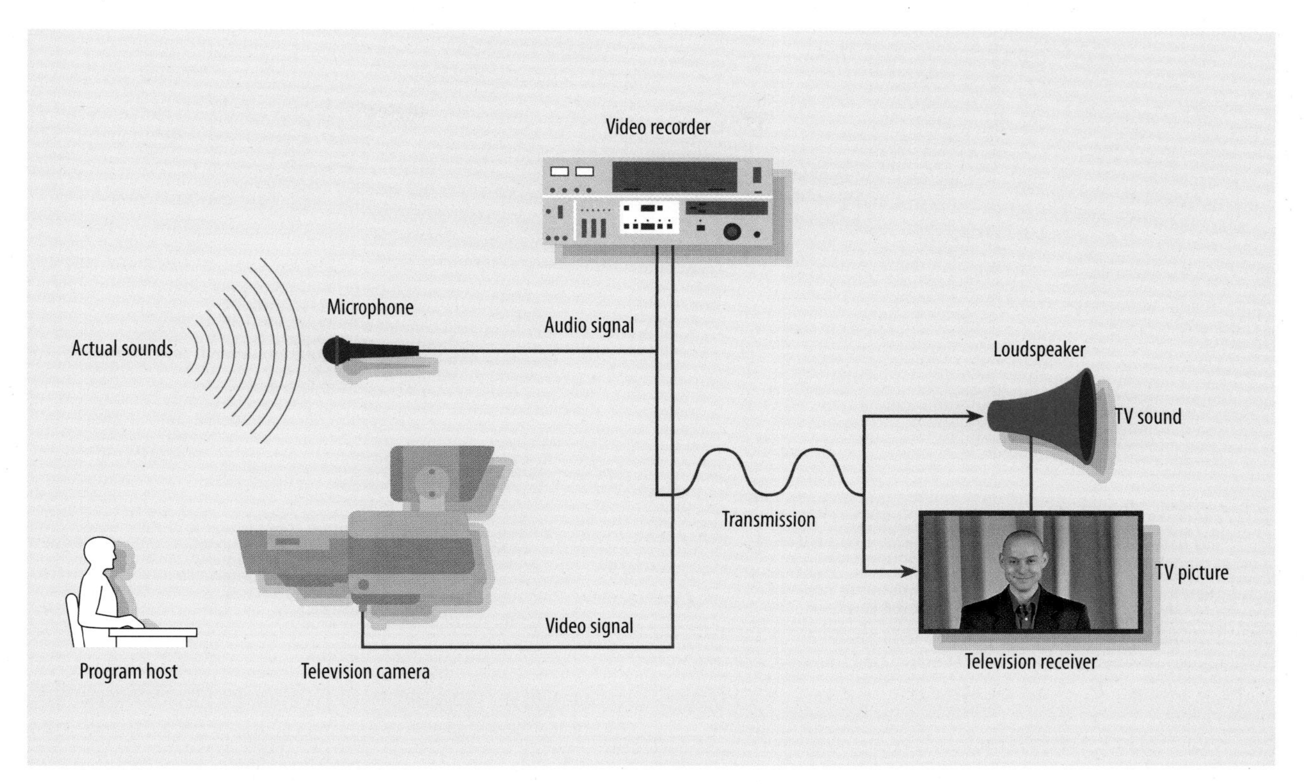

1.5 PHOTO BY EDWARD AIONA

1.5 BASIC TELEVISION SYSTEM
The basic television system converts light and sounds into electrical video and audio signals that are transmitted (wirelessly or by cable) and reconverted by the television receiver and loudspeaker into television pictures and sound.

small box. It contains all the elements needed to capture and record an event. The expanded system uses a second microphone in addition to the built-in one and may use a small transmitter that routes the signal to the television station or an ENG van. **SEE 1.7**

EFP System

The EFP system normally consists of a single portable EFP camera and an external recording device or camcorder to record various segments of an event for postproduction editing. In more elaborate productions, several cameras or camcorders are used simultaneously to capture an event from various viewpoints. **SEE 1.8**

MAJOR EQUIPMENT

With the expanded television system in mind, we briefly explore six basic production elements: camera, audio, lighting, switcher, video recorder, and postproduction editing. When learning about television production equipment, always try to see each piece and its operation within the larger context of the television system, that is, in relation to all the other equipment. Then tie the equipment to the people who operate it—the technical personnel. It is, after all, the skilled and prudent use of the television equipment by the whole production team, and not simply the smooth interaction of the machines, that gives the system its value.

Camera

The most obvious production element—the camera—comes in all sizes and configurations. Some cameras are so small that they fit easily into a coat pocket; others are so heavy that you have to strain yourself to lift them onto a camera mount. The camera mount enables the operator to move a heavy camera/lens/teleprompter assembly on the studio floor with relative ease. **SEE 1.9**

ENG/EFP camcorders are portable cameras that use a variety of recording media—videotape, hard drives, optical discs, and memory cards (also called flash drives). They operate much like consumer models except that they have

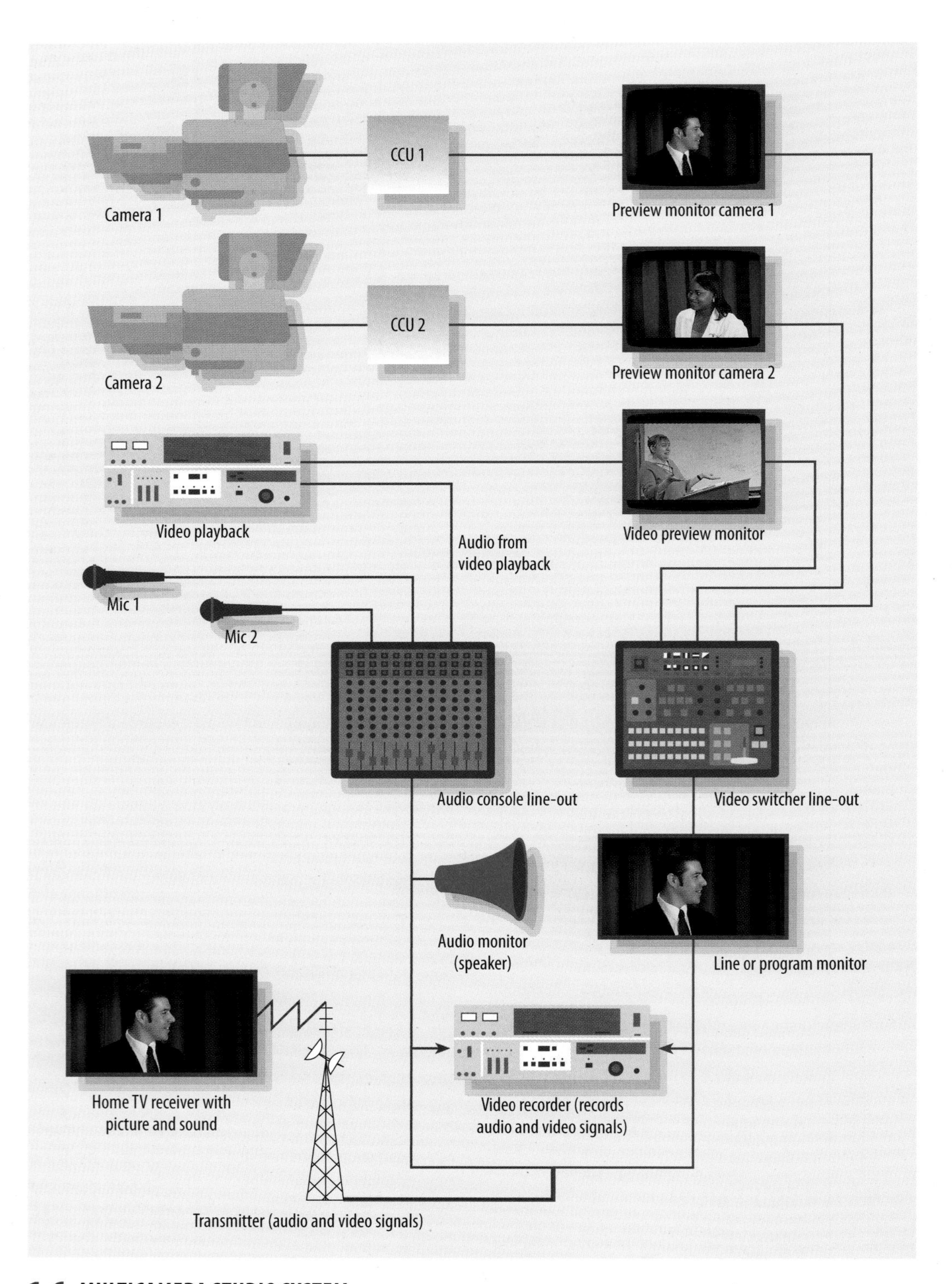

1.6 PHOTOS BY EDWARD AIONA

1.6 MULTICAMERA STUDIO SYSTEM
The multicamera studio system contains quality controls (CCU and audio console), selection controls (switcher and audio console), and monitors for previewing pictures and sound.

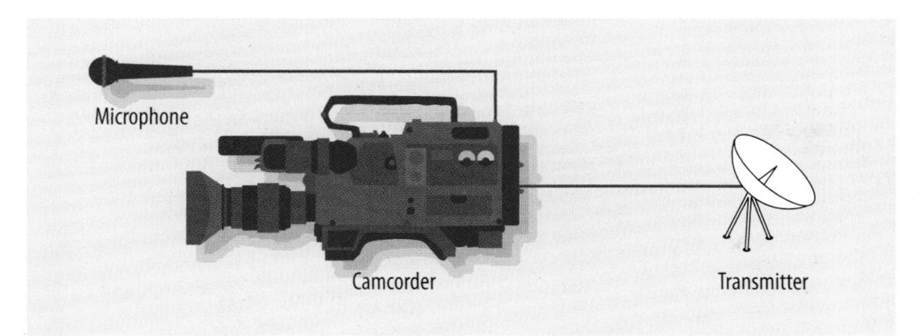

1.7 ENG SYSTEM

The ENG system consists of a camcorder and a microphone. The camcorder includes all video and audio quality controls as well as video- and audio-recording capabilities. A portable transmitter is necessary to send a live field pickup to the studio.

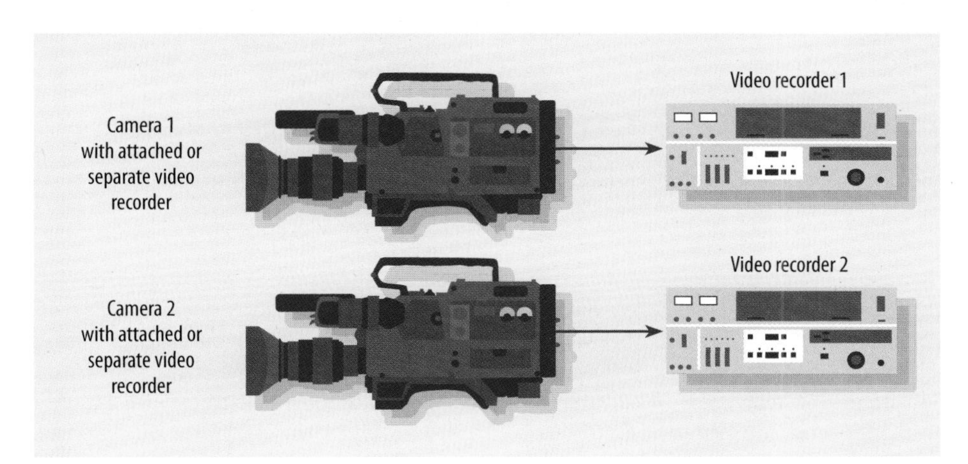

1.8 EFP SYSTEM

The EFP system is similar to that for ENG, but it may use more than one camera to feed the output to separate video recorders.

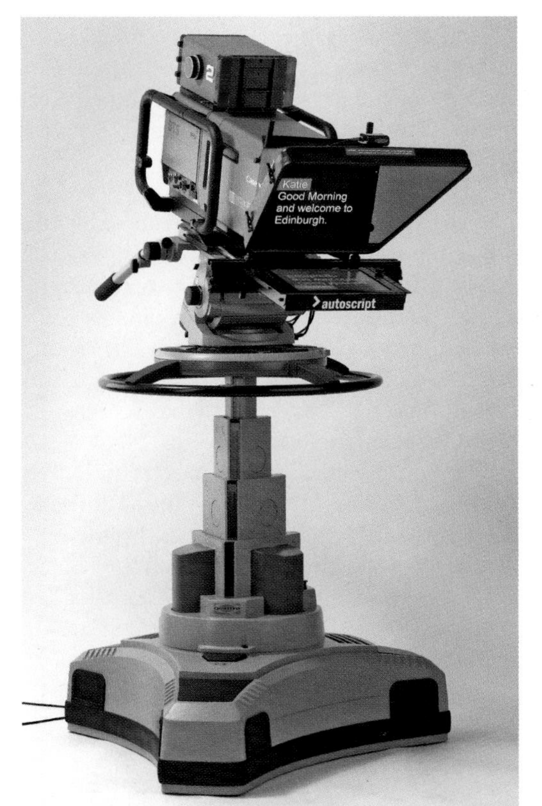

1.9 STUDIO CAMERA WITH PEDESTAL

High-quality studio cameras are mounted on a studio pedestal for smooth and easy maneuverability.

1.9 PHOTO COURTESY AUTOSCRIPT

1.10 SMALL HDV CAMCORDER
This high-definition video camcorder delivers video that comes close to that of the much more expensive HDTV camcorders. It records on MiniDV cassette tapes.

1.11 HIGH-END PROFESSIONAL HDTV CAMCORDER
This high-end camcorder can record on a hard-drive field pack or a memory card. Both can be transferred directly to an HDTV nonlinear editing system.

1.10 PHOTO COURTESY PANASONIC
1.11 PHOTO COURTESY SONY ELECTRONICS INC.

better lenses (which can be exchanged), better imaging devices (which transduce the light coming from the lens into video signals), and more controls that help produce optimal pictures even under less-than-ideal conditions. In fact, some of the new HDTV (high-definition television) camcorders are simply upgrades of high-end consumer models. **SEE 1.10**

Many high-end ENG/EFP camcorders use tapeless recording devices, which store pictures and sound on hard drives, optical discs, or memory cards. The advantages of such tapeless recording media are that they have no moving parts and they can be directly transferred into a digital editing system. **SEE 1.11** ZVL7 CAMERA→ Camera introduction ZVL8 CAMERA→ Camera moves

Audio

Although the term *television* does not include audio, the sound portion of a television show is nevertheless one of its most important elements. Television audio not only communicates precise information but also contributes greatly to the mood and the atmosphere of a scene. If you were to turn off the audio during a newscast, even the best news anchors would have difficulty communicating their stories through facial expressions, graphics, and video images alone.

The aesthetic function of sound (to make us perceive an event or feel in a particular way) becomes obvious when you listen to the background sounds of a crime show. The squealing tires during a high-speed chase are real enough, but the exciting, rhythmically fast background music that accompanies the scene is definitely artificial. We have grown so accustomed to such devices, however, that we would probably perceive the scene as less exciting if the music were missing. In fact, some crime shows and commercials carry a continuous music track with a highly rhythmic beat even through the dialogue. Frequently, sound communicates the energy of an event more readily than pictures do.

Even if you don't intend to become a sound designer, you need to learn as much as possible about the major sound production elements: microphones, sound control equipment, and sound recording and playback devices. ZVL9 AUDIO→ Audio introduction

Microphones All microphones convert sound waves into electric energy—the audio signals. The sound signals are amplified and sent to the loudspeaker, which reconverts them into audible sound. The myriad microphones available today are designed to perform different tasks. Picking up a newscaster's voice, capturing the sounds of a tennis match, and recording a rock concert—all may require different microphones or microphone sets.

Some microphones, called lavalier mics (pronounced "mikes"), are quite small and are clipped to the performer's clothing. Hand mics are larger and carried by the performer or attached to a mic stand. Boom, or long-distance, mics are either suspended from a small boom (called a fishpole, which is carried by the operator) or from a large boom, whose operator sits on a movable platform. **SEE 1.12** ZVL10 AUDIO→ Microphones→ mic types

1.12 FISHPOLE MICROPHONE
This highly directional shotgun mic is suspended from a fishpole by the boom operator.

Sound control equipment In studio productions the most important piece of sound control equipment is the audio console. At the audio console, you can select a specific microphone or other sound input, amplify a weak signal from a mic or other audio source for further processing, control the volume and the quality of the sound, and mix (combine) two or more incoming sound sources. In relatively simple productions, such as a newscast or an interview, you are mostly concerned with keeping the audio within a certain volume level. If it is too low, the viewer/listener can't hear the sound very well; if it is too high, the sound is not only hard on the ears but distorts so much that it may be impossible to fix in postproduction. **SEE 1.13**

1.13 AUDIO CONSOLE
Even a relatively simple audio console has many controls to adjust the volume and the quality of incoming sound signals and to mix them in various ways.

In ENG and EFP, the sound is normally controlled by the camera operator, who wears a small earphone that carries the incoming sound. Because the camera operator is busy running the camera, the sound controls on the camcorder are often switched to the automatic setting. In the more critical EFP, the volume of incoming sounds is usually controlled by a portable mixer. **SEE 1.14** ZVL11 AUDIO→ Consoles and mixers

1.14 AUDIO MIXER
The portable mixer has a limited number of inputs and volume controls.

Sound recording and playback devices When an event is recorded for postproduction, most of the dialogue and environmental sounds are recorded simultaneously with the picture.

In large and complex studio productions in which a single camera shoots a scene piecemeal, much in the way films are made, the audio track is subjected to much manipulation in postproduction. The sounds of explosions, sirens, and car crashes, for example, are normally dubbed in (added) during the postproduction sessions. Even parts of the original dialogue are occasionally re-created in the studio, especially when the dialogue occurs outdoors. As you undoubtedly know and have probably experienced, wind is a constant hazard to clean sound pickup. ZVL12 AUDIO→ Systems

Lighting

Like the human eye, the camera cannot see well without a certain amount of light. Because it is actually not objects we see but the light that is reflected off the objects, manipulating the light falling on objects influences the way we perceive them on-screen. The purposeful control of light and shadows is called lighting.

Types of illumination All television lighting basically involves two types of illumination: directional and diffused. Directional light has a sharp beam and produces harsh shadows. You can aim the light beam to illuminate a precise area. A flashlight and car headlights produce directional light. In television and motion pictures, these lights are called spotlights. Diffused light has a wide, indistinct beam that illuminates a relatively large area and produces soft, translucent shadows. The fluorescent lamps in a department store produce diffused lighting. Television and motion pictures use floodlights to achieve such a general nondirectional lighting. ZVL13 LIGHTS→ Light introduction

Lighting instruments In the television studio, the various types of spotlights and floodlights are usually suspended on battens that can be raised close to the ceiling and lowered close to the floor. This enables the lighting people to place the instruments in the desired positions on the battens. When the battens are raised, the cameras and crewmembers can move freely about the studio floor without interfering with the lighting. **SEE 1.15**

ENG and EFP use much smaller, portable instruments that can be set up quickly and plugged into ordinary household outlets. ZVL14 LIGHTS→ Instruments→ studio | field

Lighting techniques As mentioned, lighting is the manipulation of light and shadows that influences the way we perceive how things on-screen look and feel. All television lighting is based on a simple principle: to illuminate specific areas, mold shadows, and bring the overall light on a scene to an intensity level at which the cameras can produce optimal pictures and create a certain mood. Optimal pictures means that the colors are faithfully reproduced even in the shadow areas, that there is a certain number of brightness steps between the darkest and the brightest spots in the scene, and that you can still see some detail in the brightest and darkest areas. For some shows the lighting is deliberately flat, which means that there is little contrast between light and shadows. Such lighting is frequently used on news and interview sets, for game shows and situation comedies, and in many field productions. Crime

1.15 STUDIO LIGHTS SUSPENDED FROM MOVABLE BATTENS
Typical studio lighting uses spotlights and floodlights. All instruments are suspended from battens that can be lowered close to the studio floor and raised well above the scenery.

1.15 PHOTO BY HERBERT ZETTL

and mystery shows often use high-contrast lighting. This creates dense shadows and intensifies the dramatic tension. ZVL15 LIGHTS→ Falloff→ fast | slow | none

Switcher

The switcher works on a principle similar to that of pushbuttons on a car radio, which allow you to choose different radio stations. The switcher lets you select various video inputs, such as cameras, video recorders, and titles or other special effects, and join them through a great variety of transitions while the event is in progress. In effect, the switcher allows you to do instantaneous editing.

Any switcher, simple or complex, can perform three basic functions: select an appropriate video source from several inputs, perform basic transitions between two video sources, and create or retrieve special effects, such as split screens. **SEE 1.16**

If you now go back to figure 1.6, you can see that three video inputs—camera 1, camera 2, and a video recorder—are routed to the switcher. From these three inputs, camera 1 is selected to go on the air. ZVL16 SWITCHING→ Switching introduction

1.16 PHOTO COURTESY ECHOLAB, LLC

1.16 VIDEO PRODUCTION SWITCHER
The production switcher has rows of buttons and other controls for selecting and mixing various video inputs and creating transitions and special effects. It then sends the selected video to the line-out.

Video Recorder

One of the unique features of television is its ability to transmit a telecast live, which means capturing the pictures and the sounds of an ongoing event and distributing them instantly to a worldwide audience. Most television programs, however, originate from playback of previously recorded material. There are two basic recording systems: videotape recorders and tapeless recorders.

Videotape recorders Despite the great progress made almost daily to develop digital recording systems that are more efficient than videotape, videotape recorders (VTRs) will still be in use for some time to come, so don't throw away your old VCR, videotape collection, or tape-based camcorder just yet. Videotape is still widely used in large and small camcorders, including high-end HDV (high-definition video) and HDTV systems. Note that you can use videotape for analog as well as digital recordings.

Videotape recorders are usually classified by the electronic system used for the recording (DVCPRO, S-VHS, or VHS) and sometimes by the tape format (the width of the videotape in the videocassette). Several VTR systems still use ½-inch videocassettes (digital Betacam SX, S-VHS, and VHS). Most digital systems use ¼-inch cassettes (6.35mm) for SDTV (standard digital television), HDV, or HDTV. They come in the standard smaller MiniDV cassettes as well as larger cassettes that allow for longer recording and playback. **SEE 1.17**

1.17 PHOTO COURTESY PANASONIC

1.17 HDTV VIDEOTAPE RECORDER
This high-definition studio VTR can use a variety of cassettes (MiniDV and the larger formats) for recording and playback of high-definition footage, including digital cinema material.

Tapeless systems Great and rapid progress is being made toward a tapeless environment wherein all video recording, storage, and playback is done with non-tape-based systems. Such tapeless systems make use of hard drives, optical discs, and memory cards.

High-capacity hard drives are used extensively for the storage, manipulation, and retrieval of video and audio information by desktop computers and dedicated editing systems (specially manufactured editors) for postproduction. Hard drives with a very large storage capacity (in the multi-terabyte range—a terabyte equals 1,000 gigabytes) are called servers and have all but replaced videotape for the storage and playback of daily programming in most television stations.

Postproduction Editing

In principle, postproduction editing is relatively simple: you select the most effective shots from the original source material and join them with transitions in a specific sequence. In practice, however, postproduction editing can be quite complicated and time-consuming, especially if it also involves extensive audio manipulation. ZVL17 EDITING→ Editing introduction

Nonlinear editing A ***nonlinear editing (NLE)*** system does not use videotape. Before editing can begin, all recorded material must first be transferred to the hard drive of an editing system, which can be a laptop, a desktop computer, or a workstation especially designed

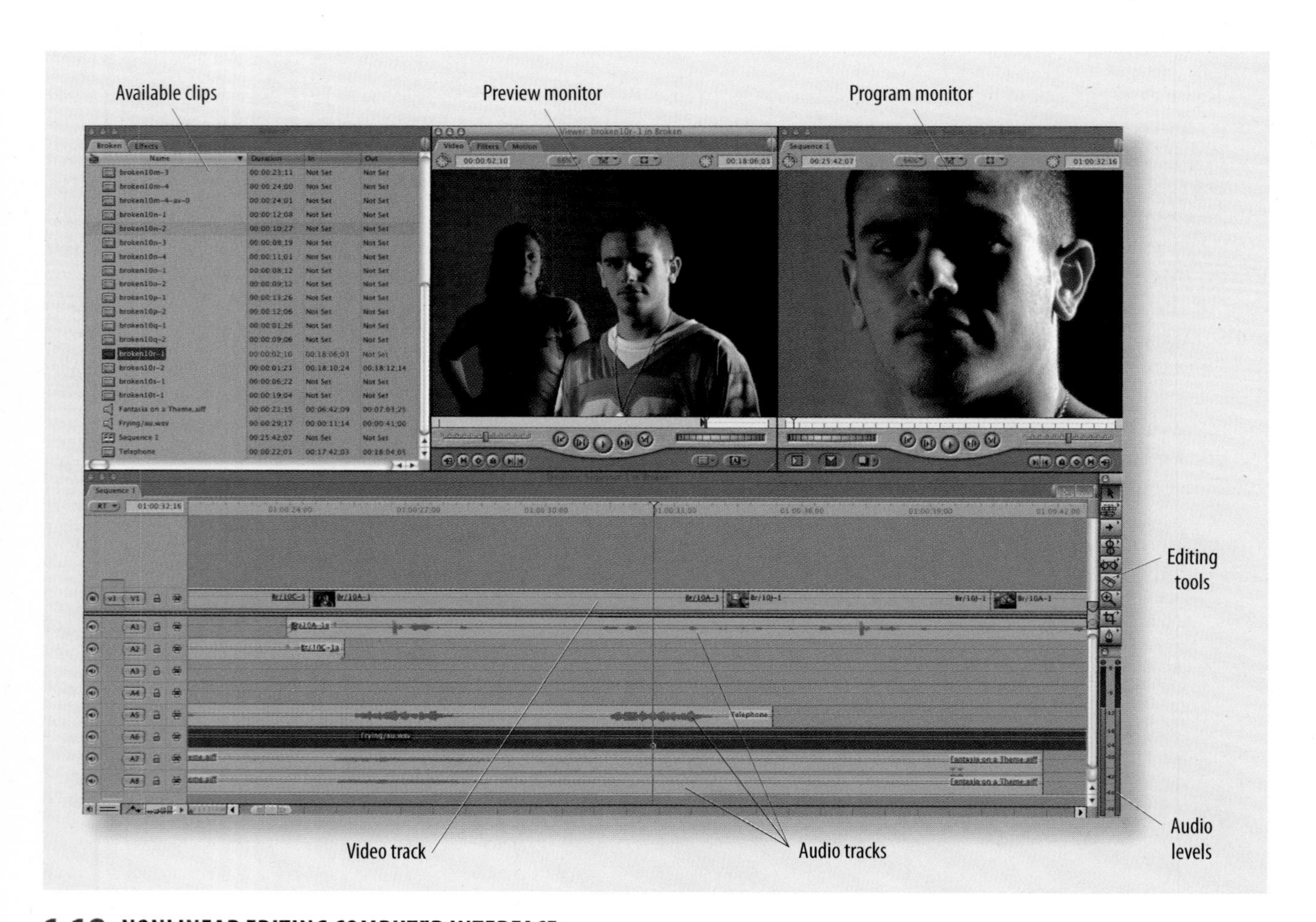

1.18 NONLINEAR EDITING COMPUTER INTERFACE
The interface of most NLE systems shows a list of available clips, a preview monitor of the upcoming shot that is to be edited to the shot shown on the program monitor, a video track (blue track with thumbnail images), two or more audio tracks (green tracks), and other information such as available transitions.

1.18 PHOTOS BY EDWARD AIONA

for postproduction editing. Once all video and audio clips are on the hard drive of the editing system, you can manipulate them pretty much as you would edit text with a word-processing program. You can call up, move, cut, paste, and join the various shots or audio segments much like words, sentences, and paragraphs when editing a document. This method is called nonlinear because you can call up any clip or frame regardless of the sequence in which it was captured.

Almost all NLE software lets you produce high-resolution full-frame, full-motion video and audio sequences. You can also decide to first produce a low-resolution preliminary rough-cut, from which you develop an EDL (edit decision list). This list is then your guide for the final high-resolution edit master recording, which is used for program duplication or broadcast. **SEE 1.18** ZVL18 EDITING→ Nonlinear editing→ system

Linear editing Whenever you use digital or analog videotape for postproduction, you are engaged in linear editing. ***Linear editing*** normally requires two source VTRs, which contain the original material that you recorded with the camera, and the record VTR, which produces the final edit master tape.

All three machines are synchronized by the edit controller, also called an editing control unit. This unit helps you find a particular scene quickly and accurately, even if it is buried midtape. It starts and stops the source and record machines and tells the record VTR to perform the edit at the precise point you have designated. ZVL19 EDITING→ Linear editing→ system

Regardless of the editing system you use, it cannot make the creative decisions for you. Thinking about postproduction as early as the preproduction stage facilitates considerably your editing chores. Always consider postproduction an extension of the creative process, not a salvage operation.

MAIN POINTS

- The basic television system consists of the equipment and the people who operate the equipment to produce specific programs. In its simplest form, the system comprises a television camera that converts what it sees into a video signal, a microphone that converts what it hears into an audio signal, and a television set and a loudspeaker that reconvert the two signals into pictures and sound.
- The expanded television system adds equipment and procedures to the basic system to make possible a wider choice of sources, better quality control of pictures and sound, and the recording and/or transmission of video and audio signals.
- The ENG (electronic news gathering) system consists basically of a camcorder and a microphone. The EFP (electronic field production) system may include multiple camcorders or field cameras, some lighting, and audio and video control equipment.
- The major production elements are the camera, audio, lighting, switcher, videotape recorder, and postproduction editing.
- There are several types of video cameras: large studio cameras that need a pedestal to be moved about the studio floor; ENG/EFP cameras that are small enough to be carried by the operator; and camcorders, which have the recording device either built into the camera or attached to it.
- Audio, the sound portion of a television show, is necessary to give specific information about what is said and to set the mood of a scene.
- Audio production elements include microphones, sound control equipment, and sound recording and playback devices.
- Lighting is the manipulation of light and shadows that influences the way we perceive objects on-screen and how we feel about the screen event.
- The two types of illumination are directional light, produced by spotlights, and diffused light, produced by floodlights.
- The switcher enables you to do instantaneous editing by selecting a specific picture from several inputs and performing basic transitions between two video sources.
- There are a variety of analog and digital tape-based recorders as well as digital tapeless video recorders.
- Non-tape-based systems include hard drives, optical discs, and memory cards or flash drives. Large-capacity hard-drive systems, called servers, are used for the recording, storage, and playback of program material.
- Postproduction editing consists of selecting various shots from the source material and putting them in a specific sequence. In nonlinear editing, the digital video and audio material is stored on a hard drive and manipulated using computer software. Nonlinear editing (NLE) systems can produce high-quality video and audio sequences for broadcast or duplication, a rough-cut of lesser video quality, or an EDL (edit decision list), which serves as a guide for the final high-quality edit master.

ZETTL'S VIDEOLAB

For your reference or to track your work, the *Zettl's VideoLab* program cues in this chapter are listed here with their corresponding page numbers.

PART

Preproduction

CHAPTER

2

The Producer in Preproduction

One morning you wake up with an idea for a documentary that will surely blow away the one you saw the night before. You grab your camera and start shooting. After a week of collecting miles of footage, you put the box with the recorded material aside until you have time for editing. When you finally look at it again, the shots don't look so good anymore, and the whole idea has somehow lost its punch. You eventually scrap the whole project.

Of course, this is a fictitious scenario. But it is used to underline the importance of carefully thinking about and preparing the whole production process before ever getting into the actual production. This step is the all-important preproduction phase.

You will notice that when studying this text you cannot simply remain a passive reader. You will be required to wear several hats and assume a variety of production roles. Sometimes you will have to act as producer and then, a few chapters later, function as a director or as a specific member of the technical crew.

In this chapter you are the producer—a person who not only comes up with a great idea but, contrary to the introductory scenario, also carries it successfully to and through the actual production phase. In section 2.1, What Producing Is All About, you are responsible for the first step of preproduction planning—to move from idea to script—and how to move from there to the production phase. Section 2.2, Information Resources, Unions, and Ratings, covers research aids, personnel unions and other legalities, as well as audience and ratings—just in case you want to be an actual producer.

KEY TERMS

demographics Audience research factors concerned with such data as age, gender, marital status, and income.

production schedule The calendar that shows the preproduction, production, and postproduction dates and who is doing what, when, and where.

program proposal Written document that outlines the process message and the major aspects of a television presentation.

psychographics Audience research factors concerned with such data as consumer buying habits, values, and lifestyles.

rating Percentage of television households tuned to a specific station in relation to the total number of television households.

share Percentage of television households tuned to a specific station in relation to all households using television (HUT); that is, all households with their sets turned on.

target audience The audience selected or desired to receive a specific message.

time line A breakdown of time blocks for various activities on the actual production day, such as crew call, setup, and camera rehearsal.

treatment A brief narrative description of a television program.

SECTION

2.1

What Producing Is All About

Producing means seeing that a worthwhile idea gets to be a worthwhile television presentation. You are chiefly responsible for all preproduction activities and for completing the various tasks on time and within budget. You are responsible for the concept, financing, hiring, and overall coordination of production activities—not an easy job by any means! But even if you happen to take on the creation of a show all by yourself, you still have to act as your own producer.

As a producer you may have to act as a psychologist and a businessperson to persuade management to buy your idea, argue as a technical expert for a certain piece of equipment, or search as a sociologist to identify the needs and the desires of a particular social group. After some sweeping creative excursions, you may have to become pedantic and double- and triple-check details, such as whether there is enough coffee for the guests who appear on your show.

Considering the painstaking work you have to do before ever getting into the production stage, you may not want to become a producer. But as a professional in television and digital cinema production, you cannot escape the all-important planning details of the preproduction phase. Even as a video journalist, you will have to make preproduction decisions while on the run.

- **PREPRODUCTION PLANNING: FROM IDEA TO SCRIPT**
  Generating program ideas, evaluating ideas, devising a program proposal, preparing a budget, and writing the script
- **PREPRODUCTION PLANNING: COORDINATION**
  People and communication, facilities request, production schedule, permits and clearances, and publicity and promotion
- **ETHICS**
  Observing the prevailing ethical standards of society

PREPRODUCTION PLANNING: FROM IDEA TO SCRIPT

Although each production has its own creative and organizational requirements, there are nevertheless techniques, or at least approaches, that you can apply to television and digital cinema production in general. Once you know the producer's basic preproduction activities, you can transfer those skills to whatever position you occupy on the production team. To help you become maximally efficient and effective in your preproduction activities, we focus here on program ideas, program proposal, budget, and script.

Generating Program Ideas

Everything you see and hear on television or in the movie theater started with an idea. As simple as this may sound, developing good and especially workable show ideas on a regular basis is not easy. As a professional television producer, you cannot wait for the occasional divine inspiration but must generate worthwhile ideas on demand.

Despite the volumes of studies written on the creative process, exactly how ideas are generated remains a mystery. Sometimes you will find that you have one great idea after another; at other times you cannot think of anything exciting regardless of how hard you try. You can break through this idea drought by engaging several people to do brainstorming or clustering.

You certainly know what group brainstorming is all about: everybody is allowed to toss out wild ideas in the hope that someone may break through the conceptual blocks and bring an end to the idea drought. The key to successful brainstorming is to not pass judgment on any comments, however far-fetched they may be. Anything goes. When playing back the audio-recorded comments, you may find that the totally unrelated comment may just trigger a new approach.

Another, more personalized and structured idea generator is clustering. This technique is a kind of brainstorming wherein you write down your ideas rather than say them aloud. To begin you write a single keyword, such as *cell phone,* and circle it. You then spin off idea clusters that somehow relate to the initial keyword. SEE 2.1

As you can see, clustering is a more organized means of brainstorming, but it is also more restrictive. Because clustering shows patterns better than brainstorming does, it serves well as a structuring technique. Although clustering is usually done by individuals, you can easily have a group of people engage in clustering and then collect the results for closer scrutiny. ZVL1 PROCESS→ Ideas

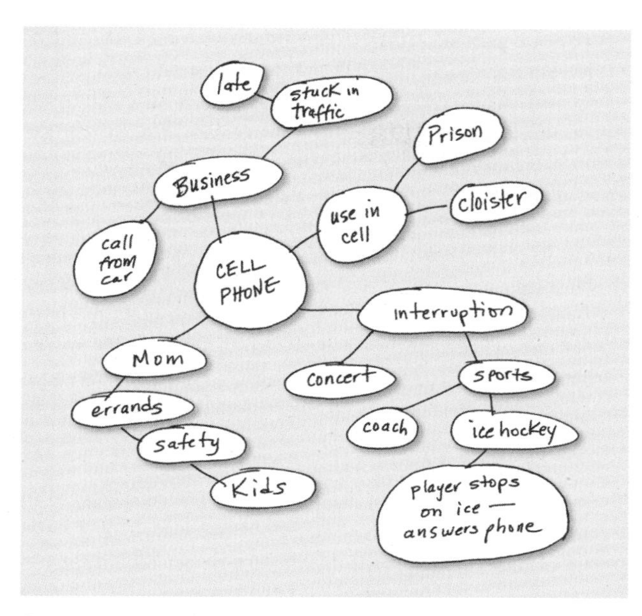

2.1 PARTIAL IDEA CLUSTER
Clustering is a form of written brainstorming. You start with a central idea and branch out to whatever associations come to mind.

Evaluating Ideas

Evaluation of the ideas is probably the most important step in the preproduction process. The two key questions you need to ask are: Is the idea worth doing? and Is the idea doable? If you can answer both with an honest "yes," proceed to the formulation of a process message.

If your answer to either or both questions is "maybe" or "no," stop right there and find a better idea. SEE 2.2

Is the idea worth doing? Whatever you do should make a difference. This means that regardless of whether you produce a brief news package, a longer feature, or a major motion picture, it should have a positive influence on somebody's life (ideally on all people watching your program).[1]

Fortunately, your idea passed the evaluation test, which means that you can progress to the formulation of the process message and the angle. Recall from chapter 1 that the process message is the basic program objective—what you would ideally like the viewers of your target audience to learn, do, and feel when watching your program. The angle is the specific focus or twist you give the story to get and keep the viewers' attention and have the actual process message come as close as possible to the defined one. The clearer you are about what your program should be and achieve, the easier it will be to write the program proposal, prepare the budget, develop the script, and decide

1. Stuart W. Hyde has lectured and written about significant vision for more than half a century. See his *Idea to Script: Storytelling for Today's Media* (Boston: Allyn and Bacon, 2003), pp. 58–64. Nancy Graham Holm, an Emmy award–winning journalist and former head of the TV Journalism Department at the Danish School of Journalism, says, "Any story worth telling has *significant vision*" [her emphasis]. She defines significant vision as "a problem to be solved, a challenge to be met, an obstacle to be overcome, a threat to be handled, a decision or choice to be made, a pressure to be relieved, a tension to be eased, a victory to be celebrated, a kindness to be acknowledged." See her *Fascination: Viewer Friendly TV Journalism* (Århus, Denmark: Ajour Danish Media Books, 2007), p. 51.

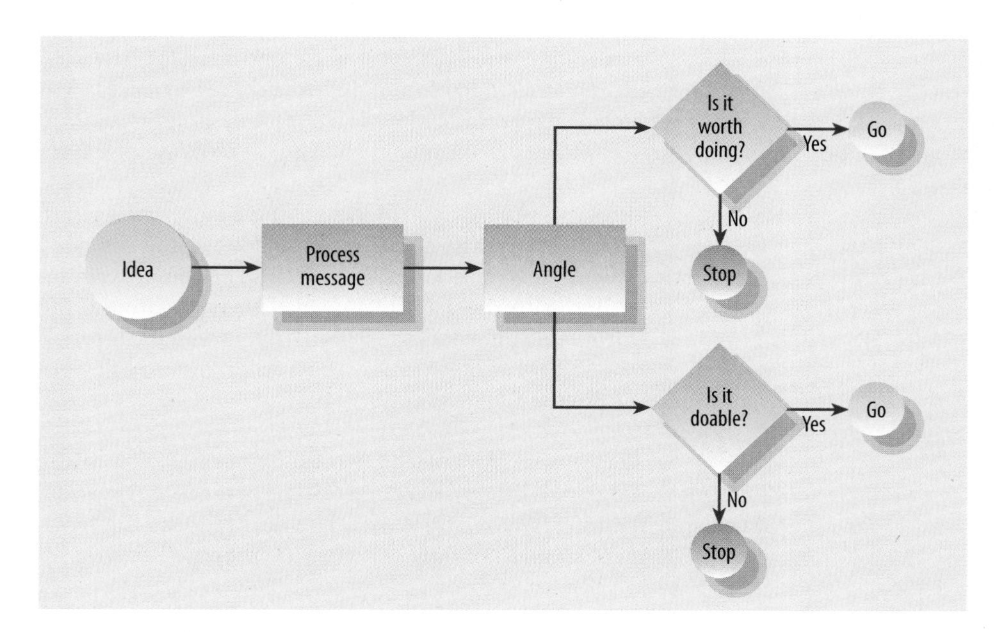

2.2 IDEA EVALUATION
Once you have translated the idea into a process message and found a useful angle, you need to ask two crucial evaluation questions: Is the idea worth doing? and Is the idea doable?

on the remaining preproduction steps—but there remains one more question to ask.

Is the idea doable? This is determined by whether you have the necessary budget, time, and facilities available to proceed to the production phase. You may have a relevant idea that has significant vision, but if you don't have the money to pay for the necessary production people, equipment, and facilities, you had better abandon the project at this point. If you have only one week to produce a story, you can't do a documentary that depicts the hard life of a Sherpa in the Himalayas. The availability of three small camcorders will not suffice for a live telecast of a high-school basketball game if you don't also have a switcher, its associated equipment, and transmission facilities. Whether a project is doable can hinge on less obvious deficiencies as well, such as bad weather, an inexperienced crew, or the lack of permits for a location or union affiliation of the talent.

Program Proposal

Once you pass the two feasibility criteria and have a clear idea of the process message and how you want to communicate it, you are ready to write the program proposal. Don't take this proposal lightly; it is a key factor in getting your program on the air as opposed to simply ending up in a "good idea" file on your hard drive.

A ***program proposal*** is a written document that stipulates what you intend to do. It briefly explains the program objective and the major aspects of the presentation. Although there is no standard format for a program or series proposal, it should at a minimum include the following information:

- Program or series title
- Program objective (process message)
- Target audience
- Show format (a single TV show or series or a digital movie)
- Show treatment (usually includes the angle)
- Production method
- Tentative budget

If you propose a series, attach a sample script for one of the shows and a list of the titles of the other shows in the series.

Program or series title Keep the title short but memorable. Perhaps it is the lack of screen space that forces television producers to work with shorter titles than do filmmakers. Instead of naming your show *The Trials and Tribulations of a University Student,* simply say *Student Pressures.*

Program objective Compared with the process message, this is a less academic description of what you want to do. For example, rather than say, "The process message is to have high-school students exposed to at least five major consequences of running a stop sign," you may write that the program's objective is "to warn teenage drivers not to run stop signs."

Target audience The ***target audience*** is whom you would primarily like to have watch the show—the elderly, preschoolers, teenagers, homemakers, or people interested in travel. A properly formulated process message will give a big clue of who the target audience is. Even when you want to reach as large an audience as possible and the audience is not defined, be specific in describing the potential audience. Instead of simply saying "general audience" for a proposed comedy series, describe the primary target audience as "the 18-to-30 generation" or the "over-60 crowd in need of a good laugh."

Once you are in the actual preproduction phase, you can define the target audience further in terms of ***demographics***—such as gender, ethnicity, education, income level, household size, religious preference, or geographical location (urban or rural)—as well as ***psychographics,*** such as consumer buying habits, values, and lifestyles. Advertisers and other video communicators make extensive use of such demographic and psychographic descriptors, but you needn't be that specific in your initial program proposal.

Show format Do you propose a single show, a new series, or part of an existing series? How long is the intended show? An example would be a two-part one-hour program dealing with the various uses of helicopters around the world. This information is vital for planning a budget or, for a station or network, to see whether the show fits into the program schedule.

Show treatment A brief narrative description of the program is called a ***treatment.*** Some of the more elaborate treatments have storyboard-like illustrations. The treatment should not only say what the proposed show is all about but also explain its angle and reflect in its writing the style of the show. The style of a treatment for an instructional series on computer-generated graphics, for example, should differ considerably from that of a drama or situation comedy. Do not include production specifics such as types of lighting or camera angles; save this information for the director. Keep the treatment brief and concise, but do not skip important aspects of a plot. A treatment should give a busy executive a fairly good idea of what you intend to do. SEE 2.3 ZVL2 PROCESS→ Proposals→ treatment

Production method A well-stated process message indicates where the production should take place and how you can do it most efficiently. Should you do a multiple- or single-camera studio production or a single-camera EFP? Is the show more effectively shot in larger segments with three or four cameras switched and in iso positions, or shot single-camera film-style for postproduction? What performers or actors do you need? What additional materials (costumes, props, and scenery) are required? ZVL3 PROCESS→ Methods→ location | studio | single-camera | multicamera

Tentative budget Before preparing the tentative budget, you must have up-to-date figures for all production services, rental costs, and wages in your area. Independent production and postproduction houses periodically issue rate cards that list the costs for services and the rental of major production items; such information is also usually available on their Web sites. Stay away from high-end services unless video quality becomes your major concern.

Preparing a Budget

When working for a client, you need to prepare a budget for all preproduction, production, and postproduction costs regardless of whether the cost is, at least partially, absorbed by the salaries of regularly employed personnel or the normal operating budget. You need to figure the costs not only for obvious items—script, talent, production personnel, studio and equipment rental, and postproduction editing—but also for items that may not be so apparent, such as videotape and/or other recording media (memory cards can be quite expensive), props, food, lodging, entertainment, transportation of talent and production personnel, parking, insurance, and clearances or user fees for location shooting.

There are many ways to present a budget, but the usual way is to divide it into preproduction (for example, script, travel to locations and meetings, location scouting, and storyboard), production (talent, production personnel, and equipment or studio rental), and postproduction (editing and sound design). Most production companies show their overall charges in this tripartite division so that the client can more easily compare your charges with those of the other bidders.

When you first present your proposal, the client may be interested not so much in how you broke down the expenses but more in the bottom-line figure. It is therefore critical that you think of all the probable expenses regardless of whether they occur in preproduction, production, or postproduction. Computer software can be of great assistance, helping you detail the various costs and recalculating them effortlessly if you need to cut expenses or if the production requirements change.

An example of a detailed budget of an independent production company is shown in the following figure. It is structured according to preproduction, production, and postproduction costs. SEE 2.4

Whenever you prepare a budget, be realistic. Do not underestimate costs just to win the bid—you will probably regret it. It is psychologically, as well as financially, easier to agree to a budget cut than to ask for more money later on. On the other hand, do not inflate the budget to ensure enough to get by even after severe cuts. Be realistic about the expenses, but do not forget to add at least a 15 to 20 percent contingency. ZVL4 PROCESS→ Proposals→ budget | try it

Writing the Script

Unless you write the script yourself, you'll need to hire a writer. Contrary to books and magazine articles, which are published in printed form, a media script is not intended as literature. Even the most literate and sophisticated script is only an intermediary in the production process. Analyzing scripts in the context of literature may be an interesting academic exercise, but it says as little about the actual television show or film as does a city map about the way the city looks and sounds. Nevertheless, a script represents an essential production element of any film and of all but the most routine television presentations.

Besides telling the talent what to say, a script indicates how a scene should be played and where and when it takes place; it also contains important preproduction, production, and postproduction information. The script is such a critical production component that if a show does not need a script, a scriptlike notice is distributed, indicating the name of the show, the date, the director, and the remark

TREATMENT FOR A ONE-HOUR SPECIAL ON THE HOMELESS

Title: HOMELESS

Proposed Actual Program Length: 45 min.

This program is intended to make the audience feel rather than watch the plight of being homeless. It will not show the customary degrading living conditions of the homeless, such as a homeless woman pushing a shopping cart past elegant shopping windows or seeking shelter in an abandoned box next to a garbage container. In fact, it will not show any footage of homeless people. Instead, it will trace a young college professor's ill-fated trip from Boston to a convention on the West Coast.

When he arrives in San Francisco, his suitcase does not show up on the baggage carousel. Only one suitcase keeps circling—and it's not his. When he gets to the baggage claims office, it is crowded. He is nervous because all the conference information and his presentation material are in the suitcase. He finally gets to the claims desk, and the stressed official asks him for the tags and hotel. Yes, his bags will be delivered to the hotel. Yes, he has the claim tags. He finds them in his wallet. Which hotel? He can't remember the exact name. It's in the folder, and the folder is in the suitcase. An impatient, not-so-friendly man behind him curses and jostles him, trying to get out of the queue.

The downtown airport bus does not go to his hotel. He takes a taxi. The cabbie circles a block, takes out his map, and finally drops off the professor in a neighborhood that doesn't seem to be an appropriate convention venue. Was he taken for a ride? When he tries to pay the expensive fare,

2.3 TREATMENT
The treatment tells the reader in narrative form what a program is all about.

his wallet is missing: cash, driver's license, credit cards, everything—gone. The cab driver radios his boss: call the cops or let him go. He lets him go.

The lobby smells. His cell phone doesn't work. Yeah, he can use the phone. But it will cost him. No money, no phone. He again interrupts the lady behind the counter with his story. "Yeah, sure! Can't you think of a better one? There is a phone outside." The graffiti-sprayed booth has only wires hanging out where the phone is supposed to be. The phone book is missing, replaced by an empty liquor bottle. Urine smell. It gets dark and starts to rain. He walks and walks and finally finds a working phone. But he has no money. The few people he approaches on the street walk faster or ask him for money. Finally, a friendly woman with a skirt that is much too tight and much too short listens to his story and gives him a dollar bill. Laughing: "I am usually the one who gets paid." He changes the bill in a bar, is allowed to use the public phone, and calls his friend on the East Coast. He reaches the answering machine.

He is outside again, in the rain, in a place he doesn't know. Hungry. Tired. He is homeless.

He finally flags down a police car and is taken to the station. No, they don't have any information about his convention. But, eventually, a police officer does a computer search and finds the name of the convention hotel. The officer, off-duty now, takes him there. Thanks! Very much! On the way in, he encounters a man, begging for coffee money. "I don't have any money." "Oh yeah?" He gets to the lobby and finds one of his colleagues with a drink in his hand and a slightly crooked nametag on his jacket. Home! Safe!

2.3 TREATMENT *(continued)*

2.4 DETAILED PRODUCTION BUDGET
These detailed budget categories are structured according to preproduction, production, and postproduction costs. Ignore the categories that are not applicable.

PRODUCTION BUDGET

CLIENT:
PROJECT TITLE:
DATE OF THIS BUDGET:
SPECIFICATIONS:

NOTE: This estimate is subject to the producer's review of the final shooting script.

SUMMARY OF COSTS	ESTIMATE	ACTUAL
PREPRODUCTION		
Personnel	______	______
Equipment and facilities	______	______
Script	______	______
PRODUCTION		
Personnel	______	______
Equipment and facilities	______	______
Talent	______	______
Art (set and graphics)	______	______
Makeup	______	______
Music	______	______
Miscellaneous (transportation, fees)	______	______
POSTPRODUCTION		
Personnel	______	______
Facilities	______	______
Recording media	______	______
INSURANCE AND MISCELLANEOUS	______	______
CONTINGENCY (20%)	______	______
TAX	______	______
GRAND TOTAL	______	______

"No script." (The basic structural ingredients of a television and film script and the various types of scripts are covered in chapter 3.)

When hiring a writer, make sure that he or she understands the program objective and, more specifically, the defined process message. If a writer disagrees with your approach and does not suggest a better one, find another writer. Agree on a fee in advance—some writers charge enough to swallow up your whole budget. ZVL5 PROCESS→ Ideas→ scripts

BUDGET DETAIL	ESTIMATE	ACTUAL
PREPRODUCTION		
Personnel		
Writer (script)		
Director (day)		
Art director (day)		
PA (day)		
SUBTOTAL		
PRODUCTION		
Personnel		
Director		
Associate director		
PA		
Floor (unit) manager		
Camera (DP)		
Sound		
Lighting		
Video recording		
C.G.		
Grips (assistants)		
Technical supervisor		
Prompter		
Makeup and wardrobe		
Talent		
Equipment and facilities		
Studio/location		
Camera		
Sound		
Lighting		
Sets		
C.G./graphics		
Video recorder		
Prompting		
Remote van		
Intercom		
Transportation, meals, and housing		
Copyrights		
SUBTOTAL		

2.4 DETAILED PRODUCTION BUDGET *(continued)*

PREPRODUCTION PLANNING: COORDINATION

Before you begin coordinating the production elements—assembling a production team, procuring studios, or deciding on location sites and equipment—ask yourself once again whether the planned production is possible within the given time and budget and, if so, whether the method (medium translation of defined process message) is indeed the most efficient.

2.4 DETAILED PRODUCTION BUDGET *(continued)*

POSTPRODUCTION	ESTIMATE	ACTUAL
Personnel		
Director	______	______
Editor	______	______
Sound editor	______	______
Facilities		
Dubbing	______	______
Window dubs	______	______
Off-line linear	______	______
Off-line nonlinear	______	______
On-line linear	______	______
On-line nonlinear	______	______
DVE	______	______
Audio sweetening	______	______
ADR/Foley	______	______
Recording media	______	______
SUBTOTAL	______	______
MISCELLANEOUS		
Insurance	______	______
Public transportation	______	______
Parking	______	______
Shipping/courier	______	______
Wrap expenses	______	______
Security	______	______
Catering	______	______
SUBTOTAL	______	______
GRAND TOTAL	======	======

For example, if you are doing a documentary on the conditions of residence hotels in your city, it is certainly easier and more cost-effective to go there and video-record an actual hotel room than to re-create one in the studio. On the other hand, if you are doing a series of interviews on what makes an effective teacher, the studio—even a small one—will serve you best. Keep in mind that the studio affords optimal control but that EFP (electronic field production) offers a limitless variety of scenery and locations at little additional cost.

Once you have made a firm decision about the most effective production approach, you must deliver what you promised to do in the proposal. You begin this coordination phase by establishing clear communication channels among all the people involved in the production, and you take care of the facilities request, the production schedule, permits and clearances, and publicity and promotion. Again, it is not your occasional flashes of inspiration that make you a good producer but your meticulous attention to detail. Preproduction is not the most exciting part of creating a show, but, from a producer's standpoint, it is the most important one.

People and Communication

Whom to involve in the post-script planning stages depends, again, on your basic objective, the process message, and whether you are an independent producer who has to hire additional personnel or whether you are working for a station or production company that has most essential creative and crew people already on its payroll and available at all times.

As a producer you are the chief coordinator among the production people. You must be able to contact every team member quickly and reliably. Your most important task is to establish a database with such essential information as names, positions, e-mail addresses, home addresses, business addresses, and various phone, pager, and fax numbers. **SEE 2.5**

Don't forget to let everyone know how you can be contacted, as well. Don't rely on secondhand information. Your communication is not complete until you hear back from the party you were trying to reach. *A good producer triple-checks everything.*

Facilities Request

The facilities request lists all pieces of production equipment and often all properties and costumes needed for a production. The person responsible for filling out such a request varies. In small-station operations and independent production companies, it is often the producer or director; in larger operations it is the production manager or director of broadcast operations.

The facilities request usually contains information concerning date and time of rehearsal, recording sessions, and on-the-air transmission; title of production; names of producer and director (and sometimes talent); and all technical elements, such as cameras, microphones, lights, sets, graphics, costumes, makeup, video recorders, video and audio postproduction facilities, and other specific production requirements. It also lists the studio and control room needed or the remote location. If you do EFP, you need to add the desired mode of transportation for yourself and the crew and the exact on-site location. If the production involves an overnight stay, communicate the

Production Personnel Contact Information
Sight Sound Motion Instructional Video
Program 4

Name E-mail	Position	Home address Work address	Home phone Work phone	Home fax Work fax	Cell phone Pager
Herbert Zettl hzettl@best.com	Producer	873 Carmenita, Forest Knolls SFSU, 1600 Holloway, SF	(415) 555-3874 (415) 555-8837	(415) 555-8743 (415) 555-1199	(415) 555-1141
Gary Palmatier bigcheese@ideas-to-images.com	Director	5343 Sunnybrook, Windsor 5256 Aero #3, Santa Rosa	(707) 555-4242 (707) 555-8743	(707) 555-2341 (707) 764-7777	 (707) 555-9873
Robaire Ream robaire@mac.com	AD	783 Ginny, Healdsburg Lightsaber, 44 Tesconi, Novato	(707) 555-8372 (415) 555-8000	 (415) 555-8080	(800) 555-8888
Sherry Holstead 723643.3722@compuserve.com	PA	88 Seacrest, Marin SH Assoc, 505 Main, Sausalito	(415) 555-9211 (415) 555-0932	(415) 555-9873 (415) 555-8383	(415) 555-0033
Renee Wong rn_wong2@earthlink.com	TD	9992 Treeview, San Rafael P.O. Box 3764, San Rafael	(415) [illegible] [illegible]	[illegible] 555-8273 [illegible]	(415) 555-3498 [illegible] 555-8988
[illegible]	Talent	253 Robert[illegible]	[illegible]	[illegible]	[illegible]

2.5 DATABASE: PRODUCTION PERSONNEL
To be able to quickly contact each production team member, the producer needs reliable contact information.

2.6 COMPUTER-BASED FACILITIES REQUEST
This facilities request lists all equipment needed for the production. Usually, the equipment permanently installed in a studio does not have to be listed again, but it must be scheduled.

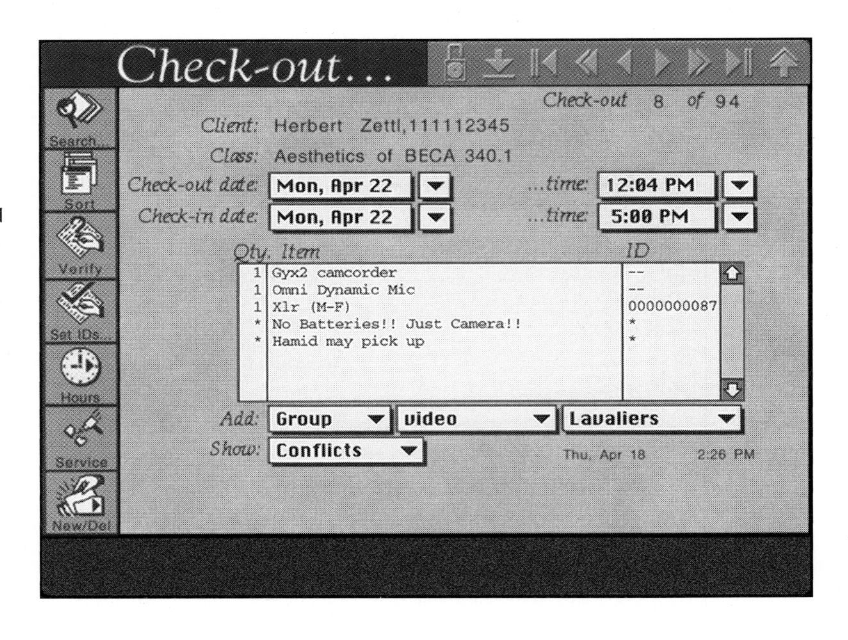

2.6 SCREEN COURTESY BROADCAST AND ELECTRONIC COMMUNICATION ARTS DEPARTMENT AT SAN FRANCISCO STATE UNIVERSITY

name and the location of the accommodations, including the customary details, such as phone numbers, when and where to assemble the next morning, and so forth.

Like the script, the facilities request is an essential communication device; be as accurate as possible when preparing it. Late changes will invite costly errors. If you have a fairly accurate floor plan and light plot, attach them to the facilities request. Such attachments will give the crew a fairly good idea of what production problems they may have to face. Facilities requests are usually distributed as "soft copy" via the internal computer system as well as on hard copy. **SEE 2.6**

Regardless of which type of production you do, always try to get by with as little equipment as possible. The more you use, the more people you need to operate it and the more that can go wrong. Do not use equipment just because it is available. Review your original process message and verify that the chosen equipment is indeed the most efficient and that the necessary equipment is actually available and within the scope of the budget. On the use of specific equipment and similar technical production items, consult the director of your show or someone on the technical staff, which may include your favorite DP or TD (director of photography or technical director, respectively).

Production Schedule

The ***production schedule*** should tell everybody involved in the production who is doing what, when, and where over the course of the three production phases. It is different from a ***time line***, which is a breakdown of time blocks for each production day. Create a realistic production schedule and stick to it. Assigning too little time will result not in a higher level of activity but usually in a higher level of anxiety and frustration. It is almost always counterproductive. On the other hand, allowing too much time for a production activity will not necessarily improve the production. Besides being costly, wasting time can make people apathetic and, surprisingly enough, often fail to meet deadlines.

One of the producer's most important responsibilities is to monitor the progress of each activity and know where everyone stands relative to the stipulated deadlines. If you don't care whether deadlines are met, you might as well do away with them. If schedules aren't kept, find out why. Again, do not rely on secondhand information. Call the people who are behind schedule and find out what the problem is. It is your job to help solve the problem and get everyone back on track or to change the schedule if necessary. Always inform all production people of all changes—even if the changes seem irrelevant or insignificant.

Permits and Clearances

Most productions involve people and facilities that ordinarily have no connection with your station or production company. These production elements need extra attention. Get the necessary permits for your crew to gain admission to a meeting or sporting event, as well as a parking permit close to the event. You may also need a permit from city hall (the mayor's media coordinator and the police department) or an insurance policy to shoot downtown. Do not ignore such requirements! "Better safe than sorry" applies to all field productions—not just to actual production activities but also to protecting yourself from legal action if a production assistant stumbles over a cable or a bystander slips on a banana peel while watching your shoot. (Copyright and union clearances are discussed in section 2.2.)

Publicity and Promotion

The best show is worthless if no one knows about it. Meet with the publicity and promotions departments (usually combined in one office or even a single person) and inform them of the upcoming production. Even if the target audience is highly specific, you must aim to reach as many viewers as possible. The job of the publicity people is to narrow the gap between the potential and the actual audience.

Regardless of what your specific producing job may be—managing ideas and production schedules or coordinating nontechnical and technical production facilities and teams—don't leave anything to chance.

Finally, you should remember and act on the following brief sermon.

ETHICS

Whatever you do as a producer, realize that your decisions, however trivial they might seem at the time, always affect a very large number of people: your audience.

Always respect and have compassion for your audience. Do not believe critics or cynics who proclaim that all television audiences have an average intelligence of a five-year-old. In fact, you and I are part of the television audience, and neither of us would appreciate having our intellect downgraded in this way by a disgruntled producer.

Never breach the prevailing ethical standards of society and the trust the audience has—and inevitably must—put in you. This doesn't mean that you should play censor to a writer who occasionally breaks the mold with frank ideas and a bold vision. It means that you and your production team should not lie to your audience. Do not fake a news story to rev up a slow news day; do not have a speech edited so that it favors your political candidate.

While realizing that producing is always connected with compromise, ask yourself if, and if so to what extent, your show will contribute to the quality of life of your audience.

I suspect you noticed that such a code of conduct concerns not just the producer but everyone working in television.

MAIN POINTS

- Producing means seeing to it that a worthwhile idea becomes a worthwhile television show. The producer manages a great number of people and coordinates an even greater number of production activities and details.
- Generating program ideas on schedule is facilitated by brainstorming and clustering.
- Preproduction includes planning how to move from the idea to the script.
- Important preproduction items are the program proposal, the treatment, the budget, and the script.
- The coordination tasks in preproduction are establishing communication among all personnel involved, filing a facilities request, creating a realistic production schedule, securing permits and clearances, and taking care of publicity and promotion.
- A good producer triple-checks everything.
- All of your actions must live up to the prevailing ethical standards.

SECTION

2.2

Information Resources, Unions, and Ratings

As a producer you need quick access to a variety of accurate information resources, an understanding of broadcast guilds and unions, and some basic knowledge of copyrights and other legal matters. Finally—like it or not—you must be conversant with the rudiments of ratings.

- **INFORMATION RESOURCES**
  The Internet, telephone directories, and other resources
- **UNIONS AND LEGAL MATTERS**
  Unions, copyrights and clearances, and other legal considerations
- **AUDIENCE AND RATINGS**
  Target audience, and ratings and share

INFORMATION RESOURCES

As a producer you must be a researcher as well as somewhat of a scrounger. On occasion you may have only a half hour to get accurate information, for example, about a former mayor who is celebrating her ninetieth birthday. Or you may have to procure a skeleton for your medical show, a model of a communications satellite for your documentary on telecommunications, or an eighteenth-century wedding dress for your history series.

The Internet Fortunately, the vast resources on the Internet put the world's information at your fingertips—almost instantly. Although Google is one of the most prominent and efficient search engines, there are other resources that can lead you to more-specialized information, including Ask.com, AltaVista, Lycos, Netscape Search, Excite, Mamma Metasearch, MetaCrawler, Wikipedia, and Yahoo!. Still others have specialized information but often rely on the larger ones. All equipment manufacturers have extensive Web sites with the most up-to-date information about their products. Amazon.com is often faster than a library for locating a specific book.

You may find, however, that the sheer volume of online information makes it difficult to find a specific item quickly. It may sometimes be faster and more convenient to use readily available printed sources or to call the local library. For example, a call to the local hospital or high-school science department may procure the skeleton more quickly than initiating a Web search. You could ask the community college science department or perhaps even the local cable company for the satellite model, and contact the historical society or the college theater arts department for the wedding dress.

Besides Internet sources, the following are some additional references and services you should have on hand.

Telephone directories There is a great deal of information in a telephone book. Get the directories of your city and the outlying areas. Also try to get the directories of the larger institutions with which you have frequent contact, such as city hall, the police and fire departments, other city and county agencies, federal offices, city and county school offices, newspapers and radio stations, colleges and universities, and museums. On the Internet you can often obtain in seconds the telephone number of practically any phone user in the world.

Other resources The local chamber of commerce usually maintains a list of businesses and community organizations. A list of the major foundations and their criteria for grants may also come in handy. When working in a television station, you can always call on the news department for quick access to a great variety of local, national, and international information sources.

UNIONS AND LEGAL MATTERS

Most directors, writers, and talent belong to a guild or union, as do almost all technical people. As a producer you must be alert to the union regulations in your production area. Most unions stipulate not only salaries and minimum fees but also specific working conditions, such as overtime, turnaround time (stipulated hours of rest between workdays), rest periods, who can legally run a studio camera and who cannot, and so forth. If you use nonunion personnel

AFTRA	**American Federation of Television and Radio Artists.** This is the major union for television talent. Directors sometimes belong to AFTRA, especially when they double as announcers and on-the-air talent. AFTRA prescribes basic minimum fees, called scale, which differ from area to area. Most well-known talent (such as prominent actors and local news anchors) are paid well above scale.
DGA	**Directors Guild of America.** A union for television and motion picture directors and associate directors. Floor managers and production assistants of large stations and networks sometimes belong to "the Guild."
WGA	**Writers Guild of America.** A union for writers of television and film scripts.
SAG	**Screen Actors Guild.** A union for all actors in television and digital cinema. Also includes talent for commercials and larger video productions.
SEG	**Screen Extras Guild.** A union for extras participating in major film or video productions.
AFM	**American Federation of Musicians of the United States and Canada.** Relevant only if live musical performances are scheduled in the production.

2.7 NONTECHNICAL UNIONS AND GUILDS

in a union station, or if you plan to air a show that has been prepared outside the station with nonunion talent, such as a play you produced with your fellow students, check with the respective unions for proper clearance.

Unions

There are two basic types of unions: those for nontechnical production personnel and those for technical personnel. Nontechnical unions are mainly those for performers, writers, and directors. **SEE 2.7** Technical unions include all television technicians, engineers, and some production personnel such as microphone boom operators, ENG/EFP camera operators, and floor personnel. **SEE 2.8**

Be especially careful about asking studio guests to do anything other than answer questions during an interview. If they give a short demonstration of their talents, they may be classified as performers and automatically become subject to AFTRA fees (see figure 2.7). Likewise, do not request the floor crew to do anything that is not directly connected with their regular line of duty, or they too may collect talent fees. Camera operators usually have a contract clause that ensures them a substantial penalty sum if they are willfully shown by another camera on the television screen. Acting students who appear in television plays produced at a high school or college may become subject to AFTRA fees if the

IBEW	**International Brotherhood of Electrical Workers.** A union for studio, master control, and maintenance engineers and technicians. May also include ENG/EFP camera operators and floor personnel.
NABET	**National Association of Broadcast Employees and Technicians.** An engineering union that may also include floor personnel and nonengineering production people, such as boom operators and dolly operators.
IATSE	**International Alliance of Theatrical Stage Employees, Moving Picture Technicians, Artists and Allied Crafts of the United States, Its Territories and Canada.** A union for stagehands, grips (lighting technicians), and stage carpenters. Floor managers and even film camera and lighting personnel can also belong.

2.8 TECHNICAL UNIONS

play is shown on the air by a broadcast station, unless you clear their on-the-air appearance with the station and/or the local AFTRA office.

Copyrights and Clearances

If you use copyrighted material on your show, you must procure proper clearances. Usually, the year of the copyright and the name of the copyright holder are printed right after the © symbol. Some photographs, reproductions of famous paintings, and prints are copyrighted as are, of course, books, periodicals, short stories, plays, and music recordings. Shows or music that you record off the air or download from the Internet as well as many CD-ROMs and DVDs also have copyright protection.

Paradoxically, when you as an artist are trying to protect your rights, you may find that the copyrights are vague; but when you as a producer use copyrighted material, you are bound by stringent laws and regulations. When in doubt, consult an attorney about copyright clauses and public domain before using other people's material in your production.

Other Legal Considerations

Check with legal counsel about up-to-date rulings on libel (written and broadcast defamation), slander (lesser oral defamation), plagiarism (passing off as one's own the ideas or writings of another), the right to privacy (not the same in all states), obscenity, and similar matters. In the absence of legal counsel, the news departments of major broadcast stations or university broadcast departments generally have up-to-date legal information available.

AUDIENCE AND RATINGS

As a producer in a television station, you will probably hear more than you care to about television audiences and ratings. Ratings are especially important for commercial stations because the cost for commercial time sold is determined primarily by the estimated size of the target audience. Even when working in public or corporate television, you will find that audience "ratings" are used to gauge the relative success of a program.

Target Audience

Broadcast audiences, like those for all mass media, are usually classified by demographic and psychographic characteristics. The standard demographic descriptors include gender, age, marital status, education, ethnicity, and income or economic status. The psychographic descriptors pertain to lifestyle, such as consumer buying habits and even personality and persuasiveness variables. When you fill out the registration card that comes with a new electronic product, you are supplying highly valuable psychographic information.

Despite sophisticated techniques of classifying audience members and determining their lifestyles and potential acceptance of a specific program or series, some producers simply use a neighbor as a model and gear their communication to that particular person and his or her habits. Don't be surprised if an executive producer turns down your brilliant program proposal with a comment such as, "I don't think my neighbor Cathy would like it." For much entertainment programming, such a subjective approach to prejudging the worth of a program might be acceptable. If you are asked to do a goal-directed program such as driver education or a commercial on the importance of water conservation, however, you need to identify and analyze the target audience more specifically. The more you know about the target audience, the more precise your objective (defined process message) and, ultimately, the more effective your actual process message will be.

Ratings and Share

An audience ***rating*** is the percentage of television households tuned to a specific station in a given population (total number of television households). You get this percentage by dividing the projected number of households tuned to your station by the total number of television households:

$$\frac{\textit{number of TV households tuned in}}{\textit{total number of TV households}} = \textit{rating figure}$$

For example, if 75 households of your rating sample of 500 households are tuned to your show, your show will have a rating of 15 (the decimal point is dropped when the rating figure is given):

$$\frac{75}{500} = 0.15 = 15 \text{ rating points}$$

A ***share*** is the percentage of television households tuned to your station in relation to all households using television (HUT). The HUT figure represents the total pie—or 100 percent. Here is how a share is figured:

$$\frac{\textit{TV households tuned to your station}}{\textit{all households using television (HUT)}} = \textit{share}$$

For example, if only 200 of the sample households have their sets actually in use (HUT = 200 = 100 percent), the 75 households tuned in to your program constitute a share of 38:

$$\frac{75}{200} = 0.375 = \text{share of } 38$$

Rating services such as A. C. Nielsen carefully select representative audience samples and query these samples through diaries, telephone calls, and meters attached to their television sets.

The problem with the rating figures is not so much the potential for error in projecting the sample to a larger population but rather that the figures do not indicate whether the household whose set is turned on has any people watching and, if so, how many. The figures also do not indicate the impact of a program on the viewers (the actual process message). Consequently, you will find that your show is often judged not by the significance of your message, the impact it has on the audience, or how close the actual effect of the process message came to the defined effect, but simply by how many people are assumed to have watched your show in relation to other shows. As frustrating as the rating system in broadcast television is, you must realize that you are working with a mass medium that by definition bases its existence on large audiences.

MAIN POINTS

- A producer needs quick and ready access to a great variety of resources and information. The Internet is an almost instantaneous and total information resource. Telephone directories and local business and community resources are also helpful.
- Most nontechnical and technical production personnel belong to guilds or unions, such as the Directors Guild of America (DGA) and the National Association of Broadcast Employees and Technicians (NABET).
- The usual copyright laws apply when copyrighted material (such as scripts, video and audio recordings, printed information, CD-ROMs, and DVDs) is used in a television production.
- An audience rating is the percentage of television households tuned to a specific station in a given sample population owning TV sets. A share is the percentage of households tuned to a specific station in relation to all other households using television (HUT).

ZETTL'S VIDEOLAB

For your reference or to track your work, the *Zettl's VideoLab* program cues in this chapter are listed here with their corresponding page numbers.

CHAPTER

3

The Script

The ***script*** is one of the most important communication devices in all three production phases; a good one tells you what the program is about, who is in it, what each person says, what is supposed to happen, and how the audience should see and hear the event. Whatever position you hold on the production team, you must be familiar with the basic structure of dramatic and nondramatic scripts and the various script formats.

Section 3.1, Basic Script Formats, covers the standard formats for film and television. Section 3.2, Dramatic Structure, Conflict, and Dramaturgy, addresses dramatic and nondramatic story structures.

KEY TERMS

A/V format Another name for the two-column AV (audio/video) script.

classical dramaturgy The technique of dramatic composition.

event order The way event details are sequenced.

fact sheet Lists the items to be shown on-camera and their main features. May contain suggestions of what to say about the product. Also called *rundown sheet.*

goal-directed information Program content intended to be learned by the viewer.

partial two-column A/V script Describes a show for which the dialogue is indicated but not completely written out.

script Written document that tells what the program is about, who says what, what is supposed to happen, and what and how the audience should see and hear the event.

show format Lists the show segments in order of appearance. Used in routine shows, such as daily game or interview shows.

single-column drama script Traditional script format for television and motion picture plays. All dialogue and action cues are written in a single column.

two-column A/V script Traditional script format with video information on page-left and audio information on page-right for a variety of television scripts, such as for documentaries and commercials. Also called *A/V format.*

SECTION

3.1 Basic Script Formats

Whether you work in television or digital movie making, you will encounter different script formats. The more common are the single-column drama script, the two-column A/V script, the partial two-column A/V script, the news script, the show format, and the fact sheet.

- **SINGLE-COLUMN DRAMA SCRIPT**
  Standard format for writing plays for television and digital cinema
- **TWO-COLUMN A/V SCRIPT**
  Fully scripted A/V format and partial two-column A/V script
- **NEWS SCRIPT**
  One of the more widely used news script formats
- **SHOW FORMAT**
  Guide to routine show segments
- **FACT SHEET**
  List of items to be mentioned and shown on the air

SINGLE-COLUMN DRAMA SCRIPT

The ***single-column drama script*** includes every word of the actors' dialogue, who is doing what when and where, and frequently how the action should play. Dramas, situation comedies, skits, and soap operas use this script format. **SEE 3.1**

TWO-COLUMN A/V SCRIPT

This popular script type is also known as the ***A/V format*** because the right column contains the audio information and the left column contains the video information. Most documentary writers prefer the convenient ***two-column A/V script*** and therefore call it the documentary format. Fortunately, all three names refer to the same thing: a script that shows what is being heard on page-right and what is being seen on page-left. The A/V format can be fully scripted, which means that everything that is spoken appears in the audio column, or partially scripted, which shows only parts of the dialogue.

Fully Scripted A/V Format

Many fully scripted documentaries use the two-column A/V format. Because a documentary is intended to record an event rather than reconstruct one, scripts are frequently written *after* the field production. Documentary scripts therefore guide the postproduction phase rather than the actual production. The major video cues are listed in the video column, and all spoken words and sound effects are written in the audio column. The script tells the editor which sound bites to use and which video to choose for the voice-over segments. One word of caution to prospective documentary writers: writing a detailed script before going out to record the event makes no sense. Instead of documenting the event, you would merely be looking for event details that fit your prejudices.

If this format is used for a studio show, it is important to indicate specific in- and out-cues that tell the director when to insert a video clip, key a title, or move the cameras to another set area. **SEE 3.2** ZVL1 PROCESS→ Ideas→ scripts

Partial Two-column A/V Script

The ***partial two-column A/V script*** indicates only part of the dialogue. In general, the opening and closing remarks are fully scripted, but the bulk of what people say is only alluded to, such as: "Dr. Hyde talks about new educational ideas. Dr. Seel replies." This kind of script is almost always used for interviews, product demonstrations, educational series, variety shows, and other program types that feature a great amount of ad-lib commentary or discussion. **SEE 3.3**

NEWS SCRIPT

Newscasts are always fully scripted. The script must include every word the anchors speak as well as instructions for what live or recorded news segments the director must call up and when. The live segments normally consist

SCENE 6

A FEW DAYS LATER. INTERIOR. CITY HOSPITAL EMERGENCY WAITING ROOM. LATE EVENING.

YOLANDA is anxiously PACING back and forth in the hospital hallway in front of the emergency room. She has come straight from her job to the hospital. We see the typical hospital traffic in an emergency room. A DOCTOR (friend of CHUCK'S) PUSHES CARRIE in a wheelchair down the hall toward YOLANDA.

CARRIE

(in wheelchair but rather cheerful)

Hi, Mom!

YOLANDA

(anxious and worried)

Carrie—are you all right? What happened?

CARRIE

I'm OK. I just slipped.

DOCTOR (simultaneously)

She has a sprained right wrist. Nothing serious...

CARRIE

Why is everybody making such a big deal out of it?

YOLANDA

(cutting into both CARRIE'S and DOCTOR'S lines)

Does it hurt? Did you break your arm?

3.1 SINGLE-COLUMN DRAMA SCRIPT

The single-column drama script contains every word of the dialogue and descriptions of primary character action. It gives minimal visualization and sequencing instructions.

GROUP 5

AIR DATE: 07/15 4:00 P.M.

VIDEO	AUDIO
OPENING MONTAGE SERV 02 CLIP 9 SOS [sound on source] **00:25**	IN-CUE: "When you drive through this valley..." OUT-CUE: "...Group 5, a remarkable union of five world-renowned artists."
JULIA in Woodacre studio **00:15**	**JULIA (on-camera)** The founding members of Group 5—a painter, a singer, a potter, a documentary videographer, and a poet—did not know one another and certainly not that they all lived in the San Geronimo Valley in Marin County.
SERV 02 CLIP 10 PACKAGE 1 Julia V/O Valley shots **3:38**	**JULIA (V/O)** They moved there to get away from city life, to trade the city's nervousness for the calm of rolling hills, ancient oaks, and redwoods. An artists' guild was definitely not on their minds.
TALIA in studio SOS	IN-CUE: "No agents, no obligations..." OUT-CUE: "...until I ran into Mr. Video at the Forest Knolls post office."
JULIA in Woodacre studio	**JULIA (on-camera)** The Mr. Video is Phil Arnone, an award-winning documentary producer and video artist who sees the world with a child's curiosity and an artist's intensity. His world consists not of spectacular vistas but, much like Talia's paintings, high-energy close-up details.
SERV 02 CLIP 11 PACKAGE 2 SOS Phil in editing room **2:47**	IN-CUE: "Yes, I am a child when it comes to looking at things, at events happening..." OUT-CUE: "...Talia and I are definitely soul brother and sister."

3.2 STANDARD TWO-COLUMN A/V SCRIPT

In a two-column A/V script, the video instructions are page-left, and the audio instructions (including the dialogue) are page-right.

VIDEO	AUDIO
	KATY:
CU of Katy	But the debate about forest fires is still going on. If we let the fire burn itself out, we lose valuable timber and kill countless animals, not to speak of the danger to property and the people who live there. Where do you stand, Dr. Hough?
	DR. HOUGH:
Cut to CU of Dr. Hough	(SAYS THAT THIS IS QUITE TRUE BUT THAT THE ANIMALS USUALLY GET OUT UNHARMED AND THAT THE BURNED UNDERBRUSH STIMULATES NEW GROWTH.)
	KATY:
Cut to two-shot	Couldn't this be done through controlled burning?
	DR. HOUGH:
	(SAYS YES BUT THAT IT WOULD COST TOO MUCH AND THAT THERE WOULD STILL BE FOREST FIRES TO CONTEND WITH.)

3.3 PARTIAL TWO-COLUMN A/V SCRIPT
This script shows the video information in the left (VIDEO) column but only partial dialogue in the right (AUDIO) column. The host's questions are usually fully scripted, but the answers are only briefly described.

of an on-location reporter describing some disaster in progress; the news "packages" make up the recorded segments. SEE 3.4

SHOW FORMAT

The ***show format*** lists only the order of particular show segments, such as "interview from Washington," "commercial 2," and "book review." It also lists the set areas in which the action takes place (or other points of origination) as well as the clock and running times for the segments. A show format is frequently used in studio productions that have established performance routines, such as a daily morning show, a panel show, or a quiz show. SEE 3.5

FACT SHEET

A ***fact sheet***, or rundown sheet, lists the items that are to be shown on-camera and indicates roughly what should be said. SEE 3.6 The fact sheet is sometimes written in the

```
Hunter's Point Package      Noon News 04/15

Studio: KRISTI       ((Kristi))

                     A LANDLORD IN HUNTER'S POINT IS UNDER FIRE
                     FOR DANGEROUS LIVING CONDITIONS IN HIS
                     BUILDINGS. RESIDENTS COMPLAIN OF RASHES...
                     HEADACHES AND NOSEBLEEDS.

                     MARTY GONZALES ASKS SEVERAL TENANTS WHO SAY
                     THAT ALL OF THIS IS DUE TO TOXIC MOLD.

                     ---------------------------------------------

Package 1            ((In-cue: "There is no official confirmation
Video and Audio      that these buildings are infested with toxic
Server 03            mold, but it sure looks like it...))
Clip 023             PACKAGE          0:42
                     ((Out-cue:  "...I wish somebody would
                     do something about it."))

Studio: KRISTI       ((Kristi))

                     THE LANDLORD DENIES THESE CHARGES AND SAYS
                     IT MUST BE THE FOGGY WEATHER. WE'LL TALK TO
                     THE LANDLORD AND HEALTH OFFICIALS RIGHT
                     AFTER THESE MESSAGES.
                     ---------------------------------------------

Server 03
Clip 112             BUMPER
Clip 005             COMMERCIAL (California Cheese)
Clip 007             COMMERCIAL (Winston Enterprises)
```

3.4 NEWS SCRIPT
The news script contains every word spoken by the anchor (Kristi), except for the occasional chitchat, and instructions for all video sources used. A package is a previously shot and edited story that contains an on-location reporter and the people interviewed.

PEOPLE, PLACES, POLITICS SHOW FORMAT (Script attached)

PRODUCTION DATE: 2/3	FACILITIES REQUEST: BECA 415
AIR DATE: 2/17	RUNNING TIME: 25:30
DIRECTOR: Whitney	HOST: Kipper

OPEN

VIDEO	AUDIO
STANDARD OPENING SERVER 1, CLIP #ST1 EFFECTS 117	SOS [sound on source] ANNOUNCER: The Television Center of the Broadcast and Electronic Communication Arts Department, San Francisco State University, presents "People, Places, Politics" --a new perspective on global events.
KEY C.G. TOPIC TITLE	Today's topic is:

SERVER 1, CLIP #033 PSAs 1 & 2

OPENING STUDIO SHOT	PHIL INTRODUCES GUESTS
KEY C.G.	NAMES OF GUESTS
CUs OF GUESTS	GUESTS DISCUSS TOPICS
CU OF PHIL	CLOSES SHOW

SERVER 1, CLIP #034 PSAs 3 & 4

CLOSE

KEY C.G. ADDRESS	ANNOUNCER: To obtain a copy of today's program, write to "People, Places, Politics," BECA Dept., San Francisco State University, San Francisco, CA 94132 E-mail: BECA@sfsu.edu
KEY C.G. NEXT WEEK	Tune in next week when we present:
	THEME MUSIC UP AND OUT

3.5 SHOW FORMAT

The show format contains only essential video information in the left (VIDEO) column and the standard opening and closing announcements in the right (AUDIO) column.

```
VIDEO PRO CD-ROM COMMERCIAL
SHOW:
DATE:

PROPS:
Desktop computer running Zettl's VideoLab 3.0. Triple-I Web page.
Video Pro poster and multimedia awards in background.
Video Pro package with disc as hand props.

1.  New multimedia product by Image, Imagination, Incorporated.
2.  Sensational success. Best Triple-I product yet.
3.  Based on ZVL 2.1, which won several awards for excellence,
    including the prestigious Invision Gold Medal.
4.  Designed for the production novice and the video professional.
5.  Truly interactive. Provides you with a video studio in your
    home. Easy to use.
6.  You can proceed at your own speed and test your progress
    after each exercise.
7.  Will operate on Windows or Macintosh platform.
8.  Special introductory offer. Expires Oct. 20. Hurry. Available
    in all major software stores. For more information or the
    dealer nearest you, visit Triple-I's Web page at
    http://www.triple-i.tv.
```

3.6 FACT SHEET
The fact sheet lists the major points to be made about the demonstrated product. No specific video or audio information is given. The talent ad-libs the demonstration, and the director follows the talent's action with the camera.

A/V format, but no specific video or audio instructions are given. It is most often supplied by a manufacturer or an advertiser who wants the talent to ad-lib about the product.

If the demonstration of the item is complicated, the director may rewrite the fact sheet and indicate key camera shots to help coordinate the talent's and the director's actions. Unless the demonstration is extremely simple, such as holding up a book by a famous novelist, directing solely from a fact sheet is not recommended. Ad-libbing by both director and talent rarely works out satisfactorily, even if the video recording is intended for postproduction editing.

MAIN POINTS

- The script is one of the most important communication devices in all three production phases.
- The common script formats are the single-column drama script, the two-column A/V (audio/video) script, the partial two-column A/V script, the news script, the show format, and the fact, or rundown, sheet.

SECTION

3.2

Dramatic Structure, Conflict, and Dramaturgy

Even if you are not a writer, you must be able to judge the quality of a script. There are several classic books on dramatic and nondramatic scriptwriting, so the information in this section is purposely kept to a minimum.[1]

▶ **STRUCTURAL COMPONENTS**
The four basic elements of drama

▶ **CONFLICT AND CLASSICAL DRAMATURGY**
Types of dramatic conflict, classical dramaturgy, and event order

▶ **NONDRAMATIC STORY STRUCTURE**
Goal-directed programs; from idea to process message: feature story; and from idea to process message: goal-directed program

▶ **BASIC DRAMATIC STORY STRUCTURE**
A good drama operates on many conscious and unconscious levels, all of which must be made explicit by the writer and, especially, the actors.

STRUCTURAL COMPONENTS

All dramas for the stage or screen have four basic elements:

- Theme—what the story is about
- Plot—how the story moves forward and develops
- Characters—how one person differs from the others and how each reacts to the situation at hand
- Environment—where the action takes place

The plot can develop from outside-in or from inside-out. A from-outside-in story has the characters react to outside influences, such as natural disasters, nasty neighbors, or the loss of a job. A from-inside-out story has the characters' behavior and choices determine the way the plot progresses. Large-screen cinema is generally driven by spectacular from-outside-in plots but has also proved that it can deal with character-based stories. Plot-based stories often emphasize outer action (such as car chases, explosions, and fights). Character-based stories thrive on inner action—the characters' psychological makeup and subsequent behavior.

Although you may not see much evidence in the daily program fare, by its very nature the television medium is ideally suited to explicate low-key, character-based, from-inside-out stories. We usually watch television alone or with family members in our home and consider the small screen a welcome and familiar—if not intimate—communication partner. This is why politicians and advertisers spend so much money on using television to deliver their messages to your home.

CONFLICT AND CLASSICAL DRAMATURGY

All drama thrives on conflict. Without conflict you have no drama. In fact, all good stories are based on some form of conflict. In a drama series, such as a soap opera or crime show, each episode solves a number of large and small problems but always ends with a new conflict to be solved in a subsequent show. Granted, this is a rather clichéd way of luring the audience back week after week, but it works.

Types of Dramatic Conflict

In the plot-based approach, the conflict grows out of circumstances that make the viewer react in a specific way.

For example, person A becomes a politician because, having been a football star, he has wide name recognition and, after marrying into a wealthy family, is urged by his father-in-law to run for an open Senate seat. His wife makes a compelling argument for how he can single-handedly change the Senate majority and bring the country back to honesty and family values. He wins but is horrified by the

1. As cited in chapter 2, the two most valuable books I've found on the art of writing for the media are: Stuart W. Hyde, *Idea to Script: Storytelling for Today's Media* (Boston: Allyn and Bacon, 2003); and Nancy Graham Holm, *Fascination: Viewer Friendly TV Journalism* (Århus, Denmark: Ajour Danish Media Books, 2007). The Hyde book contains an exceptionally lucid approach to conceptualizing a story and the various structures of dramatic and nondramatic scripts. The small Graham Holm book is a journalistic jewel. Written with conviction and wit, it conveys an amazing amount of solid information that goes way beyond the average "how-to-do-news" approach.

way the Senate makes decisions. When trying to clean up their act, he runs into a series of conflicts, gets into a mysterious car accident, and loses his arm, job, wife, friends, and self-confidence—in that order.

In the character-based approach, the character's initial decisions create the conflict with which he or she must deal. When looking at a script of a good teleplay or movie, you should be able to identify several layers of conflict—some more subtle, others quite conspicuous.

For example, person B, a physician, becomes more and more frustrated with the way the Senate ignores the wishes of its constituency, and she decides to do something about it. She quits her hospital job, starts a grassroots senatorial election campaign, and gives speeches about universal healthcare. She wins by a slim margin and, like person A, is horrified by the way the Senate makes decisions. She is immediately chastised for voting not by party lines but by what she believes to be best for people. She too has a mysterious car accident and finds herself in her former hospital, this time as a patient. She is confronted in various ways by former colleagues, who suspect that her motivation for changing professions was power rather than altruism. When she is finally well again, she wants to return to the medical profession, but her loyal husband convinces her that she is needed in the Senate and that she has already made a big difference for the people she represents. So she reenters politics, uses the media to expose some of the ways senators operate, cleans up their act, introduces a new national healthcare system, and lives happily ever after.

As you can see, person A's conflicts were basically caused by external circumstances (what other people told him to do or arranged for him), whereas person B's conflicts arose from her own decisions and convictions.

When you look at television plays and movies more carefully and identify the conflicts and how they come about, you will discover that the conflicts and how they are introduced are quite similar. You can probably come up with a few pages of conflicts with little effort, so consider the following list of possible dramatic conflicts merely examples. SEE 3.7

Classical Dramaturgy

Effective storytelling hasn't really changed much over the centuries. We still use all or most of the ingredients

3.7 DRAMATIC CONFLICTS

Though not exhaustive by any means, this list of conflicts is intended as an aid in recognizing possible conflict formulas in scripts and as a stimulus for coming up with similar lists.

GENERAL AREAS OF CONFLICT

- Cultural and ideological differences
- Ideal versus rational worldviews
- Feeling versus reason
- Wishful thinking versus reality
- Society versus the individual
- Honesty versus corruption
- Rich and poor

SPECIFIC AREAS OF CONFLICT

The conflict between the protagonist (who could be you) and:

- Various circumstances
- Environment
- Nature
- Himself/herself (yourself)
- Family members
- Relatives
- Friends
- Neighbors
- Peer pressure
- Classmates
- Conformity
- Lifestyle
- Authority
- Bureaucracy
- Power
- Faith
- Principles
- Morals
- Beliefs
- Religion
- Social system
- Government

of the ***classical dramaturgy***—the technique of dramatic composition:

- Exposition—sets the context in which the action occurs
- Point of attack—the first crisis
- Rising action—more conflicts and trouble brewing
- Climax—major crisis or turning point
- Falling action and resolution (hero runs into more trouble and either wins or loses) SEE 3.8

Go back to the treatment of the story in chapter 2 (figure 2.3) and see how, if at all, this treatment conforms to the classical dramatic structure.

Exposition (context): airport, professor on way to conference

Point of attack (first crisis): suitcase missing

Rising action (series of minor conflicts): forgets name of hotel, has to take a cab to hotel in bad neighborhood

Climax (major crisis): wallet missing

Falling action: several more conflicts—difficulty getting help, trouble finding a telephone, reaches only friend's answering machine, lost in strange city

Resolution: police come to the rescue, find colleague in lobby of conference hotel

As you can see, the five major criteria of the classical dramaturgy are included in the proposed dramatic program—and it has a good ending (the hero is rescued). If you wanted to make a tragedy out of it, you could have him get mugged and then beaten up because the villains could not find any money on him, miss his conference, and, as a result, lose his promotion. For good measure, his new marriage is also getting stressed. The hero is doomed.

This classical dramaturgy applies to good storytelling in all media—television and movies included. These media, however, have drastically altered the traditional idea of event order.

Event Order

The ease with which we can shift past, present, and future through editing has obliterated the myth that a good story must have a neat beginning, middle, and end and has redefined the concept of ***event order*** in storytelling.

3.8 CLASSICAL DRAMATURGY

The classical dramaturgy moves from the exposition (context) to the point of attack (first crisis) through rising action and a climax (major crisis) to falling action and the resolution.

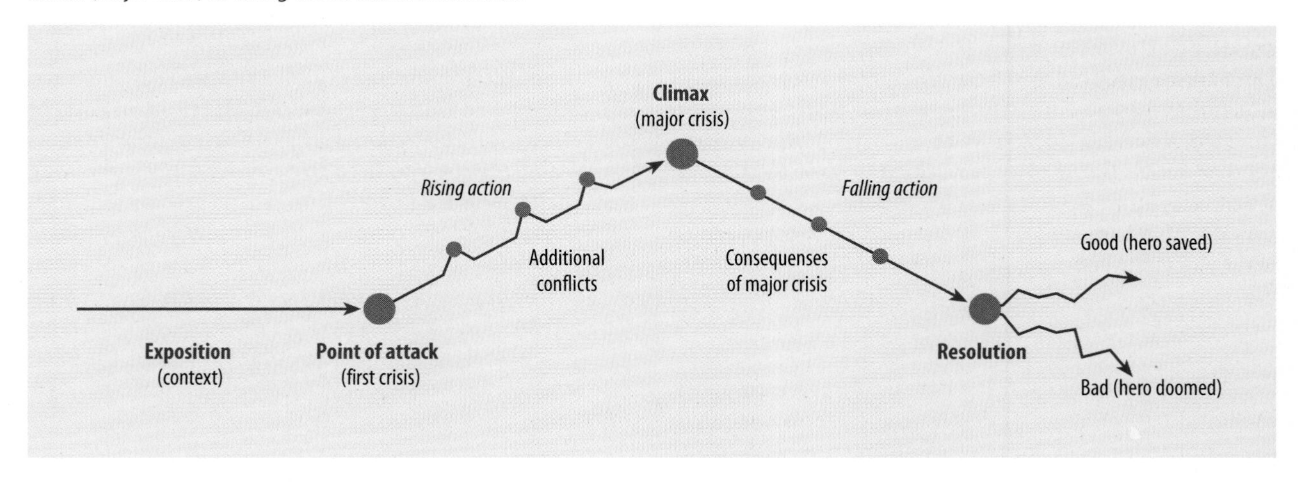

Through flashbacks you can "flash" past events onscreen while showing the present; and with flash-forwards, you can begin a story with how it ended and then follow up with what led to the conclusion. Even during a live telecast of a baseball game, you can interrupt the present and recall past moments with instant replays. Such instant replays have become an important dramaturgical ingredient in television drama. And, as you well know, soap operas and crime series never have a clean ending but entice you with an unresolved crisis to tune in again.

NONDRAMATIC STORY STRUCTURE

Nondramatic stories include everything from a news report and a documentary to a product commercial and a complex instructional video on some medical procedure. All are nonfiction and principally informational.

Goal-directed Programs

Whereas news stories and documentaries are based on the canons of storytelling and a simplified dramaturgy, commercials and instructional programs purposely comprise ***goal-directed information*** that is not bound by storytelling conventions. They are constructed by motivational and learning objectives and are calculated to have an observable, if not measurable, effect on the viewer. All programs benefit from a process message, but goal-directed programs depend on one.

From Idea to Process Message: Feature Story

Here are two examples of how different process messages can influence the concept of a feature that is not goal-directed. The basic idea is to do a story on your favorite football team. But exactly how do you want to show it? What is it you want viewers to see and feel? Let's write two process messages that will streamline your approach to what you show and how you show it.

Process message 1: *The program should make viewers feel the immense physical power of a football game.*

When translating this process message into actual medium requirements, you will need to show very tight shots of tackles. Can you do this during a regular game? Probably not. You would be better off doing it during a practice session, where you can use portable camcorders and work at close distances. Sound becomes very important. Hearing the impact will make us feel the power more than just seeing the players crash into each other. The event order is unimportant—you simply show highlights.

Process message 2: *The program should make the viewer admire the inherent beauty of a football game.*

Now you need to concentrate on the agility of well-trained athletes. Again, you probably need to record a practice session rather than an actual game, unless you have the time and the opportunity to video-record several actual games. Shots will certainly include the quarterback throwing the ball and the wide receiver catching it. Now you can even use special effects in postproduction, such as slow motion, double exposures of the receiver extending his reach to catch the game-winning pass, or, if you want to go all out, a superimposition of ballet dancers doing similar jumps.

From Idea to Process Message: Goal-directed Program

This type of program needs a highly precise process message that, much like instructional objectives, must be stated so that the results can be observed and verified.

Here is a process message for a fictional *Sesame Street* episode:

Process message: *After having watched this program, preschool children will be able to count from 1 to 5.*

The program must now fulfill a specific learning task. You can proceed with preproduction and think of content that fulfills the three credos of advertising: attention, redundancy, and variety. You need to be brief, show something that catches immediate attention (such as a cartoon character), and repeat the numbers in a variety of ways—have somebody scratch them in the sand, have them appear in the squares of a sidewalk game, have five toy cars pop up sequentially, and so forth.

In an actual instructional program in school, the message received by the students can be verified through a test and then compared with the defined process message. The closer they match, the more successful the program.

MAIN POINTS

- The four basic components of a drama are theme, plot, characters, and environment.
- All drama thrives on conflict. Conflict can be primarily outer-oriented, or plot-based, in which the circumstances prompt the character's reaction; or it can be inner-oriented, or character-based, in which the character causes the circumstances.
- The five components of the classical dramaturgy are exposition (context), point of attack (first crisis), rising action, climax (major crisis), and falling action and resolution.
- The event order of a story—beginning, middle, and end—can be changed in recorded television and film.
- Nondramatic goal-directed programs require a clear process message for production and evaluation.

ZETTL'S VIDEOLAB

For your reference or to track your work, the *Zettl's VideoLab* program cue in this chapter is listed here with its corresponding page number.

ZVL1 PROCESS→ Ideas→ scripts 44

CHAPTER

4

The Director in Preproduction

Now that you have a workable script, it is time for you as the producer to involve the director in the preproduction phase. In our simulated production environment, this means you must now switch hats and temporarily become the director.

As a director you tell talent and the entire production team what to do before, during, and after the actual production. But before you can tell *them* what to do, you obviously need a clear idea of what *you* need to do: think about what the program should look like and how to get from the idea to the television image.

It is usually the script that starts such a preproduction process in earnest; but regardless of what the script says for you to do—cover a basketball game, do a documentary on college football, or restage a play by the drama department for the local cable channel—your job as director will inevitably start with a tentative visualization process, various key images and sounds that pop into your head while reading the script. Your next steps will be to establish a process message that will make that visualization less arbitrary and to translate your visualization into the necessary medium requirements. All of this may initially seem like a waste of time, but only through effective preproduction will you be able to coordinate consistently the great variety of production elements with confidence, authority, and style. ZVL1 PROCESS→ Process introduction

Section 4.1, How a Director Prepares, looks at how to ensure that you as the director know what the show is all about, analyzing and visualizing the script, and some essential steps in preparing for the show. Section 4.2, Communication and Scheduling, addresses the functions of the support staff, the need to double-check the producer's facilities requests and schedules, and how to communicate with talent and crew.

The director's activities in the production and postproduction phases are the focus of chapter 17.

KEY TERMS

AD Stands for *associate director* or *assistant director.* Assists the director in all production phases.

DP Stands for *director of photography.* In major motion picture production, the DP is responsible for the lighting (similar to the LD in television). In smaller motion picture productions and in EFP, the DP will operate the camera. In television it refers to the camera operator, or shooter.

facilities request A list that contains all technical facilities needed for a specific production.

floor plan A diagram of scenery and properties drawn on a grid pattern.

location sketch A rough map of the locale of a remote shoot. For an indoor remote, it shows the room dimensions and the furniture and window locations. For an outdoor remote, it indicates the location of buildings, the remote truck, power sources, and the sun during the time of the telecast.

locking-in An especially vivid mental image—visual or aural—during script analysis that determines the subsequent visualizations and sequencing.

medium requirements All content elements, production elements, and people needed to generate the defined process message.

process message The message actually perceived by the viewer in the process of watching a television program. The program objective is the defined process message.

production schedule The calendar that shows the preproduction, production, and postproduction dates and who is doing what, when, and where.

sequencing The control and the structuring of a shot series during editing.

storyboard A series of sketches of the key visualization points of an event, with the corresponding audio information.

time line A breakdown of time blocks for various activities on the actual production day, such as crew call, setup, and camera rehearsal.

visualization Mentally converting a scene into a number of key video images and sounds, not necessarily in sequence. The mental image of a shot.

SECTION

4.1

How a Director Prepares

Although some directors think that their profession requires a divine gift, most good directors acquired and honed their skills through painstaking study and practice. One of the more difficult ideas to get across to the aspiring television director is the need for meticulous preproduction. Unless you act as a team of one, in which you take on the roles of producer, director, crew, as well as postproduction editor, you need careful preparation of your production before you ever set foot in the studio or drive to the location shoot. Such preproduction can be as or even more demanding than directing the production itself. Diligent preproduction minimizes—and often eliminates—the chance of failure.

The actual preproduction activities vary greatly, depending on the specific show you are asked to direct, but they all include your understanding what the show is all about, translating the script into specific visual and aural images, preparing for the show, and communicating your creative vision to the rest of the production team.

- **WHAT THE SHOW IS ALL ABOUT**
  Process message and production method
- **SCRIPT ANALYSIS**
  Locking-in point and translation
- **VISUALIZATION AND SEQUENCING**
  Formulating the process message and determining the medium requirements
- **PREPARING FOR THE SHOW**
  Floor plan, location sketch, storyboard, and marking the script

WHAT THE SHOW IS ALL ABOUT

Before you can shout "lights, camera, action" or even begin to mark the script for your live or live-recorded show, you need to find out what the show you are to direct is all about. As obvious as this may sound, you will find that as a staff director (who directs a daily variety of shows that happen in a particular time slot) you don't always have the time to carefully inquire about every detail of each show. You normally don't have time to do much preparing for a daily news show, except for marking the script at the last minute. Nor do you have to do a thorough script analysis of a daily postgame sports segment that is part of a routine morning show. This does not mean, however, that you should "wing" a show or step into the control room or a remote truck unprepared.

If you have to direct a show that is not part of a daily routine, you must try to formulate the process message and decide on the most effective production method.

Process Message

Before you do anything, revisit the ***process message***—the purpose of the show and its intended effect on a specific audience (see chapter 1). If you are not quite sure what the show is to accomplish, check with the producer. Only then can you make all the other personnel understand what the show is about and the expected outcome of the production. An early agreement between producer and director about specific communication goals and production type and scope can prevent many frustrating arguments and costly mistakes. Keep the producer apprised of your plans, even if you have been given responsibility for all creative decisions. Keep a record of telephone calls, save your e-mail, and follow up on major verbal decisions with memoranda.

Production Method

If you thoroughly understand the process message, the most appropriate production method becomes clear—that is, whether the show is best done in the studio or in the field, live or live-recorded, single-camera or multicamera, and in a sequential or nonsequential event order.

If, for example, the process message is to help the viewer participate in the excitement of watching a Thanksgiving Day parade, you need to do a live, multicamera remote in the field. A traffic safety segment on observing stop signs may require a single-camera approach and plenty of postproduction time. To help the audience gain a deeper insight into the thinking and the work habits of a famous painter, you might observe the painter in her studio over several days with a single camcorder and then edit the video-recorded material in postproduction. If the viewers are to share in the excitement of the participants and the studio audience in a new talent show and are encouraged to call in while the show is in progress, you must obviously do a live, multicamera studio production. Also the producer must provide extensive telephone feedback opportunities for the viewers.

But a clear process message might also suggest whether your approach should be a "looking-at" or "looking-into" one. Looking *at* an event means you observe it as best you can, such as objectively covering a city council meeting; looking *into* an event, on the other hand, means having the camera reveal the emotional impact through a series of close-ups and a close sound presence.[1]

If you are also the writer, such considerations must happen while writing the script. If you get a script for a show from the producer, you need to learn what to look for to make your directing reflect the intended essence of the show.

SCRIPT ANALYSIS

Because explaining the intricacies of analyzing and interpreting nondramatic and dramatic scripts is beyond the scope of this book, the following list offers some basic guidelines on reading a script as a director. This technique is especially useful when reading and visualizing a dramatic script.

Locking-in Point and Translation

Locking-in means that you conjure up a vivid visual or aural image while reading the script. This locking-in may well occur at the very opening scene, at the closing scene, or at any particularly striking scene in between. Do not try to force this locking-in process. It may well occur as an audio rather than a video image. If the script is good, the locking-in is almost inevitable. Nevertheless, there are a few steps that will expedite the process.

- Read the script carefully—don't just skim it. The video and audio information provide an overview of the show and how complex the production will be.

- Try to isolate the basic idea behind the show. Better yet, try to formulate an appropriate process message if it isn't already stated in the original proposal (usually as the "program objective.")

- Try to lock-in on a key shot, a key action, or some key technical maneuver. This may give you some idea of how to translate the images into concrete production requirements, such as camera positions, specific lighting and audio setups, video recording, and postproduction activities.

Analyzing a dramatic script is, of course, more complicated than translating the video and audio instructions of a nondramatic script into the director's production requirements. A good dramatic script operates on many conscious and unconscious levels, all of which need to be interpreted and made explicit. Above all, you should be able to define the theme of the play (the basic idea—what the story is all about), the plot (how the story moves forward and develops), the characters (how one person differs from the others and how each reacts to the situation at hand), and the environment (where the action takes place). In general, television drama emphasizes theme and character rather than plot, and inner, rather than outer, environment. Isolate all points of conflict.[2]

After the locking-in, further analysis depends greatly on what production method you choose: whether you shoot the play in sequence with multiple cameras and a switcher or with a single camera in discontinuous, out-of-sequence takes. Regardless of the method, you will need to visualize the script before deciding on location and equipment.

VISUALIZATION AND SEQUENCING

Directing starts with the visualization of the key images. ***Visualization*** means seeing the script in pictures and hearing the accompanying sounds. There is no sure-fire formula for this translation process; it requires a certain amount of imagination and artistic sensitivity and lots of practice. The best way to practice is to carefully observe the events around you—how people behave in a classroom or a restaurant or on a bus or an airplane—and mentally note what makes one event so different from the others. When you read a description of some happening in a newspaper, magazine, or novel, try to visualize it as screen images and sound.

Because the viewer sees only what the camera sees, you need to carry the initial visualization further and translate it into such directing details as where people and things should be placed relative to the camera and where the camera should be positioned relative to the event (people and things). You must then consider the ***sequencing*** of the

1. See Herbert Zettl, *Sight Sound Motion: Applied Media Aesthetics*, 6th ed. (Boston: Wadsworth, 2011), pp. 208–12.

2. See Zettl, *Sight Sound Motion*, pp. 278–79.

portions of this visualized event through postproduction editing or switching. Concurrently, you must hear the individual shots and the sequence.

Formulating the Process Message

As mentioned, a carefully defined process message facilitates the visualization process and, especially, makes it more precise. After deciding on what the target audience is to see, hear, feel, or do, you can follow the effect-to-cause model and determine just how the key shots should look and how to accomplish them.

Here is an example: You are to direct three segments of a program series on teenage driving safety. The first assignment is an interview, consisting of a female interviewer who regularly hosts the weekly half-hour community service show, a male police officer who heads the municipal traffic safety program, and a female student representative of the local high school. The second assignment is an interview with a male high-school student who has been confined to a wheelchair since a serious car accident. The third is a demonstration of some potential dangers of running a stop sign.

The scripts available to you at this point are very sketchy and look more like fact sheets than partial two-column A/V scripts. SEE 4.1–4.3

Because the producer has an unusually tight deadline for the completion of the series, she asks that you get started with preproduction despite the lack of more detailed scripts. She can give you only a rough idea of what each show is supposed to accomplish: segment 1 should inform the target audience (high-school and college students) of the ongoing efforts by the police department to cooperate with schools to teach traffic safety to young drivers; segment 2 should shock the viewers into an awareness of the consequences of careless driving; and segment 3 should make the audience aware of the potential dangers of running a stop sign.

Let's apply the effect-to-cause model and see how these scripts can be translated into video programs. (To refresh your memory of the effect-to-cause model, see chapter 1.) ZVL2 PROCESS→ Effect-to-cause→ basic idea | desired effect | cause

Despite the sketchy scripts and process messages, many images have probably entered your head already: the police officer in his blue uniform sitting next to the high-school student; a young man straining to move his wheelchair up a ramp to his front door; a car almost hit in an intersection by another car running a stop sign. Before going any further, you may want to define more precise process messages.

Process message 1: *The interview with the traffic safety officer and the student representative should demonstrate to high-school and college students a 10-point traffic safety program to help teenagers become more responsible drivers. It should also demonstrate how police and students could cooperate in this effort.*

Process message 2: *The interview with the student in the wheelchair should make viewers (in the desired target audience) gain a deeper insight into his feelings and attitudes since his accident and empathize with him.*

Process message 3: *The program should show viewers at least four different accidents caused by running a stop sign and demonstrate how to avoid them.*

A careful reading of these objectives should make your visualization a little more precise. For example, just how do you see the three people (host, officer, and student) interacting in the interview? What shots and shot sequences do you feel would best communicate the interview to the audience? Do you visualize a different approach to the interview with the student in the wheelchair? The demonstration of running a stop sign probably triggers some stereotypical Hollywood video and audio images (such as glass shattering, tires squealing, cars spinning, some subjective driver's points-of-view, and so forth).

Determining the Medium Requirements

Without trying to be too specific, you can now proceed from some general visualizations to the ***medium requirements:*** certain key visualizations and sequencing, production method (multicamera studio show or single-camera EFP), necessary equipment, and specific production procedures.

Here is how you might arrive at specific medium requirements for each segment (process message).

Segment 1 The interview is strictly informational. What the people say is more important than getting to know them. The student may not always agree with the officer's views, so the two may not only answer the interviewer but also talk to each other. According to the sketchy script, the officer's 10-point program on traffic safety and other items

TRAFFIC SAFETY SERIES

Program No: 2 Interview (Length: 26:30)
VR Date: Saturday, March 16, 4:00–5:00 P.M. STUDIO 2
Air Date: Tuesday, March 19

Host: Yvette Sharp
Guests: Lt. John Hewitt, head of traffic safety program, City Police Department
Rebecca Child, senior and student representative, Central High School

---

VIDEO	AUDIO
STANDARD OPENING	
3-shot host & guests	
CU of host (faces camera)	HOST INTRODUCES SHOW
2-shot of guests	HOST INTRODUCES GUESTS
CU of host	HOST ASKS FIRST QUESTION

INTERVIEW: Lieutenant John Hewitt is the officer in charge of the traffic safety program. Is a 20-year veteran of the City Police Department. Has been in traffic safety for the past eight years.

NOTE: HE WILL REFER TO A 10-POINT PROGRAM (DISPLAY VIA C.G.).

Rebecca Child is the student representative of Central High. She is an A student, on the debate team, and on the championship volleyball team. She is very much in favor of an effective traffic safety program but believes that the city police are especially tough on high-school students and are out to get them.

STANDARD CLOSE

CU of host	HOST MAKES CLOSING REMARKS
LS of host and guests	THEME
C.G. credits	

4.1 TRAFFIC SAFETY STUDIO INTERVIEW
This script is written as a show format. It is intended to give only basic information about the guests appearing on the show.

TRAFFIC SAFETY SERIES

Program No: 5 Location Interview (Length: 26:30)
EFP Date: Friday, March 26, 9:00 A.M.–all day
Postproduction to be scheduled
Air Date: Tuesday, April 6

Interviewer: Yvette Sharp
Interviewee: Jack Armstrong
Address: 49 Baranca Road, South City
Tel.: 990 555-9990

OPENING AND CLOSING ARE TO BE DONE ON-LOCATION

Jack is a high-school senior. He has been confined to a wheelchair since he was hit by a car running a stop sign. The other driver was from his high school. Jack was an outstanding tennis player and is proud of the several trophies he won in regional tournaments.
He is a good student and coping very well. He is eager to participate in the traffic safety program.

NOTE: EMPHASIS SHOULD BE ON JACK. GET GOOD CUs.

4.2 TRAFFIC SAFETY FIELD INTERVIEW
The location interview is also sketched out in a show format, which simply gives information about the guest to be interviewed.

TRAFFIC SAFETY SERIES

Program No: 6 Running Stop Signs (Length: 26:30)
EFP Date: Wednesday, April 7, 7:00 A.M.–all day
VR Date: Saturday, May 1, 4:00 P.M.–4:30 P.M.
Postproduction to be scheduled
Air Date: Saturday, May 15

EFP Location: Intersection of West Spring Street and Taraval Court

Contact: Lt. John Hewitt, head of traffic safety program, City Police Department
Tel.: 990 555-8888

OPENING AND CLOSING (YVETTE) ARE TO BE DONE ON-LOCATION

EFP: Program should show car running a stop sign at intersection and the consequences: almost hitting a pedestrian, jogger, bicycler; running into another car, etc. Detailed script will follow.

STUDIO: Lt. Hewitt will briefly demonstrate some typical accidents, using toy cars on a magnetic board.

NOTE: LT. HEWITT WILL PROVIDE ALL VEHICLES AND DRIVERS AS WELL AS TALENT. HE WILL TAKE CARE OF ALL TRAFFIC CONTROL, VEHICLE PARKING, AND COMMUNICATIONS. CONFIRM EFP APRIL 5.

ALTERNATE POLICE CONTACT: Sgt. Fenton McKenna (same telephone)

4.3 TRAFFIC SAFETY STOP SIGN SCENE

This show format gives basic information about the major events to be recorded at the electronic field production (EFP). All three traffic formats must eventually be scripted in a partial A/V format before the actual studio and field productions.

should be shown on-screen as C.G. (character generator) graphics, unless he brings an easel card.

The show is obviously best done as a live recording in the studio. There you can put the guests in a neutral environment, have good control over the lighting and the audio, switch among multiple cameras, and use the C.G.

The lighting should be normal; there is no need for dramatic shadows. Perhaps you can persuade the police officer to take off his cap to avoid annoying shadows on his face. How about cameras? Three or two? Even a lively exchange of ideas between the officer and the student will not require terribly fast cutting. Assuming that the host and the guests sit across from one another, you really need only two cameras. We will guide you through a possible camera setup as soon as we get a rough sketch of the set from the art director.

Segment 2 In contrast to segment 1, the segment 2 interview is much more private. Its primary purpose is not to communicate specific information but to have an emotional impact on the audience. The communication is intimate and personal; viewers should strongly empathize with the young man in the wheelchair. These aspects of the process message suggest quite readily that we visit him in his own environment—his home—and that, except for the opening shots, we see him primarily in close-ups and extreme close-ups rather than in less intense medium and long shots. Again, you will inevitably visualize certain key shots that you have called up from your personal visual reservoir.

Considering the major aspects of the process message (revealing the student's thoughts and feelings and having the audience develop empathy), the general production type and the specific medium requirements become fairly apparent. This is best done single-camera style in the student's home. First, the single camera and the associated equipment (lights and mics) cause a minimum intrusion into the environment. Second, the interview itself can be unhurried and stretch over a considerable period of time. Third, the interview does not have to be continuous; it can slow down, be briefly interrupted, or be stopped and then picked up at any time. The production can be out of sequence. Compared with segment 1, this production needs considerably more editing time.

To facilitate your visualization and sequencing, try to visit the student in his home prior to the EFP. Meeting the student and getting to know him will give you a sense of the atmosphere, enable you to plan the shots more specifically, and more accurately determine the specific medium requirements.

Segment 3 This production is by far the most demanding of you as a director. It requires the coordination of different people, locations, and actions. Start with some key visualizations. Running a stop sign is obviously best shown by having a car actually do it. To demonstrate the consequences of such an offense, you may need to show the car going through the stop sign, barely missing a pedestrian and a bicyclist who happen to be in the intersection, or even crashing into another car.

Now is the time to contact the producer again and ask her some important questions: Who will provide the vehicles for this demonstration? Who drives them? Who will be the stunt people acting as pedestrian and bicyclist? What about insurance?

At this point you would be better off abandoning the project and asking the producer to have it done by a company that specializes in such productions. In any case, it looks as though this is something you and your station may not be prepared to do. You could, however, suggest simulating these close-call actions by concentrating on the reaction of an onlooker rather than by showing the actual close calls or the crash itself. You can build up to the crash by showing the car going through the stop sign (provided the police are controlling the traffic) and then showing the frightened pedestrian jumping back onto the curb and the bicyclist trying to get out of the way (of the imagined oncoming car). A fast zoom-in on the car while it is moving toward the camera will definitely lead to an intensification of the shot and an exciting sequence when intercut with progressively closer shots of the pedestrian's frightened face. The crash itself is implied by the close-up (CU) of the frightened onlooker and a good dose of crash sounds from a sound-effect library.

In any case, you would need several camcorders capturing the same event from different points of view. Although most of the sound effects can be added in postproduction, you should still record the ambient sounds with the camera mic during the outdoor shoot.

As you can see, even the "simple" version of the crash is quite complicated. Perhaps you should simply skip this portion and ask the producer to get some stock footage of a car accident. If not, you can intercut the police officer's magnetic board demonstrations with portions of Jack's description of what he felt during the crash.

PREPARING FOR THE SHOW

Once you know what the show is all about and have a fairly well-defined process message, you can move to the next production step for the director: interpreting the floor plan or location sketch. Studying the floor plan or location sketch will help you determine the talent's principal movements, place the cameras and the microphones, and mark the script accordingly.[3]

Interpreting a Floor Plan

Unless you direct a routine studio production that occurs on the same set, such as a news, interview, or game show, you need a ***floor plan***—a diagram of scenery and properties drawn on a grid—to visualize the shots and translate them into camera positions and traffic patterns. It also influences, and sometimes dictates, how you block the talent.

With some practice you can do almost all the talent blocking and the camera positioning simply by looking at the floor plan or location sketch. You will also be able to spot potential production problems. Let's look at a simple interview floor plan and the location sketch for the stop sign segment and see whether we can isolate some potential production problems before they actually occur.

Interview floor plan Let's go back to the first segment—the studio interview with the police officer and the student representative. The art director has given you a rough sketch of the interview set and has asked you to approve it before he has his assistant prepare the final version. SEE 4.4

This interview setup seems quite workable. You can have camera 2 take care of the establishing three-shot and then move it to the right for CUs of the host. Camera 1 can be used on the guests. The line monitor is oriented toward the host so that the guests are not tempted to admire themselves on TV during the recording.

When the assistant art director hands you the final, neat version of the floor plan and the attached prop list, however, you detect some serious production problems. SEE 4.5 What are they?

Take another look at the floor plan and try to visualize some of the key shots, such as opening and closing three-shots, two-shots of the guests talking to the host and to each other, and individual CUs of the three people. Visualize the foreground as well as the background because the camera sees both. You may surmise that there are some definite camera problems with this floor plan. SEE 4.6

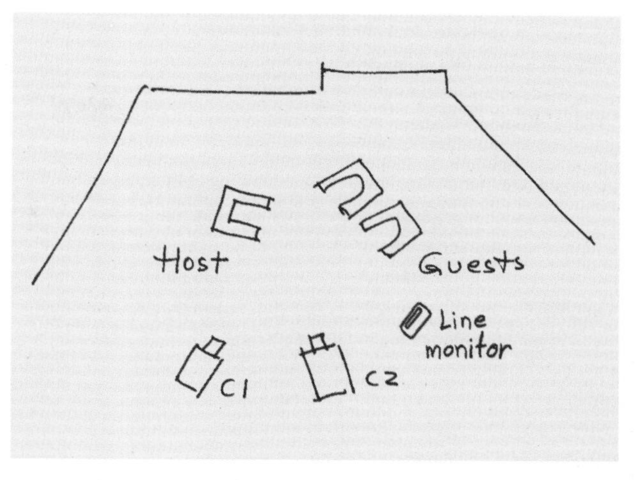

4.4 TRAFFIC SAFETY INTERVIEW: ROUGH SKETCH
This rough sketch for a studio interview set shows the approximate locations of the chairs and the cameras.

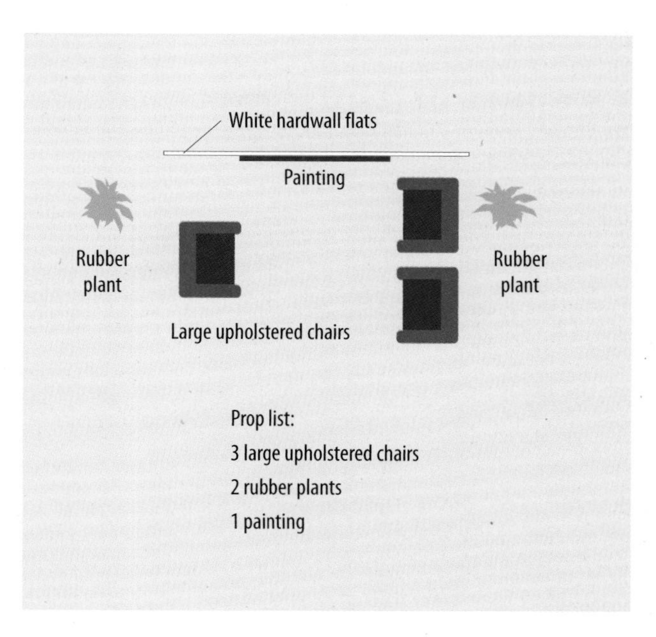

4.5 INTERVIEW SET: FLOOR PLAN AND PROP LIST
This floor plan and prop list, based on the rough sketch of an interview set, reveal serious production problems.

3. For more information about floor plans and scene design, see chapter 15.

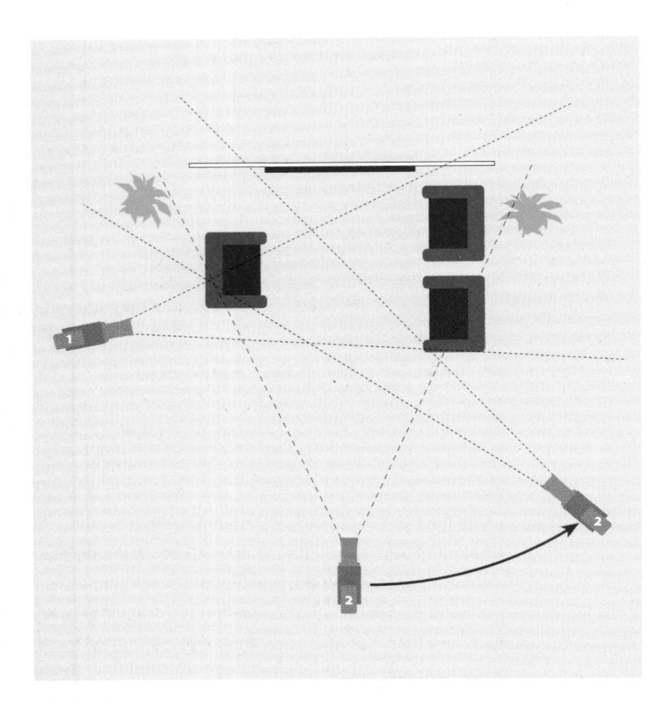

4.6 INTERVIEW SET: CAMERA POSITIONS
The camera positions reveal some of the production problems caused by this setup.

- Given the way the chairs are placed, an opening three-shot would be difficult to achieve. If camera 2 shoots from straight on, the chairs are probably too far apart. Even when using a wide-screen format, the host and the guests would seem glued to the screen edges, placing undue emphasis on the painting in the middle. Also, you would probably overshoot the set on both ends. The guests would certainly block each other in this shot.

- When you shoot from the extreme left (camera 1) to get a shot of the guests over the host's shoulder, you will overshoot the set. On a close-up you would run the risk of the rubber plant's seeming to grow out of the guest's head.

- When switching to camera 2, you will again overshoot the set, and the second rubber plant would most likely appear to grow out of the host's head (see figure 4.6).

- If you pulled the cameras more toward the center to avoid overshooting, you would get nothing but profiles.

Aside from issues with camera shots, there are additional production problems:

- White hardwall panels hardly create the most interesting background. The surface is too plain, and its color is too bright for the foreground scene, rendering skin tones unusually dark. Because the host is an African-American woman, the contrast problem with the white background is even more extreme—and you cannot correct the problem by putting more light on her.

- See how close the chairs are to the background flats? Such a setup would make proper lighting quite difficult. Once you have read chapter 12 on lighting, you may want to revisit this section and see why the setup makes lighting so difficult. Here are some clues: Any key light and fill light will inevitably strike the background too, adding to the silhouette effect. The back lights would also function as front (key) lights, causing fast falloff (dense attached shadows) toward the camera side. If you were now to lighten up the shadows on the faces with additional fill light coming from the front of the set (roughly from camera 2's position), it would inevitably hit the white flats, further aggravating the silhouette effect.

- The acoustics may also prove to be less than desirable because the microphones are very close to the sound-reflecting hardwall flats.

- The prop list signals yet more problems. The large upholstered chairs are definitely not appropriate for an interview. They look too pompous and would practically engulf their occupants.

- The painting on the back wall is seen only briefly in the opening shot and will not help make the background for the people shots more interesting.

- Finally, with the chairs directly on the studio floor, the cameras would have to look down on the performers, or the camera operators would have to pedestal all the way down and stoop for the entire interview.

We went into this detail to illustrate how a careful reading of a floor plan can uncover production problems that, if not discovered until the actual recording, would eat up valuable studio time. As you can see, even this simple floor plan and prop list reveal important clues to a variety of production problems.

The figure at the top of the facing page shows one possible solution to the aforementioned problems. You can probably come up with several more suggestions for an improved interview set. SEE 4.7

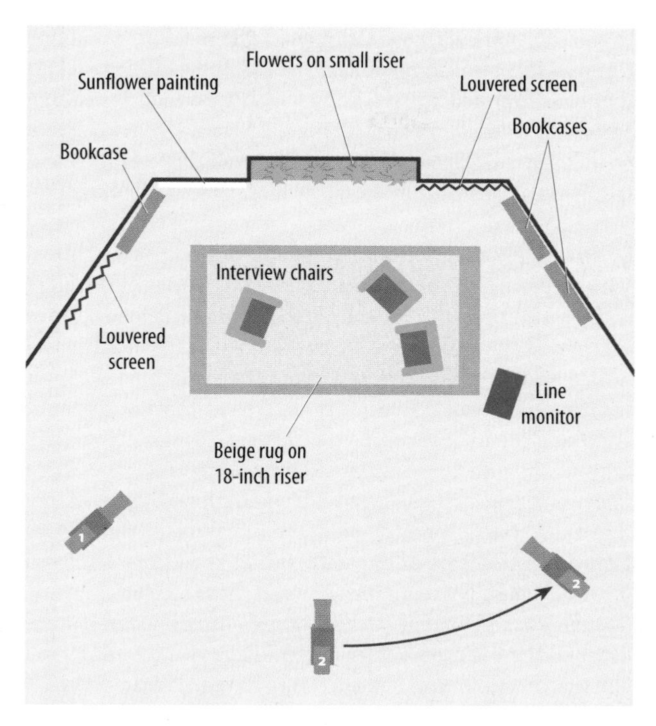

4.7 REVISED INTERVIEW SET
The revised floor plan for the interview provides for adequate background cover and interesting shots.

INTERPRETING A LOCATION SKETCH

There is no time to rest on your laurels. Just when you have fixed the interview set, the associate director comes back from a location survey for the segment on running the stop sign and shows you her ***location sketch***—a rough map of the locale of the remote shoot. **SEE 4.8** She feels that there may be several potential production problems. See if you can identify some of them.

- The intersection is obviously in the middle of downtown. You can therefore expect a great deal of traffic to pass through, and the police would not close such a busy intersection for anything but a real accident.
- Even if the intersection were not downtown, the proximity of the bank and the supermarket would make closing the intersection, even for a little while, infeasible.
- A schoolyard is very noisy during recess. Unless you do not mind the laughing and yelling of children during the production, every school recess means a forced recess for the production crew.
- The four-way stop signs make the intersection less hazardous, even if someone runs one of them. The demonstration is much more effective if one of the streets has through traffic.

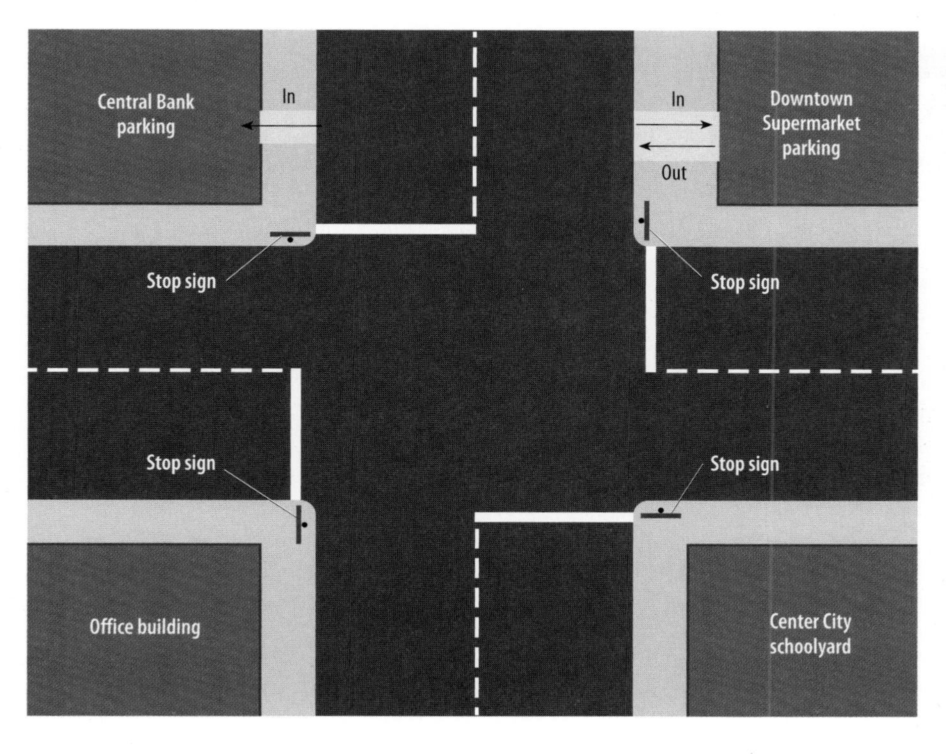

4.8 LOCATION SKETCH FOR STOP SIGN SEGMENT
This location sketch points to several major problems that make the field production infeasible.

The solution to these problems is relatively simple: have the producer contact the police department and find an intersection in a quiet neighborhood that has very little traffic. There should be sufficient alternate routes so that a temporary closure of the intersection will not cause traffic delays or prevent neighbors from getting to and from their homes.

Using a Storyboard

In a complicated assignment like the one just described, or in any directing undertaking that goes beyond a routine show, you will be greatly aided in your visualization by a good storyboard. A ***storyboard*** is a sequence of visualized shots that contains key visualization points and audio information. **SEE 4.9**

A storyboard is usually drawn on preprinted storyboard paper, which has areas that represent the video screen. Another area, usually below the screen, is dedicated to audio and other information. A storyboard can also be drawn on plain paper or created by computer. Storyboard software programs offer a great many stock images (houses, streets, cars, living rooms, and offices, for example) into which you can place figures and move them into position in the storyboard frame. **SEE 4.10**

Most commercials are carefully storyboarded shot-by-shot before they ever go into production. Storyboards help people who make decisions about the commercial see the individual shots and their intended sequence.

Storyboards are also used for other types of single-camera productions that contain a great number of especially complicated discontinuous shots or shot sequences. A good storyboard offers immediate clues to certain production requirements, such as general location, camera position, approximate focal length of the lens, method of audio pickup, cutaways, amount and type of postproduction, talent actions, set design, and hand props.

Marking the Script

Proper marking of a script will aid you greatly in multicamera directing from the control room or on-location. In control room directing, you need to coordinate many people and machines within a continuous time frame. The marked script becomes a road map that guides you through the intricacies of a production. Although there is no single correct way to mark a script, certain conventions and standards have been established. Obviously, a fully scripted show requires more precise cueing than does an interview that is directed from a simple show format. Live productions or uninterrupted recordings of live productions (live-recorded) directed from the control room in a continuous time frame need a more carefully marked script than do shows that are recorded for postproduction editing. But even in discontinuous single-camera productions, a well-marked script will help you remember camera and talent positions and make your directing more exacting.

Script marking for instantaneous editing (switching) Whatever system of script marking you choose or develop, it must be clear, readable, and above all consistent. Once you arrive at a working system, stick with it. As in musical notation, where you can perceive whole passages without reading each individual note, the script-marking system enables you to interpret and react to the written cues without having to consciously read each one. The three figures on pages 71 to 73 provide examples of script marking. **SEE 4.11–4.13** Take a look at the markings in figure 4.11 again and compare them with those in figures 4.12 and 4.13. Which script seems cleaner and more readable to you?

The script in figure 4.11 shows information that is more confusing than helpful. By the time you've read all the cue instructions, you will certainly have missed part or all of the action and perhaps even half of the talent's lines. You do not have to mark all stand-by cues or any other obvious cues that are already implied. For example, "ready" cues are always given before a cue, so they need not be part of your script markings.

In contrast, the markings in figures 4.12 and 4.13 are clean and simple. They are kept to a minimum, and there is little writing. You are able to grasp all the cues quickly without actually reading each word. As you can see, the cues in figure 4.12 provide the same information as those in figure 4.11, but they allow you to keep track of the narration, look ahead at upcoming cues, and especially watch the action on the preview monitors. Let us now highlight some of the qualities of a well-marked script from a director's point of view (refer to figure 4.12).

- All action cues are placed before the desired action.

- If the shots or camera actions are clearly described in the video column (page-left), or the audio cues in the audio column (page-right), simply underline or circle the printed instructions. This keeps the script clean and uncluttered. If the printed instructions are hard to read, do not hesitate to repeat them with your own symbols.

- If the script does not indicate a particular transition from one video source to another, it is always a cut. A large handwritten 2 next to a cue line means that the upcoming

THE RETURN OF AGENT 12
DETONATION FILMS

Shot: 327

CU – Agent 12
AGENT 12
Jetpack! Maximum burn!

Shot: 328

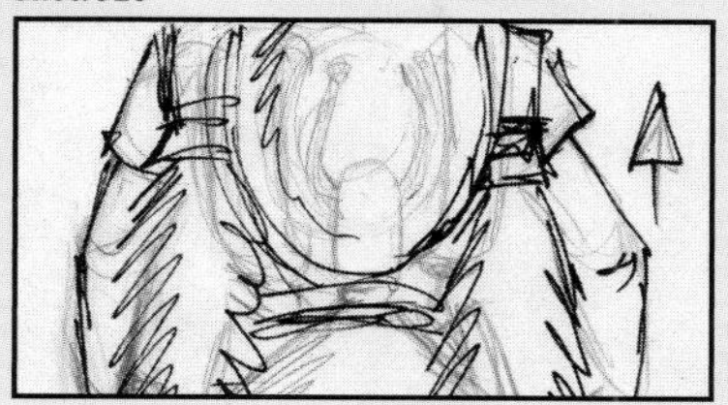

FAST CU – JETPACK rises
As the thruster IGNITES!
Audio: ROAR of jets.

Shot: 329

DRAMATIC DOWNSHOT ON AGENT 12
As he ROCKETS into the air, the ground dropping away below as he streaks UP and PAST CAM, trailing flame.
Audio: ROAR of jets, pitch drops in Doppler effect.

Shot: 330

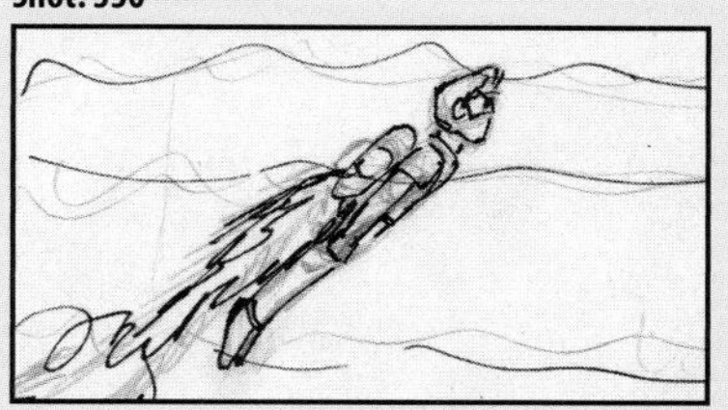

TRACKING LS – Agent 12 continues ascent.
This shot appears on-screen in following shot.

Shot: 331

ON DR. VENGEANCE
Reacting to what he's seeing on-screen.
He can't believe it.
DR. VENGEANCE
Awww, MAN! Who gave him a *jetpack?*

Shot: 332

CU
SCOTT about to speak up . . .

Shot: 333

CU
. . . but thinks better of it.

Shot: 334

MLS TIGHTEN and CANT to CU
Dr. Vengeance shouts into radio handset.
Audio: KLAXON horn sounds.
DR. VENGEANCE
Tiger! Gameboy! *Scramble!*

Shot: 335

ANGLE ON THE ROOF – Showing TIGER and GAMEBOY sprinting for the howitzer controls. The massive cannons overshadow the entire roof.
DR. VENGEANCE (ON RADIO)
Ready the roof guns!

4.9 STORYBOARD COURTESY BOB FORWARD, DETONATION FILMS

4.9 HAND-DRAWN STORYBOARD

The hand-drawn storyboard shows the major visualization points and sometimes lists the key audio sections or the shot sequence. The tilted rectangle in shot 334 indicates a canted (sideways tilted) camera.

Careful Roadways PSA

1 00:00:05:00

VO: The amazing function between the eyes and the brain is called the "sense of vision."

2 00:00:10:00

VO: The human vision system processes information faster than a Cray supercomputer.

3 00:00:13:00

VO: The human eye outperforms our best video.

4 00:00:20:00

VO: Human vision updates images, including the details of motion and color, on a time scale so rapid . . .

5 00:00:23:00

VO: . . . that a "break in the action" is almost never perceived.

6 00:00:28:00

VO: The Cray supercomputer can process more than a hundred million arithmetic operations per second.

4.10 STORYBOARD COURTESY POWERPRODUCTION SOFTWARE

4.10 COMPUTER-GENERATED STORYBOARD
The computer-generated storyboard has 3D graphics that can be used to create a variety of exterior and interior environments in which images of people can be placed.

VIDEO	AUDIO	
Effects		Ready on effects Take effects
Wipe to: VTR (SOT) (showing a series of paintings from realism to expressionism)	AUDIO IN-CUE: "ALL THE PAINTINGS WERE DONE BY ONE ARTIST... PICASSO"	Ready to wipe to VTR Roll VTR and take VTR4
	OUT-CUE: "...PHENOMENAL CREATIVE FORCE"	Track up on VTR4 Ready camera 2
MS Barbara by the easel	But even Picasso must have had some bad days and painted some bad pictures. Take a look. The woman's hands are obviously not right. Did Picasso deliberately distort the hands to make a point? I don't think so.	Cue Barbara and take camera 2
	Ready camera 3 on the easel – closeup	
CU of painting Key effects	Look at the outline. He obviously struggled. The line is unsure, and he painted this section over at least three times. Because the rest of the painting is so realistically done, the distorted hands seem out of place. This is quite different from his later period, when he distorted images to intensify the event.	Take camera 3
VTR SOT Insert time	IN-CUE: "DISTORTION MEANS POWER. THIS COULD HAVE BEEN PICASSO'S FORMULA..."	Ready to roll VTR 4 Segment 2
4:27 min	OUT-CUE: "...EXPRESSIVE POWER THROUGH DISTORTION IN HIS LATER PAINTINGS."	Roll VTR 4 and take VTR 4
CU Barbara	But the formula "distortion means power" does not always apply. Here again it seems to weaken the event. Take a look at...	Ready camera 2 Cue Barbara and take camera 2

4.11 BAD SCRIPT MARKING
This script is marked with too much unnecessary information that makes it hard to read.

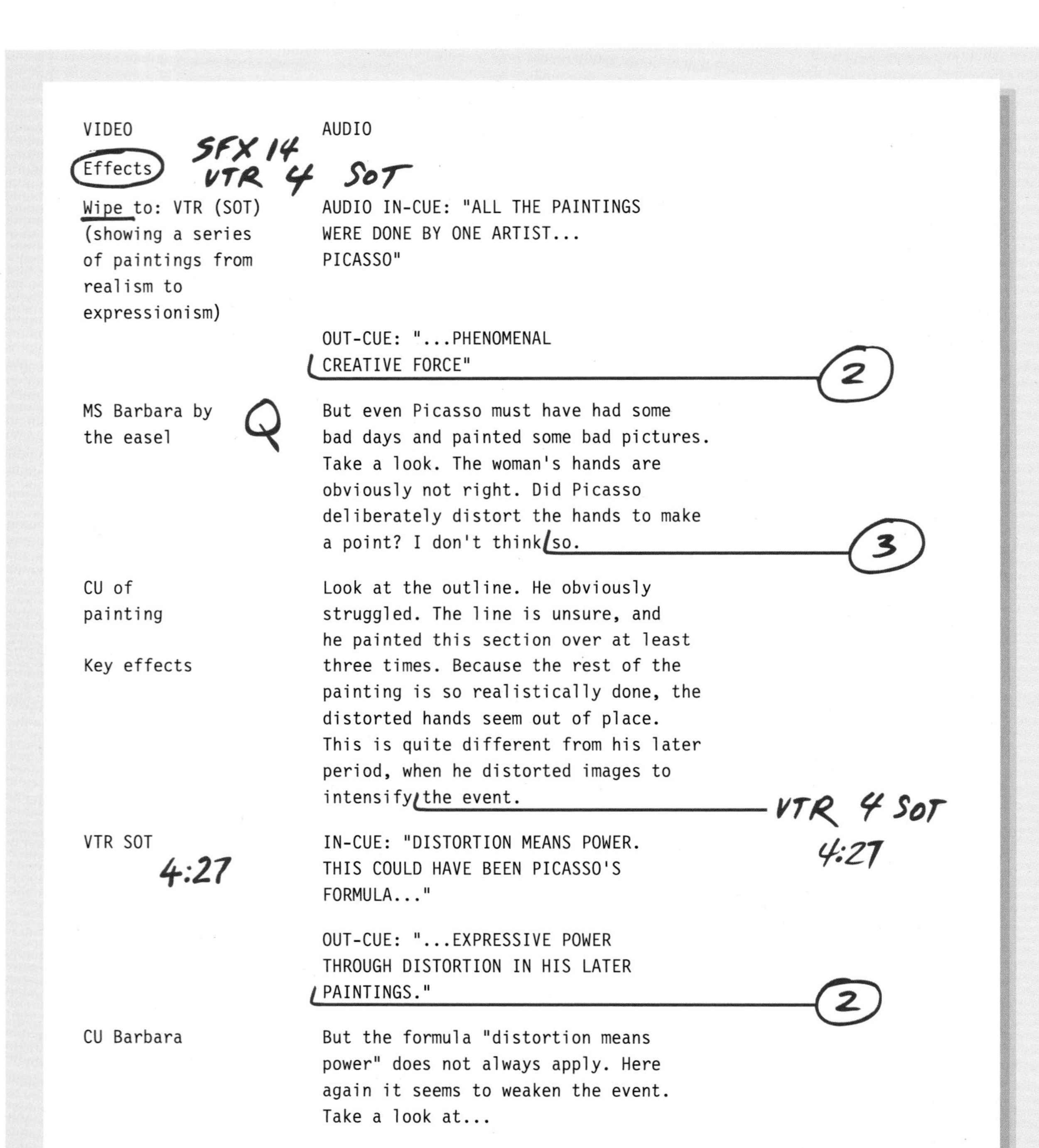

VIDEO	AUDIO
Effects	
Wipe to: VTR (SOT) (showing a series of paintings from realism to expressionism)	AUDIO IN-CUE: "ALL THE PAINTINGS WERE DONE BY ONE ARTIST... PICASSO"
	OUT-CUE: "...PHENOMENAL CREATIVE FORCE"
MS Barbara by the easel	But even Picasso must have had some bad days and painted some bad pictures. Take a look. The woman's hands are obviously not right. Did Picasso deliberately distort the hands to make a point? I don't think so.
CU of painting Key effects	Look at the outline. He obviously struggled. The line is unsure, and he painted this section over at least three times. Because the rest of the painting is so realistically done, the distorted hands seem out of place. This is quite different from his later period, when he distorted images to intensify the event.
VTR SOT	IN-CUE: "DISTORTION MEANS POWER. THIS COULD HAVE BEEN PICASSO'S FORMULA..."
	OUT-CUE: "...EXPRESSIVE POWER THROUGH DISTORTION IN HIS LATER PAINTINGS."
CU Barbara	But the formula "distortion means power" does not always apply. Here again it seems to weaken the event. Take a look at...

4.12 GOOD SCRIPT MARKING OF A/V FORMAT
This script is clearly marked and can be easily read by the director.

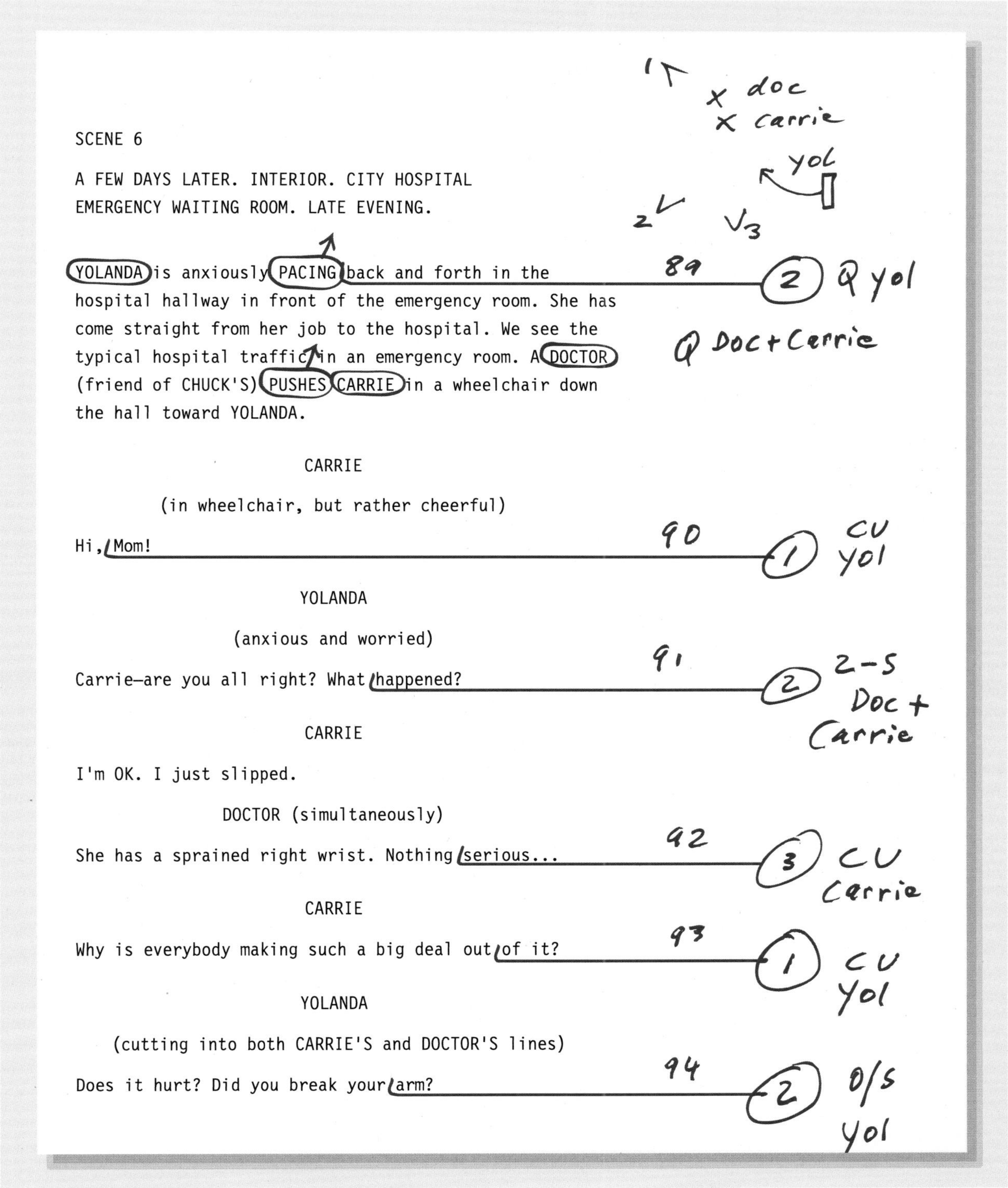

SCENE 6

A FEW DAYS LATER. INTERIOR. CITY HOSPITAL EMERGENCY WAITING ROOM. LATE EVENING.

YOLANDA is anxiously PACING back and forth in the hospital hallway in front of the emergency room. She has come straight from her job to the hospital. We see the typical hospital traffic in an emergency room. A DOCTOR (friend of CHUCK'S) PUSHES CARRIE in a wheelchair down the hall toward YOLANDA.

CARRIE

(in wheelchair, but rather cheerful)

Hi, Mom!

YOLANDA

(anxious and worried)

Carrie—are you all right? What happened?

CARRIE

I'm OK. I just slipped.

DOCTOR (simultaneously)

She has a sprained right wrist. Nothing serious...

CARRIE

Why is everybody making such a big deal out of it?

YOLANDA

(cutting into both CARRIE'S and DOCTOR'S lines)

Does it hurt? Did you break your arm?

4.13 GOOD SCRIPT MARKING OF SINGLE-COLUMN FORMAT
The marking on this multicamera single-column drama script shows the cameras used, the shot number, the type of shot, and the major actions. Note the blocking sketch at the beginning of this scene.

transition is a cut to camera 2. It also implies a "ready 2" before the "take 2" call.

■ If the show requires rehearsals, do preliminary script marking in pencil so that you can make quick changes without creating a messy or illegible script. Once you are ready for the dress rehearsal, however, you should have marked the script in bold letters. Have the associate director and the floor manager copy your markings onto their own scripts.

■ Mark the cameras by circled numbers all in one row. This allows you to see quickly which camera needs to be readied for the next shot.

■ In addition to the camera notation, number each shot in consecutive order, starting with 1, regardless of the camera you use for the shot. These numbers will not only help you ready the various shots for each camera but also make it easy to create a shot sheet for each camera. **SEE 4.14** You can now easily delete or add shots for the cameras. All you need to do is say, "delete shot 89," and camera 2 will delete the shot with Yolanda pacing back and forth.

■ You may want to devise a symbol that signifies action, such as someone coming through the door, walking over to the map, sitting down, or getting up. In figure 4.13 this cue is a handwritten arrow (↗).

■ If there are several moves by the talent, draw little maps (see figure 4.13). Such blocking sketches are usually more helpful for recalling talent moves, camera positions, and traffic than are storyboard sketches of shot compositions.

Script marking for postproduction editing The marking of the script for discontinuous takes consists of a careful breakdown and indication of the various scenes, their locations (hospital, front entrance), and principal visualizations (camera point of view, field of view). You then number the scenes in the proposed production sequence, ending up with a list of scenes that refers to the original script by page number. Here is an example:

LOCATION	SCENE	SCRIPT PAGES
Hospital	6	41–47
corridor	7	48–55
Emergency	3	5–7
room entrance	14	102–10
Yolanda's kitchen	14	2–4

In the script itself, you are free to use whatever markings you prefer. When video-recording discontinuous takes for postproduction, you obviously have more time to consult the script than during a live or live-recorded production. For discontinuous video recording, it may help to mark the talent movements on the script as well as draw next to the dialogue small storyboard sketches that show unusual shot framings. Such sketches assist in recalling what you had in mind when preparing the script. Some film directors (such as Steven Spielberg) storyboard almost every shot of the entire movie before ever shooting a single frame of film. Cinema directors often use a special and more detailed shot and scene breakdown, called mark-ups, which they can use for the actual production and especially for postproduction editing.[4]

As you can see, the more time and effort you devote to such preproduction details, the less time and effort you will have to expend during the actual video-recording sessions. Production efficiency does not mean to hurry through a production regardless of quality; it means extensive preparation.

Camera 2	
Shot #	Hospital scene 6
89	MS Yolanda follow
91	2-Shot Carrie & Doctor
94	O/S Yolanda

4.14 SHOT SHEET FOR CAMERA 2

4. You can find such mark-ups in film production books. For example, see Michael Rabiger, *Directing: Film Techniques and Aesthetics* (Burlington, Mass.: Focal Press, 2007).

MAIN POINTS

- A clear understanding of the process message (desired effect) will help you decide on the most appropriate type of production (single-camera or multicamera, studio or field, recorded or live, continuous or discontinuous takes for postproduction) and the medium requirements.
- A careful script analysis should lead to a locking-in point—an especially vivid visual or aural image—that determines the subsequent visualizations and sequencing.
- Visualization (mentally seeing and hearing key images) is crucial for the successful translation of script to screen event.
- The floor plan or location sketch enables the director to plan camera and talent positions and traffic.
- The storyboard shows drawings or computer-generated images of key visualization points of an event with accompanying audio information as well as the proper sequencing of shots.
- When preparing the show for the actual production day, you must interpret the floor plan or location sketch and mark the script.
- The important aspects of good script marking are readability and consistency.
- Precise and easy-to-read script markings help you and other production personnel anticipate and execute a great variety of cues.

SECTION

4.2 Communication and Scheduling

This section deals with the functions of the director's support staff and the necessity of working with the producer on facilities requests, schedules, and communication with talent and production personnel. ZVL3 PROCESS→ People→ nontechnical | technical

▶ **SUPPORT STAFF**
Floor manager, production assistant, and associate director

▶ **FACILITIES REQUESTS, SCHEDULES, AND COMMUNICATION**
Facilities request, production schedule, time line, and the director's communication

SUPPORT STAFF

The director's support staff depends on the size of the production company. Some documentaries are done by a three-person team: the producer/director; the ***DP*** (director of photography), who runs the camera and does the lighting; and the audio person. Many field productions are accomplished with only a two-person team—the videographer and the talent—whereby either one or both act as director. In a television station or a larger independent production company, a director will have the support of at least two or three people: the floor manager, the production assistant, and, in larger productions, an associate or assistant director.

Floor Manager

The floor manager is also called the floor director, stage manager, or unit manager, even though the unit manager functions more like a production manager or line producer who takes care of the daily production and budgetary details. The floor manager's primary functions are to coordinate all activities on the "floor" (studio or on-location site) and relay the cues from the director to the talent.

Before the production the floor manager needs to oversee and help the floor crew set up scenery, place set and hand props, dress the set, and put up displays. During rehearsals and the production, the floor manager must coordinate the floor crew and the talent and relay the director's talent cues. After the production he or she is responsible for striking the set and the props or restoring the remote production site to its original condition.

The following are some points to keep in mind when managing the floor.

- Unless you are doing a routine show that is produced on a permanent set (one that is not struck after each show), as floor manager you need to obtain a detailed floor plan and prop list. Check with the art director and the director about any specific features or changes. Get a marked script from the director so that you can anticipate talent and camera traffic. Have the director look at the set before fine-tuning the lighting. Once the lighting is complete, even minor set changes can require major lighting adjustments. When the set is put up and dressed, take a digital photo of it. Such a record is much more readily accessible than a video recording.

- You are responsible for having all hand props on the set and in operating condition. For example, if the show involves a studio demonstration of a new laptop computer, run the specific series of computer programs a few times to see how it works. Hard-to-open jars or bottles are a constant challenge to the performer. Twist the lid of a jar slightly or loosen the bottle cap so that the talent can remove it without struggling. This small courtesy can prevent retakes and frayed nerves and is usually a quick way of establishing trust between you and the talent.

- Check that the teleprompter works and that the correct copy is displayed.

- If you use an on-camera slate or clapboard in the field, have it ready and filled out with the essential information. Have several pens available as well as a rag to erase the writing.

- For complex productions study the marked script before the rehearsal and add your own cues, such as talent entrances and exits and prop, costume, or set changes. In case of doubt, ask the director for clarification.

■ Introduce yourself to the talent and the guests and have a designated place for them to sit while waiting in the studio. Because most production people are quite busy (including the director and the producer), you are the one who must establish and maintain a rapport with the talent and the guests throughout the production. Verify that they have signed the proper release forms and other necessary papers. Ask them periodically if they would like some water or coffee, whether they are comfortable, and if you can be of assistance. When working with outside talent, review your major cues with them (see chapter 16).

■ When using a teleprompter, ask the performers whether the font size is big enough and if its distance from the camera is tolerable.

■ During the rehearsal of a fully scripted show, follow the script as much as possible and anticipate the director's cues. Keep notes on especially difficult camera travels and talent actions. If the production is shot in segments for postproduction editing, pay particular attention to continuity of the talent's appearance, positions, and major moves.

■ Always carry a pen or pencil, a broad marking pen, rolls of masking and gaffer's tape, and a piece of chalk (for taping down props and equipment and for spiking—marking—talent and camera positions). Also have a large pad ready so that you can write out messages for the talent in case the I.F.B. (interruptible foldback) system breaks down or is not used.

■ During camera rehearsal deliver all cues as though you were on the air, even if the director stands right next to you. You do not always have to remain next to the camera when cueing. As much as possible, position yourself so that you can see the talent's eyes. This is one of the reasons why you should not be tied to a studio camera's intercom outlet.

■ During the show do not cue on your own, even if you think the director has missed a cue. Rather, ask the director on the intercom whether you should give the cue as marked and rehearsed. If there are interruptions in the video recording because some technical problems are being discussed in the control room, inform the talent about what is going on. Tell them they did a good job but that the director has to work out some technical details. If there are extended problem-solving interruptions, invite the talent to get out from under the lights and relax in the break area you have set up for them in the studio—but don't let them wander off.

■ After the show thank the talent or guests and show them out of the studio. You then need to supervise the strike of the set in the studio or of the items set up on-location. Be careful not to drag scenery or prop carts across cables that might still be on the studio floor. Locate objects that were brought in by a guest and see to it that they are returned. If you shot indoors on-location, put things back as you found them. A small location sketch or digital photo will be of great help when trying to return things to the way they were. When shooting on-location do not forget that you are a guest operating in someone else's space.

Production Assistant

As a production assistant (PA), you must be prepared to do a variety of jobs—from duplicating and distributing the script, looking for a specific prop, and welcoming the talent to getting coffee, calling a cab, and taking notes for the producer and the director (unless the AD is taking notes). Usually, note taking is the PA's most important assignment. You simply follow the producer and/or director with a pad and pen and record everything they mumble to themselves or tell you to write down. During the "notes" breaks, you read back your notes item by item. When in the field, you may also have to keep a field log of all the production takes, which helps the postproduction editor locate shots on the recording media. ZVL4 EDITING→ Production guidelines→ field log

Associate, or Assistant, Director

As an ***AD*** (associate, or assistant, director), you mainly assist the director in the production phase—the rehearsals and the on-the-air performance or recording sessions. In complex studio shows, a director may have you give all standby cues (for example: "Ready to cue Mary, ready 2 CU of John") and preset the cameras by telling the camera operators on the intercom the upcoming shots or camera moves. This frees up the director somewhat to concentrate more on the preview monitors or the talent's performance. Once the action is preset by you, the director then initiates it with the various action cues: "Ready 2, take 2," or, in fast dialogue, simply by snapping fingers.

In elaborate EFPs or complicated studio productions, you may have to direct the run-throughs (rehearsals) for each take, which enables the director to stand back and observe the action on the field or line monitor.

The AD in digital cinema productions may also function as a second-unit director. The second unit consists of a smaller crew whose job it is to capture various inserts, such as establishing shots, certain close-ups, special-effect sequences, or other small scenes that do not involve major dialogue.

As an AD you are also responsible for the timing of the show segments and the overall show during rehearsals as well as during the actual production.

FACILITIES REQUESTS, SCHEDULES, AND COMMUNICATION

Now is the time for you, as the director, to double-check with the producer about various important tasks that should have been done in the preproduction period, such as requesting the necessary facilities and scheduling.

Facilities Request

Even if in your operation it is the producer who requests and schedules the equipment and the facilities, you still need to verify with him or her that the ***facilities request*** actually includes all the equipment and technical facilities you may have added during your production preparations. Ask the TD (technical director) whether the final facilities request is meeting all the show's technical demands.

Production Schedule

Unless you are directing a routine show, such as the morning news, check with the producer about the ***production schedule***—the calendar that shows the preproduction, production, and postproduction dates and who is doing what, when, and where. Most likely, the producer will have taken care of the production schedule, but you as the director will have to prepare the time line.

Time Line

In larger operations the daily ***time line*** is worked out by the production manager or the line producer. In smaller production companies, the director establishes the time line for a specific production day. Assuming that you as the director are responsible for the time line, verify with the producer whether the production schedule was actually delivered to the talent and all members of the production team. A simple telephone call by the PA will confirm this distribution and make you sleep better. ZVL5 PROCESS→ Phases→ preproduction | production

As with every other aspect of television production, each production day is governed by strict time limits. Time lines differ considerably, depending on the complexity of the studio show or remote telecast. On a difficult remote, such as a parade through a narrow downtown street, you may need a whole day for the setup. The following examples are typical time lines for an interview and a one-hour soap opera segment.

Time line: interview This time line is prepared for a half-hour interview (actual length: 23 minutes), featuring two folk singers who have gained world fame because of their socially conscious songs. The singers, who accompany themselves on acoustic guitars, are scheduled to give a concert the following day on the university commons. Their contract does not allow the presence of television cameras during the actual concert, but they, their manager, and AFTRA (the talent union) agreed that the singers could come to the studio for a brief interview and play a few short selections.

The process message is relatively simple: to give viewers an opportunity to meet the two singers, learn more about them as artists and concerned human beings, and watch them perform.

To save time and money, the show is scheduled for a live-recording session. This means that the director will direct the show as though it were going on the air live, or at least with as few stop-downs (interruptions whereby the video recorder is stopped) as possible.

TIME LINE: INTERVIEW (JULY 15)

11:00 a.m.	Crew call
11:10–11:30 a.m.	Tech meeting
11:30 a.m.–1:00 p.m.	Setup and lighting
1:00–1:30 p.m.	Lunch
1:30–1:45 p.m.	Production meeting: host and singers
1:45–2:15 p.m.	Run-through and camera rehearsal
2:15–2:25 p.m.	Notes and reset
2:25–2:30 p.m.	Break
2:30–3:15 p.m.	Recording
3:15–3:30 p.m.	Spill
3:30–3:45 p.m.	Strike

As you can see from this time line, a production day is divided into blocks of time during which certain activities take place.

11:00 a.m. **Crew call** This is the time the crew must arrive at the studio.

11:10–11:30 a.m. **Tech meeting** You start the day with a technical meeting during which you discuss with the crew the process message and the major technical requirements. One of these requirements is the audio setup because the singers are obviously interested in good stereo sound. You need different mics for the interview and for the performance area. You should also explain what camera shots you want. The sincerity of the artists and their guitar-playing skills are best conveyed by CUs and ECUs (extreme close-ups), and you may want to shift the attention from one singer to the other through a rack focus effect. The audio technician may want to discuss the specific mic setup with you, such as stand mics for the performance and wireless lavaliers for the singers' crossover. The TD (acting as studio crew chief) may ask about the desired lighting and the simultaneous recording of the show on DVD. If all goes well, you can hand the guests the DVDs right after the taping as a small thank-you gesture.

11:30 a.m.–1:00 p.m. **Setup and lighting** This should be sufficient time to set up the standard interview set, place the mics, and light the interview and performance areas. Although as director you are not immediately involved in this production phase, you might want to keep an eye on the setup so that you can make minor changes before the lighting is done.

1:00–1:30 p.m. **Lunch** Tell everyone to be back by 1:30 sharp—not 1:32 or 1:35—which means that everyone has to be able to leave the studio at exactly 1:00, even if there are still some technical details left undone. Minor technical problems can be solved during the production meeting with the host and the singers.

1:30–1:45 p.m. **Production meeting: host and singers** When the singers and their manager arrive at this meeting, they have already been introduced to the host by the producer or PA. Nevertheless, check that they have signed all the necessary papers. In this meeting confirm their musical selections and the running time for each. Discuss the opening, the closing, and the crossover to the performance area. Tell them some of your visualization ideas, such as shooting very tightly during especially intense moments in their songs and for intricate guitar sections.

1:45–2:15 p.m. **Run-through and camera rehearsal** Although the setup is relatively simple and there will be little camera movement during the songs, you need to rehearse the crossovers from the interview set to the performance area and back. You may also want to rehearse some of the ECUs or the rack focus shots from one singer to the other. Then go through the opening and the closing with all facilities (theme music, credits, and name keys). Dictate to the PA any production problems you may discover during this rehearsal for the notes segment. This rehearsal is especially important for the audio engineer, who is trying to achieve an optimal sound pickup. Do not get upset when the audio technician repositions mics during the camera rehearsal. If all goes well, you may be done before 2:15 p.m.

2:15–2:25 p.m. **Notes and reset** You now gather the key production people—producer, AD, TD, audio technician, LD (lighting director), floor manager, and host—to discuss any production problems that may have surfaced during the camera rehearsal. Ask the PA to read the notes in the order written down. Direct the production team to take care of the various problems. At the same time, the rest of the crew should get the cameras into the opening positions, reset the pages of the C.G., ready the video recorder and the DVD recorder, and make minor lighting adjustments.

2:25–2:30 p.m. **Break** This short break will give everyone a chance to get ready for the recording. Don't tell the crew to "take five" but rather when to be back in the studio (at exactly 2:30).

2:30–3:15 p.m. **Recording** You should be in the control room and start both recorders no later than 2:35 p.m.—not 2:40 or 2:45. If all goes well, the half-hour show should be "in the can," or finished, by 3:15, including the stop-down time for the first crossover.

3:15–3:30 p.m. **Spill** This is a grace period because we all know that television is a complex, temperamental machine that involves many people. For example, you may have to redo the opening or the closing because the C.G. did not deliver the correct page for the opening credits or because the host gave the wrong time for the upcoming concert.

3:30–3:45 p.m. **Strike** During the strike time, you can thank the singers and their manager, the host, and the crew. Arrange for a playback in case they want to see and especially listen to the video recording right away. Play back the audio track through the best speaker system you have. All the while keep at least one eye on the strike, but do not interfere with it. Trust the floor manager and the crew to take down the set and clean the studio for the next production in the remaining 15 minutes.

One of the most important aspects of a time line is sticking to the time allotted for each segment. You must learn to get things done within the scheduled time block and, more importantly, to jump to the next activity at the precise time shown on the schedule regardless of whether you have finished your previous chores. Do not use up the time of a scheduled segment with the preceding activity. A good director terminates an especially difficult blocking rehearsal at midpoint to meet the scheduled notes and reset period. Inexperienced directors often spend a great amount of time on the first part of the show or on a relatively minor detail, then go on the air without having rehearsed the rest of the show. The time line is designed to prevent such misuse of valuable production time.

Time line: soap opera Here is an example of a time line for a more complicated one-hour soap opera. Assume that the setup and the lighting were accomplished the night before (from 3:00 to 6:00 a.m.) and that some set changes will happen after 6:00 p.m. for the following day.

AS THE SUN RISES	SEGMENT 987
6:00–8:00 a.m.	Dry run—rehearsal hall
7:30 a.m.	Crew call
8:00–8:30 a.m.	Tech meeting
8:30–11:00 a.m.	Camera blocking
11:00–11:30 a.m.	Notes and reset
11:30 a.m.–12:30 p.m.	Lunch
12:30–2:30 p.m.	Dress rehearsal
2:30–3:00 p.m.	Notes and reset
3:00–5:30 p.m.	Recording
5:30–6:00 p.m.	Spill

As you can see, this time line leaves no time for you to think about what to do next. You need to be thoroughly prepared to coordinate the equipment, technical people, and talent within the tightly prescribed time frame. There is no time allotted for striking the set because the set stays up for the next day's production.

Director's Communication

Although the producer is responsible for maintaining the contact information of the talent and the technical and nontechnical production personnel, you as the director are responsible for getting the show done on time. You should therefore establish a routine procedure for communicating efficiently with all personnel involved in a day's production. Note that such a double-check is necessary only when you work on onetime studio and field productions. If you do a routine show, you need to trust the producer or production manager to handle crew changes or substitutions for talent. Still you should have ready access to the same personnel information so that your PA can verify last-minute crew and talent substitutions. If you use e-mail for communication, request an immediate response from the recipient and always copy the producer.

Such communication verification is especially important if you are to direct a field production that needs access to restricted areas, such as a sports stadium. ZVL6 PROCESS→ People→ nontechnical

MAIN POINTS

- The director's immediate support staff normally comprises a floor manager, a PA (production assistant), and, in larger productions, an AD (associate or assistant director).
- The facilities request is an essential communication device for procuring the necessary production facilities and equipment.
- The production schedule shows the preproduction, production, and postproduction dates and who is doing what, when, and where.
- The time line shows a breakdown of time blocks for various activities on the actual production day.
- To facilitate communication between the director and the technical and nontechnical personnel, the director must establish a specific routine and stick to it. E-mail messages must be immediately acknowledged by the recipient.

ZETTL'S VIDEOLAB

For your reference or to track your work, the *Zettl's VideoLab* program cues in this chapter are listed here with their corresponding page numbers.

PART

III

Production

CHAPTER

5

Analog and Digital Television

Digital television (DTV) has led to not only dramatically improved television pictures and sound but also a convergence of various media: television is becoming interactive; large, centralized digital databases offer television news organizations instant access to news files; and computers stream audio and video content over the Internet. You can use your cell phone to text or video-record your friends, to find a nearby restaurant, or to watch television programs while walking or riding the bus to campus (assuming you will turn it off when entering the classroom).

More and more large-scale motion pictures use ***high-definition television (HDTV)*** cameras and computers rather than traditional film cameras for their studio and field productions. But even if the original footage is captured on photographic film stock rather than videotape or some other digital recording media, all subsequent production steps involve similar (or the same) equipment and processes you encounter in major television productions. Once transformed into digital data, movies can be just as easily distributed via cable, microwave, or satellite as HDTV signals are. As you can see, learning about digital processes is no longer a luxury; it has become essential for anyone involved in television and even film production.

Section 5.1, Analog and Digital Processes, explains what digital processes are all about and how they differ from analog systems. Section 5.2, Scanning Systems, introduces you to the basics of the creation of a color television image, interlaced and progressive scanning, the current digital scanning systems, and flat-panel displays.

KEY TERMS

480i The scanning system for standard analog television. Each complete television frame consists of 525 lines, of which only 480 are visible on the screen. The *i* stands for *interlaced scanning.*

480p The lowest-resolution scanning system of digital television. The *p* stands for *progressive,* which means that each complete television frame consists of 480 visible, or active, lines that are scanned one after the other (out of 525 total scanning lines). Used for standard digital television transmission.

720p A progressive scanning system of high-definition television. Each frame consists of 720 visible, or active, lines (out of 750 total scanning lines).

1080i An interlaced scanning system of high-definition television. The *i* stands for *interlaced,* which means that a complete frame is formed from two fields, each carrying half of the picture information. Each field consists of 540 visible, or active, lines (out of 1,125 total scanning lines). Just like the standard NTSC television system, the digital HDTV 1080i system produces 60 fields, or 30 complete frames, per second.

1080p A progressive scanning system. All visible lines (1,080 of 1,125) are scanned for each frame. Although the acquisition rate for digital cinema may be 24 frames per second, to avoid flicker during playback the refresh rate is generally boosted to 60 fps.

analog A signal that fluctuates exactly like the original stimulus.

aspect ratio The width-to-height proportions of the standard television screen and therefore of all standard television pictures: 4 units wide by 3 units high. For HDTV the aspect ratio is 16 × 9. The small mobile media (cell-phone) displays have various aspect ratios.

ATSC Stands for *Advanced Television Systems Committee.* Sets the standards for digital television, which largely replaces the analog NTSC standard in the United States.

ATSC signals Video signals transported as packets, each packet holding only part of a television picture and the accompanying sound plus instructions for the receiver on how to reassemble the parts into complete video pictures and sound. This system is not compatible with NTSC.

binary A number system with the base of 2.

binary digit (bit) The smallest amount of information a computer can hold and process. A charge is either present, represented by a 1, or absent, represented by a 0. One bit can describe two levels, such as on/off or black/white.

codec Short for *compression/decompression.* A specific method of compressing and decompressing digital data.

coding To change the quantized values into a binary code, represented by 0's and 1's. Also called *encoding.*

compression Reducing the amount of data to be stored or transmitted by using coding schemes (codecs) that pack all original data into less space (lossless compression) or by throwing away some of the least important data (lossy compression).

decoding The reconstruction of a video or audio signal from a digital code.

digital Usually means the binary system—the representation of data in the form of binary digits (on/off pulses).

digital television (DTV) Digital television systems that generally have a higher image resolution than standard television (STV). Also called *advanced television (ATV).*

downloading The transfer of files that are sent in data packets. Because these packets are often transferred out of order, the file cannot be seen or heard until the downloading process is complete.

field One-half of a complete scanning cycle, with two fields necessary for one television picture frame. There are 60 fields, or 30 frames, per second in standard NTSC television.

frame A complete scan from top to bottom of all picture lines by the electron beam, or one single frame of a motion series.

high-definition television (HDTV) Has at least twice the picture detail of standard (NTSC) television. The 720p uses 720 visible, or active, lines that are normally scanned progressively each $\frac{1}{60}$ second. The 1080i standard uses 60 fields per second, each field consisting of 540 visible, or active, lines. In interlaced scanning, a complete frame consists of two scanning fields of 540 visible lines. In the 1080p standard, all 1,080 lines are scanned for each frame. The refresh rate (complete scanning cycle) for HDTV systems is usually 60 fps but can vary.

interlaced scanning In this system the electron beam skips every other line during its first scan, reading only the odd-numbered lines. After the beam has scanned half of the last odd-numbered line, it jumps back to the top of the screen and finishes the unscanned half of the top line and continues to scan all even-numbered lines. Each such even- or odd-numbered scan produces a field. Two fields produce a complete frame. Standard NTSC television operates with 60 fields per second, which translates to 30 frames per second.

progressive scanning In this system the electron beam starts with line 1, then scans line 2, then line 3, and so forth, until all lines are scanned, at which point the beam jumps back to its starting position to repeat the scan of all lines.

quantizing A step in the digitization of an analog signal. It changes the sampling points into discrete values. Also called *quantization.*

refresh rate The number of complete digital scanning cycles (frames) per second.

RGB Stands for *red, green, and blue*—the basic colors of television.

sampling The process of reading (selecting and recording) from an analog electronic signal a great many equally spaced, tiny portions (values) for conversion into a digital code.

streaming A way of delivering and receiving digital audio and/or video as a continuous data flow that can be listened to or watched while the delivery is in progress.

SECTION

5.1 Analog and Digital Processes

Before you submerge yourself in the world of digital television and digital cinema, you should know the basics of analog and digital systems and how the standard scanning process works. You will also get some answers to why we converted to digital processes, especially because analog television had been doing a pretty good job all along.

- **WHAT DIGITAL IS ALL ABOUT**
  Why digital?
- **DIFFERENCE BETWEEN ANALOG AND DIGITAL**
  The digitization process: anti-aliasing, sampling, quantizing, and coding
- **BENEFITS OF DIGITAL TELEVISION**
  Quality, computer compatibility and flexibility, signal transport, and compression
- **ASPECT RATIO**
  The 4 × 3, 16 × 9, and mobile media aspect ratios
- **DIGITAL SCANNING STANDARDS**
  480p, 720p, and 1080i

WHAT DIGITAL IS ALL ABOUT

All ***digital*** computers and digital video are based on a ***binary*** code that uses the either/or, on/off values of 0's and 1's to interpret the world. The ***binary digit**, or **bit***, acts like a light switch: it is either on or off. If it is on, it is assigned a 1; if it is off, it is assigned a 0.

Why Digital?

At first glance this either/or system of binary digits may seem clumsy. For example, the simple and elegant decimal number 17 translates into the cumbersome 00010001 in binary code. The binary system uses the base-2 numbering system.[1]

The overwhelming advantage of the binary system is that it has great resistance to data distortion and error. In binary code, a light is either on or off. If the light flickers or burns at only half intensity, the digital system simply ignores such aberrations and reacts only if the switch triggers the expected on/off actions. It also permits any number of combinations and shuffling around—an extremely important feature when manipulating pictures and sound. We elaborate on the benefits of digital signals later in this chapter.

DIFFERENCE BETWEEN ANALOG AND DIGITAL

Before getting too technical, let's use a simple metaphor to explain the difference between analog and digital signal processing. The analog signal is like a ramp that leads continuously from one elevation to another. When walking up this ramp, it matters little whether you use small or big steps; the ramp gradually and inevitably leads you to the desired elevation. **SEE 5.1**

To carry on the metaphor, in the digital domain you would have to use steps to get to the same elevation. This is much more an either/or proposition. The elevation has now been quantized (divided) into a number of discrete units—the steps. You either get to the next step or you don't. There is no such thing as standing on a half or quarter step. **SEE 5.2**

The ***analog*** system processes and records a continuous signal that fluctuates exactly like the original signal (the line of the ramp). Digital processing, however, changes the ramp into discrete values. This process is called digitization. In the digitization process, the analog signal is continuously sampled at fixed intervals; the samples are then quantized (assigned a concrete value) and coded into 0's and 1's.

Digitization Process

Digitizing an analog video signal is a four-step process: anti-aliasing, sampling, quantizing, and coding. **SEE 5.3**

Anti-aliasing In this step extreme frequencies of the analog signal that are unnecessary for its proper sampling are filtered out.

1. The number 17 is represented by an 8-bit binary code. All values are mathematically represented by either 0's or 1's. An 8-bit representation of a single color pixel or sound has 2^8, or 256, discrete values.

5.1 ANALOG SIGNAL
The analog signal can be represented by a ramp that leads continuously to a certain height.

5.2 DIGITAL SIGNAL
The digital signal can be represented by a staircase that leads in discrete steps to a certain height.

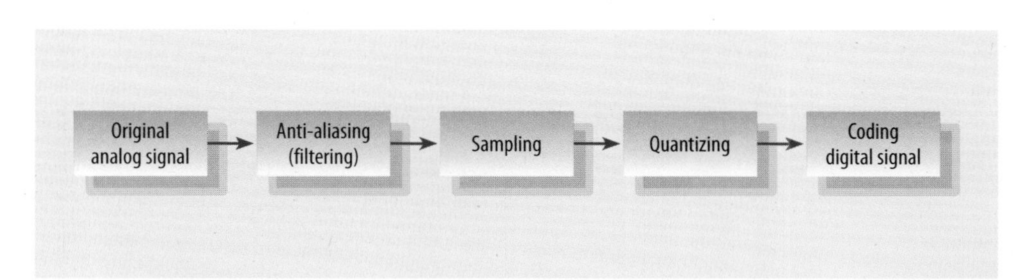

5.3 DIGITIZATION
The digitization of an analog signal is a four-step process: anti-aliasing, sampling, quantizing, and coding.

Sampling In the ***sampling*** stage, the number of points along the ramp (analog signal) is selected for building the steps (digital values). The higher the sampling rate, the more steps chosen and the more they will look like the original ramp (analog signal). Obviously, a high sampling rate (many smaller steps) is preferred over a low one (fewer but larger steps). SEE 5.4 AND 5.5 The sampling rate of a video signal is usually expressed in megahertz (MHz).

Quantizing At the ***quantizing*** digitization stage, we are actually building the steps so that we can reach the top of the staircase (which was previously the predetermined high end of the ramp) and assigning them numbers. Technically, *quantizing* means to separate a continuously variable signal into defined levels (steps) and fitting them into the desired sample range (the height of the ramp). For example, an 8-bit quantizing has a maximum number of 256 (2^8) levels. (In our metaphor we cannot use more than 256 steps.) SEE 5.6

Coding The process of ***coding*** (also called encoding) changes the quantization numbers of each step to binary numbers, consisting of 0's and 1's, and the various grouping of the bits (for us, steps). SEE 5.7

BENEFITS OF DIGITAL TELEVISION

Why go through all these processes? Wouldn't it be easier to simply walk up the (analog) ramp instead of climbing thousands or even millions of (digital) steps per second? Why do we prefer a partial (sampled) signal if we have the real thing—the complete analog signal—readily available? The simple answer is that the digital format has major advantages over the analog one: quality, computer compatibility and flexibility, signal transport, and compression.

Quality

Since long before the advent of digital video and audio systems, picture and sound quality have been a major concern

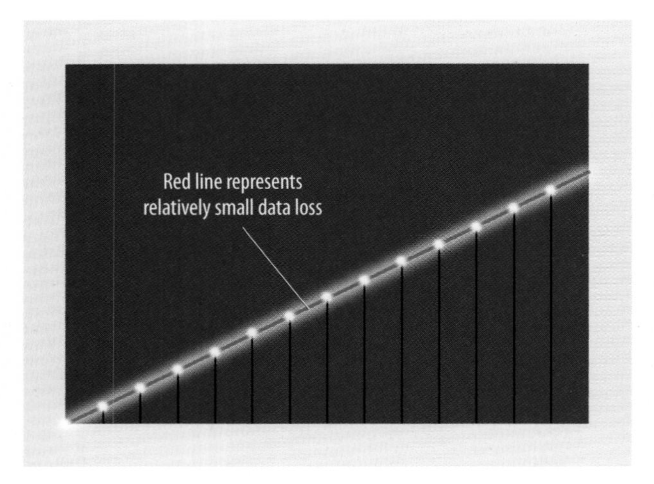

5.4 HIGH SAMPLING RATE

Sampling selects points of the original analog signal. A high sampling rate selects more points of the original signal. The digital signal comprises a higher number of smaller steps, making it look more like the original ramp. The higher the sampling rate, the higher the signal quality.

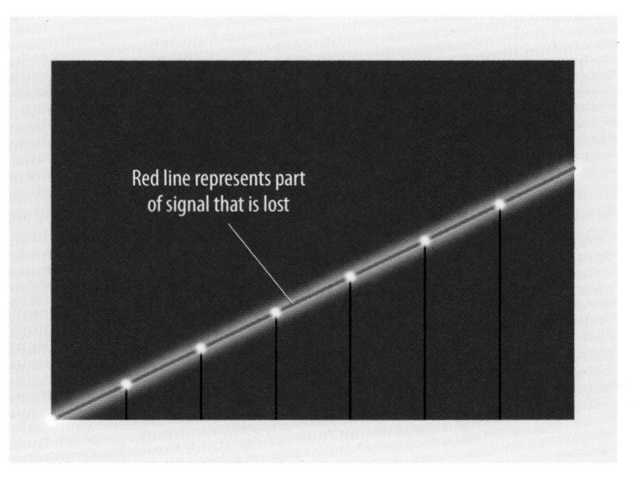

5.5 LOW SAMPLING RATE

A low sampling rate selects fewer points of the original analog signal. The digital signal is composed of a few large steps. Much of the original signal is lost.

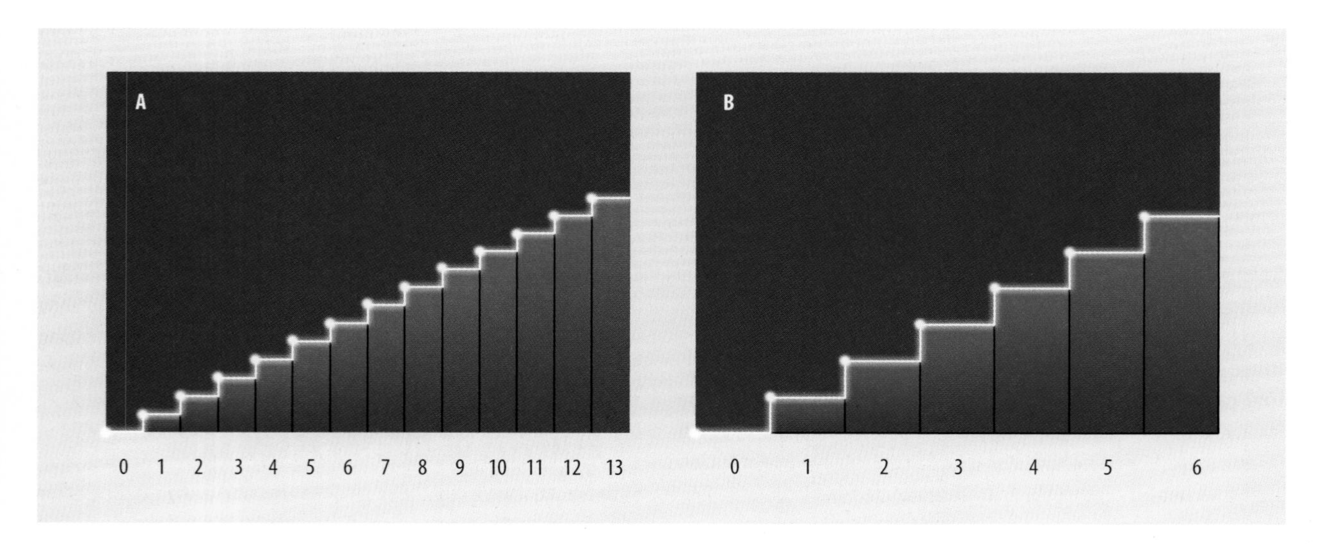

5.6 QUANTIZING

Quantizing assigns fixed positions to the selected signal samples. This is the step-building stage. Each step gets a particular number assigned. **A** High sampling rate: many small steps. **B** Low sampling rate: fewer large steps.

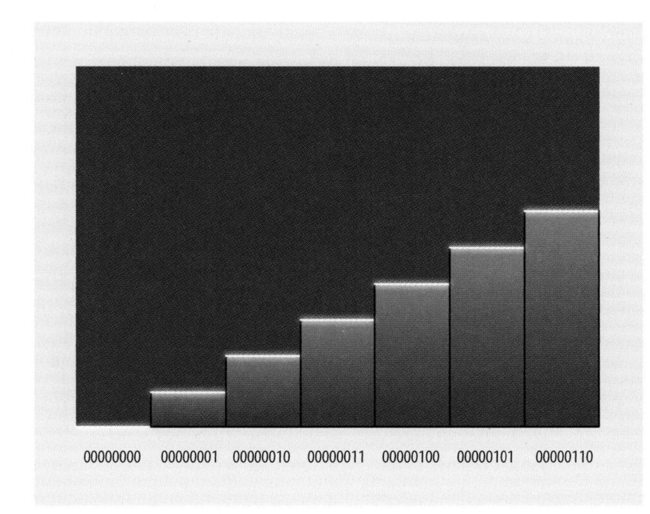

5.7 CODING
Coding, or encoding, assigns each step a binary number and groups the steps in a specific way.

of equipment manufacturers and production personnel. Such high-resolution picture quality is especially important for extensive postproduction. In the days of analog video production, complex editing and the rendering of special effects required many tape generations (the number of dubs—copies—away from the original). Unfortunately, the higher the number of generations in analog recordings, the greater the loss of quality. This is not unlike making progressive copies of a letter by photocopying each previous copy. Before long the print has deteriorated so much that you can hardly read it.

But this is where digital recordings shine. First, even small inexpensive camcorders deliver high-quality pictures and sound. Second, and more important, with digital recordings there is hardly any noticeable quality loss even after dozens of generations; the fiftieth generation looks as sharp as the original recording. In fact, through some digital wizardry you can make a copy look even better than the original. Another important quality factor is that the simple binary code is relatively immune to extraneous electronic signals—noise or artifacts—that infiltrate and distort analog signals. With digital signal processing, electronic noise is held to a minimum if not eliminated altogether.

There is a trade-off, however. High-fidelity video is often criticized for its sharpness and "in-your-face" quality by film people, who lament the lack of the softer "film look." Similarly, digital music recordings sometimes sound so crisp and clean that they lack the warmth and the texture of the original piece—or even of an analog recording. You may remember the monotone sounds of synthesized computer speech; it was missing all the complexity and subtleties (overtones) of actual speech. Audio professionals are using higher sampling rates and more complex digital signal combinations to make up for this deficiency. Paradoxically, video filters are available to make HDTV pictures less high-definition, and in digital audio a certain amount of noise is reintroduced to contribute to the "warmth" of sound. By now, however, the cameras used for digital movies and their theater projectors have matched, or even exceeded, the subtle brightness steps of film.

Computer Compatibility and Flexibility

Another big advantage of digital television is that its signals can be transferred directly from the camera to the computer without the need for digitization. The elimination of this step is especially welcome to news departments, whose members work under tight deadlines. It is also a great relief to postproduction editors, who can devote more time to the art of editing rather than sitting idle during the digitizing process.

The flexibility of the digital signal is especially important for creating special effects and computer-generated images. Even a simple weathercast or a five-minute newscast features a dazzling display of digital effects that was all but impossible with analog equipment. The opening animated title, the scene that expands full-screen from the box over the newscaster's shoulder, and the graphical transition from one story to the next where one picture peels off to reveal another underneath—all show the variety and the flexibility of digital effects. The multiple screens-within-the-screen and the various lines of text that run along the bottom, sides, or top of the main television frame are possible only through digital video effects (DVE). Computer software that allows the alteration or even the synthetic creation of audio and video images has become an essential digital production tool. Some of the spectacular movies are ample proof of such digital wizardry.

Signal Transport

Digital television, such as HDTV, differs from analog television not only by how the pictures can be manipulated but also by how they are transported. Much like the NTSC (which stands for *National Television System Committee*) standard for analog television, the ***ATSC***—the Advanced Television Systems Committee—developed the standards for digital television. Whereas NTSC analog television continuously sends picture halves (scanning fields) that form complete pictures (frames), ***ATSC signals*** are composed of picture and sound information in digital packets, each of which contains only part of a television picture and the accompanying sound plus instructions for the receiver on how to reassemble the parts into complete video pictures and sound. These two systems are obviously not compatible.

Such digital signal packets can be distributed through a variety of wires (coaxial and fiber-optic cables) or wirelessly (through a broadband router or via wireless fidelity, or WiFi). If you get impatient while waiting for a large computer file to up- or download, think about the incredible amount of data that needs to be pushed through the conduit you are using. A download speed of only 1 megabyte per second (MBps) means the computer is processing well over 8 million on/off choices (8,388,608 bits) per second.

Some of you may still be a little confused about the difference between downloading and data streaming. When you are ***downloading***, you are receiving data that are sent in packets. Because these data packets are usually sent out of order to make full use of the available transmission line, you cannot call up the entire file until the downloading process is complete. With ***streaming***, you receive digital audio and/or video data as a continuous data flow. Because the data stream is continuous and not converted into out-of-order packets, you can listen to the music or watch the initial video frames while the files for the subsequent frames are still being transferred.

But even the most efficient transmitters—broadband (multichannel) and Ethernet (single-channel) coaxial conduits—lack the capacity to stream fast enough the huge amount of data of HDTV and, even more so, interactive digital television. This is where compression comes in.

Compression

Compression is the temporary rearrangement or elimination of redundant information for easier storage and signal transmission. Remember the interlaced scanning process? Cutting each picture (frame) into two incomplete halves (fields) is done because half of a picture can be transported more easily than a whole one. Technically, a field takes up less bandwidth (data transmission capacity) than a frame. This is a crude but ingenious way of compressing an otherwise untouchable analog signal or the superlarge files of high-end HDTV for quicker transport.

Fortunately, digital signals are much more flexible in this respect and can be compressed in many different ways, called ***codecs*** (an acronym of *co*mpression/*dec*ompression). One such system does it by regrouping the original data without throwing any away. Once at the destination, the data can be restored to their original positions—a process called ***decoding***—for an output that is identical to the original input. We do this frequently when "zipping" or "stuffing" large computer files for storage and transmission and then unzipping or unstuffing them when opening the file. Or you can simply delete all data that are redundant.

Compression that results from rearranging or repackaging data is called lossless—the regenerated image has the same number of pixels and values as the original. When some pixels are eliminated in some frames because they are redundant or beyond our ordinary perception, the compression is called lossy. Even if the lost pixels are not essential for the image creation, the regenerated image is nevertheless different from the original. The obvious advantage of lossless compression is that the original image is returned without diminished quality. The disadvantage is that it takes more storage space and usually takes more time to transport and bring back from storage. Most image compression techniques are therefore the lossy kind.

To accommodate the need for transporting larger and larger files at higher and higher speeds, new codecs are constantly being developed and old ones refined. To list even the major new ones would be a futile exercise, but you should nevertheless be familiar with some of the standard ones that have been in use for quite some time.

One of the most common digital compression standards for still images is JPEG ("jay-peg"), named for the organization that developed the system—the Joint Photographic Experts Group; motion-JPEG is for moving computer images. Although a lossless JPEG technique exists, to save storage space most JPEG compressions are lossy.

Another compression standard for high-quality video is MPEG-2 ("em-peg two"), named and developed by the Moving Picture Experts Group. MPEG-2 is also a lossy compression technique, based on the elimination of redundant information from frame to frame. MPEG-4 was originally intended for streaming data. It is such an efficient codec, however, that it was modified in several ways to serve also as an effective codec for other digital processes.[2]

For example, besides offering highly efficient compression, MPEG-4 allows you to compress and encode mixed media (video, audio, and speech) all in one operation, which is especially important when using the relatively

2. Several of the major codecs and many other types of digital high-definition video and audio technology have been standardized by the ATSC. For a lucid and concise discussion of analog and digital systems as well as transmission standards, see Graham Jones, *A Broadcast Engineering Tutorial for Non-Engineers*, 4th ed. (Burlington, Mass.: Focal Press, 2011).

limited storage capacity of memory cards in your video recorder. It is also highly robust, which means that its signals are not apt to distort during various kinds of transport.

The downside of having so many codecs in operation is that they are not all compatible. For example, if your camcorder uses MPEG-2 as the recording codec, you may not capture or play back the recorded material with your editing software. Some popular nonlinear editing programs simply refuse to accept MPEG-4 footage. Some codecs are designed for PCs but will not run on Macs and vice versa. Before you buy a new camcorder, make sure its codec is compatible with your computer software. For more information about incompatible codecs, see chapters 13 and 19.

ASPECT RATIO

One of the most visible differences between analog and digital television systems is the horizontally stretched television picture of HDTV. The HDTV ***aspect ratio***—the width-to-height proportions of the screen, also standardized by the ATSC—resembles more a small motion picture screen than the traditional television screen. Aspect ratios are explored extensively in chapter 8, but here we'll take a brief look at the main characteristics of the two principal aspect ratios.

4 × 3 Aspect Ratio

Dating back to the earliest motion picture screens, the aspect ratio of the traditional television screen and some computer screens is 4 × 3, which means that its frame is 4 units wide by 3 units high, regardless of whether the units are inches or feet. This aspect ratio is also expressed as 1.33:1. For every unit in screen height, there are 1.33 units in width. **SEE 5.8**

The advantage of this classic aspect ratio is that the difference between the screen width and the screen height is not pronounced enough to unduly emphasize one dimension over the other. A close-up or an extreme close-up of a face fits well in this aspect ratio as does a horizontally stretched landscape.[3] The disadvantage is that it does not accommodate wide-screen movies that have the much more horizontally stretched aspect ratio of 1.85:1.

16 × 9 Aspect Ratio

The horizontally stretched aspect ratio of HDTV systems is 16 × 9; that is, the screen is 16 units wide by 9 units high, or 1.78:1. As you can see, this aspect ratio resembles that of a movie screen. **SEE 5.9**

Mobile Media Aspect Ratios

The small displays of mobile media, such as cell phones, iPhones, and iPods, vary from the traditional 4 × 3 format to wide-screen, square, and even vertical. On some devices you can vary the aspect ratio, depending on whether you hold it vertically or sideways. **SEE 5.10** More significant is its small size, which requires a new approach to visualization. We will discuss this subject in more detail in section 8.2.

3. See Herbert Zettl, *Sight Sound Motion: Applied Media Aesthetics,* 6th ed. (Boston: Wadsworth, 2011), pp. 84–92.

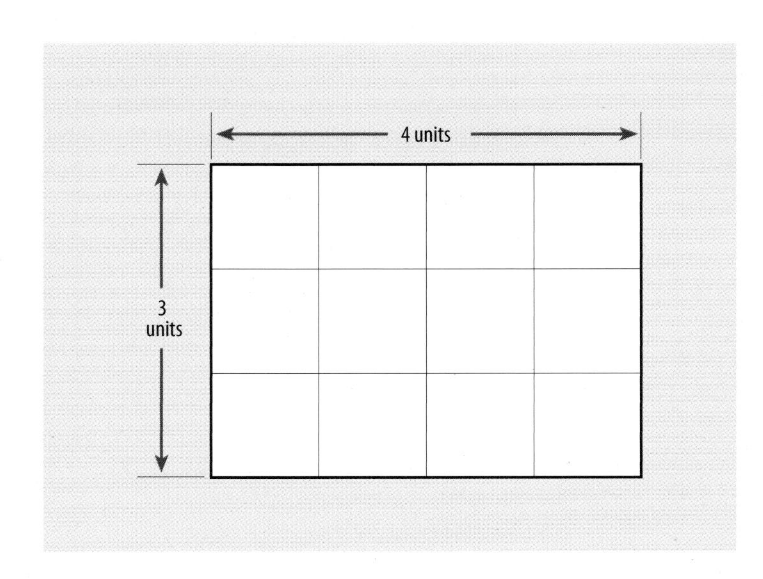

5.8 4 × 3 ASPECT RATIO
The traditional aspect ratio of the television screen is 4 × 3 (4 units wide by 3 units high). It can also be expressed as 1.33:1 (1.33 units in width for each unit of height).

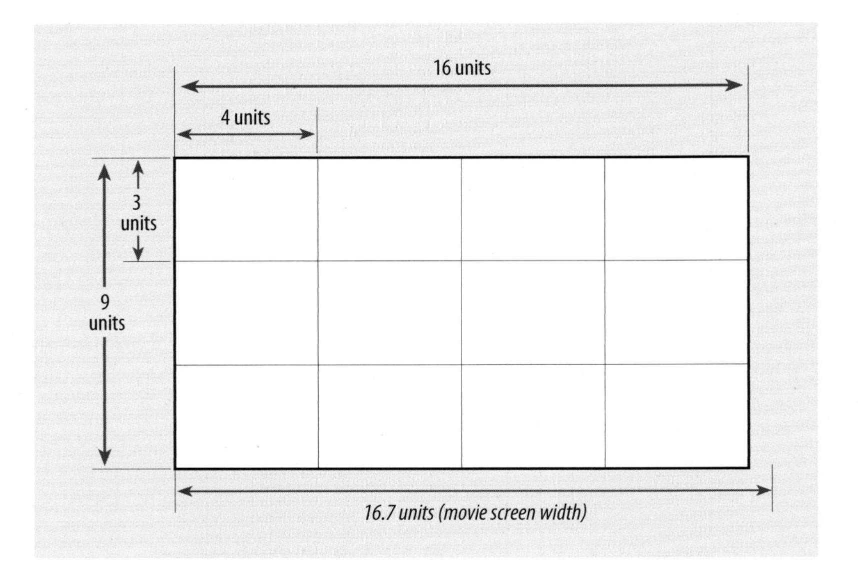

5.9 16 × 9 ASPECT RATIO
The aspect ratio of HDTV is 16 × 9 (16 units wide by 9 units high), which is a multiple of the 4 × 3 ratio ($4^2 \times 3^2$). Its horizontally stretched aspect ratio of 1.78:1 resembles that of the movie screen (1.85:1).

5.10 MOBILE MEDIA ASPECT RATIO
This cell phone has a vertical aspect ratio, which can be masked when showing television shows or films.

5.10 CELL PHONE PHOTO COURTESY NOKIA; NEWSCASTER PHOTO BY EDWARD AIONA

DIGITAL SCANNING STANDARDS

The digital scanning standard most frequently used for SDTV (standard-definition television) is 480p (480 visible lines progressively scanned); for HDTV they are 720p (720 visible lines progressively scanned) and 1080i (1080 visible interlaced scanning lines). Section 5.2 explores these standards in more detail.

MAIN POINTS

- Digital computers use binary code, consisting of 0's and 1's. This code resists data error.
- In the digital process, the analog signal is continuously sampled at specific intervals. The samples are then quantized (assigned a discrete value) and coded into groups of 0's and 1's.
- Digital television produces pictures and sound of superior quality, allows many copies with virtually no signal deterioration, provides great flexibility in image creation and manipulation, and permits data compression codecs for efficient signal transport and storage.
- The main difference between NTSC analog and ATSC digital TV signal transport is that the NTSC system sends continuously signals that make up television frames (pictures), whereas the ATSC system sends packets of data that must be interpreted and assembled by the television receiver for complete pictures and sound.
- Compared with the traditional television aspect ratio of 4 × 3 (1.33:1), HDTV systems have a wider aspect ratio of 16 × 9 (1.78:1). Mobile media displays have a variety of horizontal and vertical aspect ratios.
- The digital scanning standards are 480p, 720p, and 1080i.

SECTION

5.2 Scanning Systems

This section takes a closer look at how the basic analog color video image is created, interlaced and progressive scanning, major digital scanning systems, and digital flat-panel displays. Why bother about analog video if everything is digital anyway? There are several reasons.

First and most importantly, the imaging device (called the sensor) in your camcorder works much like an old-fashioned television set. Its surface consists of many red, green, and blue pixels that are scanned line after line by an electron beam. This scanning translates the light image as delivered by the lens into analog electrical signals. These analog signals are then digitized and further processed for digital video. Second, the analog image creation offers an easier way to understand the basic video image formation and color mixing than if you were to be confronted at this point with the complex technical workings of flat-panel video displays. Third, there are still many CRT (cathode ray tube) displays in use, especially in the high-quality (and initially very expensive) monitors in television control rooms. And, fourth, it will give you an important historical perspective of the development of video.

BASIC IMAGE CREATION

To demonstrate how the basic analog video image is created, we will put the cart before the horse and show how the video image is re-created on the cathode ray tube of a standard (old-fashioned) black-and-white television set during playback.

The video image is literally drawn onto the television screen by an electronic pencil—the electron beam. Emitted by the electron gun, the electron beam scans the inside surface of the television screen line by line, from left to right, much as we read. The inside of the television screen is dotted with thousands of light-sensitive picture elements, or pixels (tiny round, rectangular, or diamond-shaped dots), that light up when hit by the beam. If the beam is powerful, the dots light up brightly. If the beam is weaker, the dots light up only partially. If the beam is really weak, the dots don't light up at all. The process is similar to the large displays that use light bulbs for outdoor advertising, except that the light bulbs on the television screen are extremely tiny. SEE 5.11

BASIC COLORS OF THE VIDEO DISPLAY

All the beautiful images that you see on television—even the black-and-white pictures—are a mixture of three basic colors: red, green, and blue. Depending on how hard the pixels are hit by an electron beam, they light up at different intensities. Mixing these intensities produces all the other colors. Each line consists of groups of ***RGB*** (red, green, and blue) pixels, which are hit by a separate electron beam for

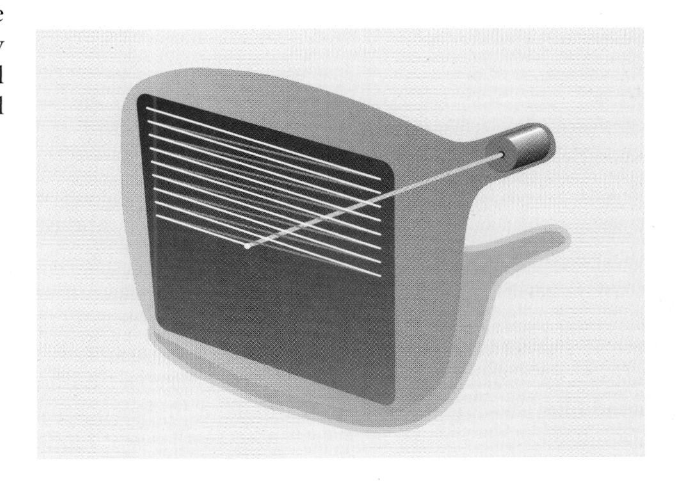

5.11 SCANNING PRINCIPLE
The electron gun in the back of the black-and-white picture tube produces an electron beam that scans the many light-sensitive pixels located on the inside of the tube's face.

Shadow mask
Electron guns
Red
Green
Blue
Blue dot
Green dot
Red dot

5.12 IMAGE FORMATION FOR COLOR TELEVISION
The color receiver has three electron guns, each responsible for either a red, a green, or a blue signal. Each of the beams is assigned to its color pixels. The shadow mask keeps the beams from spilling into the adjacent pixels.

each basic color: one for the red dots, a second for the green dots, and a third for the blue ones. Each electron beam hits its assigned color with various intensities, producing the different color mixes. Just how these three colors create all the others is explored in chapter 6. SEE 5.12

INTERLACED AND PROGRESSIVE SCANNING

As mentioned earlier in this chapter, there are basically two ways of creating a video image: one is through interlaced scanning, and the other is through progressive scanning.

Interlaced Scanning

To produce an image through ***interlaced scanning***, the electron beam scans the odd-numbered lines first; then it jumps back to the top of the screen and scans the even-numbered lines. The complete scan of all odd-numbered *or* even-numbered lines, which takes ¹⁄₆₀ second, is called a ***field***. A complete scan of all odd- *and* even-numbered lines is called a ***frame***. In the traditional NTSC system, there are 30 frames per second. The two fields are combined—interlaced—for each frame. In effect, interlaced scanning combines low-quality pictures to make a higher-quality one. Because the beam is such a speed-reader and lights up the pixels at a pretty fast clip, we perceive the two scans as a single video image. Interlaced scanning is still used for digital video because it cuts the huge amount of information in half for signal transport. SEE 5.13

5.13 INTERLACED SCANNING

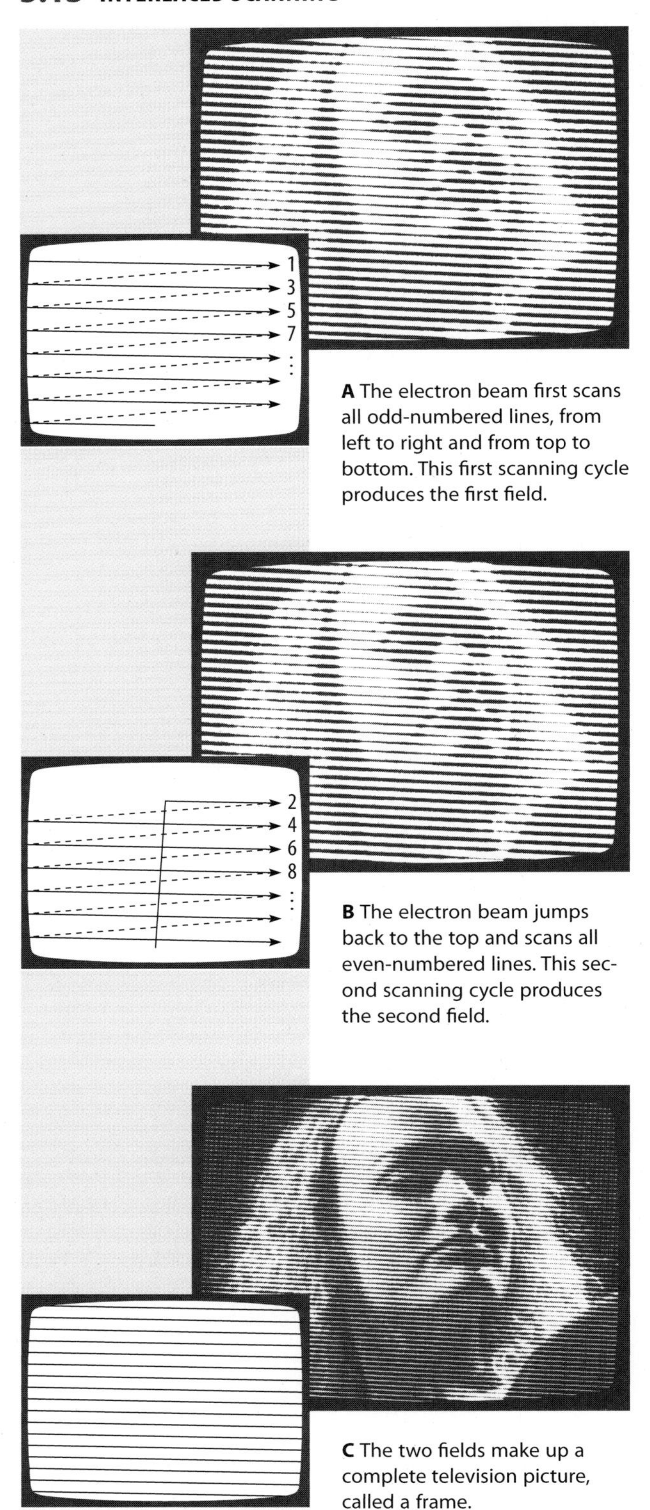

A The electron beam first scans all odd-numbered lines, from left to right and from top to bottom. This first scanning cycle produces the first field.

B The electron beam jumps back to the top and scans all even-numbered lines. This second scanning cycle produces the second field.

C The two fields make up a complete television picture, called a frame.

5.13 PHOTO BY EDWARD AIONA

Progressive Scanning

The ***progressive scanning*** system scans every line from top to bottom just as we read. A single scan produces a complete frame. The electron beam in the progressive system starts at the top left of the screen and scans the first line, then jumps back to the left at the start of the second line, scans the second line, jumps back to the third line, scans the third line, and so on. As soon as the scanning of a frame is complete, the beam jumps back to its original starting point at the top left and starts scanning the second frame, and so forth. As you can see, the beam scans all lines progressively, hence the name of the system. **SEE 5.14**

In the progressive system, the ***refresh rate***—how often the beam jumps back to scan another frame—can vary from 24 fps (frames per second) to 60 fps or more during video recording. Whatever frame rate you choose during production, it is usually standardized to 60 fps by the television receiver to avoid visible flicker.

Retrace and Blanking

Both the interlaced and progressive scanning systems use retrace and blanking. The repositioning of the beam from the end of the scanned line to the starting point of the next is called horizontal retrace. When the beam reaches the end of the last line and jumps back to the starting point of line 1 on top, it is referred to as vertical retrace. To avoid any picture interference during the horizontal and vertical retraces, the beam is automatically "starved" so that it won't light up any pixels that might interfere with the original scan; this process is called blanking. Hence, horizontal blanking occurs during the horizontal retrace, and vertical blanking is during the vertical retrace. Let's apply the two scanning systems to DTV and see how they fare.

DIGITAL SCANNING SYSTEMS

After years of wrangling over the former ATV (advanced television) and DTV (digital television) scanning standards, the ATSC has set four system standards: 480p, 720p, 1080i, and 1080p.

480 System

The ***480i*** system was the standard of STV (standard analog television). This means that all video consisted of 525 lines, of which only 480 or so were visible on the television screen. The *i* referred to interlaced scanning, meaning that each frame consisted of two fields, each composed of 252½ scanning lines. The 480i system produced 60 fields per second, which combined into 30 frames per second.

In the digital format, the 480i system changes into a ***480p*** system. This means that all 480 active lines are scanned progressively every 1/60 second. Let's take a closer look at these numbers.

As you can see, the 480p system has the same number of scanning lines as does standard television, but because the beam in progressive scanning reads all the lines before it jumps back to begin reading the next screen, progressive scanning generates a complete frame in each scanning cycle. Instead of the 60 fields, or 30 frames, per second of standard analog television, the 480p system generates 60 complete frames per second. Because you see 60 frames per second instead of 30, the pictures look better in DTV than in analog STV. In technical lingo, the increased number of frames per second results in a higher *temporal resolution.*

720p System

The true high-definition television images in the ***720p*** system are a product of both the 720 visible, or active, lines (of 750 actual scanning lines) that are scanned progressively

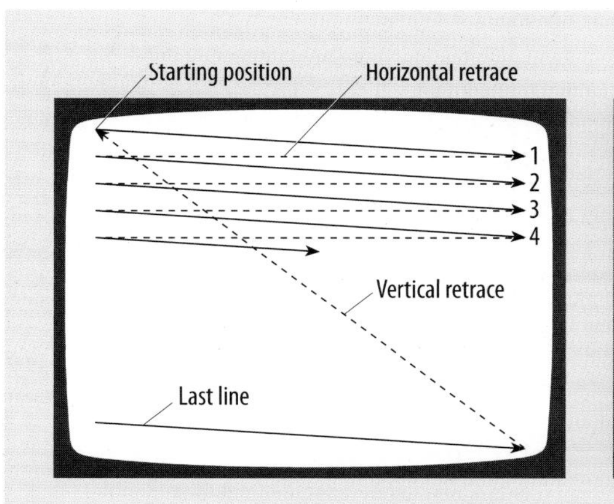

5.14 PROGRESSIVE SCANNING
In progressive scanning, the beam reads every line from top to bottom. Each complete scan produces a television frame. Retrace lines (shown as dashed in the figure) are blanked so that they don't appear on-screen.

5.14 PHOTO BY EDWARD AIONA

and the refresh rate of 60 fps (all lines are scanned every 1⁄60 second). Even though the number of frames per second is identical to that of STV (60 fps), the increased number of lines in the 720p system contributes to a superior picture. Technically, it represents a vastly improved *spatial resolution* of each frame.

1080i System

The ***1080i*** system (1,080 visible lines of 1,125 total lines) uses interlaced scanning. Much like with standard NTSC, each field of 540 visible lines is scanned every 1⁄60 second, producing 30 frames per second. The high number of scanning lines of the 1080i system dramatically improves the (spatial) resolution of the television picture—at the cost of requiring a large bandwidth for signal transport. As with the standard NTSC signal, the interlacing is primarily used as a compression device. To push half of a picture through the transmission pipeline takes up less space (bandwidth) than does a complete picture.

1080p System

The ***1080p*** system (1,080 visible lines of 1,125 total lines) uses progressive scanning. All visible lines are scanned for each frame. 1080p video cannot be broadcast because of its excessively high bandwidth. It is usually reserved for digital cinema, which uses the cinematic 24 frames per second; but this low refresh rate is generally boosted to 60 fps during playback (but not for transmission) to avoid flicker.

Maintaining Quality

In the end, it depends on how much of the original picture quality is maintained during the entire production process and, especially, during signal transmission. The problem with all these scanning standards is that what you get is not necessarily what you see. Because the signals are digital, the digital display (what the television receiver shows) does not have to mirror exactly what was sent. For example, a DTV set may receive an interlaced frame but show it as a progressive scan. It can also show the frames at a different refresh rate from what was delivered to the receiver.

To make things even more complicated, each of the scanning formats (480p, 720p, and 1080i) can have a variety of refresh rates. You may, for example, assign an HDTV camera to shoot at a frame rate of 24p (24 fps) but send it as a 60i (60 fields, or 30 fps) sequence. To fool you into an even higher resolution, the receiver may decide to double the refresh rate and show the sequence at 60 frames instead of 60 fields per second. It all boils down to giving you as sharp a picture as possible without taking up too much transmission space and time to deliver it.

Regardless of the relative picture quality of the three scanning standards, like any other system all are ultimately dependent on the program content. A bad program remains bad even when received in digital HDTV; a good program is good even if the picture quality is slightly inferior. Note, however, that picture quality becomes a real issue when using an HDTV system for instructional or training purposes, such as medical programs, and, not too surprisingly, commercials.

FLAT-PANEL DISPLAYS

Because there is a limit to the size of the cathode ray tube of the traditional television receiver, the industry has turned to flat-panel displays, such as the one on a laptop computer. The advantage of flat-panel displays over regular television receivers is that flat panels can be made very large without getting thicker or losing their resolution. In fact, a flat-panel display resembles a large painting with a modest frame. Even large flat panels can be hung on a wall like a picture. Their major disadvantages are their relatively high power consumption and cost.

As always with video technology, there are two different, incompatible types of flat-panel displays that can reproduce high-definition video images: the plasma display panel and the liquid crystal display. Because any accurate technical explanation of how flat-panel displays create video images would go beyond the intended scope of this book, we will simply explain the basic principles and differences of the two types.

Although flat panels work technically in totally different ways from the traditional CRT television system, they nevertheless maintain the basic principle of video image creation: activating a great number of specific red, green, and blue pixels in various intensities so that they form a color video image. The traditional CRT television set makes the screen light up by hitting groups of pixels with sharp electron beams; the flat-panel displays let the light of various pixels shine through the front surface.

Both types of flat-panel displays are capable of reproducing all STV and HDTV scanning systems. They can upgrade low scanning standards to higher ones and downgrade higher standards to lower ones if necessary. For example, some displays can simulate a 720p resolution by doubling some lines of a 480i input, or they can reduce a 1080i input to 720p if the set has a 720-line display limit.

Plasma Display Panel

The plasma display panel (PDP) uses two transparent (usually glass) wired plates that sandwich a thin layer of gas.

When the gas receives the voltages of the video signal, it activates the RGB pixels that are arranged very much like those of the standard television receiver and lets them shine through the top glass plate.

The advantages of a PDP are that it has an excellent contrast ratio (more than 1,000:1, which means the brightest picture spot can be at least a thousand times brighter than the darkest spot) and superior color renditions. And you can watch a plasma screen even when standing way to the side without any color or contrast deterioration. The PDP shows little or no lag when quick motion is displayed.

One of the disadvantages of a plasma display is that still pictures have a tendency to "burn in" when left on the screen too long. This means that the static image will remain on the screen as a faint image ("burn" or "ghost") even when other images are displayed. Depending on the manufacturer, some burn in relatively quickly (in less than an hour); others take several hours. If left on even longer, such a burn becomes permanent—in effect ruining the display. To ensure that a PDP maintains its quality, a static image should not be left on-screen for more than 15 minutes. Over time the pixels will not light up as brightly as before, which will cause the whole display to look dimmer. You can prevent this premature aging by not setting the brightness and contrast levels too high and by turning off the set when not in use.

Liquid Crystal Display

The liquid crystal display (LCD) panels also use two transparent plates, but instead of gas the panels sandwich a liquid whose crystal molecules change when an electric current is applied. Depending on the electric charge, the crystals either block the light shining through the glass or twist to let certain colors shine through. Laptop computers, digital clocks and telephones, and many other consumer electronics use an LCD.

The advantages of LCD panels are that the display is very bright, they do not deteriorate as fast as plasma screens, they do not burn in, and they are more economical than PDPs in power consumption. If you look really closely, you may see that LCDs have a slightly sharper image than plasma screens of the same size.

A disadvantage is a possible limited viewing angle when using cheap displays: you cannot stand or sit too far to the side of the panel or tilt it back too far without having the dark areas of the picture appear washed out. Other minor disadvantages are that very-fast-moving objects seem slightly blurred (aesthetically not really a disadvantage) and that, at least technically, the colors are not quite as subtle and accurate as in a plasma display.

Despite the slight differences between the two flat-panel display systems, both are capable of producing stunning high-definition pictures. Unfortunately, both types have proved to be anything but economical in power consumption. So, turn off your set if you are no longer watching.

MAIN POINTS

- The traditional television set is a good vehicle for explaining how to create a video image. On a television set that uses a CRT (cathode ray tube), the electron beam scans the color pixels on the inside of the CRT. The harder the pixels are hit, the brighter they light up.
- The basic colors used in television are red, green, and blue (RGB). Each of the 480 visible lines on the face of the display tube consists of groups of red, green, and blue pixels (very tiny round, rectangular, or diamond-shaped dots). Three electron beams activate these dots—one beam for the red pixels, one for the green, and one for the blue. The varying intensities of the three beams produce the colors we see on television.
- With interlaced scanning, the beam skips every other line during its first scan, reading only the odd-numbered lines. After the beam has scanned half of the last odd-numbered line, it jumps back to the top of the screen and finishes the unscanned half of the top line and continues to scan all the even-numbered lines. Each such even- or odd-numbered scan produces a field. Two fields produce a complete video frame.
- In the progressive scanning system, the electron beam scans each line, starting with line 1, then line 2, then line 3, and so on. When all lines have been scanned, the beam jumps back to its starting point to repeat the sequential scanning of all lines. Each scan of all lines results in a video frame.
- The most common refresh rate of the 480p and 720p systems is 60 frames per second (fps), whereas for the 1080i system it is 30 fps. Many high-definition television (HDTV) systems have a variable frame rate but are usually displayed at 60 fps.
- Digital television (DTV) employs four principal scanning formats: 480p (480 lines progressively scanned), 720p (720 lines progressively scanned), and 1080i or 1080p (1,080 lines with interlaced or progressive scanning, respectively). All have a 16 × 9 aspect ratio but can be switched to the traditional 4 × 3 ratio.
- The two types of flat-panel video displays are the plasma display panel (PDP), which sandwiches gas between two transparent plates, and the more common liquid crystal display (LCD), which sandwiches a liquid between two transparent plates.
- The PDP activates RGB pixels; the LCD activates a number of tiny transistors that let light shine through according to the charge they receive.

CHAPTER

6

The Television Camera

The television camera is still the single most important piece of production equipment. In fact, you can produce and show an impressive short video piece solely with a camcorder that is small enough to fit in your shirt pocket. Almost all other production equipment and techniques revolve around the camera or are greatly influenced by its technical and performance characteristics.

For example, to develop the ubiquitous camcorder, the video recorder had to be made so small that it would fit inside a camera. Because all video cameras and camcorders have become more light sensitive, we can now get by with fewer and smaller studio and field lighting instruments. The studio lights are suspended from the ceiling to facilitate the unrestricted travel of cameras during a multicamera show. Microphones are so small that they are hardly visible when clipped on people's clothing. Scenery must be constructed and makeup applied with more care because high-definition (HD) cameras deliver a sharper picture than their ***standard television (STV)*** cousins.

To maximize the full potential of the camera, you need to know how it works.

Section 6.1, How Television Cameras Work, identifies the parts, types, and electronic and operational features of cameras. Section 6.2, Resolution, Contrast, and Color, elaborates on image resolution, image contrast, and the basics of color.

KEY TERMS

beam splitter Compact internal optical system of prisms within a television camera that separates white light into the three primary colors: red, green, and blue. Also called *prism block.*

brightness The color attribute that determines how dark or light a color appears on the monochrome television screen or how much light the color reflects. Also called *lightness* and *luminance.*

camcorder A portable camera with the videotape recorder or some other recording device built into it to form a single unit.

camera chain The television camera (head) and associated electronic equipment, including the camera control unit, sync generator, and power supply.

camera control unit (CCU) Equipment, separate from the camera head, that contains various video controls, including color fidelity, color balance, contrast, and brightness. The CCU enables the video operator to adjust the camera picture during a show.

charge-coupled device (CCD) The imaging sensor in a television camera. It consists of horizontal and vertical rows of tiny image-sensing elements, called pixels, that translate the optical (light) image into an electric charge that eventually becomes the video signal.

CMOS A camera imaging sensor similar to a CCD but which operates on a different technology. It translates light into an electronic video charge that eventually becomes the video signal.

contrast ratio The difference between the brightest and the darkest portions in the picture (often measured by reflected light in foot-candles). The contrast ratio for low-end cameras and camcorders is normally 50:1, which means that the brightest spot in the picture should be no more than 50 times brighter than the darkest portion without causing loss of detail in the dark or light areas. High-end digital cameras can greatly exceed this ratio and can tolerate a contrast ratio of 1,000:1 or more.

digital cinema camera A high-definition television camera with sensors that can produce extremely high-resolution pictures exceeding 4,000 (4K) pixels per line. It records on videotape or memory cards, with a variable frame rate for normal, slow, and accelerated motion capture.

EFP camera High-quality portable, shoulder-mounted field production camera that must be connected to an external video recorder.

ENG/EFP camcorder High-quality portable field production camera with the recording device built-in.

grayscale A scale indicating intermediate steps from TV white to TV black. Usually measured with a nine-step scale for standard television. HDTV and digital cinema cameras deliver many more steps.

high-definition television (HDTV) camera Video camera that delivers pictures of superior resolution, color fidelity, and light-and-dark contrast; uses high-quality imaging sensors and lenses.

high-definition video (HDV) A recording system that produces images of the same resolution as high-definition television (720p and 1080i) with equipment that is similar to standard digital video camcorders. The video signals are much more compressed than those of HDTV, however, which results in lower overall video quality.

hue One of the three basic color attributes; hue is the color itself—red, green, yellow, and so on.

imaging device The imaging element in a television camera. Its sensor (CCD or CMOS) transduces light into electric energy that becomes the video signal. Also called *chip* and *sensor.*

pixel Short for *picture element.* (1) A single imaging element (like the single dot in a newspaper picture) that can be identified by a computer. The more pixels per picture area, the higher the picture quality. (2) The light-sensitive elements on a CCD that contain a charge.

resolution The measurement of picture detail, expressed in the number of pixels per scanning line and the number of visible scanning lines. Resolution is influenced by the imaging device, the lens, and the television set that shows the camera picture. Often used synonymously with *definition.*

saturation The color attribute that describes a color's richness or strength.

sensor The CCD or CMOS imaging device in a video camera.

standard television (STV) A system based on the NTSC scanning system of 525 (480 visible) interlaced lines.

studio camera High-quality camera with a large zoom lens that cannot be maneuvered properly without the aid of a studio pedestal or other camera mount.

sync generator Part of the camera chain; produces an electronic synchronization signal, which keeps all scanning in step.

white balance The adjustments of the color circuits in the camera to produce a white color in lighting of various color temperatures (relative reddishness or bluishness of white light).

SECTION

6.1

How Television Cameras Work

Although the electronics of the television camera have become increasingly complex, the cameras have also become more user-friendly. As you probably know from operating your own camcorder, you don't have to be a skilled electronics engineer to produce an optimal image—all you need to do is press the right buttons. Nevertheless, pressing the right buttons implies that you know how a television camera basically works. This section will help you maximize the potential of the camera and understand the specific functions of various camera types.

- **PARTS OF THE CAMERA**
  The lens, the camera itself, and the viewfinder
- **FROM LIGHT TO VIDEO SIGNAL**
  Imaging device, beam splitter, and color filter array
- **CAMERA CHAIN**
  Camera control unit, sync generator, and power supply
- **TYPES OF CAMERAS**
  Studio cameras, EFP cameras, ENG/EFP camcorders, and digital cinema cameras
- **ELECTRONIC FEATURES**
  Imaging device, scanning, resolution, aspect ratio and essential area, gain, electronic shutter, white balance, and audio channels
- **OPERATIONAL FEATURES**
  Power supply, camera cables and connectors, filter wheel, viewfinder, tally light, and intercom

PARTS OF THE CAMERA

When you take vacation pictures with your camcorder, probably the last thing on your mind is what makes a video camera work. But if you were to open up a camera (not recommended) and see the myriad electronic elements and circuits, you would probably wonder how it functions at all. Despite their electronic complexity, all television cameras (including consumer video cameras) consist of three main parts.

The first is the lens, which selects a certain field of view and produces an optical image of it. The second part is the camera itself, with its imaging device that converts into electrical signals the optical image as delivered by the lens. The third is the viewfinder, which shows a small video image of what the lens is seeing. Many cameras have an additional foldout monitor that enables you to forgo looking through an eyepiece to see the camera picture. SEE 6.1 ZVL1 CAMERA→ Camera introduction

FROM LIGHT TO VIDEO SIGNAL

All television cameras, big or small, work on the same basic principle: the conversion of an optical image into electrical signals that are reconverted by the viewfinder and the television set into visible screen images. SEE 6.2

Specifically, the light that is reflected off an object is gathered by a lens and focused on the imaging device. The imaging device, also called the sensor, is the principal camera element that transduces (converts) the light into an electric charge, which, when amplified and processed, becomes the video signal. The viewfinder and the monitor use this signal to produce the video image. With these basic camera functions in mind, we can examine the major elements and the processes involved in the transformation of light images into color television pictures. Specifically, we look at the imaging device, the beam splitter, and the color filter array.

Imaging Device

The principal electronic component that converts light into electricity is called the ***imaging device*** or ***sensor***. In tech lingo it is commonly known as the chip. There are two types of sensors in use: a ***charge-coupled device (CCD)*** and a ***CMOS*** (complementary metal-oxide semiconductor). Although technically different, they look alike and perform the same function: both convert the optical image into electrical impulses. SEE 6.3

Both the CCD and the CMOS contain hundreds of thousands or, for a high-quality chip, millions of

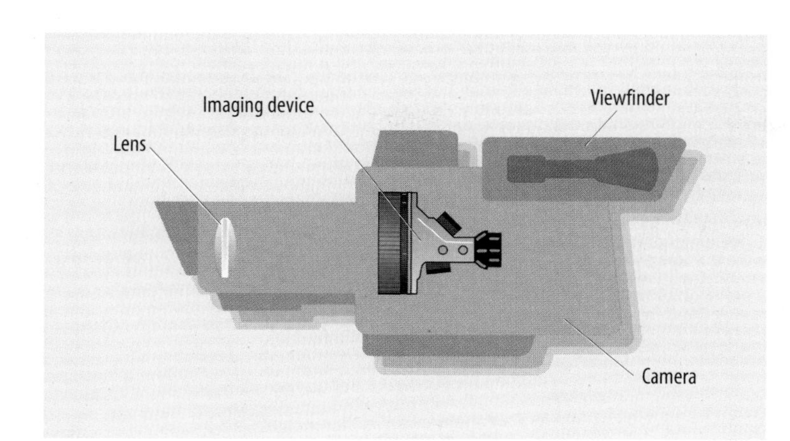

6.1 PARTS OF THE CAMERA
The main parts of a television (video) camera are the lens, the camera itself with the imaging device, and the viewfinder.

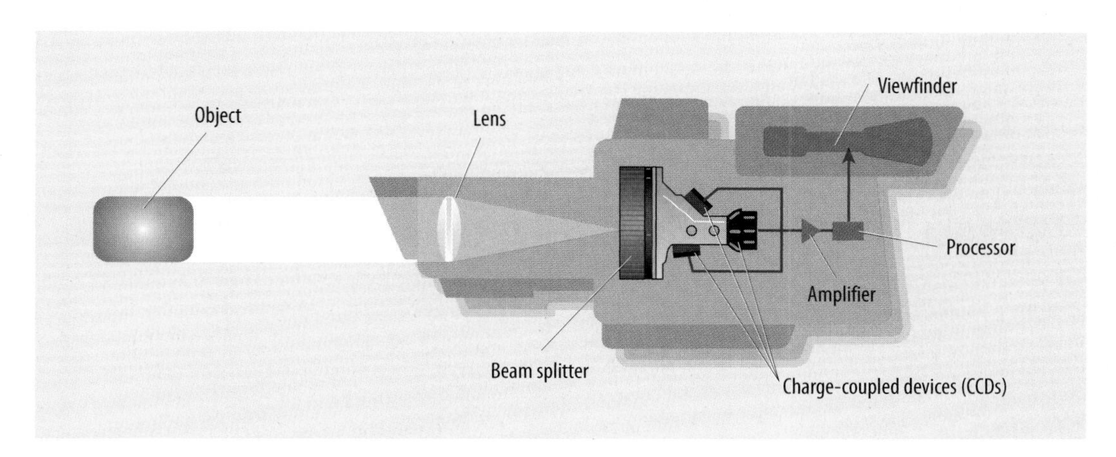

6.2 BASIC CAMERA FUNCTIONS: LIGHT TO VIDEO SIGNAL
The light reflected off the object is gathered by the lens and focused on the beam splitter, which splits the white light of the image into red, green, and blue pictures. These beams are directed toward their respective sensors (CCDs or CMOS), which transform the RGB light into electrical RGB charges; these are amplified, processed, and then reconverted by the viewfinder into video pictures.

image-sensing elements, called ***pixels*** (a compound of *pix*, for picture, and *els* for elements) that are arranged in horizontal and vertical rows. Each pixel is a discrete image element that transforms its color and brightness information into a specific electric charge. In digital cameras each pixel has a unique computer address. The electric charges from all the pixels eventually become the video signal.

Pixels function very much like tiles that compose a complete mosaic image. A certain number of such elements is needed to produce a recognizable image. If there are relatively few mosaic tiles, the object may be recognizable but the picture will contain little detail. SEE 6.4 The more and the smaller the tiles in the mosaic, the more detail the picture will have. The same is true for the sensor: the

6.3 CHARGE-COUPLED DEVICE
The CCD sensor holds many rows of thousands of pixels, each of which transforms light that enters through the window into an electric charge. The CMOS sensor looks similar but works on a different electronic principle.

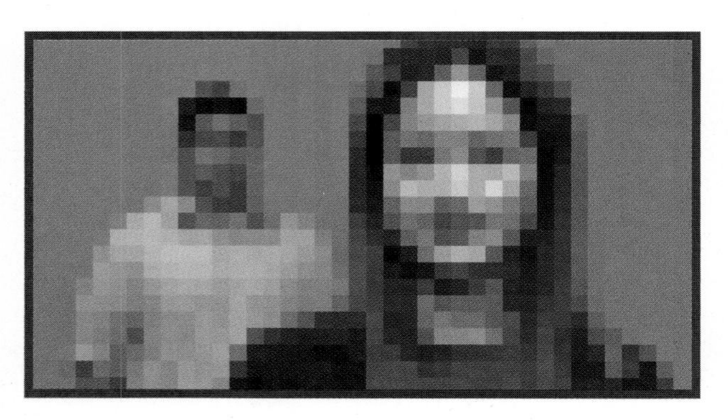

6.4 PIXELIZED SUBJECTS
Pixels function much like tiles that make up a complete mosaic image. An image comprising relatively few mosaic tiles—pixels—contains little detail. The more and the smaller the tiles, the sharper the image will be.

6.4 PHOTO BY EDWARD AIONA

more pixels it contains, the higher the resolution of the video image. Larger sensors can contain more pixels and therefore produce a sharper image than small ones. This is not unlike the negative of a photo. The larger the negative, the sharper the printed image.

The relatively large CMOS censor is why some DPs (directors of photography) prefer to shoot video with an SLR (single-lens reflex) camera, which looks and operates like a normal professional still camera.

Beam Splitter and Color Filter Array

Before the pixels on the sensors can do their translating from light to electricity, the light image coming through the lens must be split into the three primary colors of light—red, green, and blue, usually referred to as RGB. This color separation is done by the beam splitter or the color filter array.

The ***beam splitter*** contains prisms that divide the light coming from the lens into the three primary light beams. The prisms guide the three RGB beams into their designated sensors, which are attached to the prisms. High-end "prosumer" and professional cameras and camcorders normally use three sensors—one for each primary light color. The three primary colors are then electronically "mixed" into the many colors you see on the television screen. Because these prisms are contained in a small block, the beam splitter is often called the prism block. SEE 6.5

Most small consumer camcorders and, paradoxically, some high-end SLR and digital cinema cameras don't have a prism block because they use a single sensor as an imaging device. The splitting of the colors of the optical lens image into RGB is done by a single filter array that is attached to the sensor. SEE 6.6

CAMERA CHAIN

When looking at a high-quality studio camera, you can see that its cable is plugged into a dedicated outlet. This cable connects the camera to a chain of equipment necessary to produce pictures. The major parts of the ***camera chain*** are the actual camera, called the camera head because it is at the head of the chain; the camera control unit; the sync generator, which provides the synchronization pulses to keep the scanning of the various pieces of television equipment in step; and the power supply. SEE 6.7

6.5 BEAM SPLITTER
The beam splitter, or prism block, splits the incoming white light (representing the picture as seen by the lens) into red, green, and blue light beams and directs them to their respective sensors (CCDs or CMOS).

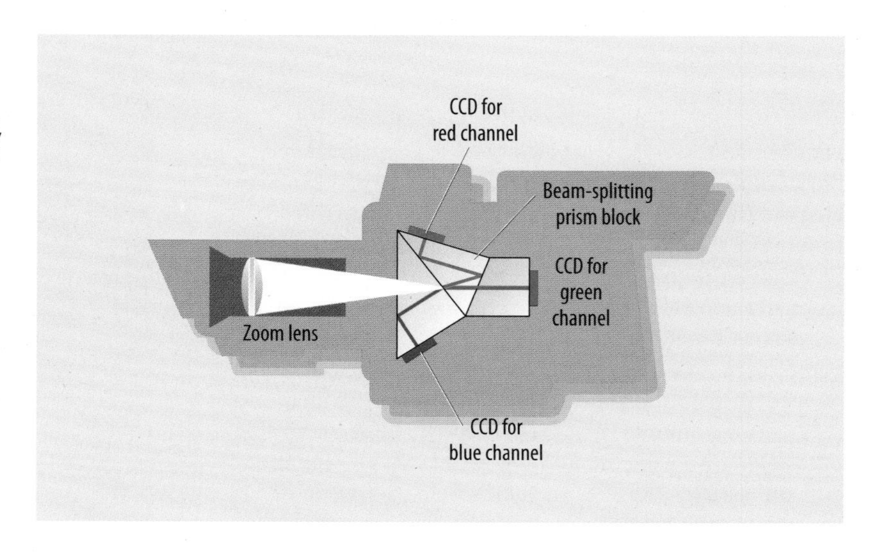

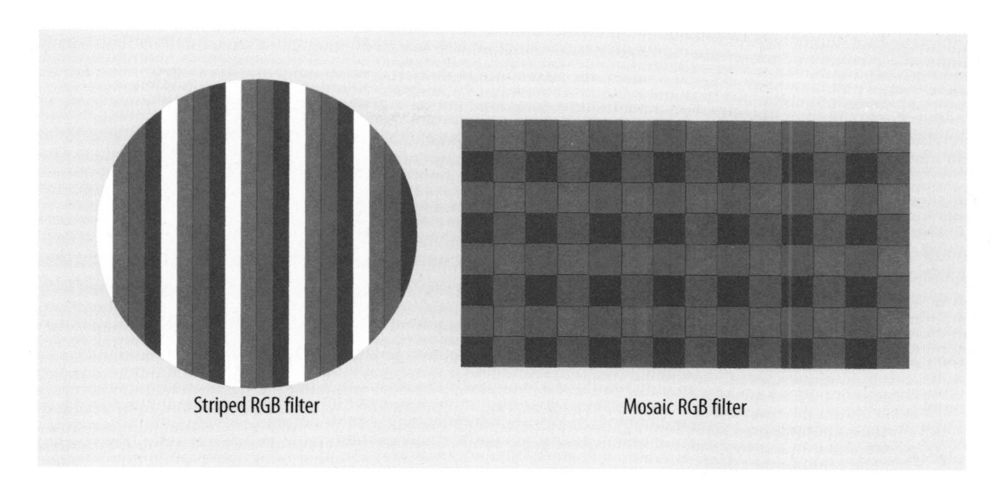

6.6 STRIPED AND MOSAIC FILTERS
Cameras with only one image sensor use a striped or mosaic-like filter instead of a prism block to divide the white light into RGB color beams. Each colored beam is then transduced by the single sensor into electric charges that eventually become the RGB signals.

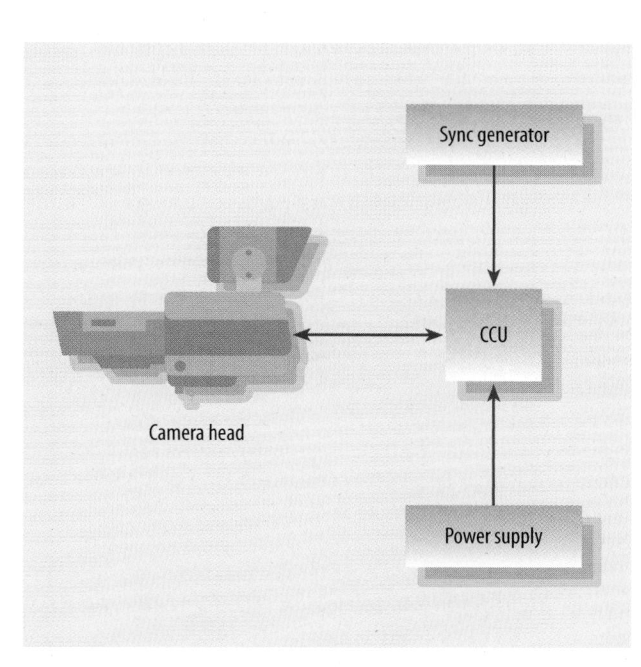

6.7 STANDARD STUDIO CAMERA CHAIN
The standard camera chain consists of the camera head (the actual camera), the camera control unit, the sync generator, and the power supply. Portable cameras contain all these functions in the camera housing.

Camera Control Unit

Each studio camera has its own ***camera control unit (CCU)***. The CCU performs two main functions: setup and control. During setup each camera is adjusted for the correct color rendition, the white balance (manipulating the three color signals so that they reproduce white correctly under a variety of lighting conditions), the proper contrast range between the brightest and the darkest areas of a scene, and the brightness steps within that range.

Assuming that the cameras are set up properly and have fair stability (which means that they retain their setup values), the video operator (VO) usually needs to control only the "master black" or "pedestal" (adjusting the camera for the darkest part of the scene) and the "white level" (adjusting the iris of the lens so that it will permit only the desired amount of light to reach the imaging device). SEE 6.8

Portable CCUs, usually called RCUs (remote control units), can be connected to cameras operating in the field (remote location). In critical field productions, RCUs allow the video operator to "shade" the pictures (maintain optimal picture quality) according not only to technical standards but also to the aesthetic requirements of the production.

Sync Generator and Power Supply

The ***sync generator*** produces electronic synchronization pulses—sync pulses—that keep in step the scanning of the various pieces of equipment (cameras, monitors, and video recorders). A genlock provides a general synchronization pulse, called house sync. Through the genlocking process, the scanning of video signals of a variety of equipment is perfectly synchronized, allowing you to switch among and intermix the video signals of various cameras and/or video recorders (VRs) without the need for additional digital equipment.

The power supply generates the electricity that drives the camera. In a studio the power supply converts AC (alternating current) to DC (direct current) power and feeds it to the cameras via the camera cable.

6.8 CAMERA CONTROL UNIT
The CCU adjusts the camera for optimal color and brightness and for varying lighting conditions.

6.8 PHOTO BY HERBERT ZETTL

Field (ENG/EFP) cameras and ***camcorders*** are self-contained, which means that the camera itself holds all the elements of the chain to produce and deliver acceptable images to the video recorder. The only part of the camera chain that can be detached from the camera head (the actual camera or camcorder) is the power supply—the battery. All ENG/EFP cameras and camcorders have built-in control equipment that can execute the CCU functions automatically. But if an ENG/EFP camera or camcorder can have a built-in control that can perform all or most CCU functions, why bother with an external CCU? Because the automated controls cannot exercise aesthetic judgment; that is, they cannot adjust the camera to deliver pictures that also suit the artistic rather than the routine technical requirements.

TYPES OF CAMERAS

Television cameras are classified by the way they are used and by their electronic makeup. Some camera types are better suited for studio use, others for the coverage of a downtown fire or the production of a documentary on pollution, and still others for taking along on vacation to record the more memorable sights. When based on their primary function, there are three types: studio cameras, EFP (electronic field production) cameras, and ENG/EFP camcorders. All three can be standard-definition television (SDTV), ***high-definition video (HDV)***, or high-definition television (HDTV, also called HD to include digital cinema). Most studio and EFP cameras are HDTV.

There is a great variety of small consumer camcorders available. Because they are not normally used in television production, they are not discussed here. You are certainly encouraged to use them as much as possible, however, which can help your understanding of how the various parts of the camcorder system interact, how to compose effective images, and how to construct meaningful picture sequences. But don't be fooled by the camera size. Some camcorders are small enough to fit into your pocket but deliver high-definition 720p video. Also, when the event is significant enough, it really doesn't matter whether you capture it with a large HD camcorder or a cell phone.

Studio Cameras

The term *studio camera* is generally used to describe high-quality ***high-definition television (HDTV) cameras***. Even when the camera itself is relatively small, the big zoom lens and the attached teleprompter make it so heavy that it cannot be maneuvered properly without the aid of a studio pedestal or other camera mount. **SEE 6.9**

Studio cameras are used for multicamera studio productions, such as news, interviews, and panel shows, and for commercials, situation comedies, daily serial dramas, and instructional shows that require high-quality video. You can also see these cameras used in such "field" locations as tennis courts, medical facilities, concert and convention halls, and football and baseball stadiums.

Considering that you can get pretty good pictures from a camera that fits in your pocket, why bother with such heavy cameras and the rest of the camera chain? The

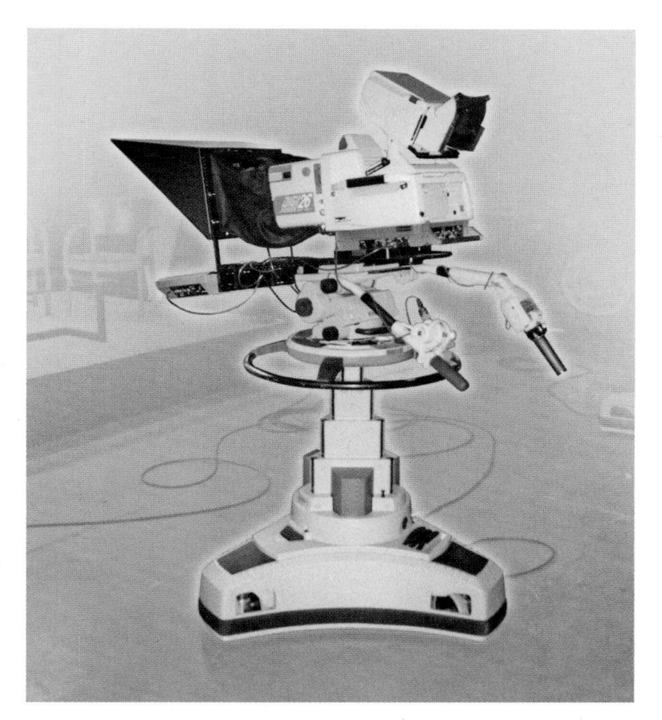

6.9 PHOTO BY HERBERT ZETTL

6.9 STUDIO CAMERA ON STUDIO PEDESTAL
Studio cameras have high-quality lenses and sensors. They are quality-controlled by the CCU. Studio cameras are too heavy to be carried and are mounted on a sturdy tripod or studio pedestal.

overriding criteria for the use of studio cameras are picture quality and control. We usually judge picture quality by the amount of sharp detail the camera and the monitor can generate, the fidelity of the colors, and the number of contrast steps from the darkest to the lightest picture area. The virtue of a studio camera is that it has superior lenses. Also, with a good video operator handling the CCU, the studio camera can deliver high-quality colors and subtle brightness steps under a variety of lighting conditions.

But *quality* is a relative term. In many productions the extra quality and control achieved with studio cameras is not worth the additional time and expense necessary for operating such equipment. For example, if you are to get a picture of an approaching tornado, you are probably not thinking about optimum picture quality. Your attention is on getting the shot and then getting out of harm's way as quickly as possible. But if picture quality is paramount, such as in the production of commercials, medical shows, and daytime serials, you would undoubtedly choose a high-end studio camera. Quality is obviously a key factor when choosing a television camera for producing digital cinema.

Other factors that ensure quality are the lens, the zoom and focus controls, and the viewfinder.

Lens The lens of the studio camera is usually larger (and often more expensive) than the ones used for ENG/EFP cameras and camcorders (see figure 6.9). The large studio lenses come with different zoom ranges (from the widest to the closest view). You can use a shorter range for studio work than for a sports pickup, where you might have to zoom in to a tight close-up of the quarterback's face with a camera located high in the stands. The quality of the lens, and especially its zoom range, are the major cost factors when buying a camcorder.

Digital cinema cameras sometimes use prime lenses, which deliver a slightly sharper picture than zoom lenses. You can't zoom with a prime lens because it has a fixed focal length. (Prime lenses are discussed in chapter 7.)

Zoom and focus controls These controls are attached to the panning handles of the studio pedestal and are driven by a servo (motorized) system. Normally, the rocker-type zoom control is on the right handle and the twist control for the focus is on the left. (Various lenses and their controls are discussed in chapter 7.)

Viewfinder When using a small camcorder, you've probably wished that you had a larger, more accurate viewfinder. Even fairly large foldout monitors wash out in bright sunlight and make it hard to see whether you are in focus. When switching to the eyepiece viewfinder, everything looks awfully small. This is why studio cameras have a large (5- to 7-inch), high-resolution viewfinder that can be tilted and swiveled sideways. Some camera operators clamp an additional flat-panel monitor onto their camcorder, especially when they can't operate the camcorder from their shoulder.

EFP Cameras

The portable ***EFP camera*** must be connected to an external video recorder. Generally, EFP cameras are high-quality cameras that can be carried on the shoulder or put on a tripod. SEE 6.10

EFP cameras are almost never used for ENG (electronic news gathering) but rather for demanding documentary productions and outdoor scenes shot single-camera film-style (where all shots are intended for postproduction editing). Because of the high quality of most EFP cameras, they are often used in the studio instead of the more expensive studio cameras.

6.10 EFP CAMERA
The EFP camera is a high-quality camera that can be carried and handled by the operator. It can also be converted into a studio camera.

6.10 PHOTO COURTESY IKEGAMI

When used in the studio, the EFP camera is placed in a specially made camera frame, a large external tally light is added, and the small eyepiece viewfinder is replaced with a larger viewfinder attached to the camera housing. The zoom and focus controls are moved to the panning handles and connected to the lens via servo cables. The ENG/EFP lens is usually replaced with a zoom lens that is more suitable to the studio environment. The whole camera assembly is then mounted on a studio pedestal. In a studio configuration, the EFP camera is usually controlled via a CCU. **SEE 6.11**

6.11 EFP CAMERA IN STUDIO CONFIGURATION
The same camera as in figure 6.10 is shown here in a studio configuration. It is placed in a housing and has a large studio lens, a 16 × 9 monitor, and external zoom and focus controls.

6.11 PHOTO COURTESY IKEGAMI

ENG/EFP Camcorders

ENG/EFP camcorders are high-quality portable field production cameras with the recording device built-in. There are basically two types: the large, high-end SDTV and HD camcorders and the much smaller HDTV camcorders.

Large SDTV and HD camcorders All full-sized standard-definition television and high-definition digital camcorders have high-quality, interchangeable zoom lenses; a high-quality imaging device with three large CCDs or a large single CMOS sensor in the megapixel range; a relatively large eyepiece viewfinder; and, on some models, an additional foldout monitor. The recording media is either a built-in high-quality videotape recorder (VTR) or a nontape recording device (hard drive, optical disc, or memory card). **SEE 6.12**

Small HDTV camcorders Smaller ENG/EFP digital camcorders operate on the same basic principle as the digital consumer models. In fact, the digital consumer models produced such high-quality video that their manufacturers upgraded them by adding a better lens, a higher-quality imaging device, and a few more video and audio controls. Because these smaller, lighter, and much less expensive models were used so frequently in professional ENG and EFP, they were called *prosumer* models (a compound of *pro*fessional and con*sumer*). Now that these smaller prosumer camcorders can produce high-definition (720p and 1080i) video, they are rapidly replacing the larger camcorders in news and documentary productions. The prosumer digital camcorder has finally shed its consumer stigma and become a professional workhorse.

Most good HDTV camcorders contain a prism block with three CCDs or high-resolution CMOS sensors attached to it. Some have a large, single sensor that divides the light coming through the lens into RGB via a striped or mosaic RGB filter array (see figure 6.6). Most small HDTV camcorders use a high-definition VTR that records on

6.12 PHOTO COURTESY PANASONIC
6.13 PHOTO COURTESY JVC

6.12 LARGE ENG/EFP CAMCORDER
This large camcorder produces high-definition images with three CMOS sensors and high-fidelity audio on four audio channels. It records video and audio on two P2 cards (high-quality solid-state memory cards) and additional data on a small SD (Secure Digital) memory card. It has a foldout color monitor and a high-resolution viewfinder for accurate focusing.

6.13 SMALL HDTV CAMCORDER
This high-end HDTV camcorder is considerably smaller and lighter than standard camcorders yet it produces excellent high-definition images and sound. It records in 1080i and 720p formats at 60, 30, and 24 frames per second.

¼-inch full-sized or mini-cassettes or, more commonly, some nontape media, such as hard drives, optical discs, or memory cards. SEE 6.13

Like a large HDTV camcorder, small HDTV camcorders can be switched between 720p and 1080i scanning lines and various frame rates, usually 24, 30, and 60 frames per second. Does this mean that the pictures of the small HDTV camcorders look as good as the large ones? Yes and no. Under normal circumstances, you will have a hard time seeing a big difference between the two, even if the high-definition images are displayed on a high-resolution monitor.

When operating under less than ideal circumstances, however, you won't be able to produce video that rivals that made with the larger camcorders. The single factor that most compromises the quality of the small camcorder video is the system's compression. The higher the compression, the more the picture quality suffers. You notice the negative compression effects especially in the color rendition, in the picture detail when shooting in high-contrast lighting, and when there is a great deal of object or camera motion. Nevertheless, you will find that in most situations a small HDTV camcorder will produce stunning images that approach and often rival the video of the much more expensive HDTV systems.

Another problem with the small HDTV camcorders is the lens. Almost all small camcorders have a built-in lens that cannot be exchanged. There are adapters that allow you to attach a lens that gives you a wider view than the regular lens, but such an arrangement is less satisfactory during production than is the slightly less flexible built-in lens. Most Canon HDTV camcorders come with interchangeable lenses.

Digital Cinema Cameras

In the context of cameras, *digital cinema* means that you use a television camera or camcorder to make movies. You can certainly use any high-end HD camera with a high-quality external video recorder or any high-end HDTV camcorder for producing digital movies, although most of today's digital cinema is shot with digital cinema cameras. SEE 6.14

The HD ***digital cinema cameras*** have large, multimillion-pixel CCDs or CMOS sensors in the 4K (4,000-plus pixels per scanning line) category. Most of these cameras use the 1,080-line standard, but some sensors have many more, thereby further increasing the definition of the video image.

Mainly because of tradition, the recording speed in digital cinema is 24 frames per second (fps), exactly like the frame rate of actual film. As pointed out earlier, most digital cinema cameras allow a variable frame rate so that you can "overcrank" or "undercrank" your shots. Overcranking in digital cinema cameras (much like in traditional filming) means that the scene is recorded at a

6.14 DIGITAL CINEMA CAMERA
This digital cinema camera (super-HD camcorder) has a high-density (4K-plus) sensor and high-quality signal processing that includes its own codec. In addition to the zoom lens, it accepts a variety of prime lenses. It records on an internal memory card and on an external hard drive or large-capacity solid-state device.

6.14 PHOTO COURTESY RED DIGITAL CINEMA

frame rate that is much higher than the playback one, so you get a slow-motion effect. Undercranking means that if the frame rate of the recording is much slower than the playback, you get accelerated motion.

An important feature of digital cinema cameras is their ability to handle high contrast and to produce a highly subtle grayscale—the many brightness steps between black and white. Until now such an expanded grayscale used to be the unique feature of traditional film.

Digital cinema cameras need special lenses to match their high-resolution sensors. These lenses can be studio zooms or a variety of prime lenses, which are similar to the fixed-focal-length lenses used with a still camera. A fixed-focal-length lens will not allow you to zoom. Additional items that facilitate the use of HDTV cameras for film use are a matte box (a lens attachment for optical special effects) and a relatively large high-resolution monochrome viewfinder.

You may hear some committed film people complain about the "in-your-face" look of HDTV images and lament the loss of the mysterious "film look." The film look has been erroneously attributed to high-resolution images, but is probably the result of the high contrast ratio and the blackouts that occur when one frame of the film changes over to the next. This constant going-to-black between frames makes us perceive a softer image. To copy this blackout sensation, some digital cinema systems use filters to reintroduce a variety of artifacts that occur in older film prints. Unfortunately, all such attempts result in a lower-quality image but not in the softer frame-by-frame cushions. As with all art, however, it is not necessarily the equipment that makes a good film but ultimately the filmmaker's creative mind and skill. In fact, some remarkable movies have been made with a small SDTV camcorder.[1]

ELECTRONIC FEATURES

When judging the quality of an HDTV camera, you need to pay attention to these electronic video features, which are typically part of every professional camera or camcorder: imaging device (sensor), scanning, resolution, aspect ratio and essential area, gain, electronic shutter, white balance, and audio channels.

Imaging Device

Most high-quality cameras and camcorders have three CCDs or CMOS sensors as the imaging device. As mentioned earlier, some camcorders and digital cinema cameras have a single high-density sensor.

Scanning

When shooting in HDTV, you should be able to switch among the 720p and 1080i scanning systems; 720p and 1080i are considered high-definition.

Resolution

Resolution refers to how sharp a picture looks. It is the measurement of picture detail—the number of distinguishable details. HDTV produces higher-resolution pictures than does STV. Video resolution is determined by the number of scanning lines (vertical resolution) and by the number of pixels that make up each line (horizontal resolution). As you recall, a 720p video has 720 visible lines that are scanned progressively (one after the other), and each line is made up of 1,280 pixels. The 1080 system uses 1,080 visible lines, with 1,920 pixels for each scanning line. This obviously produces a sharper picture than the standard TV resolution of about 480 visible lines, each of which is composed of 640 pixels. (See section 6.2 for more on scanning and resolution.)

---

1. Some big-screen documentaries and major motion pictures have been shot with a small digital camcorder. For example, David Lynch shot almost all the footage for *Inland Empire* (2006) with a Sony DSR-PD 150 with most of the controls set on automatic. See Jon Silberg, "Inland Empire—David Lynch's DV Dream," *Digital Cinematography* 2, no. 6 (2006), 26–28.

Aspect Ratio and Essential Area

All HDTV viewfinders have an aspect ratio of 16 × 9, which means they are 16 units wide by 9 units high. Most cameras that allow you to switch from a 16 × 9 to a 4 × 3 aspect ratio have provisions to indicate in the viewfinder the 4 × 3 area. Most viewfinders of HDTV camcorders have markers to help you frame a shot, such as the safe title area (the area that will definitely show on a home television set), a center cross, and the 4 × 3 area within the 16 × 9 frame. Even without such markers, it is a good idea to keep the main event within the 4 × 3 area on a 16 × 9 screen. That way you don't lose the essential parts of the event when it is shown on a standard 4 × 3 monitor.

Gain

When there is not enough light for the imaging device to produce optimal video, the gain comes to the rescue. Gain is an electric charge that tricks the sensors into believing that they are getting more light than they actually receive from the lens. The higher the gain, however, the lower the picture quality. Gain is measured in decibels (dB). Most cameras can be switched from 0 dB (no gain) to 18 dB or more (high gain). SEE 6.15

Electronic Shutter

The electronic shutter in a television camera functions just like the mechanical one in an old still camera that uses actual film. The faster the object moves, the faster the shutter speed must be to avoid a blurred image. But instead of the highest shutter speed of about $1/1,000$ second in the mechanical shutter, the electronic shutter can go from $1/8$ second to $1/10,000$ second. The reason for the higher shutter speed is that the CCDs and the CMOS sensors take some time to get fully charged and therefore blur a fast-moving object unless the shutter speed is quite high. The downside, as with a regular still camera, is that the faster the shutter speed, the more light the camera requires. For example, if a bright yellow tennis ball moves from camera-left to camera-right at high speed and the camera is set at a slow shutter speed, it looks blurred and even leaves a trail. When you increase the shutter speed so that the blur is virtually eliminated, the ball looks considerably darker.

6.17 PHOTO BY EDWARD AIONA

6.15 MANUAL GAIN CONTROL
The gain control compensates for low light levels. The higher the gain, the lower the light level can be. High gain causes video noise.

White Balance

To guarantee that a white object looks white under slightly reddish (candle or normal light bulb) or bluish (bright sunlight or fluorescent lamp) light, you need to tell the camera to compensate for the reddish or bluish light and to operate as if it were dealing with perfectly white light. This compensation in the camera is called ***white balance***. When a camera engages in white-balancing, it adjusts the RGB video signals in such a way that the white object looks white on-screen regardless of whether it is illuminated by reddish or bluish light (see chapter 11).

Audio Channels

Good HDTV cameras or camcorders provide two audio channels for high-quality (16 bit/48 kHz [kilohertz]) sound, or four audio channels of lower-quality (12 bit/32 kHz) sound. They should have XLR jacks (inputs), or at least adapters, for balanced audio cables.

OPERATIONAL FEATURES

At first glance a camcorder with the fewest buttons and a large menu of available electronic functions on its foldout monitor seems to be the most desirable—until you use it outdoors in bright sunlight. It is hard enough to dig for a specific menu item when it is buried under two or three layers of submenus; but when the sun washes out the screen display, you are in real trouble. This is why you still see some switches, buttons, and dials on the outside of the camcorder. Unfortunately, not all camcorders have the switches in convenient places.

For example, you should be able to press the *record* button with the same hand that operates the zoom control; and the switch that puts the camera into manual or autofocus should be near the manual focus ring on the lens. Experienced videographers usually prefer a camera that has the switches in the right place over one whose controls are buried in multi-tiered menus, even if the latter's video quality is slightly superior.

But there are a few more operational features of studio cameras and camcorders that need your attention: power supply, cables and connectors, filter wheel, viewfinder and monitor, and tally light and intercom.

Power Supply

When operating a studio camera, you don't have to worry about the power supply. The necessary electricity to power the camera is supplied by the camera cable. When using an ENG/EFP camera or camcorder in the field, however, the power supply becomes a major concern. All ENG/EFP cameras and camcorders use battery power. While we herald the progress in digital technology, we tend to overlook the equally dramatic development of the batteries that drive portable cameras and their associated equipment.

Most large camcorders are powered by a 12V (volt) DC battery; small camcorders use batteries that generate between 7V and 8V DC. Batteries can be clipped to the back of the camera or, in some cases, to the bottom. How long a battery lasts depends on how much recording you do and what else you are powering besides the camera/recording unit. For example, a battery will last longer if you refrain from powering with it such items as the camera light (the small light on top of the camera), the foldout monitor, the steady-shot feature, the playback of the video recorder through the foldout display, and the teleprompter.

Don't keep the camera turned on between takes; switch it to the standby mode as much as possible. This allows you to have everything up and running at the flip of a switch while preserving the battery between takes. (CMOS sensors, incidentally, draw much less power than the customary CCDs.) Always carry a fully charged spare battery with you, and exchange it a few minutes before the expiration of its life span as advertised by the manufacturer. The most reliable indicator of the remaining battery charge is the one in the camera viewfinder or menu.

Substitute power supplies are household AC current and car batteries, both of which require adapters. Use a car battery only in an emergency; car batteries are hazardous to the operator as well as to the equipment.

Cables and Connectors

Camera cables and connectors differ significantly in how they carry electronic signals to and from the camera.

Cables When requesting cable runs, you need to know which cable the camera can accept and, especially, how long a run you need.

Triaxial (triax) cables are thin and have one central wire surrounded by two concentric shields. Fiber-optic cables contain thin, flexible, glass strands and can transport a great amount of information over relatively long distances. A triax cable allows a maximum distance of almost 5,000 feet (about 1,500 meters); a fiber-optic cable can reach twice as far, to almost 2 miles (up to 3,000 meters).[2] Such a reach is adequate for most remote operations. Some cameras use a multicore cable, which contains a great number of thin wires. Multicore cables are relatively heavy and have a limited reach, but they are extremely reliable.

Connectors Before going to the field location, carefully check that the connectors on the cables actually fit into the camera jacks (receptacles) and the jacks of the auxiliary equipment. Little is more annoying than having the whole production held up for an hour or more simply because a connector on a cable does not match the jack on the camera.

Short-run video connections are done with cables that use S-video, RCA phono, USB 2.0 or 3.0, FireWire (IEEE 1394), eSATA, and HDMI (high-definition multimedia interface) connectors; BNC connectors are for longer-run coaxial cables. All but S-video cables and connectors can also be used for audio signals. USB 3.0 and eSATA connectors and interface cables are generally used for high-speed transfer of digital data, such as feeding the video and audio information from your camcorder or server to the editing system. Don't be surprised if the capture speed of the hard drive in the editing system can't quite keep up with the delivery speed of the USB 3.0 interface. Yes, it is that fast. The USB 3.0 interface is also downward compatible with the standard USB connector, which means you can use the USB 3.0 cable to plug into normal USB ports (but you then lose the USB 3.0 transfer speed).

Professional audio cables, which are especially resistant to interference, use XLR connectors. Other audio cables may use ¼-inch phone plugs or mini plugs. Although there are adapters for all plugs (whereby you can change a BNC connector into an RCA phono plug, for example), avoid them when possible. Every adapter introduces a potential trouble spot. SEE 6.16

When used in the studio, the camera cable is generally left plugged into the camera and the camera wall jack (outlet). When using studio or EFP cameras in the field, you need to carefully check whether the cable connectors fit the receptacles of the remote truck or the RCUs.

---

2. One meter (m) is a little more than 3 feet. More accurately, 1 meter = 3.28 feet. If, for example, you need a rough idea of how long a 1,000-meter cable run is in feet, multiply 1,000 by 3, which is approximately 3,000-plus feet. If you need an accurate foot reading, multiply 1,000 by 3.28, which comes to 3,280 feet.

6.16 PHOTO BY EDWARD AIONA

6.16 VIDEO AND AUDIO CONNECTORS
The most common connectors in television production are pictured here. All have adapters that let you change the plugs to fit other types of jacks. Other efficient interfaces in use are the new USB 3.0 and the eSATA cables and connectors, but the video equipment has to catch up before those are widely used. Note, however, that many pieces of equipment have unique cables and connectors.

Filter Wheel

The filter wheel is located between the lens and the imaging device. It normally holds two or three neutral density (ND) filters (referred to as ND-1, ND-2, and ND-3) and two or more color-correction filters.

The ND filters function like good sunglasses: they reduce the amount of light transmitted to the imaging device without affecting the color of the scene too much. You use them for shooting in overly bright sunlight when the smallest aperture (highest *f*-stop) of the lens still admits too much light or when the contrast is too high for the camera to handle.

The color-correction filters compensate for the relative bluishness of outdoor and fluorescent light and the relative reddishness of indoor and candlelight (see chapter 11). You can activate these filters with an ND switch near the lens barrel of the camcorder.

Viewfinder and Monitor

A good viewfinder is obviously important because it shows you the exact pictures the camera is taking. Studio cameras have a relatively large (5- or 7-inch) viewfinder that can be swiveled and tilted so that you can see what you are doing even if you are not standing directly behind the camera. SEE 6.17

Unless converted to the studio configuration, all EFP cameras and camcorders have a small (1½- to 2½-inch) high-resolution monochrome (black-and-white) or color viewfinder. It is shielded from outside reflections by a flexible rubber eyepiece that you can adjust to your eye. As pointed out earlier, most ENG/EFP cameras and all small camcorders have an additional foldout flat-panel display, called a monitor. Yes, this may be a little confusing because we also call the many screens in a television control room "monitors."

Apparently, confusing terminology has yet to deter its usage. Just remember that when you look through the tube with the shielded eyepiece when framing a shot, you are

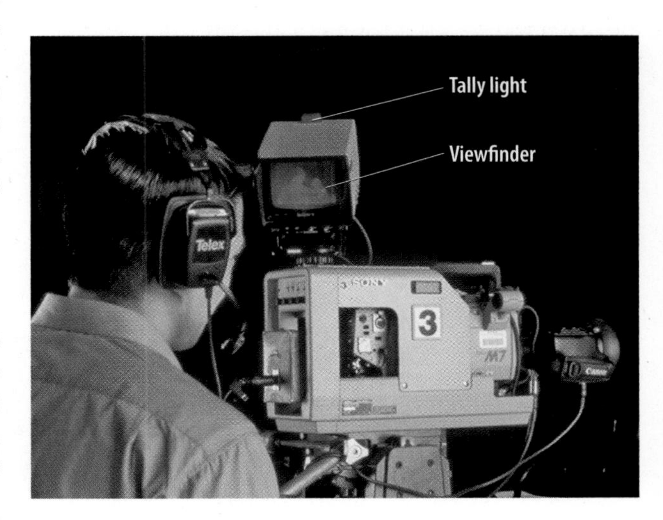

6.17 PHOTO BY EDWARD AIONA

6.17 STUDIO CAMERA VIEWFINDER
The 7-inch studio camera viewfinder can be swiveled and tilted so that the screen faces the camera operator regardless of the camera position.

looking through the viewfinder; when watching the larger foldout display, you are watching your shots on a monitor. The monitor is handy for menu displays and for framing shots when in a hurry or under pressure. Unfortunately, in bright sunlight such monitors wash out and make it hard, if not impossible, to see enough of an image for focusing or even framing a shot. You are definitely in trouble when you can't see the menu any longer during a location shoot.

External monitor Despite the viewfinder and the foldout monitor on the camera, you will find that in some situations they still seem to be in the wrong place. This is why many videographers use an additional, larger LCD (liquid crystal display) monitor that is mounted on top of the camera handle via a flexible arm. This way you can clearly see what you are shooting even if the camera sits on the ground. Similarly, if the camera is on a tripod and locked into a specific location (such as for an interview or a wedding), you can connect it to an external monitor next to the tripod and operate the camera by watching the monitor rather than the small viewfinder.

Color and monochrome All LCD monitors display a color picture, but some eyepiece viewfinders of camcorders and even some large viewfinders on studio or digital cinema cameras are often monochrome. The reason for using black-and-white viewfinders is that they can display a higher-resolution image than can most color viewfinders and especially foldout monitors. Such high-resolution video is obviously essential when working with digital cinema. Fortunately, there are high-resolution color viewfinders available, but they are very expensive. The advantages of color viewfinders and monitors are, of course, that you see the colors the camera delivers, and you can compose your shots not only by placing objects in the right position in the frame but also by seeing how the colors are distributed.

Focus-assist Even with a monochrome viewfinder, HDTV and digital cinema cameras are hard to focus: the image is so high-definition that everything looks sharp and clear even if the picture is slightly out of focus. To help you focus, most HDTV and digital cinema cameras have a focus-assist feature in the viewfinder, which simply enlarges the center section of the image. Once the enlarged portion is in focus, the entire image will be in focus. SEE 6.18

Communications center The viewfinder and/or the foldout monitor of EFP cameras and camcorders serves as an important communications center. It displays the menu for various camera and video-recording functions. When it's superimposed over the viewfinder image, you can monitor the control functions in use. SEE 6.19 Because menus differ from camera to camera, you need to study the manuals and test the options before taking the camera on an important shoot.

Try to activate the setup functions when you are indoors and the menu is clearly visible on the foldout monitor. Once you are outside in bright sunlight (or at night), searching the menu for a specific function is difficult; you will appreciate the buttons and the switches that control the most essential camera operations.

Tally Light and Intercom

All studio cameras have a tally light, which is a small red light on top of the viewfinder (see figure 6.17). Tally lights are especially important for multicamera productions. When the tally light is on, you know that that camera is "hot," that is, on the air. This signals the other cameras

6.18 FOCUS-ASSIST
The focus-assist feature enlarges the center of the image to facilitate focusing. The enlarged picture segment lets you see more readily whether the whole scene as displayed in the viewfinder is in sharp focus.

Focus-assist off

Focus-assist on

6.18 PHOTO BY GARY PALMATIER

6.19 PHOTO BY GARY PALMATIER

6.19 FOLDOUT MONITOR FUNCTION DISPLAY
The flat-panel display on a camcorder serves as a communications center, displaying a menu of vital camera functions. During camera operation, specific information is superimposed on the viewfinder picture.

(whose tally lights are off) that they are free to line up their next shots. It also helps the talent address the correct camera. As an aid to the camera operator, a small tally light inside the viewfinder hood also lights up. When two cameras are used simultaneously, such as for a split-screen effect or a superimposition (see chapter 14), the tally lights of both cameras are on. When operating a single camera, you don't need a tally light. In this case, a flashing red light inside the viewfinder and sometimes on the front of the camera indicates when it is actually recording the event.

The intercom, or intercommunication system, is especially important for multicamera productions because the director and the technical director have to coordinate the cameras' operations. All studio cameras and several high-end field cameras have at least two channels for intercommunication—one for the production crew and the other for the technical crew. Some studio cameras have a third channel that carries the program sound. (Intercoms are explained further in chapter 18.)

MAIN POINTS

- The television camera is the single most important piece of production equipment. Other production equipment and techniques are influenced by what the camera can and cannot do.
- The major parts of the camera are the lens, the camera itself with the imaging device (sensor), and the viewfinder.
- The beam splitter separates the entering white light into the three additive light primaries: red, green, and blue (RGB).
- The imaging device converts the light entering the camera into electric energy, which when amplified and processed becomes the video signal. This transducing is done by the sensor—a charge-coupled device (CCD) or a CMOS sensor—both of which consist of a solid-state chip containing rows of a great many light-sensitive pixels.
- The standard studio camera chain consists of the camera head (the actual camera), the CCU (camera control unit), the sync generator, and the power supply.
- When classified by function, the four types of television cameras are studio cameras, EFP cameras, ENG/EFP camcorders, and digital cinema cameras. EFP cameras and ENG/EFP camcorders can be SDTV (standard digital television) or HDTV (high-definition television), although most new cameras and even small camcorders are HDTV. Studio and electronic cinema cameras are exclusively HDTV.
- All electronic cinema cameras have large high-density sensors and special lenses that produce super-high-definition images. They are also equipped with various attachments that facilitate specific requirements of the filmic production process.
- The electronic features of every professional camera or camcorder include the imaging device, scanning, resolution, aspect ratio and essential area, gain, electronic shutter, white balance, and XLR jacks for balanced audio cables.
- The operational features of studio cameras, EFP cameras, and ENG/EFP camcorders include power supply, cables and connectors, filter wheel, viewfinder and monitor, tally light, and intercom.
- Studio cameras are powered through the camera cable. ENG/EFP camcorders normally operate off a battery. Large EFP cameras and camcorders use a 12V (volt) DC battery; small camcorders use batteries that can supply between 7V and 8V DC.
- Certain features of a camera that put an additional drain on the battery include the camera light, the foldout monitor, the steady-shot feature, video-recording playback through the foldout display, and the teleprompter.
- Triaxial (triax) and fiber-optic cables can carry video and audio information over relatively long distances.
- The filter wheel has one or more neutral density (ND) filters to cut excess light and reduce contrast as well as color-correction filters to facilitate white-balancing.
- Black-and-white viewfinders show more detail than normal color viewfinders and are therefore preferred for critical focusing. The viewfinder and foldout monitor can display menus of camera functions.
- Tally lights indicate when a camera is "hot" (on the air). This is important only in multicamera productions.

SECTION

6.2

Resolution, Contrast, and Color

When you listen to experienced videographers or eager equipment salespeople, you always hear them debate the relative merits of a camera's resolution, contrast, and color attributes. This section explains how these three concepts influence the video image.

- **IMAGE RESOLUTION**
  Spatial and temporal resolution
- **IMAGE CONTRAST**
  Contrast ratio and contrast control
- **COLOR BASICS**
  Color attributes (hue, saturation, and brightness) and additive and subtractive color mixing
- **LUMINANCE AND CHROMINANCE CHANNELS**
  The essential components of a color video signal

IMAGE RESOLUTION

In case you are wondering why you have to read about resolution again, it is summarized here because it is, after all, the main reason for switching to HDTV. Although *resolution* is technically the measurement of picture definition, it is frequently used to mean the same thing as picture definition—how sharp the picture looks. As you recall, the resolution of a video image is determined by how many pixels there are in each frame and how many frames per second we see. The pixel count is called the spatial resolution; the frames-per-second count is the temporal resolution.

Spatial Resolution

The spatial resolution of a video image is determined by how many scanning lines are used for each frame and how many pixels are squeezed on a single scanning line. The stack of scanning lines makes up the vertical detail, and the pixels of each scanning line make up the horizontal detail. **SEE 6.20 AND 6.21**

The more scanning lines the vertical stack contains and the more pixels each line contains, the higher the resolution. As you can see, the vertical resolution is pretty much determined by the scanning system. If you have 480 visible lines (as in the SDTV system), you have 480 lines of vertical resolution. The 720 and 1080 scanning systems have a vertical stack of 720 and 1,080 lines, respectively, and therefore a vertical resolution of 720 and 1,080. The confusion is brought about when we omit the vertical stack of scanning lines and simply refer to the number of scanning lines as "vertical lines of resolution" even though each line clearly runs horizontally.

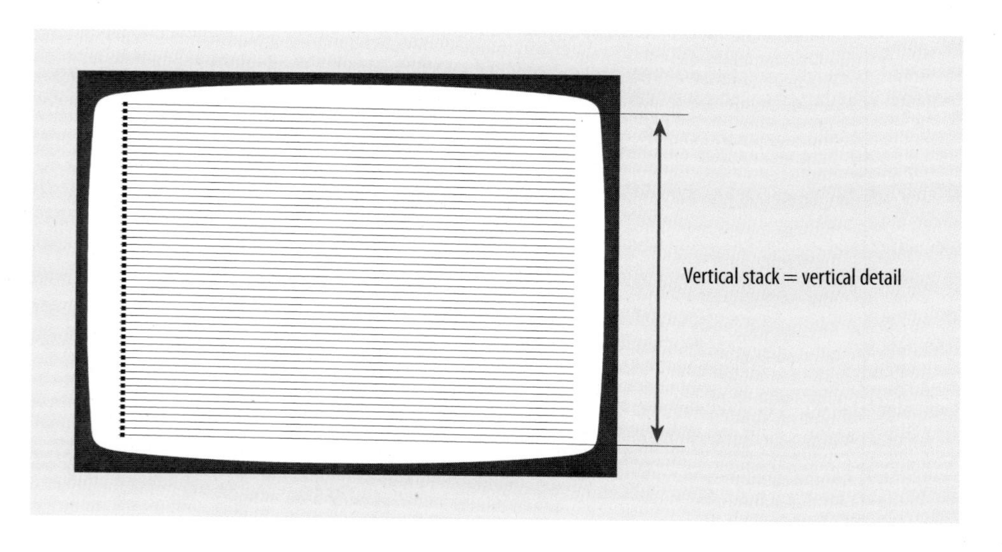

6.20 VERTICAL DETAIL (LINES OF RESOLUTION)
To measure vertical detail, we count the first pixel of each vertically stacked scanning line. The more lines the vertical stack contains, the more pixels and the higher the resolution. The number of lines is fixed by the system: the NTSC system has 525 lines, of which only about 480 are visible on-screen; HDV and HDTV systems can display 720 and 1,080 active (visible) lines; and ultra HD can display 4,000 or more lines.

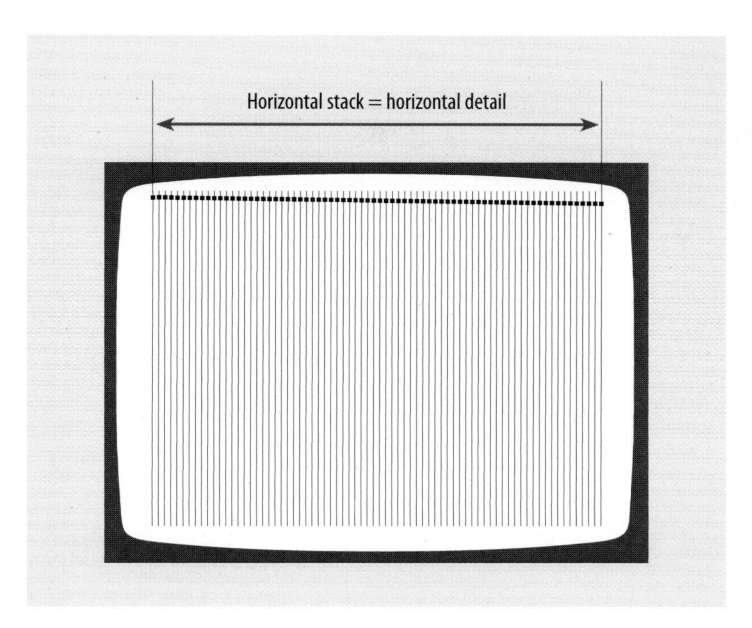

6.21 HORIZONTAL DETAIL (NUMBER OF PIXELS PER LINE)
To measure horizontal detail, we count the pixels of each horizontal line. When we connect vertically the horizontal dots of each line, we get a horizontal stack of lines, hence the horizontal "lines" of resolution. For the 1080 HDTV system, each line comprises 1,920 pixels, or 1,920 horizontal lines of resolution. Ultra HD has more than 7,000 pixels per line.

Temporal Resolution

The temporal resolution becomes an issue when something moves on-screen. A high temporal resolution makes movement look smooth without blurring the moving object; a low resolution does the opposite. In video the temporal resolution is measured by the relative frame density, that is, the number of frames per second. When the video frame is interlaced (i), it takes two scanning cycles for each complete frame. When it is scanned progressively (p), each scan produces a complete frame. This is why a progressively scanned image has a higher temporal resolution than an interlaced one.

For example, an SDTV 480p image looks much better than an STV 480i image even though the spatial resolution (number of scanning lines and number of pixels per line) is identical. The following table lists the spatial and temporal resolutions of the most common SDTV, HDTV, and digital cinema systems. SEE 6.22 Of course, you should

6.22 DTV RESOLUTION TABLE

TYPE	SPATIAL RESOLUTION		TEMPORAL RESOLUTION		
	Vertical (number of scanning lines)	**Horizontal** (number of pixels per line)	**Complete Frames per Second** i = interlaced scanning p = progressive scanning		
SDTV	480	704	24p	30p, 30i	60p
HDTV	720	1,280	24p	30p	60p
HDTV	1,080	1,920	24p	30p, 30i	60p
Digital cinema	2,304	4,096 (4K)	1–30p variable		
Super-Hi vision (ultra HD)	4,096	7,680 (8K)			60p

realize that the figures in this table are what the camera system produces, not necessarily what you get on your home screen. Most HDTV flat-panel television sets must down-convert the high-definition signal to fit the pixel array of their displays.

IMAGE CONTRAST

The range of contrast between the brightest and the darkest picture areas that the video camera and the flat-panel display can accurately reproduce is limited. That range is expressed as the ***contrast ratio***, and the grayscale shows the brightness steps between television white and television black.

Contrast Ratio

Despite their manufacturers' overly optimistic claims, even the better cameras have trouble handling high contrast in actual shooting conditions. You will run into this problem every time you video-record a scene in bright sunlight. When you adjust the camera for the extremely bright sunlit areas, the shadow becomes uniformly dark and dense. When you then adjust the lens (open its iris), you will promptly overexpose—or, in video lingo, "blow out"—the bright areas. It is best to limit the contrast and stay within a contrast ratio of about 50:1, meaning that for optimal pictures the brightest picture area can be only 50 times brighter than the darkest area. Digital cameras with high-quality sensors can tolerate much higher contrast ratios (1,000:1 or even higher), but don't be misled by the camera's specifications. It is always a good idea to avoid extreme contrasts.

Contrast Control

By watching a waveform monitor, which graphically displays the white and black levels of a picture, the video operator uses the CCUs to control contrast and adjusts the picture to the optimal contrast range, an activity called shading. SEE 6.23

To adjust a less-than-ideal picture, the VO tries to "pull down" the excessively bright values to make them match the established white level (which represents a 100 percent video signal strength). But because the darkest value cannot get any blacker and move down with the bright areas, the darker picture areas are "crushed" into a uniformly muddy, noisy dark color. If you insist on seeing detail in the dark picture areas, the video operator can "stretch the blacks" toward the white end, but, except for top-of-the-line cameras, stretching the shadow areas causes the bright areas to lose their definition and take on a uniformly white and strangely flat and washed-out color. In effect, the pictures look as though the contrast is set much too low with the brightness turned too high. Again, before the VO can produce optimal pictures through shading, you must try to reduce the contrast in the scene to tolerable limits.

In digital cinema, the contrast ratio is less a concern than the brightness steps between the limits of the contrast range. This is explained in the following section.

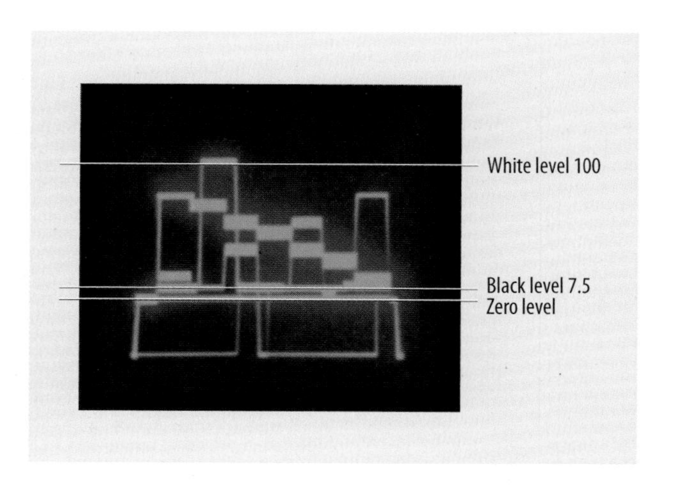

6.23 PHOTO BY HERBERT ZETTL

6.23 WAVEFORM MONITOR
The waveform monitor shows a graph of the luminance (black-and-white) portion of the video signal. It also shows the white level (the upper limit of the signal) and the black level (the lower limit of the signal).

COLOR BASICS

When you look at a red ball, its color is not part of the ball but simply light reflected off of it. The red paint of the ball acts as a color filter, absorbing all colors except red, which it bounces back. Thus the ball is stuck with the only color it has rejected: red.

Color Attributes

When you look at colors, you can easily distinguish among three basic color sensations, called attributes: hue, saturation, and brightness. SEE 6.24

Hue ***Hue*** describes the color itself, such as a red ball, a green apple, or a blue coat.

Saturation ***Saturation*** indicates the richness or strength of a color. The bright red paint of a sports car is highly

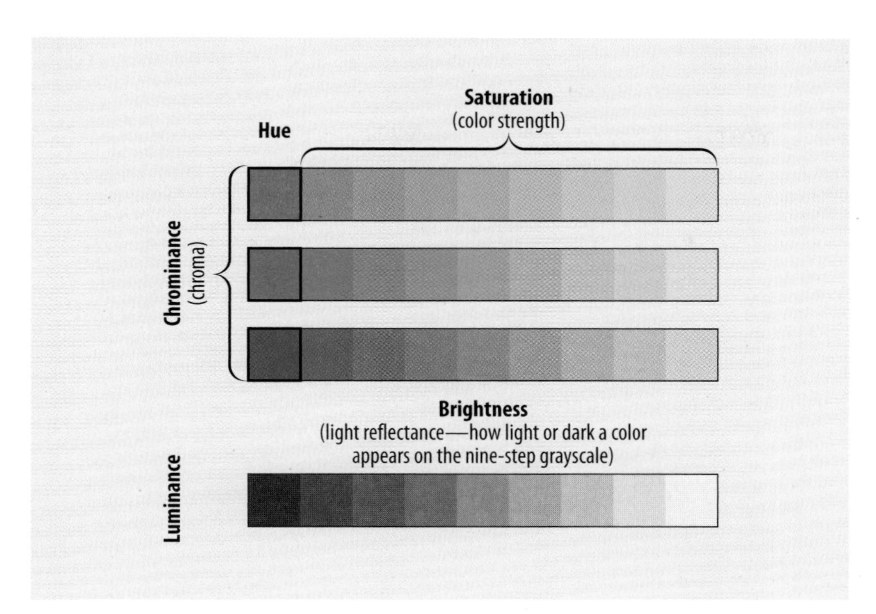

6.24 COLOR ATTRIBUTES
Hue is the term used for the base color—blue, green, yellow, and so on. *Saturation* refers to the purity and the intensity of the color. *Brightness, lightness,* or *luminance* describes the degree of reflectance—how light or dark a color appears on the grayscale.

saturated, whereas the washed-out blue of your jeans or the beige of the sand on a beach are of low saturation.

Brightness ***Brightness*** (also called lightness or luminance) is how dark or light a color appears on a black-and-white monitor or, roughly, how dark or light a color appears. When you see black-and-white television pictures on-screen, you see brightness variations only; the pictures have no hue or saturation. In television the hue and saturation properties of color are sometimes named chrominance (from *chroma,* Greek for "color"), and the brightness properties are called luminance (from *lumen,* Latin for "light"). The chrominance, or C, signals and the luminance, or Y, signal are discussed later in this section.

Brightness steps: grayscale The ***grayscale*** divides the brightness range between TV white and TV black into distinct steps. Black-and-white images are mainly discernible because of their brightness differences—their grayscale (the other factor being the outline of objects); color images get their definition not just from the color itself but also from how light or dark they are. The most common number of brightness steps between TV white and TV black on a grayscale is nine (see figure 6.24).

Not only do the new HDTV systems produce superior resolution (picture sharpness) but their brightness range is considerably larger than that of STV. Digital cinema can display many more subtle steps of gray between TV white and TV black—one of the major factors in achieving the "film look" of high-definition digital video images. This extended brightness scale is especially important for seeing detail in the darker picture areas. You will learn more about some technical details of the grayscale steps in chapter 15.

Color Mixing

When you think back to your finger-painting days, you probably had three pots of paint: red, blue, and yellow. When mixing blue and yellow, you got green; when mixing red and blue, you got purple; and when smearing red and green together, you got at best a muddy brown. An expert finger painter could achieve almost all colors by simply mixing the primary paint colors of red, blue, and yellow. Not so when mixing colored light. The three primary light colors are not red, blue, and yellow but rather red, green, and blue—in television language, RGB.

Additive mixing Assume that you have three individual slide projectors with a clear red slide (filter) in the first projector, a clear green one in the second, and a clear blue one in the third. Hook up each projector to a separate dimmer. When the three dimmers are up full and you shine all three light beams together on the same spot of the screen, you get white light (assuming equal light transmission by all three slides and projector lamps). Because you add various quantities of colored light in the process, it is called additive color mixing. This is not surprising because we can split white light into these three primaries. When you turn off

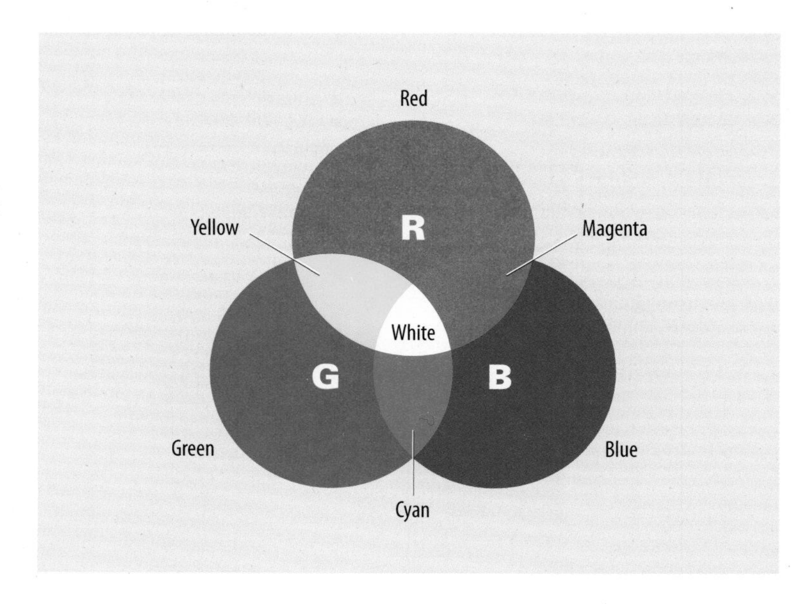

6.25 ADDITIVE COLOR MIXING
When mixing colored light, the additive primaries are red, green, and blue (RGB). All other colors can be achieved by mixing certain quantities of red, green, and blue light. For example, the additive mixture of red and green light produces yellow.

the blue projector and leave on the red and green ones, you get yellow. If you then dim the green projector somewhat, you get orange or brown. If you turn off the green one and turn on the blue one again, you get a reddish purple, called magenta. If you then dim the red projector, the purple becomes more bluish. SEE 6.25

Because the color camera works with light rather than paint, it needs the three additive color primaries (red, green, and blue) to produce all the colors you see on the television screen. You can make all other colors by adding the three light beams—primaries—in various proportions, that is, in various light intensities.

Subtractive mixing When using paint instead of colored light, the primary colors are red, blue, and yellow, or, more accurately, magenta (a bluish red), cyan (a greenish blue), and yellow. In subtractive mixing, the colors filter each other out. We skip subtractive mixing at this point because we are primarily concerned with television images.

LUMINANCE AND CHROMINANCE CHANNELS

We tend to see more detail when the picture is in black-and-white than in color. Recall that this is one of the reasons why HDTV cameras have an additional black-and-white viewfinder even if they have a high-quality color one. As stated earlier, chrominance deals with the hue and saturation attributes of a color, luminance with its brightness information. The chrominance channel in a camera and a video recorder carries the color signals, and the luminance channel carries the black-and-white signals.

Luminance Channel

The luminance channel, or Y channel, is responsible for the brightness information of the color pictures. To get the Y signal, we combine a portion of the RGB signals again, which we had just so laboriously split in the prism block. But, as you just read, when mixed in equal proportions, we get white (and a series of steps of progressively darker grays all the way to black).

The luminance signal fulfills two basic functions: it translates the brightness variations of the colors in a scene into black-and-white pictures for black-and-white receivers, and it provides color pictures with the necessary crispness and definition, just like the black dots in a four-color print. Because it has such a great influence on the sharpness of the picture, the Y signal is favored in the digital domain. Even in high-end digital HDTV cameras, the color signals are normally sampled only half as often as the luminance signal—or only one-fourth as often in HDV camcorders.[3]

3. You may hear engineers talk about 4:4:2 or even 4:4:4 versus 4:4:1 sampling ratios when talking about the quality of HDTV systems. All this means is that good cameras sample the luminance signal four times and the two Y/color difference signals only two times in the same time interval. Lesser-quality cameras sample luminance four times but color only once. Some top digital cinema cameras use 4:4:4 sampling, which means that all three RGB signals are sampled equally. For a more in-depth discussion of sampling the different types of video signals, see chapter 13.

Chrominance Channel

The chrominance channel, or C channel, transports all hue attributes. It consists of the three "slide projectors" that produce red, green, and blue light beams of varying intensities, except that in the television camera the "slide projectors" consist of the sensors that produce an electrical signal of varying intensity (voltage) for each of the three primary RGB colors. The RGB signals are usually modified and combined with the Y signal for the color video you see on the TV screen. How they are modified and combined is explained in chapter 13.

MAIN POINTS

- The resolution of video is determined by how many scanning lines there are, how many pixels there are in each line, and how many frames per second we see. The pixel count is called the spatial resolution; the frames-per-second count is the temporal resolution.
- The range of contrast between the brightest and the darkest picture areas that the video camera and a flat-panel display can accurately reproduce is limited. That limit, called the contrast range, is expressed as a ratio. It is best to limit the contrast and stay within a contrast ratio of about 50:1, meaning that for optimal pictures the brightest picture area can be only 50 times brighter than the darkest area. HDTV and digital cinema cameras have a much higher contrast ratio.
- By watching a waveform monitor, which graphically displays the white and black levels of a picture, the video operator (VO) uses the CCUs to control contrast and adjust the picture to the optimal contrast range, an activity called shading.
- The three basic color sensations, called attributes, are hue (the color itself), saturation (the purity and the intensity of the color), and brightness (the degree of reflectance, also called lightness and luminance).
- The more brightness steps between television white and black there are, the better the picture will look. The major brightness steps are displayed as a grayscale.
- Color television operates on additive mixing of the three color primaries of light: red, green, and blue (RGB).
- Color cameras contain a chrominance and a luminance channel. The chrominance channel processes the color signals—the C signals—and the luminance channel processes the black-and-white (brightness) signal, called the Y signal. The Y signal contributes to the sharpness of the picture.

ZETTL'S VIDEOLAB

For your reference or to track your work, the *Zettl's VideoLab* program cue in this chapter is listed here with its corresponding page number.

CHAPTER

7

Lenses

Lenses are used in all fields of photographic art. Their primary function is to project a clear image of the viewed scene on the film or, in the case of digital photography and television, on the electronic imaging device. As discussed in chapter 6, the lens is one of the three major parts of the camera. In studio cameras the lens is often considerably larger than the camera itself. Section 7.1, What Lenses Are, covers the basic optical characteristics of lenses and their primary operational controls. The performance characteristics of lenses, that is, how they see the world, are explored in section 7.2, What Lenses See.

KEY TERMS

aperture Iris opening of a lens, usually measured in *f*-stops.

auto-focus Automated feature whereby the camera focuses on what it senses to be the target object.

calibrate To preset a zoom lens to remain in focus throughout the zoom.

compression The crowding effect achieved by a narrow-angle (telephoto) lens wherein object proportions and relative distances seem shallower.

depth of field The area in which all objects, located at different distances from the camera, appear in focus. Depth of field depends on the focal length of the lens, its *f*-stop, and the distance between the object and the camera.

digital zooming Simulated zoom that crops the center portion of an image and electronically enlarges the cropped portion. Digital zooms lose picture resolution.

digital zoom lens A lens that can be programmed through a built-in computer to repeat zoom positions and their corresponding focus settings.

fast lens A lens that permits a relatively great amount of light to pass through at its maximum aperture (relatively low *f*-stop number at its lowest setting). Can be used in low-light conditions.

field of view The portion of a scene visible through a particular lens; its vista. Expressed in abbreviations, such as CU for close-up.

focal length The distance from the optical center of the lens to the front surface of the camera's imaging device at which the image appears in focus with the lens set at infinity. Focal lengths are measured in millimeters or inches. Short-focal-length lenses have a wide angle of view (wide vista); long-focal-length (telephoto) lenses have a narrow angle of view (close-up). In a variable-focal-length (zoom) lens, the focal length can be changed continuously from wide-angle (zoomed out) to narrow-angle (zoomed in) and vice versa. A fixed-focal-length (prime) lens has a single designated focal length.

focus A picture is in focus when it appears sharp and clear onscreen (technically, the point at which the light rays refracted by the lens converge).

***f*-stop** The calibration on the lens indicating the aperture, or iris opening (and therefore the amount of light transmitted through the lens). The larger the *f*-stop number, the smaller the aperture; the smaller the *f*-stop number, the larger the aperture.

iris Adjustable lens-opening that controls the amount of light passing through the lens. Also called *diaphragm* and *lens diaphragm.*

macro position A lens setting that allows it to be focused at very close distances from an object. Used for close-ups of small objects.

minimum object distance (MOD) How close the camera can get to an object and still focus on it.

narrow-angle lens Gives a close-up view of an event relatively far away from the camera. Also called *long-focal-length lens* and *telephoto lens.*

normal lens A lens or zoom lens position with a focal length that approximates the spatial relationships of normal vision.

rack focus To change focus from one object or person closer to the camera to one farther away or vice versa.

range extender An optical attachment to the zoom lens that extends its focal length. Also called *extender.*

selective focus Emphasizing an object in a shallow depth of field through focus while keeping its foreground and/or background out of focus.

servo zoom control Zoom control that activates motor-driven mechanisms.

slow lens A lens that permits a relatively small amount of light to pass through at its maximum aperture (relatively high *f*-stop number at its lowest setting). Can be used only in well-lighted areas.

wide-angle lens A short-focal-length lens that provides a broad vista of a scene.

z-axis An imaginary line representing an extension of the lens from the camera to the horizon—the depth dimension.

zoom lens A variable-focal-length lens. It can gradually change from a wide shot to a close-up and vice versa in one continuous move.

zoom range The degree to which the focal length can be changed from a wide shot to a close-up during a zoom. The zoom range is often stated as a ratio; a 20:1 zoom ratio means that the zoom lens can increase its shortest focal length 20 times.

SECTION

7.1

What Lenses Are

The lens determines what the camera can see. One type of lens can provide a wide vista even though you may be relatively close to the scene; another type may provide a close-up view of an object that is quite far away. Different types of lenses also determine the basic visual perspective—whether you see an object as distorted or whether you perceive more or less distance between objects than there really is. They also contribute to a large extent to the quality of the picture—its definition—and how much you can zoom in or out on an object without moving the camera. This section examines what lenses can do and how to use them.

▶ **TYPES OF ZOOM LENSES**
Studio and field lenses, zoom range, prime lenses, and lens format

▶ **OPTICAL CHARACTERISTICS OF LENSES**
Focal length, focus, light transmission (iris, aperture, and f-stop), and depth of field

▶ **OPERATIONAL CONTROLS**
Zoom control and focus control

TYPES OF ZOOM LENSES

When listening to production people talk about ***zoom lenses***, you will most likely hear one person refer to a studio rather than a field zoom, another to a 20× (20 times) lens, and yet another to a zoom lens that fits a ⅔-inch image format—and all may be talking about the same zoom lens. This section looks at these classifications.

Studio and Field Lenses

As the name indicates, studio zoom lenses are normally used with studio cameras. Field zooms include large lenses mounted on high-quality cameras that are used for remote telecasts, such as parades and sporting events. They also include the zoom lenses attached to ENG/EFP cameras. The lenses of small camcorders usually come with the camera and cannot be exchanged. Some high-end models, however, allow you to attach a variety of zoom lenses, or they have adapters for mounting various lenses. Because you can, of course, use a field lens in the studio and vice versa, a better and more accurate way to classify zoom lenses is by their zoom range and lens format, that is, what cameras they fit. **SEE 7.1 AND 7.2**

Zoom Range

If a zoom lens provides an overview, for example, of the whole tennis court and part of the bleachers when zoomed all the way out and (without moving the camera closer to the court) a tight close-up of the player's tense expression when zoomed all the way in, the lens has a good zoom range. The ***zoom range*** is the degree to which you can change the focal length of the lens (and thereby the field of view, or vista) during the zoom.

The zoom range of a lens is often stated as a ratio, such as 10:1 or 40:1. A 10:1 zoom means that you can increase the shortest focal length 10 times; a 40:1, 40 times. To make things easier, these ratios are usually listed as 10× (10 times) or 40× (40 times), referring to the maximum magnification of the image of which the lens is capable. **SEE 7.3**

The large (studio) cameras that are positioned on top of the bleachers for sports coverage may have zoom ranges of 40× and even 70×. In the studio the cameras are well served by a 20× zoom lens. The smaller and lighter ENG/EFP camcorder lenses rarely exceed a 15× zoom range.

Optical and digital zoom ranges You may have noticed that the optical zoom range on a small camcorder is rather limited; an optical zoom range of 15× is considered excellent even for high-end camcorders. This is why most small camcorders offer the option of increasing the zoom range digitally. During an optical zoom to a tighter shot, the image magnification is achieved by moving elements within the lens. In effect, you are continually changing the focal length during the zoom-in or zoom-out. In a digital zoom, such a change in focal length does not occur.

In ***digital zooming*** the electronics of the camera simply select the center portion of the long shot and enlarge the cropped area to full-screen size. The problem with digital zooming is that the enlarged picture portions noticeably reduce the image resolution (recall the mosaic tiles in figure 6.4). At one point in digital zooming, the

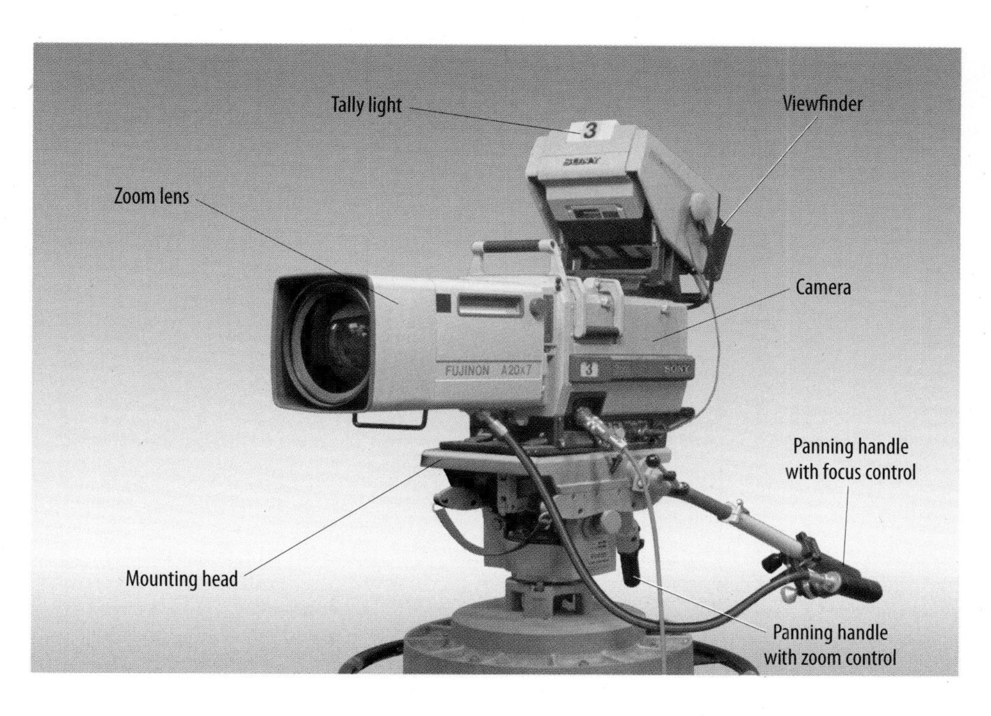

7.1 STUDIO ZOOM LENS
High-quality studio lenses are heavy and often larger than the camera itself.

7.1 PHOTO BY EDWARD AIONA

"mosaic tiles" can get so large that they look more like a special effect than a magnification of the original image. Some cameras have a digital zoom that is remarkably good, although even the best digital zoom does not achieve the crispness of the optical zoom.

All studio, field, and ENG/EFP lenses of large camcorders and some high-end small camcorders are detachable and interchangeable with lenses of a different zoom range. Many small camcorders, however, have a built-in lens that cannot be detached. This is a major handicap, especially because the wide-angle positions of these lenses are usually not wide enough.

Studio and large field lenses Note that a 20× studio lens becomes a field lens if it is used "in the field," that is, for a production that occurs outside the studio. Generally, field lenses have a much greater zoom range (from 40× to 70×) than studio cameras. Some field lenses have an even greater zoom range, allowing the camera operator to zoom from a wide establishing shot of the football stadium to a tight close-up of the quarterback's face. Despite the great zoom range, these lenses deliver high-quality pictures even in relatively low light levels. For studio use such a zoom range would be unnecessary and even counterproductive.

ENG/EFP lenses These lenses are much smaller, to fit the portable cameras. Their normal zoom range varies between 11× and 20×. A 15× zoom lens would be sufficient for most ENG/EFP assignments, but sometimes you might want a closer view of an event that is relatively far away. You would then need to exchange the 15× zoom lens for one with a higher zoom range—such as 20× or even 30×.

7.2 ENG/EFP ZOOM LENS
The ENG/EFP camera lens is considerably lighter and smaller than a studio zoom lens. Although ENG/EFP lenses are not as high quality as studio lenses, they nevertheless have many of the studio zoom's features, such as servo and manual zoom controls, automatic iris control, and sometimes an auto-focus feature.

7.2 PHOTO COURTESY PANASONIC.

7.3 PHOTOS BY EDWARD AIONA

7.3 MAXIMUM ZOOM POSITIONS OF A 10× LENS

The 10× zoom lens can increase its focal length 10 times. It magnifies a portion of the scene and seems to bring it closer to the camera and ultimately the viewer.

You can also use a range extender (discussed later in this chapter), which would let you zoom beyond the normal zoom range to a tighter shot.

A more important consideration for ENG/EFP lenses is whether they have a wide enough angle of view (a very short focal length), which would allow you to shoot in highly cramped quarters, such as in a car, a small room, or an airplane. Also, the wide-angle view is important for shooting in the wide-screen 16 × 9 format.

Many lenses have digital or mechanical stabilizers that absorb at least some of the picture wiggles resulting from camera operation, especially when in a narrow-angle (zoomed-in) position. Realize, however, that such stabilizers cause an additional drain on the battery. Use this feature only when you don't have a tripod or are unable to stabilize the camera in any other way.

Range extenders If a zoom lens does not get you close enough to a scene from where the camera is located, you can use an additional lens element called a ***range extender***, or simply an extender. This optical element, usually available only for lenses on professional cameras, does not actually extend the range of the zoom but rather shifts the magnification—the telephoto power—of the lens toward the narrow-angle end of the zoom range. Most lenses have 2× extenders, which means that they double the zoom range in the narrow-angle position, but they also reduce the wide-angle lens position by two times. With such an extender, you can zoom in to a closer shot, but you cannot zoom back out as wide as you could without the extender. There is another disadvantage to range extenders: they cut down considerably the light entering the camera, which can be problematic in low-light conditions.

Prime Lenses

You have probably used prime lenses on a single-lens reflex (SLR) still camera. Prime lenses are also called fixed-focal-length lenses because their focal length cannot be changed. They are more apt to be used on digital cinema cameras than on television camcorders. Prime lenses are also used on digital SLR cameras that are used for video capture. You cannot zoom in or out with a prime lens, so you have to physically move the camera closer or farther away from the object if you want a closer or wider shot of it.

Why use a relatively clumsy lens when you have a much more versatile zoom lens available? Because the light has to travel through less glass in a prime lens than in a zoom lens (see figure 7.6). As a consequence, a prime lens delivers a sharper image than a zoom lens, especially under low-light conditions. You won't see this difference on a small screen (such as a regular TV set), but you can probably see it when the shot is projected on a large motion picture screen.

Lens Format

Because camera lenses are designed to match the size of the imaging device, you may hear about a lens format or an image format of ⅓-inch, ½-inch, or ⅔-inch. This means that you can use only a lens that fits the corresponding size of the sensor. Most large camcorders use the ½-inch format, and small camcorders use a ⅓-inch format. Digital cinema cameras must use lenses that fit their unique larger

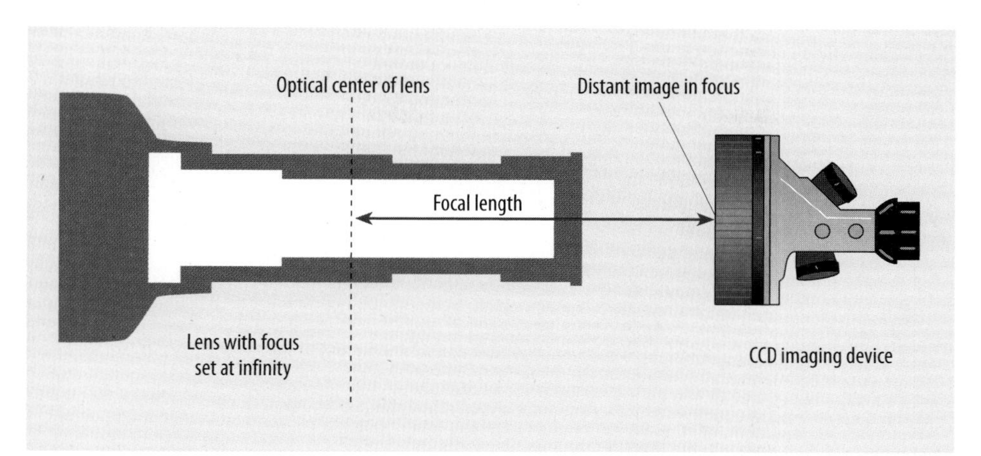

7.4 FOCAL LENGTH
The focal length is the distance from the optical center of the lens to the front surface of the imaging device.

sensors. The term *lens format* may also refer to whether a lens is used for standard NTSC cameras or high-definition (HD) cameras.

OPTICAL CHARACTERISTICS OF LENSES

Effective use of a camera depends to a great extent on your understanding of four optical characteristics of lenses: focal length; focus; light transmission—iris, aperture, and *f*-stop; and depth of field.

Focal Length

Technically, ***focal length*** refers to the distance from the optical center of the lens to the point where the image the lens sees is in focus. This point is the camera's imaging device. SEE 7.4 Operationally, the focal length determines how wide or narrow a vista a particular camera can display and how much and in what ways objects appear magnified.

When you zoom all the way out, the focal length of the lens is short and at the maximum wide-angle position; the camera will provide a wide vista or field of view. When you zoom all the way in, the focal length is long and at the maximum narrow-angle (telephoto) position; the camera will provide a narrow vista—a close-up view of the scene. SEE 7.5 When you stop the zoom approximately halfway between these extreme positions, the lens has the normal focal length. This means that you will get a "normal" vista that approximates your actually looking at the scene. Because the zoom lens can assume all focal lengths from its maximum wide-angle position (zoomed all the way out) to its maximum narrow-angle position (zoomed all the way

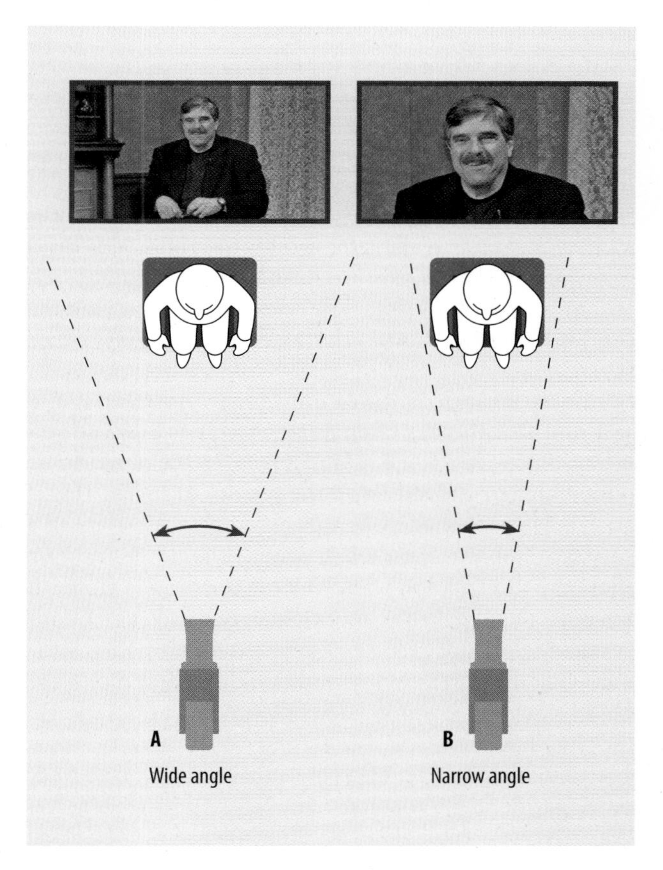

7.5 WIDE-ANGLE AND NARROW-ANGLE ZOOM POSITIONS
A The wide-angle zoom position (zoomed out) has a wider vista (field of view) than **B**, the narrow-angle zoom position (zoomed in). Note that zooming in magnifies the subject.

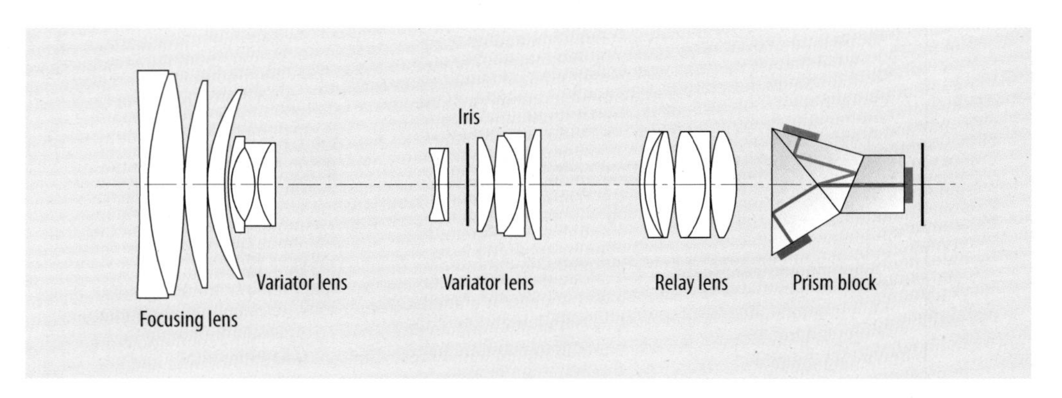

7.6 ELEMENTS OF A ZOOM LENS
A zoom lens consists of many sliding and stationary lens elements that interact to maintain focus throughout the continuous change of focal length. The front elements control the focus; the middle elements control the zoom.

in), it is called a variable-focal-length lens. ZVL1 CAMERA→ Zoom lens→ normal | wide | narrow | try it

On the television screen, a zoom-in appears as though the object is gradually coming toward you. A zoom-out seems to make the object move away from you. Actually, all that the moving elements within the zoom lens do is gradually magnify (zoom-in) or reduce the magnification (zoom-out) of the object while keeping it in focus, but the camera remains stationary during both operations. SEE 7.6

Minimum object distance and macro position You will find that there is often a limit to how close you can physically move a camera (and a lens) to the object to be photographed and still keep the picture in focus. This is especially problematic when trying to get a close-up of a very small object. Even when zoomed in all the way, the shot may still look too wide. Moving the camera closer to the object will make the shot tighter, but you can no longer get the picture in focus. Range extenders help a little; but while they provide you with a tighter close-up of the object, they force you to back off with the camera to get the shot in focus.

One way to solve this problem is to zoom all the way out to a wide-angle position. Contrary to normal expectations, the wide-angle zoom position often allows you to get a tighter close-up of a small object than does the extended narrow-angle zoom position (zoomed all the way in with a 2× extender). But even with the lens in the wide-angle position, there is usually a point at which the camera will no longer focus when moved too close to the object. The point where the camera is as close as it can get and still focus on the object is called the ***minimum object distance (MOD)*** of the lens.

Many field lenses on ENG/EFP cameras have a ***macro position***, which lets you move the camera even closer to an object without losing focus. When the lens is in the macro position, you can almost touch the object with the lens and still retain focus; you can no longer zoom, however. The macro position changes the zoom lens from a variable-focal-length lens to a fixed-focal-length, or prime, lens. The fixed focal length is not a big disadvantage in this case because the macro position is used only in highly specific circumstances. For example, if you need to get a screen-filling close-up of a postage stamp, you would switch the camera to the macro position, but then you cannot use the camera for zooming until you switch back to the normal zoom mechanism.

Focus

A picture is "in focus" when the projected image is sharp and clear. The ***focus*** depends on the distance from the lens to the film (as in a nondigital still or movie camera) or from the lens to the camera's imaging device (beam splitter with sensors). Simply adjusting the distance from the lens to the film or imaging device brings a picture into focus or takes it out of focus. In television zoom lenses, this adjustment is accomplished not by moving the lens or the prism block (beam splitter) but by moving certain lens elements relative to each other through the zoom focus control (see figure 7.6).

Focus controls come in various configurations. Portable cameras have a focus ring on the lens that you turn;

studio cameras have a twist grip attached to the panning handle (see figure 7.18). Many camcorders have an automatic focus feature, called auto-focus, which is discussed in the context of operational controls later in this section.

If properly preset, a zoom lens stays in focus during the entire zoom range, assuming that neither the camera nor the object moves very much toward or away from the other. But because you walk and even run while carrying an ENG/EFP camera, you cannot always prefocus the zoom. In such cases you would do well by zooming all the way out to a wide-angle position, considerably reducing the need to focus.

Presetting (calibrating) the zoom lens There is a standard procedure for presetting, or ***calibrating***, the zoom lens so that the camera remains in focus throughout a zoom: Zoom all the way in on the target object, such as a newscaster on a news set. Focus on the newscaster's face (the eyes or the bridge of the nose) by turning the focus control. When zooming back out to a long shot, you will notice that everything remains in focus. The same is true when you zoom in again. You should now be able to maintain focus over the entire zoom range. If you move the camera, however, or if the talent moves after you preset the zoom lens, you need to calibrate the lens again.

For example, if you had preset the zoom on the news anchor but then the director instructs you to move the camera a little closer and to the left so that the anchor could more easily read the teleprompter, you would not be able to maintain focus without presetting the zoom again from the new position. If, after presetting the zoom, you were asked to zoom in on the map behind the anchor, you would have to adjust the focus while zooming past her—not an easy task for even an experienced camera operator.

If the camera moves are predetermined and repeated from show to show, as in a daily newscast, you can use the preset features of the digital zoom lens. The lens then remembers the various zoom positions and performs them automatically with the push of a button. This is especially important when using robotic cameras that are run by computers rather than human beings.

Unless you have an automatic focus control, you must preset the zoom lens even when covering a news event in the field. You may have noticed that unedited video of a disaster (such as a tornado or fire) often contains brief out-of-focus close-ups followed by focusing and quick zoom-outs. Despite the precarious situations, the camera operator was calibrating the zoom lens so that it would stay in focus during subsequent zooms.

Light Transmission: Iris, Aperture, and f-Stop

Like the pupil in the human or cat eye, all lenses have a mechanism that controls how much light is admitted through them. This mechanism is called the ***iris***, or lens diaphragm. The iris consists of a series of thin metal blades that form a fairly round hole—the ***aperture***, or lens opening—of variable size. **SEE 7.7**

If you "open up" the lens as wide as it will go, or, technically, if you set the lens to its maximum aperture, it admits the maximum amount of light. **SEE 7.8A** If you close the lens somewhat, the metal blades of the iris form a smaller hole and less light passes through the lens. If you close

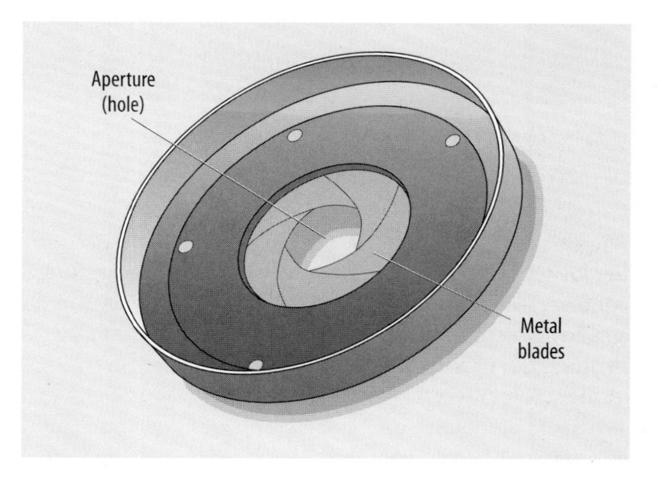

7.7 LENS IRIS

The iris, or lens diaphragm, consists of a series of thin metal blades that form, through partial overlapping, an aperture, or lens opening, of variable size.

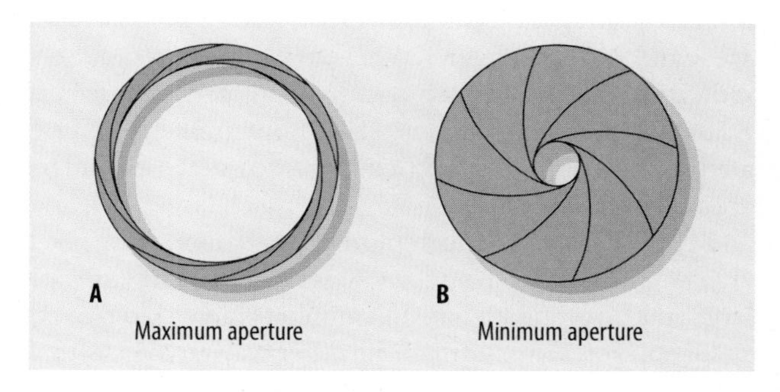

7.8 MAXIMUM AND MINIMUM APERTURES

A At the maximum aperture, the iris blades form a large opening, permitting a great amount of light to enter the lens. **B** At the minimum setting, the blades overlap to form a small hole, admitting only a small amount of light.

7.9 *f*-STOP SETTINGS
The *f*-stop is a calibration that indicates the size of the aperture.

7.9 PHOTO COURTESY NIKON, INC.

the lens all the way—that is, if you set it to its minimum aperture—very little light is admitted. **SEE 7.8B** Some irises can be closed entirely, which means that no light at all goes through the lens.

f-stop The standard scale that indicates how much light goes through a lens, regardless of the lens type, is the ***f*-stop**. **SEE 7.9** If, for example, you have two cameras—a camcorder with a 10× zoom lens and a field camera with a large 50× lens—and both lenses are set at *f*/5.6, the imaging devices in both cameras will receive an identical amount of light.

Regardless of camera type, *f*-stops are expressed in a series of numbers, such as *f*/1.7, *f*/2.8, *f*/4, *f*/5.6, *f*/8, *f*/11, and *f*/16 (see figure 7.9). The *lower* *f*-stop numbers indicate a relatively *large* aperture or iris opening (lens is relatively wide open). The *higher* *f*-stop numbers indicate a relatively *small* aperture (lens is closed down considerably). A lens that is set at *f*/1.7 has a much larger aperture and therefore admits much more light than one that is set at *f*/17. (The reason why the low *f*-stop numbers indicate large apertures and high *f*-stop numbers indicate relatively small apertures, rather than the other way around, is that the *f*-stop numbers actually express a ratio. In this sense *f*/4 is actually *f*/¼; that is, *f* one over four.) Most lenses produce the sharpest pictures between *f*/5.6 and *f*/8. Some lenses extend the optimal focus to *f*/11.

Lens speed The "speed" of a lens has nothing to do with how fast it transmits light but with how much light it lets through. A lens that allows a relatively great amount of light to enter is called a ***fast lens***. Fast lenses go down to a small *f*-stop number (such as *f*/1.4). Most good studio zoom lenses open up to *f*/1.6, which is fast enough to make the camera work properly even in low-light conditions.

A lens that transmits relatively little light at the maximum iris aperture is called a ***slow lens***. A studio lens whose lowest *f*-stop is *f*/2.8 is obviously slower than a lens that can open up to *f*/1.7. Range extenders render the zoom lens inevitably slower. A 2× extender can reduce the lens speed by as much as two "stops" (higher *f*-stop numbers), for instance, from *f*/1.7 to *f*/4 (see figure 7.9). This reduction in light transmission is not a big handicap because range extenders are normally used outdoors, where there is enough light. The more serious problem is a slight deterioration of the original picture resolution.

Remote iris control Because the amount of light that strikes the camera's imaging device is so important to picture quality, the continuous adjustment of the iris is a fundamental function of video control. Studio cameras have a remote iris control, which means that the aperture can be continuously adjusted by the video operator (VO) from the camera control unit (CCU). If the set is properly lighted and the camera properly set up (electronically adjusted to the light/dark extremes of the scene), all that the VO has to do to maintain good pictures is work the remote iris control—open the iris in low-light conditions and close it down somewhat when there is more light than needed.

Auto-iris switch Most cameras, especially ENG/EFP camcorders, can be switched from the manual to the auto-iris mode. **SEE 7.10** The camera then senses the light entering the lens and automatically adjusts the iris for optimal camera performance. This auto-iris feature works well so long as the scene does not have too much contrast.

There are circumstances in which you may want to switch the camera over to manual iris control. For example, if you took a loose close-up shot of a woman wearing a white hat in bright sunlight, the automatic iris would adjust to the bright light of the white hat, not to the darker (shadowed) face under the hat. The auto-iris control would therefore give you a perfectly exposed hat but an underexposed face. In this case you would have to switch to manual iris control, zoom in on the face to eliminate most of the white

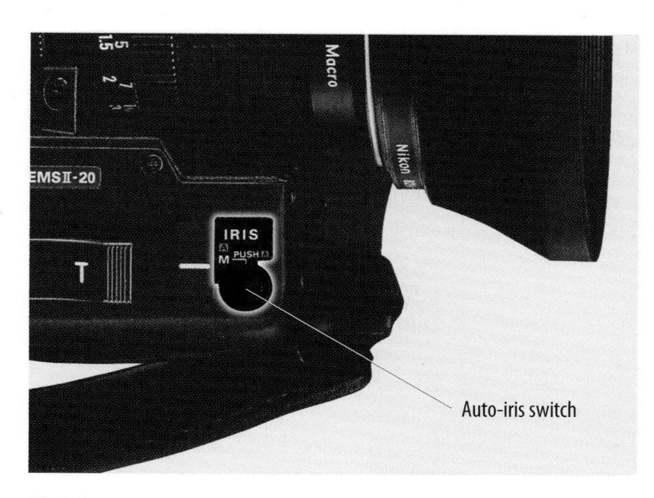

7.10 AUTO-IRIS SWITCH
The auto-iris switch lets you change the aperture control from manual to automatic. You can quickly change back to manual simply by pressing the auto-iris switch without interrupting your shot.

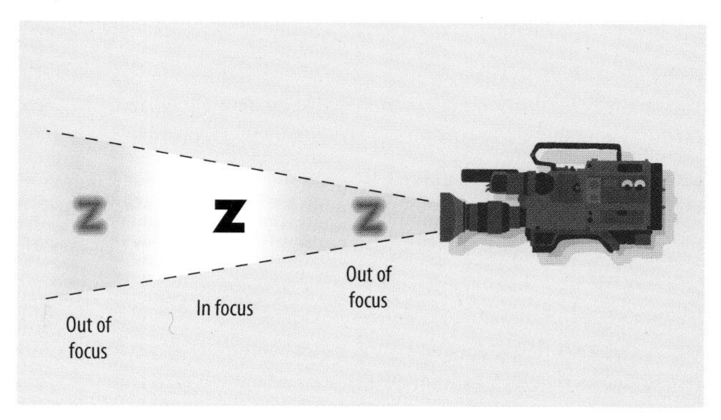

7.11 DEPTH OF FIELD
The depth of field is the area within which all objects, although located at different distances from the camera, are in focus.

hat, then adjust the iris to the light reflecting off the face rather than the hat. When switching to manual iris control, however, you will find that even a fairly good ENG/EFP camera can't handle such an extreme contrast. In this case you might try a neutral density (ND) filter, which would lower the extreme brightness without making the dense shadow areas any darker. (Other ways to handle extreme contrast are explained in chapter 12.) ZVL2 CAMERA→ Exposure control→ aperture | *f*-stop | auto-iris | try it

Depth of Field

If you place objects at different distances from the camera, some will be in focus and others will be out of focus. The area in which the objects are in focus is called ***depth of field***. The depth of field can be shallow or great, but it is always greater behind the object than in front of it. **SEE 7.11**

If you have a shallow depth of field and you focus on an object in the foreground, the middleground and background objects will be out of focus. **SEE 7.12** If the depth of field is great, all objects (foreground, middleground, and background) will be in focus, even though you focus on the middleground object only. **SEE 7.13**

With a great depth of field, there is a large "sharp zone" in which people or objects can move toward or away from the camera without going out of focus and without any need for adjusting the camera focus. If they move in a shallow depth of field, however, they can quickly become

7.12 SHALLOW DEPTH OF FIELD
With a shallow depth of field, the area in which an object is in focus is limited.

7.13 GREAT DEPTH OF FIELD
With a great depth of field, almost everything in the camera's field of view appears in focus.

7.10 PHOTO COURTESY NIKON, INC.
7.12 PHOTO BY HERBERT ZETTL
7.13 PHOTO BY HERBERT ZETTL

blurred unless you adjust the focus. A similar thing happens when you move the camera. A great depth of field makes it relatively easy to move the camera toward or away from the object because you do not have to work any controls to keep the picture in focus. If you move the camera similarly in a shallow depth of field, you must adjust the focus continuously to keep the target object sharp and clear.

Why use a shallow depth of field when a great depth of field makes it so much easier to keep an object in focus? Because a shallow depth of field helps define picture depth. It is also a simple yet highly effective way to draw attention to people or objects that are positioned along the ***z-axis***—an imaginary line that extends from camera to horizon. Finally, a shallow depth of field has become an important element in the director's aesthetic arsenal.

Operationally, the depth of field depends on the coordination of three factors: the focal length of the lens, the aperture, and the distance between the camera and the object.

Focal length The focal length of the lens is the factor that most influences the depth of field. In general, wide-angle lenses and, of course, wide-angle (short-focal-length) zoom positions (zoomed out) have a great depth of field. Narrow-angle lenses and narrow-angle (long-focal-length) zoom positions (zoomed in) have a shallow depth of field. You may want to remember a simple rule of thumb:

- ***Depth of field increases as focal length decreases.***

When running after a fast-moving news event, should you zoom all the way in or all the way out? All the way out. Why? Because, first, the wide-angle position of the zoom lens will at least show the viewer what is going on. Second, and most importantly, the resulting great depth of field will help keep most of your shots in focus, regardless of whether you are close to or far from the event or whether you or the event is on the move.

Aperture Here are rules of thumb for apertures and depth of field:

- ***Large f-stop numbers (such as f/16 or f/22) contribute to a great depth of field.***
- ***Small f-stop numbers (such as f/1.7 or f/2) contribute to a shallow depth of field.***

The following is an example of how everything in television production seems to influence everything else: If you have to work in low-light conditions, you need to open up the iris to get enough light for the camera. But this large aperture (low *f*-stop number) reduces the depth of field. Thus, if you are to cover a news story when it is getting dark and you have no time or opportunity to use artificial lighting, focus becomes critical—you are working in a shallow depth of field. This problem is compounded when zooming in to tight close-ups. On the other hand, in bright sunlight you can stop down (decrease the aperture) and thereby achieve a great depth of field. Now you can run with the camera or cover people who are moving toward or away from you without too much worry about staying in focus—provided the zoom lens is in a wide-angle position.

Camera-to-object distance The closer the camera is to the object, the shallower the depth of field. The farther the camera is from the object, the greater the depth of field. Camera-to-object distance also influences the focal-length effect on depth of field. For example, if you have a wide-angle lens (zoom lens in a wide-angle position), the depth of field is great. But as soon as you move the camera close to the object, the depth of field becomes shallow. The same is true in reverse: If you work with the zoom lens in a narrow-angle position (zoomed in), you have a shallow depth of field. But if the camera is focused on an object relatively far from the camera (such as a field camera located high in the stands to cover an automobile race), you work in a fairly great depth of field and do not have to worry too much about adjusting focus, unless you zoom in to an extreme close-up. **SEE 7.14** ZVL3 CAMERA→ Focusing→ focus ring | depth of field | great depth | shallow | rack focus | auto-focus | try it

- ***Generally, the depth of field is shallow when you work with close-ups and low-light conditions. The depth of field is great when you work with long shots and high light levels.***

OPERATIONAL CONTROLS

You need two basic controls to operate a zoom lens: the zoom control, which lets you zoom out to a wide shot or zoom in to a close-up, and the focus control, which slides the lens elements that lie close to the front of the zoom lens back and forth until the image or a specific part of the image is sharp. Both controls can be operated manually or through a motor-driven servo control mechanism.

Zoom Control

Zoom lenses are equipped with a servo mechanism whose motor activates the zoom, but they may also have a mechanical zoom control that can override the servo zoom at any time.

7.14 DEPTH-OF-FIELD FACTORS

DEPTH OF FIELD	FOCAL LENGTH	APERTURE	*f*-STOP	LIGHT LEVEL	SUBJECT-TO-CAMERA DISTANCE
Great	Short (wide-angle)	Small	Large *f*-stop number (*f*/22)	High (bright light)	Far
Shallow	Long (narrow-angle)	Large	Small *f*-stop number (*f*/1.4)	Low (dim light)	Near

Servo zoom control All types of professional cameras (studio and ENG/EFP) have a ***servo zoom control*** for their lenses. The servo zoom control for studio cameras is usually mounted on the right panning handle, and you zoom in and out by moving the thumb lever, similar to a rocker switch. When pressing the right side of the lever, you zoom in; when pressing the left side, you zoom out. The farther you move the lever from the central position, the faster the zoom will be. **SEE 7.15**

The automation lets you execute extremely smooth zooms. Most servo mechanisms for studio cameras offer a choice of at least two zoom speeds: normal and fast. The fast zoom setting is used when fast zoom-ins are required for emphasis. For example, the director may call for a very fast zoom-in on a ringing telephone or a contestant's face. Normal zoom speeds are simply not fast enough to highlight such events.

The servo zoom control for ENG/EFP camcorders is directly attached to the lens; for small camcorders it is sometimes built into the camera housing. The rocker switch, which controls the zoom, is usually marked with a *W* (for wide) and a *T* (for tight or telephoto). To zoom in press the *T* side of the switch; to zoom out press the *W* side. **SEE 7.16** The servo control housing has a strap attached, which lets you support the shoulder-mounted or handheld camcorder while operating the zoom control. This way your other hand is free to operate the manual focus control.

Manual zoom control ENG and EFP often require extremely fast zoom-ins to get fast close-ups or to calibrate the zoom lens as quickly as possible. Even fast servo settings are usually too slow for such maneuvers. Camcorder lenses therefore have an additional manual zoom control, which is activated by a ring on the lens barrel. **SEE 7.17** By moving the ring clockwise (to zoom in) or counterclockwise (to zoom out), you can achieve extremely fast zooms not possible with the servo control. Some zoom rings have a lever

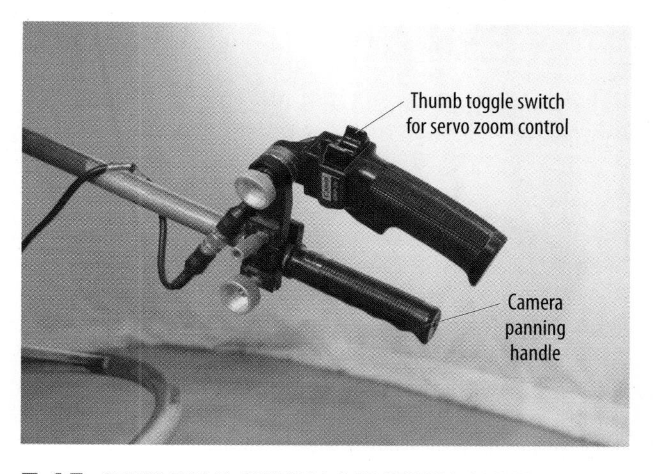

7.15 SERVO ZOOM CONTROL FOR STUDIO CAMERA
This zoom control is attached to the camera panning handle. By moving the rocker switch with your thumb to the right or left, you zoom in or out, respectively. The farther you move the lever from the central position, the faster the zoom will be.

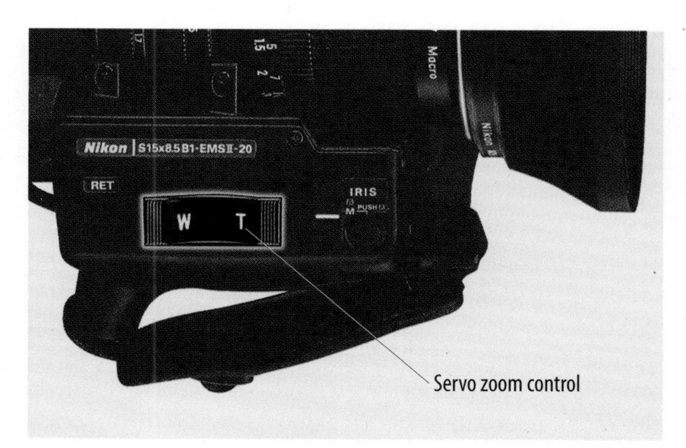

7.16 SERVO ZOOM CONTROL FOR ENG/EFP CAMERA
For ENG/EFP cameras and camcorders, the servo zoom control is part of the lens assembly.

7.15 PHOTO BY EDWARD AIONA

7.16 PHOTO COURTESY NIKON, INC.

7.17 MANUAL ZOOM CONTROL ON LENS
The ring behind the focus control on ENG/EFP and prosumer camera lenses activates a manual zoom control.

attached to facilitate mechanical zooming (see figure 7.9). In addition to news coverage, this manual zoom option is especially important for sports, where getting quick close-ups is the rule rather the exception.

Digital zoom lens Do not confuse a digital zoom lens with digital zooming. A ***digital zoom lens*** is a fancy (and often confusing) name for a lens that can preset certain zoom positions and then trigger the operation with the push of a button. This preset device, which also remembers focus calibration, is highly accurate, provided the camera and the subject are in exactly the same positions as during setup. It is most practical when using robotic cameras (cameras whose movements are controlled by computer and not by an operator), such as during studio newscasts.

Focus Control

The focus control activates the focus mechanism in a zoom lens. The focus mechanism on studio cameras differs from those on ENG/EFP cameras or camcorders.

Studio cameras For studio cameras the focus control ordinarily consists of a twist grip similar to a motorcycle throttle, or a knob, usually mounted on the left panning handle. Two or three turns are sufficient to achieve focus over the full zoom range. As with the servo zoom control, the focus operations are transferred by the drive cable from the panning-handle control to the lens, but the lens executes the focusing though an electric motor. **SEE 7.18**

ENG/EFP cameras EFP cameras and all camcorders have a focus ring near the front of the zoom lens, much like a prime lens for a regular still camera (see figure 7.9). You focus the lens by turning the focus ring clockwise or counterclockwise until the viewfinder shows the image sharply and clearly.

Digital cinema cameras When using prime lenses on digital cinema cameras, a flexible cable is attached to the focus ring on the lens. A twist-knob at the end of the cable allows the assistant camera operator to "pull focus," which means to stay in focus when somebody approaches the camera or when the camera moves toward a person or an event.

Remote control Robotic or fixed cameras, such as the ones mounted in classrooms, have various types of focus controls that can be operated from a remote control station.

Focus-assist As mentioned, focusing a high-definition television (HDTV) image is not always easy because the high resolution can fool you into believing that the picture is in focus when it is not. In fact, the problem with using HDTV for movie making is that the sharpness always simulates a great depth of field, much to the dismay of directors whose style depends on a shallow depth of field.

To help HDTV camera operators focus and stay in focus, some cameras enlarge part of the viewfinder image. Once you get the enlarged portion into focus, the whole image will be crisp. The extralarge sensors in digital

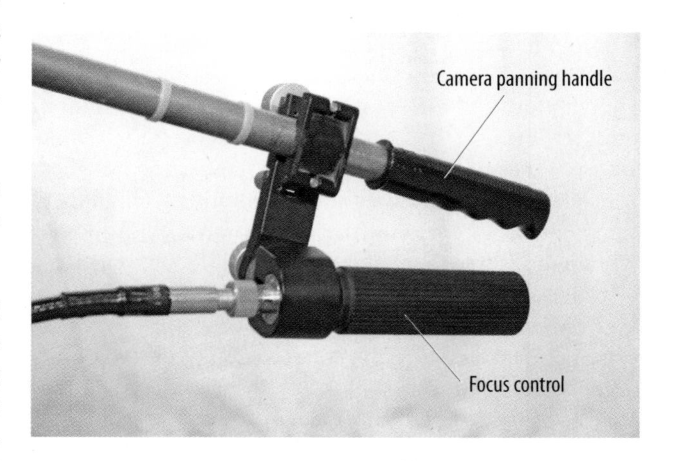

7.18 SERVO FOCUS CONTROL FOR STUDIO CAMERA
The twist grip of the servo focus control for a studio zoom lens turns clockwise and counterclockwise for focusing.

7.18 PHOTO BY EDWARD AIONA

cinema cameras allow the customary manipulation of the depth of field.

Servo focus The servo focus control lets you preset the lens so that it keeps focus during carefully rehearsed camera and/or subject movements. Because even the smartest servo focus control will not help you stay in focus if the camera or subject movements have not been carefully rehearsed, most camera operators prefer to use the manual focus controls.

Auto-focus The ***auto-focus*** feature essentially tries to sense which part of the image you want in focus. It works well in most conventional shots but will cause a problem when the principal object to be focused on is not obvious. The auto-focus usually settles for the object that is more or less in the center of the frame and closest to the camera. If you want to focus on part of a scene that is farther in the background and off to one side, the auto-focus will not comply. Unless you are running after a news story, switch to the manual focus. Don't be swayed by "professionals" who reject outright any auto-focus feature, however. So long as the camera gives you the desired focus, use the automatic feature. In many cases the auto-focus is faster and more accurate than a hasty manual one.

MAIN POINTS

- The primary function of the lens is to produce a sharp optical image on the front surface of the camera's imaging device.
- A 15× zoom lens can extend the focal length by 15 times.
- A range extender (an additional lens element) extends the telephoto power of a zoom lens (permits a closer shot) but reduces the range at the wide-angle end.
- All television cameras are equipped with zoom (variable-focal-length) lenses.
- All digital cinema cameras and SLR cameras that are used for video can accept prime lenses.
- The major optical characteristics of lenses are focal length, focus, light transmission (iris, aperture, and *f*-stop), and depth of field.
- The focal length of a lens determines how wide or narrow a vista the camera can show and how much and how close or far away the object seems to be from the camera (viewer). Zoom lenses have a variable focal length, whose major positions are wide-angle, normal, and narrow-angle (telephoto).
- A wide-angle lens (zoomed out) gives a wide vista. A narrow-angle lens (zoomed in) gives a narrow vista but magnifies the object so that it appears closer to the camera than it actually is. A normal lens (zoom position toward the midrange of the zoom) approximates the angle of human vision.
- A picture is in focus when the projected image is sharp and clear.
- A zoom lens needs to be preset (calibrated) so that focus is maintained over the zoom range. If a lens is properly focused when zoomed in, it should remain in focus when zoomed out and in again.
- The lens iris, or diaphragm, controls the amount of light passing through the lens. It consists of a series of thin metal plates that form a hole known as the aperture, or lens opening.
- The *f*-stop is a standard scale indicating how much light passes through the lens. Low *f*-stop numbers indicate large apertures; high *f*-stop numbers indicate small apertures.
- Studio cameras have a remote iris control, which is operated by the VO (video operator) from the CCU (camera control unit).
- The area in which objects at different distances from the camera are in focus is called depth of field. The depth of field depends on the focal length of the lens, the aperture (*f*-stop), and the distance from camera to object.
- The two basic operational controls for the zoom lens are the zoom control and the focus control. On ENG/EFP cameras and camcorders, both can be operated either manually or automatically by servo control.
- Auto-focus is an automated feature whereby the camera focuses on what it senses to be the target object. High-definition television (HDTV) lenses have a focus-assist feature whereby the camera operator selects the target area.

SECTION

7.2

What Lenses See

The performance characteristics of a lens refer to its vista, what it can and cannot do, and how it generally behaves in common production practice. Because the camera normally processes only visual information that the lens can see, knowledge of the performance characteristics—how it sees the world and how it influences the aesthetic elements of a picture—will aid you greatly in composing effective shots and in many other production tasks. This section explores these concepts.

▶ **HOW LENSES SEE THE WORLD**
Field of view, object and distance distortion, movement, and depth of field

▶ **IMAGE STABILIZATION**
Reducing or eliminating minor image wobbles

HOW LENSES SEE THE WORLD

Although all television cameras use zoom lenses, it might be easier to learn how various zoom positions influence what you see in the viewfinder by describing three zoom positions as though they were fixed-focal-length (prime) lenses. Recall that prime lenses have a specific focal length that cannot be changed. They are normally classified as wide-angle, or short-focal-length, lenses; normal, or medium-focal-length, lenses; and narrow-angle, or long-focal-length, lenses, also called telephoto lenses.

Let's adjust a zoom lens to correspond to these standard focal lengths and observe their performance characteristics, which include field of view, object and distance distortion, movement, and depth of field.

Wide-angle Lens

As you recall, you need to zoom all the way out to achieve the maximum short focal length, or wide angle, of the zoom lens.

Field of view The ***wide-angle lens*** affords a wide vista. You can have a relatively wide ***field of view***—the portion of a scene visible through the lens—with the camera rather close to the scene. When you need a wide vista (long shot) or, for example, when you need to see all five people on a panel in a relatively small studio, a wide-angle lens (wide-angle zoom position) is mandatory. The wide-angle lens is also well suited to provide pictures that fit the horizontally stretched 16 × 9 HDTV aspect ratio. SEE 7.19

Object and distance distortion A wide-angle lens makes objects relatively close to the camera look large and objects only a short distance away look small. This distortion—large foreground objects, small middleground, and even smaller background objects—seems to pull objects apart and stretch the perceived distance. The wide-angle lens also influences our perception of perspective. Because parallel lines seem to converge faster with this lens than you ordinarily perceive, it gives a forced perspective that aids the illusion of exaggerated distance and depth (see 7.25a). With a wide-angle lens, you can make a small room appear spacious or a hallway seem much longer than it really is. SEE 7.20–7.23

Such distortions can also work against you. If you take a close-up of a face with a wide-angle lens, the nose, or whatever is closest to the lens, will look unusually large

7.19 PHOTO BY EDWARD AIONA

7.19 WIDE-ANGLE LONG SHOT: NEWSROOM
The wide-angle lens (zoom position) gives you a wide vista. Although the camera is relatively close to the news set, we can see the whole set.

7.20 PHOTO BY EDWARD AIONA

7.20 WIDE-ANGLE DISTORTION: LINEAR PERSPECTIVE
The length of this hallway is greatly exaggerated.

7.21 PHOTO BY EDWARD AIONA

7.21 WIDE-ANGLE DISTORTION: TRUCK
The wide-angle lens intensifies the raw power of this truck. Note that the apparent size of the front grill is greatly exaggerated by the wide-angle lens.

7.22 PHOTO BY EDWARD AIONA

7.22 WIDE-ANGLE DISTORTION: EMPHASIS ON FOREGROUND OBJECT
The importance of this telephone call in the video adaptation of a famous crime novel is emphasized by the wide-angle distortion.

7.23 PHOTO BY EDWARD AIONA

7.23 WIDE-ANGLE DISTORTION: DEPTH ARTICULATION
Shooting through a permanent foreground piece with the wide-angle lens creates a spatially articulated, forceful picture.

compared with the other parts of the face. **SEE 7.24** Such facial distortions are quite appropriate, however, when the objective is to emphasize stress or unstable psychological conditions or when creating stylistic effects in crime shows, for example.

Movement The wide-angle lens is also a good dolly lens. Its wide field of view de-emphasizes camera wobbles and bumps during dollies, trucks, and arcs (see chapter 8), but because the zoom lens makes it so easy to move from a long shot to a close-up and vice versa, dollying with a zoom lens has almost become a lost art.

Most of the time, a zoom will be perfectly acceptable as a means of changing the field of view (moving to a wider or closer shot). You should be aware, however, that there

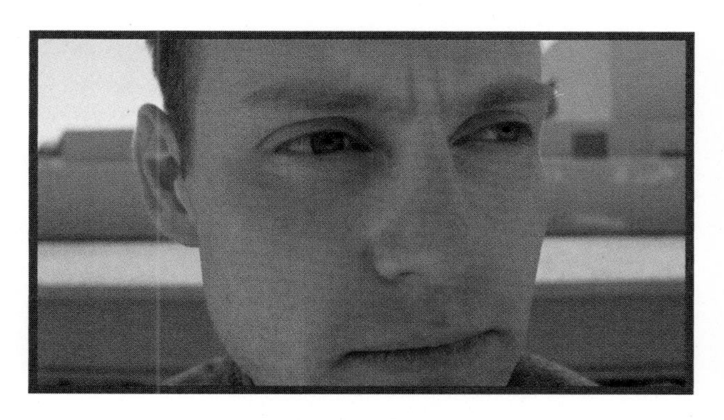

7.24 PHOTO BY EDWARD AIONA

7.24 WIDE-ANGLE DISTORTION: FACE
This face is greatly distorted because the shot was taken with a wide-angle lens at a close distance.

is a significant aesthetic difference between a zoom and a dolly. Whereas the zoom seems to bring the scene to the viewer, a dolly seems to take the viewer into the scene. Because the camera does not move during the zoom, the spatial relationships among objects remain constant. The objects appear to be glued into position—they simply get bigger (zoom-in) or smaller (zoom-out). In a dolly, however, the relationships among objects change constantly. You seem to move past them when dollying in or out.[1] Be sure to recalibrate the zoom lens when you reach the end of the dolly so that you can zoom in and out from the new position without losing focus. ZVL4 CAMERA→ Camera moves→ dolly | zoom | try it

When people or objects move toward or away from the camera, their speed appears greatly accelerated by the wide-angle lens. The wide-angle zoom position is often used to accelerate the speed of a car or a dancer moving toward or away from the camera.

When covering a news event that exhibits a great deal of movement or that requires you to move rapidly, you should put the zoom lens in its extreme wide-angle position, as this will reduce camera wobbles to a great extent and make it much easier to keep the event in the viewfinder. Also, the great depth of field helps you keep the pictures in focus. The disadvantage of the extreme wide-angle position is that you need to move the camera quite close to the action if you want a closer look.

Depth of field The wide-angle lens generally has a great depth of field. When zoomed all the way out, you should have few focus problems, unless you work in low-light conditions (which require a large aperture) or are extremely close to the object.

Normal Lens

The zoom position for a normal focal length lies somewhere in the midrange of a zoom lens, perhaps a little more toward the wide-angle position (see 7.25b).

Field of view The ***normal lens*** offers a field of view (focal length) that approximates that of normal vision. It gives you the perspective between foreground and middleground that you actually see.

Object and distance distortion Whereas the wide-angle lens makes objects seem farther apart and rooms seem larger than they actually are, the normal lens or the midrange zoom positions make objects and their spatial relationships appear more like our normal vision. When shooting graphics such as charts that are positioned on an easel, you should put the zoom lens in the midrange position.

Movement With the normal lens (midrange zoom positions), you have a much more difficult time keeping the picture in focus and avoiding camera wobbles, even when the camera is mounted on a studio pedestal. When carrying an ENG/EFP camera or camcorder, this lens position makes it hard to avoid camera wobbles even when standing still. If you must have such a field of view, put the camera on a tripod.

Because the distance and the object proportions approximate our normal vision, the dolly speed and the speed of objects moving toward or away from the camera also appear normal. But again, such movement may cause focus problems, especially when the object gets fairly close to the camera.

Depth of field The normal lens has a considerably shallower depth of field than the wide-angle lens under similar conditions (same *f*-stop and object-to-camera distance). As mentioned before, a medium depth of field is often preferred in studio work and electronic field production (EFP) because the in-focus objects are set off against a slightly out-of-focus background. The objects are emphasized, and a busy background or the inevitable smudges on the television scenery receive less attention. Most importantly, foreground, middleground, and background are better defined.[2]

Of course, a large depth of field is necessary when there is considerable movement of camera and/or subjects. Also, when two objects are located at widely different distances from the camera, a great depth of field enables you to keep both in focus simultaneously. Most outdoor telecasts, such as sports remotes, require a large depth of field.

Narrow-angle, or Telephoto, Lens

When you zoom all the way in, the lens is in the maximum narrow-angle, long-focal-length, or telephoto, position.

Field of view The ***narrow-angle lens*** not only reduces the vista but also magnifies the background objects. Actually,

1. See Herbert Zettl, *Sight Sound Motion: Applied Media Aesthetics,* 6th ed. (Boston: Wadsworth, 2011), pp. 282–84.

2. See Zettl, *Sight Sound Motion,* pp. 168–70.

7.25 PHOTOS BY HERBERT ZETTL

7.25 FIELD OF VIEW AND PERSPECTIVE: COMPARISON OF WIDE-ANGLE, NORMAL, AND NARROW-ANGLE LENSES

A The wide-angle lens expands space, stretching the row of columns.

B The normal lens gives a field of view that approximates normal vision.

C The narrow-angle (telephoto) lens compresses space.

when you zoom in, all the zoom lens does is magnify the image. You get a view as though you were looking through binoculars, which, in effect, act as telephoto lenses. **SEE 7.25**

Object and distance distortion Because the enlarged background objects look big in comparison with the foreground objects, an illusion is created that the distance between foreground, middleground, and background has decreased. The long lens seems to compress the space between the objects, in direct contrast to the effect created by the wide-angle lens, which exaggerates object proportions and therefore seems to increase relative distance between objects. A narrow-angle lens, or telephoto zoom position, crowds objects on-screen. This crowding effect, called aesthetic ***compression***, can be positive or negative. If you want to show how crowded the freeways are during rush hour, for example, use the zoom lens in the telephoto position. The long focal length shrinks the perceived distance between the cars and makes them appear to be bumper-to-bumper. **SEE 7.26**

But such depth distortions by the narrow-angle lens also work to a disadvantage. You are certainly familiar with the deceptive closeness of the pitcher to home plate on the television screen. Because television cameras must remain at a considerable distance from the action in most sporting events, the zoom lenses usually operate at their extreme telephoto positions (zoomed in all the way) or with powerful range extenders. The resulting compression effect makes it difficult for viewers to judge actual distances. **SEE 7.27**

Movement The narrow-angle lens gives the illusion of reduced speed of an object moving toward or away from the camera. Because the narrow-angle lens changes the size of an object moving toward or away from the camera

7.26 PHOTO BY ALEX ZETTL
7.27 PHOTO BY HERBERT ZETTL

7.26 POSITIVE AESTHETIC COMPRESSION WITH NARROW-ANGLE LENS

With a narrow-angle lens, the background is greatly enlarged and the distance between the cars seems reduced. The feeling of a traffic jam is intensified.

7.27 NEGATIVE AESTHETIC COMPRESSION WITH NARROW-ANGLE LENS

This shot was taken with a zoom lens in an extreme long-focal-length position. Note how the pitcher, batter, catcher, and umpire all seem to stand only a few feet apart from one another. The actual distance between the pitcher and the batter is 60½ feet.

much more gradually than does the wide-angle lens, the object seems to move more slowly than it actually does; in fact, an extreme narrow-angle lens virtually eliminates such movement. The object does not seem to change size perceptibly even when traveling a considerable distance relative to the camera. Such a slowdown is especially effective if you want to emphasize the frustration of someone running toward the camera but not getting any closer. Added to the compression effect (figure 7.26), the drastic reduction of the perceived speed of traffic will certainly emphasize the congestion. ZVL5 CAMERA→ Picture depth→ perspective and distortion | try it

You cannot dolly with a narrow-angle lens or with a zoom lens in the telephoto position (zoomed in); its magnifying power makes smooth camera movement impossible. If you work outdoors, even wind can be a problem. A stiff breeze may shake the camera to such a degree that the greatly magnified vibrations become clearly visible on-screen.

In the studio the telephoto position may present another problem. The director may have you zoom in on part of an event, such as the lead guitarist in a rock performance, and then, after you have zoomed in, ask you to truck (move the camera sideways) past the other members of the band. But the telephoto position makes any wobble-free camera movement extremely difficult, if not impossible. What you can do to minimize the wobbles is zoom out before trucking.

Depth of field and selective focus Unless the object is far away from the camera, long-focal-length lenses have a shallow depth of field. You can use a shallow depth of field effectively to draw attention to certain parts of an event that is staged along the z-axis—an imaginary line representing an extension of the lens from the camera to the horizon (the depth dimension). One of the most popular shallow-depth-of-field effects is selective focus.

Let's assume that you are about to take a quick close-up of a medium-sized object, such as a can of soup. You do not have to bother putting up a background for it—all you need to do is move the camera back and zoom in on the display. With the zoom lens in a telephoto (narrow-angle) position, decreasing the depth of field to a large extent, the background is sufficiently out of focus to prevent undesirable distractions. This technique is called ***selective focus***, meaning you can focus either on the foreground, with the middleground and the background out of focus; on the middleground, with the foreground and the background out of focus; or on the background, with the foreground and the middleground out of focus. SEE 7.28 AND 7.29

By now you know that you can use selective focus to easily shift emphasis from one object to another. For example, you can zoom in on a foreground object, thus reducing the depth of field, and focus on it with the zoom lens in the telephoto position. Then, by refocusing on the person behind it, you can quickly shift the emphasis from the foreground object to the person in the middleground. This technique is called racking focus or, simply, ***rack focus***. ZVL6 CAMERA→ Focusing→ rack focus | try it

The advantage of a shallow depth of field also applies to unwanted foreground objects. In a high-school baseball pickup, for example, the camera behind home plate may have to shoot through the chain-link backstop. But because the camera is most likely zoomed in on the pitcher, or on other players a considerable distance from the camera, you work with a relatively shallow depth of field. Consequently,

7.28 SELECTIVE FOCUS: FOREGROUND IN FOCUS
In this shot the camera-near person is in focus, drawing attention away from the two people in the background.

7.29 SELECTIVE FOCUS: BACKGROUND IN FOCUS
Here the focus and the attention are shifted from the camera-near person in the foreground to the two people farther away.

everything fairly close to the camera, such as the chain-link fence, is so out of focus that it becomes virtually invisible. The same principle works for shooting through birdcages, prison bars, or similar foreground objects.

IMAGE STABILIZATION

The best way to keep an image steady is to put the camera on a tripod or studio pedestal. There are also software programs that allow you to make a shot jitter-free in postproduction, but this process is quite slow, even if you have a superfast processor in your computer. If you are operating a shoulder-mounted camera or carrying it in your hands, the image stabilizer may come to your rescue.

Image Stabilizer

Some lenses have an image stabilization device built into the lens or the camera. The lens stabilization is mostly mechanical; the stabilization inside the camera is usually electronic. Both systems reduce and often eliminate subtle image shifts caused by minor camera shakes. But even with the best image stabilizer, the pictures will be rendered useless by the inevitable camera wobbles when moving the camera in the telephoto position.

MAIN POINTS

- The performance characteristics of wide-angle, normal, and narrow-angle lenses (zoom lenses adjusted to these focal lengths) include field of view, object and distance distortion, movement, and depth of field.

- A wide-angle lens (zoom lens in the wide-angle position, zoomed all the way out) offers a wide vista. It gives a wide field of view with the camera relatively close to the scene. It also distorts objects close to the lens and exaggerates proportions. It is ideal for camera movement because it minimizes camera wobbles and makes it easy to keep the picture in focus during camera movement.

- The normal lens gives a field of view that approximates that of normal vision. The normal lens (zoom lens in the midrange position) does not distort objects or the perception of distance.

- When a camera is moved with the lens in the midrange zoom position, camera wobbles are emphasized considerably more than with a wide-angle lens. The shallower depth of field makes it harder to keep the picture in focus.

- A narrow-angle lens (zoom lens in the telephoto position, zoomed all the way in) has a narrow field of view and enlarges the objects in the background. Exactly opposite of the wide-angle lens, which increases the perceived distance between objects, the narrow-angle lens seems to compress the space between objects at different distances from the camera. It slows the perception of object speed toward and away from the camera.

- The magnifying power of a narrow-angle lens prevents any camera movement while on the air. Narrow-angle lenses have a shallow depth of field, which makes keeping in focus more difficult but allows for selective focus.

- Selective focus, which puts only a specific area along the z-axis (an imaginary line from camera to horizon) in focus, with all others out of focus, is possible only in a shallow depth of field. Rack focus means changing emphasis from one z-axis area to another though a focus shift.

ZETTL'S VIDEOLAB

For your reference or to track your work, the *Zettl's VideoLab* program cues in this chapter are listed here with their corresponding page numbers.

CHAPTER

8

Camera Operation and Picture Composition

Digital cameras with their automated features have become so user-friendly that anybody can run a camera, right? Wrong! Although you may see news programs or even entire television series that are shot by amateurs, professional camera work still requires practice and a basic knowledge of how to move the camera with or without a camera support—and especially how to compose effective pictures.

There is nothing wrong with using the automated features of an electronic camera so long as the conditions allow it. But even the smartest automated camera has no way of knowing what part of the event you consider important and how best to clarify and intensify the selected event details through maximally effective shots. Nor can a camera exercise aesthetic judgment—how to frame an extreme close-up, for example. This is why it is important to learn as much as possible about camera operation before trying to do your blockbuster documentary.

Section 8.1, Working the Camera, discusses the basic camera movements, the standard mounting equipment, and the do's and don'ts of camera operation. Section 8.2, Framing Effective Shots, focuses on some of the aesthetic aspects of picture composition in various aspect ratios.

KEY TERMS

arc To move the camera in a slightly curved dolly or truck.

aspect ratio The width-to-height proportions of the standard television screen and therefore of all standard television pictures: 4 units wide by 3 units high. For HDTV the aspect ratio is 16 × 9. The small mobile media (cell-phone) displays have various aspect ratios.

camera stabilizing system Camera mount whose mechanism holds the camera steady while the operator moves.

cant To tilt the shoulder-mounted or handheld camera sideways.

close-up (CU) Object or any part of it seen at close range and framed tightly. The close-up can be extreme (extreme or big close-up—ECU) or rather loose (medium close-up—MCU).

closure Short for *psychological closure.* Mentally filling in spaces of an incomplete picture.

crane (1) Motion picture camera support that resembles an actual crane in both appearance and operation. The crane can lift the camera from close to the studio floor to more than 10 feet above it. (2) To move the boom of the camera crane up or down. Also called *boom.*

cross-shot (X/S) Similar to the over-the-shoulder shot except that the camera-near person is completely out of the frame.

dolly (1) Camera support that enables the camera to move in all horizontal directions. (2) To move the camera toward (dolly in) or away from (dolly out or back) the object.

extreme close-up (ECU) Shows the object with very tight framing.

extreme long shot (ELS) Shows the object from a great distance. Also called *establishing shot.*

field of view The portion of a scene visible through a particular lens; its vista. Expressed in symbols, such as *CU* for close-up.

headroom The space left between the top of the head and the upper screen edge.

jib arm Similar to a camera crane. Permits the jib arm operator to raise, lower, and tongue (move sideways) the jib arm while tilting and panning the camera.

leadroom The space left in front of a person or an object moving toward the edge of the screen.

long shot (LS) Object seen from far away or framed loosely. Also called *establishing shot* and *full shot.*

medium shot (MS) Object seen from a medium distance. Covers any framing between a long shot and a close-up. Also called *waist shot.*

monopod A single pole onto which you can mount a camera.

mounting head A device that connects the camera to the tripod or studio pedestal to facilitate smooth pans and tilts. Also called *pan-and-tilt head.*

noseroom The space left in front of a person looking or pointing toward the edge of the screen.

over-the-shoulder shot (O/S) Camera looks over a person's shoulder (shoulder and back of head included in shot) at another person.

pan To turn the camera horizontally.

pedestal (1) Heavy camera dolly that permits raising and lowering the camera while on the air. (2) To move the camera up and down via a studio pedestal.

quick-release plate A mounting plate used to attach camcorders and ENG/EFP cameras to the mounting head.

robotic pedestal Remotely controlled studio pedestal and mounting head. It is guided by a computerized system that can store and execute a great number of camera moves. Also called *robotic.*

studio pan-and-tilt head A camera mounting head for heavy cameras that permits extremely smooth pans and tilts.

tilt To point the camera up or down.

tongue To move the boom or jib arm with the camera from left to right or right to left.

tripod A three-legged camera mount. Can be connected to a dolly for easy maneuverability.

truck To move the camera laterally by means of a mobile camera mount. Also called *track.*

two-shot Framing of two people.

z-axis An imaginary line representing an extension of the lens from the camera to the horizon—the depth dimension.

zoom To change the lens gradually to a narrow-angle position (zoom-in) or to a wide-angle position (zoom-out) while the camera remains stationary.

SECTION

8.1

Working the Camera

Although it may be fun to hold your small camcorder out the window of a moving car or swing it through the air like a butterfly net, such moves rarely produce satisfactory pictures. Like painting, good camera work is more difficult than it looks, but the learning curve can be considerably flattened by mastering a few basics of camera operation.

▶ **STANDARD CAMERA MOVEMENTS**
Pan, tilt, pedestal, tongue, crane or boom, dolly, truck or track, arc, cant, and zoom

▶ **CAMERA SUPPORTS**
Monopod, tripod and tripod dolly, studio pedestal, mounting (pan-and-tilt) heads, and special camera mounts

▶ **WORKING THE CAMCORDER AND THE EFP CAMERA**
Some basic camera "don'ts"; and camera setup, operation, and care—the basic operational steps before, during, and after a field production

▶ **WORKING THE STUDIO CAMERA**
Camera setup, operation, and care—the basic operational steps before, during, and after a studio production

STANDARD CAMERA MOVEMENTS

Before learning to operate a camera, you should become familiar with the most common camera movements. *Left* and *right* always refer to the camera's point of view. The camera mounting equipment has been designed solely to help you move the camera smoothly and efficiently. The major camera movements are pan, tilt, pedestal, tongue, crane or boom, dolly, truck or track, arc, cant, and zoom. SEE 8.1

- **Pan** means turn the camera horizontally, from left to right or from right to left. When the director tells you to "pan right," which means point the lens and the camera to the right (clockwise), you must push the panning handles to the left. To "pan left," which means swivel the lens and the camera to the left (counterclockwise), you push the panning handles to the right.

- **Tilt** means point the camera up or down. When you "tilt up," you make the camera lens point up gradually. When you "tilt down," you make the camera lens point down gradually.

- **Pedestal** means elevate or lower the camera on a studio pedestal. To "pedestal up" you raise the camera; to "pedestal down" you lower the camera.

- **Tongue** means move the whole camera from left to right or from right to left with the boom of a camera crane. When you tongue left or right, the camera usually points in the same general direction, with only the boom moving left (counterclockwise) or right (clockwise).

- **Crane**, or boom, means move the whole camera up or down on a camera crane or jib arm. The effect is somewhat similar to an up or down pedestal except that the camera swoops over a much greater vertical distance. You either "crane [or boom] up" or "crane [or boom] down."

- **Dolly** means move the camera toward or away from the scene in more or less a straight line by means of a mobile camera mount. When you "dolly in," you move the camera closer to the scene; when you "dolly out" or "dolly back," you move the camera farther away from the scene.

- **Truck**, or track, means move the camera laterally by means of a mobile camera mount. To "truck left" means to move the camera mount to the left with the camera pointing at a right angle to the direction of travel. To "truck right" means to move the camera mount to the right with the camera pointing at a right angle to the direction of travel.

- **Arc** means move the camera in a slightly curved dolly or truck movement with a mobile camera mount. To "arc left" means to dolly in or out in a camera-left curve or to truck left in a curve around the object; to "arc right" means to dolly in or out in a camera-right curve or to truck right in a curve around the object.

- **Cant** means tilt the shoulder-mounted or handheld camera sideways. The result, called a canting effect, is

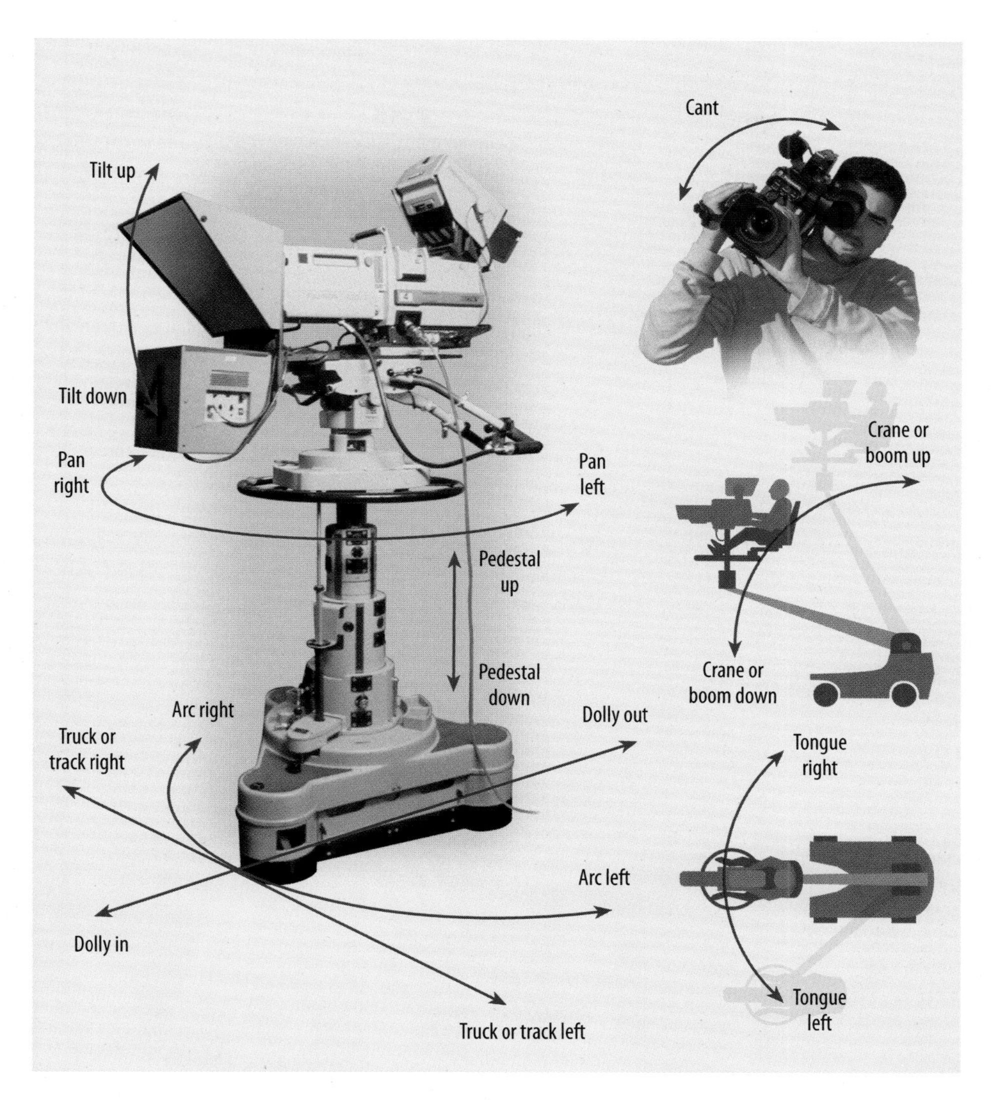

8.1 PHOTOS BY EDWARD AIONA

8.1 CAMERA MOVEMENTS
Major camera movements include pan, tilt, pedestal, tongue, crane or boom, dolly, truck or track, arc, and cant.

a slanted horizon line, which puts the scene on a tilt. Through the skewed horizon line, you can achieve a highly dynamic scene.

■ **Zoom** means change the focal length of the lens through the use of a zoom control while the camera remains stationary. To "zoom in" means to change the lens gradually to a narrow-angle position, thereby making the scene appear to move closer to the viewer; to "zoom out" means to change the lens gradually to a wide-angle position, thereby making the scene appear to move farther away from the viewer. Although not a camera movement per se, the zoom effect looks similar to that of a moving camera and is therefore classified as such. ZVL1 CAMERA→ Camera moves→ dolly | zoom | truck | pan | tilt | pedestal | try it

CAMERA SUPPORTS

Even if your camera is small and lightweight enough to carry in your hands, you should mount it on a camera support whenever possible. Using a camera support will reduce fatigue and especially prevent unnecessary and distracting camera motion. Unless motivated, as in some commercials and music shows, wild and rapid camera movement draws too much attention to itself and is one of the sure signs of amateur camera handling.

Monopod

The ***monopod*** is a single pole, or a single "pod," onto which you can mount a camera. When using a monopod, you must balance the support with one hand and operate the camera with the other. You can use a monopod to support some of the weight of a shoulder-mounted camera.

Tripod and Tripod Dolly

The most common mounting device for camcorders is the ***tripod***. You certainly know what a tripod is, but, like cameras, tripods can be good, passable, and useless. The most important criteria for a good tripod are that it is sturdy, easy to set up and level on any type of terrain, and, ideally, not too heavy. The other major feature of a good tripod is its mounting head, which we discuss later in this section.

Leveling All good tripods have an air bubble that tells you when the tripod is level. If the ground is very uneven, you can first adjust the legs so that the tripod looks relatively even, then do the fine-tuning with the leveling bowl on the mounting head.

Spreader and dolly The tips of the tripod legs are equipped with spikes and/or rubber cups that keep them from slipping. A further safeguard that prevents the legs from spreading is, paradoxically, the spreader. Spreaders are important for tripods that support heavy camcorders, but you don't need them for small camcorders. SEE 8.2

Tripod dolly You can also place a tripod on a three-caster dolly base, which is simply a spreader with wheels. Because the tripod and the dolly are collapsible, they are ideal for fieldwork. The wheels of the dolly should have cable guards that prevent the camera cable from getting caught under the dolly base or run over by the dolly wheels. SEE 8.3

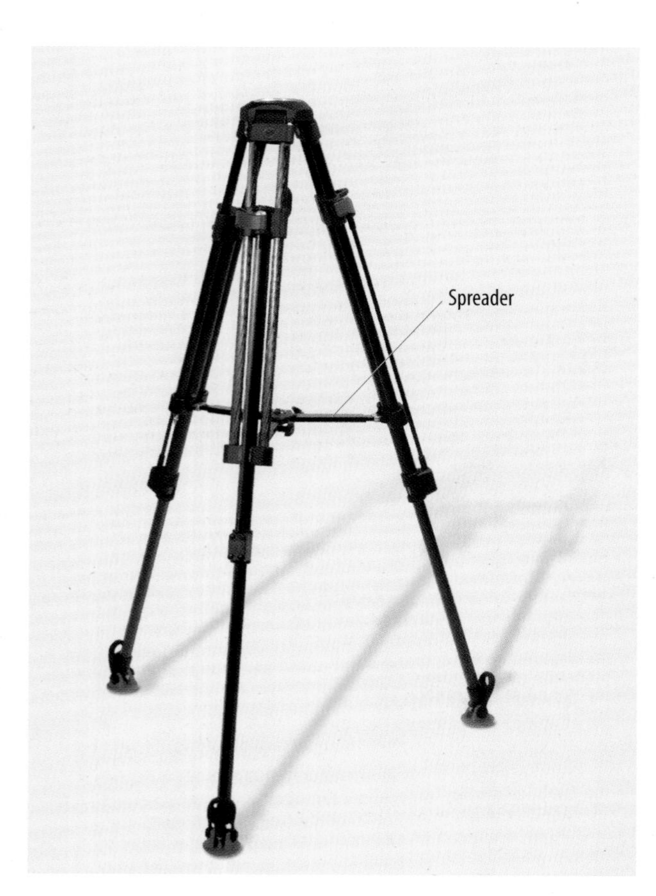

8.2 TRIPOD WITH BUILT-IN SPREADER
The tripod is one of the most basic camera supports and is used extensively in field productions. This tripod has a built-in spreader at midlevel.

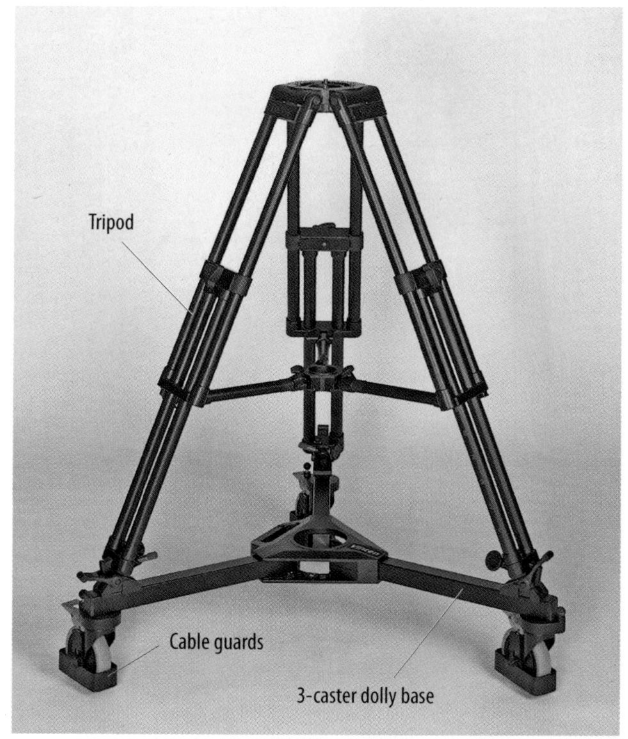

8.3 COLLAPSIBLE TRIPOD MOUNTED ON DOLLY BASE
The tripod can be mounted on a tripod dolly, which permits quick repositioning of the camera.

8.2 PHOTO COURTESY SACHTLER
8.3 PHOTO COURTESY VINTEN, INC.

Studio Pedestal

With a studio pedestal, you can move a camera in all directions (assuming there is a smooth floor) and elevate and lower the camera while on the air. This up-and-down movement adds an important dimension to the art of television and cinema photography. You can not only adjust the camera to a comfortable working height but also change the eye level from which you look at an event. For example, if you are in danger of overshooting the set, you can always pedestal up (raise the camera) and look down on the scene. To make a person look more imposing, you can pedestal down (lower the camera) and look up at him or her. This also heightens the perceived energy of the event, such as looking up at the lead singer of a rock group or at a truck.

All studio pedestals have similar operating features. You can pedestal up and down as well as steer the pedestal smoothly in any direction with a large horizontal steering wheel. The telescoping pedestal column can be locked at any vertical position.

Like tripod dollies, studio pedestals need a cable guard to keep from running over cables. Always check that the adjustable skirt of the pedestal base is low enough to push a cable out of the way rather than roll over it. **SEE 8.4**

Generally, you work the pedestal in the parallel, or crab, steering position, which means that all three casters point in the same direction. **SEE 8.5A** If you want to rotate the pedestal itself, to move it closer to a wall or piece of scenery for example, you can switch it from the crab to the tricycle steering position. **SEE 8.5B**

Mounting (Pan-and-tilt) Heads

The ***mounting head*** connects the camera to the tripod or studio pedestal. The mounting head (not to be confused with the camera head, which represents the actual camera) allows you to tilt and pan the camera extremely smoothly. This is why the mounting heads are also called pan-and-tilt heads.

Most professional ***studio pan-and-tilt heads*** have four controls: a tilt and pan drag and a tilt and pan lock. The drag controls give various degrees of resistance to panning and tilting to make the camera movements optimally smooth. The lock controls immobilize the pan-and-tilt mechanism to keep the camera from moving when left unattended. **SEE 8.6**

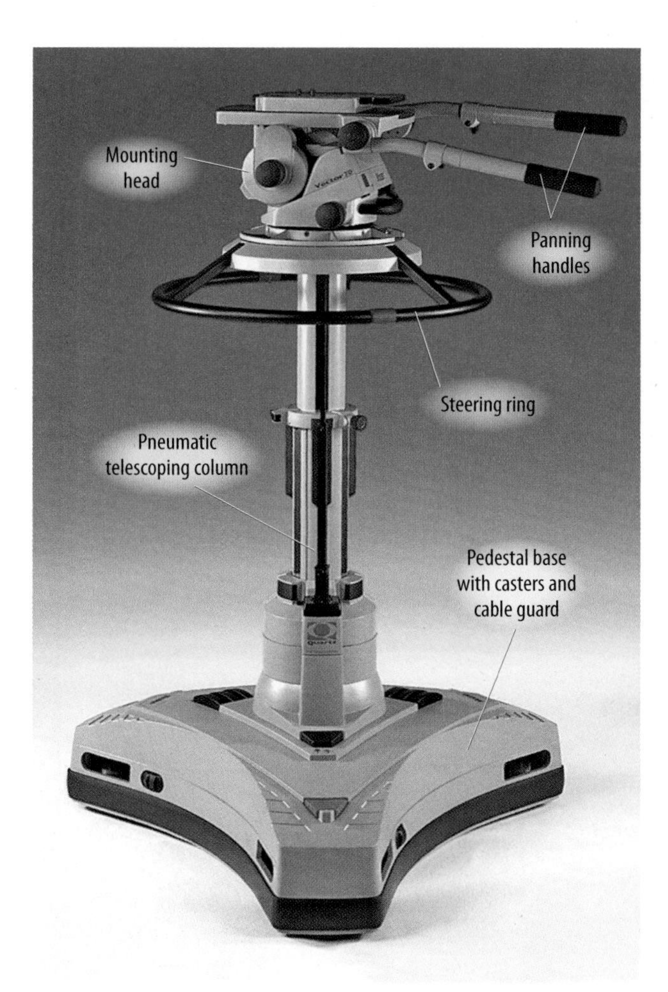

8.4 STUDIO PEDESTAL
The studio pedestal permits smooth dollies and trucks and has a telescoping center column that pedestals the camera from a low of 2 feet to a maximum height of about 6 feet above the studio floor.

8.4 PHOTO COURTESY VINTEN, INC.

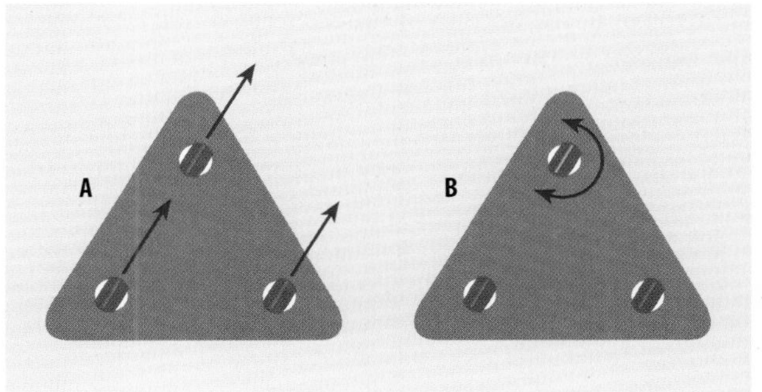

8.5 PARALLEL (CRAB) AND TRICYCLE STEERING
A In the parallel, or crab, position, all three casters point in the same direction. **B** In the tricycle position, only one wheel is steerable. A foot pedal allows a quick change from parallel to tricycle steering.

8.6 ENG/EFP CAMERA MOUNTING HEAD
This mounting head is designed for mounting and operating ENG/EFP cameras and camcorders on tripods. It has a limited weight capacity.

Pan-and-tilt heads, designed to connect heavy studio or field cameras, have an adjustable counterbalance mechanism that permits smooth pans and keeps the heavy load balanced during tilts so that the camera does not tip forward or backward. SEE 8.7

Whatever mounting head you use, do not tighten the drag control to lock the pan-and-tilt mechanism or use the lock control to adjust the drag. Using the drag control to lock the camera will ruin the mounting head in a very short time; and trying to use the locking device for tilt and pan drag controls will almost always result in jerky and uneven camera movements.

Quick-release plate and wedge mount Attaching the camera to the mounting head is done with a quick-release plate for most camcorders and with a wedge mount for heavy studio cameras. Both are metal plates that are screwed onto the bottom of the camera. When attached to the bottom of the camera, the ***quick-release plate*** allows you to take the camera off the tripod and then reattach it in a balanced position very quickly (hence its name). You will find this feature very convenient, especially in a production where you have to switch frequently from a tripod-mounted camera to a handheld one. SEE 8.8

The wedge mount helps you properly balance the heavy studio camera and lens. Once balanced, the wedge will keep the camera in a balanced position on subsequent mountings (see figure 8.7).

Special Camera Mounts

These camera mounts are designed for specific operations. They include camera stabilizing systems, SLR and pocket-sized-camera supports, jib arms, and robotics.

8.7 STUDIO CAMERA MOUNTING HEAD
This studio mounting head is designed for heavier cameras. It is normally used for mounting studio cameras with big lenses and teleprompters onto studio pedestals. Note the wedge mount, which makes it easy to connect the studio camera to the mounting head in a balanced position.

8.6 PHOTO COURTESY MANFROTTO
8.7 PHOTO COURTESY VINTEN, INC.

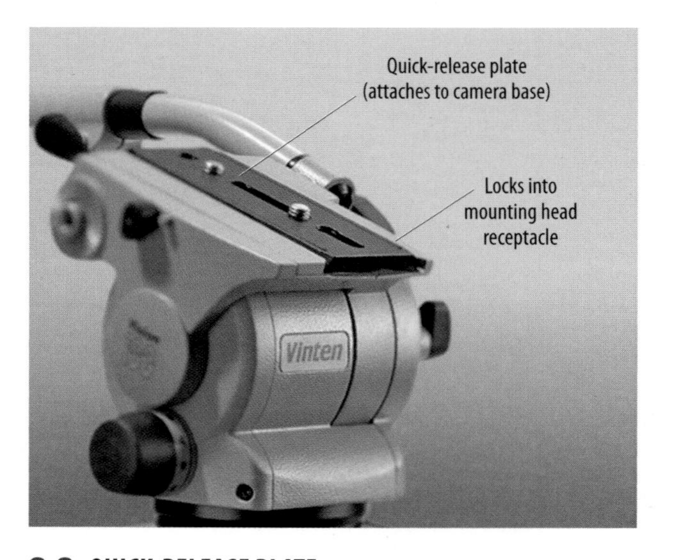

8.8 QUICK-RELEASE PLATE
The quick-release plate enables you to reattach the camera to the mounting head in a balanced position without time-consuming readjustment.

8.8 PHOTO COURTESY VINTEN, INC.

8.9 STABILIZER FOR HEAVY CAMCORDERS
This type of stabilizing system, called a Steadicam, allows you to walk or run with the camera while keeping the pictures perfectly steady. The heavy spring-balanced mechanism is connected to a body harness.

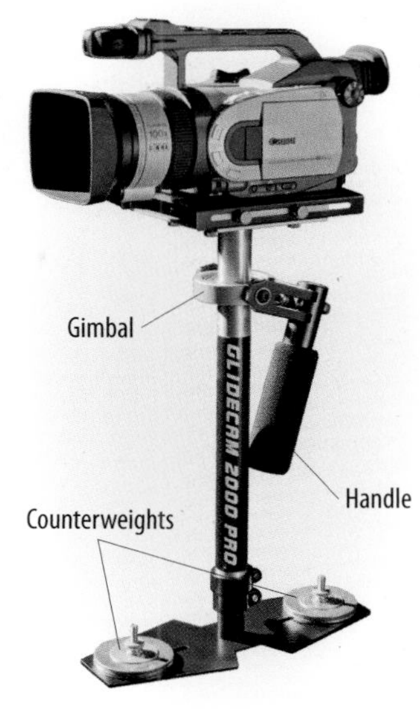

8.10 STABILIZER FOR LIGHTWEIGHT CAMCORDERS
Stabilizer systems such as this one are designed so that the operator can carry the mount with the camera in one hand.

8.11 BEANBAG
This canvas bag filled with synthetic material adjusts to any camera and any object on which it is mounted. Both bag and camera can be easily secured with nylon rope.

Camera stabilizing systems A ***camera stabilizing system*** is a mount that allows the camera operator to walk, run, and even jump while keeping the image steady. One of the most popular camera stabilizing systems is the Steadicam.

For heavy camcorders, the spring-loaded and gimbals-equipped mechanism (which allows the camera to remain level even when its support is tipped) is attached to a harness that supports the vertical monopod and the camera. SEE 8.9 For small and lightweight camcorders, the camera is connected via gimbals to a handgrip and a counterweight below the camera. SEE 8.10

But don't expect to get perfectly smooth pictures the first time you try using such a device. All these stabilizing systems require practice before you can make full use of their potential.

When mounting a camcorder to a moving car, you can use a simple beanbag. This camera mount consists of a canvas bag filled with high-tech foam that molds itself to the shape of any camcorder. All you do is set the camera on the bag and then strap the bag with the camera to the object that acts as a camera mount. You can use this bag mount on cars, boats, mountain ledges, bicycles, or ladders. SEE 8.11

8.12 JIB ARM
The jib arm lets the camera operator dolly, truck, and boom the camera up and down and simultaneously pan, tilt, focus, and zoom.

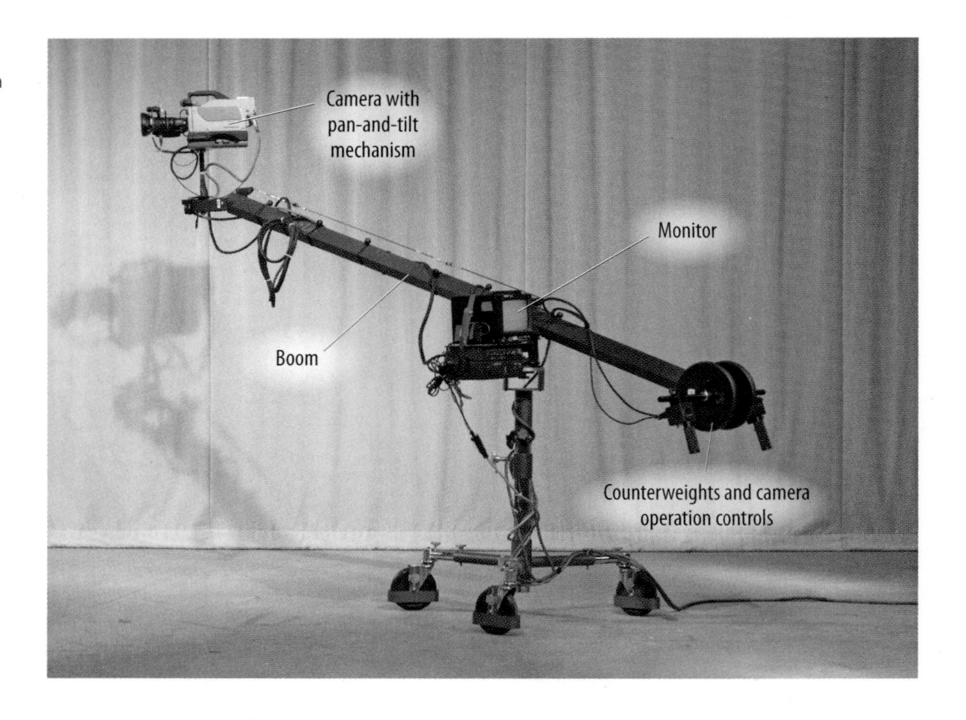

8.12 PHOTO BY EDWARD AIONA

SLR and pocket-sized-camera supports There are a variety of supports available for trucking a single-lens reflex (SLR) camera sideways or dollying it in and out for short distances. Because the camera is relatively small and light, you can use a small bracket to attach it quite easily to a moving car or even a bicycle. Pocket-sized camcorders (such as the Flip or cell phones) can be mounted on tiny tripods whose legs can be bent to adjust the camcorder's height and level.

Jib arm The ***jib arm*** is a cranelike device that lets you—by yourself—lower the camera practically to the studio floor, raise it to 12 feet or even higher, tongue the jib arm and swing it a full 360 degrees, dolly or truck the whole assembly, and, at the same time, tilt, pan, focus, and zoom the camera. Obviously, all these movements require practice if they are to look smooth on the air. The camera and the jib arm are balanced by a monitor, the battery pack, remote camera controls, and, for good measure, actual counterweights. SEE 8.12

Robotic camera mounts Automated pedestals and mounting heads, usually called robotics, are used more and more for shows with rigid production formats, such as newscasts, teleconferences, and certain instructional programs. The ***robotic pedestal*** consists of a remotely controlled studio pedestal and a mounting head. SEE 8.13

Several studio pedestals can be programmed so that they move their cameras into position for specific shots, such as a two-shot and then close-ups (CUs) of the news anchors followed by a wide shot of the meteorologist and the weather map—all without camera operators. For teleconferencing and classrooms, smaller robotic camera systems often operate from fixed positions, with their pan-and-tilt mechanisms controlled by a single operator in the control room. SEE 8.14

WORKING THE CAMCORDER AND THE EFP CAMERA

Similar to learning how to ride a bike, working the camcorder or studio camera cannot be learned from a book (even this one!). You simply need to do it. Knowing some camera "don'ts," however, as well as what to do before, during, and after the shoot can greatly speed up your learning how to operate the camera. When you are caught up in a large studio production or covering a hot news story in the field, it's easy to forget that the camera is an extremely complex piece of machinery. Although it may not be as precious or fragile as your grandmother's china, it still needs careful handling and a measure of respect. The following "don'ts" before learning the "do's" of camera operation are to prevent you from damaging or losing the equipment before

8.13 PHOTO COURTESY VINTEN, INC.
8.14 PHOTO COURTESY VINTEN, INC.

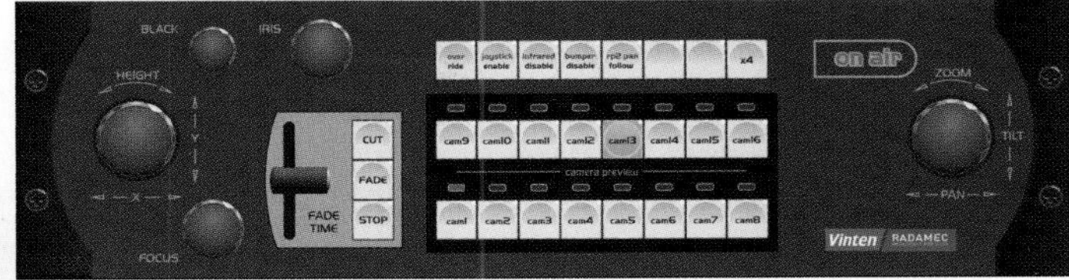

8.14 OPERATOR CONTROL PANEL FOR ROBOTIC PEDESTAL
With such computer control panels, you can preprogram and control several robotic cameras.

8.13 ROBOTIC PEDESTAL
The robotic pedestal is fully automated and needs no camera operator. All necessary camera movements and functions are computer-controlled.

you ever get to use it. In this light these taboos represent a positive beginning.

Some Basic Camera "Don'ts"

■ Do not leave a camcorder in a car—even in the trunk—for an extended period of time unless the car is safely locked in a garage. Like people and animals, electronic equipment tends to suffer from excessive heat. More importantly, keeping the camera gear with you as much as possible is a fairly simple way of preventing theft.

■ Do not leave a camcorder unprotected in the rain, hot sun, or extreme cold. When you must use a camcorder in the rain, protect it with a "raincoat"—a prefabricated plastic hood—or at least a plastic sheet. A simple but effective means of keeping rain away from a camera is a large umbrella. Some zoom lenses stick in extremely wet or cold weather. Test the lens before using it on-location. Never use wet tapes or other recording media.

■ Do not point the lens for an extended period of time at the midday sun. Although the imaging device will not be damaged by the intensity of the sunlight, it may suffer from the heat generated by the focused rays. The same goes for the viewfinder: don't leave it pointed at the sun for an extended period of time; the viewfinder's magnifying lens can collect the sun's rays, melting its housing and electronics.

■ Do not leave camcorder batteries in the sun or, worse, drop them. Although a battery may look rugged from the outside, it is actually sensitive to heat and shock. Some batteries should not be charged in extremely cold temperatures.

■ Do not lay a camcorder on its side. You run the risk of damaging the viewfinder or the clipped-on microphone on the other side. When finished shooting, cap the camera with the external lens cover and, just to make sure, close the aperture to the *C* (cap) or *close* position.

Before the Shoot

■ Before doing anything else, count all the pieces of equipment and mark them on an equipment checklist. If you need auxiliary equipment, such as external microphones, camera lights, a power supply, or field monitors, check on the right connectors and cables. Take some extra adapters along.

■ Unless you are running after hot news, first set up the tripod and see whether it is level. Do some panning and tilting to determine the optimal pan and tilt drags. Check the pan and tilt locks. Insert the battery or connect the camcorder to its alternate power supply (AC/DC converter and transformer) and do a brief test recording before taking the camcorder into the field. Make sure the camcorder records audio as well as video.

■ If you are engaged in elaborate field productions using high-quality EFP cameras and separate video recorders, check the connecting cables and the power supplies (usually batteries). You may need a video feed from the camera (or recorder) to a battery-powered field monitor for the director. Do all the plugs fit the intended jacks? In EFP the wrong, or even a loose, connector can mean a lost production day. As with the camcorder, hook up all the equipment you will use in the field and do a test recording before going on-location. Never assume that everything will work merely because it worked in the past.

■ Check that the external microphone (usually a lavalier or hand mic) and the camera mic are working properly. Most camera mics need to be switched on before they become operational. Scratch the top of the mic and listen for whether it makes any noise. Is there sufficient cable for the external mic so that the reporter can work far enough away from the camera? If you are primarily doing news that requires an external mic for the field reporter, you may want to keep the external mic plugged in to save time and minimize costly mistakes.

■ Does the portable camera light work? Don't just look at the lamp. Turn on the light to verify that it works. When using a separate battery for the light, make sure that it is fully charged. If you have additional lights, are they all operational? Do you have enough AC extension cords to power the additional lights?

■ When using a separate video recorder for EFP, do a test recording to ensure that the recorder is in good working order. (See chapter 13 for details on video recording.)

■ Double-check that the recording media (videotape, P2 cards or other flash drives, hard disks, or optical discs) work in the camcorder you are using. Some small camcorders cannot accept the larger tapes and use only mini-cassettes. Before putting the recording media into the camcorder, do a quick check of whether the record-protect tab is in the *off* position (see chapter 13). Always take along more recording media than you think you will need.

■ Although you are not a maintenance engineer, carry some spare fuses for the principal equipment. Some ENG/EFP cameras and camcorders have a spare fuse right next to the active one. Note, however, that a blown fuse indicates a malfunction in the equipment. Even if the camcorder works again with the new fuse, have it checked when you return from the shoot.

■ Like carrying a medical first-aid kit, always have a field production kit that contains the following necessities: several extra videocassettes or appropriate recording media, an audio recorder and several of its recording media, an additional microphone and a small mic stand, one or more portable lights and stands, additional lamps for all lighting instruments, AC cords, spare batteries for all battery-powered equipment, various clips or wooden clothespins, gaffer's tape, a small reflector, a roll of aluminum foil, a small white card for white-balancing, light-diffusing material, light blue gels for boosting color temperature, various effects filters, a can of compressed air for cleaning lenses, and a camera raincoat.

You should also carry such personal survival items as water and energy bars, a working flashlight, an umbrella, some spare clothes, and, yes, toilet paper. Once you have worked in the field a few times, you will know how to put together your own field production kit.

During the Shoot

After getting some field production experience, you will probably develop your own techniques for carrying and operating a camcorder or ENG/EFP camera. There are obvious differences between the small handheld camcorders and the larger ones that are shoulder-mounted. But most of the general operating procedures are the same.

■ First and foremost, put the camera on a tripod whenever possible. You will have more control over framing and steadying the shot. You will also get less tired during a long shoot.

■ Unless you want to use the camcorder's fully automated white-balance feature, you must white-balance before starting to shoot. Be sure to white-balance the camera in the same light that illuminates the scene you are shooting. If you don't have a white card, focus the camera on anything white, such as somebody's shirt or the back of a script. Repeat the white balance each time you encounter new lighting conditions, such as when moving from an interview on a street corner to the interior of a new restaurant. Careful white-balancing may save you hours of color correction in postproduction.

■ Keep the camera as steady as possible. This is especially important when the zoom lens is in the telephoto position. When operating a small camcorder, support it in the palm of your hand and use the other hand to support the camera arm or the camcorder itself. Whenever possible, press your elbows against your body, inhale, and hold your breath during the shot. Bend your knees slightly when shooting, or lean against a sturdy support to increase the stability of the camera. Such camera handling is recommended even if you have the image stabilizer turned on. Note that image stabilizers correct only minor camera wobbles and drain the battery in the process. **SEE 8.15 AND 8.16**

8.15 PHOTO BY EDWARD AIONA
8.16 PHOTO BY EDWARD AIONA

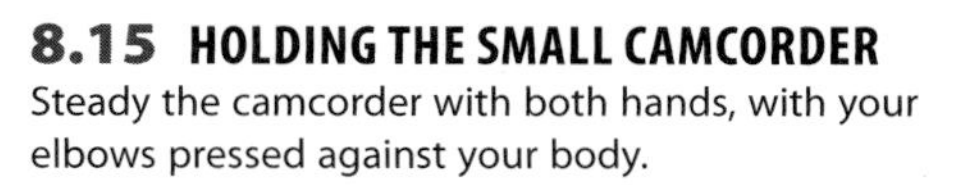

8.15 HOLDING THE SMALL CAMCORDER
Steady the camcorder with both hands, with your elbows pressed against your body.

8.16 STEADYING THE CAMERA OPERATOR
Lean against a tree or wall to steady yourself and the camcorder.

- Use the viewfinder rather than the foldout monitor to compose your shots. The viewfinder makes focusing more accurate (you get a sharper image) and is a better guide to proper exposure (*f*-stop) than the foldout display. When shooting outdoors, you will find that the foldout monitor is often rendered useless by the sun shining on it, obliterating the image. Use the monitor only if the camera is in the automatic mode and you need only a rough guide to framing a shot.

- Large and some midsized camcorders are designed to be shoulder-mounted. There are shoulder pods available for the midsized camcorders. Assuming that you are right-handed, carry the camera on your right shoulder and slip your right hand through the support strap on the zoom lens. This helps you steady the camera while allowing you to operate the zoom and auto-focus controls. Your left hand is free to operate the manual focus ring. If you are left-handed, reverse the procedure and switch the lens grip and the viewfinder to the left side. SEE 8.17

8.17 PHOTO BY EDWARD AIONA

8.17 SHOULDER-MOUNTED CAMCORDER
Carry the larger camcorder on your shoulder. One hand slips through the strap on the lens to steady the camcorder and operate the zoom. The other hand is free to operate the focus ring at the front of the zoom lens and provide further support for the camcorder.

- When moving the camcorder, zoom all the way out. Because of the great depth of field, you have fewer problems keeping the event in focus even if you or the subject moves. Even in the wide-angle position, move the camera as smoothly as possible.

- Before zooming, calibrate the zoom lens by zooming in all the way on the target object, focusing, and then zooming back out to the desired shot. Your subsequent zooms will remain in focus unless the camcorder or the object gets closer or farther away. If this happens, you need to recalibrate the lens.

- If the target object is obvious, put the camcorder on auto-focus. When in a hurry, the auto-focus gives you better-focused pictures than if you are not careful focusing with the manual control.

- Achieving optimal focus is especially difficult with HDTV (high-definition television) because the high-resolution picture looks in focus even if it is slightly out of focus. If the scene is steady, put the camera on auto-focus. Otherwise, rack through the focus a few times to see where the optimal focus lies. Look through the viewfinder rather than at the foldout monitor when focusing. Switch to the focus-assist feature if the camcorder has one.

- To pan the small—and the large—camcorder, move it with your whole body rather than just your arms. First, point your knees in the direction of the *end* of the pan. Then twist your body with the camera aimed toward the beginning of the pan. During the pan you are like a spring that is uncoiling from the start of the action to the finish. Always bend your knees slightly when shooting; as in skiing, your knees act as shock absorbers. Don't panic if you lose the subject temporarily in the viewfinder. Keep the camera steady, look up to see where the subject is, and aim the camera smoothly in the new direction. **SEE 8.18**

When moving with people who are walking, get in front of them with the camera and walk backward at the same speed. This way you can see their faces rather than their backs. Moving backward also forces you to walk on the balls of your feet, which are better shock absorbers than your heels. Watch that you do not bump into or stumble over something while walking backward—a quick check of your proposed route can prevent unexpected mishaps. With the zoom lens in the wide-angle position, you are often closer to the object than the viewfinder image indicates. Be careful not to hit something or someone with the camera, especially if you walk forward with it into a crowd. **SEE 8.19**

8.18 PHOTO BY EDWARD AIONA

8.18 PANNING THE CAMCORDER
Before panning, point your knees in the direction of the end of the pan, then uncoil your upper body during the pan.

- Under normal conditions put the camera in the auto-iris mode. "Normal conditions" means that the lighting does not produce a high contrast (such as the midday sun) or that you don't have to create special effects, such as a very shallow depth of field. Despite the objections of some especially critical camera operators, using auto-iris on an overcast day will often yield better-exposed video than doing it manually, especially during electronic news gathering (ENG).

- Don't forget the audio. Use earbuds or headphones to listen to the sound. All camcorders record on two audio tracks. Whenever possible, check the audio level before and during recording. When working in relatively quiet

8.19 PHOTO BY EDWARD AIONA

8.19 WALKING BACKWARD
When moving with something or somebody, walk backward rather than forward. The balls of your feet act like shock absorbers.

surroundings, record with the automatic gain control (AGC). Otherwise, you need to switch to manual gain control, take a sound level, and record. (See chapters 9 and 10 for more on ENG sound.)

■ Always record sound with the camera mic, even if you are using an external mic for the primary sound pickup. This ambient sound is important for achieving continuity in postproduction editing. When the reporter is holding the external mic, do not start to run away from him or her to get a better shot of the event. Either you both run together, or you must stay put.

■ Whenever you use a camcorder in the field, use common sense. Be mindful of your and other people's safety. Always ask yourself whether the risk is worth the story. In ENG reliability and consistency are more important than sporadic feats, however spectacular. Do not risk your neck and the equipment to get a shot that would simply embellish a story. Leave that type of shooting to the gifted amateurs.

After the Shoot

■ Unless you have just shot a really hot story that must air immediately, even unedited, take care of the equipment before delivering the recording media. If you are properly organized, it should take just a few minutes.

■ Close the lens iris to the *C* or *close* position and replace the lens cap.

■ Put all the switches in the *off* position, unless you are running after another story, in which case put the camera on standby.

■ Extract the recording media (tape, disc, flash drive) from the camera, label it and its box, and put it in a safe place.

■ As soon as you return from the assignment, recharge all batteries. If the camcorder got wet, wait until everything has dried out before putting the camcorder in its case. Moisture is one of the most serious threats to camcorders. If you have time, check all the portable lights so that they will work for the next assignment. Coil all AC extension cords—you will not have time to untangle them during an ongoing event.

WORKING THE STUDIO CAMERA

In one way a studio camera is easier to operate than a portable camera: it is mounted on some kind of camera support, and the video operator (VO) makes all the electronic adjustments, "shading" the camera at the CCU (camera control unit). In another way, however, you may find operating the studio camera more difficult because you have to steer the heavy pedestal to truck, arc, dolly, and pedestal up and down. You have to zoom in and out and adjust the focus while composing effective pictures, listening to the director's commands, and watching the studio traffic around you. Obviously, a smooth coordination of all these factors takes practice. The following steps for before, during, and after a show will facilitate your operating the studio camera.

Before the Show

■ Put on your headset and check that the intercom system is functioning. You should hear at least the director, the technical director (TD), and the video operator.

■ Unlock the pan-and-tilt mechanism on the mounting head and adjust the horizontal and vertical drag, if necessary. Check that the camera is balanced on the mounting head (the camera should not dip forward or tilt back when unlocked). Unlock the pedestal, then pedestal up and down.

■ See how much camera cable you have and whether there are any obstacles that may interfere with the cable run. Check that the pedestal skirt or other type of cable guard is low enough to move the cable out of the way rather than roll over it.

■ Ask the VO to uncap the camera from the CCU, and ask if you can remove the lens cap. You can then see in the viewfinder the pictures the camera actually takes. Is the viewfinder properly adjusted? Like a home television set, the viewfinder can be adjusted for brightness and contrast. If you need framing guides, flip the switch that shows the essential area (safe title area) and the screen-center mark. The center mark is especially important if you have to keep the principal action in the 4 × 3 screen area on a 16 × 9 viewfinder (see chapter 15).

■ Check the zoom lens. Zoom in and out. Get a feel for how close you can get to the main event from a certain position. Is the lens clean? If it is dusty, use a fine camel-hair brush and carefully clean off the larger dust particles. With a small rubber bulb or a can of compressed air, blow off the finer dust. Do not blow on the lens with your mouth: the moisture will fog it up and get it even dirtier.

■ Rack through focus: move the focus control from one extreme position to the other. Can you move easily and smoothly into and out of focus, especially when in a narrow-angle, zoomed-in position?

■ Calibrate the zoom lens by zooming all the way in and focusing on the target object. You should now remain in focus throughout the zoom, provided neither the object nor the camera moves toward or away from each other.

■ If you have a shot sheet, this is a good time to practice the more complicated zoom and dolly or truck shots. A shot sheet is a list of every shot a particular camera has to get. It is attached to the camera to help the operator remember the shot sequence.

■ If a teleprompter is attached to the camera, check all the connections.

■ Lock the camera again (the pedestal and the pan-and-tilt mechanism) before leaving it. Don't ever leave a camera unlocked, even for a short while. Some pedestals have a parking brake. Set the brake on the pedestal before leaving the camera.

■ Cap the camera if you leave it for a prolonged period of time.

During the Show

■ Put on the headset and establish contact with the director, TD, and VO. Unlock the camera and recheck the pan and tilt drags and the pedestal movement.

■ Calibrate the zoom at each new camera position. Ensure that you can stay in focus over the entire zoom range.

■ When checking the focus for a new shot, rack through focus a few times to determine at which position the picture is the sharpest. When focusing on a person, the hairline usually gives you enough detail to determine the sharpest focus. In extreme close-ups (ECUs), focus on the bridge of the nose. Achieving and remaining in proper focus is more difficult with an HDTV camera. If available, use the focus-assist feature.

■ If you anticipate a dolly, set the zoom lens to the wide-angle position. With the zoom lens in the extreme wide-angle position, the depth of field should be large enough so that you need to adjust focus only when you are very close to the object or event.

■ If you have a difficult truck or arc to perform, have a floor person help you move and steer the camera. You can then concentrate on the camera operation.

■ In a straight dolly, you can keep both hands on the panning handles. If you have to steer the camera, steer with your right hand, keeping your left hand on the focus control.

■ When you pedestal up or down, try to brake the camera before it hits the stops at the extreme pedestal positions. Generally, keep your shots at the talent's eye level unless the director instructs you to shoot from either a high (pedestal up and looking down) or a low (pedestal down and looking up) angle.

■ When you operate a freewheel camera dolly, always preset the wheels toward the intended camera movement to prevent the dolly from starting off in the wrong direction. Check that the cable guards are low enough to prevent the camera from running over the cables on the studio floor instead of pushing them out of the way.

■ Use masking tape on the studio floor to mark the critical camera positions.

■ At all times during the show, be aware of the activity around you. Where are the other cameras? Where is the microphone boom? The fishpole operator? The floor monitor? It is your responsibility to keep out of the view of the other cameras and not hit anything (including floor personnel and talent) during your moves. Rugs are a constant hazard to camera movement. When dollying into a set that has a rug, watch the floor so that you do not suddenly dolly up onto the rug.

■ Watch for the tally light to go out before calibrating the zoom or moving the camera to a new shooting position. This is especially important if your camera is engaged in special effects. With effects involving two cameras (such as a superimposition), the tally lights of both cameras are on (see chapter 14).

■ In general, keep your eyes on the viewfinder. If the format allows, look around for something interesting to shoot between shots. The director will appreciate good visuals in an ad-lib show (in which the shots have not been rehearsed). If you have a shot sheet, though, stick to it, however tempting the shot possibilities may be. Do not try to outdirect the director. Mark all shot changes on the shot sheet.

■ If you work without a shot sheet, try to remember the type and the sequence of shots from the rehearsal. A good camera operator has the next shot lined up before the director calls for it. If you work from a shot sheet, go to the next shot immediately after the preceding one—don't wait until the last minute.

■ Listen carefully to what the director tells all camera operators (not just you) so that you can coordinate your shots with those of the other cameras. Also, you can avoid wasteful duplication of shots by knowing approximately what the other cameras are doing.

■ Avoid unnecessary chatter on the intercom.

After the Show

■ At the end of the show, wait for the "all clear" signal before you lock the camera. Ask the VO whether you can cap the lens.

■ Lock the mounting head and the pedestal in a low position. Push the camera to its designated parking place in the studio. Coil the cable as neatly as possible on the cable hanger or, when left on the studio floor, in the customary figure-eight loops.

MAIN POINTS

- The standard camera movements are pan, turning the camera horizontally; tilt, pointing the camera up or down; pedestal, lowering or elevating the camera on a studio pedestal; tongue, moving the whole camera from left to right or from right to left with the boom of a jib arm; crane or boom, moving the whole camera up or down on a camera jib arm; dolly, moving the camera toward or away from the scene; truck or track, moving the camera laterally; arc, moving the camera in a slightly curved dolly or truck movement; cant, tilting the camera sideways; and zoom, changing the focal length of the lens while the camera is stationary.
- The standard camera mounts are handheld and shoulder-mounted, the monopod, the tripod and the tripod dolly, the studio pedestal, and special camera mounts, which include stabilizing systems, SLR (single-lens reflex) camera and pocket-sized-camera supports, the jib arm, and the robotic pedestal.
- The mounting head connects the camera to the tripod or studio pedestal and allows the camera to be smoothly tilted up and down and panned horizontally.
- The quick-release plate helps to repeatedly attach and detach a camcorder to and from its mounting head. The wedge mount facilitates attaching the heavier cameras to the mounting head in a balanced position.
- Before using a camcorder, check that the batteries are fully charged and that you have enough videotape or other recording media for the assignment. Do an audio check with the camera mic and the external mic.
- When working a camcorder or portable camera, handle it with the utmost care. Do not leave it unprotected in the sun or uncovered in the rain.
- At all times pay particular attention to white balance, presetting the zoom, and recording ambient sound. Switch to automatic controls if the conditions are right.
- After the production put everything back carefully so that the equipment is ready for the next assignment.

SECTION

8.2 Framing Effective Shots

The basic purpose of framing a shot is to show images as clearly as possible and to present them so that they convey meaning and energy. Essentially, you clarify and intensify the event before you. When working a camcorder, you are the only one who sees the images before they are video-recorded. You therefore cannot rely on a director to tell you how to frame every picture for maximum effect.

The more you know about picture composition, the more effective your clarification and intensification of the event will be. But even if you are working as a camera operator during a multicamera studio show or a large remote where the director can preview all the camera pictures, you still need to know how to compose effective shots. The director might be able to correct some of your shots, but he or she will certainly not have time to teach you the fundamentals of good composition.

This section describes the major compositional principles and explains how to frame a shot for maximum clarity and impact.

▶ **SCREEN SIZE AND FIELD OF VIEW**
Operating with close-ups and medium shots rather than long shots and extreme long shots

▶ **FRAMING A SHOT: STANDARD TV AND HDTV ASPECT RATIOS**
Dealing with height and width, framing close-ups, headroom, noseroom and leadroom, and closure

▶ **DEPTH**
Creating the illusion of a third dimension in both aspect ratios: choice of lens, positioning of objects, depth of field, and lighting and color

▶ **SCREEN MOTION**
Z-axis motion (movement toward and away from the camera) and lateral movement in both aspect ratios

SCREEN SIZE AND FIELD OF VIEW

Screen size and field of view are closely related. On the large movie screen, you can show a relatively large vista with a great amount of event detail. When the same scene is shown on television or on a mobile media display, however, not only will you have difficulty making out the smaller event details but, more importantly, you will lose the shot's aesthetic impact. This is why some film critics suggest seeing a particular film "on the big screen."

Screen Size

To reveal event details on the relatively small television screen, you must show them in close-ups rather than long shots. In other words, the field of view must generally be tighter on television than on the motion picture screen. Such a close-up approach necessitates choosing and emphasizing those details that contribute most effectively to the overall event and that add aesthetic energy.

Field of View

Field of view refers to how far or how close the object appears relative to the camera, that is, how close it will appear to the viewer. It is basically organized into five steps: ***extreme long shot (ELS)***, also called establishing shot; ***long shot (LS)***, also called full shot or establishing shot; ***medium shot (MS)***, also called waist shot; ***close-up (CU)***; and ***extreme close-up (ECU)***. SEE 8.20 ZVL2 CAMERA→ Composition→ field of view

Four other ways of designating conventional shots are: bust shot, which frames the subject from the upper torso to the top of the head; knee shot, which frames the subject from just above or below the knees; ***two-shot***, with two people or objects in the frame; and three-shot, with three people or objects in the frame.

Although more a blocking arrangement than a field of view, you should also know two additional shots: the over-the-shoulder shot and the cross-shot. In the ***over-the-shoulder shot (O/S)***, the camera looks at someone over the shoulder of the camera-near person. In a ***cross-shot (X/S)***, the camera looks alternately at one or the other

Extreme long shot (ELS), or establishing shot

Long shot (LS), or full shot

Medium shot (MS), or waist shot

Close-up (CU)

Extreme close-up (ECU)

8.20 FIELD-OF-VIEW STEPS

The shot designations range from ELS (extreme long shot) to ECU (extreme close-up).

8.20 PHOTOS BY EDWARD AIONA

person, with the camera-near person completely out of the shot. **SEE 8.21**

Of course, exactly how to frame such shots depends not only on your sensitivity to composition but also on the director's preference.

FRAMING A SHOT: STANDARD TV AND HDTV ASPECT RATIOS

Many high-end studio cameras, EFP cameras, and most camcorders have a switch for changing the ***aspect ratio*** from the standard 4 × 3 format to the HDTV 16 × 9. Although the aspect ratios of STV (standard television) and HDTV are quite different and require different technical manipulations, many of the aesthetic principles of good picture composition apply to both. This section takes a closer look at dealing with height and width, framing close-ups, headroom, noseroom and leadroom, and closure.

Dealing with Height and Width

You will find that the 4 × 3 aspect ratio is well suited to framing a vertical scene, such as a high-rise building, as well as a horizontally oriented vista. **SEE 8.22 AND 8.23** It is

Bust shot

Knee shot

Two-shot (two persons or objects in frame)

Three-shot (three persons or objects in frame)

Over-the-shoulder shot (O/S)

Cross-shot (X/S)

8.21 OTHER SHOT DESIGNATIONS

Other common shot designations are the bust shot, knee shot, two-shot, three-shot, over-the-shoulder shot, and cross-shot. Note that the bust shot is similar to the MS and that the knee shot is similar to the LS.

8.21 PHOTOS BY EDWARD AIONA

8.22 FRAMING A VERTICAL VIEW
The 4 × 3 aspect ratio allows you to frame a vertical scene without having to use extreme camera distance or angles.

8.23 FRAMING A HORIZONTAL VIEW
The 4 × 3 aspect ratio readily accommodates a horizontal vista.

8.24 FRAMING HEIGHT AND WIDTH IN A SINGLE SHOT
The 4 × 3 aspect ratio easily accommodates both horizontal and vertical vistas.

8.22 PHOTO BY HERBERT ZETTL
8.23 PHOTO BY HERBERT ZETTL
8.24 PHOTO BY HERBERT ZETTL

8.25 FRAMING A HORIZONTAL VIEW IN THE HDTV ASPECT RATIO
The 16 × 9 format is ideal for framing wide horizontal vistas.

8.26 FRAMING A VERTICAL VIEW IN THE HDTV ASPECT RATIO
The 16 × 9 format makes it difficult to frame a vertical object. One way to frame a tall object is to shoot it from below and cant the camera.

8.25 PHOTO BY HERBERT ZETTL
8.26 PHOTO BY GARY PALMATIER

8.27 NATURAL MASKING OF THE SCREEN SIDES IN THE HDTV ASPECT RATIO
You can use parts of the natural environment to block the sides of the wide 16 × 9 screen to create a vertical space in which to frame the vertical object. In this shot the foreground buildings create a vertical aspect ratio for the high-rise building.

8.27 PHOTO BY HERBERT ZETTL

also relatively easy to accommodate a scene that has both wide and high elements. **SEE 8.24**

Although the horizontally stretched 16 × 9 aspect ratio makes horizontal scenes look spectacular, it presents a formidable obstacle to framing a vertical view. **SEE 8.25 AND 8.26** You can either tilt the camera up to reveal the height of the object or shoot from below and cant the camera to make the subject fit into the diagonal screen space. Another frequently used film technique for dealing with vertical objects is to have other picture elements block the sides of the screen and, in effect, give you a vertical aspect ratio in which to frame the shot. **SEE 8.27**

Framing Close-ups

Close-ups and extreme close-ups are common elements in the visual language of television because, compared with the large motion picture screen, even large television

8.28 PHOTO BY EDWARD AIONA

8.28 FRAMING A CLOSE-UP
The normal close-up shows the head of the person and part of the shoulders.

8.29 PHOTO BY EDWARD AIONA

8.29 FRAMING AN EXTREME CLOSE-UP
In an extreme close-up, you should crop the top of the head while keeping the upper part of the shoulders in the shot.

8.30 PHOTO BY EDWARD AIONA

8.30 FRAMING A CLOSE-UP IN THE HDTV ASPECT RATIO
When framing the same close-up in the 16 × 9 format, both screen sides look conspicuously empty.

8.31 PHOTO BY EDWARD AIONA

8.31 FRAMING AN EXTREME CLOSE-UP IN THE HDTV ASPECT RATIO
In the 16 × 9 format, the extreme close-up of the person seems oddly squeezed between the upper and lower screen edges.

screens are relatively small. The 4 × 3 aspect ratio and the small screen of the standard television receiver are the ideal combination for CUs and ECUs of people's heads. SEE 8.28

As you can see, the normal CU shows the customary headroom and part of the upper body. The ECU is somewhat trickier to frame: the top screen edge cuts across the top part of the head, and the lower edge cuts just below the top part of the shoulders. SEE 8.29 ZVL3 CAMERA→ Composition→ close-ups

When you try to frame the same shot in the HDTV 16 × 9 aspect ratio, however, you are left with a great amount of space on both sides of the subject's face. The close-up looks somewhat lost in the wide-screen format, and the extreme close-up looks as though it is squeezed between the top and bottom screen edges. SEE 8.30 AND 8.31 You can solve this problem relatively easily by including some visual elements in the shot that fill the empty spaces on either side. SEE 8.32 Some directors simply tilt the camera or the talent somewhat so that the shot occupies more of the horizontal space.

One of the advantages of the HDTV aspect ratio is that it lets you easily frame close-ups of two people face-to-face. Such an arrangement is difficult in the traditional format because the two dialogue partners must stand uncomfortably close to each other. SEE 8.33

Headroom

Because the edges of the television frame seem to attract like magnets whatever is close to them, leave some space above people's heads—called ***headroom***—in normal long shots, medium shots, and close-ups. SEE 8.34 Avoid having the head "glued" to the upper edge of the frame. SEE 8.35

8.32 NATURAL MASKING OF A CLOSE-UP IN THE HDTV ASPECT RATIO
To avoid excessive empty space when framing a screen-center close-up of a person in the 16 × 9 format, you can mask the sides with objects from the actual environment.

8.33 FACE-TO-FACE CLOSE-UPS IN THE HDTV ASPECT RATIO
The 16 × 9 format makes it relatively easy to have two people face each other on a close-up without having to stand uncomfortably close together.

8.32 PHOTO BY EDWARD AIONA
8.33 PHOTO BY EDWARD AIONA

8.34 NORMAL HEADROOM
Headroom counters the magnetic pull of the upper frame. The person appears comfortably placed in the frame.

8.35 TOO LITTLE HEADROOM
With no, or too little, headroom, the person looks cramped in the frame and the head seems to be glued to the upper screen edge.

8.36 TOO MUCH HEADROOM
With too much headroom, the pull of the bottom edge makes the picture bottom-heavy and strangely unbalanced.

8.34 PHOTO BY EDWARD AIONA
8.35 PHOTO BY EDWARD AIONA
8.36 PHOTO BY EDWARD AIONA

Because you lose a certain amount of picture area in video recording and transmission, you need to leave a little more headroom than feels comfortable. Leaving too much headroom, however, is just as bad as having too little. SEE 8.36 If the camera is so equipped, you can use the frame guide in the viewfinder to see the picture area that actually appears on the television screen. The headroom rule applies equally to both aspect ratios. ZVL4 CAMERA→ Composition→ headroom

Noseroom and Leadroom

Somebody looking or pointing in a particular direction other than straight into the camera creates a screen force called an index vector. You must compensate for this force by leaving some space in front of the vector. When someone looks or points screen-left or screen-right, the index vector needs to be balanced with ***noseroom***. A lack of noseroom or leadroom makes the picture look oddly out of balance; the person seems to be blocked by the screen edge. SEE 8.37 AND 8.38

All motion creates a motion vector. When someone or something moves in a screen-right or screen-left direction, you must leave ***leadroom*** to balance the force of the motion vector. SEE 8.39 Even in a still photo, you can see that without proper leadroom the cyclist seems to be crashing into the left screen border. SEE 8.40

To avoid such crashes, you must always lead the moving object with the camera rather than follow it. After all, we want to see where the moving object is going, not where it has been. Note, however, that neither of the leadroom examples here represents actual motion vectors because they are still pictures. In a still picture, they do not move but simply point in a specific direction and are, therefore, index vectors. ZVL5 CAMERA→ Composition→ leadroom

8.37 PROPER NOSEROOM
To absorb the force of the strong index vector created by the person's looking toward the screen edge, you need to leave some noseroom.

8.38 LACK OF NOSEROOM
Without noseroom the person seems to be blocked by the screen edge, and the picture looks unbalanced.

8.37 PHOTO BY EDWARD AIONA
8.38 PHOTO BY EDWARD AIONA

8.39 PROPER LEADROOM
Assuming that the cyclist is actually moving, his motion vector is properly neutralized by the screen space in front of him. We like to see where the person is heading, not where he has been. Note that a still picture cannot show a motion vector. What you see here is an index vector.

8.40 LACK OF LEADROOM
Without leadroom the moving person or object seems to be hindered or stopped by the screen edge.

8.39 PHOTO BY EDWARD AIONA
8.40 PHOTO BY EDWARD AIONA

Closure

Closure, short for *psychological closure,* is the process by which our minds fill in information that we cannot actually see. Take a look around you: you see only parts of the objects that lie in your field of vision. There is no way you can ever see an object in its entirety unless the object moves around you or you move around the object. Through experience we have learned to mentally supply the missing parts, which allows us to perceive a whole world although we actually see only a fraction of it. Because close-ups usually show only part of an object, your psychological closure mechanism must work overtime.

Positive closure To facilitate closure you should always frame a shot in such a way that the viewer can easily extend the figure beyond the screen edges into off-screen space and perceive a sensible whole. **SEE 8.41** To organize the visual world around us, we also automatically group things together so that they form a pattern. **SEE 8.42 AND 8.43** You would be hard-pressed not to perceive figure 8.42 as a triangular pattern and figure 8.43 as a semicircle.

Negative closure This closure automation can also work against good composition. For example, when framing a close-up of a face without giving prominent visual clues to

8.41 FACILITATING CLOSURE BEYOND THE FRAME
In this shot we perceive the whole figure of the person and her guitar although we see only part of them. This shot gives us sufficient clues to project the figure beyond the frame and apply psychological closure in the off-screen space.

8.41 PHOTO BY EDWARD AIONA

8.42 TRIANGLE CLOSURE
We tend to organize things into easily recognizable patterns. This group of similar objects forms a triangle.

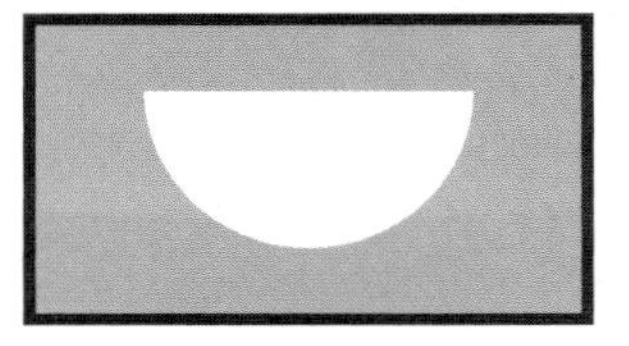

8.43 SEMICIRCLE CLOSURE
These objects organize the screen space into a semicircle.

8.42 PHOTO BY HERBERT ZETTL
8.43 PHOTO BY HERBERT ZETTL

help viewers project the image beyond the screen edges, the head seems oddly cut off from its body. **SEE 8.44** You therefore need to provide enough visual clues to lead the viewers' eyes beyond the frame so they can apply closure and perceive the complete person in off-screen space. **SEE 8.45**

Our mechanism to organize our environment into simple patterns is so strong that it often works against reason. In the excitement of getting a good story and an interesting shot, it is easy to forget to look *behind* the object of attention, but it is often the background that spoils a good picture composition. **SEE 8.46** As you can see in figure 8.46, we tend to perceive the background as part of the foreground. The reporter seems to be balancing a street sign on his head. Once you are aware of the background, it is relatively easy to avoid illogical closure. ZVL6 CAMERA→ Composition→ closure

DEPTH

Because television and movie screens are flat, we must create the illusion of a third dimension. Fortunately, the principles for creating the illusion of depth on a two-dimensional surface have been amply explored and established by painters and photographers over the years. The camera lens will do much of the work for you. To create and intensify the illusion of depth on the most basic level, try to establish a clear division of the image into foreground, middleground, and background. **SEE 8.47** To do this you need to consider four factors: choice of lens, positioning of objects, depth of field, and lighting and color.

Choice of lens A wide-angle zoom position exaggerates depth. Narrow-angle positions reduce the illusion of a third dimension.

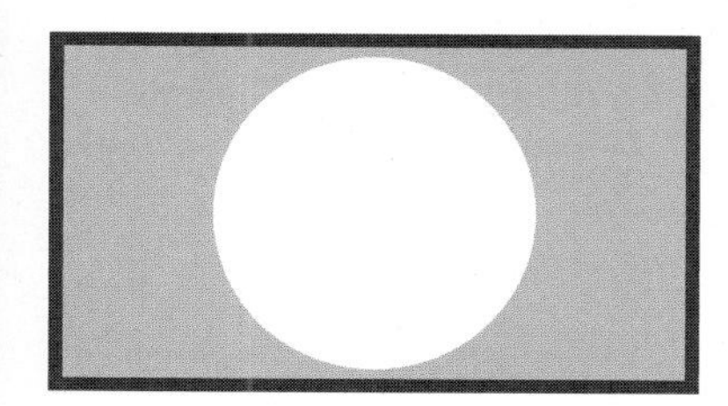

8.44 PHOTO BY EDWARD AIONA

8.44 UNDESIRABLE CLOSURE WITHIN THE FRAME
This shot is badly framed because we apply closure within the frame without projecting the rest of the person into off-screen space.

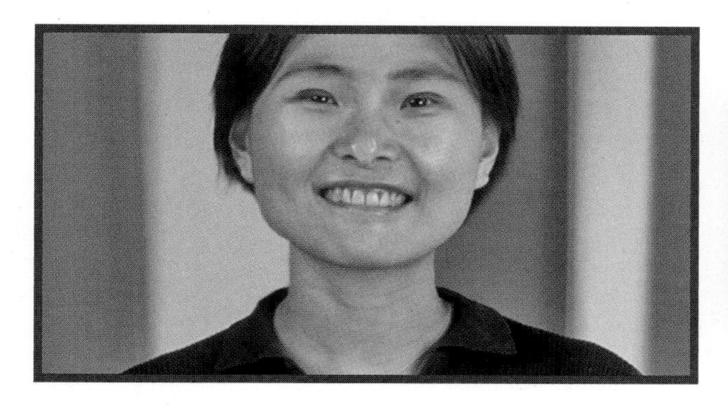

8.45 PHOTOS BY EDWARD AIONA

8.45 DESIRABLE CLOSURE IN OFF-SCREEN SPACE
In this ECU there are enough on-screen clues to project the rest of the person's head and body into off-screen space, thus applying closure to the total figure.

8.46 PHOTO BY HERBERT ZETTL
8.47 PHOTO BY HERBERT ZETTL

8.46 ILLOGICAL CLOSURE
Although we know better, we perceive this reporter as balancing a street sign on his head.

8.47 FOREGROUND, MIDDLEGROUND, AND BACKGROUND
In general, try to divide the z-axis (depth dimension) into a prominent foreground (tree trunks), middleground (embankment of the creek and the bushes), and background (mountains). Such a division helps create the illusion of screen depth.

Positioning of objects The ***z-axis***—the imaginary line representing an extension of the lens from the camera to the horizon—has significant bearing on perceiving depth. Anything positioned along the z-axis relative to the camera will create the illusion of depth. To emphasize depth, always try to include a foreground object in the shot. This inevitably pushes everything else farther back.

Depth of field A slightly shallow depth of field is usually more effective to define depth because the in-focus foreground object is more clearly set off against the out-of-focus background.

Lighting and color A brightly lighted object with strong (highly saturated) color seems closer than one that is dimly lighted and has washed-out (low-saturation) colors. ZVL7 CAMERA→ Picture depth→ z-axis | lens choice | perspective and distortion

SCREEN MOTION

Contrary to the painter or the still photographer, who deals with the organization of static images within the picture frame, the television camera operator must almost always cope with framing images in motion. Composing moving images requires quick reactions and full attention throughout the telecast or recording. The study of the moving image is an important part of learning the fine art of television and film production; here we look at some of its most basic principles.[1]

When framing for the traditional 4 × 3 aspect ratio and small screen, movements along the z-axis (toward or away from the camera) are stronger than any type of lateral motion (from one screen edge to the other). Fortunately, they are also the easiest to frame: you simply keep the camera as steady as possible and make sure that the moving object does not go out of focus as it approaches the camera. Remember that a wide-angle zoom position gives the impression of accelerated motion along the z-axis, whereas a narrow-angle zoom position slows z-axis motion for the viewer.

When working in the 16 × 9 HDTV aspect ratio, lateral movement takes on added prominence. The stretched screen width gives you a little more breathing room when two people face each other (see figure 8.33), and a laterally moving object has more time to travel from one screen edge to the other.

If you are on a close-up and the subject shifts back and forth, don't try to follow each minor wiggle. You might run the risk of making viewers seasick; at the very least, they will not be able to concentrate on the subject for very long. Keep the camera pointed at the major action area or zoom out (or pull back) to a slightly wider shot.

When one of the persons in a two-shot moves out of the frame, do not try to keep both people in the frame—stay with just one of them. SEE 8.48 AND 8.49 ZVL8 CAMERA→ Screen motion→ z-axis | lateral | close-ups

If even after extensive rehearsals you find that in an over-the-shoulder shot the person closer to the camera blocks the other person, who is farther away from the camera, you can solve the problem by trucking or arcing to the right or left. SEE 8.50 AND 8.51

1. For an extensive discussion of screen forces and how they can be used for effective picture composition, see Herbert Zettl, *Sight Sound Motion: Applied Media Aesthetics,* 6th ed. (Boston: Wadsworth, 2011), pp. 103–25, 253–73.

8.48 TWO PERSONS SAYING GOOD-BYE
If in a two-shot the people walk away from each other toward the screen edges, don't try to keep both people in the shot.

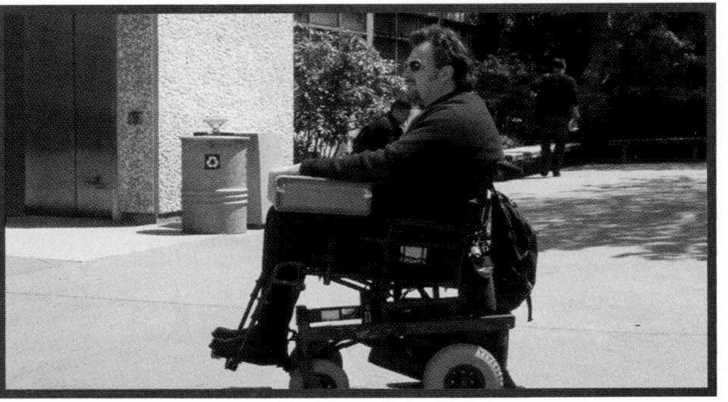

8.49 CAMERA STAYS WITH ONE OF THEM
You must decide which person you will keep in the frame and let the other move off-camera.

8.48 PHOTO BY EDWARD AIONA
8.49 PHOTO BY EDWARD AIONA

8.50 CAMERA-FAR PERSON BLOCKED
In an over-the-shoulder shot, you may find that the camera-near person blocks the camera-far person.

8.51 CAMERA TRUCKS TO CORRECT
To correct this over-the-shoulder shot so that the camera-far person can be seen, simply truck or arc the camera to the right.

8.50 PHOTO BY EDWARD AIONA
8.51 PHOTO BY EDWARD AIONA

Whatever you do to organize screen motion, do it smoothly. Try to move the camera as little as possible unless you need to follow a moving object or dramatize a shot through motion. Because you can move a camcorder so easily, it may be tempting to animate a basically static scene by moving the camera with great fervor. Don't do it. Excessive camera motion is a telltale sign of an amateur camera operator.

MAIN POINTS

- Because the television screen is relatively small, we use more close-ups and medium shots than long shots. When shooting for wide-aspect-ratio, large-screen HDTV, you can use more medium shots and long shots.
- Field of view refers to how much of a scene you show in the viewfinder, that is, how close the object appears relative to the viewer. The field of view is organized into five steps: ELS (extreme long shot, or establishing shot), LS (long shot, full shot, or establishing shot), MS (medium shot, or waist shot), CU (close-up), and ECU (extreme close-up).
- Alternate shot designations include the bust shot, the knee shot, the two-shot, the three-shot, the over-the-shoulder shot (O/S), and the cross-shot (X/S).
- In organizing the screen area for the traditional 4 × 3 and the HDTV 16 × 9 aspect ratios, the major considerations are dealing with height and width, framing close-ups, headroom, noseroom and leadroom, and closure.
- A simple and effective way to organize screen depth is to establish a distinct foreground, middleground, and background.
- In creating the illusion of a third dimension (depth), you need to consider the choice of lens, positioning of objects, depth of field, and lighting and color.
- In organizing screen motion for the 4 × 3 aspect ratio, z-axis motion (movement toward or away from the camera) is stronger than lateral movement (from one side of the screen to the other). When working in the 16 × 9 aspect ratio, lateral movement becomes more prominent.

ZETTL'S VIDEOLAB

For your reference or to track your work, the *Zettl's VideoLab* program cues in this chapter are listed here with their corresponding page numbers.

ZVL1 CAMERA→ Camera moves→ dolly | zoom | truck | pan | tilt | pedestal | try it 141

ZVL2 CAMERA→ Composition→ field of view 154

ZVL3 CAMERA→ Composition→ close-ups 157

ZVL4 CAMERA→ Composition→ headroom 158

ZVL5 CAMERA→ Composition→ leadroom 158

ZVL6 CAMERA→ Composition→ closure 160

ZVL7 CAMERA→ Picture depth→ z-axis | lens choice | perspective and distortion 162

ZVL8 CAMERA→ Screen motion→ z-axis | lateral | close-ups 162

CHAPTER

9

Audio: Sound Pickup

We are usually so engrossed in the barrage of colorful pictures when watching television that we are often totally unaware of the sound—unless there is an audio problem. All of a sudden we realize that without sound we have a hard time following what is going on. So long as we can hear the sound track, we can turn away from the TV and still know pretty much what's happening on-screen. But isn't a picture worth a thousand words? Apparently not in television. Because so much information is transmitted by someone talking, the infamous "talking head" is not such a bad production technique after all, provided the person talking has something worthwhile to say.

Sound is also important for establishing mood and intensifying an event. A good chase sequence invariably has a barrage of sounds, including agitated music and squealing tires. Some of the crime shows carry some kind of rhythmic thumping throughout the dialogue. It is so cleverly done that most of us are hardly aware of the nonverbal track. Another important function of rhythmic sounds is to help structure the quick cuts and the visual fragments of a string of close-ups to form a meaningful whole.

If sound is indeed such an important production element, why do we fail to have better sound on the average television show? Even when you produce a short scene as an exercise in the studio, you will probably notice that although the pictures may look acceptable or even spectacular, it is usually the sound portion that could stand some improvement. It is often assumed, unfortunately, that by sticking a microphone into a scene at the last minute the audio requirements have been satisfied. Don't believe it. Good television audio needs at least as much preparation and attention as the video portion. And, like any other production element, television audio should not simply be added—it should be integrated into the production planning from the very beginning.

Section 9.1, How Microphones Hear, covers the sound pickup portion of ***audio*** (from the Latin verb *audire,* "to hear"), including the sound-producing and operational types of microphones. In Section 9.2, How Microphones Work, you learn about the more technical aspects of sound-generating elements and the various microphone uses in electronic news gathering (ENG), electronic field production (EFP), and music setups.

KEY TERMS

audio The sound portion of television and its production. Technically, the electronic reproduction of audible sound.

cardioid Heart-shaped pickup pattern of a unidirectional microphone.

condenser microphone A mic whose diaphragm consists of a condenser plate that vibrates with the sound pressure against another fixed condenser plate, called the backplate. Also called *electret microphone* and *capacitor microphone.*

direct insertion Recording technique wherein the sound signals of electric instruments are fed into an impedance-matching box and from there into the mixing console without the use of a speaker and a microphone. Also called *direct input.*

dynamic microphone A mic whose sound pickup device consists of a diaphragm that is attached to a movable coil. As the diaphragm vibrates with the air pressure from the sound, the coil moves within a magnetic field, generating an electric current. Also called *moving-coil microphone.*

fishpole A suspension device for a microphone; the mic is attached to a pole and held over the scene for brief periods.

flat response Measure of a microphone's ability to hear equally well over its entire frequency range. Is also used as a measure for devices that record and play back a specific frequency range.

foldback The return of the total or partial audio mix to the talent through headsets or I.F.B. channels. Also called *cue-send.*

frequency response Measure of the range of frequencies a microphone can hear and reproduce.

headset microphone Small but good-quality omni- or unidirectional mic attached to padded earphones; similar to a telephone headset but with a higher-quality mic.

impedance Type of resistance to electric current. The lower the impedance, the better the signal flow.

lavalier microphone A small mic that can be clipped onto clothing. Also called *lav.*

omnidirectional Pickup pattern in which the microphone can pick up sounds equally well from all directions.

phantom power The power for preamplification in a condenser microphone, supplied by the audio console rather than a battery.

pickup pattern The territory around the microphone within which the mic can "hear equally well," that is, has optimal sound pickup.

polar pattern The two-dimensional representation of a microphone pickup pattern.

ribbon microphone A mic whose sound pickup device consists of a ribbon that vibrates with the sound pressures within a magnetic field. Also called *velocity mic.*

shotgun microphone A highly directional mic for picking up sounds from a relatively great distance.

system microphone Mic consisting of a base upon which several heads can be attached that change its sound pickup characteristic.

unidirectional Pickup pattern in which the microphone can pick up sounds better from the front than from the sides or back.

wireless microphone A system that transmits audio signals over the air rather than through mic cables. The mic is attached to a transmitter, and the signals are received by a receiver connected to the audio console or recording device. Also called *RF (radio frequency) mic* and *radio mic.*

SECTION

9.1 How Microphones Hear

The pickup of live sounds is done through a variety of microphones. How good or bad a particular microphone is depends not only on how it is built but especially on how it is used. Section 9.1 focuses on the specific types and uses of microphones.

- **MICROPHONE TYPES BY HOW THEY HEAR**
  Sound-generating elements (dynamic, condenser, and ribbon), pickup patterns (omnidirectional and unidirectional), and additional microphone features

- **MICROPHONE TYPES BY HOW THEY ARE USED**
  Mobile microphones (lavalier, hand, boom, headset, and wireless) and stationary microphones (desk, stand, hanging, hidden, and long-distance)

MICROPHONE TYPES BY HOW THEY HEAR

Choosing the most appropriate microphone, or mic (pronounced "mike"), and operating it for optimal sound pickup requires that you know the three basic types of sound-generating elements, their pickup patterns, and some additional microphone features.

Sound-generating Elements

All microphones *transduce* (convert) sound waves into electric energy, which is amplified and reconverted into sound waves by the loudspeaker. The initial conversion is accomplished by the microphone's generating element. There are three major types of sound-converting systems, which are used to classify microphones: dynamic, condenser, and ribbon.

Dynamic microphones These are the most rugged. ***Dynamic microphones***, also called moving-coil microphones, can tolerate reasonably well the rough handling that television mics frequently (though unintentionally) receive. They can be worked close to the sound source and withstand high sound levels without damage to the microphone or excessive input overload (distortion of very high-volume sounds). They can also withstand fairly extreme temperatures. As you can probably guess, they are an ideal outdoor mic.

Condenser microphones Compared with dynamic mics, ***condenser microphones*** are much more sensitive to physical shock, temperature change, and input overload, but they usually produce higher-quality sound when used at greater distances from the sound source. Unlike dynamic mics, the condenser has a built-in preamplifier. This preamp strengthens the sound signal so that it does not get lost on the way to the camcorder, mixer, or console. The power supply for condenser mics is usually a battery. Although these batteries last for about a thousand hours, you should always keep spares on hand, especially if you are using condenser mics for ENG/EFP. Many times condenser mic failures can be traced to a dead or wrongly inserted battery. **SEE 9.1**

Condenser mics can also be powered through voltage supplied by the audio console or mixer through the audio

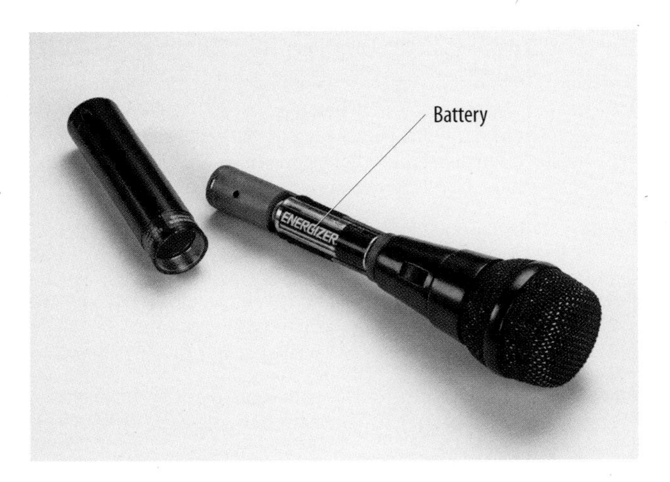

9.1 PHOTO BY EDWARD AIONA

9.1 POWER SUPPLY BATTERY FOR CONDENSER MICROPHONE
Many condenser microphones are powered by a battery rather than from the console (phantom power); be sure to observe the + and – poles as indicated on the battery housing.

cable. This method of supplying power to the mic's preamp is called ***phantom power***.

Ribbon microphones Similar in sensitivity and quality to condenser mics, ***ribbon microphones*** produce a warmer sound, frequently preferred by singers. Unlike condenser mics, which you may use outdoors under certain circumstances, ribbon mics are strictly for indoor use. They are sometimes called velocity mics. ZVL1 AUDIO→ Microphones→ mic choice | transducer

Pickup Patterns

Whereas some microphones, like our ears, hear sounds from all directions almost equally well, others hear sounds better when they come from a specific direction. The territory within which a microphone can hear equally well is called its ***pickup pattern;*** its two-dimensional representation is called the ***polar pattern***, as shown in figures 9.2 through 9.4.

In television production you need to use both omnidirectional and unidirectional mics, depending on what and how you want to hear. The ***omnidirectional*** microphone hears sounds from all (*omnis* in Latin) directions more or less equally well. SEE 9.2 The ***unidirectional*** microphone hears better in one (*unus* in Latin) direction—the front of the mic—than from its sides or back. Because the polar patterns of unidirectional microphones are roughly heart-shaped, they are called ***cardioid***. SEE 9.3

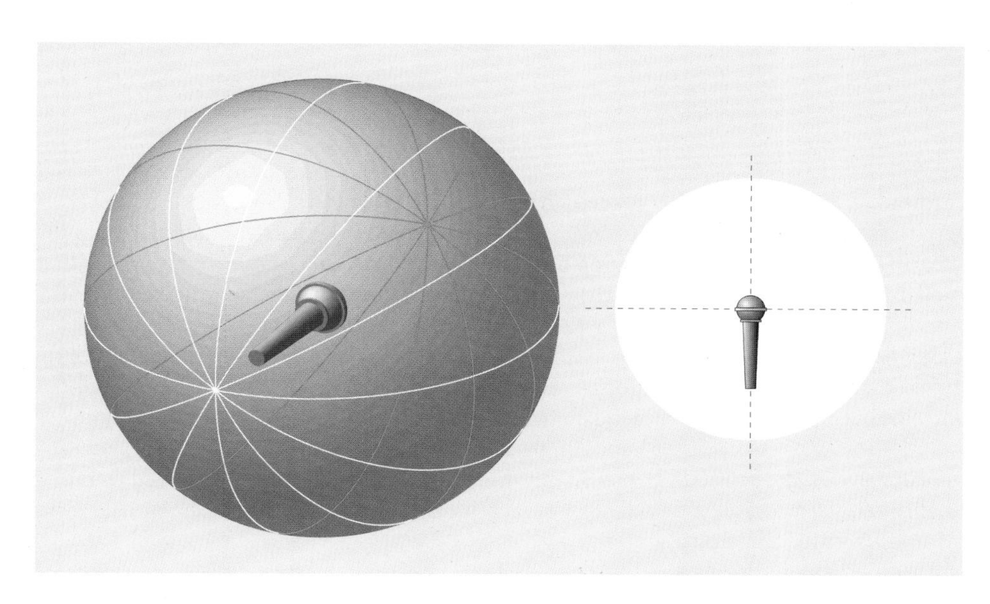

9.2 OMNIDIRECTIONAL PICKUP AND POLAR PATTERNS
The omnidirectional pickup pattern is like a sphere with the mic in its center. All sounds that originate within its pickup pattern are heard by the mic without marked difference.

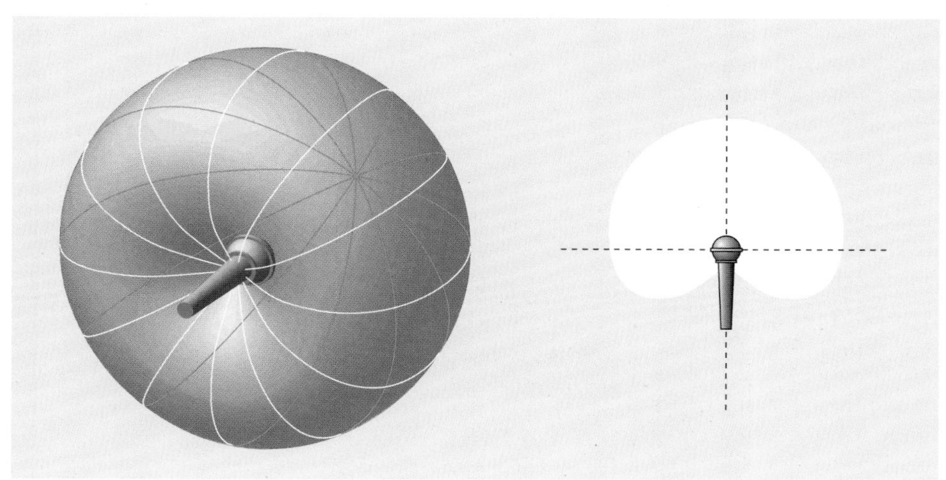

9.3 CARDIOID PICKUP AND POLAR PATTERNS
The heart-shaped pickup pattern makes the mic hear better from the front than from the sides. Sounds to its rear are suppressed.

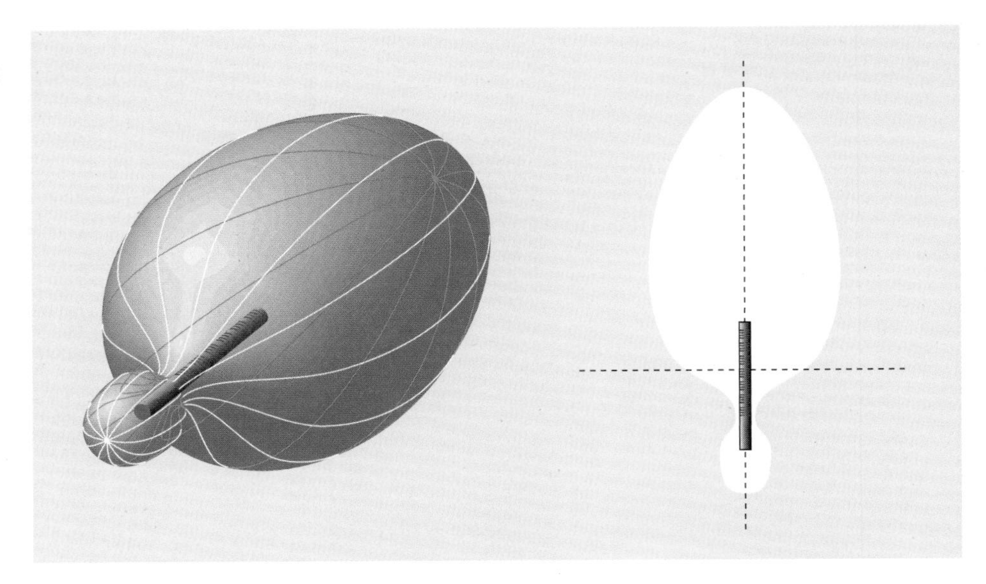

9.4 HYPERCARDIOID PICKUP AND POLAR PATTERNS
The supercardioid and hypercardioid pickup patterns narrow the sound pickup. They have a long but narrow reach in front and eliminate most sounds coming from the sides. They also hear sounds coming from the rear.

The supercardioid, hypercardioid, and ultracardioid microphones have progressively narrower pickup patterns, which means that their hearing is more and more concentrated in the front. Their claim to fame is that they can hear sounds from far away and make them appear to be relatively close. These mics also hear sounds that are behind them; but because they excel in hearing in one direction (a narrow path in front), they still belong to the unidirectional group. **SEE 9.4**

Which type to use depends primarily on the production situation and the sound quality required. If you are doing a stand-up report (standing in front of the actual scene) on conditions at the local zoo, you would want a rugged, omnidirectional mic that not only favors speech but also picks up some of the animal sounds for authenticity. If, on the other hand, you are video-recording a singer in the studio, you should probably choose a high-quality mic with a more directional cardioid pickup pattern. To record an intimate conversation between two soap opera characters, a hypercardioid shotgun mic is probably best. Unlike the omnidirectional mic, the shotgun mic can pick up their conversation from relatively far away without losing sound presence (the perceived closeness of the sound) while ignoring to a large extent many of the other studio noises, such as people and cameras moving about, the humming of lights, or the rumble of air conditioning. A table of the most common microphones and their characteristics is included in section 9.2 (see figure 9.32). ZVL2 AUDIO→ Microphones→ pickup patterns

Microphone Features

Microphones that are held close to the mouth have a built-in pop filter, which eliminates the sudden breath pops that might occur when someone speaks directly into the mic. **SEE 9.5** When used outside, all types of microphones are susceptible to wind, which they reproduce as low rumbling noises. To reduce wind noise, put a windscreen made of acoustic foam rubber over the mic. The popular name is zeppelin because it resembles an airship. **SEE 9.6**

To cut the wind noise even more, you need to pull a windsock, or wind jammer, over the windscreen. The wind

9.5 POP FILTER
The built-in pop filter eliminates breath pops.

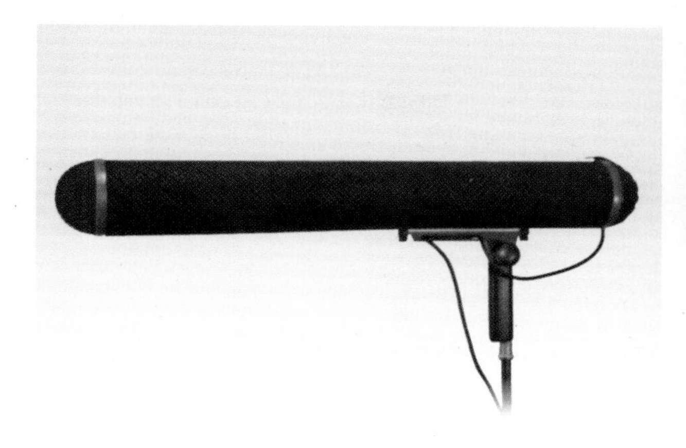

9.6 WINDSCREEN
The windscreen, commonly known as a zeppelin, is normally made of acoustic foam rubber. It covers the whole microphone to reduce the low rumble of wind noise.

9.7 LAVALIER MICROPHONE
This lavalier mic is properly attached for optimal sound pickup.

jammer is made from synthetic material and resembles more a mop than a sophisticated audio device (see figure 9.14). Whatever you use, bear in mind that the rumble of wind noise cannot be totally eliminated. The only way to have no wind noise on a video recording is to shoot when there is no wind. You can, however, use filters in postproduction that can reduce or eliminate some wind noise, or you can cover it up with sound effects or music.

To reduce the need for microphones with different pickup patterns, you can use a ***system microphone,*** which consists of a base upon which several "heads" can be attached. These heads can change the pickup pattern from omnidirectional to hypercardioid. As convenient as this may be, you will find that most audio engineers favor the individual mics built for specific applications.

MICROPHONE TYPES BY HOW THEY ARE USED

Some microphones are designed and used primarily for sound sources that are moving, whereas others are used for stationary sound sources. When grouped according to how they are used, there are mobile and stationary microphones (see figure 9.32). Of course, any of the mobile mics can be used in a stationary position, and the stationary mics can be moved about if the production situation so requires.

The mobile microphones include lavalier, hand, boom, headset, and wireless mics. The stationary microphones include desk, stand, hanging, hidden, and long-distance mics.

Lavalier Microphones

The first of the mobile type, the ***lavalier microphone,*** usually referred to as a lav, is probably the most frequently used on-camera mic in television. The high-quality lavaliers, which range in size from a small earbud to the eraser on the end of a pencil, can be clipped to clothing. Because of their size, they are unobtrusive and look more like jewelry than a technical device. **SEE 9.7**

Lavalier microphones are dynamic or condenser, with an omnidirectional or cardioid pickup pattern. They are designed primarily for voice pickup. The quality of even the smallest one is amazingly high. Once the lav is properly attached to the performer (approximately 5 to 8 inches below the chin, on top of the clothes, and away from anything that could rub or bang against it), sound pickup is no longer a worry. The audio engineer also has less difficulty riding the gain (adjusting the volume) of the lavalier than the boom mic or hand mic. Because the distance between the mic and the sound source does not change during the performance, an even sound level can be achieved more easily than with other mobile microphones.

The use of lavalier mics frees the lighting people from "lighting around the boom" to avoid shadows. They can concentrate more on the aesthetic subtleties of lighting as required by the scene.

Although the action radius of performers is still limited by the lavalier cable, the cable is flexible enough that they can move quickly and relatively unrestricted in a limited studio area. For greater mobility you can plug the lavalier into a small cell-phone-sized transmitter, which you can clip on a belt or put in a coat pocket, and use it as a wireless mic (see figure 9.21).

Despite their small size and high-quality sound pickup characteristics, lavs are durable and relatively immune to physical shock. Because they are so small and lightweight, some production people unfortunately take much less care when handling a lav than with other, larger mics. If you happen to drop a lavalier, or any other mic, check it immediately to see if it is still operational.

When to use lavalier microphones The following are some typical productions that use lavs as the primary microphone.

News The lavalier is the most efficacious sound pickup device for all types of indoor news shows and interviews. You can also use it outdoors with a windscreen attached for ENG/EFP.

Interviews So long as the interview takes place in one location, the wearing of lavalier mics by the interviewer and each guest ensures good, consistent voice pickup.

Panel shows Rather than use desk mics, which are apt to pick up the unavoidable banging on the table, you can achieve good audio with individual lavaliers. But note that each panel member needs his or her own lavalier mic. With several people on a panel, however, using lavaliers gets unwieldy. Desk mics are, therefore, the better choice.

Instructional shows In shows with a principal performer or television teacher, the lavalier is ideal. The sound pickup is the same whether the instructor speaks to the class or turns to the whiteboard.

Dramas Some multicamera studio productions, such as soap operas, use wireless lavaliers for audio pickup. In such productions the lavs are hidden from camera view. If properly attached to the talent's clothing so that the voices do not sound muffled, lavalier mics seem the ideal solution to a traditionally difficult sound pickup problem. Once the levels are set, the audio engineer needs to do very little to keep the voices balanced. As mentioned earlier, the lighting director (LD) can design the lighting without worrying about boom or camera shadows. As you will read later, the constant presence of the fixed distance between mic and mouth, however, can be a problem.

Music The lavalier has been successfully used for the pickup of certain instruments, such as a string bass, with the mic taped below the fingerboard. In the realm of music, there is always room for experimentation; do not be too limited by convention. If the lavalier sounds as good as or better than a larger, more expensive mic, stick with the lavalier.

ENG/EFP The lav is often used for field reports, in which case you need to attach the little windscreen. Wireless lavs are used when the field reporter needs a great deal of mobility. For example, if you talk with a farmer about the drought while walking with him in the parched field, two wireless lavs will solve the audio problem. Wireless lavs can also save you many headaches when picking up the school principal's comments while conducting a tour through the newly completed computer lab.

Disadvantages of lavalier mics Despite their versatility and high-quality pickup, there are some disadvantages to the lavalier:

- The lavalier can be used for only one sound source at a time—that of the wearer. Even for a simple interview, each participant must wear his or her own mic.
- The wearer cannot move the mic any closer to the mouth; consequently, if there is extraneous noise, it is easily picked up by an omnidirectional mic, although a unidirectional lavalier will usually make such unwanted ambient sounds less of a problem.
- The lav is prone to distort very loud input levels. This is especially important when you use a lav for sound inputs other than speech.
- Although the lav allows considerable mobility, a wired lavalier can limit the performer's action radius.
- Because it is attached to clothing, the lav tends to pick up occasional rubbing noises, especially if the performer moves around a great deal.
- The voice sounds muffled when the mic is concealed underneath clothing.
- If the performer's clothes generate static electricity, the mic may pick up the discharge as loud, sharp pops.
- If two lavaliers are at a certain distance from each other, they may cancel out some frequencies and make the voices sound strangely thin (see figure 9.26).
- The main problem with using lavs for drama is not operational but aesthetic. Because the lavalier mic is always at the same distance from its sound source, long shots sound exactly the same as close-ups. The unchanging presence does not contribute to a credible sound perspective (wherein close-ups sound closer and long shots sound farther away). This is why most productions of television dramas use boom mics rather than lavaliers (see chapter 10).

How to use lavalier microphones Lavaliers are easy to use, but there are some points you need to consider:

- Be sure to put it on. You would not be the first performer to be discovered sitting on, rather than wearing, the microphone by airtime when getting cued.
- To put on the mic, bring it up underneath the blouse or jacket and attach it on the outside. Clip it firmly to the clothing so that it does not rub against anything. Do not wear jewelry in proximity to the mic. If you get rubbing noises, put a piece of foam rubber between the mic and the clothing.
- Tape the cable to part of the clothing so that it cannot pull the microphone sideways.
- Loop the cable or even make a loose knot just below the clip to block unwanted pops and rubbing noises.
- If you hear electrostatic pops, treat the clothes with antistatic laundry spray, available at supermarkets.
- If you must conceal the mic, do not bury it under layers of clothing; keep it as close to the surface as possible.
- If you use the dual-redundancy microphone system (which uses two identical mics for the sound pickup in case one fails), fasten both mics securely to a clip designed to hold two lavaliers so that they do not touch each other. Don't use two lavs without a clip, and don't activate both at the same time.
- Avoid hitting the microphone with any object you may be demonstrating on-camera.
- If the lav is a wireless and/or condenser mic, check that the battery is in good condition and correctly installed.
- When using a wireless lav, double-check that the transmitter is turned on (there are normally two switches—one for power and one for the mic) and that it is turned off when leaving the set.
- If your lavalier was used as a wireless mic, don't walk off with the mic still clipped to your clothing. Turn off the transmitter, take off the mic, and remove the cable from under the clothing before leaving the set. Put the mic down gently.
- When using a lavalier outdoors, attach the windscreen. You can also make a windscreen by taping a piece of acoustic foam or cheesecloth over the mic. Experienced EFP people claim that by wrapping the mic in cheesecloth and covering it with the tip of a child's woolen glove, wind noise is virtually eliminated.

Hand Microphones

As the name implies, the hand microphone is handled by the performer. It is used in all production situations in which it is most practical, if not imperative, that the performer exercise some control over the sound pickup. Hand mics are used extensively in ENG, where the reporter often works in the midst of noise and commotion. In the studio or on-stage, hand mics are used by singers and by performers who do audience participation shows. With the hand mic, the performer can approach and talk at random to anyone in the audience.

For singers the hand mic is part of the act. It gives them freedom of movement, especially if it is wireless; they can work the mic during a song, switching it from one hand to the other or holding it close to the mouth to increase the intimacy during tender passages or farther away during louder, more external ones. You will find that most singers have a preference for the type of mic they use, so you don't have to worry about just which mic to give them.

The wide variety of uses makes heavy demands on the performance characteristics of a hand mic. Because it is handled so much, it must be rugged and capable of withstanding physical shock. And because it is often used extremely close to the sound source, it must be insensitive to plosive breath pops and input overload distortion (see section 9.2). When used outdoors on remote locations, it must withstand rain, snow, humidity, heat, and extreme temperature changes and yet be sensitive enough to pick up the full range and the subtle tone qualities of a singer's voice. Finally, it must be small enough to be handled comfortably by the performer.

Of course, no single mic can fulfill all these requirements equally, which is why some hand mics are built for outdoor use whereas others work best in the controlled studio environment. Normally, you should use dynamic mics for outdoor productions. Their built-in pop filter and sometimes even built-in windscreen produce acceptable audio even in bad weather. **SEE 9.8** Condenser or ribbon mics do not fare as well outdoors but are excellent for more demanding sound pickup, such as of singers.

The major disadvantage of the hand mic is what we just listed as one of its advantages: the sound control by the performer. If a performer is inexperienced in using a hand mic, he or she might produce more pops and bangs than intelligible sounds or may, much to the dismay of

9.8 DYNAMIC HAND MICROPHONE FOR OUTDOOR USE
The hand mic is rugged, has a built-in windscreen, and is insulated to prevent rubbing sounds from the talent's hands.

9.9 POSITION OF DIRECTIONAL HAND MIC DURING SONG
For optimal sound pickup, the singer holds the microphone close to her mouth, at approximately a 45-degree angle.

9.8 PHOTO BY EDWARD AIONA
9.9 PHOTO BY EDWARD AIONA

the camera operator, cover the mouth or part of the face with the mic.

Another disadvantage of most hand mics is that their cables can restrict movement, especially in ENG, when a field reporter is tethered to the camcorder. Although wireless hand mics are used successfully in the studio, try to stay away from them when working outdoors. As more and more devices become wireless, there is the ever-increasing problem of signal interference. A cable is still the most reliable connection between the mic and the audio mixer or camcorder.

How to use hand microphones Working with the hand mic requires dexterity and foresight. Here are some helpful hints:

- Although the hand mic is fairly rugged, treat it gently. If you need both hands during a performance, do not just drop the mic; put it down gently or wedge it under your arm. If you want to impress on someone the sensitivity of a microphone, especially that of the hand mic, turn up the volume level and feed the clanks and bangs back into the studio for all to hear.

- Before the telecast check your action radius to see if the mic cable is long enough for your actions and laid out for maximum mic mobility. The action radius is especially important in ENG, where you may be closely tied to the camcorder. If you have to move about a great deal, use a wireless hand mic or lavalier.

- Always test the microphone before the show or news report by speaking into it or lightly scratching the pop filter or windscreen. Do not blow into it. Have the audio engineer or camcorder operator confirm that the mic is working properly.

- When using an omnidirectional mic, speak across rather than into it. With a directional hand mic, hold it close to your mouth at approximately a 45-degree angle to achieve optimal sound pickup. Unlike the reporter, who speaks *across* the omnidirectional hand mic, the singer sings *into* the directional mic. SEE 9.9

- If the mic cable gets tangled, do not yank on it. Stop and try to get the attention of the floor manager.

- When walking a considerable distance, do not pull the cable with the mic. Tug the cable gently with one hand while holding the microphone with the other.

- When in the field, always test the microphone before the show or news report by having the camcorder operator record some of your opening remarks and then play them back for an audio check. Insist on a mic check, especially if the crew tells you not to worry because they've "done it a thousand times before"!

- When doing a stand-up news report in the field under normal conditions (no excessively loud environment, no strong wind), hold the microphone at chest level. Speak toward the camera, across the microphone. SEE 9.10 If the background noise is high, raise the mic closer to your mouth while still speaking across it. SEE 9.11

- When interviewing someone, hold the microphone to your mouth whenever you speak and to the guest's whenever he or she answers. Unfortunately, this obvious procedure is sometimes reversed by nervous performers.

9.10 HAND MIC POSITION: CHEST
When used in a fairly quiet environment, the hand mic should be held chest high, parallel to the body.

9.11 HAND MIC POSITION: MOUTH
In a noisy environment, the hand mic must be held closer to the mouth. Note that the talent is still speaking across the mic rather than into it.

9.10 PHOTO BY HERBERT ZETTL
9.11 PHOTO BY HERBERT ZETTL

- Do not remain standing when interviewing a child. Crouch down so that you are at the child's eye level; you can then keep the mic close to the child in a natural way. You become a psychological equal to the child and also help the camera operator frame an acceptable picture. SEE 9.12

- Always coil the mic cables immediately after use to protect them and have them ready for the next job.

9.12 USE OF HAND MIC WITH CHILD
When interviewing a child, crouch down to the child's eye level. The child is more at ease, and the camera operator is able to frame a better shot.

9.12 PHOTO BY HERBERT ZETTL

Boom Microphones

When a production, such as of a dramatic scene, requires that you keep the microphone out of camera range, you need a mic that can pick up sound over a fairly great distance while making it seem to come from close up and which keeps out most of the extraneous noises surrounding the scene. The ***shotgun microphone*** fills that bill. It is highly directional (supercardioid or hypercardioid) and has a far reach with little loss of presence. SEE 9.13

The directionality of the shotgun mic is achieved by allowing extraneous sounds to enter the mic through the many slots on the shotgun barrel. Once trapped, they get canceled. Because it is usually suspended from some kind of boom, or is handheld with your arms acting as a boom, we call it a boom microphone.

How to use shotgun microphones The external microphones you see attached to camcorders are small shotgun mics. They operate like a lens except that they capture sounds instead of light. Most commonly, shotgun mics are handheld on a mount or suspended from a fishpole boom. Both methods work fairly well for short scenes, where the mic is to be kept out of camera range. The advantages of holding it are threefold: the microphone is extremely

9.13 SHOTGUN MIC
The shotgun mic has a highly directional (super- or hypercardioid) pickup pattern and a far reach, permitting the pickup of sounds that are relatively far away.

flexible—you can carry it into the scene and aim it in any direction without any extraneous equipment; by holding the shotgun, or by working the fishpole, you take up very little production space; and you can easily work around the existing lighting setup to keep the mic shadows out of camera range.

There are, however, disadvantages: you can cover only relatively short scenes without getting tired; you have to be relatively close to the scene to get good sound pickup, which is often difficult, especially if the set is crowded; if the scene is shot with multiple cameras (as in a studio production), you are often in danger of getting in the wide-shot camera view; and the mic is apt to pick up some handling noises, even if you carry it by the shock mount (a suspension device that prevents transmitting handling noises to the mic).

Because there is a great likelihood that you will be asked to handle a shotgun mic, here are some tips:

- Always wear good-quality earphones or earbuds so that you can hear what the mic is actually picking up. Listen not only to the sound quality of the dialogue but also for unwanted noise. If you hear sounds that are not supposed to be there, tell the director about the interference immediately after the take (from start to stop of the show segment being video recorded). You can plug the earphones into the camera audio jack that usually takes a mini plug.
- Always carry the shotgun mic by the shock mount. Do not carry it directly or you'll end up with more handling noises than actors' dialogue. SEE 9.14
- Do not cover the ports (openings) at the sides of the shotgun with anything but the windscreen. These ports must be able to receive sounds to keep the pickup pattern directional. Holding the mic by the shock mount minimizes the danger of covering the ports.
- Watch that you do not hit anything with the mic and that you do not drop it. A short cable can yank it right out of your hands.
- Aim the mic as much as possible toward whoever is speaking, especially if the mic is fairly close to the sound source.
- Watch for unwanted mic shadows.

9.14 PHOTO BY EDWARD AIONA

9.14 HANDHELD SHOTGUN MIC
Always hold the shotgun mic by its shock mount. When outdoors a windscreen is mandatory. This mic has an additional wind jammer attached.

How to use fishpole microphones A ***fishpole*** is an extendable metal pole that lets you mount a shotgun mic. It is used mostly outdoors for ENG/EFP but can, of course, be applied to brief scenes in the studio in place of the big perambulator boom. You will find that a short fishpole is relatively easy to handle, whereas working a long or fully extended fishpole can be quite demanding and tiring, especially during long, uninterrupted takes.

Short shotgun mics, called pencil mics by audio people, are frequently preferred over the more directional long shotguns when picking up sounds indoors. The obvious advantage of a pencil mic is that it is smaller and therefore lighter; it is also not quite so directional as the long shotgun, so you don't always need to point it directly at the person who is speaking. If, for example, you cover a conversation between two people, you can often position the pencil mic between them for a balanced pickup. To make your job even easier, most pencil mics let you exchange the unidirectional head with one that has a slightly wider cardioid pattern.

When using the fishpole, many of the foregoing points apply. Here are some additional ones:

- Check that the mic is properly shock-mounted so that it does not touch the pole or the mic cable. The normal shock mount consists of an outside ring within which the mic barrel is suspended with rubber bands.
- Fasten the mic cable to the pole. Some fishpoles let you feed the cable through the pole, like through a conduit.
- Hold the fishpole from either above or below the sound source. SEE 9.15 AND 9.16
- If you are recording two people talking to each other, point the mic at whoever is speaking.

9.15 FROM-ABOVE MIC POSITION
The short fishpole is usually held as high as possible and dipped into the scene from above.

9.16 FROM-BELOW MIC POSITION
The fishpole can also be held low, with the mic aimed at the sound source from below.

9.15 PHOTO BY EDWARD AIONA
9.16 PHOTO BY EDWARD AIONA

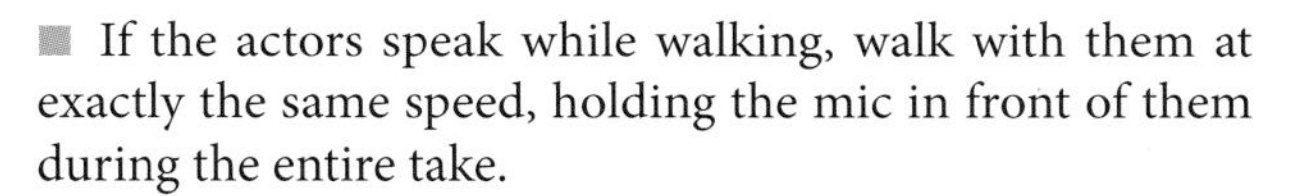

- If the actors speak while walking, walk with them at exactly the same speed, holding the mic in front of them during the entire take.
- Watch for obstacles that may block your way, such as cables, lights, cameras, trees, or pieces of scenery. Because you usually walk backward while watching the actors, rehearse your route a few times.
- Before each take check that you have enough mic cable for the entire walk.
- If you have a long fishpole, anchor it in your belt and lower it into the scene as though you were "fishing" for the appropriate sound. **SEE 9.17**

9.17 HANDLING THE LONG FISHPOLE BOOM
The long fishpole can be anchored in a belt and raised and lowered similar to an actual fishing pole.

9.17 PHOTO BY EDWARD AIONA

Big perambulator boom When working with large, multicamera studio productions, such as situation comedies and soap operas, you will find that despite the presence of lavalier mics, the big perambulator boom is very much alive and well. In the controlled environment of the studio, the big boom is still one of the most effective ways of getting a high-quality mic close to the talent while keeping it out of camera view. **SEE 9.18**

There are several reasons why the big boom has not achieved great popularity in routine studio productions: it takes up a large amount of space, it makes lighting difficult, and it is not easy to operate. That said, we haven't come up with anything better for picking up actors' dialogue during long takes. You can extend or retract the boom, simultaneously pan it horizontally, move it up and down vertically, and rotate and tilt the mic toward the sound source. During all these operations, the boom assembly can be moved when the fully extended boom cannot reach the sound source.

How to use boom microphones However and wherever you move the boom, do it smoothly. It is better to be a little off while following the actors than to lose them entirely because you can't stop your wild boom swing in time. Here are some more operational tips:

9.18 BIG PERAMBULATOR BOOM
The big boom can extend to a 20-foot reach, pan 360 degrees, and tilt up and down. The microphone itself can be rotated by about 300 degrees—almost a full circle.

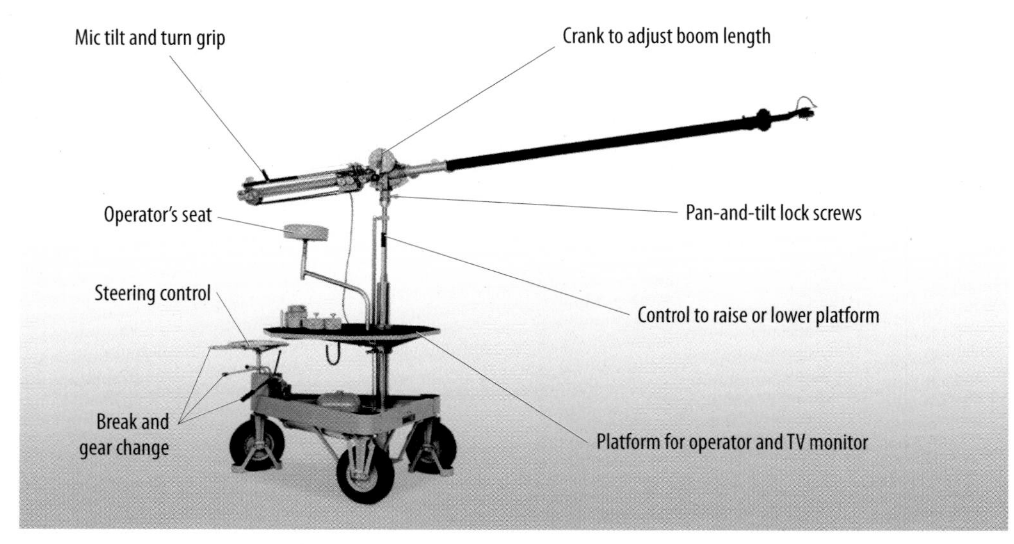

9.18 PHOTO BY HERBERT ZETTL

- Try to keep the mic in front of the sound source and as low as possible without getting it in the picture.

- Do not ride the mic directly above the talent's head—the actor speaks from the mouth, not the top of the head.

- Watch the studio line monitor or the monitor on your dolly (which shows the picture that goes on the air or is video recorded). Try to ascertain during rehearsal how far you can dip the mic toward the sound source without getting it or the boom in the picture. The closer the mic, the better the sound. In boom mic operation, you will rarely get close enough to violate the minimum distance required of shotgun mics to avoid the proximity effect (breath pops and boomy bass). The optimum distance for boom mics is when the talent can almost touch the mic by reaching up at about a 45-degree angle.

- If the boom gets in the picture, it is better to pull it back than to raise it. By retracting the boom, you pull the mic out of the camera's view and at the same time keep it in front of, rather than above, the sound source.

- Watch for shadows. Even the best LD cannot avoid shadows but can only redirect them. If the major boom positions are known before the show, work with the LD to light around them. You may sometimes have to sacrifice audio quality to avoid boom shadows.

- If you discover a boom shadow when the camera is already on the air, do not suddenly move the mic—everyone will see the shadow travel across the screen. Try to sneak it out of the picture very slowly or, better, just keep the mic and the shadow as steady as possible until a relief shot permits you to move into a more advantageous position.

- Listen for good audio balance. With a highly directional shotgun mic, you normally must rotate it toward whoever is talking. In fully scripted shows, the audio board operator in the booth may follow the scripted dialogue and signal you (the boom operator) whenever the mic needs to be rotated from one actor to the other. ZVL3 AUDIO→ Microphones→ mic types | placement

Headset Microphones

The ***headset microphone*** consists of a small but good-quality omni- or unidirectional mic attached to earphones. One of the earphones carries the program sound (whatever sounds the headset mic picks up or is fed from the station), and the other carries the I.F.B. (interruptible foldback or feedback) cues and instructions of the director or producer. Headset mics are used in certain EFP situations, such as sports reporting, or in ENG from a helicopter or convention floor. The headset mic isolates you sufficiently from the outside world so that you can concentrate on your specific reporting job in the midst of much noise and commotion while at the same time keeping your hands free to track players' statistics on a laptop or to buttonhole someone for an interview. SEE 9.19

Wireless Microphones

In production situations in which complete and unrestricted mobility of the sound source is required, ***wireless microphones*** are used. If, for example, you are recording a singer who is also doing some dance moves, or if you

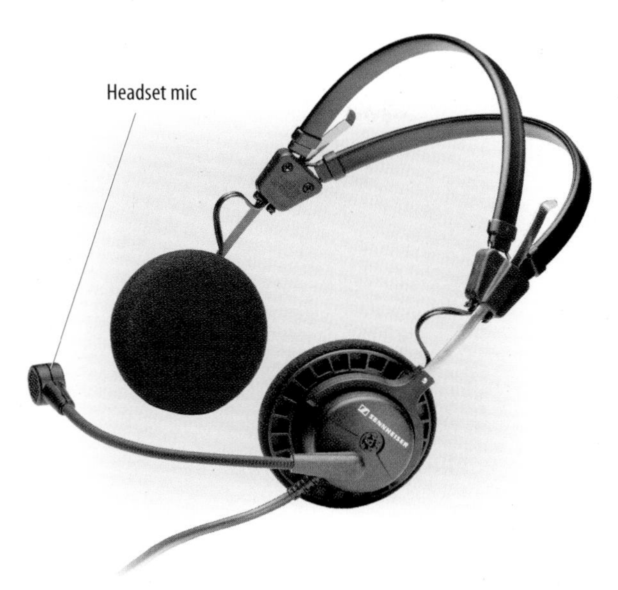

9.19 HEADSET MICROPHONE
The headset mic is similar to an ordinary telephone headset except that it has bigger, padded earphones and a higher-quality microphone.

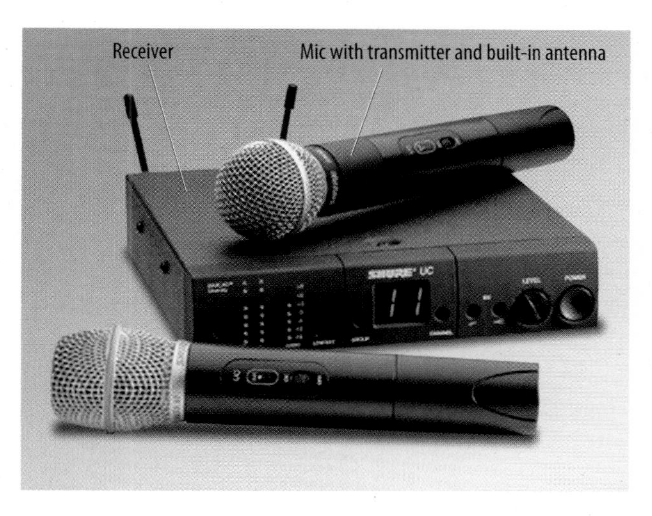

9.20 WIRELESS HAND MIC AND RECEIVER
The wireless hand mic normally has the transmitter built into the housing. The antenna either is built into the mic or sticks out at the bottom. The receiver, which is tuned to the frequency of the hand mic's transmitter, picks up the signal and sends it via mic cable to the audio console or camcorder.

are asked to pick up the clatter of a racer's skis on an icy downhill course, the wireless mic is the obvious choice. Wireless mics are also used extensively for newscasts, for EFP, and occasionally for multicamera studio productions of dramatic shows. Wireless mics actually broadcast their signals. They are therefore also called RF (radio frequency) mics or radio mics. Most wireless microphones are used as either hand or lavalier mics.

In wireless hand mics, the battery-powered transmitter is built into the microphone itself. Some models have a short antenna protruding from the bottom of the mic, but in most the antenna is incorporated into the microphone housing or cable. SEE 9.20

The wireless lavalier mic is connected to a battery-powered transmitter that is either worn in the hip pocket or taped to the body. The antenna is usually tucked into the pocket or strung inside the clothing. SEE 9.21

The other important element of the wireless microphone system is the receiver (see figure 9.20). The receiver tunes in to the frequency of the wireless transmitter and can receive the signal from as far away as 1,000 feet (approximately 330 meters) under favorable conditions. When conditions are more adverse, the range may shrink to about 100 feet (about 33 meters). To ensure optimal signal reception, you can set up several receiving stations in the studio as well as in the field. When the signal gets too weak for one of the receivers, the other or others will take over. This is called diversity reception.

9.21 WIRELESS LAVALIER MIC AND TRANSMITTER
The wireless lavalier mic has a separate transmitter that is worn by the talent. The receiver picks up the mic's signal and routes it via mic cable to the audio mixer, console, or camcorder.

How to use wireless microphones The wireless mic works best in the controlled environment of a studio or stage, where you can determine the precise range of the performer's movements and find the optimal position for the receiver(s). Most singers prefer working with the wireless hand mic because it affords them unrestricted movement. It is also useful in audience participation shows, where the performer walks into the crowd for brief, unscripted interviews. The wireless lavalier mic has been used successfully for musicals and dramatic shows and, of course, in many ENG/EFP situations.

When using any wireless mic, you should watch for these problems:

- If the transmitter is taped to the body, the performer's perspiration can reduce signal strength, as does, of course, the increasing distance from transmitter to receiver.
- Large metal objects, high-voltage lines and transformers, X-ray machines, microwave transmissions, and cell phones—all can interfere with the proper reception of the wireless mic signal.
- Although most wireless equipment offers several frequency channels, there is still some danger of picking up extraneous signals, especially if the receiver operates in the proximity of someone else's wireless signals or other strong radio signals. Interference is evident by pops, thumps, signal dropouts, and even the pickup of police band transmissions. As pointed out before, the increase of wireless communication also increases the likelihood of signal interference.

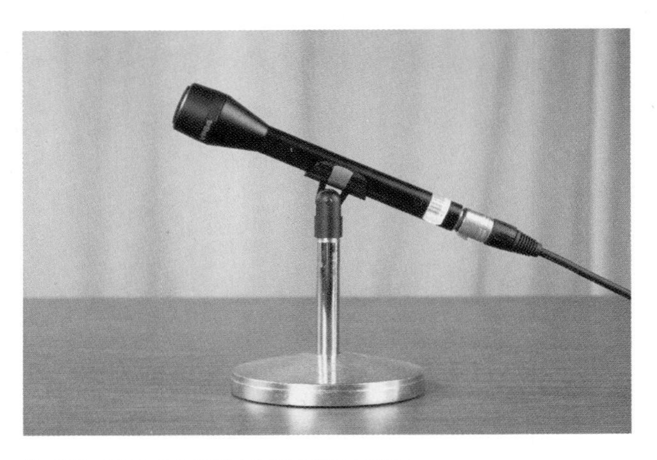

9.22 MICROPHONE ON DESK STAND
In television production desk mics are usually hand mics clipped to a desk stand.

9.22 PHOTO BY EDWARD AIONA

Desk Microphones

As the name implies, desk microphones are usually put on tables or desks. These stationary mics are widely used in panel shows, public hearings, press conferences, and other programs where the performer is speaking from behind a desk, table, or lectern. These mics are used for voice pickup only. Because the performer is usually doing something—shuffling papers, putting things on the desk, accidentally bumping the desk with feet or knees—desk microphones must be rugged and able to withstand physical shock.

Dynamic, omnidirectional mics are generally used as desk mics. If a high separation of sound sources is desired, however, unidirectional mics are an option. When you use a desk mic for a speaker at a lectern, use a unidirectional rather than an omnidirectional mic. The unidirectional mic will make it easier for you to avoid feedback through the speakers in the same room. Once properly positioned the unidirectional mic is less apt to cause feedback. Generally, most hand mics double as desk mics—all you do is place them in a desk stand and position them for optimal sound pickup. **SEE 9.22**

Boundary microphone One type of desk mic is the boundary microphone or, as it is commonly called, the pressure zone microphone (PZM). These mics don't look like ordinary microphones and operate on a different principle. **SEE 9.23**

The boundary microphone is mounted or positioned close to a reflecting surface, such as a table or a plastic plate accessory. When placed into this sound "pressure zone," the mic receives both direct and reflected sounds at the same time. Under optimal conditions the boundary microphone produces a clearer sound than do ordinary mics. Its chief advantage is that it can be used for the simultaneous voice pickup of several people with equal fidelity. Boundary mics have a wide, hemispheric pickup pattern and are therefore well suited for large group discussions and audience

9.23 BOUNDARY MICROPHONE
This mic must be mounted or put on a reflecting surface to build up the "pressure zone" at which all sound waves reach the mic at the same time.

9.23 PHOTO COURTESY AKG ACOUSTICS

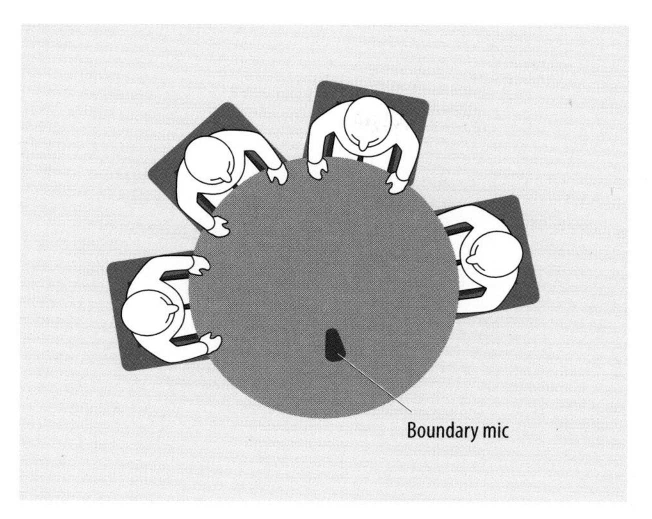

9.24 BOUNDARY MIC USED FOR MULTIPLE-VOICE PICKUP
With the boundary mic on a table, the sound pickup is equal for all people sitting around it.

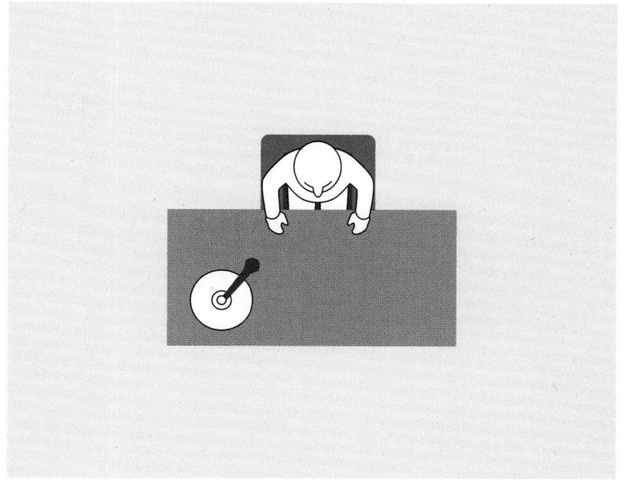

9.25 DESK MIC PLACEMENT FOR SINGLE PERFORMER
The desk mic should be placed to the side of the talent and aimed at the talent's collarbone so that he or she speaks across rather than into it. In this way the mic is not obstructing the shot of the talent.

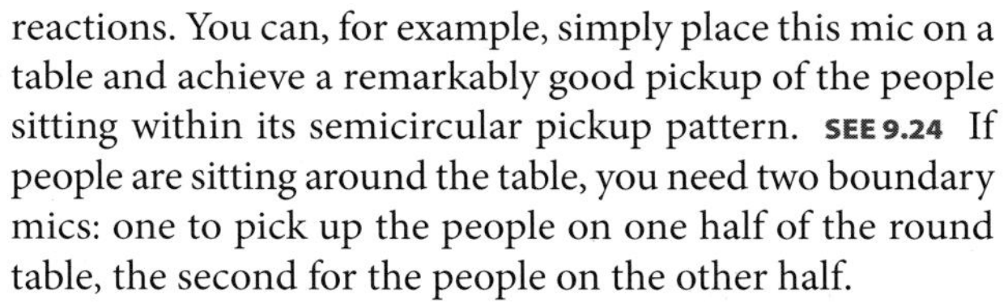

reactions. You can, for example, simply place this mic on a table and achieve a remarkably good pickup of the people sitting within its semicircular pickup pattern. **SEE 9.24** If people are sitting around the table, you need two boundary mics: one to pick up the people on one half of the round table, the second for the people on the other half.

Unfortunately, when used as a table mic, the boundary mic also picks up paper rustling, finger tapping, and the thumps of people knocking against the table; pads for the mic minimize or virtually eliminate such problems. Note that the boundary mic can also be used for miking musical instruments, such as a piano.

How to use desk microphones Desk mics, like peanuts, seem to be irresistible—not that performers want to eat them, but when sitting or standing behind a desk mic they feel compelled to grab it and pull it toward them no matter how carefully you might have positioned it. Polite or not-so-polite requests not to touch the mic seem futile. Sooner or later the talent will move the mic. To counter this compulsion, consider taping the mic stand to the table, or at least tape the microphone cable securely and unobtrusively so that the mic can be moved only a short distance.

- As with the hand mic, no attempt is made to conceal the desk mic from the viewer. Nevertheless, when placing it on a desktop or lectern, consider the camera picture as well as optimal sound pickup. Performers certainly appreciate it if the camera shows more of them than the microphone. If the camera shoots from straight on, place the mic somewhat to the side of the performer and point it at his or her collarbone rather than mouth, giving a reasonably good sound pickup while allowing the camera a clear shot of the performer's face. **SEE 9.25**

- When integrating the mic unobtrusively in the picture, do not forget about concealing the mic cable as best as you can. Even if the director assures you that the mic cable on the floor will never be seen, don't bet on it.

- When using two desk mics for the same speaker as a dual-redundancy precaution, use identical mics and place them as close together as possible. As noted, *dual-redundancy* is the rather clumsy term for using two mics for a single sound source so that you can switch from one to the other in case one fails. Do not activate them at the same time unless you are feeding separate audio channels. If both mics are on at the same time, you may experience multiple-microphone interference, which means that when the mics receive the same sound at slightly different times, they tend to cancel each other's frequencies. To avoid such interference, place the mics so that they are at least three times as far apart as any mic is from its sound source. **SEE 9.26**

- Although almost a lost cause, remind panel members—or anyone working with a desk mic—not to reposition

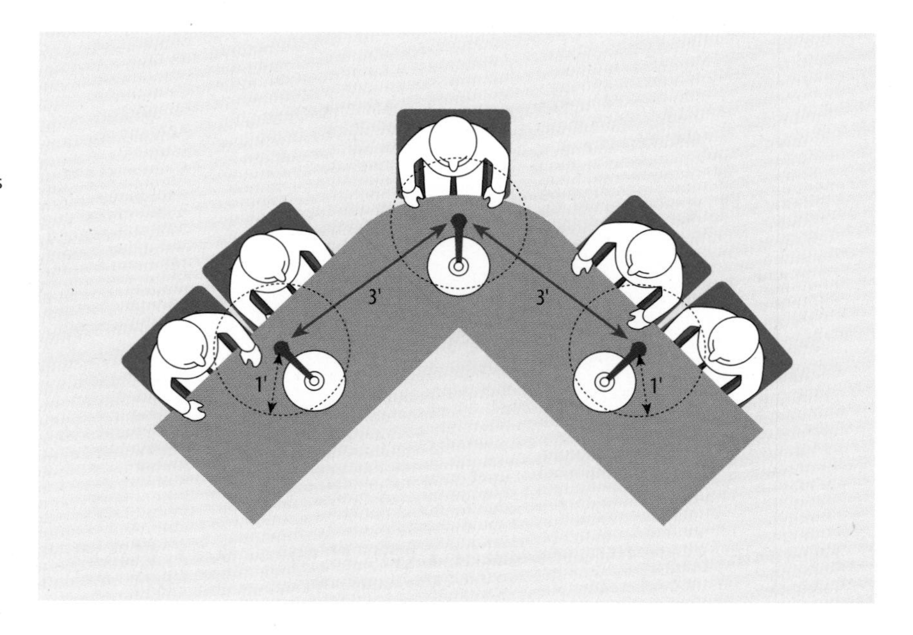

9.26 MULTIPLE-MICROPHONE SETUP
When using a multiple-microphone setup, keep the individual mics at least three times as far apart as the distance any mic is from its sound source.

the mic once it is set and to avoid banging on the table or kicking the lectern, even if the discussion gets lively. Tell participants not to lean into the mics when speaking.

Stand Microphones

Stand microphones are used whenever the sound source is fixed and the type of programming permits them to be seen. For example, there is no need to conceal the mics of a rock group; on the contrary, they are an important show element. You are certainly familiar with the great many ways rock performers handle the stand mic. Some tilt it, lift it, lean against it, hold themselves up by it, and, when the music rocks with especially high intensity, even swing it through the air like a sword (not recommended, by the way).

The quality of stand mics ranges from dynamic hand mics clipped to a stand to highly sensitive ribbon or condenser mics used exclusively for music recording sessions.

How to use stand microphones Stand mics are usually placed in front of the sound source—whether it is a singer or the speaker of an amplified electric guitar. SEE 9.27 Many singers prefer to unclip the mic from the stand for some especially energetic or intimate passages, so you need to ensure that the mic can be easily detached and then reattached to the stand. For the pickup of a singer using an acoustic guitar, you may attach two mics to a single stand: one for the singer and one a little lower for the guitar.

9.27 STAND MIC FOR SINGER
The singer stands in front of the stand mic and sings directly into it.

9.27 PHOTO COURTESY BROADCAST AND ELECTRONIC COMMUNICATION ARTS DEPARTMENT AT SAN FRANCISCO STATE UNIVERSITY

Hanging Microphones

Hanging microphones are used whenever any other concealed-microphone method (boom or fishpole) is impractical. You can hang the mics (high-quality cardioid condenser mics but also lavaliers) by their cables over any fairly stationary sound source. Most often, hanging mics are used in dramatic presentations where the action is fully blocked so that the actors are in a precise location for each delivery of lines. A favorite spot for hanging mics is the upstage door (at the back of the set), from which the actors deliver their hellos and good-byes when entering or leaving the major performance area. The boom can generally not

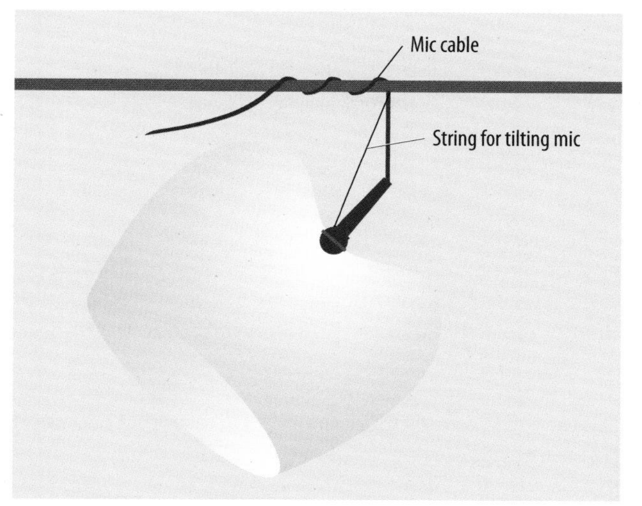

9.28 AUDIO POOL OF HANGING MICROPHONE
Hanging microphones are high-quality unidirectional mics that are normally suspended by their cables from the lighting grid. The talent must remain in the audio pool to be properly heard.

reach that far to adequately pick up voices. The actors have to take care to speak only within the "audio pool" of the hanging mic. Similar to the spotlight pool, where the actors are visible only so long as they move within the limited circle of light, the audio pool allows actors to be heard only when they are within the limited range of the mic. **SEE 9.28**

The sound quality from hanging mics is not always the best. The sound source is often at a distance from the mic, and if the actor is not precisely within the audio pool, his or her voice is off-mic. In the case of the upstage door, such quality loss is actually an asset because it underscores the physical and psychological distance of the departing person. Unfortunately, hanging mics have the annoying tendency to pick up the shuffling of feet and the rumbling of moving camera pedestals almost as well as the voices. A further disadvantage is that when positioned close to the studio lights, the hanging mic might pick up and amplify their hum. Hanging mics are nevertheless popular in dramas, studio productions, and audience participation shows. They are easy to set up and take down and, when in the right positions, produce acceptable sound.

How to use hanging microphones You may find that a single suspended boundary mic will meet the audio requirements better than several regular hanging mics. Mount the boundary mic on a sound-reflecting board (such as a 3-by-4-foot piece of plastic or plywood), suspend it above and in front of the general sound-generating area (such as an audience area), and angle the reflecting board for optimal pickup. **SEE 9.29** Regardless of whether the sound source is near the mic or farther away, the sounds still have good presence. This positive aspect turns negative in dramatic productions, where sound perspective is an important factor. This is one of the reasons why in complex productions the boom is still preferred over the boundary mic.

When suspending a microphone from the lighting grid, be careful not to place the mic next to a hot lighting

9.29 BOUNDARY MIC USED AS HANGING MICROPHONE
When using a boundary mic as a hanging microphone, mount it on an additional sound-reflecting board and angle it toward the sound source for optimal pickup. The shotgun mics are for the host and guests' audio pickup.

instrument. It will inevitably pick up its hum. Separate the mic cables from the studio lights or the AC cables to minimize electronic interference. If that is not possible, cross the mic and power cables at right angles rather than having them run parallel.

Hidden Microphones

You may sometimes find that you need to hide a small lavalier mic in a bouquet of flowers, behind a centerpiece, or in a car to pick up a conversation during studio productions or in EFP where microphones should be out of camera range. SEE 9.30 Realize that it is time-consuming to place a hidden mic so that it yields a satisfactory pickup. Often you get a marvelous pickup of various noises caused by people hitting the table or moving their chairs but only a poor pickup of their conversation.

Again, the boundary mic can serve as an efficient "hidden" mic. Especially because it looks nothing like an ordinary mic, you may get away with not hiding it at all; simply place it on a table among other eclectic objects.

How to use hidden microphones Hiding mics presents unexpected problems. These tips may minimize or eliminate some of them:

- Do not try to conceal the mic completely unless there is an extreme close-up of the object to which it is attached.
- Realize that you must hide not only the microphone but also the cable. If you use a wireless setup, you must hide the transmitter as well.
- Secure the mic and the cable with tape so that they do not come loose. The setup must withstand the rigors of the rehearsals and the video-recording sessions.
- Do not hide a mic in enclosed spaces such as empty drawers or boxes. The highly reflecting enclosure will act as a reverberation chamber and make the voices sound as though the actors themselves were trapped in the drawer.

Long-distance Microphones

It is often the sounds more than the pictures that carry and communicate the energy of an event. The simplest way to pick up the sounds of a sporting event, for example, is to place normal shotgun mics at strategic positions and aim them at the main action. The sounds of the spectators are picked up by additional omnidirectional mics. Coverage of a single tennis match may involve six or more microphones to pick up the sounds of the players, the judges, the ball, and the crowd. Place a fairly dense windscreen on every mic to eliminate wind noise as much as possible.[1]

1. You will find highly useful suggestions for how to mic a variety of sports for ambient sound in Stanley R. Alten, *Audio in Media*, 9th ed. (Boston: Wadsworth, 2011), pp. 257–76.

9.30 LAVALIER AS HIDDEN MIC
This "hidden" lavalier microphone is attached to the rear-view mirror to pick up the conversation inside the car. Note that the mic is not covered, to ensure optimal sound pickup.

9.31 PARABOLIC REFLECTOR MICROPHONE
The parabolic reflector mic is used primarily for sound pickup over long distances, such as crowd noises in a stadium.

9.30 PHOTO BY HERBERT ZETTL
9.31 PHOTO BY HERBERT ZETTL

An old-fashioned but successful means of picking up distant sounds is the parabolic reflector microphone, which consists of a parabolic dish (similar to a small satellite dish) that has an omnidirectional mic facing inward at its focal point. All incoming sounds are reflected toward and concentrated at the mic. **SEE 9.31** A popular use of the parabolic mic is to pick up the sounds of the bands during a parade, the collisions of football players, and the enthusiastic chanting of a group of home-team fans. Because the parabolic reflector directs the higher sound frequencies to the mic better than the lower ones, the sounds take on a slight telephonic tone. We tend to ignore this impaired sound quality, however, when the mic is used primarily for ambient (environmental) sounds that communicate the feel of an event (such as a football game) rather than precise information.

MAIN POINTS

- Audio is the sound portion of a television show. It transmits information (such as a news story), helps establish the specific time and locale of the action, contributes to the mood, and provides continuity for the picture portions.
- The three major types of microphones are dynamic, condenser, and ribbon. Each type has a different sound-generating element that converts sound waves into electric energy—the audio signal.
- Some microphones can hear sounds equally well from all directions (omnidirectional); others hear better from a specific direction (unidirectional or cardioid).
- Microphones are classified according to their operation and are either mobile or stationary. The mobile types include lavalier, hand, boom, headset, and wireless mics. The stationary types include desk, stand, hanging, hidden, and long-distance mics.
- The lavalier microphone, or lav for short, is most common in normal television operations. It is usually clipped to clothing. Although it is extremely small, it provides high-quality sound reproduction.
- Hand microphones are used when the performer needs to exercise some control over the sound pickup.
- When the microphone must be kept out of camera range, it is usually handheld or mounted on and operated from a fishpole or microphone boom. All boom mics are highly directional.
- The headset mic is used when the talent needs both hands free to take notes or work with scripts or a keyboard. Headset microphones are especially practical for sportscasting and for ENG (electronic news gathering) from a helicopter or convention floor.
- When unrestricted mobility of the sound source is required, a wireless, or RF (radio frequency), microphone is used. Also called radio mics, wireless mics need a transmitter and a receiver.
- Desk microphones are simply hand mics clipped to a desk stand. They are often used for panel discussions.
- Stand microphones are employed whenever the sound source is fixed and the type of programming permits the mics to be seen by the camera, such as in rock concerts.
- Hanging microphones are popular in some studio productions because the mics are kept out of camera range without using booms.
- Hidden microphones are small lavaliers concealed behind or within set dressings.
- Long-distance mics are shotgun or parabolic reflector mics that pick up sound over relatively great distances.

SECTION

9.2

How Microphones Work

Section 9.1 examined sound pickup and the electronic and operational characteristics of microphones. Although you may never see the inside of a microphone (unless you break one), you should have some idea of how they work. Knowing the differences among the sound-generating elements helps you choose the right mic for a specific sound pickup. This section also includes a list of popular mics and their primary uses, some additional mic features, examples of mic setups for music groups, and microphone use specific to ENG/EFP.

- ▶ **SOUND-GENERATING ELEMENTS**
  The diaphragm and the sound-generating element—and sound quality—of dynamic, condenser, and ribbon microphones

- ▶ **SPECIFIC MICROPHONE FEATURES**
  High and low impedance, frequency response, flat response, balanced and unbalanced mics and cables, and audio connectors

- ▶ **MIC SETUPS FOR MUSIC PICKUP**
  Possible setups for various musical events

- ▶ **MICROPHONE USE SPECIFIC TO ENG/EFP**
  Ambient sounds and line-out tie-in

SOUND-GENERATING ELEMENTS

Simply speaking, microphones convert one type of energy—sound waves—to another—electric energy. All microphones have a diaphragm, which vibrates with the sound pressures, and a sound-generating element, which transduces the physical vibrations of the diaphragm into electric energy; but the particular process each mic uses to accomplish this conversion determines its quality and use.

Dynamic Microphones

In the dynamic microphone, the diaphragm is attached to a coil—the voice coil. When someone speaks into the mic, the diaphragm vibrates with the air pressure from the sound and makes the voice coil move back and forth within a magnetic field. This action produces the audio signal. Because of this physical process, dynamic mics are sometimes called moving-coil microphones.

The diaphragm–voice coil element is physically rugged, so the mic can withstand and accurately translate high sound levels and other air blasts close to it with little or no sound distortion.

Condenser Microphones

In the condenser microphone, also called electret or capacitor microphone, the movable diaphragm constitutes one of the two plates necessary for a condenser to function; the other, called the backplate, is fixed. Because the diaphragm moves with the air vibrations against the fixed backplate, the capacitance of this condenser is continuously changed, thus creating the audio signal. The condenser mic hears better than the dynamic mic (wider frequency response) but is also more sensitive to high-intensity sound sources (prone to overload distortion).

Ribbon Microphones

In the ribbon, or velocity, microphone, a very thin metal ribbon vibrates within a magnetic field. These vibrations create the electrical audio signal. The ribbon mic hears extremely well and produces warm sounds, but its "ears" are extremely sensitive. The ribbon is so fragile that even moderate physical shocks to the mic, or sharp air blasts close to it, can damage and even destroy the instrument.

Sound Quality

High-quality microphones can hear higher and lower sounds better than the less expensive models can. Other, less definable quality factors are whether a mic produces especially warm or crisp sounds, but don't be misled by specifications or the often very strong personal preferences of singers or sound engineers. What is a good mic? It's the one that produces the sounds you want. To help you get started on your mic selection for a particular assignment, the table in figure 9.32 lists some of the more popular mics and their most common uses. **SEE 9.32**

9.32 TABLE OF MICROPHONES

MICROPHONE	ELEMENT TYPE / *PICKUP PATTERN*	CHARACTERISTICS	USE
SHOTGUN MIC — LONG			
Sennheiser MKH 70	Condenser / *Supercardioid*	Excellent reach and presence, therefore excellent distance mic. Extremely directional. Quite heavy when held on extended fishpole.	Boom, fishpole, handheld. Best for EFP and sports remotes to capture sounds over considerable distances.
SHOTGUN MICS — SHORT			
Sennheiser MKH 60	Condenser / *Supercardioid*	Good reach and wider pickup pattern than long shotguns. Less presence over long distances but requires less precise aiming at sound source. Lighter and easier to handle than long shotgun mics.	Boom, fishpole, handheld. Especially good for EFP indoor use.
Neumann KMR 81	Condenser / *Supercardioid*	Slightly less reach than the MKH 60 but has warmer sound.	Boom, fishpole, handheld. Especially good for EFP. Excellent dialogue mic.
Sony ECM 672	Condenser / *Supercardioid*	Highly focused but slightly less presence than long shotguns.	Boom, fishpole, handheld. Especially good for EFP indoor use.
HAND, DESK, AND STAND MICS			
Electro-Voice 635N/D	Dynamic / *Omnidirectional*	An improved version of the classic 635A. Has good voice pickup that seems to know how to differentiate between voice and ambience. Extremely rugged. Can tolerate rough handling and extreme outdoor conditions.	Excellent mic (and therefore standard) for all-weather ENG and EFP reporting assignments.

9.32 PHOTOS COURTESY SENNHEISER ELECTRONIC CORPORATION; GEORG NEUMANN GMBH; SONY ELECTRONICS, INC.; AND ELECTRO-VOICE

9.32 TABLE OF MICROPHONES *(continued)*

MICROPHONE	ELEMENT TYPE / *PICKUP PATTERN*	CHARACTERISTICS	USE
HAND, DESK, AND STAND MICS (continued)			
Electro-Voice RE50	Dynamic / *Omnidirectional*	Similar to the Electro-Voice 635N/D. Rugged. Internal shock mount and blast filter.	Good, reliable desk and stand mic. Good for music pickup, such as vocals, guitars, and drums.
Beyerdynamic M58	Dynamic / *Omnidirectional*	Smooth frequency response, bright sound. Rugged. Internal shock mount. Low handling noise.	Good ENG/EFP mic. Especially designed as an easy-to-use hand mic.
Shure SM57	Dynamic / *Cardioid*	Good-quality frequency response. Can stand fairly high input volume.	Good for music, vocals, electric guitars, keyboard instruments, and even drums.
Shure SM58	Dynamic / *Cardioid*	Rugged. Good for indoors and outdoors.	Standard for vocals and speech.
Shure SM81	Condenser / *Cardioid*	Wide frequency response. Also good for outdoors.	Excellent for miking acoustic instruments.
Beyerdynamic M160	Double ribbon / *Hypercardioid*	Sensitive mic with excellent frequency response. Can tolerate fairly high input volume.	Especially good for all sorts of music pickup, such as strings, brass, and piano. Also works well as a stand mic for voice pickup.
AKG D112	Dynamic / *Cardioid*	Rugged. Specially built for high-energy percussive sound.	For close miking of kick drum.

9.32 TABLE OF MICROPHONES *(continued)*

MICROPHONE	ELEMENT TYPE / *PICKUP PATTERN*	CHARACTERISTICS	USE
LAVALIER MICS			
Sony ECM 44	Condenser / *Omnidirectional*	Excellent presence. Produces close-up sound. But, because of this excellent presence, does not mix well with boom mics, which are normally farther away from the sound source.	Excellent for voice pickup in a controlled environment (studio interviews, studio news, and presentations).
Sennheiser MKE 102	Condenser / *Omnidirectional*	Mixes well with boom mics. Excellent, smooth overall sound pickup. Very sensitive to clothes noise and even rubbing of cable, however. Must be securely fastened to avoid rubbing noises.	Excellent for most lavalier uses. Works well as a concealed mic.
Sony ECM 77B	Condenser / *Omnidirectional*	Has wide frequency response. Smooth overall sound pickup. Excellent quality. Blends well with larger mics. Optimal positioning not too critical. Works with battery or phantom power. Very reliable.	Widely used in TV for news, interviews, and documentaries.
Professional Sound PSC MilliMic	Condenser / *Omnidirectional*	Extremely small yet has excellent pickup quality. Blends well with boom mics. Well shielded against electromagnetic interference.	Excellent as a concealed mic for interviews, dramas, and documentaries. Works well outdoors.

9.32 PHOTOS COURTESY SONY ELECTRONICS, INC.; SENNHEISER ELECTRONIC CORPORATION; AND PROFESSIONAL SOUND CORPORATION

SPECIFIC MICROPHONE FEATURES

When working with audio equipment, you will probably hear some terms that are not self-explanatory: high- and low-impedance mics, frequency response, flat response, and balanced and unbalanced mics and cables. Although these features are quite technical in nature, you need to know at least their operational requirements.

Impedance

Impedance is the total resistance of an electric circuit to the flow of AC current. Technically, impedance is also AC frequency-dependent. In practical terms this means that you need to match the impedance of the source with the impedance of the equipment to which the source is connected. If, for example, you plug in a high-Z (high-impedance) source, such as an electric bass guitar, into the low-Z (low impedance) mic input of an audio mixer, you will end up with a distorted sound. Fortunately, most audio equipment has a built-in device that will do the matching of impedances for you. If not, there are adapters available that will take care of impedance matching.

Frequency Response

The ability of a microphone to hear extremely high and low sounds is known as the ***frequency response***. A good microphone hears better than most humans and has a frequency range of 20 Hz (*hertz*, which measures cycles per second) to a very high 20,000 Hz. Many high-quality mics are built to hear equally well over the entire frequency range, a feature called ***flat response***. High-quality mics should therefore have a great frequency range and a relatively flat response.

Balanced and Unbalanced Mics and Cables, and Audio Connectors

All professional microphones have a balanced output that is connected by three-wire microphone cables with three-pronged connectors, called XLR connectors, to recorders and mixers. The balanced line rejects hum and other electronic interference.

Unbalanced mics and cables may work just as well so long as the cables are quite short and there is no interference. Unbalanced cables carry the audio signal on only two wires. The well-known two-wire connectors include the ¼-inch phone plug, the RCA phono plug, and the mini plug. **SEE 9.33** ZVL4 AUDIO→ Connectors→ overview

With adapters you can connect an XLR to the unbalanced connectors and vice versa. Note, however, that every adapter is a potential trouble spot. Always try to find a mic cable with the appropriate connector already attached.

MIC SETUPS FOR MUSIC PICKUP

The following suggestions of how to mike musical events should be taken with a grain of salt. It's hard to find two audio experts who agree on just how a musical event should be miked and what mics to use. Nevertheless, the suggested setups will help you get started.

The sound pickup of an instrumental group, such as a rock band, is normally accomplished with several stand mics. These are placed in front of each speaker that emits the amplified sound of a particular instrument as well as in front of unamplified sound sources, such as singers and drums. The mic to use depends on such factors as studio acoustics, the type and the combination of instruments, and the aesthetic quality of the desired sound.[2]

Generally, the rugged dynamic (omnidirectional or cardioid) mics are used for high-volume sound sources, such as drums, electric guitar speakers, and some singers, whereas condenser or ribbon mics are used for such sound sources as singers, strings, and acoustic guitars. Although

2. For a detailed discussion of music recording, see Alten, *Audio in Media*, pp. 371–418.

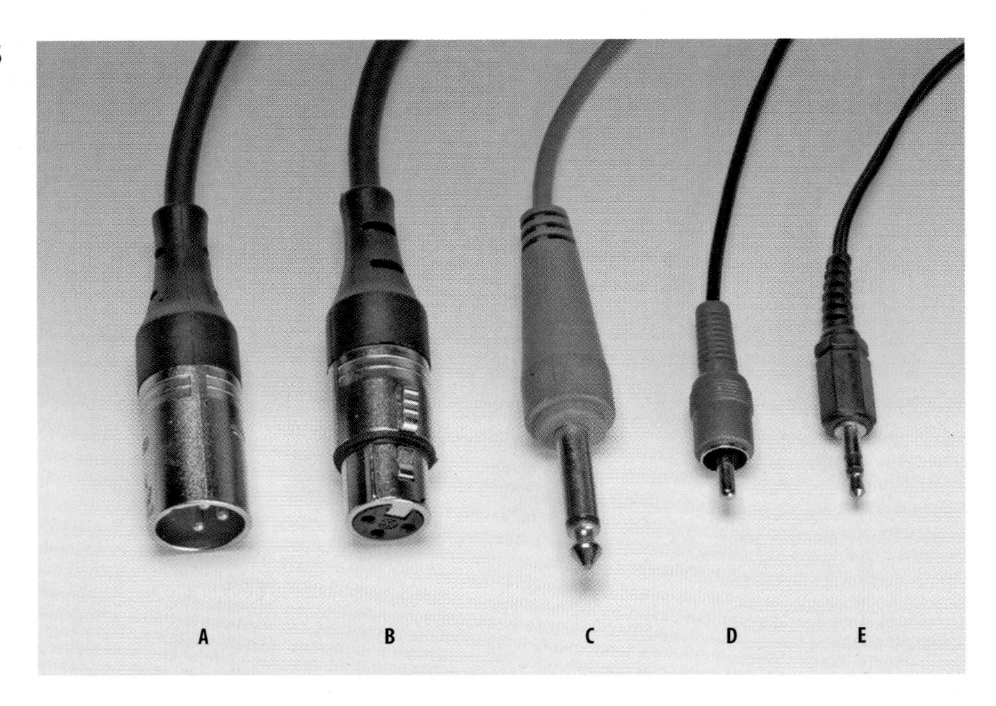

9.33 AUDIO CONNECTORS
Balanced audio cables use XLR connectors (**A** and **B**); unbalanced cables use the ¼-inch phone plug (**C**), the RCA phono plug (**D**), and the mini plug (**E**).

9.33 PHOTO BY EDWARD AIONA

many factors influence the type of microphone used and its placement, the figures in this section give some idea of how three different yet typical musical events may be miked. Again, the final criterion is not what everybody tells you but whether the playback loudspeakers give you the sounds you expected.

Microphone Setup for Singer and Acoustic Guitar

For a singer accompanying himself or herself on an acoustic guitar, one cardioid condenser or ribbon mic is sufficient in most cases. Try a Beyerdynamic M160 (ribbon) or a Shure SM81 (condenser). You may also try to attach two microphones on a single mic stand—one for the singer, pointing just below the mouth, and another pointing at the guitar. **SEE 9.34**

Microphone Setup for Singer and Piano

If the concert is formal, with the vocalist singing classical songs with piano accompaniment, you should keep the mics out of the picture. You may want to try a Beyerdynamic M160 mic suspended from a small boom. For the piano tape a boundary mic on the lid in the low-peg position or directly on the sounding board. **SEE 9.35** Another way of miking a piano is to have one Shure SM81 mic pointing at the lower half of the strings and another at the upper half, or pointing toward each other from opposite sides of the keyboard. Good results have also been achieved by putting the mic underneath the piano close to the sounding board and about a foot behind the pedals.

If the concert comprises popular songs, such as light classics or rock, a hand mic, such as a Shure SM58, may be a more appropriate choice for the singer. The miking of the piano does not change.

Microphone Setup for Small Rock Group and Direct Insertion

When setting up for a rock group, you need microphones for the singers, the drums, and the speakers that carry the sound of amplified instruments (such as electric guitars and keyboards). The sound signals of electric instruments, such as the bass, are often fed directly to the mixing console without the use of a speaker and a microphone. This technique is called ***direct insertion*** or direct input. Because

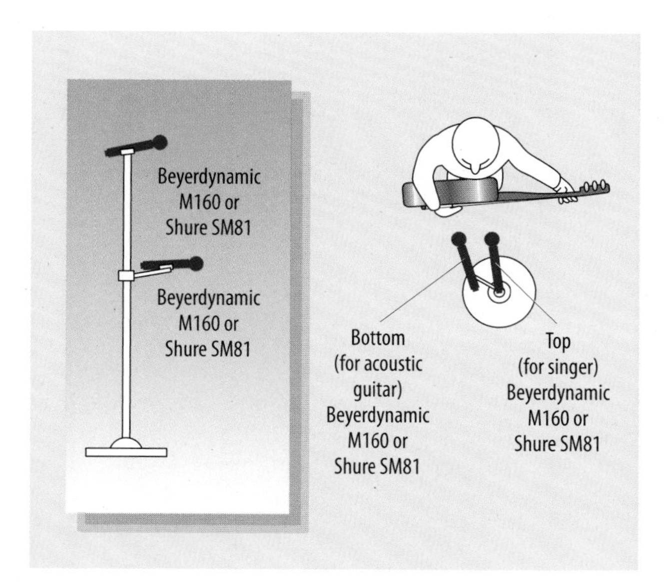

9.34 MICROPHONE SETUP FOR SINGER AND ACOUSTIC GUITAR
The mic setup for a singer with an acoustic guitar is to have one mic for the voice and another lower on the same mic stand for the guitar.

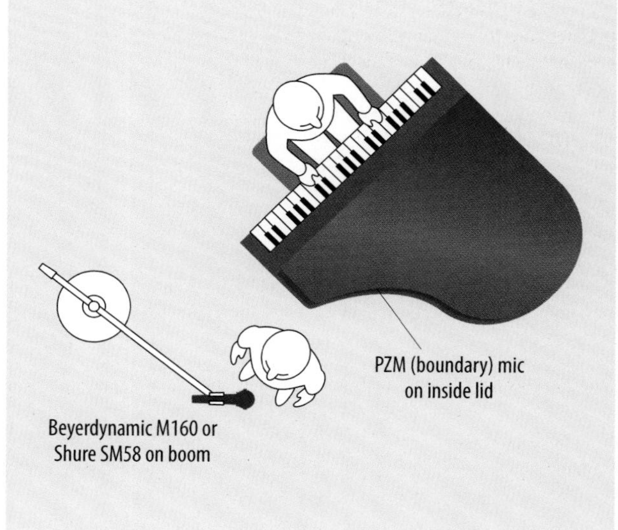

9.35 MICROPHONE SETUP FOR SINGER AND PIANO
If the singer's mic is to be out of camera view, it should be suspended from a boom. The piano is miked separately. For an on-camera mic, the singer can use a hand mic.

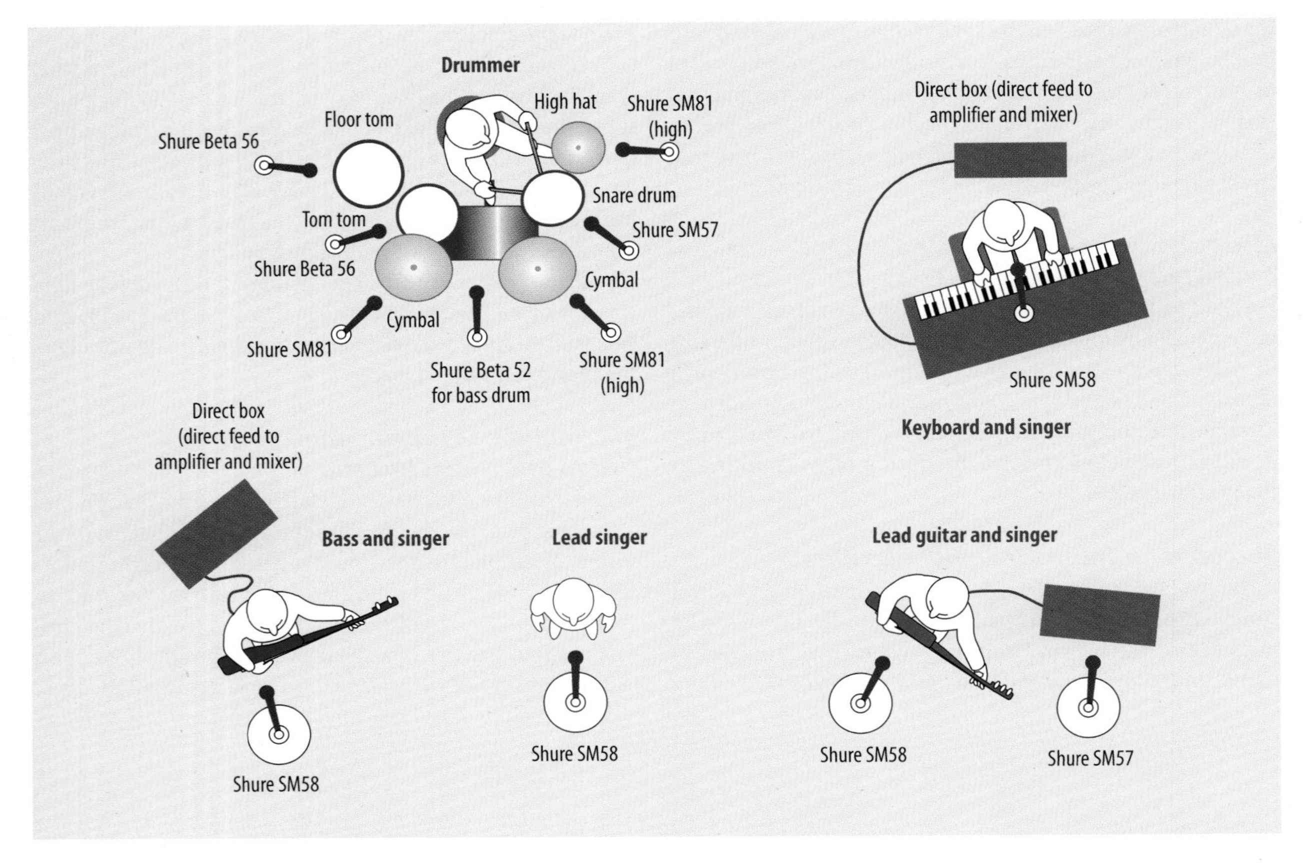

9.36 MICROPHONE SETUP FOR SMALL ROCK BAND
The types and the placement of microphones in this illustration are a general suggestion for how you may start with the mic setup. The Shure microphones are suggested here because they are high-quality mics at a reasonable price. The final criterion for a successful setup is when the sounds coming out of the control room speakers are satisfactory.

most electric instruments are high impedance and all other professional sound equipment is low impedance, you need to match impedances through the direct box (unless the input equipment does it for you).

When setting up mics and speakers, watch for feedback and multiple-microphone interference. For the band members to hear themselves, you must supply the foldback sound mix through either earphones or speakers. ***Foldback***, also called cue-send, is the return to the musicians of the total or partial audio mix from the mixing console. Again, note that the microphones in figure 9.36 are merely suggestions. By all means, try the mics you have and listen to the mix.

Generally, use dynamic or ribbon mics for singers, dynamic mics for instruments, and dynamic mics for high-output sound sources, such as the bass drum. **SEE 9.36**

MICROPHONE USE SPECIFIC TO ENG/EFP

The sound pickup requirements in ENG/EFP do not differ significantly from those in studio operation. In the field as in the studio, your ultimate objective is optimal sound. You will find, however, that sound pickup in the field is much more challenging than in the studio. When outdoors there is the ever-present problem of wind noise and other unwanted sounds, such as airplanes or trucks passing by during a critical scene. The best way to combat wind noise is to use a highly directional mic, cover it with an effective windscreen and a wind jammer, and hold it as close to the sound source as possible without causing a proximity effect (distorted sound when the shotgun mic is too close to the source). Contrary to most studio shows, ambient (environmental) sounds are often needed to support the video.

After you are done with the video recording of an indoor event, you need to record room tone—the "silence" of an empty room without speaker or audience and the ambient sounds of the room with the audience present but without the speaker's voice. These ambient sounds are essential for masking edits in postproduction.

When on an ENG assignment, always have the camera microphone (the shotgun mic, which is built into, or clipped to, the camera) open to record ambient sounds, even when shooting "silent" footage. In fact, when using a hand mic for a stand-up report (with the reporter describing a news event on-location), you must also turn on the camera mic for the ambient sounds. Feed the two mics into separate audio tracks during the recording. As with room tone, such ambient sounds are essential for sound continuity in postproduction editing. The split tracks allow the video editor to control at least somewhat the mix between the reporter's voice and the ambient sounds.

If you have only one microphone, which you must use for voice pickup, record the ambient sounds on a portable audiocassette recorder or on the recording media you are using after finishing the voice work. Again, the editor will appreciate some authentic sounds with which to bridge the edits.

You may find that a seemingly simple audio pickup, such as a speech in a large conference room, can pose a formidable audio problem, especially if you cannot get close enough in the crowded and noisy room for a clean voice pickup. In this case it may be easier to ask the engineer in charge (usually the audiovisual manager of the hotel or convention center) to assist you with a line-out tie-in. In such a setup, you do not need a microphone to pick up the speaker's sound but simply a direct feed from the audio control board of the in-house audio system to the audio input of your camcorder. In effect, you "tie in" to the audio feed from the audio system of the conference room.

MAIN POINTS

- All microphones have a diaphragm, which vibrates with sound pressure, and a generating element, which transduces the physical vibrations of the diaphragm into electric energy.
- In the dynamic, or moving-coil, mic, the diaphragm is attached to the voice coil. The air pressure moves the voice coil back and forth within a magnetic field. This type of generating element makes the mic quite rugged.
- The condenser, or electret, mic has a condenser-like generating element. The movable diaphragm constitutes one of the two condenser plates; a fixed backplate is the other. The moving condenser plate changes the capacitance of the condenser, creating the audio signal. Condenser mics have a wide frequency response.
- In the ribbon, or velocity, mic, a thin metal ribbon vibrates within a magnetic field. Because the ribbon is fragile, such mics are generally used indoors under controlled conditions.
- Impedance is the total resistance of an electric circuit to the flow of AC current. The impedance of the signal source (mics and electric instruments) must be matched with the equipment to which it is connected (mixer or console).
- High-quality microphones have a wide frequency range and pick up sounds equally well over the total range—called a flat response.
- Microphones can be balanced or unbalanced. Most professional mics have a balanced output. Balanced microphones and mic cables have three-pronged XLR connectors. The balanced audio cable prevents external signals from causing a hum in the audio track. Unbalanced (two-wire) cables have phone plug, RCA phono plug, and mini plug connectors.
- Direct insertion, or direct input, involves feeding the sound signal of an electric instrument into a direct box that matches impedances with the input of the console. This eliminates the use of a speaker and a microphone.
- Foldback is the return to the musicians of the total or partial audio mix from the mixing console.

ZETTL'S VIDEOLAB

For your reference or to track your work, the *Zettl's VideoLab* program cues in this chapter are listed here with their corresponding page numbers.

CHAPTER

10

Audio: Sound Control

The previous chapter dealt mostly with sound pickup—the types of microphones and their uses. This chapter explores the equipment and the techniques of controlling sound and sound recording in television studio and field production. Section 10.1, Sound Controls and Recording, identifies the major equipment and production techniques for mixing and recording sound in the studio and the field. Section 10.2, Stereo, Surround Sound, and Sound Aesthetics, highlights the principal aesthetic factors of sound.

You should realize that audio production is a highly specialized field in its own right and that this chapter is limited to the major equipment, the basic production techniques, and some fundamental aesthetic considerations. Even if you don't intend to become a sound designer, you need to know what good audio is all about. Whatever you do, the most important prerequisite to successful audio for television is, and will always be, a good pair of ears. ZVL1 AUDIO→ Audio introduction

KEY TERMS

ambience Background environmental sounds.

audio control booth Houses the audio, or mixing, console; analog and digital playback machines; a turntable; a patchbay; a computer display; speakers; intercom systems; a clock; and a line monitor.

automatic gain control (AGC) Can regulate sound volume or video brightness and contrast levels.

calibrate To make all VU meters (usually of the audio console and the video recorder) respond in the same way to a specific audio signal.

digital audiotape (DAT) Recording system in which the sound signals are encoded on audiotape in digital form. Includes digital recorders as well as digital recording processes.

digital cart system Digital audio system that uses built-in hard drives, removable disks, or read/write optical discs to store and access audio information almost instantaneously. It is normally used for the playback of brief announcements and music bridges.

equalization Controlling the quality of sound by emphasizing certain frequencies while de-emphasizing others.

figure/ground Emphasizing the most important sound source over the general background sounds.

memory card A read/write solid-state storage device for large amounts of digital video and audio data. Also called *flash drive* and *flash memory card.*

mini disc (MD) Optical 2½-inch-wide disc that can store one hour of CD-quality audio.

mixing Combining two or more sounds in specific proportions (volume variations) as determined by the event (show) context.

mix-minus Type of multiple audio feed missing the part that is being recorded, such as an orchestra feed with the solo instrument being recorded. Also refers to a program sound feed without the portion supplied by the source that is receiving the feed.

MP3 A widely used lossy compression system for digital audio. Most Internet-distributed audio is compressed in the MP3 format.

peak program meter (PPM) Meter in an audio console that measures loudness. Especially sensitive to volume peaks, it indicates overmodulation.

sound perspective Distant sound must go with a long shot, close sound with a close-up.

surround sound Sound that produces a soundfield in front of, to the sides of, and behind the listener by positioning loudspeakers either to the front and the rear or to the front, sides, and rear of the listener.

sweetening Variety of quality adjustments of recorded sound in postproduction.

volume-unit (VU) meter Measures volume units, the relative loudness of amplified sound.

SECTION

10.1

Sound Controls and Recording

When watching a television program, we are generally not aware of sound as a separate medium. Somehow it seems to belong to the pictures, and we become aware of the audio portion only when it is unexpectedly interrupted. But in your own video recordings, you probably notice that there are always some minor or even major audio problems that tend to draw attention away from your beautiful shots. Although audio is often treated casually, you quickly realize that the sound portion is, indeed, a critical production element that requires your full attention.

▶ **MAJOR AUDIO PRODUCTION EQUIPMENT**
The audio console, the audio mixer, the patchbay, and tapeless and tape-based audio-recording systems

▶ **BASIC OPERATION FOR STUDIO AUDIO**
The audio control booth and basic audio operation

▶ **BASIC OPERATION FOR FIELD AUDIO**
Keeping sounds separate, using the automatic gain control in ENG and EFP, using an XLR pad, and ENG/EFP mixing

MAJOR AUDIO PRODUCTION EQUIPMENT

The major components of audio equipment are the audio console, the patchbay, and analog and digital tapeless and tape-based audio-recording systems.

Audio Console

Regardless of individual models—analog or digital—all audio consoles, or audio control boards, are built to perform five major functions:

- Input: to preamplify and control the volume of the various incoming signals
- Mix: to combine and balance two or more incoming signals
- Quality control: to manipulate the sound characteristics
- Output: to route the combined signals to a specific output
- Monitor: to listen to the sounds before or as their signals are actually recorded or broadcast SEE 10.1

Input Studio consoles have multiple inputs to accept a variety of sound sources. Even small studio consoles may have 16 or more inputs. Although that many inputs are rarely used in the average in-house production or broadcast day, they must nevertheless be available for the program you may have to do the next day.

Each input module requires that you select either the mic-level or the line-level input. Mic-level inputs are for sound sources that need to be preamplified before they are sent to the input controls. All microphones need such preamplification and are therefore routed to the mic input. Line-level inputs, such as CD players, DVD players, and digital audiotape (DAT) recorders, have a strong enough signal to be routed to the line input without preamplification. All incoming audio signals must reach line-level

10.1 PHOTO BY EDWARD AIONA

10.1 AUDIO CONSOLE
Each module of this audio console contains a volume control (slide fader), various quality controls, and assignment switches. It can route several mixes to different destinations.

strength before they can be further adjusted or mixed at the audio console.

Because not all input levels of microphones or line signals are the same, they run the risk of becoming overamplified. To prevent this you can manipulate the signals individually with the trim control, which adjusts the input strength of the mic signals so that they won't become distorted during further amplification.

Regardless of input, the audio signals are then routed to the volume control, a variety of quality controls, switches (mute or solo) that silence all the other inputs when you want to listen to a specific one, and assignment switches that route the signal to certain parts of the audio console and to signal outputs. SEE 10.2

Volume control All sounds fluctuate in volume (loudness). Some sounds are relatively weak, so you have to increase their volume to make them perceptible. Other sounds come in so loud that they overload the audio system and become distorted or they outweigh the weaker ones so much that there is no longer proper balance between the two. The volume control that helps you adjust the incoming sound signals to their proper levels is usually called a pot (short for potentiometer) or a fader (also called attenuator or gain control). SEE 10.3

Mix The audio console lets you mix, or combine, the signals from various inputs, such as two lavalier mics, the

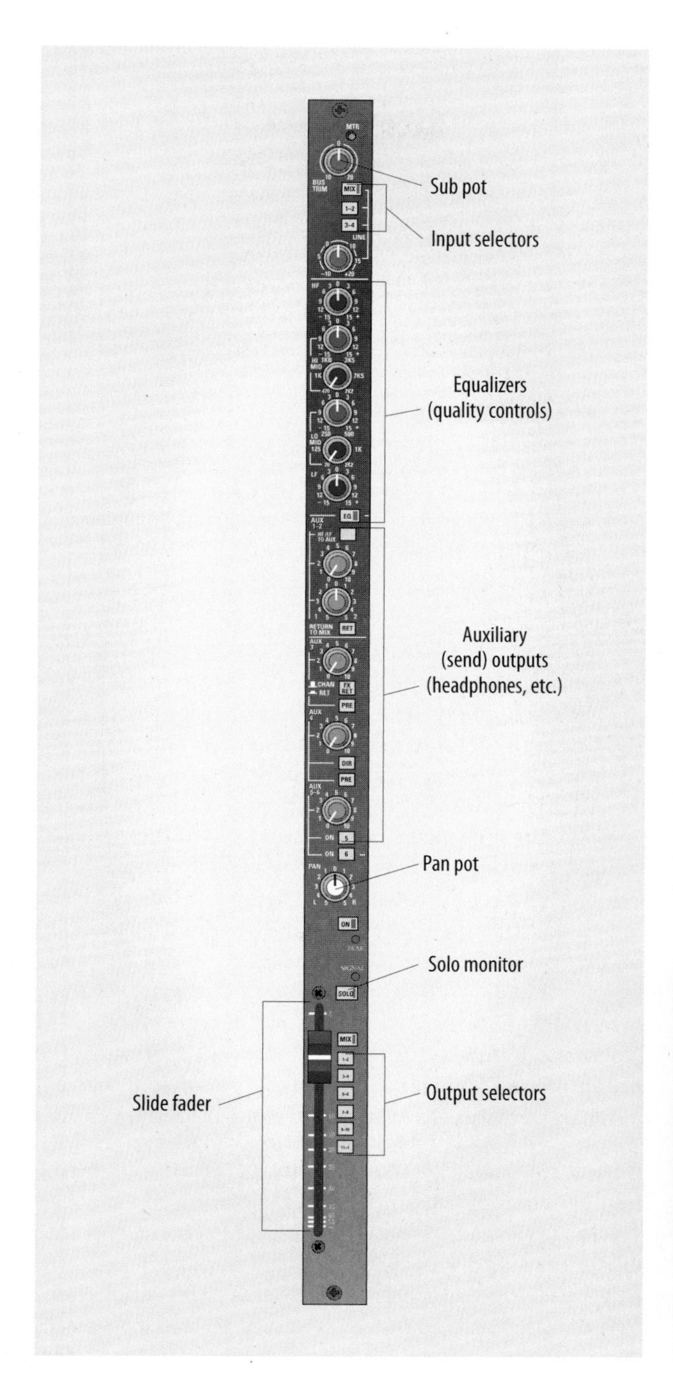

10.2 AUDIO CONSOLE MODULE
The major controls on this module are the slide fader volume control, the solo monitor (turns off automatically all other channels except the one on this module), the pan pot (moves the sound horizontally from one stereo speaker to the next), quality controls such as the equalizers, and various input and output selectors.

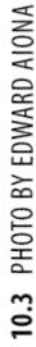

10.3 SLIDE FADERS
Pushing the fader up increases the volume; pulling it down decreases the volume.

background music, and the sound effect of a phone ring. The mix bus combines these various audio signals with the specific volume that you assign. Without the ***mixing*** capability of the board, you could control only one input at a time. The completed mix is then fed to the line-out.

A mix bus is like a riverbed that receives the water (signals) from several different streams (inputs). These streams (various sound signals) converge (mixed sound signal) and finally flow downstream along the riverbed (mix bus) to their destination (recorder). ZVL2 AUDIO→ Consoles and mixers→ parts | signals | control | try it

Solo switch and pan pot The solo switch lets you listen to a particular incoming sound while silencing all the other sound signals you are mixing. Every console module has a solo switch. The pan pot shifts the sound or sounds you select horizontally from left to right or vice versa (clockwise or counterclockwise when mixing surround sound).

10.4 ANALOG VU METER
The VU meter indicates the relative loudness of a sound. The upper figures ranging from –20 to +3 are the volume units (decibels). The lower figures represent a percentage scale, ranging from 0 to 100 percent signal modulation (signal amplification). Overmodulation (too much signal amplification) is indicated by the red line on the right (0 to +3 VU).

10.4 PHOTO COURTESY SELCO PRODUCTS COMPANY

Quality control All audio consoles have controls that let you shape the character of a sound (see figure 10.2). Among the most important are equalization, filters, and reverberation (reverb) controls.

The process of controlling the audio signal by emphasizing certain frequencies and de-emphasizing or eliminating others is called ***equalization***. It can be accomplished manually or automatically through an equalizer, which works like the tone control on a home stereo receiver. It can boost or reduce selected frequencies and thereby influence the character of a sound. For example, you can make a sound more brilliant by boosting the high frequencies or more solid by boosting the lows, or you can eliminate a low-frequency hum or a high-frequency hiss. Filters eliminate automatically all frequencies above or below a certain point. The reverb controls can add an increasing amount of reverberation to each of the selected inputs.

Output The mixed and quality-processed signal is then routed to the output, called the line-out. To ensure that the mixed signals stay within the acceptable volume limits, they are regulated by final volume controls—the master pots—and metered by volume indicators, the most common of which is the ***volume-unit (VU) meter***. As the volume varies, the needle of the VU meter oscillates back and forth along a calibrated scale. **SEE 10.4**

If the volume is so low that the needle barely moves from the extreme left, you are riding the gain (or volume) "in the mud." If the needle oscillates around the middle of the scale and peaks at the red line on the right, you are riding the gain correctly. If the needle swings in the red on the right side of the scale, and even occasionally hits the right edge of the meter, the volume is too high—you are "bending the needle," "spilling over," or "riding in the red."

Much like the volume indicator of home stereo systems, the VU meter in many audio consoles consists of light-emitting diodes (LEDs), which show up as thin, colored light columns that fluctuate up and down a scale. When you ride the gain too high, the column shoots up on the scale and changes color. **SEE 10.5**

Some audio consoles have an additional ***peak program meter (PPM)***, which measures loudness peaks. A PPM reacts more quickly to the volume peaks than does the needle of the VU meter and clearly shows when too strong a signal overloads the system and distorts loudness.

Output channels We often classify audio consoles by the number of output channels. Older television consoles had several input channels but only one output channel because television sound was monophonic. Today, however, even small television consoles have at least two output channels to handle stereophonic sound or to feed two pieces of equipment (such as headphones and a video recorder) simultaneously with two identical but independent signals.

With high-definition television (HDTV), the sound requirements also change. Very much like motion pictures, large-screen TV displays require surround sound, which involves multiple discrete output channels and a variety of speakers that are strategically placed in front and in back

10.5 DIGITAL LED VU METER
The light-emitting diode meters indicate input overload by lighting up in a different color (usually red). When the meter column turns red, the signal is overloaded and being clipped.

10.5 PHOTO BY EDWARD AIONA

of the display screen (see figure 10.18). This increasing demand for high-quality audio has led to greater use of multichannel (output) consoles in the audio control booth and especially in the audio postproduction room.

To identify how many inputs and outputs a specific console has, they are labeled with the number of input and output channels, such as an 8 × 1 or a 32 × 4 console. This means that the small 8 × 1 console has eight inputs and one output; the larger 32 × 4 console has 32 inputs and four outputs. With a single output channel, the 8 × 1 board obviously is monophonic.

Most larger television audio consoles have eight or more output channels (with eight master pots and eight VU meters), each of which can carry a discrete sound signal or mix. The advantage of multiple outputs is that you can feed the individual signals onto a multitrack audio recorder for postproduction mixing. If, for example, there are 24 inputs but only two outputs, you need to mix the input signals down to two, which you can then feed to the left and right channels of a stereo recorder. But if you want to keep the various sounds separated to exercise more control in the final postproduction mix, or if you want to feed separate surround-sound speakers, you need more outputs. Even when covering a straightforward rock concert, for example, you may have to provide one mix for the musicians, another for the audience, one for the video recorder, and yet another for the multitrack audiotape recorder (ATR). You will be surprised by how fast you run out of available inputs and outputs even on a big console.

In-line consoles Some of the more elaborate consoles have input/output, or I/O, modules, which means that each input has its own output—all in one line. If, for example, there are 24 inputs and each receives a different sound signal, you could send each signal directly to a separate track of a 24-track recorder without feeding them through any of the mix buses. That way you use the console to control the volume of each input, but the console does not function as a mixing or quality-control device. In fact, the sound is sent to the audio recorder in its raw state. The mixing and quality control of the various sounds are done in the postproduction and mixdown sessions. The I/O circuits let you try out and listen to all sorts of mixes and sound manipulations without affecting the original signals sent to the recorder.

Phantom power Don't let the name scare you. As mentioned earlier, the "phantom" in phantom power is more like "virtual." All it means is that the audio console or some other source, rather than a battery, supplies the preamplification power to condenser mics.

Monitor and cue All consoles have a monitor system, which lets you hear the final sound mix or allows you to listen to and adjust the mix before switching it to the line-out. A separate audition or cue return system lets you hear a particular sound source without routing it to the mix bus. This system is especially important when you want to cue a digital audio recording or check a specific track of a compact disc (CD) or a digital versatile disc (DVD) while on the air with the rest of the sound sources.

Digital consoles All digital consoles have centralized controls that trigger sound control and routing functions for each input module. These controls are not unlike the delegation controls of a video switcher (see figure 14.7). The advantage is that this routing architecture keeps the console relatively small and workable. Most larger digital consoles let you preset, store, recall, and activate many of the audio control functions. For example, you can try out a particular mix with specific volume, equalization, and reverberation

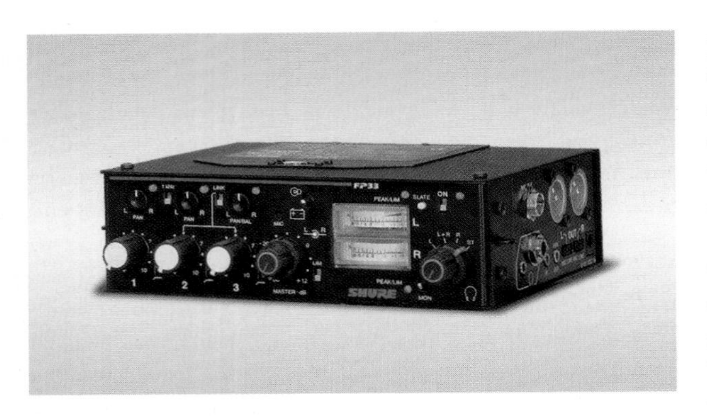

10.6 PORTABLE MIXER
This portable mixer has three inputs and two outputs. The volume controls are rotary knobs. Such big knobs and switches are especially convenient in the field, where digital menus are often hard to see and activate.

10.6 PHOTO COURTESY SHURE, INC.

values for each of the individual sounds, store it all in the computer's memory, try something else, and then recall the original setup with the press of a button.

Audio Mixer

An audio mixer differs from an audio console in that it normally serves only the input (volume control) and mixing (combining two or more signals) functions. **SEE 10.6**

Most portable mixers have only three or four input channels and one or two outputs. Even then the small mixers require that you distinguish between mic-level (low signal strength) and line-level (high signal strength) input sources. A switch above or below each sound input must be set either to *mic* for low-level inputs, such as all microphones, or to *line* for high-level sources, such as the output of a CD player. Because most of the time you will use the field mixer for mixing microphones, double-check that the input switch is set to *mic*. If you are not sure whether a particular piece of audio equipment produces a mic-level or a line-level signal, do a brief test recording. Don't rely on the VU meter when playing back the test recording—you should listen to it with headphones. The VU meter might show the recording to be in the acceptable volume range, but it will not reflect sound distortions.

Even though some digital mixers have more inputs as well as some quality controls, elaborate mixing in the field is not recommended unless you're doing a live telecast.

Patchbay

The primary function of the patchbay, or patch panel, is connecting and routing audio signals to and from various pieces of equipment. You can accomplish this by using actual wires that establish specific connections or with a computer that rearranges the signals and sends them according to your instructions. Whichever method you use, the principle of patching is the same. Here we use wires, called patch cords, to explain a simple patching procedure.

Assume that during a newscast you want to operate two microphones, a remote feed from a field reporter, and a CD. Lav 1 and lav 2 are the newscasters' lavaliers. The remote feed comes from the field reporter with a live story. The CD contains the opening and closing theme music for the newscast.

Any one of these audio sources can be patched to an individual volume control (pot or fader) in any desired order. Suppose you want to operate the volume controls in the following order, from left to right: CD, lav 1, lav 2, remote feed. You can easily patch these inputs to the audio console in that order. If you want the inputs in a different order, you need not unplug the equipment; all you do is pull the patch cords and repatch the inputs in the different order. **SEE 10.7**

Wired patchbay All wired patch panels contain rows of holes, called jacks, which represent the outputs and the inputs. The upper rows of jacks are normally the outputs (which carry the output signals from mics, CDs, and so forth). The rows of jacks immediately below the output jacks are the input jacks, which are connected to the audio console. The connection between output and input is made with the patch cord, a wire that has the same plugs on either end.

To accomplish a proper patch, you must plug the patch cord from one of the upper output jacks into one of the lower input jacks. **SEE 10.8** Patching output to output (upper-row jack to another upper-row jack) or input to input (lower-row jack to lower-row jack) will give you nothing but headaches.

To reduce the number of patch cords, certain frequently used connections between outputs (a specific mic, digital recorder, or CD) and inputs (specific volume controls assigned to them) are directly wired, or normaled, to one another. This means that the output and the input of a circuit are connected without a patch cord. By inserting a patch cord into one of the jacks of a normaled circuit, you *break,* rather than establish, the connection.

Although patching helps make the routing of an audio signal more flexible, it can also cause some problems. Patching takes time; patch cords and jacks get worn out from frequent use, which can cause a hum or an intermittent

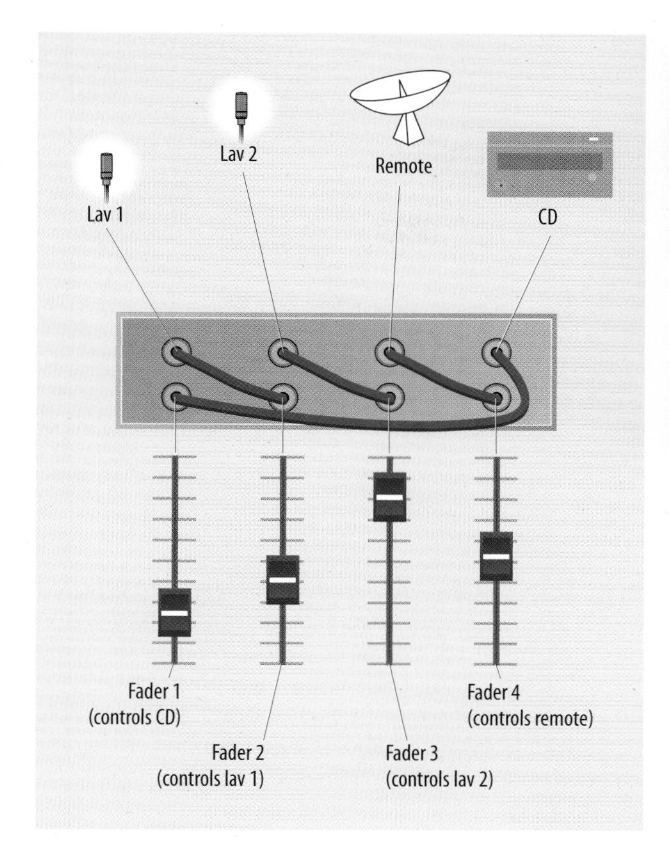

10.7 PATCHING
This patching shows that the signal outputs (audio sources) of two microphones, a remote feed, and a CD are grouped in the following order of fader inputs: CD, lavalier 1, lavalier 2, remote feed.

10.8 PHOTO BY EDWARD AIONA

10.8 PATCHBAY WITH PATCHES
All patchbays connect the signal outputs (mics, CDs, video recorder) to specific input modules of the audio console. The patching is accomplished by connecting the audio outputs (top row) to the inputs (bottom row) with patch cords.

connection; and many patch cords crisscrossing each other are confusing and look more like spaghetti than orderly connections, making individual patches difficult to trace. Also, when patching with a corresponding fader still set at a reasonably high volume, the pop caused by plugging or unplugging the patch cord can blow even the most robust speaker. Once again, although physical connections are important because you can see which signal goes where, the computer can perform many of the routine patching functions more efficiently.

Computer patching In computer patching, the sound signals from the many sources, such as mics, direct boxes, CDs, DVDs, or video recordings, are routed to the patch panel programmer, which assigns the multiple signals to specific fader modules of the audio console for further processing. To route lavalier 1 to pot 2, and the CD to pot 1, for example, you don't need any physical patches; you simply enter the routing information into the patch panel programmer, which tells the electronic patch panel to connect the inputs to the desired faders on the console, show the information on the display, and store your patching commands for future use. Computer patching is as easy as pasting words with a word processor. If something goes wrong, however, it is usually much harder to trace the incorrect computer patch than locating physically a specific patch cord on the patchbay.

Tapeless Audio-recording Systems

All audio recording in professional television is done digitally. As with video, digital audio recordings excel not only in sound quality but also in maintaining that quality throughout extensive postproduction editing. Because digital systems allow you to see a visual display of the recorded sounds, they make editing much more precise than with analog methods. Digital recording can be done with tape as a recording media but is more often done with tapeless media, such as computer hard drives, mini discs, or memory cards.

In television production the more popular tapeless audio-recording systems include the digital cart system, mini disks and memory cards, hard drives with removable or fixed disks, and optical disc systems with a variety of CD and DVD formats.

10.9 DIGITAL CART SYSTEM
This digital cart system can record, store on its built-in hard drive, and play back instantly a great amount of audio data. You can also edit audio files with it, interface it via Ethernet with digital audio workstations, and exchange audio data with a removable 250 MB (megabyte) Zip disk.

10.9 PHOTO COURTESY 360 SYSTEMS

10.10 PROFESSIONAL MINI DISC RECORDER/PLAYER
Mini disc recorders can store and provide random access to more than an hour of high-quality digital stereo audio. The disc cassette measures only about 2¾ inches and operates much like a small removable disk in a hard drive.

10.10 PHOTO COURTESY SENNHEISER ELECTRONIC CORPORATION

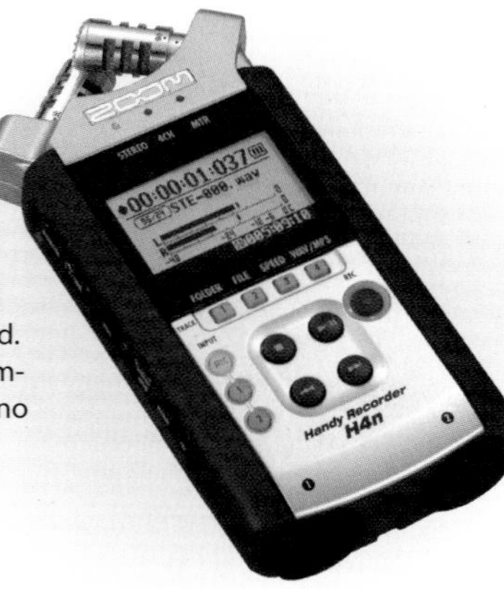

10.11 SD MEMORY CARD RECORDER
This digital stereo recorder uses an SD flash drive that can hold up to three hours of high-quality stereo sound. The advantage of flash memory cards is that they have no moving parts.

10.11 PHOTO COURTESY SAMSON TECHNOLOGIES CORP.

Digital cart system A ***digital cart system*** lets you record a great amount of audio data (literally hundreds of hours) on a built-in hard drive, cue up almost instantly any audio file stored on the drive, edit audio files, construct a playlist for an automated playback sequence, operate with a remote control, and exchange audio data that were recorded on a separate removable disk. **SEE 10.9**

Mini discs and flash memory cards The ***mini disc (MD)*** is a read-only or read/write optical disc that can store more than an hour of high-quality digital stereo audio. Its small size (about 2¾ inch), large storage capacity, and easy cueing make it a useful playback device for television production. **SEE 10.10**

The ***memory card***, or flash drive, is used extensively in digital audio recording. One such media is the SD (Secure Digital format) memory card, the same as, or very much like, the one you may be using in a digital still camera or camcorder. There are small and relatively inexpensive digital recorders that have two remarkably good condenser mics built-in and can store on a 2 GB (gigabyte) SD flash drive a little more than three hours of high-quality stereo sound. When compressed with the ***MP3*** system, the SD card can store about 30 hours of music. The advantage of flash memory cards is that they have no moving parts. **SEE 10.11**

Hard drives There are large-capacity systems built specifically for audio production and postproduction that store audio information just like you would on a computer hard drive. **SEE 10.12** For example, with this recorder you

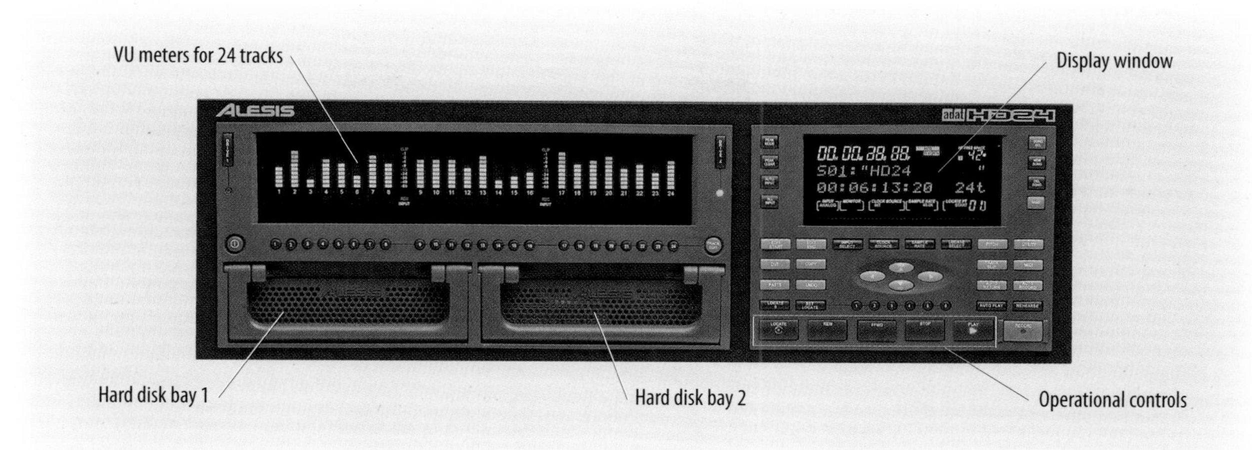

10.12 DIGITAL MULTITRACK RECORDER/PLAYER
This digital recorder can record 24 tracks on two high-capacity hard drives. Both hard disks are removable and can be swapped from bay to bay.

10.12 PHOTO COURTESY ALESIS

can record 24 tracks of high-quality audio on two hard drives; do editing with common cut, copy, and paste commands; and move entire tracks from one position to another. You can randomly access and play any audio file almost instantly.

As you know, the ubiquitous Apple iPod classic is a powerful recording device. The 160 GB model can store thousands of songs and photos or store and display 200 hours of video. If you have the right software, you can transfer the recorded information to your laptop for further processing.

CDs and DVDs The professional compact disc and digital versatile disc players are often used in television (and radio) stations for playing back commercially produced music and other audio material. The rewritable CDs and DVDs are used for multiple recording and playback. There are several different CD and DVD formats on the market, all of which perform similar production functions: the storage and the playback of a variety of audio material. Some of the more elaborate models let you load several CDs at the same time and select and program separate tracks for automatic playback. SEE 10.13

Although CDs and DVDs can theoretically withstand unlimited playback without deterioration, they are nevertheless vulnerable. If you scratch the shiny side or even the label side, the disc won't play past the scratch. And if there are fingerprints on the disc, the laser may try to read the fingerprints instead of the imprinted digits. When handling CDs and DVDs, try to keep your hands off the surface and always lay the disc down on its label side—not its shiny side.

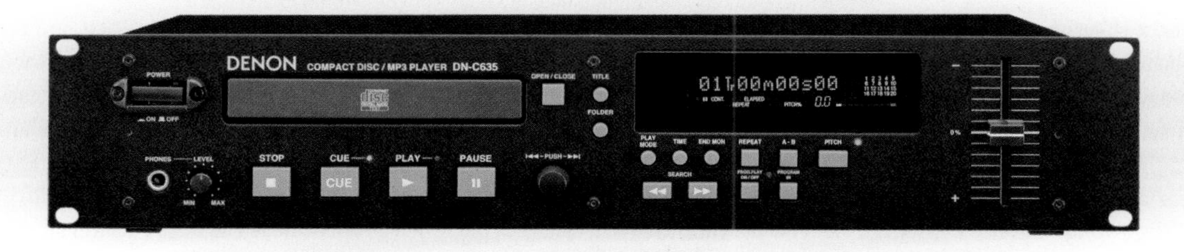

10.13 PROFESSIONAL CD PLAYER
Professional CD and DVD players allow instant random access to various tracks. The play sequence can be stored and displayed on playback.

10.13 PHOTO COURTESY DENON BRAND COMPANY

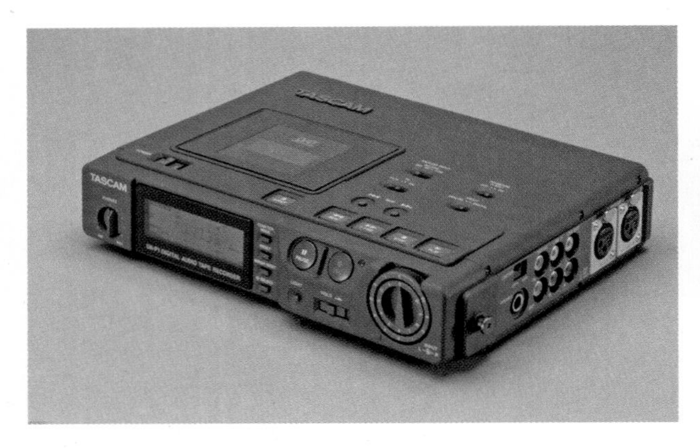

10.14 PORTABLE DAT RECORDER
This portable DAT recorder can record up to two hours on a single battery charge. It has one balanced stereo input (two XLR jacks) and four unbalanced inputs (RCA phono jacks). Its excellent frequency response lets you make high-fidelity recordings of speech and music.

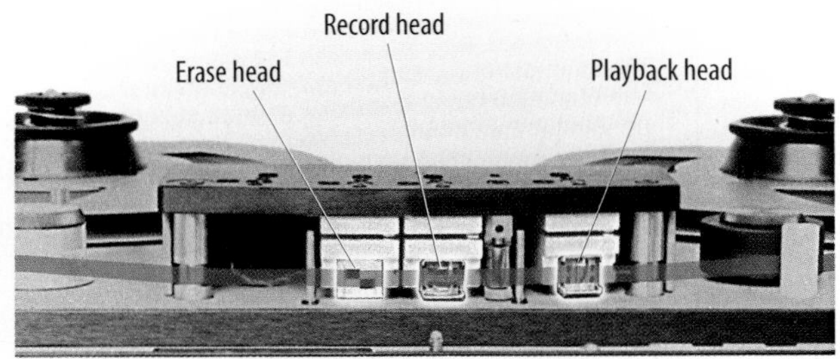

10.15 ANALOG AUDIOTAPE HEAD ASSEMBLY
The head assembly of an analog reel-to-reel ATR consists of an erase head, a record head, and a playback head.

10.14 PHOTO COURTESY TASCAM CORPORATION
10.15 PHOTO BY WILLIAM STORM

Tape-based Audio-recording Systems

Contrary to hard drives, optical discs, and all kinds of flash memory cards, tape-based systems can record digital or analog information.

Tape-based recorders: videotape The tape-based audio-recording devices include videotape used by standalone or camcorder videotape recorders (VTRs). Most digital audio recording for video is done simultaneously with the picture on one or two tracks of the videotape (see figure 13.4). High-end camcorders offer four tracks, but if you need multiple tracks, you should feed them into a mixer and route them to a separate multitrack recorder for postproduction editing.

Tape-based digital recorders: DAT Although mostly replaced by the more efficient tapeless recording systems, you may still find digital tape recorders in operation. Normally, a ***digital audiotape (DAT)*** recorder refers to a digital audiocassette recorder. **SEE 10.14** These initially quite expensive recorders, whose mechanics and electronics were adapted from videotape machines, can record virtually noise-free high-fidelity sound, record time code with the audio (for synchronizing with other tracks or video), and provide high-speed cueing information. The problem with these recorders is that they are quite sensitive, especially to moisture, and need constant maintenance to work reliably.

Tape-based analog recorders Don't dismiss analog audio just yet. Many older camcorders and VHS recorders are analog, and there are extensive analog sound archives that will most likely remain analog even in the digital age. Even you may have a collection of analog equipment that most likely includes an analog cassette machine. Some audio purists have returned to analog sound systems because, according to them, analog recordings have a warmer sound than digital ones.

All analog audio recorders are tape-based. Some die-hard analog audio fans continue to use the two types of analog systems: audiocassette recorder/players and open-reel audiotape recorders.

Professional audiocassette systems are similar to the one you may have at home except that they have slightly more sophisticated electronics to reduce noise, and more durable tape transports that allow faster and smoother fast-forward and rewind speeds. Despite the narrow tape, audiocassettes produce good enough sound for casual listening.

The open-reel audiotape system is generally used for multitrack recording or for playing back longer pieces of audio material. A look at the open-reel head assembly will give you a good idea of how the system works. **SEE 10.15** The tape moves from a supply reel to a takeup reel over at least three heads: the erase head, the record head, and the playback head. When the ATR is being used for recording, the erase head clears the portions of the tape of all audio material that might have been left on the tape from a previous recording; the record head then puts the new audio material on the tape. When the tape is played back, the playback head reproduces the audio material previously recorded. The erase and record heads are not activated during playback.

BASIC OPERATION FOR STUDIO AUDIO

All television control rooms are divided into an audio control section and a program control section. The audio control booth contains all the necessary equipment for studio audio operation.

Audio Control Booth

The ***audio control booth*** is a soundproof room adjacent to the larger program control room. Most audio booths provide visual access to the control room. Ideally, the audio technician should be able to see the preview monitors in the program control room so that he or she can anticipate the audio cues.

Generally, the audio control booth houses the audio, or mixing, console and the recording and playback equipment. There is also a physical patchbay (despite the presence of computer patching) and one or more desktop computers. You will also find cue and program speakers, intercom systems, a clock, and a line monitor. The audio technician (or audio operator or audio engineer) operates the audio controls during a show. **SEE 10.16**

Learning to operate all this equipment takes time and practice. Fortunately, in most studio productions your audio tasks consist mostly of making sure that the voices of the news anchors or panel guests have acceptable volume levels and are relatively free of extraneous noise and that the sound appears with the pictures when video recordings are played back. Most likely you will not be asked to do intricate sound manipulations during complex recording sessions—at least not right away. Consequently, the focus here is on the basic audio control factors: audio system calibration, volume control, and live studio mixing.

Audio system calibration Before you send your mix from the console to the equipment that records the output, you need to make sure that the input volume (recording level) of the recording equipment (video or audio recorder) matches that of the console output (line-out signal). This process is called audio system calibration or simply calibration. (Note that audio calibration has nothing to do with zoom lens calibration, whereby you adjust the zoom lens so that it stays in focus during the entire zoom range.)

To ***calibrate*** the audio system, you normally send a 0 VU tone, which is generated in the console, from the console to the video recorder (VR). The VR operator adjusts the input level on the video recorder so that it also reads 0 VU. When dealing with digital audio, it is a good idea

10.16 AUDIO CONTROL BOOTH
The television audio control booth contains a variety of audio control equipment, such as the audio console with computer display; the patchbay; CD, DVD, and mini disc players; DAT recorder/players; loudspeakers; intercom systems; and a video line monitor.

10.16 PHOTO BY EDWARD AIONA

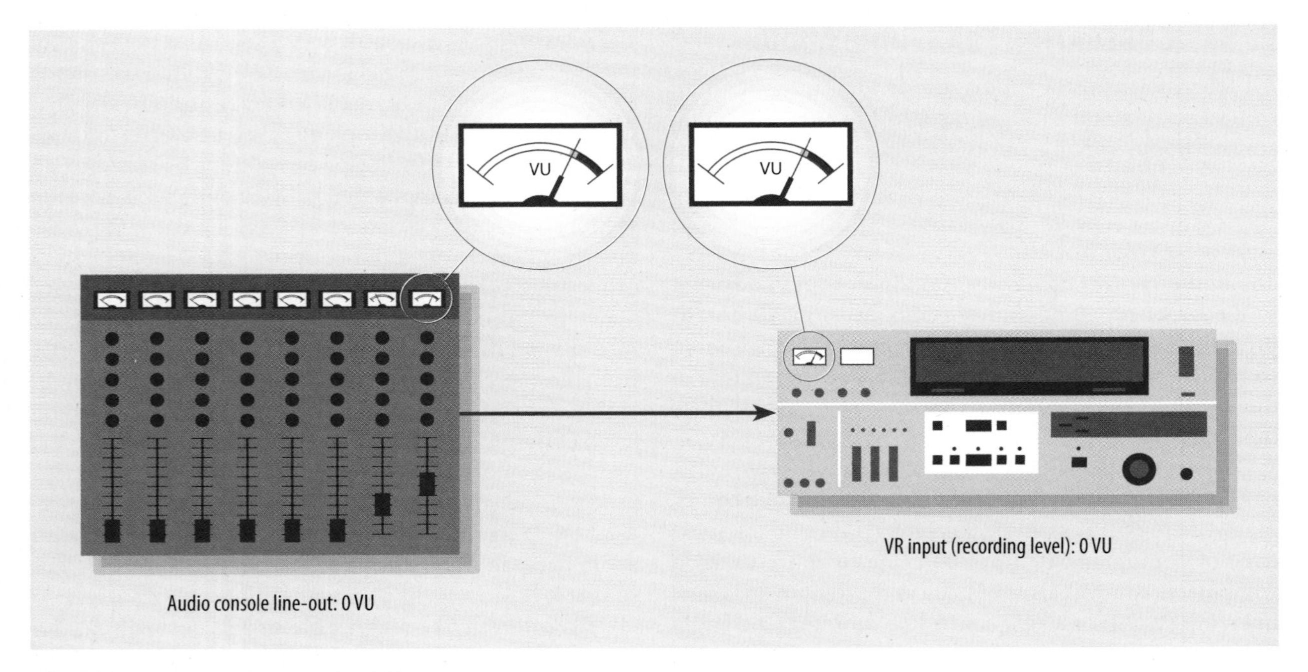

10.17 AUDIO SYSTEM CALIBRATION
An audio system is calibrated when all VU meters respond in the same way to a specific audio signal—the control tone. Here the line-out of the audio mixer is calibrated with the input (recording level) of the VR. Both VU meters show the same value.

to set the recording level somewhat below 0 VU to avoid distortion of some of the louder bursts; if this is done, however, the inputs of all other audio-related equipment must also be calibrated in the same way. **SEE 10.17** ZVL3 AUDIO→ Consoles and mixers→ calibration

Volume control Once the system is calibrated, you can pay attention to the finer points of adjusting the volume of the incoming sound sources. Before starting the video recording, you should always take a level—that is, adjust the input level so that the talent's speech falls more or less within the tolerable volume range (not riding in the mud and not bending the needle). Ask the talent to talk long enough for you to see where the lower and upper limits of the speech volume are, then place the fader between these two extremes. An experienced performer will stay within this volume range even in subsequent takes.

Unfortunately, when asked to give a level, most performers consider it an intrusion on their concentration and simply count rapidly to three or four; then, when they are on the air, their voices rise to the occasion—and also in volume. Always be prepared for this sudden volume increase. Experienced performers give a few of their opening remarks in about as loud a voice as they will use when on the air. But don't bet on it.

Clipping in digital sound When riding the gain consistently at too high a level so that the VU meter occasionally spills into the red zone, or the digital LED meter goes so high that the green column changes to red (see figures 10.4 and 10.5), you end up not with a recording that is slightly too loud but with "clipping" the peaks, which results in distorted sound. Although it is relatively easy to boost sound that was recorded at a slightly lower-than-normal level (even at the risk of amplifying some of the noise with the low-level sounds), it is difficult and more often impossible to fix clipped sound in postproduction. You should therefore always leave some "headroom," that is, set the recording levels somewhat lower than for analog recording to absorb the occasional volume peaks. The warning levels are sometimes colored orange or pink before they turn into red. When calibrating the system, audio people recommend setting the maximum level for digital audio between –6 dB [decibels] and –12 dB rather than at 0 dB. In practice, just stay at least –6 dB below the clipping threshold.

Live studio mixing Live mixing means that you combine and balance sounds while the production is in progress. Studio mixing may range from the relatively simple task of riding gain for the newscaster's lavalier mic or balancing the voices of several panel members during a discussion to the more complicated job of switching among multiple audio sources during a newscast or recording a rock band or even a dramatic scene for an interactive multimedia program on how to recognize potential shoplifters.

As with the setup of mics for a complex production, there is no formula for how an optimal mix is achieved. When riding gain for the single mic of the news anchor, simply keep his or her level within the acceptable audio range and watch that the anchor is clearly heard. When controlling the audio of the panel discussion, riding gain is easiest if every member wears a lavalier. Once the levels are set, you have little to do except bring down the fader somewhat if one of the members gets excited and starts talking much louder than normal, or bring it up when somebody drifts off mumbling.

When using desk mics, the most important audio job is before the show even starts—the mic setup. Remember to place the mics at least three times as far apart as the distance of any mic to the panel member (as described in chapter 9) to eliminate possible multiple-microphone interference. After taking preliminary levels, adjust the mics for optimal positions and tape the stands to the table. Take another level, adjust the faders for all mics, and hope that the panel members' kicking and banging on the table will be minimal.

The multisource newscast is more challenging. For example, you may need to switch quickly from the anchor's introduction to SOT (sound on tape) or SOS (sound on source) of a video from the video server, and from there to the co-anchor, to the guest in London (remote source), back to the co-anchor, to another SOS clip from the video server, back to the anchor, to a commercial, and so forth. You will find that labeling each audio input will greatly facilitate your audio control: simply put a strip of masking tape below the faders and mark them with a grease pencil. As for volume control, you have to watch the remote sources and the SOT/SOS clips more than the mics of the anchors and the weathercaster (whose voice levels you have set before the newscast).

The mixing for the rock band or dramatic scene for the multimedia project can be quite complicated and is best left to an audio expert. Again, the initial choice of mics and their proper placement is more challenging than the mixing itself. The recording engineer may also have to patch the mics for multiple audio feeds, such as foldback, mix-minus, audience feed, and video feed. A ***mix-minus*** feed is a type of foldback in which you send into the studio a complete mix (usually the entire band) minus the sound generated in the studio (such as the lead singer's voice).

Surprisingly enough, the recording of a symphony orchestra is generally simpler. All you have to do is hang two good condenser mics close together in front of and high above the orchestra (about 10 feet or higher). Point them in a V to the left and the right of the orchestra, with the tip of the V facing the audience. Most of the balancing of instruments and volume will be done for you by the conductor.

BASIC OPERATION FOR FIELD AUDIO

As with all audio, the better the sound pickup, the easier the sound control during the production or in postproduction. (Refer to chapter 9 for advice on what mics to use outdoors and how to achieve optimal sound under various field conditions.)

Unless you are engaged in a big remote (see chapter 18), the audio equipment in the field is much less elaborate than its studio counterparts. This is not because you don't need to produce optimal audio in the field but simply because in electronic news gathering (ENG) the audio requirements are usually more modest. Similarly, in electronic field production (EFP) most of the quality control is done in postproduction in the studio. But don't be fooled into thinking that field audio is somehow easier than studio audio. On the contrary—sound pickup and recording in the field are actually more difficult. In the field you have to worry about wind noise, barking dogs, traffic sounds, airplanes overhead, chattering onlookers, or rooms that produce the dreaded inside-a-barrel sounds.

Keeping Sounds Separate

The key to good field audio is keeping the primary sounds as separate from the environmental sounds as possible. For example, you usually want to record the reporter's mic input on one audio track and the camera mic's input of primarily ambient sounds on the second audio track. There will nevertheless be circumstances in which you need to mix and balance several sound sources in the field. For example, if you have to cover an interview of four people in somebody's living room, you should balance their voices right then and there.

Whichever pickup technique you choose, be sure to record the sound at 16 bits/48 kHz (kilohertz), especially if you use a small digital camcorder. If you record at a lower bit rate (such as 12 bits), the audio will drift away from the video and remain out of sync when you transfer the footage to a DVD or hard drive.

Using the AGC in ENG and EFP

When you are on an ENG assignment and cannot watch the VU meter on the camcorder, switch on the ***automatic gain control (AGC)***, which boosts low sounds and reduces high-volume sounds so that they conform to the acceptable volume range. Note, however, that the AGC does not discriminate between wanted and unwanted sounds; it faithfully boosts the sound of the passing truck and the coughing crewmember and even the noise of the pauses when the field reporter is thinking of something clever to say. Whenever possible, and especially when in noisy surroundings, switch off the AGC, take a level, and try to watch the audio levels in the camcorder viewfinder or on the VU meter of the mixer.

Using an XLR Pad

An XLR pad is a simple and reliable way of preventing overload distortion or clipping when the incoming signal is too strong. For all practical purposes, it substitutes for the console's trim control, which you don't find on a portable mixer or camcorder. It looks like an XLR connector, and you use it in a similar way: you plug one end into the XLR connector of the (balanced) cable coming from the sound source; you plug the other end into the XLR connector of the cable going to the mixer, the audio input of the camcorder, or a separate audio recorder. This pad introduces a little bit of noise but is not nearly so detrimental to the field audio as clipping.

ENG/EFP Mixing

In ENG/EFP mixing there are always assignments for which you have to control more audio sources than two microphones. Even a simple assignment such as covering the opening of the local elementary school's new gym will most likely require that you mix at least three microphones: the field reporter's mic, the lectern mic for the speeches, and a mic to pick up the school show choir. If you run out of mic inputs on the field mixer, you can always cover the choir with the camera mic.

Despite the number of mics, the mixing itself is relatively simple. Once you have set the level for each input, you probably need to ride gain only for the reporter's mic during interviews and for the various speakers at the lectern. You may also want to bring up (increase the gain of) the choir mic during the performance. Although in an emergency you could try to pick up most of these sounds with the camera mic or by pointing a shotgun mic at the various areas, the multiple-mic setup and the portable mixer afford you better control.

Here are a few basic guidelines for live ENG/EFP mixing:

- Even if you have only a few inputs, label each one with what it controls, such as field reporter's mic, audience mic, and so forth. You would be surprised at how quickly you forget whose mic corresponds to which pot. If you have to turn over the audio control to someone else, he or she can take over without long explanations.

- If you do a complicated mix in the field, protect yourself by feeding it not only to the camcorder but also to a separate audio recorder for probable remixing in postproduction.

- If you work with a separate audio recorder, calibrate the audio output of the camera with the audio input of the audio recorder.

- If recording for postproduction, try to put distinctly different sound sources on separate audio tracks, such as the reporter's and guests' voices on one track and the speaker's lectern mic and the choir on the other. That way it will be easier during postproduction ***sweetening*** (getting rid of unwanted noises and improving the sound quality) to balance the reporter's voice with the other sounds.

It is usually easier to do complicated and subtle mixing in postproduction rather than live in the field. This does not mean that you should forgo filtering out as much unwanted sound as possible during the on-location pickup, assuming that the mixer has some basic quality controls available. If it doesn't, don't worry. If any sweetening is to be done, do it in postproduction. Remember that the more attention you pay to good sound pickup in the field, the less time you need in postproduction. A good wireless lavalier is often the best solution to good field audio.

MAIN POINTS

- The major production equipment for studio audio includes the audio console, the audio mixer, the patchbay, and tapeless and tape-based audio-recording systems.
- Audio consoles perform five major functions: input—select, preamplify, and control the volume of the various incoming signals; mix—combine and balance two or more incoming signals; quality control—manipulate the sound characteristics; output—route the combined signal to a specific output; and monitor—route the output or specific sounds to a speaker or headphones so that they can be heard.
- The audio control area of a television studio includes the audio control booth, a soundproof room that is used for the sound control of daily broadcasts.
- The basic audio operation includes: the audio system calibration, which means that all volume-unit (VU) meters or digital volume meters in the system must respond in the same way to a specific control tone; volume control; and live studio mixing.
- Digital sound signals have no headroom, which means that any signal that peaks above the maximum level will be clipped, causing irreparable damage to the recorded sound.
- Live studio mixing usually involves combining and balancing sounds while the production is in progress.
- In EFP and ENG, the key to good field audio is keeping the various sound sources reasonably separate so that they can be properly mixed in postproduction.
- The automatic gain control (AGC) is a convenient means of keeping the volume within acceptable limits, but in its automatic amplification it will not distinguish between desired sounds and unwanted sounds.
- The XLR pad is a device that, much like the trim control on the audio console, prevents input overload.

SECTION

10.2

Stereo, Surround Sound, and Sound Aesthetics

This section introduces stereo and surround sound and the principal aesthetic sound factors.

- **STEREO AND SURROUND SOUND**
  Making sound multidimensional

- **BASIC AESTHETIC SOUND FACTORS**
  Environment, figure/ground, perspective, continuity, and energy

STEREO AND SURROUND SOUND

As you read this brief discussion of stereo and surround sound, imagine the sound not as the only means of expression, such as the songs to which you may be listening, but as part of an audio/video structure.

Stereo Sound

Stereo sound, which defines especially the horizontal audio field (left-right or right-left positioning of the major audio source) is of little use when played back on a standard television set. Because the horizontal dimension of the screen is so small, any panning (horizontal positioning) of sound will inevitably lead to off-screen space, even if you sit in the sweet spot (the center where you perceive the two or more channels as one). But hearing a conversation happening in off-screen space when seeing both dialogue partners on-screen makes little sense. At best, stereo for standard television will enrich the general shape of the sound, that is, make it more spacious.

With large-screen, home-theater HDTV video projections, however, stereo sound becomes extremely important for keeping up with and balancing the high-energy video. In fact, the movielike experience when watching large-screen video projections will be greatly intensified by a surround-sound system.

Surround Sound

Surround sound is a technology that produces a soundfield in front of, to the sides of, and behind the listener, enabling one to hear sounds from the front, sides, and back. Developed originally for film reproduction, it is now used for HDTV and other large-screen home-theater arrangements. The most prevalent surround-sound system is still Dolby 5.1, which positions three speakers in front and two in back for sound reproduction. These five speakers are supported by an additional subwoofer that can be positioned anywhere between the speakers, sometimes in the rear, and sometimes between the center and left speakers. This .1 speaker can reproduce especially low-frequency, thunderous sounds. Because low frequencies are omnidirectional, the exact placement of the subwoofer is not critical. SEE 10.18

Just in case you want even more interference by the constant low-frequency sound dribble that accompanies the clipped police-speak dialogue in crime shows, you can install the 7.1 system, which has three speakers in front, two on the sides, and two in the rear. The .1 subwoofer can go anywhere. SEE 10.19

Good surround-sound mixing generally restricts on-screen dialogue to the center-front speaker and spreads action close to the actors to the three front speakers. But if the video shows the hero standing amid downtown traffic, playing in an orchestra, or dodging bombs, all five or seven speakers are active as well as the subwoofer.[1]

BASIC AESTHETIC SOUND FACTORS

As reiterated throughout this chapter, the bewildering array of audio equipment is of little use if you cannot exercise some aesthetic judgment—make decisions about how to work with television sound artistically and not just technically. Yet aesthetic judgment is not arbitrary or totally personal; there are some common aesthetic factors to which we all react similarly.

When dealing with television sound, you should pay particular attention to five basic aesthetic factors: environment, figure/ground, perspective, continuity, and energy.

1. See Stanley R. Alten, *Audio in Media*, 9th ed. (Boston: Wadsworth, 2011), pp. 54–58.

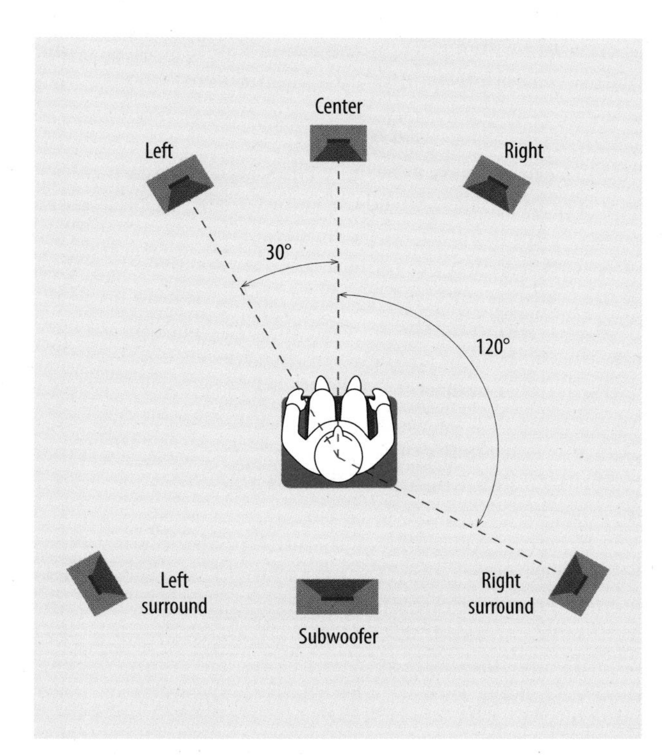

10.18 5.1 SURROUND SOUND
The 5.1 Dolby surround-sound system uses six speakers—three in front and two in back, plus a (.1) subwoofer, which can go just about anywhere on the floor.

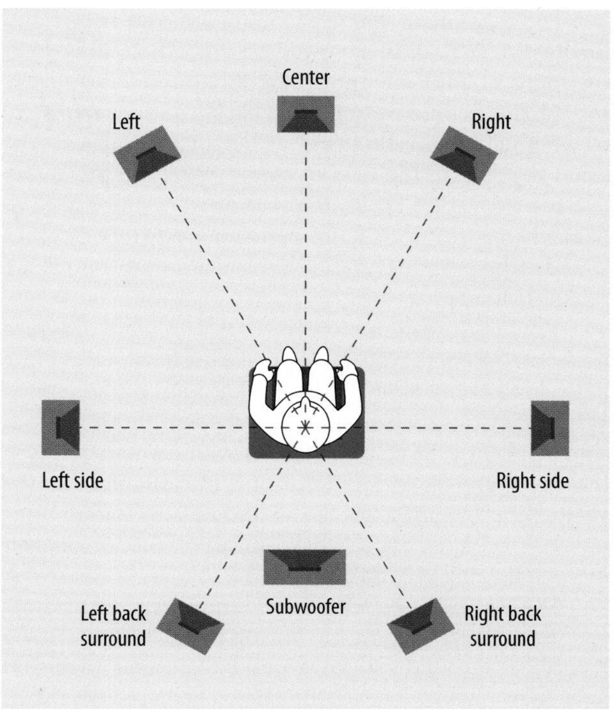

10.19 7.1 SURROUND SOUND
The 7.1 surround-sound system uses eight speakers—three in front, two on both sides of the listener, two in the rear, and a subwoofer. As with 5.1 Dolby surround sound, the placement of the nondirectional (.1) subwoofer is not critical.

Environment

In most studio recordings, we try to eliminate as much ambient sound as possible. In the field these sounds, when heard in the background of the main sound source, are often important indicators of where the event takes place or even how it feels. Such sounds help establish the general environment of the event.

For example, when covering a downtown fire, the sirens, the crackling of the fire, the noise of the fire engines and the pumps, the tense commands of the firefighters, and the agitated voices of onlookers are significant in communicating to the television viewer some of the excitement and apprehension on the scene. Now consider the recording of a small orchestra. In a studio recording, the coughing of a crewmember or musician during an especially soft passage would certainly prompt a retake. Not so in a live concert. We have learned to identify occasional coughing and other such environmental sounds as important indicators of the immediacy of the event.

Environmental sounds are especially important in ENG. By using an omnidirectional mic, you pick up the ambient sounds automatically with the main audio source. But, as mentioned before, if you intend to do some postproduction, use a more directional (cardioid) mic for recording the main sound source, such as the reporter or the guest, on audio track 1; use another mic (usually the camera mic) for recording the ambient sounds on audio track 2.

Figure/Ground

One important perceptual factor is the ***figure/ground*** principle, whereby we tend to organize our visual environment into a relatively mobile figure (a person or car) and a relatively stable background (a wall, houses, or mountains). We can expand this principle a little and say that we single out an event that is important to us and make it the foreground while relegating all other events to the background—the environment.

For example, if you are looking for your friend and finally discover her in a crowd, she immediately becomes the focus of your attention—the foreground—while the rest of the people become the background, regardless of whether they walk ahead of her or behind her. The same happens in the field of sound. We have the ability to perceive, within limits, the sounds we want or need to hear (the figure) while ignoring to a large extent all other sounds (the ground), even if the background sounds are relatively louder.

When showing a close-up (CU) of someone in a noisy environment, you should make the figure (CU of the person talking) louder and the background sounds softer. When showing the person in a long shot, however, you should increase the volume of the environmental sounds so that the figure/ground relationship is more in line with the video. When emphasizing the foreground, the sounds must not only be louder but also have more presence (explained in the following section).

You can now see why it is so important to separate sounds as much as possible during the recording. If you record background and foreground all on one track, you have to live with whatever the mic picked up; manipulating the individual sounds would be very difficult if not impossible. With the figure sounds on one track and the background sounds on the other, the manipulation is relatively easy.

Perspective

Sound perspective means that close-up pictures are matched with relatively nearby sounds, and long shots correspond with sounds that seem to come from farther away. Close sounds have more presence than distant sounds—a sound quality that makes us feel in proximity to the sound source. Generally, background sounds have less presence, and close-ups have more presence. Experienced singers hold the mic close to the mouth during intimate passages but pull it back a little when the song becomes less personal.

Such a desirable variation of sound presence is virtually eliminated when using lavalier mics in a drama. Because the distance between mic and mouth is about the same for each actor, their voices exhibit the same presence regardless of whether they are seen in a close-up or a long shot. The necessary presence must then be achieved in time-consuming and costly postproduction. This is why boom mics are still preferred in many multicamera productions of television plays such as soap operas. The boom mic can be close to an actor during a close-up and moved somewhat farther away during a long shot to stay out of the picture—a simple solution to a big problem.

Continuity

Sound continuity is especially important in postproduction. You may have noticed the sound quality of a reporter's voice change depending on whether he was speaking on- or off-camera. When on-camera the reporter used one type of microphone and was speaking from a remote location; then he returned to the acoustically treated studio to narrate the off-camera segments of the video-recorded story, using a high-quality mic. The change in microphones and locales gave the recordings distinctly different qualities. This difference may not be too noticeable during the actual recordings, but it becomes obvious when they are edited together in the final show.

How can you avoid such continuity problems? First, have the reporter record the narration on-site. Second, use identical mics (or mics that produce a similar sound quality) for the on- and off-camera narrations. Third, always record some ***ambience*** at the on-camera location. You can then mix these sounds with the off-camera, voice-over narration. Fourth, if you have time for a sweetening session, try to match the on-camera sound quality through equalization and reverberation. When producing this mix, feed the ambient sounds to the reporter through earphones while he or she is doing the voice-over narration; this will help recapture the on-site energy.

Sometimes you may hear the ambience punctured by brief silences at the edit points. The effect is as startling as when an airplane engine changes its pitch unexpectedly. The easiest way to restore the background continuity is to cover up these silences with prerecorded ambience. Always record a few minutes of "silence" (room tone or background sounds) before and after video-recording or whenever the ambience changes decisively (such as a concert hall with and without an audience). ZVL4 EDITING→ Continuity→ sound

Sound is also a chief element in establishing visual continuity. A rhythmically precise piece of music can help a disparate series of pictures seem continuous. Music and sound are often the important connecting link among abruptly changing visual sequences.

Energy

Unless you want to achieve a special effect through contradiction, you should match the general energy of the pictures with a similar sound intensity. Energy refers to all the factors in a scene that communicate a certain degree of aesthetic force and power. Obviously, high-energy scenes, such as a series of close-ups of an ice-hockey game or a rock band in action, can stand higher-energy sounds than can a more tranquil scene, such as lovers walking through a meadow. Good television audio depends a great deal on your ability to sense the general energy of the pictures or sequences and to adjust the volume and the sound presence accordingly. ZVL5 AUDIO→ Aesthetics→ continuity | environment | sound perspective | try it

MAIN POINTS

- Surround-sound technology uses several speakers that surround the listener. The customary 5.1 system places three speakers in front of the listener and two in back to produce a soundfield that surrounds the listener. The .1 woofer can be placed anywhere. The 7.1 system has three speakers in front, two to the side, and two in back. The .1 subwoofer can go anywhere.
- The five major aesthetic factors in sound control are: environment—sharpening an event through ambient sounds; figure/ground—emphasizing the most important sound source over the general background sounds; perspective—matching close-up pictures with nearby sounds, and long shots with distant sounds; continuity—maintaining the quality of sound when combining various takes; and energy—matching the force and the power of the pictures with a similar intensity of sound.

ZETTL'S VIDEOLAB

For your reference or to track your work, the *Zettl's VideoLab* program cues in this chapter are listed here with their corresponding page numbers.

CHAPTER

11

Lighting

Why bother with lighting when even a small and inexpensive camcorder can see virtually in the dark? This is a valid point so long as the available light in which you are video recording is the right type and in the right place to produce good pictures. Unfortunately, this is rarely the case, and we have to give nature a little help. This "little help" is called lighting. Specifically, lighting helps the video camera see well, that is, produce technically optimal pictures. It also helps the viewer see well—to recognize what things and people look like and where they are in relation to one another and to their immediate environment. Finally, it helps establish for the viewer a specific mood that intensifies the feeling about the event.

Section 11.1, Lighting Instruments and Lighting Controls, describes the tools you need to accomplish effective lighting. Section 11.2, Light Intensity, Lamps, and Color Media, introduces a few more elements about light, how to control and measure it, and how to use colored light.

KEY TERMS

barn doors Metal flaps mounted in front of a lighting instrument that control the spread of the light beam.

baselight Even, nondirectional (diffused) light necessary for the camera to operate optimally. Normal baselight levels are 150 to 200 foot-candles (1,500 to 2,000 lux) at *f*/8 to *f*/16. Also called *base.*

broad A floodlight with a broadside, panlike reflector.

clip light Small internal reflector spotlight that is clipped to scenery or furniture with a gator clip. Also called *PAR (parabolic aluminized reflector) lamp.*

color temperature The standard by which we measure the relative reddishness or bluishness of white light. It is measured on the Kelvin (K) scale. The standard color temperature for indoor light is 3,200K; for outdoor light it is 5,600K. Technically, the numbers express Kelvin degrees.

cookie A popularization of the original term *cucoloris.* Any pattern cut out of thin metal that, when placed inside or in front of an ellipsoidal spotlight (pattern projector), produces a shadow pattern. Also called *gobo.*

dimmer A device that controls the intensity of light by throttling the electric current flowing to the lamp.

ellipsoidal spotlight Spotlight that produces a very defined beam, which can be shaped further by metal shutters.

flag A thin, rectangular sheet of metal, plastic, or cloth used to block light from falling on specific areas. Also called *gobo.*

floodlight Lighting instrument that produces diffused light with a relatively undefined beam edge.

fluorescent Lamps that generate light by activating a gas-filled tube to give off ultraviolet radiation, which lights up the phosphorous coating inside the tube.

foot-candle (fc) The amount of light that falls on an object. One foot-candle is the amount of light from a single candle that falls on a 1-square-foot area located 1 foot away from the light source.

Fresnel spotlight One of the most common spotlights, named after the inventor of its lens. Its lens has steplike concentric rings.

gel Generic term for color filters put in front of spotlights or floodlights to give the light beam a specific hue. *Gel* comes from *gelatin,* the filter material used before the invention of more durable plastics. Also called *color media.*

HMI light Stands for *hydragyrum medium arc-length iodide light.* Uses a high-intensity lamp that produces light by passing electricity through a specific type of gas. Needs a separate ballast.

incandescent The light produced by the hot tungsten filament of ordinary glass-globe or quartz-iodine lamps (in contrast to fluorescent light).

incident light Light that strikes the object directly from its source. An incident-light reading is the measure of light in foot-candles (or lux) from the object to the light source. The light meter is pointed directly into the light source or toward the camera.

Kelvin (K) Refers to the Kelvin temperature scale. In lighting it is the specific measure of color temperature—the relative reddishness or bluishness of white light. The higher the K number, the more bluish the white light. The lower the K number, the more reddish the white light.

LED light Stands for *light emitting diode light.* Its light source is an array of semiconductors (a solid-state electronic device) that emits light when electricity passes through. Can produce different-colored light.

lumen The light intensity power of one candle (light source radiating isotropically, i.e., in all directions).

luminaire Technical term for a lighting instrument.

luminant Lamp that produces the light; the light source.

lux European standard unit for measuring light intensity. 11.75 lux = 1 fc; usually roughly translated as 10 lux = 1 fc.

neutral density (ND) filter Filter that reduces the incoming light without distorting the color of the scene.

patchboard A device that connects various inputs with specific outputs. Also called *patchbay.*

pattern projector An ellipsoidal spotlight with a cookie (cucoloris) insert, which projects the cookie's pattern as a cast shadow.

quartz A high-intensity incandescent light whose lamp consists of a quartz or silica housing (instead of the customary glass) that contains halogen gas and a tungsten filament. Produces a very bright light of stable color temperature (3,200K). Also called *TH (tungsten-halogen) lamp.*

reflected light Light that is bounced off of the illuminated object. A reflected-light reading is done with a light meter held close to the illuminated object.

scoop A scooplike television floodlight.

scrim A spun-glass material that is put in front of a lighting instrument as an additional light diffuser or intensity reducer.

softlight Television floodlight that produces extremely diffused light.

spotlight A lighting instrument that produces directional, relatively undiffused light with a relatively well-defined beam edge.

SECTION

11.1

Lighting Instruments and Lighting Controls

When you turn on the light in a room, you are concerned primarily with having enough illumination to see well and get around. Contrary to the lighting in your home, however, lighting for television and digital cinema must also please the camera and fulfill certain aesthetic functions, such as simulating outdoor or indoor light or creating a happy or sinister mood. Studio lighting requires instruments that can simulate bright sunlight, a street lamp at a deserted bus stop, the efficiency of a hospital operating room, and the horror of a medieval dungeon. It must also reflect the credibility of a news anchor, the high energy of a game show, and the romantic mood in a love scene.

When in the field, you need lighting instruments that are easy to transport and set up and flexible enough to work in a great variety of environments for a multitude of lighting tasks. This section describes the major studio and field lighting instruments and the types of lighting controls. Section 11.2 addresses light intensity, types of lamps, and color media. The techniques of lighting are covered in chapter 12.

- **STUDIO LIGHTING INSTRUMENTS**
  Spotlights and floodlights
- **FIELD LIGHTING INSTRUMENTS**
  Portable spotlights, portable floodlights, and camera lights
- **LIGHTING CONTROL EQUIPMENT**
  Mounting devices, directional controls, intensity controls, and the basic principle of electronic dimmers

STUDIO LIGHTING INSTRUMENTS

All studio lighting is accomplished with a variety of spotlights and floodlights. These instruments, technically called ***luminaires***, are designed to operate from the studio ceiling or from floor stands.

Spotlights

Spotlights produce directional, well-defined light that can be adjusted from a sharp light beam like the one from a focused flashlight or a car headlight to a softer beam that is still highly directional but that lights up a larger area. All studio spotlights have a lens that helps sharpen the beam. Most studio lighting uses two basic types of spotlights: the Fresnel and the ellipsoidal spot.

Fresnel spotlight Named for the early-nineteenth-century French physicist Augustin Fresnel (pronounced “fra-nel”) who invented the lens used in it, the ***Fresnel spotlight*** is widely used in television studio and film production. SEE 11.1 It is relatively lightweight and flexible and has a high output. The spotlight can be adjusted to a “flood” beam position, which gives off a widespread light beam; or it can be “spotted,” or focused to a sharp, clearly defined beam.

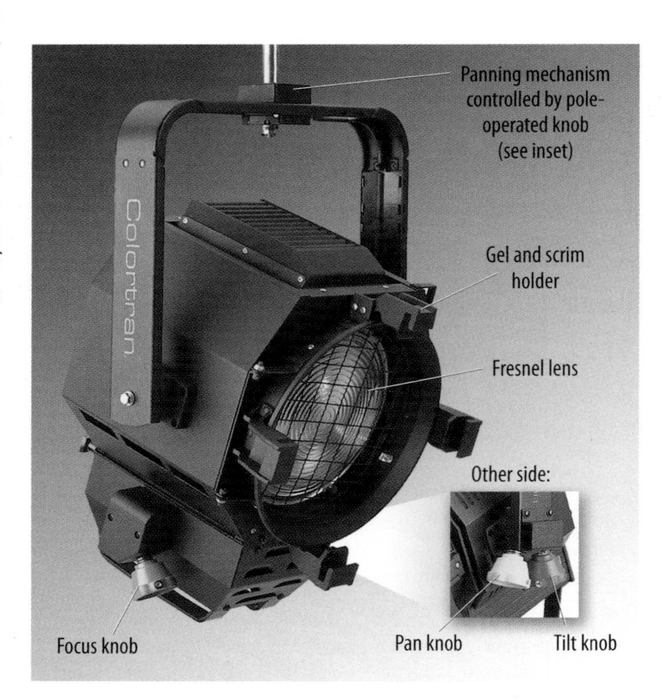

11.1 FRESNEL SPOTLIGHT
This spotlight is one of the most useful lighting instruments in the studio. Note the pan, tilt, and focus controls, which can be operated from the studio floor with a lighting pole.

11.1 PHOTO COURTESY LEVITON, COLORTRAN DIVISION

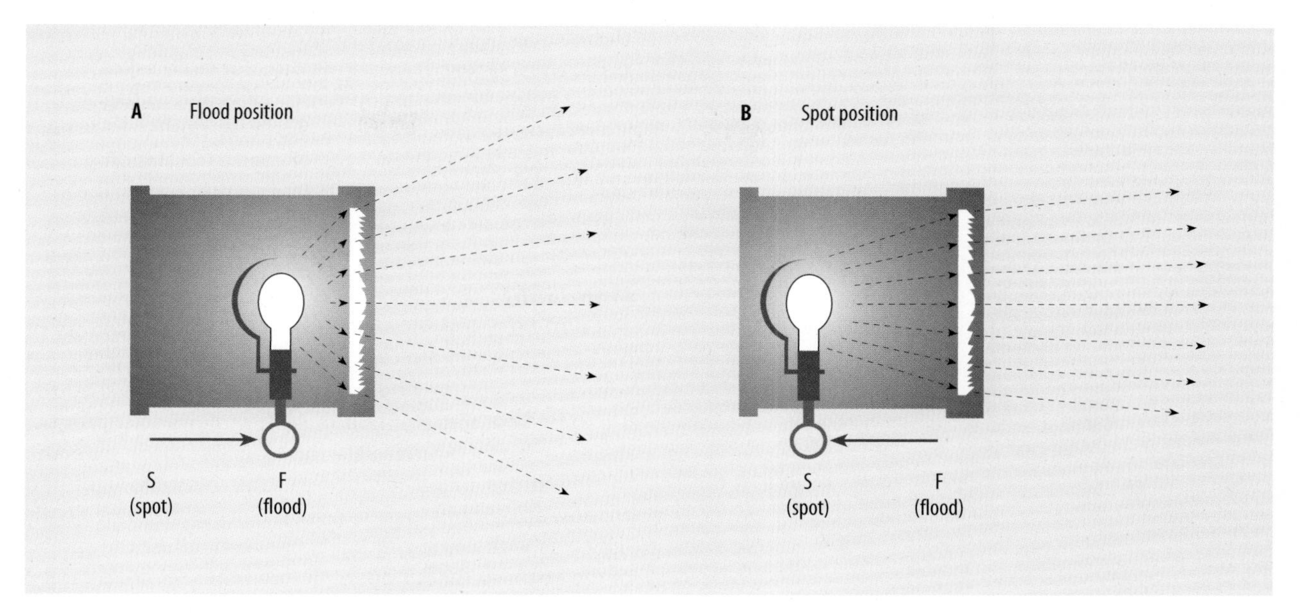

11.2 BEAM CONTROL OF FRESNEL SPOTLIGHT
A To flood (spread) the beam, turn the focus spindle, ring, or knob so that the lamp-reflector unit moves toward the lens.
B To spot (focus) the beam, turn the focus spindle, ring, or knob so that the lamp-reflector unit moves away from the lens.

You manipulate the relative spread of the beam with a focus control that changes the distance between the lamp and the lens. Most Fresnel spotlights have a lamp-reflector unit inside the lighting instrument that slides toward or away from the lens. To spot, or focus, the beam, turn the control so that the lamp-reflector unit moves *away* from the lens. To flood, or spread, the beam, turn the control so that the lamp-reflector unit moves *toward* the lens. Even in the flood position, the beam of the spotlight is still directional and much sharper than that of a floodlight. The flood position softens the beam (and with it the shadows) and simultaneously reduces the amount of light falling on the object. Always adjust the beam gently. When the lamp is turned on, its hot filament is highly sensitive to shock. **SEE 11.2**

Some Fresnel spots have external knobs with which you can also control the pan and the tilt of the instrument without climbing a ladder and doing it manually (see figure 11.1).

Fresnel spotlights come in different sizes, depending on how much light they produce. Obviously, the larger instruments produce more light than the smaller ones. The size of Fresnel spotlights is normally given in the wattage of the lamp. For example, you might be asked to rehang the "1K" (1 kilowatt [kW] = 1,000 watts) Fresnel or change the lamp in the "2K" Fresnel.

In most television studios, the most common Fresnels are the 1kW and the 2kW instruments. For maximum lighting control, technicians usually prefer to operate with as few (yet adequately powerful) lighting instruments as possible. The increased sensitivity of cameras has made the 1kW Fresnel the workhorse in average-sized studios.

Ellipsoidal spotlight A favorite for theater lighting, the ***ellipsoidal spotlight*** produces a sharp, highly defined beam. Even when in a flood position, the ellipsoidal beam is still sharper than the focused beam of a Fresnel spot. Ellipsoidal spots are generally used when specific, precise lighting tasks are necessary. For example, if you want to create pools of light reflecting off the studio floor, the ellipsoidal spot is the instrument to use.

As with the Fresnel, you can spot and flood the light beam of the ellipsoidal. Instead of sliding the lamp inside the instrument, however, you focus the ellipsoidal spot by moving its lens in and out. You can even shape the beam into a triangle or rectangle by adjusting the four metal shutters inside the instrument. **SEE 11.3**

Ellipsoidal spotlights come in sizes from 500W (watts) to 2kW, but the most common is 750W. Some ellipsoidal spotlights can also be used as ***pattern projectors***. These instruments are equipped with a slot next to the beam-shaping shutters, which can hold a metal pattern called a

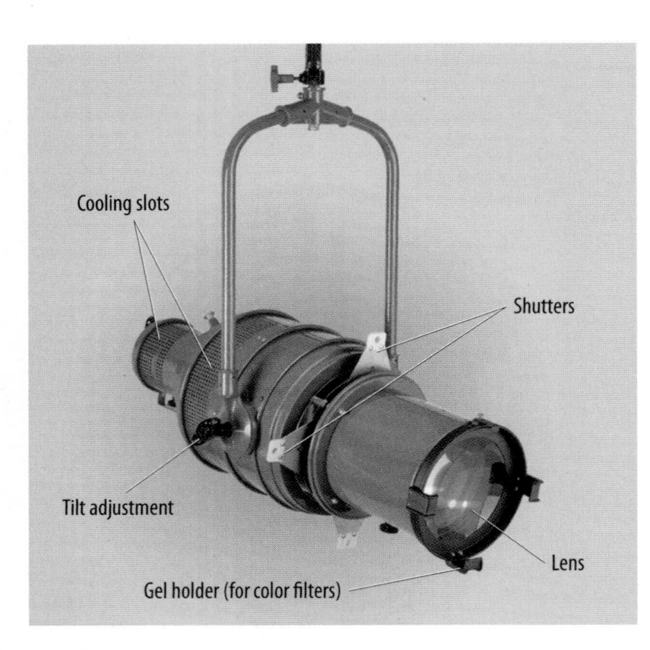

11.3 ELLIPSOIDAL SPOTLIGHT
The highly focused beam of the ellipsoidal spotlight can be further shaped by shutters. It produces the most directional beam of all the spotlights.

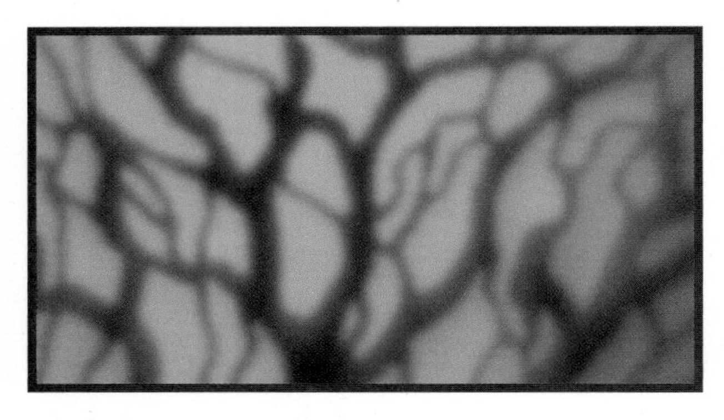

11.4 COOKIE PATTERN ON CYCLORAMA
The cookie pattern is projected by an ellipsoidal spotlight (pattern projector) in which you can insert a variety of metal templates, called cookies. Because the spotlight can be focused, you can make the projected pattern look sharp or soft.

cucoloris, or ***cookie*** for short. When the spotlight is turned on, the cookie causes a shadow pattern on any surface. Most often it is used to break up flat surfaces, such as the cyclorama (large cloth drape used for backing of scenery) or the studio floor. **SEE 11.4**

To make lighting terminology a little more confusing, some lighting people call these metal templates "gobos." Unfortunately, *gobo* seems to have as ambiguous a meaning as the word *spring*. If a lighting director (LD) asked you to fetch a gobo, he or she could mean a cookie; a flag, which is a rectangular piece of metal, plastic, or cloth to keep light from falling onto certain areas; or even a freestanding piece of scenery, such as prison bars or a picture frame, through which the camera can shoot a related scene.

Floodlights

Floodlights are designed to produce great amounts of highly diffused light. They are often used as principal sources of light (key lights) in situations where shadows are to be kept to a minimum, such as news sets, product displays, and commercials for skin lotion or makeup; to slow down falloff (reduce contrast between light and shadow areas); and to provide baselight. With some floodlights, as with some spotlights, you can adjust the spread of the beam so that undue spill into other set areas can be minimized. You can also create a floodlight effect by flooding the beam of a spotlight and diffusing it further with a ***scrim***—a spun-glass material held in a metal frame—in front of the instrument.

There are four basic types of studio floodlights: the scoop; the softlight and the broad; the fluorescent floodlight bank; and the strip, or cyc, light.

Scoop Named for its peculiar scooplike reflector, the ***scoop*** is one of the more popular floodlights. Although it has no lens, it nevertheless produces a fairly directional but diffused light beam. **SEE 11.5**

The standard scoop emits a fixed diffused beam of light. You can increase the diffusion of the beam by attaching a scrim (see figure 11.5). Although the light output through the scrim is considerably reduced, some lighting people put scrims on all scoops, not only to produce highly diffused light but also to protect studio personnel in case the hot lamp inside the scoop shatters. Most scoops operate with a 1,000W or 1,500W lamp. ZVL1 LIGHTS→ Instruments→ studio | field

Softlight and broad ***Softlights*** are used for even, extremely diffused lighting. They have large tubelike lamps, a diffusing reflector in the back of the large housing, and a diffusing material covering the front opening to further scatter the light. Softlights are often used for flat (virtually shadowless) lighting setups. You can also use softlights to increase the baselight level without affecting specific

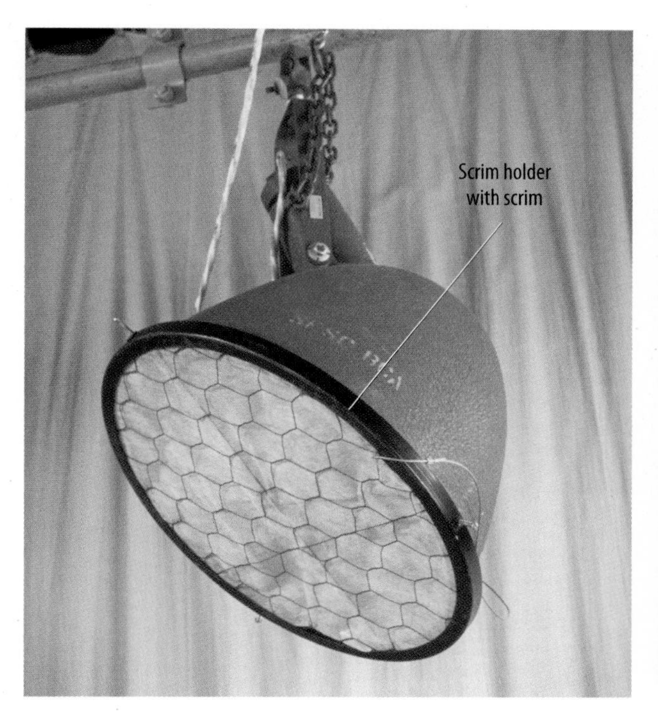

11.5 SCOOP
The scoop is a rugged, all-purpose floodlight. Its scooplike reflector gives its beam some directionality. This scoop has a scrim attached to soften the beam.

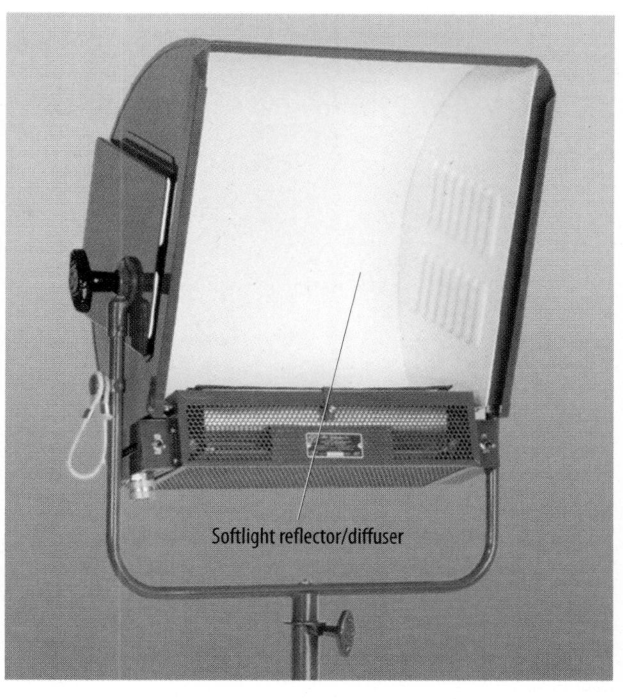

11.6 INCANDESCENT SOFTLIGHT
The softlight produces extremely diffused light and is used for illumination with slow falloff. It renders a scene almost shadowless.

11.5 PHOTO BY EDWARD AIONA
11.6 PHOTO COURTESY MOLE-RICHARDSON COMPANY

lighting where highlights and shadow areas are carefully controlled. For example, if a scene calls for a hallway with alternating bright and dark areas, you can lighten up the dark areas with softlights to provide enough baselight for the camera to see well even in the dark areas. Softlights come in various sizes and use incandescent or HMI lamps, which are discussed in section 11.2. **SEE 11.6**

The ***broad*** (from *broadside*) is similar to a softlight except that it has a higher light output that causes more distinct shadows. Broads also have some provision for beam control. They are generally used in digital cinema production to evenly illuminate large areas with diffused light. **SEE 11.7** Smaller broads emit a more directional light beam than do the larger ones, for evenly illuminating smaller areas. To permit some directional control over the beam, some broads have barn doors—movable metal flaps—to block gross light spill into other set areas.

Fluorescent floodlight bank The fluorescent floodlight bank goes back to the early days of television lighting, when the banks were large, heavy, and inefficient. Today's

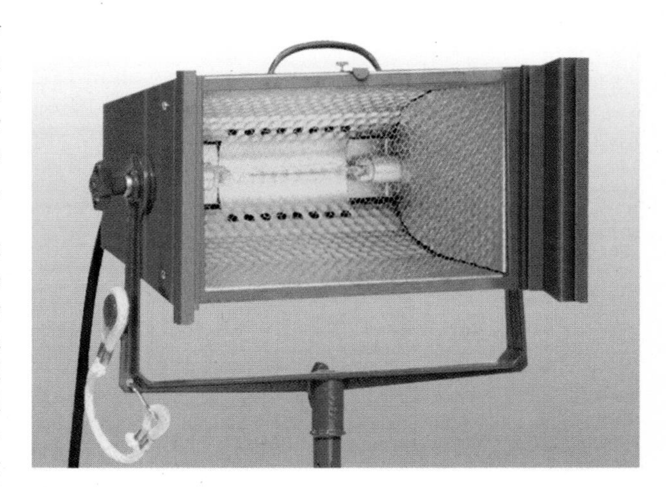

11.7 LARGE BROAD
This instrument illuminates a fairly large area with diffused light. Its light output is normally greater than that of a softlight of equal size.

11.7 PHOTO COURTESY MOLE-RICHARDSON COMPANY

11.8 FLUORESCENT FLOODLIGHT BANK
These floodlight banks act like softlights except that they do not get as hot as incandescent floodlights of equal output. Some floodlight banks use lamps that operate on fluorescent-like principles.

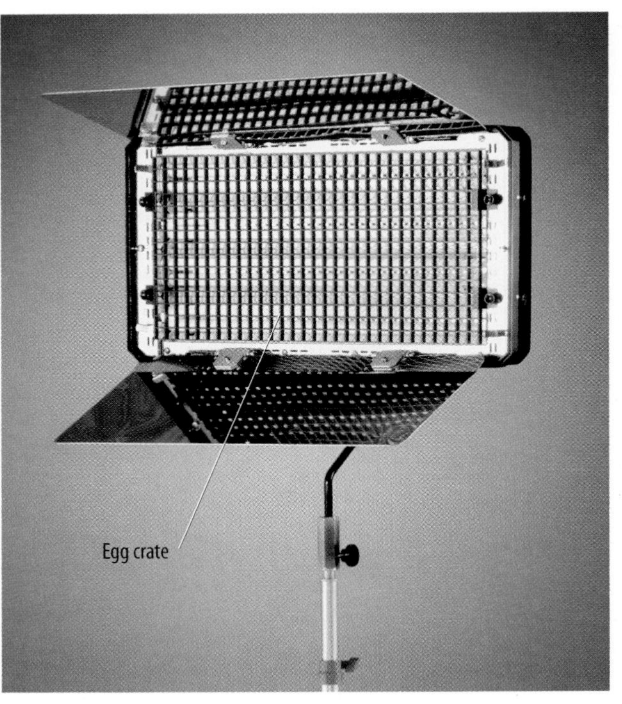

11.9 EGG CRATE ON FLUORESCENT FIXTURE
The egg crate makes the floodlight from a fluorescent fixture more directional without influencing the light's softness.

11.8 PHOTO COURTESY LOWEL-LIGHT MFG., INC.
11.9 PHOTO COURTESY LOWEL-LIGHT MFG., INC.

fluorescent banks are relatively lightweight, are much more efficient, and can burn close to the standard indoor color temperature (3,200K), giving off a warmer (more reddish) light. By simply changing the tubes, you can approximate the standard outdoor color temperature (5,600K) or achieve even higher ones (more-bluish light) that resemble the extremely bluish midday sunlight filtered by a hazy sky. (Color temperature is explained in detail in section 11.2 and chapter 12. For now it should suffice to know that a high color temperature refers to white light with a slight bluish tint; a low color temperature refers to white light with a slight reddish tint. Color temperature has nothing to do with how hot a lamp gets.)

Other advantages of fluorescent banks are that they use less power than incandescent lamps and they burn much more coolly—a definite advantage when lighting interiors with poor ventilation. The disadvantages are that fluorescent banks are still quite large and bulky and their color spectrum is sometimes uneven. This means that the light emitted does not reproduce all colors faithfully. Some lamps still cause a persistent and noticeable greenish tint.

Floodlight banks have rows of low-powered fluorescent lamps inside a housing that looks similar to a softlight. **SEE 11.8** Some fluorescent studio lighting fixtures have a gridlike contraption, called an egg crate, attached to make the light beam more directional without losing its softness. **SEE 11.9**

Strip, or cyc, light This type of instrument is commonly used to achieve even illumination of large set areas, such as the cyclorama (cyc) or some other uninterrupted background. Similar to the border, or cyc, lights of the theater, television strip lights consist of rows of three to 12 quartz lamps mounted in long, boxlike reflectors. The more sophisticated strip lights have, like theater border lights, colored-glass frames for each of the reflector units so that the cyc can be illuminated in different colors. **SEE 11.10**

11.10 STRIP, OR CYC, LIGHT
Strip lights are used to illuminate cycloramas and other large areas that need even illumination.

11.10 PHOTO COURTESY MOLE-RICHARDSON COMPANY

You can also use strip lights as general floodlights by suspending them from the studio ceiling, or you can place them on the studio floor to separate pillars and other set pieces from the lighted background. Strip lights are sometimes used for silhouette lighting (where the background is evenly illuminated and the foreground objects remain unlit) and special-effects chroma-key lighting (see chapter 12). ZVL2 LIGHTS→ Design→ silhouette

For relatively static scenes, such as news or interviews, you will find that it is often easier to use the much lighter and more flexible field lighting instruments, even if there are a great variety of studio lights hanging from the lighting grid. There are several advantages to using these lighter instruments instead of those on the grid: you can place the small instruments anywhere in the studio with a minimum of effort, they can be easily repositioned to get the desired lighting effect, they draw considerably less power than the larger instruments, and they generate less heat. The following section highlights some of the principal portable field lighting instruments.

FIELD LIGHTING INSTRUMENTS

You can use studio lighting instruments on remote locations, but you'll find that most of them are too bulky to move around easily, their large plugs do not fit the normal household receptacles, and they draw too much power. Once in place and operating, they may not provide the amount or type of illumination you need for good field lighting. Unless you do big remotes or a movie scene where the lighting requirements rival studio lighting, you need instruments that are easy to transport and quick to set up and that give you the lighting flexibility you need in the field.

Although many portable lighting instruments fulfill dual spotlight and floodlight functions, you may still find it useful to group them, like studio lights, into those categories. Note, however, that by putting a diffusing scrim in front of the spotlight, or by bouncing its beam off the ceiling, a wall, or a large foam-core board, the spotlight will take on the function of a floodlight. On the other hand, you can use a small floodlight and control its beam with barn doors so that it illuminates a relatively limited area, operating as a spot.

Portable Spotlights

Portable spotlights, often called hard lights (as the opposite of softlights), are easy to set up and transport and small enough to be effectively hidden from camera view even in cramped interiors. They come in different sizes and with or without a lens. Some portable spots have a Fresnel lens, just like the Fresnel spots hanging from the studio lighting grid, and some have no lens and are, therefore, called open-face spots.

Spotlights with lenses These include low-powered (up to 750W) Fresnel spots, the much smaller (125W to 250W) spotlight with a prismatic lens or simply a glass cover, and the HMI spots.

The portable Fresnels are identical to the ones hanging in the studio except that they operate with lower-wattage lamps. They are usually mounted on a tricaster light stand. SEE 11.11

The smaller spotlights with a plain (prismatic) lens emit enough light to be used as a key light for an interview in a small area. The beam of these small spotlights can be

11.11 SMALL FRESNEL SPOTLIGHT
This low-powered (300W to 650W) Fresnel spotlight is especially effective in lighting for electronic field production (EFP). You can focus or diffuse its beam and attach four-way barn doors and color media (gels).

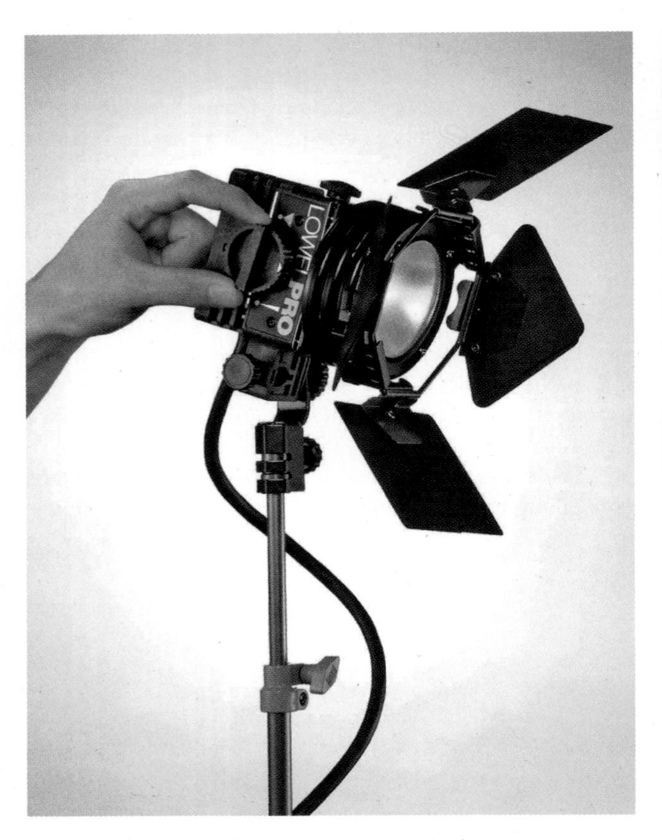

11.12 SMALL SPOTLIGHT WITH PRISMATIC LENS
This low-powered (125W and 250W) spotlight has an efficient reflector and a plain lens (not a Fresnel) that, despite its small size, make it into a highly effective lighting instrument. It is primarily used for EFP and electronic news gathering (ENG).

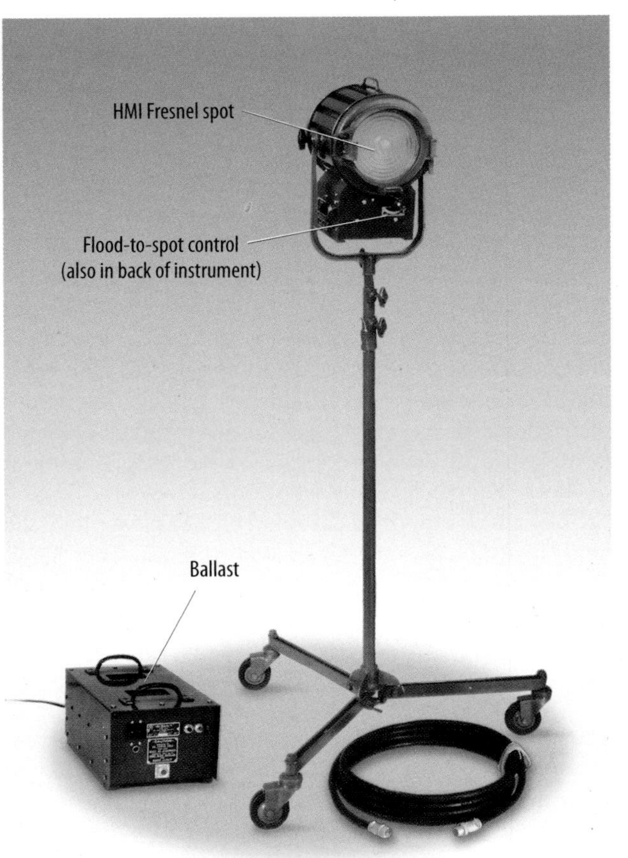

11.13 PORTABLE HMI FRESNEL SPOT WITH BALLAST
The HMI Fresnel spotlight burns with the daylight standard (5,600K). It needs considerably less power than does an incandescent light of equal intensity.

spread or focused just like a larger Fresnel, which makes them ideal instruments in tight quarters. SEE 11.12

When working in digital cinema, you will undoubtedly use the portable HMI (200W or higher) Fresnel spots. ***HMI lights*** are highly efficient, which means that they use very little power to produce a great amount of light. One of the major advantages of these superefficient HMI lights is that you can use up to five 200W instruments simultaneously without overloading a single circuit, assuming that nothing else is plugged into the same circuit. Because you plug most of the lights into household outlets, you can light most interiors with minimal time and effort.

HMI lights also have some serious drawbacks, however. They are expensive and need a separate ballast unit to run the lamp. They can cause an audio hum or, when shooting at specific frame rates, a noticeable flicker. The larger instruments are normally used outdoors to fill in dense shadows caused by the supreme spotlight—the sun. In any case, test them out with the camera before using them on-location. SEE 11.13

Open-face spotlights Mainly because of weight considerations and light efficiency, the open-face spotlight has no lens. This permits a higher light output, but the beam is less even and precise than that of the spots with a lens. In most remote lighting tasks, however, a highly defined beam offers no particular advantage. Because you usually have to work with a minimum of lighting instruments, a fairly general illumination is often better than a highly defined one. SEE 11.14

You can spot or spread the beam of the high-efficiency quartz lamp through a focus control lever or knob on the

11.12 PHOTO COURTESY LOWEL-LIGHT MFG., INC.
11.13 PHOTO COURTESY MOLE-RICHARDSON COMPANY

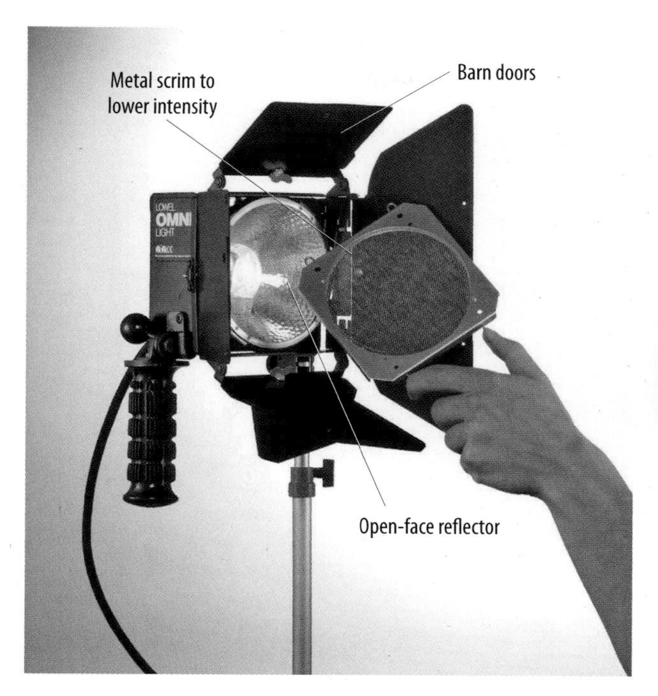

11.14 OPEN-FACE SPOT
The open-face (external reflector) spotlight has no lens. Its beam spread can be adjusted to a spot or moderate flood position. It is one of the most versatile lighting instruments in field production.

11.15 CLIP LIGHT
The clip light, or PAR lamp, consists of a normal internal reflector bulb (such as a PAR 38), a socket with an on/off switch, and a clip for fastening the lamp to a support. You can also get a small metal housing with a two-way barn door for the clip light.

11.14 PHOTO COURTESY LOWEL-LIGHT MFG., INC.
11.15 PHOTO BY HERBERT ZETTL

back. Unfortunately, the focused beam is not always even. When you place the spot close to an object, you may notice (and the camera surely will) that the rim of the beam is intense and "hot" while the center of the beam has a "hole"—a low-intensity dark spot. If you place the instrument too close when lighting a face, for example, the hot spot may cause a glowing white area surrounded by red on the lighted face or, at best, a distinct color distortion. But by spreading the beam a little and pulling the instrument farther away from the person, or by placing a scrim in front of the spotlight, you can usually correct the problem. In fact, when using a spot for general indoor lighting, you should routinely place a scrim or similar light-diffusing material in front of the instrument.

Most open-face spots use 300W to 500W lamps and can therefore be plugged into a regular household receptacle without risking a circuit overload. They can also be adapted to run off of 12V (volt) batteries. Most of these instruments have a power switch close to the lamp so that you can turn it off anytime it is not in use. All these small spotlights come as part of a lighting kit—a suitcase containing several such instruments and light stands.

Open-face lights get extremely hot: don't touch the front of the instrument when handling it, and place it far enough away from curtains and other combustible materials to prevent fires. With any lighting instrument, always be careful not to overload the circuit; that is, do not exceed the circuit's rated amperage by plugging in too many instruments per outlet. Extension cords also add their own resistance to that of the lamp, especially when they get warm. Ordinary household outlets can tolerate a load of up to 1,200 watts. You can therefore plug four 300W spots or two 500W instruments into a single circuit without risking an overload (see chapter 12).

Internal reflector spot Although these lights are not part of the usual arsenal of field lighting equipment, they have been used successfully in many small-area lighting situations. This small spotlight is also known as a ***clip light*** because it is usually clipped onto things. These lamps are also called PAR lamps for their parabolic aluminized reflector, which is the inside coating of the lamp. SEE 11.15

Internal reflector spots come in a variety of beam spreads, from a soft, diffused beam to a hard, precisely shaped beam. For even better beam control, as well as for the protection of the internal reflector bulb, the lamp can be used in a metal housing with barn doors attached.

Portable Floodlights

Most ENG/EFP lighting requires a maximum amount of even illumination with a minimum of instruments and

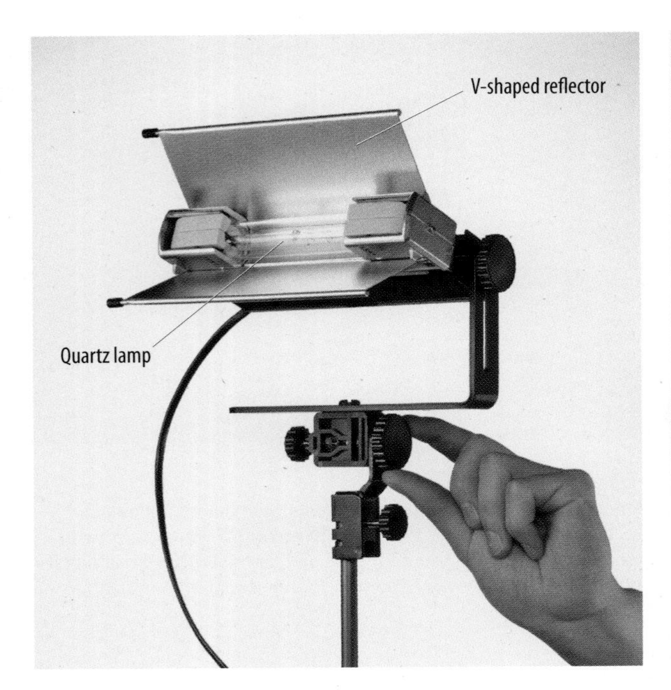

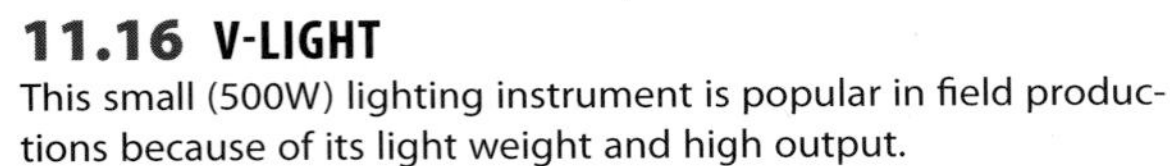
11.16 V-LIGHT
This small (500W) lighting instrument is popular in field productions because of its light weight and high output.

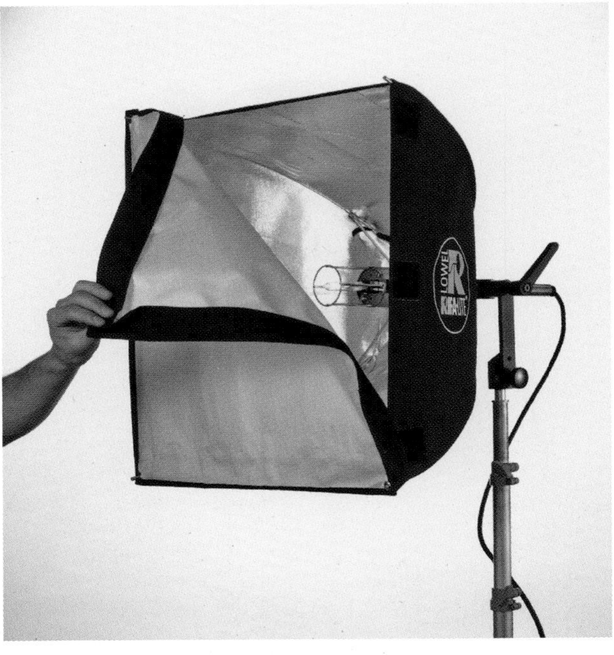

11.17 SOFT-BOX
The softlight comes as a single unit comprising the lamp and the diffusion tent. It can be folded up for easy transport.

11.16 PHOTO COURTESY LOWEL-LIGHT MFG., INC.
11.17 PHOTO COURTESY LOWEL-LIGHT MFG., INC.

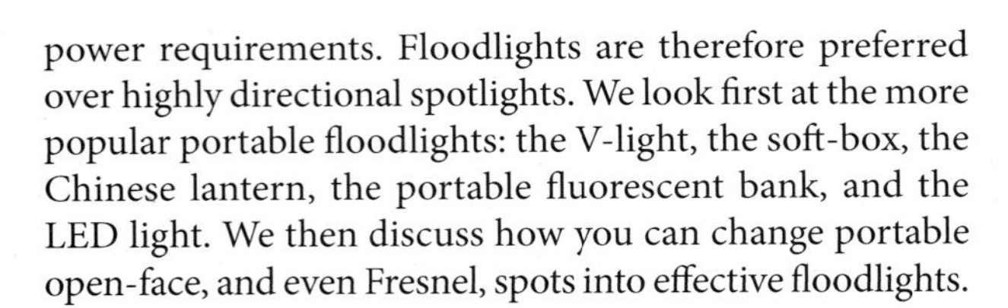

power requirements. Floodlights are therefore preferred over highly directional spotlights. We look first at the more popular portable floodlights: the V-light, the soft-box, the Chinese lantern, the portable fluorescent bank, and the LED light. We then discuss how you can change portable open-face, and even Fresnel, spots into effective floodlights.

V-light One of the more popular floodlights is the V-light. Although the V-light was originally a specific floodlight manufactured by the Lowel-Light Manufacturing company, it has become the generic name for any small instrument that consists of a large (500W) incandescent quartz lamp mounted in a V-shaped metal reflector. **SEE 11.16** The V-light is highly portable and easy to set up and can light up large areas relatively evenly. Be careful when handling such lights—they get very hot. Don't touch them when they are switched on, and keep them away from combustible materials.

Soft-box The soft-box (250W to 1kW), also called light box or tent, is simply a black heat-resistant cloth bag with a scrim at its opening. It has a built-in quartz (tungsten-halogen) lamp. You can also attach a small Fresnel spot with a ring adapter or use a variety of fluorescent lamps. Because it is lightweight and can be folded up, it makes an ideal portable softlight. **SEE 11.17**

Chinese lantern A highly effective portable softlight is the Chinese lantern. This softlight is a more durable version of an actual round or bulb-shaped Chinese lantern. It is usually suspended from a mic stand or a microphone fishpole (see chapter 9). You can put various types of low-powered lamps inside the same lantern, such as a 250W clip light, a 200W household light bulb, or even a daylight (5,600K) lamp if you want to match outdoor light. **SEE 11.18**

The Chinese lantern gives off a very soft yet noticeable light that is especially useful for close-ups. The advantage of using the lantern as a principal light source is that you can follow the subject as you would with a microphone (but you have to secure the lantern to the pole so that it doesn't swing). Because the lantern has an opening on the bottom to vent the heat, keep the bright spot coming from this hole out of the picture. For brief takes, covering it with a light scrim works well. There are huge Chinese lanterns that are used primarily for providing even light for large, reflective objects, such as automobiles or major appliances.

11.18 CHINESE LANTERN
This floodlight is modeled after an actual Chinese lantern. It can be suspended from a fishpole and illuminates a relatively large area with extremely soft light.

11.19 PORTABLE FLUORESCENT BANK
The portable fluorescent bank can be mounted on a light stand. It has great light output and emits no heat.

11.18 PHOTO COURTESY CHIMERA
11.19 PHOTO COURTESY LOWEL-LIGHT MFG., INC.

Portable fluorescent bank Even small portable fluorescent floodlights are considerably bulkier and heavier than comparable incandescent instruments. But because fluorescent floodlights use much less power and generate practically no heat, they are frequently used for indoor EFP lighting. As mentioned before, the problem with fluorescent lights is that they do not always accurately reproduce all colors, even if the camera has been properly white-balanced. Because fluorescent instruments do not always burn at the standard Kelvin ratings of 3,200K and 5,600K, pay particular attention to white-balancing the camera. Check the colors and especially the skin tones on a well-adjusted field monitor before starting to video record. If highly accurate color reproduction is not a major concern, however, the small fluorescent unit is a valuable EFP lighting tool. SEE 11.19

11.20 LED LIGHT
This small LED (light-emitting diode) panel can illuminate a limited area close to it with varying color temperatures and different colors.

11.20 PHOTO COURTESY LOWEL-LIGHT MFG., INC.

LED light *LED* stands for *light-emitting diode.* The light source is not a lamp but an array of semiconductors (a solid-state electronic device) that light up when electricity passes through. The ***LED light*** looks like a small computer screen or a stretched foldout monitor, but instead of displaying an image it simply emits soft white light. You can make it produce various colors as well as white light of different color temperatures (see chapter 12). Although the light output of the small LED panel (up to 7 inches wide) is quite modest, it generates enough light to illuminate an object sufficiently for acceptable video images, provided it is fairly close to the object. SEE 11.20 The LED panel makes a good camera light because it does not cause hot spots. When it's not mounted on the camera, you can use it to light up small areas, such as a car interior.

Although they are quite expensive, LED lights have several advantages over standard incandescent lights: they last longer, are more durable, consume less energy, radiate less heat, and can produce various colors without the need for filters. A major operational disadvantage is that their light output has a severely limited distance from source to object, is sometimes difficult to focus, and makes some colors look bad.

Diffusing Portable Spotlights

The open-face instruments (discussed at the beginning of this section) can also be used as floodlights: just change the light from a spot to a flood position. You will find that despite the flood control, however, you will not always get the even diffusion you may need. Fortunately, there are several ways to achieve a more diffused light with these instruments.

Bouncing the light The simplest way to diffuse the light is to bounce it off of the wall, the ceiling, or a piece of foam core. Unfortunately, bouncing light drastically reduces its intensity, even if the walls are painted a light color. To salvage maximum light intensity, try to get the instrument as close to the wall or ceiling as possible without charring the paint.

Attaching a scrim The most popular diffusers are scrims and frosted gels. As mentioned, scrims are spun-glass diffusers that you can put in front of small spotlights, floodlights, or open-face spots to achieve maximum diffusion of the light. The simplest way to attach a scrim to an open-face instrument is to clip it to the barn doors with wooden clothespins. Don't use plastic ones: open-face lights get very hot and will melt plastic within a few minutes. **SEE 11.21**

Scrims come in various thicknesses; the thinner ones absorb less light, and the thicker ones absorb more light. You can also convert a scoop into a softlight by attaching a scrim that is trimmed to fit a scrim holder (see figure 11.5). Some lighting people prefer frosted gels as diffusers. Frosted gels are white translucent sheets of plastic that have a semiopaque surface. Like scrims, they come in different densities that diffuse and therefore reduce the intensity of the light beam by varying degrees.

Using a diffusion umbrella Another highly effective diffusion device is the umbrella. The small, silvery, heat-resistant umbrella is not to protect you from the rain but to reflect and diffuse the light source that shines into it. You can attach the scooplike umbrella to the lighting instrument and/or the light stand and then aim the umbrella's opening in the general direction of illumination. You need to shine the light into the umbrella opening, not on the rounded surface. **SEE 11.22**

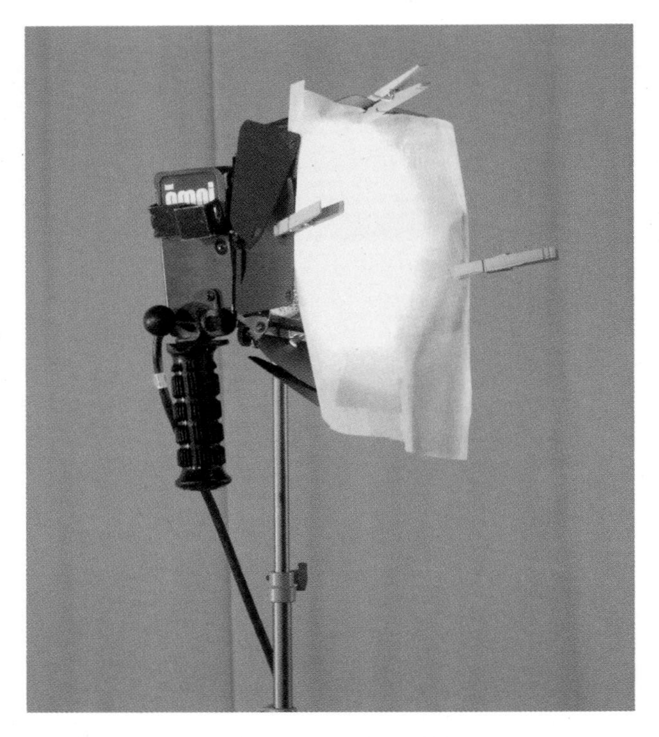

11.21 PHOTO BY EDWARD AIONA

11.21 SCRIM ATTACHED TO BARN DOORS
To further diffuse the beam of an open-face instrument, you can attach a scrim to the barn doors with wooden clothespins.

Attaching a soft-box As mentioned earlier, you can use a soft-box rather than an umbrella to change an incandescent spotlight into an effective softlight. **SEE 11.23** ZVL3 LIGHTS→ Instruments→ field

When doing elaborate field productions, such as covering a high-school basketball game, you can try to use high-powered V-lights and umbrellas, but you may need larger floodlights, such as scoops, or floodlight banks. If available, HMI floodlights would probably be the most efficient instruments. A few 1kWs or even 575W HMI instruments in the flood position are all you need to light up a gymnasium—but then you must white-balance the cameras for the 5,600K outdoor color temperature.

11.22 DIFFUSION UMBRELLA
The umbrella reflector is a popular diffusion device. Note that the lighting instrument shines into, not away from, the inside of the umbrella.

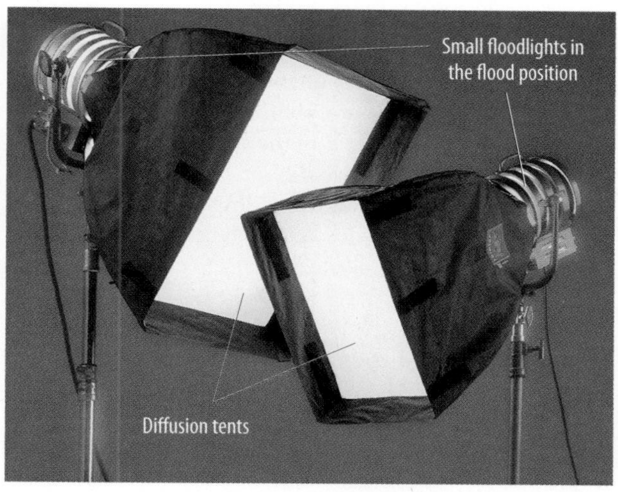

11.23 SOFT-BOX ON FRESNEL SPOTLIGHT
You can turn a small Fresnel spot into a softlight by diffusing its beam with a portable soft-box.

11.22 PHOTO COURTESY LOWEL-LIGHT MFG., INC.
11.23 PHOTO COURTESY CHIMERA

Camera Lights

Electronic news gathering requires yet another type of light, which can be mounted on top of the camera or handheld by the camera operator or an assistant. SEE 11.24 Incandescent camera lights have a high light output. They are open-faced and relatively small and have an assortment of diffusion filters and a daylight filter (5,600K), which you can flip over the lens of the light. The advantages of using an LED panel as a camera light are that it draws very little power and it can adjust to practically any color temperature. The disadvantage is that it has very limited throw. Camera lights draw their power either from the camera battery or from a larger battery that can be attached to a tripod or carried by the camera operator.

11.24 CAMERA LIGHT
This small light is mounted on the camera and powered by the camcorder battery or a separate battery pack. Its beam can be diffused by flipping up a diffusion filter.

11.24 PHOTO COURTESY FREZZI ENERGY SYSTEMS

LIGHTING CONTROL EQUIPMENT

To understand lighting control, you need to be familiar with some specific equipment: mounting devices, directional controls, intensity controls, and electronic dimmers.

Mounting Devices

Mounting devices let you safely support a variety of lighting instruments and aim them in the desired direction. Good mounting devices are as important as the instruments themselves. The major devices specially designed and intended for studio lights are: the pipe grid and the counterweight batten, the C-clamp, the sliding rod and the pantograph, and floor stands. Portable lights are mounted primarily on collapsible stands. For on-location lighting there are additional mounting devices available, such as small booms, cross braces, and braces that fit over doors and furniture.

Pipe grid and counterweight batten Studio lights are hung either from a fixed pipe grid or from a counterweight batten. The pipe grid consists of heavy steel pipe strung either crosswise or parallel. The height of the grid is determined by the height of the studio ceiling, but even in rooms with low ceilings the pipe should be mounted approximately 2 feet below the ceiling so that the lighting instruments or hanging devices can be easily attached. The space above the grid is also necessary to dissipate the heat generated by the lights. **SEE 11.25**

Unlike the pipe grid, which is permanently mounted below the ceiling, a counterweight batten can be raised and lowered to any desired position and locked firmly in place. **SEE 11.26** The battens and the instruments are counterweighted by heavy iron weights and moved by means of a rope-and-pulley system or by individual motors. **SEE 11.27** Before unlocking a counterweight rope to move a batten up or down, always check that the batten is properly weighted. You can do this by counting the weights and comparing them with the type and the number of instruments mounted on the batten. The counterweights and the instruments should roughly balance each other.

The obvious advantage of the counterweight battens over the pipe grid system is that the instruments can be hung, adjusted, and maintained from the studio floor. You will find, however, that you cannot do without a ladder entirely. Although you can initially adjust the instruments to a rough operating position, you need to re-aim them once the battens are locked at the optimal height.

By that time, however, the studio floor is generally crowded with sets, cameras, and microphones, which prevent lowering the battens to a comfortable working height, so you use a lighting pole (a long wooden pole with a hook at one end) to aim the lighting instrument at the desired target and to focus its beam.

C-clamp The lighting instruments are directly attached either to the batten by a large C-clamp or to hanging devices (discussed next). You need a wrench or key to securely fasten the C-clamp to the round metal batten. The lighting instrument is attached to the C-clamp and can be swiveled

11.25 PIPE GRID
This simple pipe grid supports all the necessary lighting for a small performance area, such as a news, interview, or kitchen set.

11.25 PHOTO BY EDWARD AIONA

11.26 COUNTERWEIGHT BATTEN
The counterweight batten can be raised and lowered and locked at a specific operating height.

11.26 PHOTO BY EDWARD AIONA

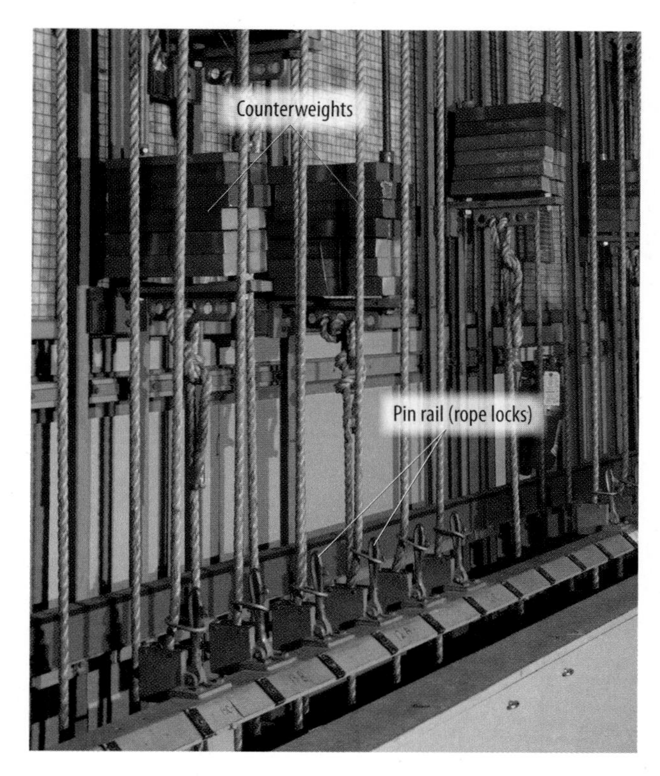

11.27 COUNTERWEIGHT RAIL
The battens and the lighting instruments attached to them are counterweighted by heavy iron weights and moved up and down by a rope-and-pulley system from a common rail.

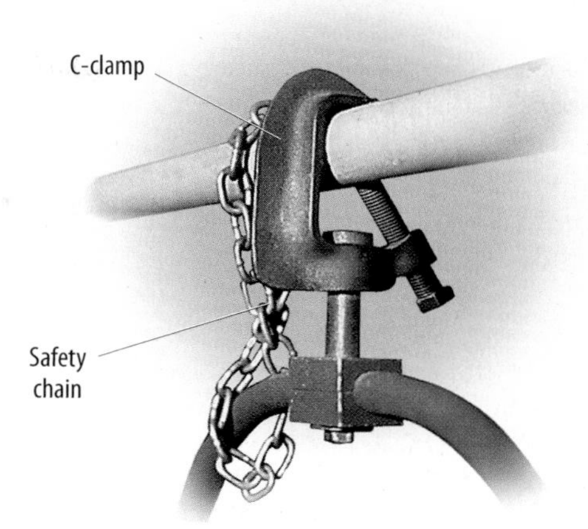

11.28 C-CLAMP
The C-clamp is the connection between the lighting instrument and the batten. Even when the C-clamp is securely tightened to the batten, you can swivel the instrument as necessary.

11.27 PHOTO BY EDWARD AIONA 11.28 PHOTO BY HERBERT ZETTL

horizontally without loosening the bolt that holds it to the batten. Although the C-clamp will support the lighting instrument and not fall off the batten even if the large bolt becomes loose, you should nevertheless check that all C-clamps on the grid are securely tightened. As an added safety measure, all lighting instruments should be chained or secured to the batten itself by a strong steel cable loop. Similarly, the barn doors must be secured to the lighting instruments. Even if you are under severe time pressure when rehanging lights, do not neglect to secure each instrument with the safety chain or cable. **SEE 11.28**

Sliding rod and pantograph If the studio has a fixed pipe grid, or if you need to raise or lower individual instruments without moving an entire batten, you can use sliding rods. A sliding rod consists of a sturdy pipe attached to the batten by a modified C-clamp; it can be moved and locked into a specific vertical position. For additional flexibility, the more expensive sliding rods have telescopic extensions. **SEE 11.29** Some high-end lighting systems have motor-driven sliding

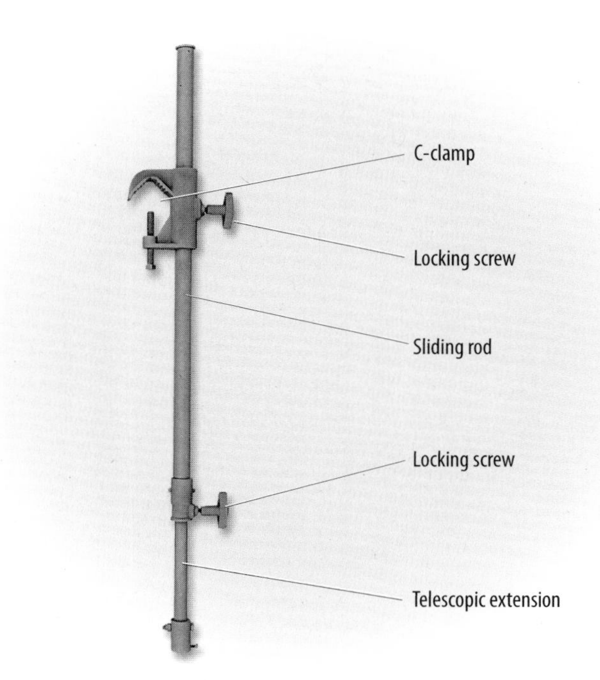

11.29 SLIDING ROD (TELESCOPE HANGER)
This sliding rod, also called a telescope hanger, allows you to move the instrument up and down and lock it into position. It is used primarily on lighting grids but also on counterweight systems when more vertical control is needed.

11.29 PHOTO BY HERBERT ZETTL

rods whose vertical movement can be remotely activated from the studio lighting control.

For some reason the pantograph, a spring-loaded hanging device that can be adjusted with a lighting pole to any vertical position within its 12-foot range, has gone out of fashion. Pantographs are actually more useful for adjusting the height of scoops and other floodlights than are sliding rods. **SEE 11.30**

Floor stands You will find that when video-recording a scene film-style with a single camera, there will be many lighting instruments on the studio floor. Mounting the lights on roller-caster floor stands certainly speeds up a lighting job. Such stands can hold all types of instruments: scoops, broads, spots, and even strip lights. Professional stands usually have a switch to turn the light on and off. **SEE 11.31**

For small lighting instruments, you can use the collapsible light stands in most lighting kits. Just be sure to secure each stand with a sandbag to prevent it and the light from tipping over. **SEE 11.32**

In a pinch, you can always make a simple lighting bridge out of 1 × 3 lumber that will hold one or two portable spotlights for back-lighting. Whatever mounting devices you use—including your own contraptions—see to it that the lighting instrument is securely fastened and sufficiently far from curtains, upholstery, and other combustible materials.

Directional Controls

You are already familiar with the spot and flood beam control on spotlights. Several other devices, such as barn doors, flags, and reflectors, can help you control the direction of the beam.

Barn doors This admittedly crude beam control method is very effective for blocking certain set areas partially or totally from illumination. ***Barn doors*** consist of two or four metal flaps that you can fold over the lens of the lighting instrument to prevent the light from falling on certain areas. For example, if you want to keep the upper part of the scenery dark without sacrificing illumination of the lower

Pantograph

11.30 PANTOGRAPH
You can adjust this spring-loaded pantograph quickly and easily by pushing it up or pulling it down with a lighting pole. The springs act as a counterweight for the lights.

11.31 FLOOR STANDS
A floor stand can support a variety of lighting instruments and can be adapted for an easel or for large reflectors.

11.30 PHOTO BY HERBERT ZETTL
11.31 PHOTO BY EDWARD AIONA

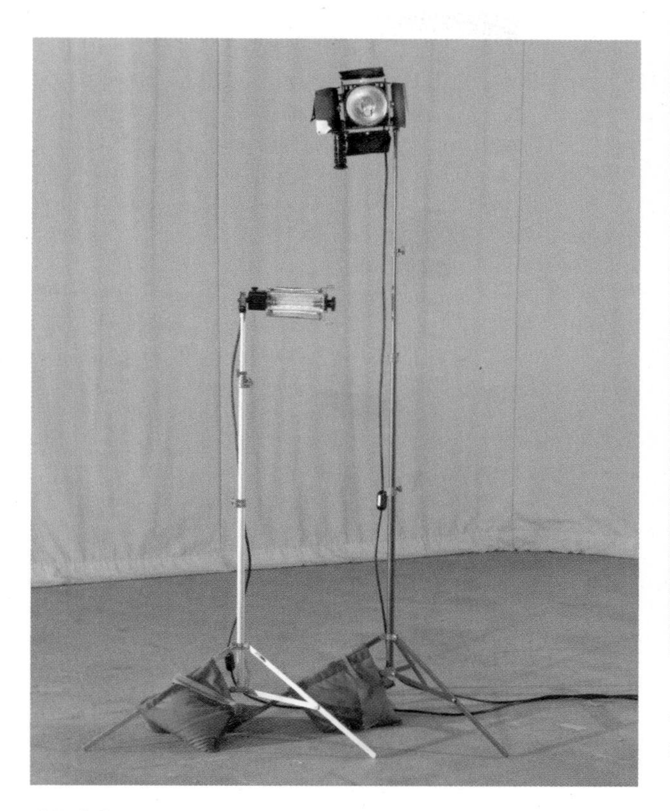

11.32 PORTABLE LIGHT STANDS
These light stands are designed for relatively lightweight portable instruments and can be extended to a height of 8 to 10 feet. Because light stands tend to tip over when fully extended, always secure them with sandbags.

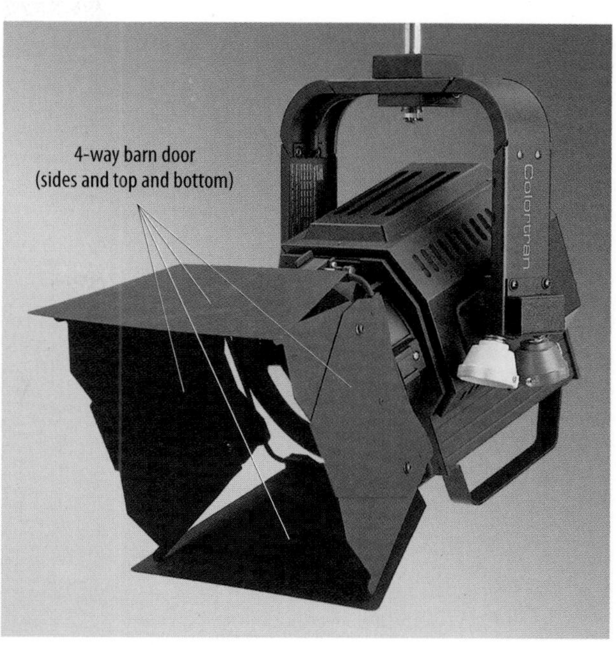

11.33 FOUR-WAY BARN DOOR
This four-way barn door allows you to control the beam spread on all four sides—top, bottom, left, and right.

11.32 PHOTO BY EDWARD AIONA
11.33 PHOTO COURTESY LEVITON, COLORTRAN DIVISION

part, you simply block off the upper part of the beam with a barn door. Or if you want to eliminate a boom shadow, you can partially close a barn door. SEE 11.33 ZVL4 LIGHTS→ Instruments→ beam control

Barn doors are also effective for preventing the back light from shining into the camera lens, which can cause lens flare (an uncontrolled light reflection inside the lens that shows up as superimposed rays of light circles). Because barn doors slide into their holders easily, they have a tendency to slide out of them just as readily. Always secure barn doors to their instruments with the safety chain or cable. Barn doors also get very hot: wear protective gloves while adjusting them when the instrument is turned on.

Flags Rectangular metal frames with heat-resistant cloth or thin metal sheets of various sizes, ***flags*** act very much like barn doors except that you don't place them directly on the lighting instrument. Flags are mounted on light stands and put anywhere they are needed to block the light from falling on a specific area without being seen by the camera. In movie lingo, flags are also called gobos—yet another definition of gobo, this time referring to a flag and not to a cookie (the metal template that is inserted into an ellipsoidal spotlight to produce a shadow pattern; see figure 11.4). Obviously, you can use flags only if the camera and talent movements have been carefully blocked and rehearsed. SEE 11.34

Reflectors Mirrors are the most efficient reflectors. You can position them to redirect a light source (often the sun) into areas that are too small or narrow for setting up lighting instruments. For example, if you had to light up a long, dark hallway that has an exterior door, you could use mirrors to redirect the sunlight into the hall and reflect it off the wall. This technique would save you setup time, equipment, and electricity. Most often, however, you use reflectors to produce highly diffused light to lighten up dense shadows (in media aesthetic language, to slow down falloff) on someone's face or on an object. You don't use mirrors to slow down falloff; rather, you use material that will reflect only a portion of the light and diffuse it at the same time.

11.34 FLAG
Flags come in various sizes and densities. You use them to prevent light from hitting specific set areas.

Most LDs prefer a large sheet of white foam core; it is lightweight, quite sturdy, simple to set up or hold, and easily replaced if it gets dirty or broken. Any large white cardboard will do almost as well. If you need a more efficient reflector (one that reflects more light), you can crumple up some aluminum foil to get an uneven surface (for a more diffused reflection) and then tape it to a piece of cardboard. **SEE 11.35** ZVL5 LIGHTS→ Field→ use of reflectors

Commercial reflectors come in white, silver, and gold and can be folded up for easy transport and setup. The silver and white models reflect a higher-color-temperature light than do the gold-colored ones. **SEE 11.36**

Intensity Controls: Instrument Size, Distance, and Beam

When you are out in the field, you probably won't carry a sophisticated digital dimmer with you, so you must find other ways of controlling light intensity. The three most common methods are selecting an instrument of the proper size, adjusting the distance of lighting instrument to object, and focusing or diffusing the light beam.

Instrument size The simplest way to control light intensity is obviously to turn on only a certain number

11.35 FOIL REFLECTOR
This homemade but highly efficient reflector uses crumpled aluminum foil taped to a piece of cardboard.

11.36 COLLAPSIBLE REFLECTOR
This reflector can be folded up for easy transport. It has a silver-colored reflector on one side and a warmer, gold-colored reflector on the other.

11.34 PHOTO BY EDWARD AIONA
11.35 PHOTO BY EDWARD AIONA
11.36 PHOTO BY EDWARD AIONA

of instruments of a specific size (wattage). Because of the light sensitivity of today's cameras and the high light output of incandescent and fluorescent lamps, you don't need the large instruments you may still see in major motion picture productions. The largest instrument used in most television studios is a 2kW Fresnel spotlight. The lights for ENG/EFP rarely exceed 650W. A small LED light draws less current than a 100W light bulb. In fact, many crime dramas are partially shot in available light—without any lighting instruments.

Distance When you move the lighting instrument closer to the object, the intensity of the light increases; if you move it farther away, the intensity decreases. You can apply this principle easily so long as the instruments are mounted on light stands and you can move them without too much effort. In many cases this is the most efficient way of controlling light intensity on an ENG/EFP shoot. You can also apply this principle in the studio if the lights are mounted on a movable batten. In general, try to position the instruments as low as possible while keeping them out of camera range. In this way you achieve maximum light intensity with minimal power. ZVL6 LIGHTS→ Instruments→ field

Beam The more focused the light beam, the higher its intensity; the more diffused the light beam is, the less intensity it has. Note that, except for a laser, even a focused light beam becomes more diffused the farther away the instrument is from its target object.

You have already learned about diffusing the beam using the focus control in the instrument as well as with scrims and reflectors. You can also use a wire-mesh screen to diffuse and block a certain amount of light: you simply slide the metal screen directly into the gel holder in front of the instrument, much like scrims and frosted gels (see figure 11.14). Depending on the fineness of the mesh, the screen dims the light without influencing its color temperature. The problem with wire-mesh screens is that the heat of the quartz lamp tends to burn up the fine metal wires within a relatively short time; the screens become brittle and eventually disintegrate.

Basic Principle of Electronic Dimmers

The most precise light control is the electronic dimmer. With a ***dimmer*** you can easily manipulate each light, or a group of lights, to burn at a given intensity, from an *off* position to full strength. Regardless of whether you use an old manual dimmer or a sophisticated digital computer dimmer, both operate on the same basic principle: by allowing more or less voltage to flow to the lamp, the lamp burns with a higher or lower intensity. If you want the lighting instrument to burn at full intensity, the dimmer lets all the voltage flow to the lamp. If you want it to burn at a lesser intensity, the dimmer reduces the voltage. To dim the light completely—called a blackout—the dimmer permits no voltage (or at least an inadequate voltage) to reach the lamp.

The downside of dimming—manual and computerized—is that lowering the voltage will cause incandescent lamps to lower their color temperature to a more reddish light. We discuss this problem in more detail in chapter 12.

Manual dimmers Although few manual dimmers are still in use, it is easier to explain basic dimmer operation with a manual dimmer than with a complex computerized one. Each individual lighting instrument has its own dimmer control, which in a manual operation is a fader with a specific calibration. As you can see in the following figure, the left fader is moved to 2, which is barely enough to make the lamp glow. The fader on the right is all the way up at 10, which means that its lamp burns at full intensity. **SEE 11.37**

Computerized dimmers Most larger studio dimmers can control hundreds of lights with a great number of intensity steps from no light to full burn. Although the dimming commands are generated by the computer, such a dimmer has nevertheless a number of manual faders that can override the stored dimming program. **SEE 11.38** You

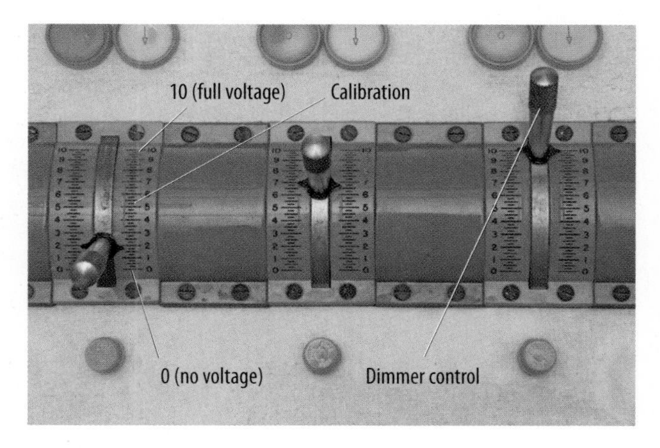

11.37 MANUAL DIMMER CALIBRATION
The higher you push the lever on this manual dimmer, the more voltage flows to the lamp. At the 0 setting, no voltage flows to the lamp; at a setting of 10, the lamp burns at full intensity. The 2 setting on the left dimmer makes the lamp barely glow; the 10 setting on the right dimmer makes the lamp burn at full intensity.

11.37 PHOTO BY EDWARD AIONA

11.38 COMPUTERIZED DIMMER CONTROL

This simple but powerful dimmer and lighting effects panel can control 48 channels in a single scene or 24 channels in two scenes. The video display shows the status of submaster settings, lighting effects, and patching data. Large dimmers can control hundreds of channels.

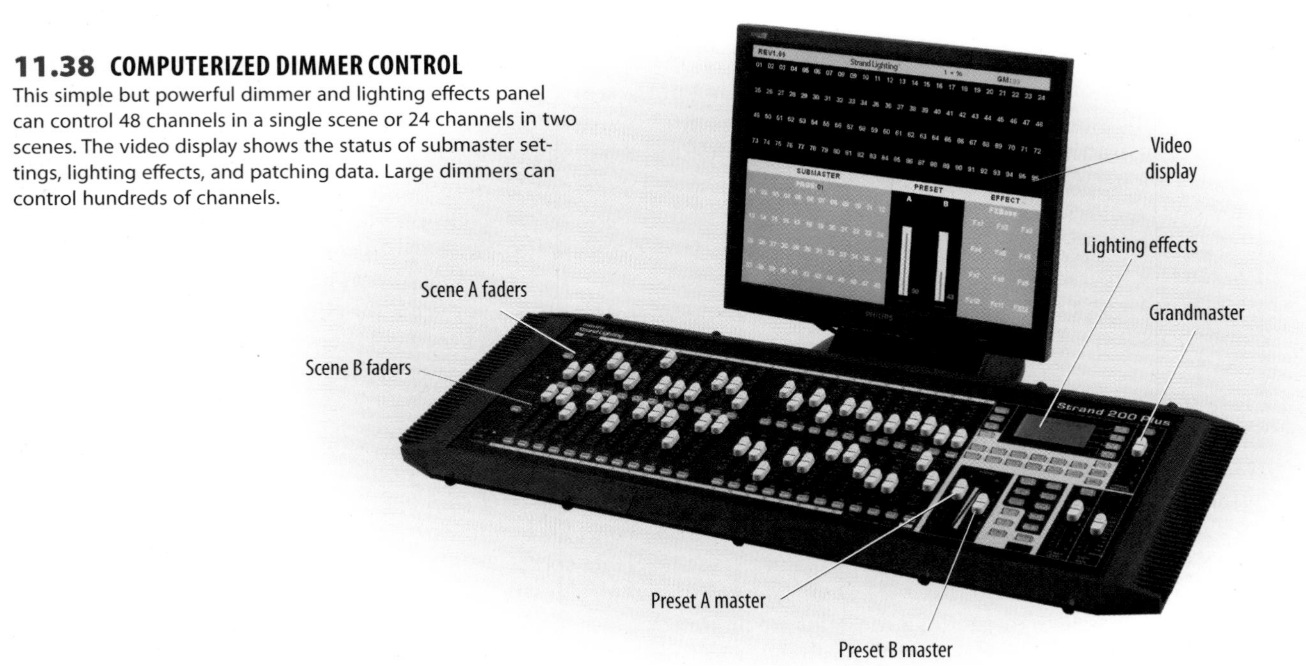

11.38 PHOTO COURTESY STRAND LIGHTING

should realize, however, that the actual dimming is done with some kind of equipment that regulates the voltage flowing to the lamps.

Patchboard The ***patchboard***, or patchbay, makes it possible to connect each lighting instrument to a specific dimmer. Let's assume that you have one lamp in your study and another lamp in your bedroom but only one dimmer. If you want to dim both lights to save electricity, you can plug lamp 1 (study) and lamp 2 (bedroom) into the same dimmer. The patchboard of an actual dimmer system works in the same way. Although the computer can patch a great variety of instruments to any one of dozens of dimmers, we again use a manual system to explain the patching principle.

To patch a lighting instrument into a dimmer, you select its designated patch cord and plug it into the dimmer receptacle (called a jack). **SEE 11.39**

11.39 MANUAL PATCHBOARD

The patchboard enables you to establish power connections between specific lighting instruments and specific dimmers.

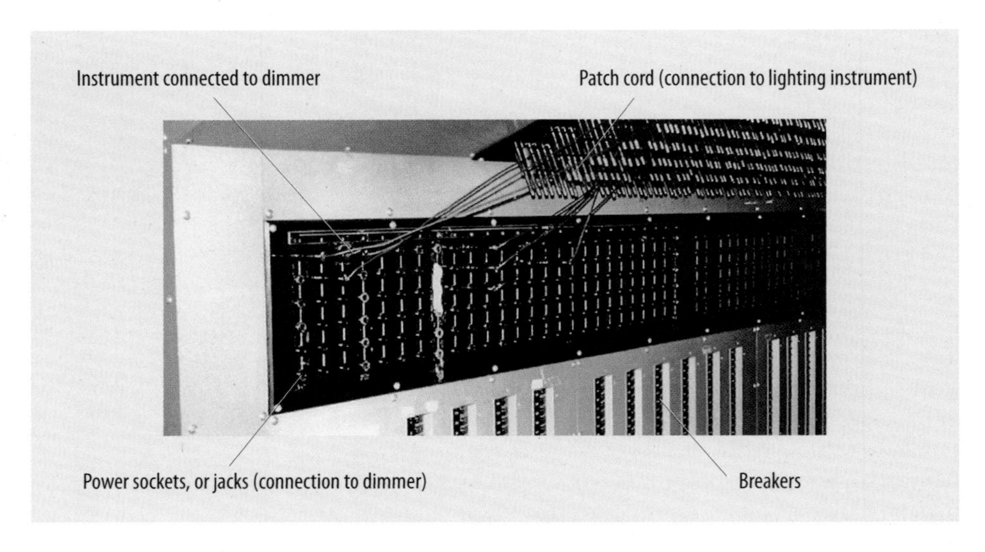

11.39 PHOTO BY HERBERT ZETTL

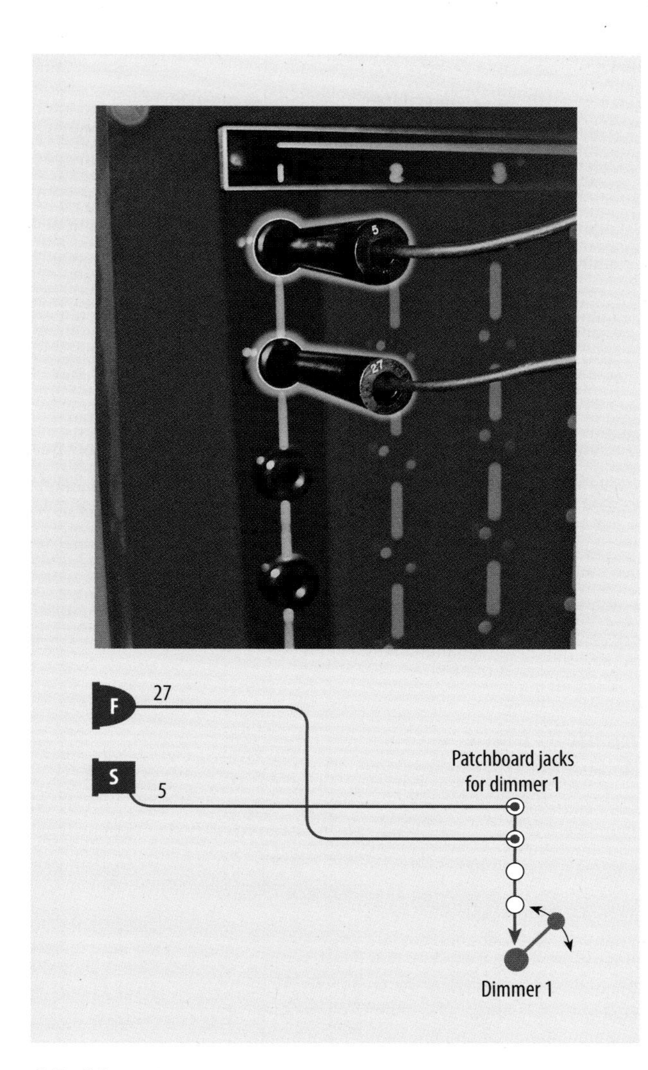

11.40 PHOTO BY HERBERT ZETTL

11.40 MANUAL PATCHING
As you can see, the patches for the lighting instruments (spotlight 5 and scoop 27) are both patched to dimmer 1. Consequently, both lighting instruments respond identically to any dimmer 1 setting.

Figure 11.40 shows lighting instruments 5 (a Fresnel spot) and 27 (a scoop) patched into dimmer 1. When you bring up dimmer 1 at the dimmer control (move the fader to a designated intensity setting—let's say 8), both instruments—spotlight 5 and scoop 27—should light up simultaneously and be dimmed equally to a number 8 intensity. SEE 11.40 If you want to control them separately, you would plug spotlight 5 into dimmer 1 and scoop 27 into dimmer 2.

Besides controlling the intensity of the light, patching enables you to change from one type of lighting setup to another quickly and easily. For example, you may change a dining room set from day to night by simply switching from the stored daylight setup to the nighttime one. In this way you can light several studio areas at once, store the lighting setup in the dimmer's memory, and activate part of or all the stored information whenever necessary. Some shows may require that you go from one background color to another, such as from red to blue. With the dimmer you can simply fade down all instruments that throw red light onto the background while at the same time bringing up the blue lights.

MAIN POINTS

- All studio lighting is accomplished by a variety of spotlights and floodlights, technically called luminaires.
- Studio spotlights include the Fresnel spot and the ellipsoidal spot.
- Studio floodlights include the scoop, the softlight and the broad, the fluorescent floodlight bank, and the strip, or cyc, light.
- Field lighting uses the small portable spotlights with lenses (small Fresnel and HMI spotlights), open-face spotlights without a lens, and the internal reflector spot (clip light).
- Most portable floodlights are open-faced, which means they have no lens. Small fluorescent banks are also used as portable floodlights. Soft-box diffusers can turn a spotlight into a floodlight.
- Simple ENG lighting is often done with small, versatile camera lights or LED panels that are mounted on the camera or handheld.
- Lighting control equipment includes mounting devices, directional controls, intensity controls, and dimmers.
- The principal mounting devices are the pipe grid and the counterweight batten, the C-clamp, the sliding rod and the pantograph, and floor stands.
- Directional controls include barn doors, flags, and reflectors.
- Basic intensity controls are the size of the instrument (lamp wattage), the relative distance of lighting instrument to target object, and the relative focus or diffusion of the beam.
- With an electronic dimmer, you can easily manipulate a light, or a group of lights, to burn at a given intensity. The patchboard, or patchbay, makes it possible to connect each lighting instrument to a different dimmer.

SECTION

11.2

Light Intensity, Lamps, and Color Media

Before learning to do actual lighting in the studio and the field, you need to study a few more elements about light, how to control and measure it, and how to produce colored light. This section adds to the technical details given in section 11.1.

- **LIGHT INTENSITY**
  Using foot-candles and lux to measure incident and reflected light
- **CALCULATING LIGHT INTENSITY**
  The lumen and the inverse square law
- **OPERATING LIGHT LEVEL: BASELIGHT**
  Establishing optimal light levels
- **COLOR TEMPERATURE**
  The reddishness and bluishness of white light and how to control color temperature
- **TYPES OF LAMPS**
  Basic luminants: incandescent, fluorescent, HMI, and LED
- **COLOR MEDIA**
  Plastic sheets (gels) that change the color of light

LIGHT INTENSITY

Although some video cameras can produce pictures in almost total darkness, most standard cameras need a certain amount of light for optimal performance. As sensitive as our eyes are, they cannot always tell accurately just how much light an instrument produces, how much light is actually on the set or on-location, how much light an object actually reflects, and how much light the camera lens actually receives. A light meter gives us a more accurate reading of light intensity.

Foot-candles and Lux

The standard units of measuring light intensity are the American ***foot-candle (fc)*** and the European ***lux***. Because ordinary television lighting doesn't require extremely precise units of intensity, you can simply figure lux by multiplying foot-candles by a factor of 10, or you can figure foot-candles by dividing lux by 10:

- To find lux when given foot-candles, multiply foot-candles by 10.
- To find foot-candles when given lux, divide lux by 10.

For example, 100 foot-candles equal about 1,000 lux (100 × 10), and 2,000 lux are about 200 foot-candles (2,000 ÷ 10). If you want to be more accurate, use a factor of 10.75 to calculate foot-candles from lux or lux from foot-candles.

Equipped with foot-candles or lux as the unit of light intensity, you can now measure the two types of light intensity: incident light and reflected light. ZVL7 LIGHTS→ Measurement→ meters

Incident Light

An incident-light reading gives you some idea of how much light reaches a specific set area. When measuring ***incident light***, you are actually measuring the amount of light that falls on a subject or a performance area but not what is reflected by it. To measure incident light, you must stand in the lighted area or next to the subject and point the incident-light meter toward the camera lens. The meter will give a quick reading of the overall light level that reaches the set area. This general light level is also called baselight. But incident light can also refer to the light produced by a particular instrument. If you want a reading of the intensity of the light coming from specific instruments, you should point the foot-candle (or lux) meter into the lights. SEE 11.41

Such measurements may come in handy, especially when you need to duplicate the illumination for a scene shot on the same set over a period of several days. For some reason duplicating the exact lighting from one day to the next is difficult, even with a computer-assisted dimmer that faithfully duplicates the settings of the previous day. An incident-light check, however, guarantees identical or fairly close intensities.

To discover possible holes in the lighting (unlighted or underlighted areas), walk around the set with the light meter pointed at the major camera positions. Watch the light meter: whenever the needle dips way down, it is indicating a hole.

Reflected Light

The reading of ***reflected light*** gives you an idea of how much light is bounced off the various objects. It is primarily used to measure contrast.

To measure reflected light, you must use a reflected-light meter (most common photographic light meters measure reflected light). Point it closely at the lighted object, such as the dancer's face or blond hair, and then at the dark background—all from the direction of the camera (the back of the meter should face the principal camera position). **SEE 11.42**

Do not stand between the light source and the subject when taking this reading or you will measure your shadow instead of the light actually reflecting off the subject. To measure contrast, point the meter first at the lighted side of the object and then move it to the shadow side. The difference between the two readings gives you the contrast ratio. (Chapter 12 describes contrast ratio and its importance in television lighting.)

Do not be a slave to these measurements and ratios, however. A quick check of the baselight is all that is generally needed for most lighting situations. In especially critical productions, you may want to check the reflectance of faces or exceptionally bright objects. Some people get so involved in reading light meters and oscilloscopes that visually display the light levels against camera tolerances that they forget to look at the monitor to see whether the lighting looks the way it was intended. If you combine your knowledge of how the camera works with artistic sensitivity and, especially, common sense, you will not let the light meter tell you how to light but rather use it as a guide to make your job more efficient.

CALCULATING LIGHT INTENSITY

Light intensity measures how much light strikes an object. One foot-candle (candlepower) is the amount of light from a single candle (1 lumen) that falls on a 1-square-foot surface located 1 foot away from the candle. One lux is the light that falls on a 1-square-meter surface (about 3 by 3 feet) generated by a single candle (1 lumen) that burns at a distance of 1 meter. The light intensity norm for both foot-candles and lux of one candle is 1 ***lumen***.

Light intensity is subject to the inverse square law. This law states that if a light source radiates isotropically (uniformly in all directions), such as a candle or a single light bulb burning in the middle of a room, the light intensity falls off (gets weaker) at $1/d^2$ where d is the distance from the source. For example, if the intensity of a light is 1 fc at a distance of 1 foot from the source, its intensity at a distance of 2 feet is ¼ fc. **SEE 11.43**

The inverse square law also applies to lux. In this case the light intensity is measured off a surface of 1 m^2 located 1 meter from the light source of 1 lumen.

This formula tells you that light intensity decreases the farther away you move the lighting instrument from the object, and it increases if you move the light closer. Otherwise, the formula does little to make television lighting more accurate. The beams of a searchlight, a flashlight, car headlights, and a Fresnel or an ellipsoidal spotlight do not

11.41 INCIDENT-LIGHT READING
To read incident light, you point the light meter at the camera or into the lights while standing next to the lighted subject or performance area. Here the LD checks the intensity of a specific light.

11.42 REFLECTED-LIGHT READING
To measure reflected light, you point the reflected-light meter (used in still photography) close to the lighted subject or object.

11.41 PHOTO BY EDWARD AIONA
11.42 PHOTO BY EDWARD AIONA

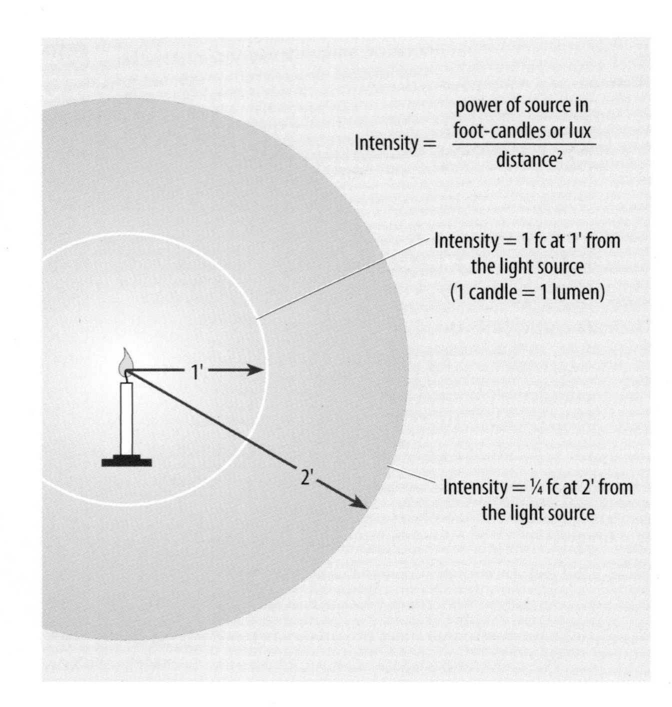

11.43 INVERSE SQUARE LAW
Note that the inverse square law applies only to light sources that radiate isotropically (uniformly in all directions). This law applies equally to lux.

radiate light isotropically (like a candle) but are collimated (the light rays are made to run parallel as much as possible) and therefore do not obey the inverse square law. Even floodlights radiate their light more in the direction of the reflector opening than its back. The more collimated the light—that is, the more focused its beam—the slower its intensity decreases with distance. This is why we "focus" a spotlight when we want more light on an object and "flood" its beam when we want less light, without changing the distance between the lighting instrument and the object. An example of an extremely well-collimated light is a laser beam, which, as you know, maintains its intensity over a great distance.

OPERATING LIGHT LEVEL: BASELIGHT

To make the camera "see well" so that the pictures are relatively free of video noise (artifacts in the picture, or "snow"), you must establish a minimum operating light level, called baselight or base. As you recall, ***baselight*** is the general, overall light level on a scene.

Baselight Levels

Many an argument has been raised concerning adequate minimum baselight levels for different cameras. The problem is that baselight levels do not represent absolute values but are dependent on other production factors, such as the sensitivity of the camera, the desired lighting contrast, the general reflectance of the scenery, and, of course, the aperture of the lens (*f*-stop). When shooting outdoors on an ENG assignment, you do not have much control over baselight levels; you must accept whatever light there is. But even there you might be able to use sunlight reflectors to lighten up shadow areas, or additional lighting instruments to boost available light. Often the problem is not inadequate baselight but, paradoxically, too much light. Let's have a brief look at both situations.

Not enough baselight Although you often hear that consumer camcorders can operate in light levels as low as 0.1 fc or even 0.02 fc (10 or even 2 lux), the light levels for optimal camera performance are much higher. Professional ENG/EFP and studio cameras normally need about 100 fc, or approximately 1,000 lux, for optimal picture quality at an aperture setting of *f*/5.6 to *f*/8.0. These *f*-stops produce the highest-resolution images. You will probably read camera specifications that use 200 fc (2,000 lux) as the standard illumination and then give the highest *f*-stop, such as *f*/11, at which the camera still delivers optimal pictures.

Most video cameras can work at baselight levels that are considerably lower without noticeable loss of picture quality. By turning on the gain control and switching to a low gain setting (which, as you recall, will electronically boost the video signal), you may get an acceptable image even in low-light conditions. Despite manufacturers' claims to the contrary, high gain can cause increased video noise and occasional color distortion. For home video and even ENG, video quality may be secondary to picture content, but it is of major concern for EFP and studio shows that must tolerate many picture manipulations in postproduction editing.

Here is the rule of thumb: in general, a camera has less trouble producing true colors and crisp, high-quality pictures when the light level is fairly high and the contrast limited than under very low light levels with high-contrast lighting.

Too much baselight Despite the validity of this general rule for baselight and picture quality, there will be instances when there is simply too much light for the camera to operate properly. You can cope with too much light by reducing the lens aperture, which translates into setting the *f*-stop to a higher number, such as *f*/22, and/or using an ND filter that is part of the filter wheel inside the camera. Much like a small aperture, a ***neutral density (ND) filter*** reduces the amount of light falling on a scene or entering the beam splitter in the camera without changing the color temperature of the light. Such ND filters will also help you control the extreme contrast between light and shadows when shooting outdoors on a sunny day. ZVL8 LIGHTS→ Measurement→ baselight

Color Temperature

As promised in previous chapters, here is an explanation of what color temperature is all about. You may have noticed that a fluorescent tube gives off a different "white" light than does a candle. The fluorescent tube actually emits a white light that has a bluish-green tinge, whereas the candle produces a more reddish-white light. The setting sun gives off a much more reddish light than does the midday sun, which is more bluish. These color variations in light are called ***color temperature***. Note that color temperature has nothing to do with physical temperature, that is, how hot the lamp actually gets; it is strictly a measure of the relative reddishness or bluishness of white light.

This reddishness and bluishness of white light can be precisely measured; it is expressed in degrees of color temperature, or ***Kelvin (K)*** degrees. In lighting lingo, the degrees are dropped and a specific color temperature is referred to only as a certain amount of K.

The color temperature standard for indoor illumination is 3,200K, which is a fairly white light with just a little reddish (warm) tinge. All studio lighting instruments and portable lights intended for indoor illumination are rated at 3,200K, assuming they receive full voltage. Lighting instruments used to augment or simulate outdoor light have lamps that emit a 5,600K light. They approximate more the bluish light of the outdoors. SEE 11.44

When you dim a lamp that is rated at 3,200K, the light becomes progressively more reddish, similar to sunlight at sunset. The color camera, when adjusted to seeing white in 3,200K light, will faithfully show this increasing reddishness. For example, the white shirt of a performer will gradually turn orange or pink, and the skin tones will take on an unnatural red glow. Some lighting experts therefore warn against any dimming of lights that illuminate performers or performance areas. Skin tones are, after all, the only real standard viewers have by which to judge the accuracy of the television color scheme. If skin colors are distorted, how can we trust the other colors to be true? So goes the argument. Practice has shown, however, that you can dim a light by 10 percent or even a little more without the color change becoming too noticeable on a color monitor. Note that dimming the lights by at least 10 percent will not only reduce power consumption but just about double the life of the lamps. ZVL9 LIGHTS→ Color temperature→ white balance | controlling | try it

How to control color temperature As you learned in chapter 6, you need to white-balance the camera to ensure the correct color reproduction even if the illumination has different color temperatures. You may find, however, that occasionally the camera will refuse to white-balance although you follow exactly the procedures outlined here.

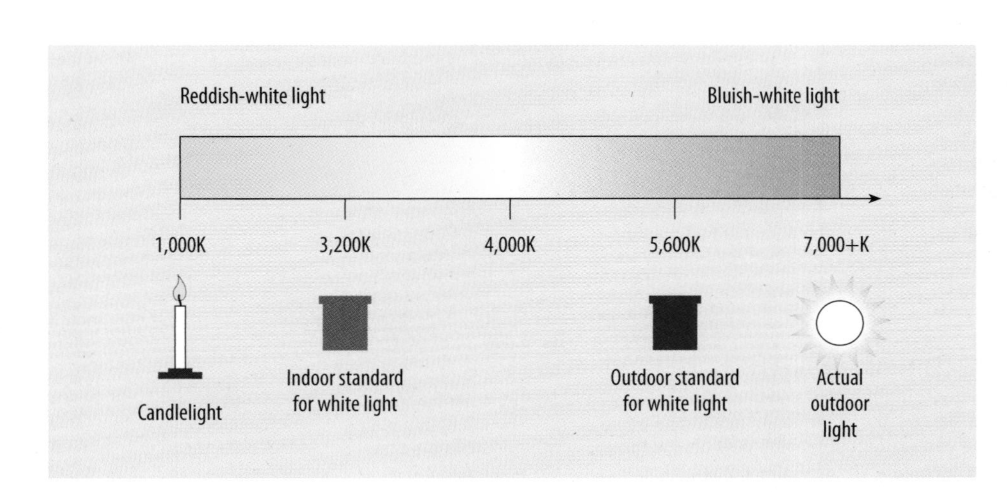

11.44 COLOR TEMPERATURE
Color temperature, measured on the Kelvin scale, measures the relative reddishness or bluishness of white light. The norm for indoor light is 3,200K; for outdoor light it is 5,600K.

This difficulty may be caused either by a color temperature that is too low (light is too reddish) or too high (too bluish) for the normal white-balance circuits to handle, or by mixing color temperatures in single shot.

To take care of the first white-balance problem, you need to choose one of the color filters on the filter wheel inside the camera (see chapter 6). Light-blue filters compensate for the reddishness of low-color-temperature light; amber or light-orange filters compensate for the bluishness of high-color-temperature light. When you have lights with two different color temperatures in one shot, you need to either raise the 3,200K indoor standard to the 5,600 outdoor one or bring down the outdoor standard to match the indoor one.

To raise the color temperature of the reddish light (to make it more bluish), you can put a light-blue gel (sheet of colored plastic) in front of the lighting instrument's lens. You can also lower the color temperature (to make it more reddish) by placing an amber or light-orange gel in front of the lighting instrument or cover the entire window with amber plastic sheets that act like gigantic filters, lowering the high outdoor color temperature to the lower indoor standard. The advantage of this method is that the whole interior is adjusted to the 3,200K standard. A quicker and cheaper way is to let the high-color-temperature outdoor light stream through the window and put bluish filters in front of the indoor lighting instruments to raise their light to the outdoor standard. **SEE 11.45** ZVL10 LIGHTS→ Color temperature→ light sources

In certain circumstances, you can get away with mixing lights of different color temperatures so long as one or the other dominates the illumination. For example, if you are in an office that is illuminated by overhead fluorescent tubes and you need to add key and back lights (see chapter 12) to give the performer more sparkle and dimension, you can most likely use normal portable lighting instruments that burn at the indoor color temperature standard. This is because the portable instruments provide the dominant light, so they overpower the overhead lights that now act as rather weak fill lights. The camera will have little trouble

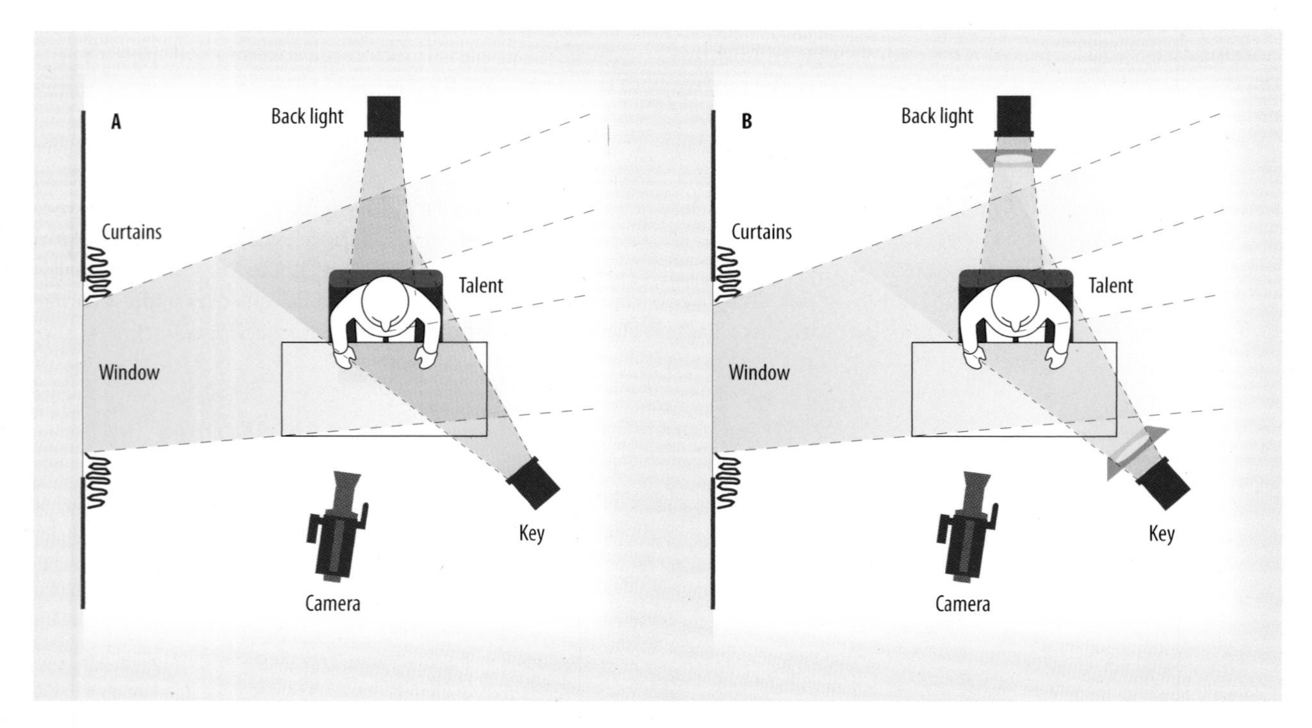

11.45 MATCHING COLOR TEMPERATURES OF DIFFERENT LIGHT SOURCES

A When illuminating an object with indoor light mixed with outdoor light coming through a window, you need to equalize the color temperatures of both light sources to ensure proper white-balancing.

B To equalize the color temperatures, you can put light-blue gels on the indoor lighting instruments to raise their 3,200K color temperature to the more prominent 5,600K daylight coming through the window.

white-balancing on the strong indoor lights while more or less ignoring the higher color temperature of the overhead fluorescent lights.

TYPES OF LAMPS

Lighting instruments are classified not only by function (spotlight or floodlight) but also by the lamp they use—technically called a ***luminant***. Television lighting generally uses four basic types of luminants: incandescent, fluorescent, HMI, and LED.

Incandescent

The ***incandescent*** lamp operates on the same principle as the ordinary household light bulb: it generates light by heating up a filament with electricity. The incandescent lamps used in television resemble the ones in home fixtures except that they usually have more wattage and therefore produce higher-intensity light. The disadvantages of regular incandescent lamps are that the higher-wattage lamps are quite large, the color temperature becomes progressively lower (more reddish) as the lamp ages, and they have a relatively short life.

Incandescent luminants include the smaller but hotter ***quartz*** lamps, also called TH (tungsten-halogen) lamps. The quartz lamp has a tungsten filament that is encased in a quartz bulb filled with halogen gas. The advantages of a quartz lamp over regular incandescent systems are that it is smaller and maintains its color temperature over its entire life. The disadvantage is that it burns at an extremely hot temperature. When changing quartz lights, *do not touch the lamp with your fingers.* The old lamp may still be hot enough to burn your skin, and your fingerprints will cause the new one to develop weak spots in the quartz housing that may cause the lamp to explode. Always use gloves, a paper towel, or a clean rag when handling lamps.

Fluorescent

A ***fluorescent*** lamp generates light by activating a gas-filled tube to give off ultraviolet radiation. This radiation in turn lights up the phosphorous coating inside the tube. Despite improved fluorescent lamps that produce a fairly even white light, many still have a tendency to give off a slightly greenish light. If accurate color rendering is critical to your assignment, test out a fluorescent light before using it. Check especially whether its color temperature blends with that of the other lighting instruments.

HMI

HMI (which stands for *hydragyrum medium arc-length iodide*) lamps generate light by moving electricity through various types of gases. This creates a sort of lightning inside the lamp, which is the discharge that creates the illumination. To create the lightning inside the lamp, you need a ballast—a fairly heavy transformer. HMI lamps produce light with a color temperature of 5,600K, the outdoor standard. (See section 11.1 for the advantages and the disadvantages of the HMI when used in production.) As with quartz lamps, do not touch HMI lamps with your hands: your fingerprints will weaken the quartz housing and cause the lamp to burn out in a relatively short time.

LED

Although not a lamp, the light-emitting-diode panel lights up when voltage is applied and then illuminates an area close to it. The panel itself can change colors without the use of color filters. Because the LED light does not "throw" a beam, it is—so far—restricted to small- and close-area lighting.

COLOR MEDIA

You can produce a great variety of colored light simply by putting different color media, or ***gels***, in front of the lighting instrument. (*Gel* is short for *gelatin,* which was the color medium used before the more heat- and moisture-resistant plastic was developed.) Color media are sheets of highly heat-resistant plastic that act as color filters. They are used extensively to color-tint scenic backgrounds or to create color special effects, such as a sunset or dark blue sky, or the color effects in dance programs, rock concerts, or some mystery or outer-space adventure shows. SEE 11.46

How to Use Color Media

You can cut the color media sheet to fit the frame of the gel holder of the lighting instrument. You then slip the gel holder into brackets in front of the lens of the lighting instrument. If the colored lighting does not have to be too precise, you can use wooden clothespins (plastic ones melt) to hang the color sheets from the barn doors like laundry on a clothesline. The advantages of this method are that it saves you from having to cut the expensive gels and they are farther away from the heat generated by the lamp. Highly focused instruments generate so much heat that they will burn out the center of even the most heat-resistant gels.

11.46 COLOR MEDIA
Color media, or gels, are colored filters that are put in front of lighting instruments to produce colored light.

11.46 PHOTO BY EDWARD AIONA

You can avoid such burns by putting the instrument into more of a flood position (by moving the lamp-reflector unit toward the lens), thereby dissipating some of the heat.

Mixing Color Gels

When using gels, you can mix the colors additively or subtractively. For example, if you put a red gel in one instrument and a green gel in the other and then partially overlap their beams, you get yellow in the overlap. Because you added one light on top of the other, this is additive mixing. If, however, you were to put both gels—the red and the green—in front of the same instrument, you would get no light from the instrument. This is because the red gel blocks (subtracts) all the green light, and the green gel negates all the red light.

A similar problem occurs if you shine colored lights on colored objects. We see an apple as red because the color filters in the apple absorb all colors of white light except red, which is reflected back to our eyes. A green apple absorbs all colors except green, which is reflected back and makes the apple look green. What would happen if you shined a red light on a green apple? Would it turn yellow? No, the apple would look dark brown or black. Why? Because the red light that shines on the green apple contains no green, the apple can't reflect any green. In the same way, you may have a problem using yellow objects under blue "night" illumination: the blue light contains no yellow, and the objects therefore have no yellow to reflect, so they turn dark gray or black. When choosing a certain color for your LED light, you may run into the same problem. Try the LED lighting out on camera before the actual recording session.

Most lighting experts advise against using colored lights to illuminate talent and performance areas unless, of course, it's for special effect, such as the greenish tint on crime shows or the multicolored lights on a rock music scene. If colors are critical, try to keep the colored light away from the faces.

MAIN POINTS

- Light intensity is measured in foot-candles (fc) or lux. To find lux when given foot-candles, multiply foot-candles by 10. To find foot-candles when given lux, divide lux by 10. If you need to be most accurate, use 10.75 instead of 10 as the conversion factor.
- To measure incident light (the light that falls on the scene), point the light meter away from the lighted scene toward the camera or into the lights that are illuminating the subject.
- To measure reflected light, use a reflected (standard) light meter and point it closely at various areas of the lighted subject or object. Reflected-light readings measure primarily contrast.
- The inverse square law in illumination applies only if the light source radiates isotropically (uniformly in all directions), such as a bare light bulb or a candle. Because most television lighting instruments collimate the light (focus the light rays), the inverse square law does not apply to the same degree. The general principle, however, still holds true: the farther away the light source is from the object, the less intense the light; the closer the light is to the object, the more intense the light.
- Baselight is the overall light level on a scene. Cameras require a minimum baselight level for optimal operation.

- Color temperature, measured in Kelvin degrees (K), is the relative reddishness or bluishness of white light. The standard color temperature for indoor light is 3,200K (warm-colored light); the outdoor standard is 5,600K (cool-colored light). Do not mix color temperatures in a single shot.
- White-balancing means to adjust the circuits of the camera to reproduce white under a various color temperatures.
- Lamps are labeled by the type of luminant: incandescent, including quartz; fluorescent; HMI; and LED.
- Color media, normally called gels, are colored plastic filters that, when put in front of the lens of a lighting instrument, give the light beam the color of the gel.
- Colored light beams mix additively, but overlaying gels mix subtractively.

ZETTL'S VIDEOLAB

For your reference or to track your work, the *Zettl's VideoLab* program cues in this chapter are listed here with their corresponding page numbers.

CHAPTER

12

Techniques of Television Lighting

When watching television you will probably notice that newscasts, situation comedies, and game shows are brightly lit with a minimum of shadows on faces. But when watching crime shows or soap operas, you will often see more deep shadows on the actors' faces, and even the colors are sometimes distorted. The techniques of television lighting suggest how to achieve such different lighting effects and more.

In most video production situations, especially electronic field production (EFP), available space, time, and people are insufficient for you to accomplish motion picture–quality lighting. The time allotted to lighting is often so short that all you can do is flood the studio or location site with highly diffused light, regardless of the nature of the event. Although such a technique may please the camera and probably the video operator (who because of the uniform light levels has little shading to do), it does not always fulfill the aesthetic requirements of the production. A dramatic studio scene that is supposed to play on a dark street corner will not look convincing if everything is brightly and evenly illuminated by softlights. But to light the street corner set so that it looks believable requires not only time but especially that you know the principles of manipulating light and shadows. Lighting for digital cinema requires even more care and time because you normally have to light for every shot or, at best, for every scene.

The ever-present time limitation should not preclude good and creative television lighting, but it does call for a high degree of efficiency. Without a thorough understanding of the basic lighting principles, you can easily spend all your allotted time, and part of the rehearsal time, trying to achieve a specific lighting effect that, in the end, might look out of place. Efficiency in lighting also means careful preparation.

This chapter will help you with such preparations. Section 12.1, Lighting in the Studio, covers basic and special-effects studio lighting techniques and principles; section 12.2, Lighting in the Field, addresses lighting techniques for electronic news gathering (ENG), electronic field production, and on-location digital cinema productions.

KEY TERMS

background light Illumination of the set, set pieces, and backdrops. Also called *set light.*

back light Illumination from behind the subject and opposite the camera.

cameo lighting Foreground figures are lighted with highly directional light, with the background remaining dark.

chroma keying An effect that uses color (usually blue or green) for the backdrop, which is replaced by the background image during a key.

contrast ratio The difference between the brightest and the darkest portions in the picture (often measured by reflected light in foot-candles). The contrast ratio for low-end cameras and camcorders is normally 50:1, which means that the brightest spot in the picture should be no more than 50 times brighter than the darkest portion without causing loss of detail in the dark or light areas. High-end digital cameras can exceed this ratio and can tolerate a contrast ratio of 1,000:1 or more.

cross-keying The crossing of key lights for two people facing each other.

diffused light Light that illuminates a relatively large area with an indistinct beam. Diffused light, created by floodlights, produces soft shadows.

directional light Light that illuminates a relatively small area with a distinct beam. Directional light, produced by spotlights, creates harsh, clearly defined shadows.

falloff (1) The speed with which light intensity decays. (2) The speed (degree) with which a light picture portion turns into shadow area. Fast falloff means that the light areas turn abruptly into shadow areas and there is a great brightness difference between light and shadow areas. Slow falloff indicates a very gradual change from light to dark and a minimal brightness difference between light and shadow areas.

fill light Additional light on the opposite side of the camera from the key light to illuminate shadow areas and thereby reduce falloff. Usually done with floodlights.

floor plan A diagram of scenery and properties drawn onto a grid pattern. Can also refer to *floor plan pattern.*

high-key Light background and ample light on the scene. Has nothing to do with the vertical positioning of the key light.

key light Principal source of illumination.

kicker light Usually directional light that is positioned low and from the side and the back of the subject.

light plot A plan, similar to a floor plan, that shows the type, size (wattage), and location of the lighting instruments relative to the scene to be illuminated and the general direction of the beams.

location survey Written assessment, usually in the form of a checklist, of the production requirements for a remote.

low-key Dark background and illumination of selected areas. Has nothing to do with the vertical positioning of the key light.

photographic lighting principle The triangular arrangement of key, back, and fill lights, with the back light opposite the camera and directly behind the object, and the key and fill lights on opposite sides of the camera and to the front and the side of the object. Also called *triangle lighting.*

side light Usually directional light coming from the side of an object. Acts as additional fill light or a second key light and provides contour.

silhouette lighting Unlighted objects or people in front of a brightly illuminated background.

SECTION

12.1

Lighting in the Studio

Lighting is the control of light and shadows. Both are necessary to show the shape and the texture of a face or an object, to suggest a particular environment, and, like music, to create a specific mood. Regardless of whether you do lighting for dramatic or nondramatic productions, you will find that there are usually many solutions to any one problem. And though there is no universal recipe that works for every possible lighting situation, there are some basic principles that you can easily adapt to a great variety of specific requirements. When faced with a lighting task, do not start with anticipated limitations. Start with how you would like the lighting to look and then adapt to the existing technical facilities and especially the available time.

▶ **SAFETY**
Basic lighting safety precautions

▶ **QUALITY OF LIGHT**
Directional and diffused light, and color temperature

▶ **LIGHTING FUNCTIONS**
Terminology and specific functions of the main light sources

▶ **SPECIFIC LIGHTING TECHNIQUES**
High- and low-key, flat, continuous-action, large-area, high-contrast, cameo, silhouette, and chroma-key-area lighting, and controlling eye and boom shadows

▶ **CONTRAST**
Contrast ratio, measuring contrast, and controlling contrast

▶ **BALANCING LIGHT INTENSITIES**
Key-to-back-light ratio and key-to-fill-light ratio

▶ **LIGHT PLOT**
Indicating the location of instruments and their beams

▶ **OPERATION OF STUDIO LIGHTS**
Preserving lamps and power, and using a studio monitor

SAFETY

In the actual operation of lighting instruments and the associated control equipment, you should heed the rule for all production activities: *safety first.*

■ As mentioned in chapter 11, always wear gloves when working with active (switched-on) lighting instruments. The gloves will protect you from burns when touching the housing and barn doors of hot instruments and will give you some protection from electric shock.

■ Always secure the lighting instruments to the battens with safety chains or cables and attach the barn doors and the scrims to the lighting instruments. Check all C-clamps periodically, especially the bolt that connects the lighting instrument to the hanging device.

■ Be careful when plugging in lights and when moving active instruments. Because the hot lamps are especially vulnerable to physical shock, try not to jolt the instrument; move it gently.

■ When replacing lamps, wait until the instrument has cooled somewhat. Always turn off the instrument before reaching in to remove a burned-out lamp. As a double protection, unplug the light at the batten. Do not touch a new quartz lamp with your fingers. Fingerprints—or anything else clinging to the quartz housing of the lamp—cause the lamp to overheat and burn out. Wear gloves or, if you have nothing else, use a paper towel or even your shirttail when handling a quartz lamp.

■ Watch for obstacles above and below when moving ladders. Do not take any chances by leaning way out to reach an instrument. Position the ladder so that you can work from behind, rather than in front of, the instrument.

■ When adjusting a light, try not to look directly into it; look instead at the object to be lighted and see how the beam strikes it. If you must look into the light, wear dark glasses and do so only briefly.

■ Before you start patching (assuming that you use a physical patchboard) have all dimmers and breakers in the *off* position. Do not "hot-patch" by connecting the power cord of the instrument to the power outlet on the batten with the breaker switched on. Hot-patching can burn your hand and also pit the patches so that they no longer make the proper connection.

QUALITY OF LIGHT

Whatever your lighting objective, you will be working with two types of light: directional and diffused. And, as you have read in chapter 11, normal white light can have different color temperatures.

Directional Light and Diffused Light

Directional light, produced by spotlights, illuminates a relatively small area with a distinct light beam and produces dense, well-defined shadows. The sun on a cloudless day acts like a giant spotlight, producing dense and distinct shadows. ***Diffused light*** illuminates a relatively large area with a wide, indistinct beam. It is produced by floodlights and creates soft, transparent shadows. The sun on a cloudy or foggy day acts like an ideal floodlight because the overcast transforms the harsh light beams of the sun into highly diffused light.

Actually, it is the density of the shadows and their falloff that indicates whether the light is directional or diffused. If you looked only at the illuminated side, you would have a hard time telling whether it was directional or diffused light.

Color Temperature

Color temperature is a light quality that influences especially the colors of the lighted object. Although our brain may compensate for the different color temperatures and sees white as white even if it is relatively reddish or bluish, the television camera does not. If you aim to approximate the screen colors of a face or an object as closely as possible to those of the real thing, you need to white-balance the camera every time you enter a specific lighting environment.

LIGHTING FUNCTIONS

You will notice that lighting terminology is based not so much on whether the instruments are spotlights or floodlights but rather on their functions and their position relative to the object being lighted.

Terminology

Although there are variations for the following terms, most lighting people in the photographic arts (including video and digital cinema) use this standard terminology.

- The ***key light*** is the apparent principal source of directional illumination falling on a subject or an area; it reveals the basic shape of the object.
- The ***back light*** produces illumination from behind the subject and opposite the camera; it distinguishes the shadow of the object from the background and emphasizes the object's outline.
- The ***fill light*** provides generally diffused illumination to reduce shadow or contrast range (to slow falloff). It can be directional if the area to be "filled in" is rather limited.
- The ***background light,*** or *set light,* is used specifically to illuminate the background or the set and is separate from the light provided for the performers or performance area.
- The ***side light*** is placed directly to the side of the subject, usually on the opposite side of the camera from the key light. Sometimes two side lights are used opposite each other, acting as two keys for special-effects lighting of a face.
- The ***kicker light*** is a directional illumination from the back, off to one side of the subject, usually from a low angle opposite the key light. Whereas the back light merely highlights the back of the head and the shoulders, the kicker light highlights and defines the entire side of the person, separating him or her from the background.

Specific Functions of Main Light Sources

How do these lights now function in basic lighting tasks? Let's take a look.

Key light As the principal source of illumination, the major function of the key light is to reveal the basic shape of the subject. **SEE 12.1** To achieve this the key light must produce some shadows. Fresnel spotlights, medium spread, are normally used for key illumination. But you can use a scoop, a broad, or even a softlight for a key if you want softer shadows or, technically, slower falloff. In the absence of expensive softlights, some lighting directors (LDs) take a cue from filmmakers and still photographers and use reflectors as key and fill lights. Instead of diffusing the key and fill lights with diffusion material, such as scrims or frosted gels, you do not aim the key light (a Fresnel spot) directly at the subject but rather bounce it off white foam core or a large white posterboard. The reflected, highly diffused light nevertheless produces distinct, yet extremely soft, slow-falloff shadows.

Because during the day we see the principal light source—the sun—coming from above, the key light is normally placed above and to the right or left front side of the object, from the camera's point of view. Look again at figure 12.1, which shows the woman illuminated with the key light only, and notice that the falloff is very fast,

12.1 KEY LIGHT
The key light represents the principal light source and reveals the basic shape of the object or person.

12.2 KEY AND BACK LIGHTS
The back light provides more definition to the actual shape of the subject (her hair on camera-left), separates her from the background, and gives her hair sparkle and highlights.

12.1 PHOTO BY EDWARD AIONA
12.2 PHOTO BY EDWARD AIONA

blending part of her hair and shoulder with the dark background. To help clarify the outline and the texture of the woman's right (camera-left) side, you obviously need light sources other than the single key light.

Back light Adding illumination from behind helps separate the subject from the background. SEE 12.2 Note how the back light helps distinguish between the shadow side of the woman and the dark background, emphasizing the outline—the contour—of her hair and shoulders. We have now established a clear figure/ground relationship, which means that we can easily perceive a figure (the woman) in front of a (dark) background. Besides providing spatial definition, the back light adds sparkle and professional polish.

In general, try to position the back light as directly behind the subject (opposite the camera) as possible; there is no inherent virtue in placing it somewhat to one side or the other unless it is in the camera's view. A more critical problem is controlling the vertical angle at which the back light strikes the subject. If it is positioned directly above the person, or somewhere in that neighborhood, the back light becomes an undesirable top light: instead of revealing the subject's contour to make her stand out from the background and giving the hair sparkle, the light simply brightens the top of her head, causing dense shadows below her eyes and chin. On the other hand, if the back light is positioned too low, it shines into the camera.

To get good back lighting on a set, you need a generous space between the performance areas (the areas in which the talent move) and the background scenery. Therefore you must place "active" furniture, such as chairs, tables, sofas, or beds actually used by the performers, at least 8 to 10 feet away from the walls toward the center of the set. If the talent works too close to the scenery, the back lights

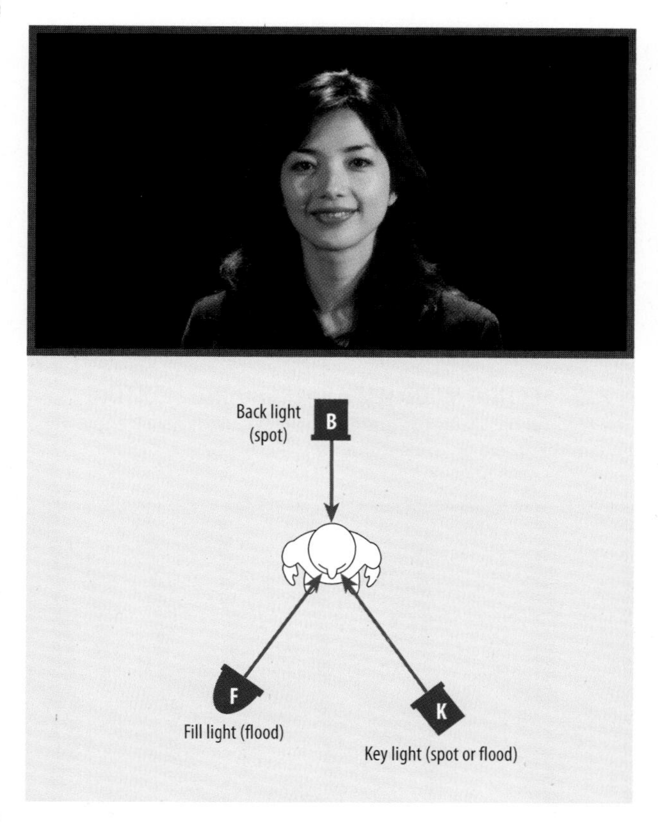

12.3 PHOTO BY EDWARD AIONA

12.3 KEY, BACK, AND FILL LIGHTS
The fill light slows falloff, making the shadow side (camera-left) more transparent and revealing details without eliminating altogether the form-revealing shadows.

must be tilted at very steep angles to reach over the scenery, and such steep angles inevitably cause undesirable top light.

Fill light Now take another look at figure 12.2. Despite the back light, the difference between the light and shadow sides is still extreme, and the light side of the face still changes abruptly to a dense shadow. This change is called falloff. ***Falloff*** refers to the speed (degree) to which a light picture portion turns into shadow area. If the change is sudden, as in figure 12.2, it is fast falloff. With fast falloff the shadow side of the subject's face is very dense; the camera sees no shadow detail. To slow down the falloff, that is, to make the shadow less prominent and more transparent, you need some fill light. **SEE 12.3**

Not surprisingly, you place the fill light on the opposite side of the camera from the key light. A highly diffused floodlight or reflected light is generally used as fill. The more fill light you use, the slower the falloff becomes. When the intensity of the fill light approaches or even matches that of the key light, the shadows, and with them the falloff, are virtually eliminated. This gives the subject a flat look—shadows no longer help define shape and texture.

When doing critical lighting in a specific area and not wanting the fill light to spill over too much into the other set areas, you can use a Fresnel spotlight as fill by spreading the beam as much as possible or by putting a scrim in front of the lens. You can then use the barn doors to further control the spill.

The Photographic Principle: Triangle Lighting

With the three main light sources (key, fill, and back) in the triangle setup, you have established the basic ***photographic lighting principle,*** also called triangle lighting (see figure 12.3). But you are not done yet! You must now fine-tune this lighting arrangement. Take a good hard look at the lighted object or, if possible, the studio monitor to see whether the scene (in our case, the close-up of the woman) needs some further adjustment for optimal lighting. Are there any undesirable shadows? Are there shadows that distort rather than reveal the face? How is the light balance? Does the fill light wash out the necessary shadows, or are the shadows still too dense? Is the back light too strong for the key/fill combination? Our lighting in figure 12.3 is good for explaining the photographic principle.

Background, or set, light To illuminate the background (walls or cyclorama) of the set or portions of the set that are not a direct part of the principal performance area, you use the background light, or, as it is frequently called, the set light. To keep the shadows of the background on the same side as those of the person or object in front of it, the background light must strike the background from the same direction as the key light. **SEE 12.4**

As you can see in the figure, the key light is placed on the camera-right side, causing the shadows on the subject to fall on the camera-left side. Consequently, the background light is also placed on camera-right to make the shadows on camera-left correspond with those of the foreground. If you place the background light on the opposite side from the key, the viewer may assume that there are two separate light sources illuminating the scene or, worse, that there are two suns in our solar system. ZVL1 LIGHTS→ Triangle lighting→ key | back | fill | background | try it

Background light frequently goes beyond its mere supporting role to become a major production element.

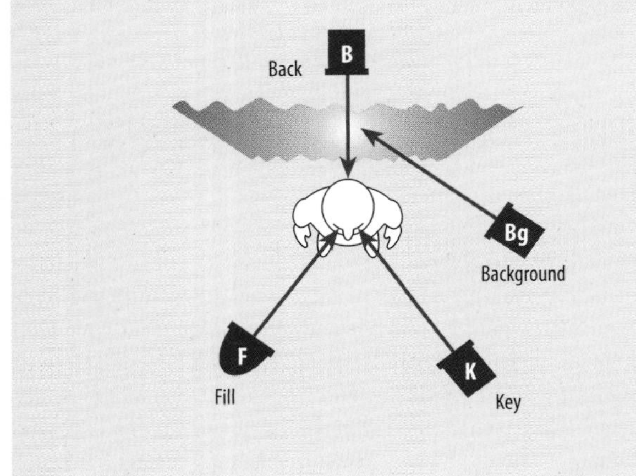

12.4 BACKGROUND LIGHT
The background light illuminates the background area. It must be on the same side of the camera as the key light to keep the background shadows (curtain) on the same side as the foreground shadows (woman).

12.5 SETTING MOOD WITH BACKGROUND LIGHTING
The colorful background lighting in this set suggests a trendy environment and an upbeat mood.

12.6 SETTING LOCALE WITH BACKGROUND LIGHTING
Background lighting can place an event in a specific locale or environment. Here the background light produces barlike shadows, suggesting that the scene takes place in a prison.

12.4 PHOTO BY EDWARD AIONA
12.5 PHOTO COURTESY BROADCAST AND ELECTRONIC COMMUNICATION ARTS DEPARTMENT AT SAN FRANCISCO STATE UNIVERSITY
12.6 PHOTO BY EDWARD AIONA

Besides accentuating an otherwise dull, monotonous background with a slice of light or an interesting cookie, the background light can be a major indicator of the show's locale, time of day, and mood. **SEE 12.5** A cookie projection of prison bars on the wall, in connection with the clanging of cell doors, immediately places the event in a prison. **SEE 12.6**

A long slice of light or long shadows falling across the back wall of an interior set suggests, in connection with other congruent production clues, late afternoon or evening. In normal background lighting of an interior setting, try to keep the upper portions of the set relatively dark, with only the middle and lower portions (such as the walls) illuminated. This will emphasize the talent's face and shoulders against a slightly darker background and will set off furniture and medium- and dark-colored clothing against the lighter lower portions of the set. Also, the dark upper portions suggest a ceiling. You can darken the upper portions of the set easily by using barn doors to block off any spotlight (including the background lights) that would hit those areas.

Side light Usually placed directly to the side of the subject, the side light can function as a key or fill light. When used as a key, it produces fast falloff, leaving half the face in dense shadow. When used as a fill, it lightens up the whole shadow side of the face. When you place side lights on opposite sides of the person, the sides of the face are bright, with the front of the face remaining shadowed. **SEE 12.7**

12.7 PHOTO BY EDWARD AIONA

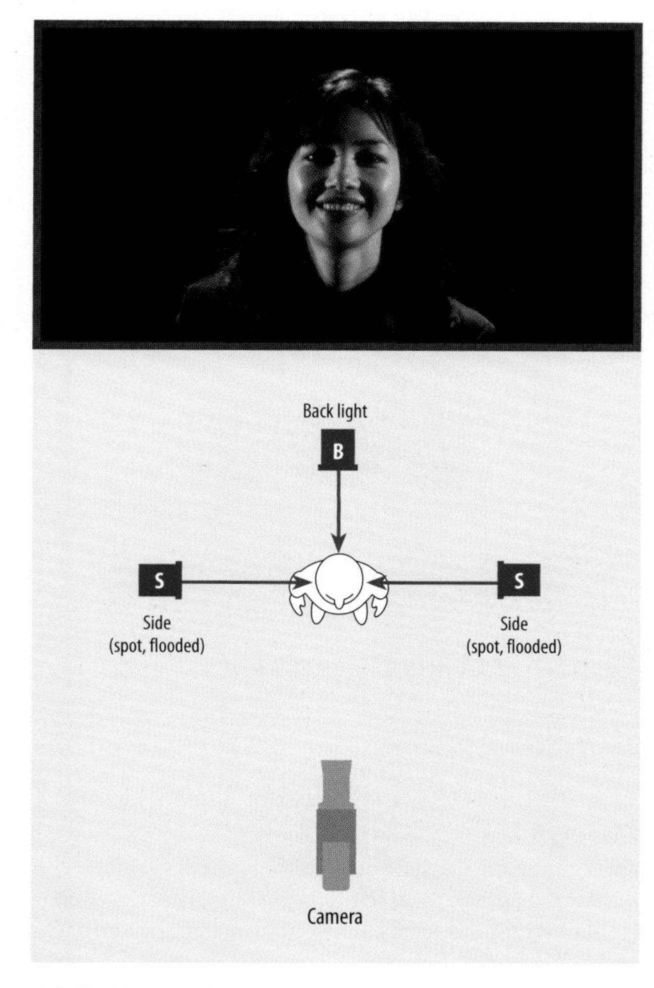

12.7 SIDE LIGHT
The side light strikes the subject from the side. It can act as key and/or fill light. In this case two opposing side lights are used as two keys.

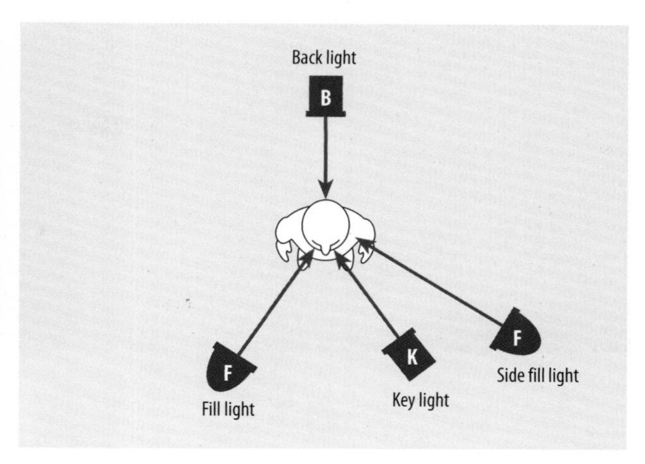

12.8 SIDE FILL-LIGHT SETUP
The side fill light provides soft illumination, with the key (spot) adding sparkle. When the key is turned off, the side fill takes over the function of the key light.

The side light becomes a major light source if the camera's shooting arc is exceptionally wide. If, for instance, the camera moves around the subject from a 6 o'clock to an 8 o'clock position, the side light takes on the function of the key light and provides essential modeling (lighting for a three-dimensional effect). Although Fresnel spots at a wide-beam setting are generally used for side lighting, using scoops or broads as side lights can produce interesting effects.

For extrabrilliant lighting, you can support the key light with side fill light. The fill light gives the key side of the subject basic illumination, with the key light providing the necessary sparkle and accent. SEE 12.8

Kicker light Generally a sharply focused Fresnel spot, the kicker light strikes the subject from behind and on the opposite side of the camera from the key light (the fill-light side). Its main purpose is to highlight the subject's contour at a place where key-light falloff is the densest and where the dense shadow of the subject opposite the key-lighted side tends to merge with the dark background. The function of the kicker is similar to that of the back light except that the kicker "rims" the subject not at the top-back but at the lower side-back. It usually strikes the subject from below eye level. Kicker lights are especially useful for creating the illusion of moonlight. SEE 12.9

SPECIFIC LIGHTING TECHNIQUES

Once you are familiar with how to apply the photographic principle in a variety of lighting situations, you can move on to a few specific lighting techniques. These include: high- and low-key lighting, flat lighting, continuous-action lighting, large-area lighting, high-contrast lighting, cameo lighting, silhouette lighting, chroma-key-area lighting, and controlling eye and boom shadows.

Accurate lighting is always done with basic camera positions and points of view in mind. It therefore helps

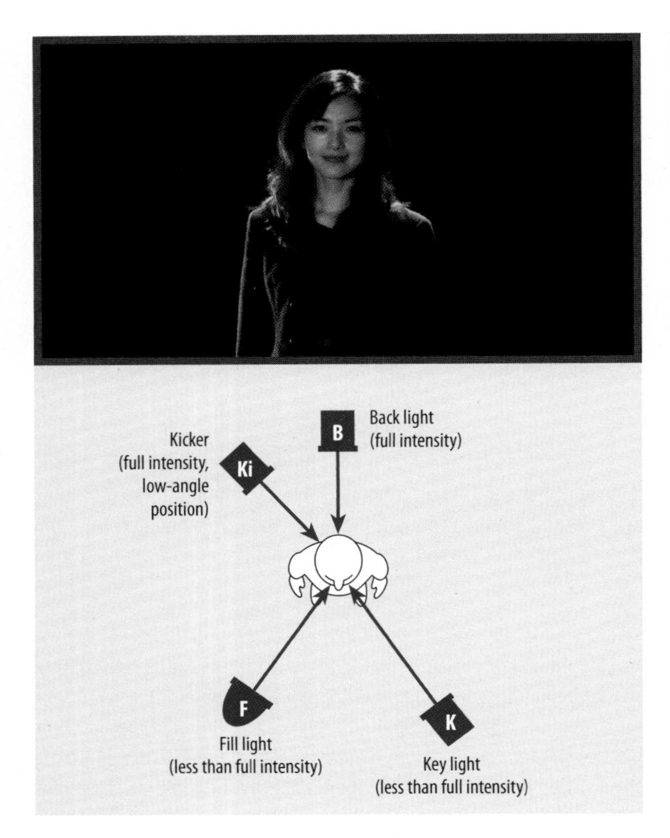

12.9 KICKER LIGHT
The kicker light rims the subject opposite the key, emphasizing contour. Like the back light, the kicker helps separate the foreground subject from the background.

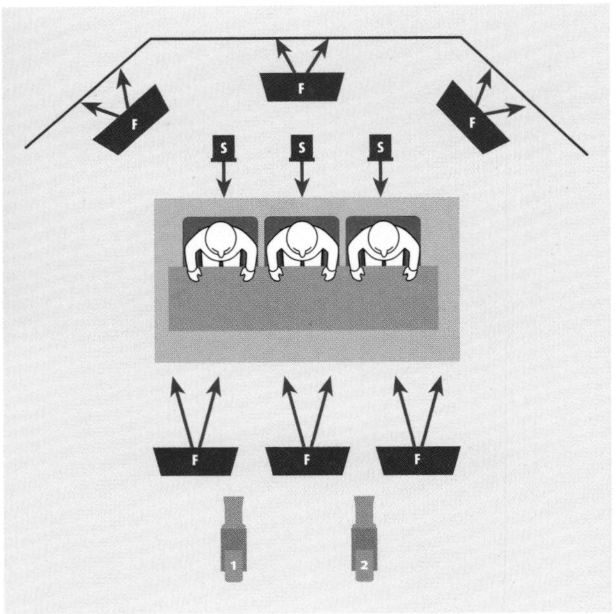

12.10 FLAT-LIGHTING SETUP FOR NEWS
This flat-lighting setup consists of three frontal softlights that act as key and fill lights, three spots or spotted floods for back lights, and three background floodlights.

12.9 PHOTO BY EDWARD AIONA

immensely to know at least the basic camera positions and the range of all major camera viewpoints before starting with the lighting.

High- and Low-key Lighting

A light background, a generous amount of light (high baselight levels), and slow falloff usually constitute a ***high-key*** scene with an upbeat, happy mood. That is why situation comedies and game shows are much more brightly lighted (higher baselight level and less contrast) than are mystery and police dramas (lower baselight level and more contrast). Selective, fast-falloff lighting with distinct shadows and normally a dark background define a ***low-key*** scene. It creates a dramatic or mysterious mood. Do not confuse *high-key* and *low-key* with high and low vertical hanging positions of the key light or with the intensity with which it burns.

Flat Lighting

Flat lighting means that you light for optimal visibility with minimal shadows. Most flat-lighting setups use floodlights (softlights or fluorescent banks) for front lighting and background lighting and more focused instruments (Fresnel spots or small broads) for back lights. This setup is the favorite lighting technique for more or less permanently installed news sets and interview areas. **SEE 12.10** As you can see in the figure, the basic lighting triangle is preserved. In effect, you have three key lights, or, if you wish, three fill lights, evenly illuminating the front area. The back lights add the sparkle and make the flatness of the lighting setup less noticeable. The additional background lights illuminate the set.

The major disadvantage of flat lighting is that it looks like what it is: flat.

Continuous-action Lighting

When watching dramas or soap operas on television, you probably notice that many of them have fast-falloff, low-key lighting, which means prominent shadows and relatively dark backgrounds. In such multicamera productions, the cameras look at a scene from different points of view, and people and cameras are always on the move. Wouldn't it be easier to light "flat," that is, to flood the whole performance area with flat light rather than with spotlights? Yes, but then the lighting would not contribute to the mood of the scene or how we feel about the persons acting in it. Fortunately, the basic lighting triangle of key, back, and fill lights can be multiplied and overlapped for each set or performance area for continuous-action lighting. Even if there are only two people sitting at a table, you have to use a multiple application of the basic lighting triangle. SEE 12.11

To compensate for the movement of the performers, you should illuminate all adjacent performance areas so that the basic triangle-lighted areas overlap. The purpose of overlapping is to give the performers continuous lighting as they move from one area to another. It is too easy to concentrate only on the major performance areas and to neglect the small, seemingly insignificant areas in between. You may not even notice the unevenness of such lighting until the performers move across the set and all of a sudden they seem to be playing a "now you see me, now you don't" game, popping alternately from a well-lighted area into dense shadow. In such situations a light meter comes in handy to pinpoint the "black holes."

If you do not have enough instruments to apply multiple-triangle lighting for several performance areas when lighting for continuous action, you must place the instruments so that each can serve two or more functions. In reverse-angle shooting, for instance, the key light for one performer may become the back light for the other and vice versa. This technique is generally called ***cross-keying***. Or you may have to use a key light to serve also as directional fill in another area. Because fill lights have a diffused beam, you can use a single fill light to lighten up dense shadows in more than one area. SEE 12.12

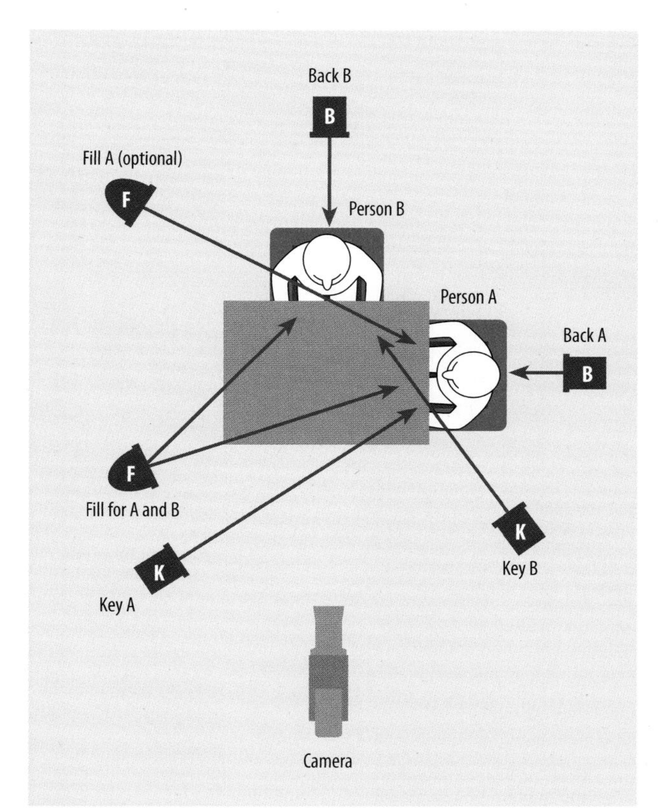

12.11 MULTIPLE-TRIANGLE APPLICATION
In this lighting setup, a separate lighting triangle with its own key, back, and fill light is used for each of the two persons (performance areas). If floodlights are used for the keys, you can probably dispense with the fill lights.

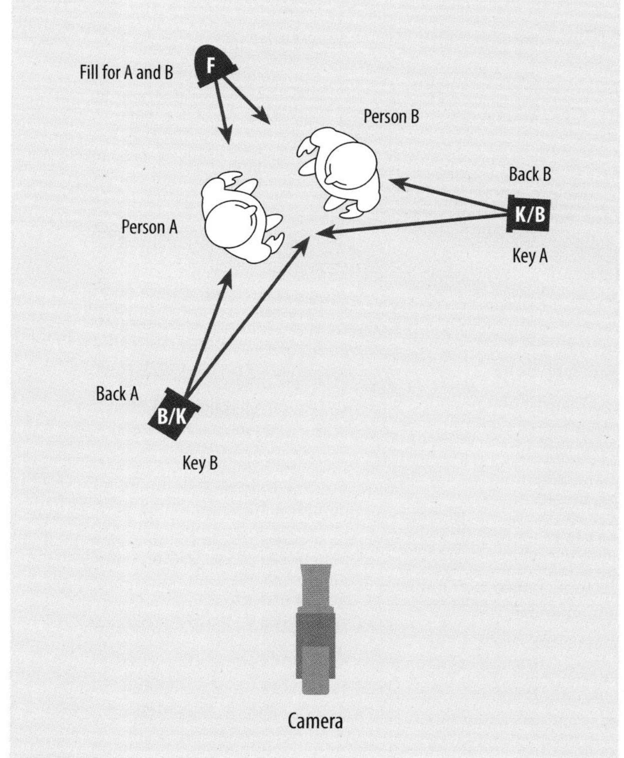

12.12 CROSS-KEYING
In this lighting setup, the key light for person A (the camera-near person) also functions as a back light for person B (the camera-far person); and the back light for person A is the key for person B.

Of course, the application of lighting instruments for multiple functions requires exact positioning of set pieces such as tables and chairs, clearly defined performance areas, and blocking (movements of performers). Directors who decide to change blocking or move set pieces after the set has been precisely lighted are not very popular with the lighting crew.

Large-area Lighting

For large-area lighting, such as for an audience or orchestra, the basic photographic principle still holds: all you do is partially overlap one triangle on another until you have adequately covered the entire area. Instead of key-lighting from just one side of the camera and fill-lighting from the other, however, key-light from both sides of the camera with Fresnel spots in the flood position. The key lights from one side act as fill for the other. If the area is really big, you can have additional sets of Fresnel spots positioned closer to the center.

The back lights are strung out in a row or a semicircle opposite the main camera position. The fill lights (broads or scoops) usually come directly from the front. If the cameras move to the side, some of the key lights also function as back lights. You can also use broads or fluorescent banks instead of Fresnel spots for this type of area lighting. SEE 12.13

For some assignments, such as lighting a school gym for a basketball game, check first whether the available light is adequate. Turn on a camera and look at the monitor. If you need more light, you can simply flood the gym with highly diffused light. As mentioned, one possibility is to use fairly high-powered open-face instruments with light-diffusing umbrellas. Watch out that you don't mix color temperatures of the house lights and your additional lighting. See to it that all cameras are properly white-balanced.

High-contrast Lighting

The opposite of flat lighting is high-contrast lighting, much of which mirrors motion picture lighting techniques. Because of the increased tolerance of today's video cameras to low light levels and higher-contrast lighting, many television plays make extensive use of fast-falloff lighting.

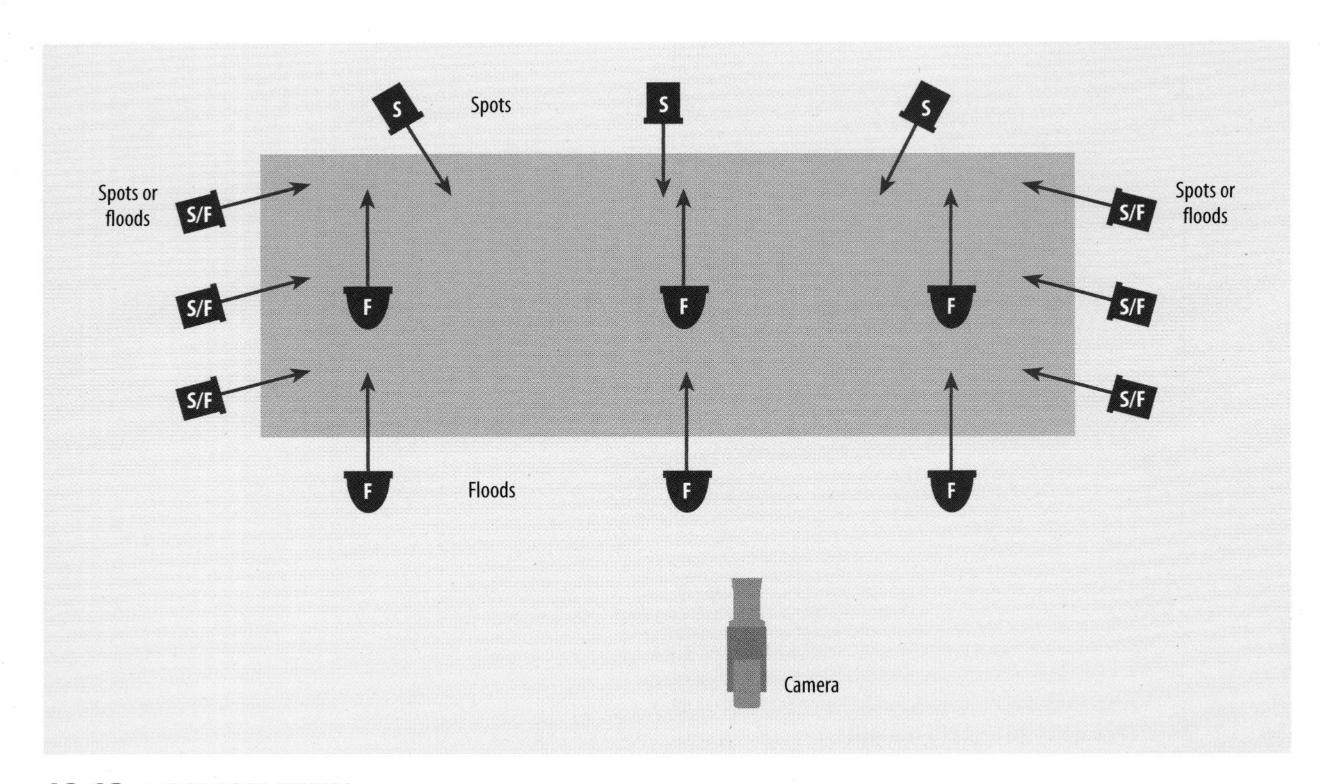

12.13 LARGE-AREA KEYING
In this lighting setup, the Fresnel spots at the left and right sides function as keys and directional fill lights. Fresnel spots are also strung out as regular back lights behind the main action area. If necessary, scoops provide additional fill light from the front.

You may have noticed that some series, such as crime or medical shows, use extremely fast-falloff lighting (harsh shadows). Extremely fast falloff on a face inevitably looks more dramatic than if it were lighted with slow falloff. **SEE 12.14** ZVL2 LIGHTS→ Design→ high key | low key

Prominent side lighting and fast-falloff lighting can be combined for dramatic effect. **SEE 12.15** And instead of always having the key-lighted side face the camera, you may show the shadow side to establish a certain mood. **SEE 12.16**

In addition to fast-falloff lighting, color distortion can add dramatic impact. In this example the scene is purposely given a green tint. **SEE 12.17**

Realize that such lighting effects require not only skill but also a lot of production time. That said, you should still

12.14 FAST-FALLOFF LIGHTING ON FACE

The fast-falloff lighting increases the dramatic impact of this close-up.

12.14 PHOTO BY EDWARD AIONA

12.15 HARSH SIDE LIGHT

The fast falloff and prominent side lighting intensify the mysterious mood of this scene.

12.15 PHOTO BY EDWARD AIONA

12.16 SHADOW SIDE TOWARD CAMERA

The camera-far person is lighted so that his shadow side, rather than key side, is seen by the camera. This shadow reversal has dramatic impact.

12.16 PHOTO BY EDWARD AIONA

12.17 COLOR DISTORTION
In addition to the fast-falloff lighting, the ominousness of this scene is further emphasized by the green tint.

12.17 PHOTO BY EDWARD AIONA

12.18 CAMEO LIGHTING
In cameo lighting, the background is kept dark, with only the foreground person illuminated by highly directional spotlights.

12.18 PHOTO BY EDWARD AIONA

try to apply some of these lighting techniques whenever possible and when appropriate to the show. If your lighting time is limited, however, stay away from such precision lighting and stick with the classic triangle-lighting approach. You might be pleasantly surprised to find that by turning on a few Fresnel spots and fill lights that are already pointed at the set area, your lighting will look quite acceptable.

Cameo Lighting

Certain television shows, especially those of a dramatic nature, are staged in the middle of an empty studio against an unlighted background. This technique, where the performers are highlighted against a plain dark background, is commonly known as ***cameo lighting*** (from the cameo art form in which a light relief figure is set against a darker background). SEE 12.18 Like the close-up, cameo lighting concentrates on the talent and not the environment.

All cameo lighting is highly directional and is achieved most effectively using spotlights with barn doors. In small studios the background areas are carefully shielded from any kind of distracting spill light with black, light-absorbing draperies.

Silhouette Lighting

Lighting for a silhouette effect is the opposite of cameo lighting. In ***silhouette lighting*** you light the background but leave the figures in front unlighted. This way you see only the contour of objects and people but not their volume and texture. To achieve silhouette lighting, use highly diffused light (usually from softlights, cyc lights, or scoops with scrims) to evenly illuminate the background. Obviously, you light in silhouette only those scenes that gain by emphasizing contour. SEE 12.19 You can also use silhouette lighting to conceal the identity of a person appearing on-camera. ZVL3 LIGHTS→ Design→ silhouette

Chroma-key-area Lighting

The chroma-key set area normally consists of a plain blue or green backdrop. It is used to provide backgrounds that are electronically generated, replacing the blue or green areas during the key—a process called ***chroma keying***. A popular

12.19 PHOTO BY EDWARD AIONA

12.19 SILHOUETTE LIGHTING
In silhouette lighting, only the background is lighted, with the figure in front remaining unlighted. It emphasizes contour.

use of the chroma key is a weather report. Although the weathercaster seems to be standing in front of a large weather map, she is in fact standing in front of an empty, evenly lighted blue or green backdrop. When the blue or green areas are electronically replaced by the weather map during the key, the weathercaster must look into a monitor to see the map. **SEE 12.20** (See chapter 14 for an in-depth explanation of the chroma-key process.)

The most important aspect of lighting the chroma-key area is even background illumination, which means that the blue or green backdrop must be lighted with highly diffused instruments, such as softlights or floodlight banks. If there are unusually dark areas or hot spots (undesirable concentrations of light in one area), the electronically supplied background image looks discolored or, worse, breaks up.

When lighting the weathercaster in the foreground, prevent any of the lights used for the foreground from hitting the backdrop. Such spill would upset the evenness of the chroma-key background illumination and lead to keying problems. In practice this means that the key and the directional fill light (a Fresnel in the flood position) must strike the subject from a steeper-than-normal angle. You may find that using softlights for the key and fill lights on the weathercaster will not affect the chroma key even if they spill onto the backdrop.

If you use a chroma-key light ring on the camera, do not place any reflective objects close to the light source. The light ring is an actual ring that fits over the lens and emits a relatively weak ray of blue or green light (see figure 14.26). This light beam sufficiently illuminates a beaded gray cloth to make it function as a blue or green chroma-key backdrop. Because the talent in front of the backdrop does not wear any reflecting beads and is sufficiently far away from the camera, he or she will not reflect any blue or green light coming from the light ring. If a highly reflecting object is placed near the camera with the light ring, however, the object will take on a blue or green sheen or, worse, let some of the keyed background image show through. We revisit the ring light in section 14.2.

Sometimes the outline of a weathercaster looks out of focus or seems to vibrate during the chroma key. One reason for such vibrations is that especially dark colors or shadows at the contour line take on a blue or green tinge, caused by a reflection from the colored backdrop. During the chroma key, these blue or green spots become transparent and let the background picture show through.

12.20 PHOTOS BY EDWARD AIONA

12.20 CHROMA-KEY EFFECT: WEATHERCAST

A In this weathercast, the blue background is evenly lighted with floodlights. The weathercaster is lighted with the standard triangle arrangement of key, back, and fill lights.

B During the chroma key, the weathercaster seems to stand in front of the satellite view.

12.21 SHADOW CAUSED BY EYEGLASSES
The steep angle of the key light causes the shadow of the woman's glasses to fall right across her eyes.

12.22 KEY LIGHT LOWERED
By lowering the key light, the shadow moves up and is hidden behind the glasses.

12.21 PHOTO BY EDWARD AIONA
12.22 PHOTO BY EDWARD AIONA

To counteract a blue reflection, try putting a light-yellow or amber gel on all the back lights or kicker lights. For green reflections, use a light-magenta or soft-pink gel. The back lights then not only separate the foreground subject from the background picture through contour illumination but also neutralize the blue or green shadows with the complementary yellow or pink filters. As a result, the outline of the weathercaster will remain relatively sharp even during the chroma key.

Controlling Eye and Boom Shadows

Two fairly persistent problems in studio lighting are the shadows caused by eyeglasses and microphone booms. Depending on the specific lighting setup, such unwanted shadows can present a formidable challenge to the lighting crew. Most often, however, you can correct such problems rather quickly.

Key light and eye shadows The key light's striking the subject from a steep angle will cause large dark shadows in any indentation and under any protrusion, such as in the eye sockets and under the nose and the chin. If the subject wears glasses, the shadow of the upper rim of the frames may fall directly across the eyes, preventing the camera (and the viewer) from seeing them clearly. **SEE 12.21**

There are several ways to reduce these undesirable shadows. First try to lower the vertical position of the light itself or move the key light farther away from the subject. When you lower it (with a movable batten or a rod), notice that the eye shadows seem to move farther up the face. As soon as the shadows are hidden behind the upper rim of the glasses, lock the key light in position. Such a technique works well so long as the subject does not move around too much. **SEE 12.22**

You can also reduce eye shadows by illuminating the person from both sides with similar instruments. Also try repositioning the fill light so that it strikes the subject directly from the front and from a lower angle, thus placing the shadows upward, away from the eyes. Annoying reflections from eyeglasses can be eliminated with the same technique.

Boom shadows Although you may not normally use a large microphone boom in the studio except for some dramatic productions, the principles of dealing with boom shadows also apply to handheld microphone booms, such as fishpoles and even handheld shotgun mics.

When you move a boom mic in front of a lighted scene—in this case a single person—and move the boom around a little, you may notice shadows on the actor or on the background whenever the mic or boom passes through a spotlight beam. (You can easily test for shadows by substituting a broomstick or the lighting pole.) You can deal with boom shadows in two ways: move the lights and/or the mic boom so that the shadow falls out of camera range, or use such highly diffused lighting that the shadow becomes all but invisible.

First you need to find the light that is causing the boom shadow. The easiest way to locate it is to move your head directly in front of the boom shadow and look at the microphone suspended from the boom. The shadow-causing light will now inevitably shine into your eyes. Be careful not to stare into the light for a prolonged

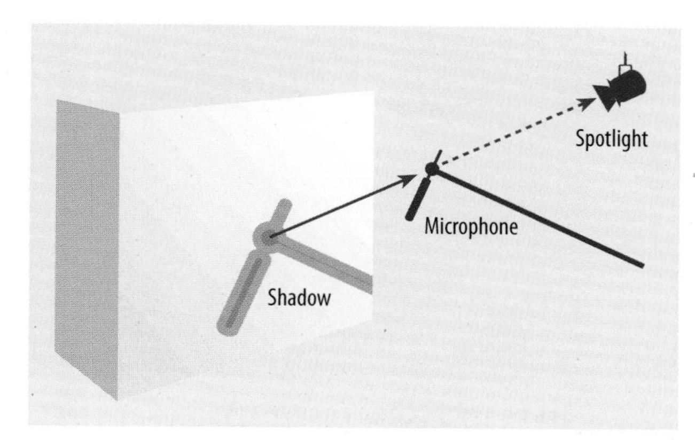

12.23 LOCATING THE SHADOW-CAUSING LIGHT
The instrument that causes the undesirable boom shadow lies at the extension of a line drawn from the shadow to the microphone causing it.

period. More precisely, the instrument lies at the extension of a line drawn from the shadow to the microphone causing it. SEE 12.23

To get rid of the shadow, simply turn off the offending instrument. You may be pleasantly surprised to find that you have eliminated the shadow without impeding the overall lighting. If such a drastic step seriously weakens the lighting setup, try to position the boom so that it does not have to travel through this light. If you use a handheld fishpole boom, walk around the set while pointing the mic toward the sound source. Watch the shadow move on the background wall until it is out of camera range. If the mic is still in a position for optimal sound pickup, you have solved the problem. You may locate such a shadow-safe spot more readily when holding or placing the boom parallel to the key-light beam rather than when crossing it. Some LDs use the key and fill lights close to side-light positions to provide a "corridor" in which to operate the boom.

Another simple way to avoid boom shadows is to light more steeply than usual. You do this by moving the key light closer to the performance area. The closer the lights are to actors, the steeper the lights will have to be angled to hit the target. The boom will now cast its shadow on the studio floor rather than on the talent's face or background scenery and thus be out of camera range. The downside to this technique (besides increasing the heat on the actors) is that the steep key lights produce dense and prominent shadows under the eyes, nose, and chin.

You can also try to use barn doors to block off the part of the spotlight that causes the boom shadow. Such a technique is especially useful when the shadow appears in the upper part of the background scenery.

CONTRAST

In chapter 6 you learned about contrast ratio and the way television cameras react to it. Now you'll get acquainted with how lighting affects contrast and how to keep it within tolerable limits. Contrast does not depend so much on how much light comes from the lighting instruments (incident-light reading) as on how much light is reflected by the illuminated objects (reflected-light reading). For example, a white refrigerator, a yellow plastic raincoat, and a polished brass plate reflect much more light than does a dark-blue velvet cloth, even if they are illuminated by the same source. If you place the brass plate on the velvet cloth, there may be too much contrast for the television camera to handle—and you have not even begun with the lighting.

What you must consider when dealing with contrast are the constant relationships among multiple factors, such as how much light falls on the subject, how much light is reflected, and how much difference there is between the foreground and the background or the lightest and darkest spots in the same picture. Because we deal with relationships rather than absolute values, we express the camera's contrast limit as a ratio.

Contrast Ratio

As you learned, ***contrast ratio*** is the difference between the brightest and the darkest spots in the picture (often measured by reflected light in foot-candles). The brightest spot, that is, the area reflecting the greatest amount of light, is called the reference white, and it determines the "white level." The area reflecting the least amount of light is the reference black, which determines the "black level." With a contrast limit of 50:1, the reference white should not reflect more than 50 times the light of the reference black.

Measuring Contrast

As pointed out in chapter 11, you measure contrast with a reflected-light reading—by first pointing the light meter close to the brightest spot (often a white card on the set, which serves as the reference white) and then to the darkest spot. Even if you don't have a light meter or waveform monitor for checking the contrast ratio, you can tell by looking at the camera viewfinder or studio monitor. When the white areas, such as the white tablecloths in a restaurant

12.24 ZEBRA STRIPES ON OVEREXPOSED MOUNTAIN
With the zebra control set at 100 percent, the zebra stripes appear on all overexposed areas.

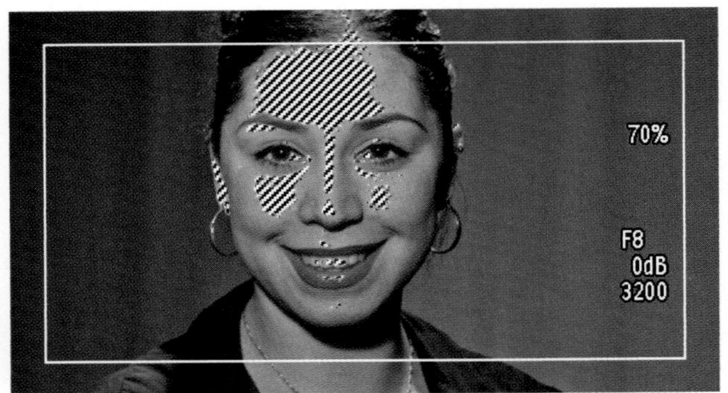

12.25 ZEBRA STRIPES FOR CORRECT EXPOSURE ON FACE
With the zebra control set at 70 percent, you've reached the proper exposure (*f*-stop) when the zebra stripes appear on the bright areas of the face.

12.24 PHOTO BY GARY PALMATIER
12.25 PHOTO BY EDWARD AIONA

set, look awfully bright, or the dark clothing of the people sitting at the tables is a dense black without any detail, the contrast is obviously great and probably exceeds the optimal ratio. With a little practice, squinting your eyes when taking a brief look at the set will give you a good idea about the contrast ratio even without using the light meter. ZVL4 LIGHTS→ Measurement→ contrast

Zebra stripes Most good camcorders show a zebra pattern when a certain white level is achieved or exceeded. If you have only one setting for zebra stripes, they will appear over the areas that exceed the brightness limits. This overexposure level for zebra stripes is sometimes labeled 100 percent zebra. **SEE 12.24**

If you have a choice between 70 percent zebra and 100 percent zebra, choose the 70 percent setting to get the proper exposure for skin tones on a face. When using manual exposure control, you should open up the iris until you see the zebra stripes appear on parts of the close-up of a face (such as the cheekbones, nose, forehead, and chin). This indicates the correct exposure. **SEE 12.25**

Controlling Contrast

The problem with contrast occurs when the bright areas reach a certain upper limit and are electronically clipped, very much like an audio signal that is too loud. When this happens, all detail in the bright areas is lost. This clipping is done either automatically or, when using a CCU (camera control unit), by the video operator (VO): when the VO "pulls down the whites," the black areas become crushed into a uniformly dark area without shadow detail. High-end cameras can tolerate a much higher contrast, however, and they show shadow detail in the dark areas even if the VO, also called the shader, brings extremely bright areas into tolerable limits.

If you feel that the contrast ratio is too high, think about what you can do to reduce it before fussing with the lighting. For example, changing the white tablecloth to a pink or light-blue one will help eliminate the contrast more readily than dimming some of the lights.

Such help is much appreciated by the video operator, who is ultimately responsible for controlling contrast. If there is too much contrast, however, even the best VOs have difficulty unless they're working with top-of-the-line cameras. For example, it is difficult for the camera to reproduce true skin color if the talent is wearing a highly reflective starched white shirt and a light-absorbing black jacket. If the camera adjusts for the white shirt by clipping the white level, the talent's face will go dark. If the camera tries to bring up the black level (making the black areas in the picture light enough to distinguish shadow detail), the face will be overexposed, or, in production slang, "blown out."

One big advantage of shooting in the studio is that you can control the light intensity and, with it, the contrast. Even if the talent wear contrasting clothes, you can always reduce the contrast by adjusting the key and fill lights so that the differences between light and dark areas are somewhat reduced.

Here are several tips for preventing an overly high contrast ratio.

- Be aware of the general reflectance of the objects. A highly reflective object obviously needs less illumination than does a highly light-absorbing one.

- Avoid extreme brightness contrasts in the same shot.
- Have the talent avoid clothes whose colors are too contrasting (such as a starched white shirt with a black suit).

Many contrast problems, however, occur when shooting outdoors on a sunny day. These problems and how to solve them are explored in section 12.2.

BALANCING LIGHT INTENSITIES

Even if you have carefully adjusted the position and the beam of the key, back, and fill lights, you still need to balance their relative intensities. For example, it is not only the direction of the lights that orients the viewer in time but also their relative intensities. A strong back light with high-key, slow-falloff front lighting can suggest the early-morning sun; a generous amount of strong back light and very low-intensity front lighting can suggest moonlight.

Because balancing the three lights of the lighting triangle depends on what you intend to convey to the viewer, you can't use precise intensity ratios among key, back, and fill lights as an absolute guide to effective lighting. Nevertheless, there are some ratios that have proved beneficial for a number of routine lighting assignments. You can always start with these ratios and then adjust them to your specific lighting task.

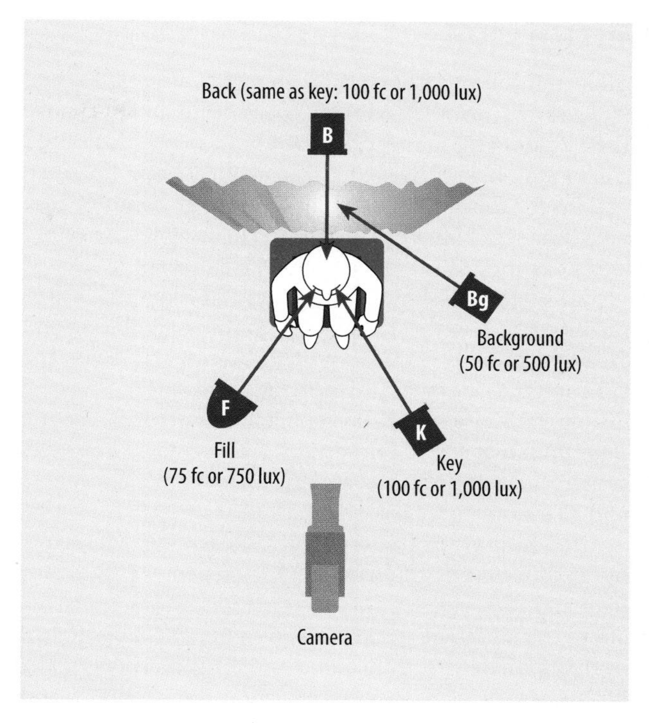

12.26 LIGHTING RATIOS
Lighting ratios differ, depending on the specific lighting task. These ratios are a good starting point.

Key-to-back-light Ratio

In normal conditions back lights have approximately the same intensity as key lights. An unusually intense back light tends to glamorize people; a back light with an intensity much lower than that of the key tends to get lost on the monitor. A television performer with blond hair and a light-colored suit will need considerably less back light than will a dark-haired performer in a dark suit. The 1:1 key-to-back-light ratio (key and back lights have equal intensities) can go as high as 1:1.5 (the back light has one and a half times the intensity of the key) if you need a fair amount of sparkle or if the talent has dark, textured, light-absorbing hair.

Key-to-fill-light Ratio

The fill-light intensity depends on how fast a falloff you want. If you want fast falloff for dramatic effect, little fill is needed. If you want very slow falloff, higher-intensity fill is needed. As you can see, there is no single key-to-fill-light ratio, but for starters you may want to try a fill-light intensity that is half that of the key and go from there. Remember that the more fill light you use, the less modeling the key light is doing but the smoother the texture becomes (such as that of a person's face). If you use almost no fill light, the dense shadows reveal no picture detail.

If you are asked to light for a high-baselight, low-contrast scene (high-key lighting), you may want to use floodlights for both the key and the fill, with the fill burning at almost the same intensity as the key. As you know by now, high-key has nothing to do with the actual positioning of the key light but rather the intensity of the overall light level. The back light should probably burn with a higher intensity than the key or the fill light to provide the necessary sparkle. In a low-key scene, the back light is often considerably brighter than the key and fill lights. SEE 12.26

Again, as helpful as light meters are in establishing rough lighting ratios, do not rely solely on them. Your final criterion is how the picture looks on a well-adjusted monitor.

LIGHT PLOT

Most routine programs, such as news, interviews, and panel shows, are relatively easy to light and do not change their lighting setup from show to show, so you don't need a light plot. If you have to light an atypical show, however, such as for the video recording of a college dance or play performance, a light plot makes the lighting less arbitrary and saves the crew considerable time and energy.

The ***light plot*** shows the location of the lighting instruments relative to the set and the illuminated objects and areas, the principal directions of the beams, and ideally the type and the size of the instruments used.

In drawing a successful light plot, you need an accurate ***floor plan*** that shows the scenery and the stage props, the principal talent positions, and the major camera positions and shooting angles (see section 15.2).

If you don't use a computer, an easy way to make a light plot is to put a transparency over a copy of the floor plan and draw the lighting information on the transparency. Use different icons for spotlights and floodlights, drawing arrows to indicate the main directions of the beams. **SEE 12.27** Some LDs use cutouts of spotlights and floodlights, which they lay on the floor plan and move to the appropriate positions.

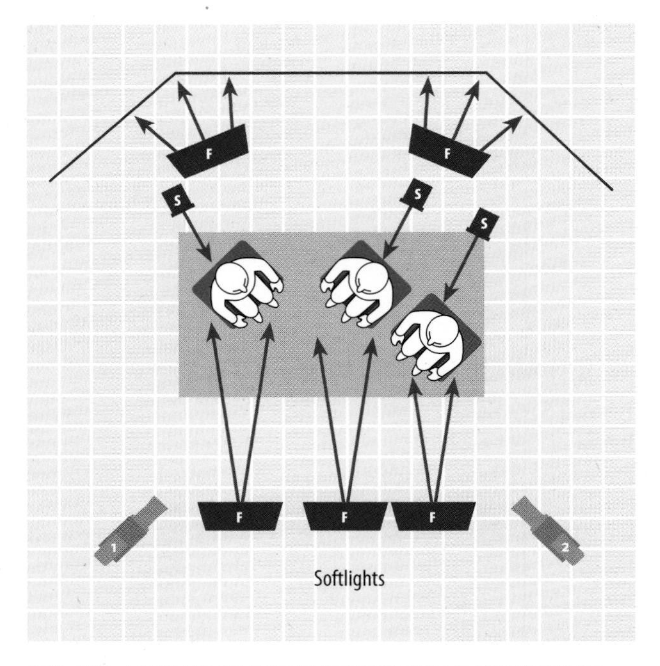

12.27 LIGHT PLOT FOR FLAT LIGHTING OF INTERVIEW
This light plot shows the slow-falloff (flat) lighting setup for a simple interview. Ordinarily, such a simple setup would not require a light plot. Note that the sketch is not to scale.

Try to work with the set designer (usually the art director) or the floor manager (who is responsible for putting up the set) as much as possible to have them place the set in the studio where you won't have to move any, or only a few, instruments to achieve the desired lighting. *Placing a small set to suit the available lighting positions is much easier than moving the lights to suit the location of a small set.*

Studio lighting is successful when you get it done on time. With due respect to creative lighting, don't fuss over a single dense shadow somewhere on the background while neglecting to light the rest of the set. If you are really pressed for time, turn on some floodlights and back lights that hang in approximate positions and hope for the best. More often than not, the lighting will look acceptable.

OPERATION OF STUDIO LIGHTS

When initially hanging lights, divide the studio into major performance areas and hang the appropriate instruments (spotlights and floodlights) in the triangular arrangements of the basic photographic principle.

Preserving Lamps and Power

Always warm up large instruments through reduced power by keeping the dimmer low for several minutes before supplying full power. This will prolong the lamp life and the Fresnel lenses, which occasionally crack when warmed up too fast. This warm-up period is essential for getting HMI lights up to full operation.

Do not waste energy. Dry runs (without cameras) can be done just as efficiently when illuminated by work lights as by full studio lighting. If you have movable battens, telescope hangers, or pantographs, try to bring the lights down as close as possible to the object or scene to be illuminated. As you know, light intensity drops off considerably the farther the light moves from the object. Bring the lights up full only when necessary.

Using a Studio Monitor

If you intend to use a well-adjusted color monitor as a guide for lighting, you must be ready for some compromise.

As noted, the lighting is correct if the studio monitor shows what you want the viewer to perceive. To get to this point, you should use the monitor as a guide to lighting, rather than the less direct light meter. Switch the camera to automatic exposure for the setup and then, if you have a VO operating a CCU, work with the VO to fine-tune the lighting.

MAIN POINTS

- Exercise caution during all lighting operations. Do not look directly into the instruments, and wear gloves when handling the hot lights.
- All lighting uses directional and/or diffused light. Color temperature is important when trying to achieve true colors on the screen. You should white-balance a camera every time you move in a specific (new) lighting environment.
- The key light is the principal source of illumination and reveals the basic shape of the object.
- The back light provides more definition to the object's outline, separates it from the background, and gives it sparkle.
- The fill light reduces falloff and makes the shadows less dense.
- Most television and film lighting setups use the basic photographic principle, or triangle lighting, of key, back, and fill lights.
- The background, or set, light illuminates the background of the scene and the set. The side light acts as additional fill or a side key. The kicker light is used to outline the contour of an object that would otherwise blend in with the background.
- Specific lighting techniques include high- and low-key lighting, flat lighting, continuous-action lighting, large-area lighting, high-contrast lighting, cameo lighting, silhouette lighting, chroma-key-area lighting, and controlling eye and boom shadows.
- Falloff indicates how fast the lighted side of a subject changes to shadow and how dense the shadows are. Fast falloff means that the light and shadow areas are distinct and that the shadows are dense. Slow falloff means that the transition from light to shadow is more gradual and that the shadows are transparent. Generally, fast falloff means high-contrast lighting; slow falloff means low-contrast, or flat, lighting.
- A high-key scene has a light background, a generally high baselight level, and usually an upbeat, happy mood. A low-key scene has a dark background with selective fast-falloff lighting and a dramatic or mysterious mood.
- The contrast ratio is the difference between the lightest and the darkest areas in a picture, often measured by reflected light in foot-candles.
- The light plot indicates the location of the lighting instruments, the principal direction of their beams, and sometimes the type and the size of the instruments used.

SECTION

12.2 Lighting in the Field

When engaged in location lighting, you are not working in the studio, where all the lighting equipment is in place and ready to go. Every piece of equipment, however large or small, must be hauled to the remote location and set up in places that always seem either too small or too large for good video lighting. You also never seem to get enough time to experiment with various lighting setups to find the most effective one. Whatever the remote lighting task, you need to be especially efficient in the choice of instruments and their use. This section explains the techniques of location lighting and describes some of its essential requirements.

- **SAFETY**
  Primary safety concerns: electric shock, cables, and fires
- **LOCATION LIGHTING**
  Shooting in bright sunlight, in overcast daylight, in indoor light, and at night
- **LOCATION SURVEY**
  Survey checklists and power supply

SAFETY

As in the studio, safety is a primary concern when lighting in the field. In fact, there are more safety hazards in the field than in the studio. No production, however exciting or difficult, excuses you from abandoning safety for expediency or effect.

Electric Shock

Be especially careful with electric power when on-location. A charge of 110V (volts) can be deadly. Secure cables so that people do not trip over them. Every connection—from cable to power outlet, from cable to cable, and from cable to lighting instrument—can cause an electric shock if not properly joined. If you use regular AC (alternating current) extension cables, tape all connections so they won't pull apart accidentally. Be especially careful when using power cords in the rain.

Cables

String the cables above doorways or tape them to the floor and cover them with a rubber mat or flattened cardboard at points of heavy foot traffic. A loose cable not only can trip someone but may also topple a lighting instrument and start a fire. See that all light stands are secured with sandbags.

Fires

As discussed in chapter 11, portable incandescent lighting instruments get very hot when turned on for only brief periods of time. Place them as far as possible from combustible materials, such as drapes, books, tablecloths, wood ceilings, and walls. It pays to double-check. If they must be close to walls or other combustibles, insulate them with aluminum foil.

LOCATION LIGHTING

There is no clear-cut division between lighting for ENG and EFP, except that in electronic news gathering you often have to shoot in available light or with the camera light. But when called upon to do an interview in a hotel room or in the office of a CEO, or when covering a ceremony on the steps of city hall, ENG and EFP lighting techniques are pretty much the same. The big difference is that in electronic field production you have enough lead time to survey the lighting requirements before the event takes place; but then you may be expected to make the office of a corporate president look like the best Hollywood can muster or to illuminate the hearing room of the board of supervisors so that it rivals a courtroom scene in the latest blockbuster movie—all without adequate time or equipment.

When lighting a remote location for digital cinema, you will proceed in pretty much the same way you would to light a set in the studio, except that you need to haul all the lighting equipment to the location and then set it up in usually cramped quarters.

When engaged in location lighting, you will find yourself confronted with problems both indoors and out.

When outdoors you have to work with available light—the illumination already present at the scene. At night you must supplement available light or provide the entire illumination. Although you have a little more time in EFP and certainly in digital cinema than in ENG, you must still work quickly and efficiently to obtain not only adequate lighting but also the most effective lighting possible under the circumstances.

Shooting in Bright Sunlight

Most lighting problems occur when you have to shoot in bright sunlight. A shooter's nightmare is having to cover a mixed choir, with the women dressed in starched white blouses and black skirts, and the men in white shirts and black jackets, with half of them standing in the sun and the rest of them in a deep shadow against a sun-flooded white building. Even a good digital ENG/EFP camcorder would have trouble handling such high contrast.

If you put the camera in the auto-iris mode, it will faithfully read the bright light of the shirts and the light background and close its iris for optimal exposure. The problem is that the drastic reduction of light coming through the lens will darken equally drastically the shadow area and the people standing in it. The black skirts and jackets will turn into a dull black and lose all detail. If you switch to manual iris to open the aperture somewhat to achieve some transparency in the shadows and the black skirts and jackets, the white shirts and blouses as well as the sunlit background will be overexposed. Worse, the highlights on the perspiring foreheads and occasional bald spots of the choir members will begin to "bloom," turning the skin color into strangely luminescent white spots surrounded by a pinkish rim.

Should you give up? No, even though your options are somewhat limited. Here are some potential remedies.

- Whenever possible, try to position the talent in a shadow area, away from a bright background. You could probably move the whole choir in the shadow and away from the sunlit building. For a single on-camera person, you can always create a shadow area with a large umbrella.

- Ask whether the male choir members can take off their black jackets. This is worth a try, even though you will probably be turned down. You will obviously be stuck with the black skirts.

- Shoot from an angle that avoids the white building in the background.

- Stretch a large piece of gauze or other diffusion material between two stands to diffuse the sunlight striking the choir, and use large reflectors to slow down the falloff. Recall that a reflector functions like an effective fill light. SEE 12.28 AND 12.29

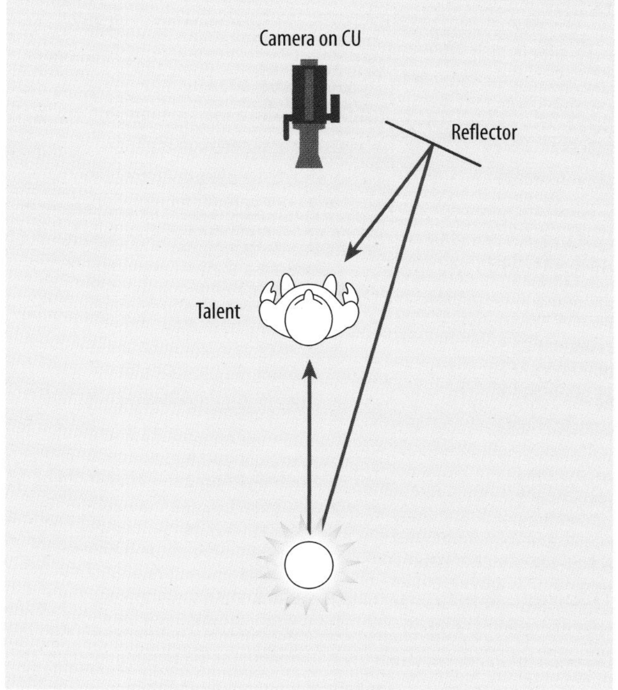

12.28 USING A REFLECTOR: SHOOTING AGAINST THE SUN
When shooting against the sun, reflect as much sunlight as possible back to the talent with a simple reflector (in this case a white card).

12.28 PHOTO BY EDWARD AIONA

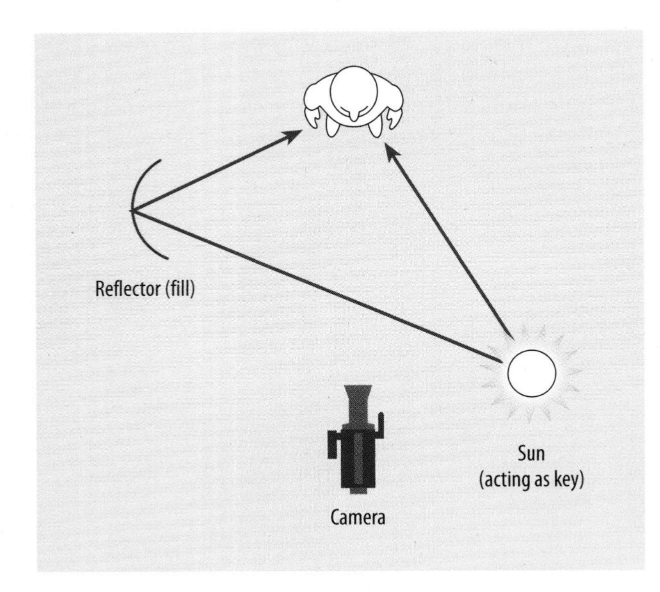

12.29 USING A REFLECTOR: SHOOTING WITH THE SUN
When shooting in bright sunlight, the dark shadows can be easily lightened with a reflector.

- Use the neutral density (ND) filters in the camera. The ND filters act like sunglasses of varying densities, reducing the amount of light that falls on the imaging device without distorting the actual colors of the scene. In fact, an ND filter seems to reduce extreme brightness while still revealing detail in the shadow areas. It will certainly eliminate the red-rimmed flares on the shirts and the perspiring foreheads of the choir without rendering the rest of the people invisible.

- Be sure the camera is switched to manual exposure control.

- Feed the camcorder into a good and shaded field monitor and adjust the iris.

- Keep your fingers crossed. ZVL5 LIGHTS→ Field→ outdoor | indoor | mixed | try it

Shooting in Overcast Daylight

The ideal light for outdoor shooting is an overcast day: the clouds or fog act as diffusers for the harsh sunlight, providing an even illumination similar to that of softlights. Do not be surprised if you have to use an ND and/or color-correction filter. The light of a cloudy day is often surprisingly bright and usually has a high color temperature.

Even in diffused lighting, avoid an overly bright background. If you must shoot against a light background, zoom in on the person, thereby avoiding as much of the background as possible. Be sure that you have manual iris control and adjust the iris to meet the light requirements of the person rather than of the background. It is better to have an overexposed background than an underexposed person.

Shooting in Indoor Light

You encounter countless amounts and types of light when shooting indoors. Some interiors are illuminated by the daylight that comes through large windows, others by high-color-temperature fluorescent banks that make up a light ceiling. Still others, such as windowless hotel rooms, have desk and floor lighting that provide a romantic mood but hardly the proper illumination for good television pictures. The major problem here is not so much how to supply additional light but how to place the instruments for optimal aesthetic effect and how to match the various color temperatures.

In all cases try first to set up the lighting triangle of key, fill, and back lights. If this isn't possible, try to adjust the setup so that you maintain at least the effect of triangle lighting. Whenever possible try to get some back-light effect; it is often the back light that distinguishes good lighting from mere illumination.

Let's assume that you are lighting an interview of the CEO of a software company. Except for some cutaway close-ups of the interviewer at the end of the show, the CEO is seen in a close-up for most of the interview. Let's put her in three different environments: in a windowless hotel room, in a hotel room with a window, and in her office with a large picture window behind her desk.

Windowless room In a room with no windows, you can simply set up portable, open-face lights in a typical triangle. Use a diffused key light, a more focused back light of the same type, and a reflector or softlight (or diffusion tent) for the fill (see figure 12.3). If you have a fourth instrument, you can use it as a background light. If only two instruments are available, use an open-face spot as a back light and use a diffused light (open-face spot with scrim, tent, or umbrella) as a key, placed so that most of the face is illuminated. The spill of the key will also take care of the background lighting. SEE 12.30

If the director insists on cross shooting with two cameras to catch the immediacy of the interviewer's asking questions or reacting to the CEO, you can still get by with two or three instruments. Place two open-face spots or small Fresnel spots with tents, scrims, or umbrella reflectors

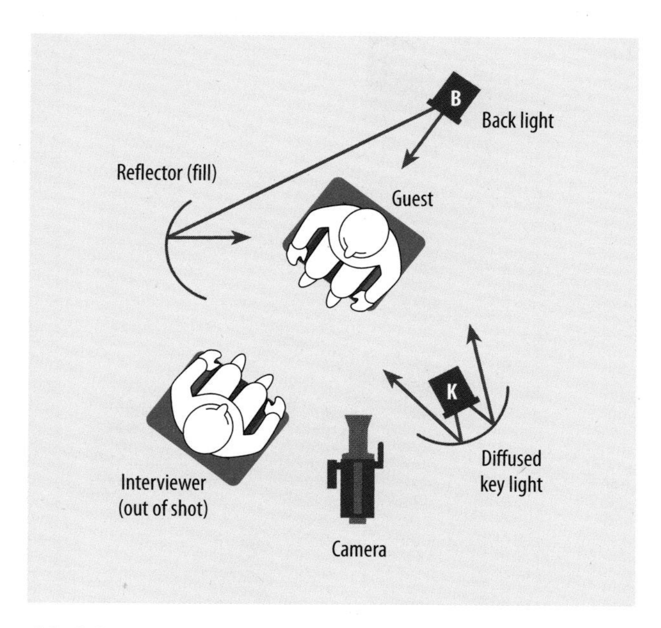

12.30 TRIANGLE LIGHTING FOR INTERVIEW
This one-person lighting setup uses two instruments. The diffused key light is an open-face spot with a scrim, a tent, or an umbrella. The back light is a spread or focused open-face spot. If fill light is necessary, it can be created with a softlight or a reflector. You can use an additional softlight as a background light. Note that the interviewee is looking at the interviewer, who is sitting next to the camera, out of the shot.

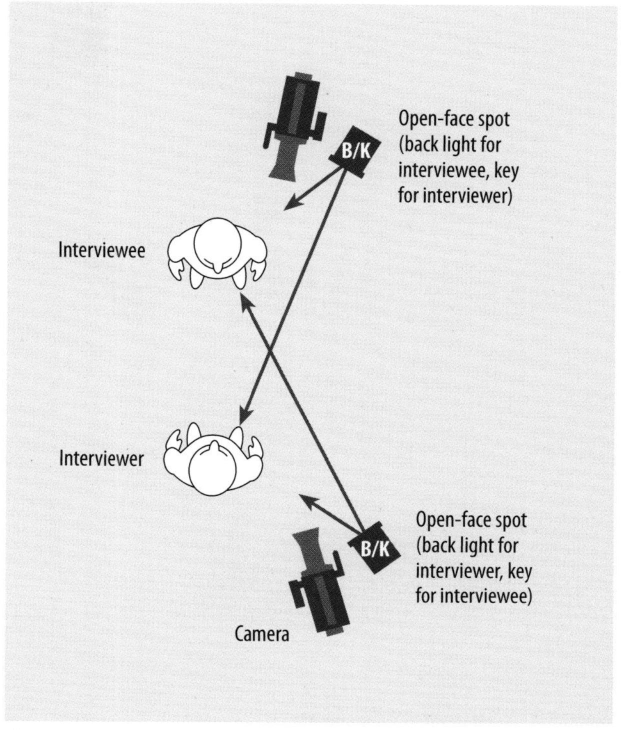

12.31 CROSS-KEYING FOR INTERVIEW
The two portable lights serve multiple functions: key and back lights for the interviewer and the interviewee. If you have a third light, use it as fill.

so that they shine over the shoulder of the participants sitting opposite each other. In this cross-keying, the two lights now serve as multifunction key and back lights. You can use the third instrument as general fill light. This lighting setup can also be used for an interview in a hallway, living room, or other such location. **SEE 12.31**

Room with window When there is a window in the room, you can use it as a key or even a back light. If you use the window as a key, you need a reflector or fill light on the opposite side. In any case, you need a strong back light. To match the outdoor color temperature of the window light, both the fill and back lights need either 5,600K lamps or 3,200K lamps with blue gels to raise their color temperature. **SEE 12.32**

The better way of lighting is to position the CEO so that the window acts as a back light—without letting it get into the shot. You can then use a single diffused 5,600K key light (an open-face spot with a 5,600K lamp or a 3,200K lamp with a blue gel) to illuminate most of her face, eliminating the need for a fill light. **SEE 12.33**

Panoramic office window A typical problem is having to shoot against a large window. If, for example, the CEO insists on making her statement from behind her desk that is located in front of a large picture window, your lighting problem is identical to that of shooting a person standing in front of a bright background: If you set the iris according to the background brightness, the person in front tends to appear in silhouette. If you adjust the iris for the person in front, the background is overexposed. Here are some possible solutions.

- Frame a fairly tight close-up and set the exposure for the face rather than the background. The little you see of the background window will be overexposed but will not render the close-up unusable.
- Draw the drapes or the blinds and light the person with portable instruments.
- Move the camera to the side of the desk and have the person face the camera. You can then shoot parallel to the

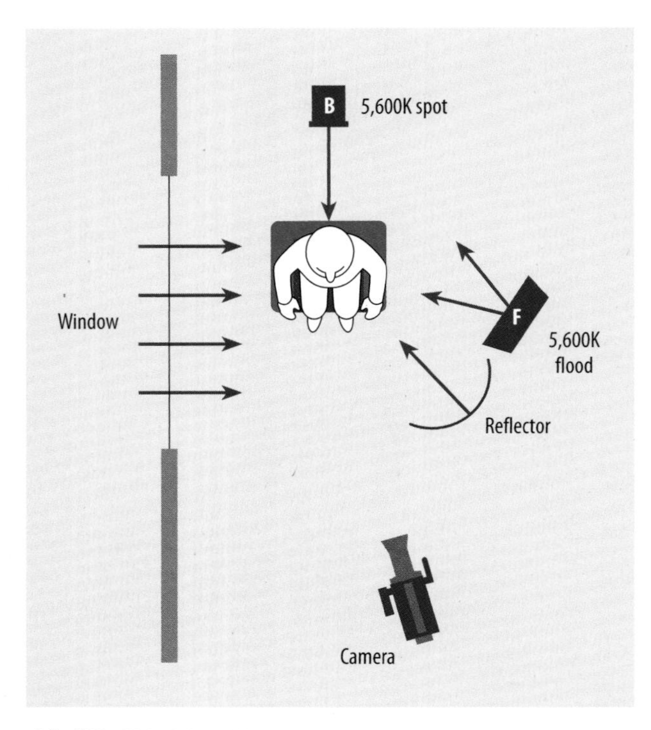

12.32 WINDOW AS KEY LIGHT
The daylight shining through a window can serve as the key light and a reflector as the fill light. If you use a portable light as the fill and/or back light, you need to bring its color temperature up to the 5,600K daylight standard.

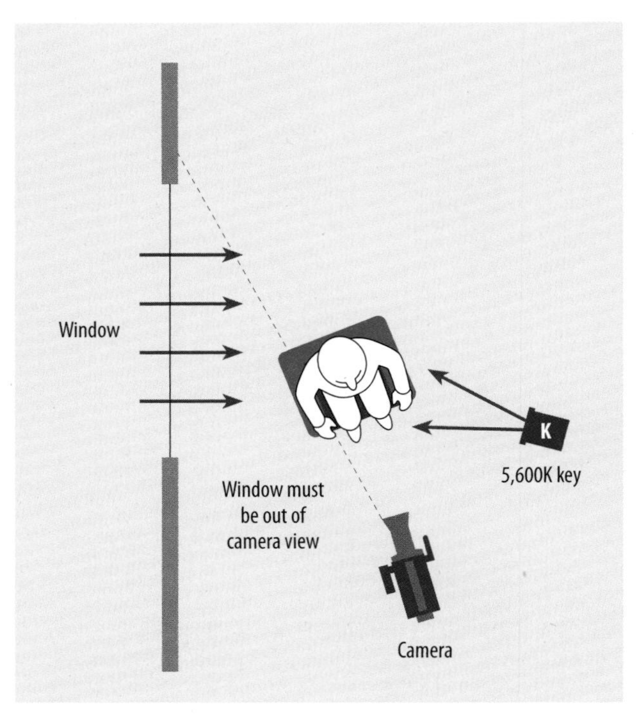

12.33 WINDOW AS BACK LIGHT
You can use a window as a back light so long as you place the talent with the window out of the shot. The key light can be a diffused open-face spot that burns at 5,600K.

window. You can use the light from the window as the key, and fill with a large reflector or an additional light on a stand (see figure 12.32).

- If the person insists on having the window in the background, you must cover it with large color temperature filters and/or ND filters (plastic sheets) of varying densities. Use two strong but diffused open-face 5,600K instruments as key and fill, or use a large, highly efficient reflector that bounces the light from the window onto the CEO's face. Bear in mind that these procedures take a great amount of time and are generally left to EFP.

- Take a picture of the window view and use it as a chroma-key video source (see chapter 14).

Large-area indoor lighting Sometimes you have to deal with groups of people who are gathered in locations with inadequate illumination. Typical examples are meeting rooms, hotel lobbies, and hallways. Most of the time, the available light or a camera light provides enough illumination to cover the speaker and individual audience members. If you are to do extensive coverage of such an event, however, you need additional illumination.

The quickest and most efficient way to light such a location is to establish a general, nondirectional baselight level. Use two or three open-face spots or V-lights and bounce the light off the ceiling or walls. If you have light-reflecting umbrellas, direct the lights into the open part of the umbrellas and place them so that you can cover the event area with the diffused light. You will be surprised by how much illumination you can get out of a single V-light when diffused by an umbrella. If that is not possible, direct the lights on the group but diffuse the beams with scrims. The most effective method is to use portable quartz, fluorescent, or HMI softlights and flood the active area. Always be sure to white-balance the camera for the light in which the event actually takes place.

As you probably noticed, all these lighting techniques aim to establish a high baselight level. Even when pressed for time, try to place one or two diffused back lights out of camera range. They will provide sparkle and professional polish to an otherwise flat scene. **SEE 12.34**

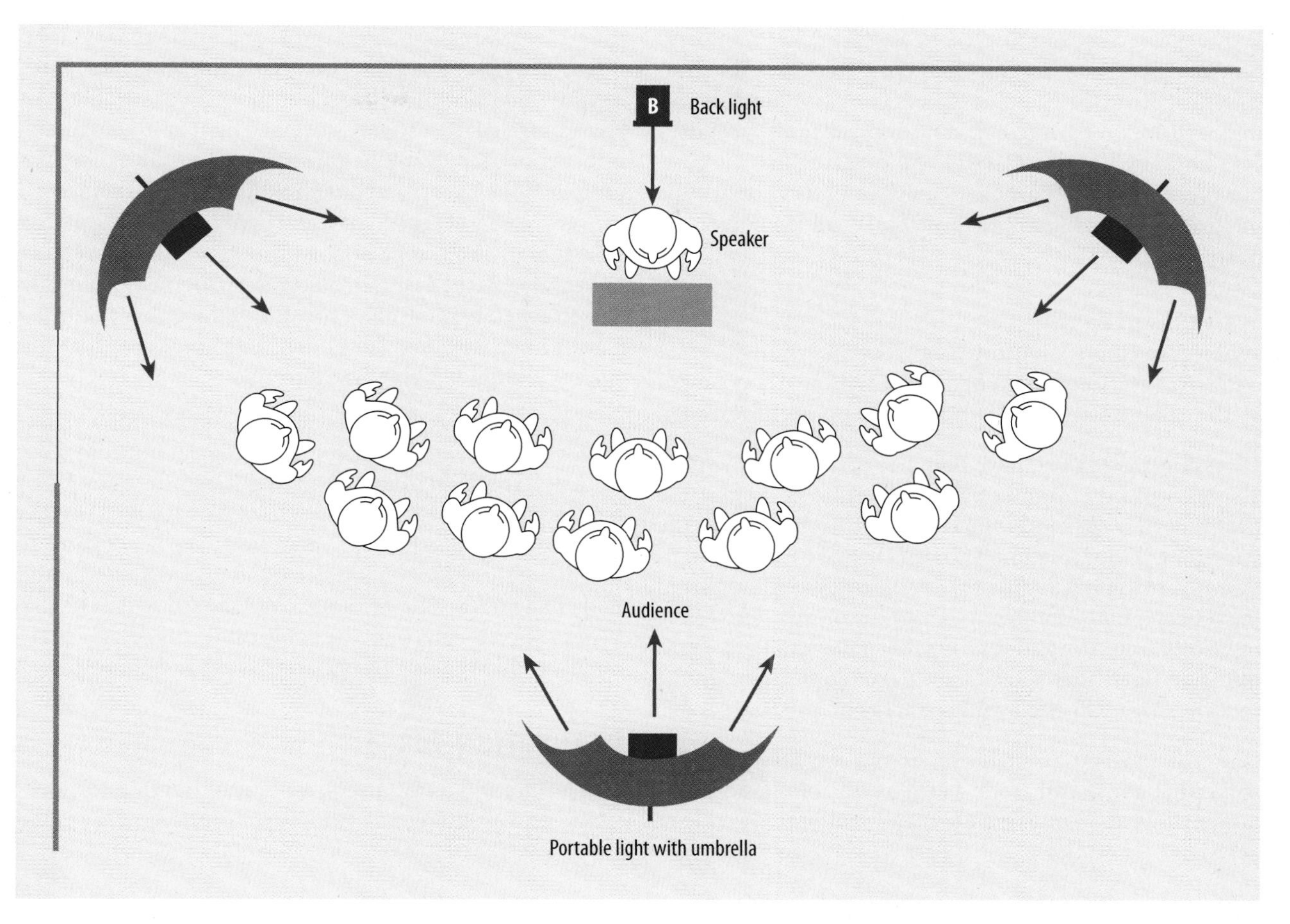

12.34 LARGE-AREA LIGHTING
To establish sufficient baselight over a large area, you need highly diffused light. Here three portable incandescent lights and light-diffusing umbrellas provide maximally diffused light over the entire area. You can, of course, use fluorescent or HMI lights in place of the quartz lights.

Working with fluorescents The basic problem of working with the fluorescent lights used in stores, offices, and public buildings is their color temperature; it is usually higher than the 3,200K indoor standard of incandescent lights. Even if some fluorescent tubes burn at the warmer indoor color temperature, they can have a strange greenish blue tint. So if you turn on the camera light for additional illumination, you are confronted with two color temperatures. Some lighting people advise turning off the fluorescents altogether when working with quartz lights (3,200K), but this is unrealistic. If you need to get a fast-breaking story and you shoot in a hallway that is illuminated by fluorescent lights, you certainly do not have time to locate and persuade the building manager to turn off the lights and then to relight the scene before starting to shoot.

If the fluorescent lights give enough illumination, simply select the appropriate color temperature filter in the camera (to bring down the high color temperature of the fluorescents) and white-balance the camera with the available light. If you have to use a camera light for additional illumination, either boost the color temperature of the camera light (by inserting a dichroic filter that often comes with the camera light) or white-balance the camera with the illumination provided by the camera light (3,200K).

As mentioned in chapter 11, portable incandescent lights—including the camera light—are strong enough to wash out the fluorescent baselight. If available, the better solution, of course, is to use floodlights that burn at the outdoor color temperature of 5,600K or floodlights whose color temperature is raised by light-blue gels.

One word of caution: Despite all the praise for fluorescent field lights, stay away from them if color reproduction is critical. Even the best fluorescent lamps do not give you the color mix for white light that you get from incandescent and HMI lamps. Careful white-balancing will help, but you may still discover a greenish or bluish tint to your pictures that is difficult, if possible at all, to correct in postproduction.

Shooting at Night

When covering a nighttime news event, you will most often use the camera light or a single light that is operated by the camera assistant. Here are some points to consider.

- Assuming that you have only one camera light and no assistant, use the camera light and aim it straight at the field reporter. The closer the reporter is to the camera, the stronger the illumination. You can change the light intensity by moving just one or two steps toward or away from the reporter and zooming in or out to compensate for the repositioning. Almost all professional camera lights have diffusion filters that you can use to soften the light on the reporter. This will also avoid hot spots on the face.

- If you have an assistant, he or she can hold the light somewhat above camera level (to avoid shining the light directly into the reporter's eyes) and a little to the side of the camera so that the single camera light acts as a key light. If you are fairly close to the event, put this single light into a semiflood position to avoid hot spots. Take advantage of any additional light source, such as a lighted store window or a street lamp, as fill by positioning the subject appropriately. Don't worry about mixing color temperatures; viewers readily accept color distortions when seeing events shot at night. You could also use the store window as a key light and have the assistant hold a reflector on the opposite side to generate some fill. If you don't have an assistant, just learn to live with fast falloff. Once again, avoid shooting against a brightly lighted background.

- If you are to cover a brief feature report outside a hospital, for example, and you are not under great time pressure, use a portable light mounted on a light stand as a key. Use a lighted hospital door or window as fill or back light. In this case position the reporter so that he or she is not directly in front of the door or window but off to one side, with the door or window out of camera range (see figure 12.33). Whenever possible, plug the lights into regular household outlets rather than using batteries as a power source.

LOCATION SURVEY

One of the most important aspects of lighting for electronic field production is a thorough ***location survey*** of the

12.35 EFP LOCATION SURVEY

INDOORS	OUTDOORS
AVAILABLE LIGHT	
Is the available light sufficient? If not, what additional lights do you need? What type of available light do you have? Incandescent? Fluorescent? Daylight coming through windows?	Do you need any additional lights? Where is the sun in relation to the planned action? Is there enough room to place the necessary reflectors?
PRINCIPAL BACKGROUND	
Is there any action planned against a white wall? Are there windows in the background? If so, do they have curtains, drapes, or venetian blinds that can be drawn? If you want to use the daylight from the window, do you have lights that match the color temperature of the daylight (5,600K)? If the window is too bright, or if you have to reduce the color temperature coming through the window, do you have the appropriate ND or color filters to attach to the window? You will certainly need some reflectors or other type of fill-light illumination.	How bright is the background? Even if the sun is not hitting the background at the time of the survey, will it be there when the actual production takes place? When shooting at the beach, does the director plan to have people perform with the ocean as the background? You will need reflectors and/or additional lights (HMIs) to prevent the people from being shown in silhouette, unless the director plans on extreme close-ups most of the time.

12.35 EFP LOCATION SURVEY *(continued)*

INDOORS	OUTDOORS
CONTRAST	
If there are dense shadows or if the event takes place in high-contrast areas (sunlight and shadows), you need extra fill light and/or ND filters to reduce the contrast.	Does the production take place in bright sunlight? Can the scene be moved into the shadow area? If not, you must provide for a generous amount of fill light (reflectors and/or HMI spotlights) to render the shadows transparent, or ND filters to reduce the glare of overly bright areas.
LIGHT POSITIONS	
Can you place the lights out of camera range? What lighting supports do you need (light stands, gaffer grip, clamps)? Do you need special mounting devices, such as battens or cross braces? Are the lighting instruments far enough away from combustible materials? Are the lights positioned so that they do not interfere with the event? People who are not used to television complain mostly about the brightness of the lights.	If you need reflectors or additional lights on stands, is the ground level enough for the stands to be securely placed? Will you need to take extra precautions because of wind? (Take plenty of sandbags along, or even some tent stakes and rope, so that you can secure the light stands in case of wind.)
POWER REQUIREMENTS	
Your main concern will be power and how to get it to the lighting instruments. Is the necessary power available nearby? Do you need a generator? If you can tap available power, make sure you can tell the engineer in charge the approximate power requirement for all lights. (Simply add up the wattage of all the lights you plan to use, plus another 10 percent to ensure enough power.) Do you have enough extension cords to reach all lighting instruments? Do you know exactly where the outlets are, what the capacities of the circuits are, and which outlets are on the same circuit? Make a rough sketch of all outlets and indicate the distance to the corresponding light or lights. What adapters do you need to plug lights into the available outlets? Do you have the necessary cables, extension cords, and power strips so that you can get by with a minimum of cable runs? In the projected cable runs, have you taken all possible safety precautions?	You do not need to use lighting instruments very often when shooting outdoors unless you shoot at night or need to fill in particularly dense shadows that cannot be reached with a simple reflector.

remote site. **SEE 12.35** The survey checklists in figure 12.35 are intended for relatively simple productions, as are all other discussions of EFP. (For more detailed information on location surveys, see "Remote Survey" in chapter 18.) The lighting for large and complex EFPs is more closely related to motion picture techniques and is not addressed here. But even in a relatively simple EFP, you will find that the power supply is one of the key elements for good remote lighting.

Power Supply

In EFP you have to work with three types of power for lighting instruments: household current (usually from 110V to 120V), generators, and 7.2V, 12V, or 30V batteries.

The most frequently used power supply is household current. When using regular wall outlets, be aware of the power rating of the circuits, which is usually 15 or 20 amps (amperes) per circuit. This rating means that you can theoretically plug in a 1,500W (or 2kW) instrument, or any combination of lights that does not exceed 1,500W (or 2,000W), without overloading the circuit, provided nothing else is on the same circuit. But that is not always wise to do. Recall the discussion about extension cords that build up additional resistance, especially when warm. To be on the safe side, do not load up a single circuit to full capacity. Otherwise you may find that the lights go out just at the most important part of the shoot.

You can find the capacity of the circuit by checking its fuse or breaker. Each breaker is labeled with the number of amps it can handle. You can now figure the total wattage capacity of each circuit: simply multiply the number of amps of the circuit (15 or 20 amps) by 100 (assuming the household current rates between 110V and 120V). This gives you an upper limit: 1,500 watts for a 15-amp breaker (15 amps × 100 volts = 1,500 total wattage) or 2,000 watts for a 20-amp breaker (20 amps × 100 volts = 2,000 watts). But don't press your luck. Try to use lower-wattage instruments per circuit to ensure that the lights will work properly during the entire production. SEE 12.36

If you need to power more lights than a single circuit can handle, plug them into different circuits. But how do you know which outlets are on separate circuits?

Determining the circuits Normally, several double wall outlets are connected to a single circuit. You can determine which outlets are on the same circuit by plugging one low-powered lamp into a particular outlet. Find the specific circuit breaker that turns off the lamp. Switch the breaker on again. The light should light up again. Now plug the light into the next convenient outlet and switch off the same circuit breaker or fuse. If the light goes out, the plugs are on the same circuit. If the light stays on, it's a different circuit and you are safe to use it.

Safe power extensions Obviously, you need enough extension cords to get from the outlets to the lighting instruments. You can minimize cable runs by using power strips (multiple-outlet boxes), especially if you use low-wattage instruments. The larger the wires in the extension cords (lower gauge ratings), the more wattage they can handle without getting unduly hot. Have enough and various kinds of adapters available so that lights can be plugged into the existing outlets.

Whenever there is doubt about the availability or reliability of power, use a generator, the responsibility of which falls to the engineering crew. The circuit ratings and the allowable combined wattage of the lights per circuit still apply.

For relatively simple on-location productions, you may power the lights with batteries. First check whether the lamps in the portable lights are appropriate for the voltage of the battery. Obviously, you cannot use a 12V lamp with a 30V battery. Then check that the batteries are properly charged and that there are enough spares for the duration of the production. Turning off the lights whenever possible saves battery power and greatly extends the life of the lamps.

12.36 CALCULATING ELECTRIC POWER REQUIREMENTS

WATTAGE OF LAMP	NUMBER OF INSTRUMENTS PER 15-AMP CIRCUIT
100	15
150	10
175	9
200	7
350	4
500	3
750	2
1,000	1
1,500	1

To find the maximum load (watts) for a single circuit, use the following formula:

amperes × volts = watts

The ampere rating of a standard household circuit is 15 amps (normally stamped on the circuit breaker). This means that the circuit can theoretically support a maximum load of 15 amps × 110 volts = 1,650 watts.

To be safe always figure 100 volts instead of 110 volts:

15 amps × 100 volts = 1,500 watts

To calculate how many instruments to plug into a single circuit, divide their total wattage into 1,500 watts (maximum load). The table at left lists the number of instruments of a certain wattage that you can safely plug into a single 15-amp circuit.

MAIN POINTS

- Safety is a primary concern when lighting in the field. Never abandon safety for expediency or effect.
- When shooting in bright sunlight, try to place the talent in the shade. If you must shoot in the sun, use a reflector and/or a neutral density (ND) filter to reduce contrast.
- The best outdoor shooting light is an overcast day. The clouds act as a giant diffusion filter.
- Use the basic photographic principle (triangle lighting) when lighting a single-person interview in a windowless room. If you have only two instruments, use a softlight from the front as key and fill and use a second instrument as a back light. When cross shooting use two instruments to fulfill key- and back-light functions.
- When a window is present, use it for fill or back light. Any indoor lights must then burn at 5,600K. Gel the 3,200K indoor lights with light-blue media or use 5,600K lamps. Use a large panoramic window for the key light or cover it with a curtain and use a triangle lighting setup. If the window is in the shot, filter the intensity of the light and lower the color temperature with gels on the window and add 3,200K key and fill lights.
- When shooting under fluorescent lights, use 5,600K lights for additional key and back lights or "wash out" the fluorescent ceiling lights with incandescent key, back, and fill lights.
- When shooting at night, use the camera light as the principal light source if no other light is available. Use a diffusion filter on the camera light and any other available light or a reflector for fill.
- Before doing any EFP lighting, conduct a location survey.
- The formula for figuring the electric power rating is amperes × volts = watts.
- When powering portable lights with household current, check the capacity of the circuits and do not overload them.

ZETTL'S VIDEOLAB

For your reference or to track your work, the *Zettl's VideoLab* program cues in this chapter are listed here with their corresponding page numbers.

CHAPTER

13

Video-recording and Storage Systems

Although one of television's great assets is its capability to transmit an event "live," that is, while the event is in progress, most programs have been prerecorded on some kind of video-recording device. There is virtually no show on the air, including live newscasts, that does not contain a preponderance of prerecorded material.

The ubiquitous camcorder, like the still camera, made the recording of various slices of life—observed or constructed—even more popular. In fact, it was principally the breakthrough of small recording devices and high-quality rechargeable batteries that made the portable camcorder possible and precipitated a revolution in video production. That you can get a high-definition television (HDTV) camcorder small enough to hold in your hand is a sheer miracle. Manufacturers are constantly striving to compress more and more video and audio information onto ever-smaller storage devices while making the retrieval of program material as quick and simple as possible. Despite the rapid move toward a totally tapeless operation, videotape is still a prominent recording media and, therefore, discussed in this chapter. Note that many of the operational principles of using videotape also apply to nontape media.

As you know by now, there is more to video recording than simply pressing the little red button on a camcorder and waiting for the *rec* light to flash. Section 13.1, Tape-based and Tapeless Video Recording, explains the principal analog and digital recording systems, and section 13.2, How Video Recording Is Done, covers basic video recording in the studio and in the field.

KEY TERMS

analog recording systems Record the continually fluctuating video and audio signals generated by the video and/or audio source.

codec Short for *compression/decompression*. A specific method of compressing and decompressing digital data.

component system A process in which the luminance (Y, or black-and-white) signals and the color (C) signals, or all three color signals (RGB), are kept separate throughout the recording and storage process.

composite system A process in which the luminance (Y, or black-and-white) signal and the chrominance (C, or color) signal as well as sync information are encoded into a single video signal and transported via a single wire. Also called *NTSC signal.*

compression Reducing the amount of data to be stored or transmitted by using coding schemes (codecs) that pack all original data into less space (lossless compression) or by throwing away some of the least important data (lossy compression).

control track The area of the videotape used for recording the synchronization information (sync pulse). Provides reference for the running speed of the videotape recorder, for the reading of the video tracks, and for counting the number of frames.

digital recording systems Sample the analog video and audio signals and convert them into discrete on/off pulses. These digits are recorded as 0's and 1's.

electronic still store (ESS) system An electronic device that can grab a single frame from any video source and store it in digital form. It can retrieve the frame randomly in a fraction of a second.

field log A record of each take during the video recording.

framestore synchronizer Image stabilization and synchronization system that stores and reads out one complete video frame at a time. Used to synchronize signals from a variety of video sources that are not genlocked.

JPEG A video compression method mostly for still pictures, developed by the Joint Photographic Experts Group.

memory card A read/write solid-state storage device for large amounts of digital video and audio data. Also called *flash drive* and *flash memory card.*

MPEG A compression technique for moving pictures, developed by the Moving Picture Experts Group.

MPEG-2 A compression standard for motion video.

MPEG-4 A compression standard for motion video. It has many codec variations for use in video compression and Internet streaming.

RGB component system Analog video-recording system wherein the red, green, and blue (RGB) signals are kept separate throughout the entire recording and storage process and are transported via three separate wires.

tape-based video recorder All video recorders (analog and digital) that record or store information on videotape. All tape-based systems are linear.

tapeless video recorder All digital video recorders that record or store digital information on media other than videotape.

time base corrector (TBC) Electronic accessory to a video recorder that helps make playbacks or transfers electronically stable.

video leader Visual material and a control tone recorded before the program material. Serves as a technical guide for playback.

video recorder (VR) All devices that record video and audio. Includes videotape, hard disks, read/write optical discs, and memory cards.

videotape recorder (VTR) Electronic recording device that records video and audio signals on videotape for later playback or postproduction editing.

videotape tracks Most videotape systems have a video track, two or more audio tracks, a control track, and sometimes a separate time code track.

Y/C component system Analog video-recording system wherein the luminance (Y) signals and the chrominance (C) signals are kept separate during signal encoding and transport but are combined and occupy the same track when actually laid down on videotape. The Y/C component signal is transported via two wires. Also called *S-video.*

Y/color difference component system Video-recording system in which three signals—the luminance (Y) signal, the red signal minus its luminance (R–Y) signal, and the blue signal minus its luminance (B–Y)—are kept separate throughout the recording and storage process.

SECTION

13.1

Tape-based and Tapeless Video Recording

Despite the great variety of video-recording devices, there are basically two types of systems: tape-based and tapeless. Tape-based systems can record analog or digital signals; tapeless systems can record only digital information. The great advantage of a tapeless operation is that you can store and archive vast amounts of information on media that take up very little space—and you can retrieve the information much faster than if it were stored on videotape.

To help you make sense of the many video-recording systems, section 13.1 gives an overview of tape-based and tapeless recording devices and how they work.

▶ **RECORDING SYSTEMS**
Tape-based systems, analog and digital systems, and tapeless systems

▶ **TAPED-BASED VIDEO RECORDING**
How videotape recording works, operational controls, and time base corrector and frame store synchronizer

▶ **TAPELESS VIDEO RECORDING**
Hard drives, read/write optical discs, and memory cards

▶ **ELECTRONIC FEATURES OF VIDEO RECORDING**
Composite and component signals, sampling, and compression

RECORDING SYSTEMS

On your visit to the video-recording section of a television station or video production company, you hear engineers discussing the advantages of tapeless video recorders when compared with videotape recorders and why some analog recorders are still in use. Then they get into such tech-talk as sampling, compression, the quality difference between component and composite video recordings, and so forth. They are basically discussing the types and the functions of video-recording systems and the major electronic features that determine the quality of the video recorders.

Tape-based Systems

This classification is based on the recording media used rather than how the video recorder works. Obviously, all ***tape-based video recorders*** use videotape as the recording media. What is not so obvious is that the ***videotape recorders (VTRs)*** can be either analog or digital.

Analog and Digital Systems

At this juncture you may ask: why bother with analog tape-recording systems when everything is digital anyway? You've got a point, but the venerable VHS (video home system) you may have is still used in millions of homes for off-the-air recording of favorite programs and playing rental movies. Besides, there are all the favorite tapes of birthdays, weddings, and off-the-air recordings that are still awaiting transfer to a less cumbersome digital media, such as digital versatile discs (DVDs). So, don't recycle your VHS recorder just yet.

In fact, the higher-quality S-VHS (Super VHS) and the high-end analog Betacam recorders are still produced by their manufacturers, and many consumer and professional analog camcorders are just too good to throw away. Their analog output can be easily digitized and imported by digital editing systems—such as a laptop computer.

Analog VTRs The basic problem with less-than-high-end video recording is that the recordings deteriorate quickly from one generation to the next (a generation is the number of dubs away from the original recording). This shortcoming is especially apparent during editing and the rendering of special effects. Both operations usually require several tape generations.

Without going into too much detail, the following table gives an overview of the major ***analog recording systems***. **SEE 13.1**

Digital VTRs The major advantages of ***digital recording systems*** are that they are more compact and that even inexpensive models produce high-quality pictures and sound. They maintain their quality through repeated dubs. For all practical purposes, the fiftieth generation looks the same as the original recording. Unlike analog recordings, digital recordings do not need to be converted for computer storage on a hard disk for nonlinear editing and special-effects manipulation. Except for a few older, high-end ½-inch

13.1 ANALOG VIDEOTAPE-RECORDING SYSTEMS

SYSTEM	CASSETTE	PRODUCTION CHARACTERISTICS
VHS	½-inch (12.7mm)	Low to fair quality for first generation. Rapid deterioration from one generation to the next.
Hi8	8mm (approx. 0.31-inch)	Sharp images. Good quality for first generation.
S-VHS	½-inch (12.7mm)	Very good quality. Will hold up fairly well for three or four generations.
Betacam SP	½-inch (12.7mm)	Superior quality. Will stand up well over multiple generations.

13.2 DIGITAL VIDEOTAPE-RECORDING SYSTEMS

SYSTEM	CASSETTE	PRODUCTION CHARACTERISTICS
DV	¼-inch (6.35mm) mini	Good digital quality.
Digital8	8mm	Good quality—about the same as DV.
DVCAM	¼-inch (6.35mm) full-sized	Excellent quality.
DVCPRO	¼-inch (6.35mm) full-sized	Excellent quality.
Betacam SX	½-inch (12.7mm) full-sized	Superior quality.
HDV	¼-inch (6.35mm) mini	Excellent quality. HDTV resolution.
HDTV	¼-inch (6.35mm) full-sized	Superior quality. High resolution and color.

digital recording systems still in use (such as the D5 VTRs), almost all new digital videotape systems use the much smaller ¼-inch cassettes. They do, however, differ in size and range from the mini-cassette you are probably using in your camcorder to larger (but not wider) cassettes that allow longer continuous recording. SEE 13.2

Tapeless Systems

Tapeless systems record only digital signals and use a variety of recording media: computer hard disks, read/write optical discs, and various memory cards, also called flash drives. The latter have no moving parts but are more limited in their recording capacity than are hard drives and optical discs. Because most professional television equipment is digital, the more flexible ***tapeless video recorders*** are more desirable.

There are four big advantages of a tapeless recording system over videotape: it allows almost instant random access to any frame in the recording, it is much less prone to wear and tear, it facilitates the entire editing process, and the archived material takes up considerably less space.

All tapeless video-recording systems operate on the same principle: they store digital video and audio data in computer files that can be identified and randomly retrieved. If that sounds familiar, it's because disk-based systems are, indeed, nothing but specialized computers.

This is why you can use a laptop or desktop computer and appropriate software as the key elements for a disk-based recording and editing system.

TAPED-BASED VIDEO RECORDING

Although there is a great difference in the electronic makeup of analog and digital video recorders, the basic recording and operational principles are quite similar and often identical.

How Videotape Recording Works

When recording video and audio signals, the videotape moves past a rotating head assembly that "writes" the video and audio signals on the tape during the recording process and "reads" the magnetically stored video and audio signals off the tape during playback. In the play mode on some VTRs, the same heads used for recording are also used to read the information off the tracks and convert it back into video signals. Others use different heads for the record and playback functions. Some analog VTRs use two or four heads for the record/play (write/read) functions.

Digital VTRs record the coded video and audio signals, which consist of combinations of on/off pulses (0's and 1's). Some digital VTRs have more read/write heads than analog VTRs for various video, audio, control, and cue tracks. For a simple explanation of how video recording works, the following discussion uses an analog VTR with two record/playback heads.

Record/playback heads As mentioned, during video recording the tape moves past a rotating head assembly that writes the video and audio signals on the tape while recording and reads the information off the tape during playback. Good VTRs use one set of heads for recording the signals and another set for playing back the information. SEE 13.3

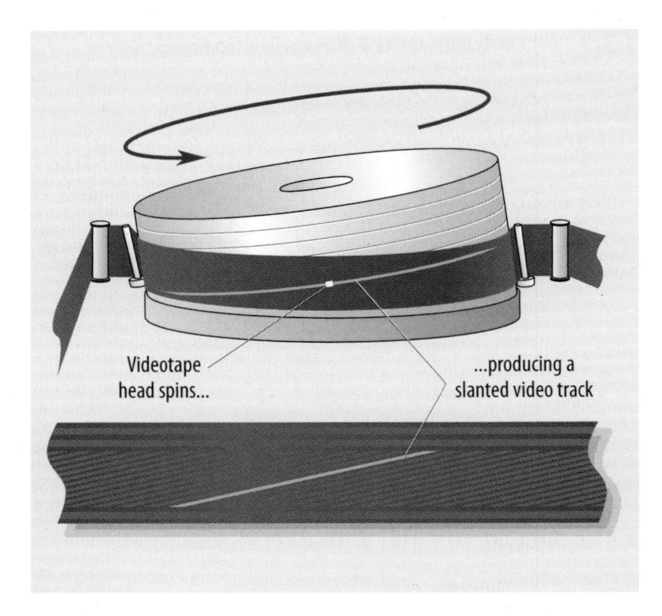

13.3 BASIC VIDEOTAPE-RECORDING METHOD
When recording, the videotape moves past a revolving head drum, which contains record and playback heads. These heads write the video and audio information onto the tape or read it from the tape.

Videotape tracks There are many ways in which video and audio signals are written to videotape, but they all use comparable ***videotape tracks:*** a video track, two or more audio tracks, one or more control tracks, and various data tracks. All video tracks are recorded as helical (slant) tracks, but the audio and control information can be on separate longitudinal tracks or incorporated in the helical tracks of the video signals. In its simplest form, the standard VHS recorder puts four separate tracks on the tape: the video track containing the picture information, two audio tracks containing all sound information, and a control track that regulates the videotape and the rotation speed of the VTR heads. SEE 13.4 ZVL1 EDITING→ Postproduction guidelines→ tape basics

Video track When you record the video signal in the normal NTSC composite configuration, one pass of the head records the first field (half a frame) of video information. The next pass of the head lays down the second field right next to it, completing a single video frame. Because two fields make up a single frame, the two heads must write 60 tracks for 60 fields, or 30 frames, for each second of NTSC video.[1] Digital systems use high-speed heads to lay down 20 or more tracks to record a single video frame.

Audio tracks Analog audio information is generally recorded on longitudinal tracks (stripes) near the edge of the tape. Because of the demand for stereo audio and for keeping certain sounds separate even in monophonic sound, all VTR systems (even VHS recorders) provide at least two audio tracks. Digital VTRs have the multitrack audio information embedded in the video track.

1. There are actually only 29.97 fps (frames per second).

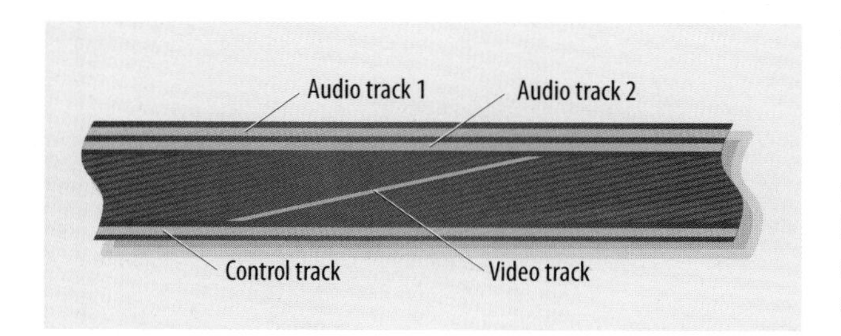

13.4 BASIC VIDEOTAPE TRACK SYSTEM
The basic videotape track system comprises a slanted video track, two or more audio tracks, and a control track.

Control and cue tracks The ***control track*** contains evenly spaced blips or spikes, called the sync pulse, which mark each complete television frame. These pulses synchronize the tape speed (the speed with which the tape passes from the supply reel to the take-up reel in the cassette) and the rotation speed of the record heads so that a tape made on a similar machine can be played back without picture breakup. The control track is also essential for linear videotape editing. Some VTRs have a separate track, called the cue track, for additional data, such as the SMPTE time code. This code marks each frame with a different address—a number that shows elapsed time and number of frames. (You will read more about time code in chapter 19.)

Digital VTR systems have a control track and a separate cue track for time code, and on each pass of the record heads they squeeze yet more coding information into a subcode area on a tape that is half the width of that in a standard VHS cassette.

Operational Controls

When merely looking at a high-end SD (standard-definition) VTR and the same-quality HDTV VTR, you will be hard-pressed to tell the difference unless you read the nameplates on the recorders. Although their electronic make-up is different, they look almost identical. SEE 13.5 AND 13.6

Even these high-end digital VTRs have the same controls as your home videocassette recorder (VCR), but the professional models have shuttle and especially additional audio controls. You know the basic VTR functions, so we address them only briefly. All VTRs (and VCRs), low-end or high-end, analog or digital, have these basic controls: the *play, stop, record, fast-forward, rewind,* and *eject* buttons and the audio volume controls. SEE 13.7

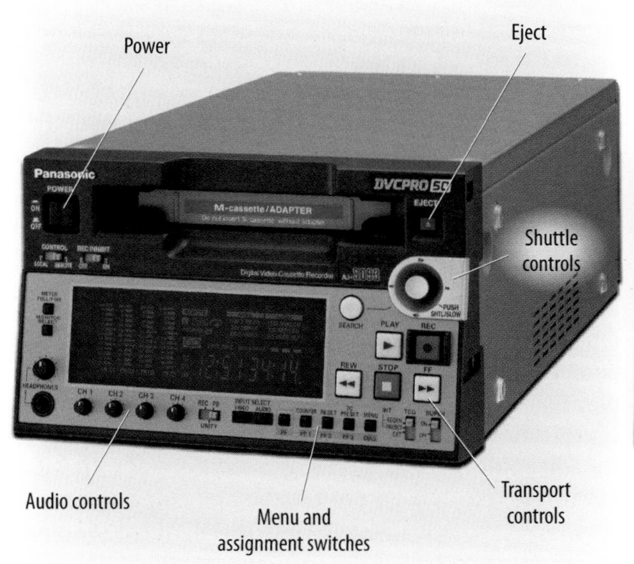

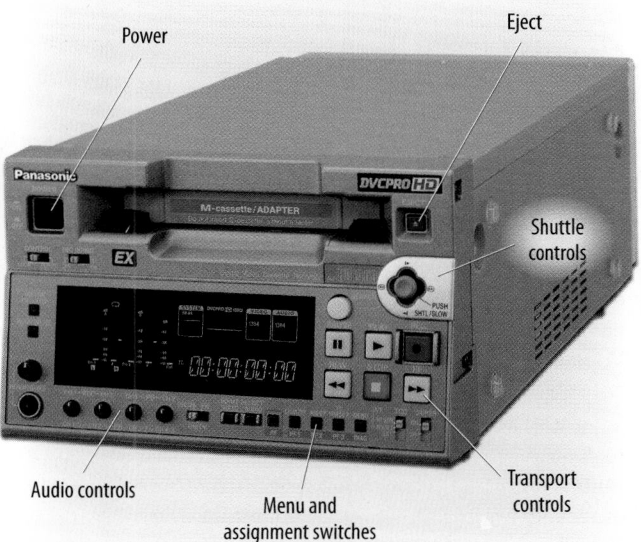

13.5 PHOTO COURTESY PANASONIC
13.6 PHOTO COURTESY PANASONIC

13.5 STANDARD-DEFINITION STUDIO VTR
This high-end digital VTR has all the controls of a standard VTR, some additional audio controls, and a variety of menu options.

13.6 HIGH-DEFINITION DIGITAL STUDIO VTR
This high-end digital HDTV VTR has the same operational controls as a SD digital or standard analog VTR.

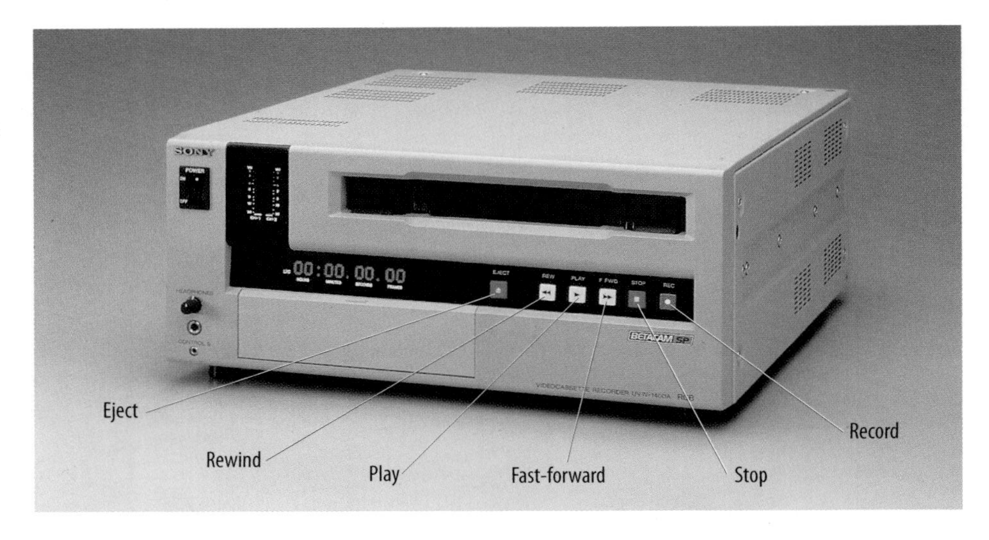

13.7 BASIC VTR CONTROLS
Standard VTR controls are similar to those on a home videocassette recorder.

13.7 PHOTO COURTESY SONY ELECTRONICS, INC.

The more sophisticated VTRs have additional controls for these operations: standby, pause or still, search or shuttle, and audio level.

Standby In the standby mode, the VTR threads the tape and rotates the video heads, but the tape remains stationary. The rotating video heads are disengaged and not in contact with the tape.

Pause or still The pause control will stop the tape with the heads still moving. In this mode the rotating video heads are in contact with the tape and continuously scan the adjacent video fields and produce a still—or freeze—frame on the video monitor or in the camera viewfinder. Do not keep a machine on pause for too long—the heads are apt to scrape the iron-oxide coating off the tape and leave you with nothing but clogged heads and video noise on the monitor. Most VTRs switch off pause if the tape has had enough abuse. Even so, don't leave a tape too long in pause mode, especially if you want to use it for editing.

Search or shuttle This control lets you advance or reverse the tape at speeds that may be much faster or slower than the normal record/play speed. The shuttle feature is especially important when searching for a particular shot or scene on the videotape. You can advance the tape frame-by-frame or rattle through a whole scene until you find the right picture. You can also slow down the shuttle enough to get a jogging effect, which shows a frame-by-frame advancement of the videotape. Some high-end recorders keep the original pitch of the recorded audio rather than raise it to the infamous chipmunk voice at high-speed shuttle or lower it to angry growls at a low-speed shuttle.

Audio level The main audio controls are the volume control and volume-unit (VU) monitoring for each audio channel. Some recorders have separate volume controls for sound recording and playback. The audio dub control lets you record sound information without erasing the pictures already recorded on the video track. Most professional analog VTRs allow to you choose between regular (analog) and hi-fi (digital) audio. Depending on how the sound on the videotape was recorded, you may have to switch to or from hi-fi audio. If the audio doesn't play back properly, try the other mode.

Time Base Corrector and Framestore Synchronizer

One of the main problems of tape-based video recorders is that you may experience some picture breakup when playing back a tape that was recorded on a different VTR. If the record and play machines are not perfectly aligned, if the edit points were not perfectly matched, or if you switch live from a remote source to a studio, you may experience a time base error, usually called a sync roll, which will show up as a glitch or a band that slowly moves up or down the screen, separating the upper and lower halves of the picture. SEE 13.8 This is why all professional videotape recording employs additional stabilizing equipment: the time base corrector or the framestore synchronizer.

13.8 PHOTO BY EDWARD AIONA

13.8 SYNC ROLL
The sync roll is caused by a time base error—a momentary picture breakup at the edit point or when two video sources that have different sync pulses are switched.

Time base corrector The ***time base corrector (TBC)*** adjusts the scanning of the play VTR to the one that did the original recording. Such synchronization is especially important in editing, where one or more source (or play) VTRs supply the material to be edited, and another VTR (the record VTR) does the actual editing. The TBC makes sure that the scanning clocks of all VTRs are in step. It can be a stand-alone piece of equipment, but most high-end VTRs have a built-in TBC to prevent picture breakup and ensure jitter-free pictures.

Framestore synchronizer This electronic marvel is an all-digital signal stabilizer that fulfills the same functions as the TBC and then some. The framestore synchronizer grabs each frame of the video signal, digitizes it, and stores it momentarily until its scanning is synchronized with the next good video frame it receives. It can be used to stabilize weak signals or signals that are temporarily interrupted. You may have seen a picture freeze momentarily when watching a live news report from a remote location. But how can a live picture freeze? Because the framestore synchronizer was doing its job. As you can see, in the digital age, even live events are temporarily recorded, however fleeting this recording may be.

TAPELESS VIDEO RECORDING

This overview looks at the recording features of hard drives, read/write optical discs, and memory cards.

Hard Drives

The most prominent hard drive used is the one in your computer. Even today's small laptops have hard drives with enough storage capacity for most simple editing jobs. But there are also specialized hard drives that serve specific recording functions.

Large-capacity hard disks The most efficient way to store and retrieve digital video and audio information for postproduction editing is with large-capacity external hard disks. Accelerated hard-drive speeds and highly efficient compression techniques enable you to store hours of video and audio information and call up any frame in a fraction of a second. Unlike videotape, which inevitably degrades after repeated use, the hard disk has no such limitation—its data remains like new even after a great many recordings and erasures.

Video servers In the ongoing move to an entirely tapeless operation, almost all television stations use video servers. If you think that servers are merely high-capacity hard drives, you are only half-right. Servers are indeed very large-capacity (in the multi-terabyte range) disk arrays that can ingest (fancy broadcast lingo for *input*) and record simultaneously multiple video and audio sources on several input channels. What makes a server so different from a standard large hard drive, however, is that it can serve several clients simultaneously. The brain of the server is a system of metadata (instructions) that lets you: store content in a variety of ways that facilitate fast search, such as name of program, date, time code, or function; order selected content for different clients; and create a playlist for automated on-the-air playback.

Best of all, a server is specifically built for multitasking. For example, a broadcast server lets you feed the newsroom with the latest footage of a breaking story, supply an editor with stored clips for an upcoming documentary, and offer a graphic designer a choice of backgrounds for the latest weather report—all while feeding the video and audio content of the daily telecast to the transmitter, some of it even in slow motion. SEE 13.9

Portable hard drives These small hard drives are usually built into camcorders, but some can also be docked with ENG/EFP cameras. Still others can be clipped onto a video camera or belt and connected to a camcorder via FireWire. Such detachable hard drives are often called field packs. All

13.9 VIDEO SERVER
The video server comprises large-capacity disk arrays (in the multi-terabyte range) that can ingest a great variety of different video and audio material, store it, and make it available for instant and simultaneous retrieval to multiple clients. The on-the-air content can be programmed for automated retrieval and broadcasting.

13.10 HARD DISK FIELD PACK
The 80 GB hard disk in this portable video recorder is capable of storing six hours of video and audio data in the DV (Digital Video) and HDTV formats.

13.9 PHOTO BY EDWARD AIONA
13.10 PHOTO COURTESY CANON USA

such portable hard drives are rated in gigabytes of storage and can hold several hours of SDTV (standard-definition television) or HDTV footage. Besides high-quality video, such hard drives can record time code and two- or four-channel audio. The USB, FireWire, eSATA, or Ethernet connections are especially handy when transferring the video recording to the editing system. SEE 13.10

Some portable hard drives allow you to do editing in the field with an externally connected laptop—a big advantage in electronic news gathering (ENG). Some camcorders have a tape-based as well as a disk-based recording system. You can use both simultaneously for recording, or use the tape as a backup in case the hard drive crashes (which is rare).

Electronic still store systems In effect a large slide collection that allows you to access any slide in a fraction of a second, the ***electronic still store (ESS) system*** can grab any frame or short clip from various video sources (camera, videotape, or computer) and store it in digital form on a hard disk. Although it used to be a stand-alone piece of equipment that was dedicated to the storage of still frames, the still- and clip-store functions have largely been incorporated into the software of video switchers, servers, or high-capacity character and graphics generators.

Read/Write Optical Discs

There are a variety of read/write optical discs that can record and play back a great amount of digital information. The optical discs most often used for storing video are DVDs, although in the true spirit of needless competition, you now must have at least two DVD systems to play all three DVD formats: the standard DVD; the HD DVD, which is downward compatible with the standard DVD (it can play the standard DVD, but the standard DVD player can't play the HD DVD); and the Blu-ray system, which can

13.11 DVD STANDARDS AND OPERATIONAL FEATURES

The maximum playing times are approximate. The capacities (in gigabytes) are given for single-side and double-side recording.

STANDARD	CAPACITY	VIDEO PLAY TIME (IN HOURS)
DVD*	4.7 GB	3.8
HD DVD	25 GB; 50 GB	8.5; 17
HDTV	15 GB; 30 GB	5.1; 10.2
Blu-ray	50 GB double-layer	4.5 (HDTV and 3D HDTV); 20 (SDTV)

*DVD has stereo sound; HD DVD and HDTV have audio tracks for surround sound.

play only its own Blu-ray discs. The table below describes the standards and the operational characteristics of each system. SEE 13.11

Some camcorders use read/write optical discs rather than hard disks as their recording media. The advantages of optical discs are that they are easy to store and they permit extremely fast access time. The disadvantage is that they must be handled very carefully. A single scratch can ruin two days' worth of shooting. Be especially careful with Blu-ray discs; they scratch rather easily despite their plastic coating.

Memory Cards

Memory cards, also called *flash drives* and *flash memory cards* (among other names), are solid-state digital storage media. They have become effective recording media for ENG/EFP camcorders. Their biggest advantage over hard drives and optical discs is that they have no moving parts. Much like the memory card in a digital still camera, video flash drives can be used over and over again without any noticeable deterioration.[2] Other advantages are that they are small and lightweight. Some can be inserted directly into the card slot of a laptop computer. The downside is that even the largest cards have a relatively limited storage capacity, especially when recording in high-definition. The real problem with using these cards is not technical but economic: when considering the cost per gigabyte of storage, the cards can be outrageously expensive. SEE 13.12

2. Note that some smaller camcorders use SD (Secure Digital) cards for their recording media. A 32 GB SD card can record almost three hours of HDTV footage.

13.12 PANASONIC P2 CARD RECORDING TIMES

This solid-state P2 memory card can be extracted from the camcorder and plugged directly into a computer slot for nonlinear editing. This table shows the approximate recording times for the normal 8 MB (megabyte) card as well as the 16 MB and 32 MB cards.

SIZE	RECORDING SYSTEM	RECORDING TIME (IN MINUTES)
8 MB	DVCPRO HD	8
	24p HD	20
	DVCPRO50	32
	DVCPRO	64
16 MB	DVCPRO HD	16
	24p HD	40
	DVCPRO50	64
	DVCPRO	128
32 MB	DVCPRO HD	32
	24p HD	80
	DVCPRO50	128
	DVCPRO	256

ELECTRONIC FEATURES OF VIDEO RECORDING

This discussion of electronic ***video recorder (VR)*** features concerns primarily the digital recorders and recording processes. If the material is getting a little too technical for your taste, at least try to remember its impact on video recording and editing. Nevertheless all this information is important if you aim to become a video professional or to hone the production skills you already possess. This section addresses composite and component systems, sampling, and compression.

Composite and Component Systems

There are basically two systems of recording video signals: composite and component. The ***composite system*** combines the black-and-white and color signals into one; the ***component system*** keeps the black-and-white and color signals separate in three different ways. The composite and component systems are not compatible with each other.

Composite system The composite system combines the color (C, or chrominance) and the brightness (Y, black-and-white, or luminance) information into a single—composite—signal. Only one wire is necessary to transport the composite signal. Because this electronic combination was standardized some time ago by the National Television System Committee (NTSC), the composite signal is also called the NTSC signal or, simply, NTSC. The NTSC system is different from other composite systems, such as the European PAL system. A standard conversion is necessary when systems don't match. Most such standard conversions are done in the satellite that distributes the signal.

The advantage of the composite system is that it is one way of compressing an analog signal to save bandwidth in the signal transport, video recording, and transmission. The major disadvantage of the composite system is that there is usually some interference between chrominance and luminance information that gets worse and therefore more noticeable with each videotape generation. SEE 13.13

Y/C component system The advantage of the ***Y/C component system*** is that it produces higher-resolution pictures that will suffer less in subsequent tape generations than do composite tapes. In the Y/C system, also called S-video, the Y (black-and-white) signal and the C (color) signals are kept separate during signal encoding and transport but are combined and occupy the same track when actually laid down on videotape. The Y/C configuration requires two wires to transport the Y/C component signal. SEE 13.14

To maintain the advantages of Y/C component recording, other equipment used in the system, such as monitors,

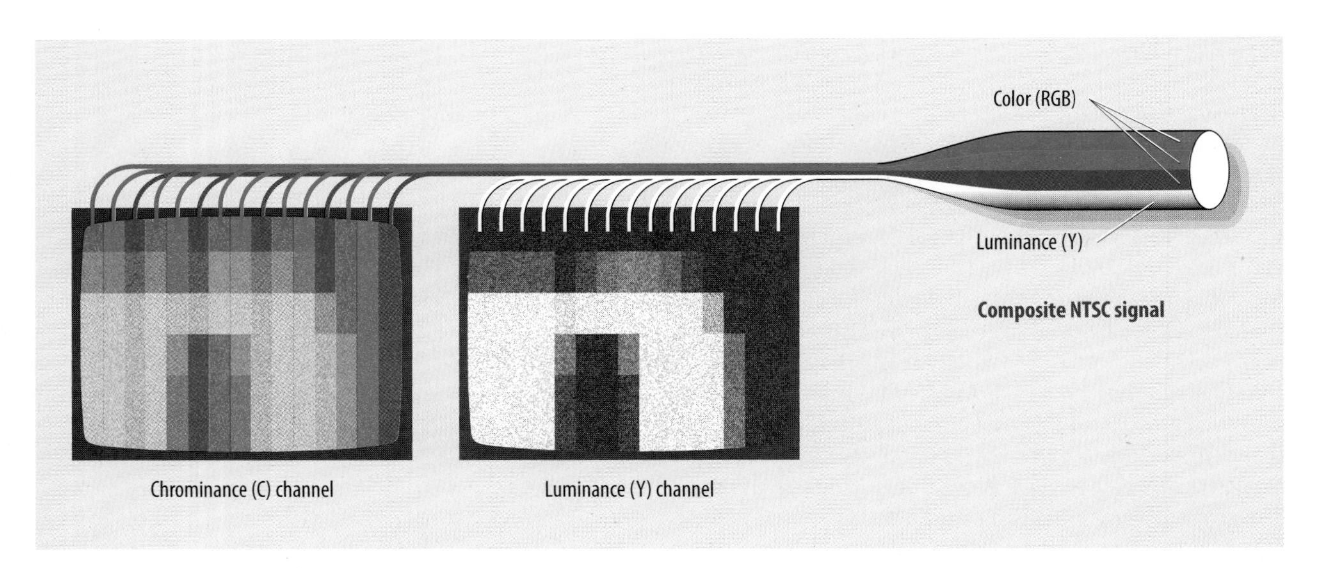

13.13 COMPOSITE SYSTEM
The composite system uses a video signal that combines the color (C, or chrominance) and luminance (Y, or brightness) information. It needs a single wire to be transported and recorded on videotape as a single signal. It is the standard NTSC system.

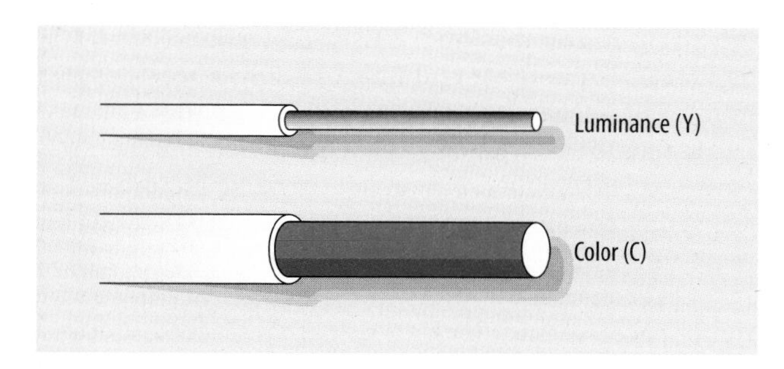

13.14 Y/C COMPONENT SYSTEM
The Y/C component system separates the Y (luminance) and C (color) information during signal encoding and transport, but it combines the two signals on the videotape. It needs two wires to transport the separate signals.

must also keep the Y and C signals separate. This means you cannot play a Y/C component videotape on a regular VHS recorder but only on an S-VHS recorder.

Y/color difference component system In the ***Y/color difference component system***, the luminance signal, the red signal minus its luminance (R–Y), and the blue signal minus its luminance (B–Y) are transported and recorded as three separate signals. The green signal is matrixed (regenerated) from these three signals.

The analog color difference signal, sometimes called the RGB signal, needs three wires to transport the three separate signals. SEE 13.15 When digitized, however, the sampled Y and color difference signals can be sent one after the other in a rapid-fire sequence. They can therefore be sent much like the composite signal in a single wire.

RGB component system In the true ***RGB component system***, the red, green, and blue signals are kept separate and treated as separate components throughout the recording and storage process. Each of the three signals remains separate even when laid down on videotape. This system is actually the most desirable for maintaining video quality. When working with a true RGB component system for HDTV, however, it takes up so much bandwidth, storage space, and retrieval time that it is used only by very high-end recorders and editing and graphics systems. The Y/color difference system was developed to overcome these problems. SEE 13.16

The Y/C, Y/color difference, and RGB component signals eventually must combine the separate parts into a single NTSC composite signal for traditional television broadcast and distribution.

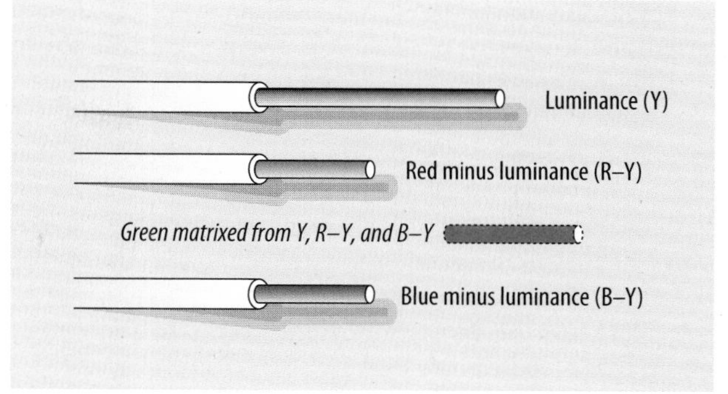

13.15 Y/COLOR DIFFERENCE COMPONENT SYSTEM
The Y/color difference component system separates the three RGB signals throughout the recording process. It needs three wires to transport the three component signals: the Y (luminance) signal, the R–Y (red minus luminance) signal, and the B–Y (blue minus luminance) signal. The green signal is then matrixed (regenerated) from these signals.

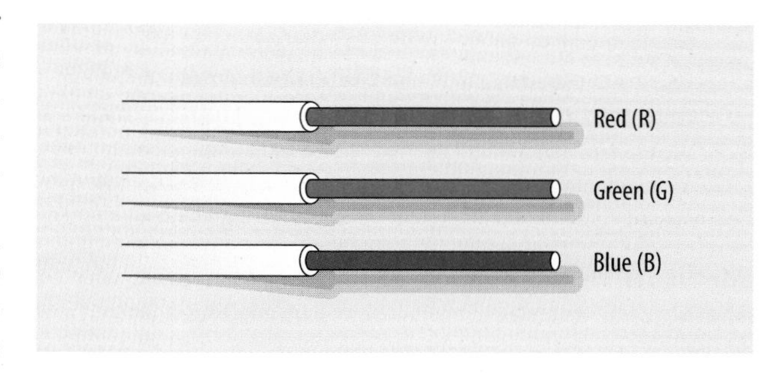

13.16 RGB COMPONENT SYSTEM
Like the Y/color difference system, the RGB component system (also called the RGB system) separates the three RGB signals throughout the recording process. It needs three wires to transport the signals. It provides the greatest color quality but takes up an inordinate amount of frequency space.

Sampling

Although you learned in chapter 5 what sampling is all about, you still may wonder what video and digital cinema experts are talking about when they discuss the relative merits of 4:2:2 over 4:1:1 sampling in video recording. All that this means is that in the digitizing process, the C (color) signals are sampled less frequently than the Y (black-and-white) signal. In 4:2:2 sampling, the Y signal is sampled twice as often as the C signals. During every period that the Y signal is sampled four times, the C signals are sampled twice. In a 4:1:1 sampling process, the Y signal is sampled four times, whereas the C signals are sampled only once. The luminance signal receives such privileged treatment because it is a major contributor to picture sharpness.

Most normal productions look stunning with 4:1:1 sampling. If, however, you require high-quality color that must withstand a variety of special effects, such as chroma-key effects (see chapter 14) or a great number of key layers, you will do well to use equipment that employs the higher 4:2:2 sampling ratio.

Here are the most important points to remember about these recording features:

- The composite video signal of the NTSC composite system is of lower quality than that of the Y/C component system, which is somewhat inferior to the Y/color difference component system and the RGB component system.
- A 4:2:2 sampling ratio produces better colors than does a 4:1:1 ratio, although the latter certainly produces great images. In fact, you would notice the difference between the two sampling ratios only when using or building complex effects or when trying to maintain color fidelity in extremely high-contrast video images.

Compression

All digital video pictures are compressed in one way or another. The specific ***codec*** (short for *compression/decompression*) system is often more important than the sampling ratio a camcorder uses. For example, a cell-phone video is much more compressed than that of a small HDTV camcorder. The less the images are compressed, the better they look.

As you recall from chapter 5, ***compression*** refers to the rearrangement or elimination of redundant picture information for more-efficient storage and signal transport. Lossless compression means that we rearrange the data so that they take up less space. This technique is similar to repacking a suitcase to make all the stuff fit into it. In lossy compression we throw away some of the unnecessary items and therefore can use a much smaller suitcase. Most codecs are of the lossy kind. The more relevant compression question, however, may well be whether a camcorder uses the intraframe or interframe compression system.

Understanding the difference between these two systems is important when you get to actual editing in the postproduction phase.

Intraframe compression This compression system was designed primarily for still images but has been adapted to moving images. Intraframe compression looks at each frame and throws away all video information that is unnecessary to perceiving pretty much the same picture as the original. In technical terms it eliminates spatial redundancy for each frame.

Let's consider our overstuffed suitcases again. To save some space, we must look at each of the packed suitcases (each frame) and ask whether we can get along with two shirts instead of six; then we move to the other suitcase to see whether we can take out five of the six sweaters we packed, especially since we are going to go to a warm, sunny location. We continue to check each suitcase (frame) to see what we can leave behind. Pretty soon we will have discarded enough unnecessary clothing (redundant pixels) to get by with much smaller suitcases.

The ***JPEG*** system—a video compression method used mostly for still pictures—employs this intraframe compression technique. What we did was apply lossy compression to each frame—or, in our example, suitcase—one by one.

The big advantage of the intraframe codec is that each frame contains a usable picture. When editing you can choose any frame as the edit-in or edit-out point and all the ones in between. The initial disadvantage of this system was that repacking each of the suitcases (frames) took time, that is, it was too slow when capturing fast-moving objects; but this drawback has largely been eliminated in most intraframe codecs.

Interframe compression This system was developed for moving images. Rather than compress each video frame independently of all the others, interframe compression looks for redundancies from one frame to the next in a motion sequence. Basically, the system compares each frame with the preceding and following ones and keeps only the pixels that constitute a change.

For example, if you see a cyclist moving against a cloudless blue sky, the system will not bother with repeating all the information that makes up the blue sky but only with the position change of the cyclist. As you can see, interframe compression looks for temporal redundancy (change from frame to frame) rather than spatial redundancy within a single frame.

Let's use the suitcase example one last time. We now have two people with suitcases. John has already packed his big suitcase, and Ellen is ready to begin packing. Before she starts, however, she checks with John to see what he has packed (full video frame 1). To her delight he has packed a lot of stuff she wanted to take along, so she needs to fit only a few more items into a very small suitcase (interframe-compressed frame 2). ***MPEG-2***, a compression standard for motion video, uses the interframe technique. (***MPEG*** stands for *Moving Picture Experts Group.*)

The problem with the interframe system is in editing. Because some of the compressed frames are incomplete, they can't be used as the start or end point of an edit. The system therefore periodically sends a full reference frame (say, every fifth or tenth frame), called an I-frame, that is compressed but, unlike the other frames, holds video information for a complete picture. Such I-frames are intraframe compressed and are therefore not influenced by the adjacent frames. Because an I-frame contains the information for a whole picture, an editor can go to the full frame to do the actual cut.

Being restricted to every fifth or tenth frame for a cut, however, does not please an editor who may need to match each frame of lip movement with the corresponding sound; but in most cases a five-frame cutting restriction is not too much of a handicap. This is why systems designed for editing include these reference I-frames as often as feasible. Sophisticated MPEG-2 editing systems can recalculate a complete frame anywhere in the compressed video for critical frame-accurate editing.

MPEG-4, another compression standard for motion video, was originally developed to facilitate Internet streaming. As you know, *streaming* means sending and receiving digital video and audio data as a continuous data flow that can be watched while the delivery is in progress. The more efficient the compression systems are, the faster and smoother the streaming process becomes. MPEG-4 proved to be such an efficient codec that software engineers started tinkering with it to make it compress video and audio even during the camcorder capture process with little damage to the pictures or sound.

To list all the codecs based on MPEG-4 would be of little help to you. Realize, however, that not all MPEG-4 variations are compatible with all the popular editing systems. Before buying a camcorder whose codec is based on MPEG-4, you should find out whether you will be able to import its recordings into your editing system.

In general, this formula applies to all codecs: the less compression, the better the image quality. But then there is another, not-so-happy formula: the less compression, the more unwieldy the huge amount of information becomes.

MAIN POINTS

- There are two basic types of recording systems: tape-based and tapeless.
- Videotape can be used for recording analog or digital information.
- All analog recording systems use videotape as the recording media.
- Tapeless recording systems can record digital information only.
- Digital recordings are virtually immune to deterioration in subsequent dubs.
- During the recording, the videotape moves past one or more rotating heads that write video tracks on the tape as it passes. During playback the same or different heads read the video and audio information off the tape.
- The basic tracks on a videotape are the video track, two audio tracks, and the control track.
- The time base corrector (TBC) and the framestore synchronizer are electronic devices that help stabilize the playback of video recorders (VRs) and synchronize the scanning from remote sources.
- Tapeless video recording uses hard drives, read/write optical discs, and memory cards as the recording media.
- There are basically two systems of recording video signals: composite and component. The composite system combines the black-and-white and color signals into one. The component system keeps the black-and-white and color signals separate in three different ways: Y/C, Y/color difference, and RGB.
- The composite signal is called NTSC. A component signal is of higher quality than a composite one. The composite and component systems are not compatible with each other.
- Sampling concerns mainly color fidelity. A sampling ratio of 4:2:2 is slightly better than a ratio of 4:1:1.
- Intraframe compression eliminates all redundant pixels in each video frame.
- Interframe compression eliminates redundant pixels from frame to frame.
- Some codecs may not be compatible with your editing software.

SECTION

13.2

How Video Recording Is Done

Now that you know all about video-recording systems, you need to know what to do with them. This section introduces you to the operational uses of video recording and the video-recording procedures in studio production and ENG/EFP.

- **USES OF VIDEO RECORDING AND STORAGE**
  Building a show, time delay, program duplication and distribution, and record protection and reference

- **VIDEO-RECORDING PRODUCTION FACTORS**
  Recording preparations (schedule and equipment checklist) and production procedures (video leader, recording checks, time code, recordkeeping, and specific aspects of disk-based video recording)

USES OF VIDEO RECORDING AND STORAGE

Video recording is used primarily for building a show, time delay, program duplication and distribution, and the creation of a protection copy for reference and study.

Building a Show

One of the major uses of video recordings is to build a television show from previously recorded video segments. This construction process is done through postproduction editing. It may include assembling multiple segments shot at different times and locations, or it may simply involve condensing a news story by cutting out the nonessential parts. It also includes stringing together longer multi-camera scenes that were switched (instantaneously edited) and video recorded. A good example of this technique is the recording of relatively long and uninterrupted studio segments of soap operas and then editing them together for the program in a postproduction session.

Time Delay

Through video recording, an event can be stored and played back immediately or hours, days, or years after its occurrence. In sports many key plays are recorded and shown right after they occur. Because the playback of the recording happens so quickly after the actual event, they are called instant replays. Network shows that you can watch at the same schedule time in each time zone are time delayed through some type of video recording. For example, by recording in San Francisco a satellite feed at 3:00 p.m. (Pacific Standard Time, or PST) of an awards show that takes place at 6:00 p.m. (Eastern Standard Time, or EST) in New York, it can be played back and aired at 6:00 p.m. (PST) in San Francisco.

Program Duplication and Distribution

Video recordings can be easily duplicated and distributed to television outlets by mail, courier, telephone line, coax or fiber-optic cable, satellite, or Internet streaming. With satellite or Internet streaming, a single video recording can be distributed simultaneously to multiple destinations around the world with minimal effort. Eventually, digital movies will be similarly distributed to movie houses by satellite or fiber-optic cable rather than by shipping the heavy 35mm film cans.

Record Protection and Reference

To preserve the recordings of important events, make protection copies of the original video recordings. Make these dubs with equipment that has the same or a better recording quality as that which was used to shoot the original footage. DVDs are an excellent archival recording media. They take up very little space, and the playback equipment is readily available.

The problem with digital recordkeeping is the rapidly changing technology that makes one recording device obsolete in just a few years. You have probably run into this problem with playbacks of DVDs and Blu-ray discs or with the ever-changing systems software. Unless you transfer the digital records periodically to the latest system, your archives are worthless once the playback system becomes obsolete.

VIDEO-RECORDING PRODUCTION FACTORS

Unless you are chasing a hot news story, video recording needs careful preparation and attention to detail. An entire production can be lost if you as the video recorder

(VR) operator forget to push the right button during the recording process. You might do well to follow some checklists for the recording preparations and the actual recording activities.

Recording Preparations

When doing video recordings of a studio show, the wrong cable, connector, or videocassette is no great disaster. But when you are out in the field, a wrong cable or connector may cancel the entire EFP. Let's take a brief look at scheduling, the equipment checklist, and general recording preparations.

Schedule Double-check that the equipment you need for the recording or dubbing is actually available. Have a backup recorder standing by in case somebody is using "your" (scheduled) machine. Be reasonable in your request. You will find that recording equipment is usually available for the actual studio or field production but not always for your playback demands.

If you are asked to simply check the order of the scenes on a video recording or to time some segments, import it to your laptop or iPod, or dub the material down to a regular ½-inch VHS format or a DVD so that you can watch the material at home. That way you free up the high-quality machines or servers for more important tasks, and you are not tied to a precise schedule when reviewing the recordings. Once the source footage is on the computer or server, you can dub it to your notebook and watch it while riding the bus.

Always follow the established routine for equipment check-out and use. If you don't, you not only jeopardize your recording session but possibly some other people's as well. When sharing a server, do you know how to access the material? What are the established safeguards from accidental erasure?

Equipment checklist Like a pilot who goes through a checklist before every flight, you should have your own equipment checklist every time you do a production. Such a list is especially important in EFP. This brief checklist is limited to video recording.

- *Dubbing.* When dubbing you need two recorders: the source machine and the record machine. Try not to use the camcorder as the play VTR. The VTR mechanism in a camcorder is not as rugged as that of a studio VTR. Even if your recording is on another media, such as a hard drive, optical disc, or memory card, transfer it to your computer before you start your postproduction activities.

- *Correct recorder.* Does the video recorder match the media used in the camcorder? If the camcorder used an optical disc for recording, you obviously can't play it back on a VTR. As obvious as this example is, you may find that some digital VTRs do not have provisions for playing mini-cassettes. Does the codec of the camcorder match that of the editing software?

- *Combine jobs.* Can you combine certain recording jobs, such as doing a window-dub when making a protection copy of the original recording? (As you will read in chapter 19, a window dub is a copy of the recording with the time code address keyed over each frame.)

- *Video recorder status.* Does the VR actually work? If possible, do a brief test recording to ensure that it functions properly.

- *Power supply.* Do you have a fully charged battery for the camcorder and/or the external video recorder? Always take an extra fully charged battery along. When using household current for the power supply, you need the appropriate transformer/adapter. Before leaving for the field location, check that the connecting cable from the power supply fits the jack on the VR or camcorder. Do not try to make a connector fit if it is not designed for that jack. You may blow more than a fuse if you do.

- *Correct recording media.* Do you have the correct recording media, that is, the videocassette, disc, or flash drive that fits the camcorder or external VR? Also check that the boxes contain the correct media. Because even same-sized cassettes can be loaded with different lengths of tape, check the supply reel to see that it contains the amount of tape indicated on the label. If, for example, the box says that it contains a 184-minute tape but your check shows only a relatively small amount of tape on the supply reel, the box is obviously mislabeled.

- *Enough media.* Do you have enough recording media for the proposed production? This is especially important when you record HDTV. High-definition data take up much more storage space than is advertised on the box (which usually shows the recording time for standard digital video).

If the largest media does not hold enough data for the entire event, you need to schedule two machines or you will lose a few minutes during the media change. Especially

13.17 DIGITAL VIDEOCASSETTE IN RECORD-PROTECT POSITION
Digital videocassettes have a movable tab that prevents accidental erasure. To record on the cassette, the tab must be in the closed position.

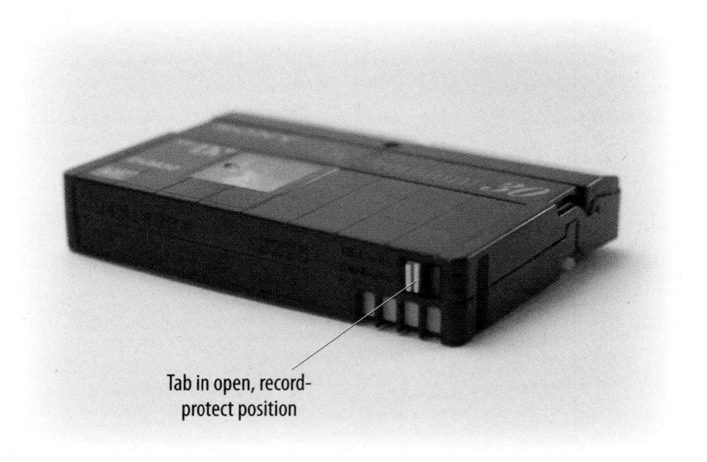

13.17 PHOTO BY EDWARD AIONA

when doing multiple recordings for instant replay, you need three or four times the normal storage capacity. Always take along more recording media than you think you'll need.

■ *Record protection.* If a VTR refuses to record despite a careful check of the connecting cables, remove the cassette to see whether its recording is enabled. All cassettes have a device, sometimes called the record inhibitor, to protect the videotape from accidental erasure. VHS and S-VHS ½-inch cassettes have a record-protect tab on the back edge at the lower left. When this tab is broken off, the cassette is record-protected. You can enable the cassette again by putting a piece of masking tape or even gaffer's tape over the tab opening.

Digital videocassettes have a tab on the back edge that you can move into or out of a record-protect position. If it is in the open position, you cannot record on the cassette. On some cassettes this tab is colored so that you can easily see whether it is in the closed (recording OK) or open (protect) position. **SEE 13.17** Even small flash drives such as SD cards have a movable record-protect tab that you can move from the closed (recording OK) to the open (protect) position. **SEE 13.18**

Routinely check the record-protect tab before using a cassette for recording. Although you cannot record on a record-protected tape, any cassette will play back with or without a record-protect device in place.

Production Procedures

If you have followed the basic preproduction steps, you should have little trouble during the actual recording, although the following elements still need attention: video

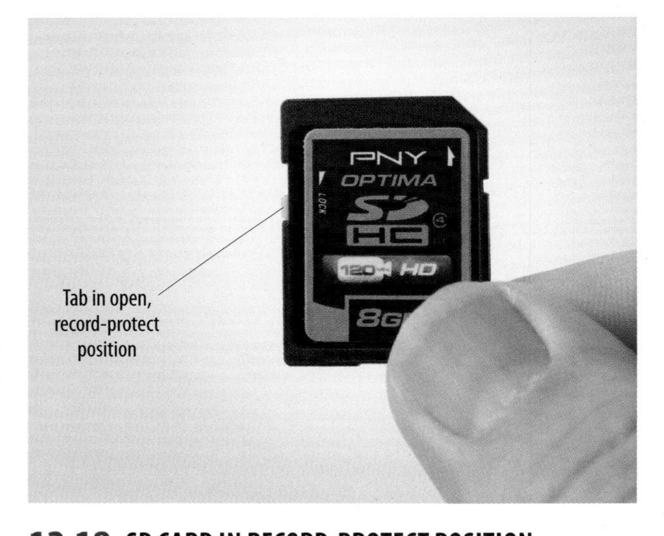

13.18 PHOTO BY EDWARD AIONA

13.18 SD CARD IN RECORD-PROTECT POSITION
Even the small SD cards have a record-protect tab. You cannot record on this SD card because its tab is in the open (protect) position.

leader, recording checks, time code, recordkeeping, and specific aspects of disk-based video recording.

Video leader When playing back a properly executed video recording, you will notice some front matter at the head of the recording: color bars, a steady tone, an identification slate, and perhaps some numbers flashing by with accompanying audio beeps. These elements, collectively called the ***video leader***, help adjust the playback and record

13.19 PHOTO BY GARY PALMATIER

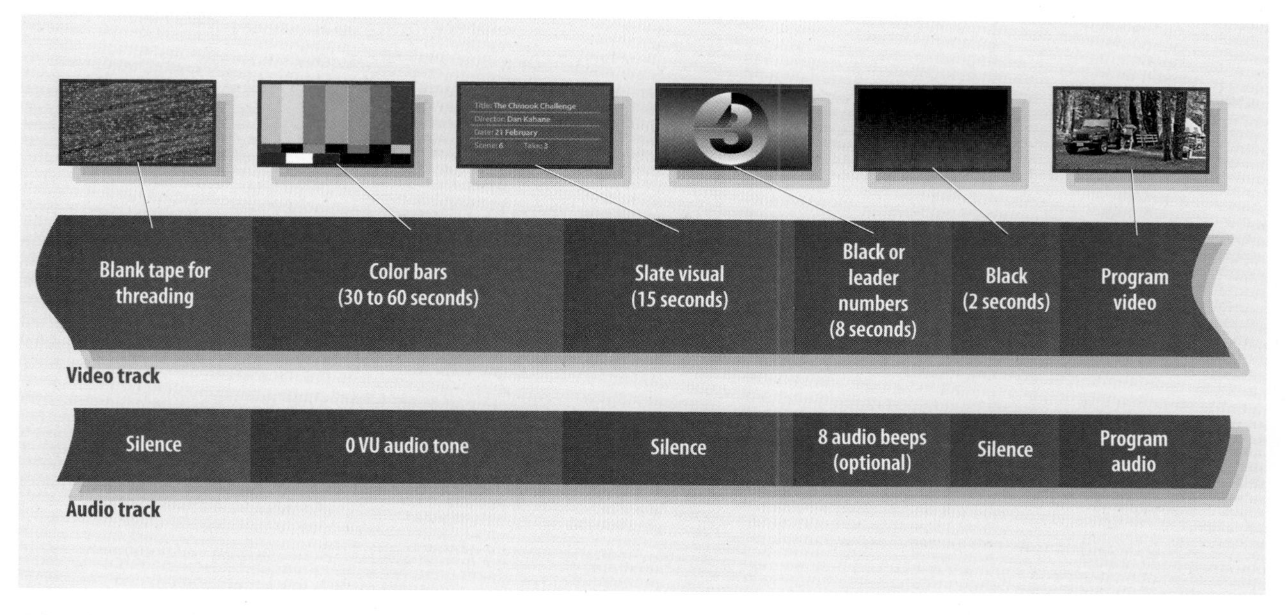

13.19 VIDEO LEADER
The video leader helps adjust the playback and record machines to standard audio and video levels.

machines to standard audio and video levels. Let us look at them one by one. SEE 13.19 ZVL2 EDITING→ Postproduction guidelines→ leader

Color bars help the VR operator match the colors of the playback machine with those of the record machine. It is therefore important that you record the color bars (fed by color-bar generators located in master control or built into ENG/EFP cameras) for a minimum of 30 seconds each time you use a new recording media or begin a new recording session. Some VR operators prefer to have the color bars run for a full minute or more so that they do not have to rerun the bars if the equipment requires further adjustment; but this may be overdoing it, especially if you have to listen to the test tone for a whole minute.

Most audio consoles and even some field mixers can generate the test tone that you need in calibrating the line-out level of the audio console or mixer with the input (record) level of the video recorder (see chapter 10). You should record this test tone along with the color bars. Obviously, these test signals should be recorded with the equipment that you use for the subsequent video recording; otherwise the playback will be referenced to the recorded color bars and test tone but not to the video-recorded material. The director refers to these test signals as "bars and tone." When doing a studio show, you will hear the director call for bars and tone right after the video recorder has stabilized. In EFP the camera or VR operator will take care of this reference recording.

The slate and the clapboard give pertinent production information along with some technical details. Normally, the slate indicates the following identification data:

- Show title and scene number (corresponding to that in the script)
- Take number (how often you record the same thing)
- Recording date

Some slates also list the director, the location (especially for EFP), and additional in-house information, such as "reel" numbers (number of the specific media used), editing instructions, name of producer, and so on. The essential information is the name of the show and the scene and take numbers.

You should use the clapboard rather than the slate when video/audio synchronization is a crucial postproduction element. It is used much like the slate, except that the clapboard is recorded until the floor manager has snapped the clapstick. This way you get a visual marker (the frame where the clapstick meets the board) and an audio marker (the clap) in a single frame. This frame gives the editor a

13.20 CLAPBOARD
The clapboard is used to identify the scene and take numbers and to help synchronize video and audio.

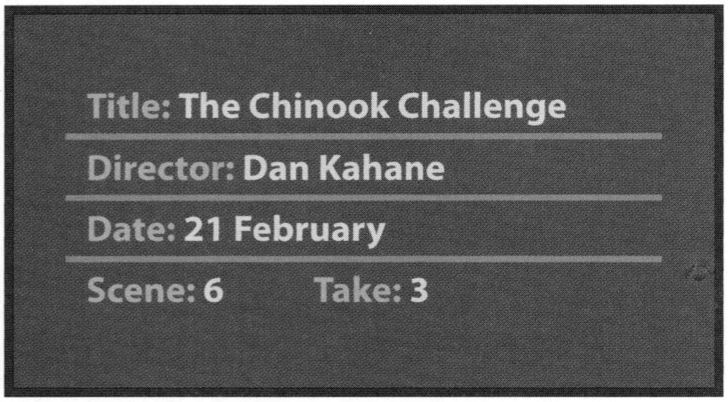

13.21 CHARACTER-GENERATED SLATE
The electronic slate gives pertinent information about the production. It is recorded at the beginning of each take.

reference point for lining up the slate frame with the audio clap on the visual waveform for exact video/audio synchronization. This frame is also a starting point for the time code if it has to be laid down in postproduction. **SEE 13.20**

In the studio the slate is usually generated by the character generator (C.G.) and recorded right after the color bars. **SEE 13.21** In the absence of a C.G., you can use a small whiteboard with a 4 × 3 or 16 × 9 aspect ratio. Because the information on the slate changes from take to take, the slate surface should be easily cleaned (chalk or dry-erase markers work well). The slate identifies the scene as well as the take, so you must use it every time you record a new take, regardless of how short or how complete the take may be. ZVL3 EDITING→ Production guidelines→ slate

Assume that you are the director of the weekly *President's Chat* production. You have just videotape-recorded about 10 seconds of the first take when the college president stumbles over the name of the new dean. You stop the tape, keep calm, roll the tape again, and wait for the "in-record" confirmation from the VTR operator. Before repeating the president's introduction, you need to record the slate again. It reads: scene 1, take 1. But shouldn't the slate read: scene 1, take 2? Yes—the C.G. operator obviously forgot to change the slate. Should you go on, or stop the tape again to correct the slate? In this case you might as well keep going. The VTR operator, who keeps the field log, can note the false 10-second start and record the second take as take 1. If, however, you are breaking up the president's "chat" into several short takes to be assembled in postproduction, the slate numbers must be accurate.

Leader numbers are used for the accurate cueing of the videotape during playback. The leader numbers flash at 1-second intervals from 10 to 3 or from 5 to 3 and are usually synchronized with short audio beeps. The last two seconds are normally kept in black and silent so that they do not accidentally appear on the air if the videotape is punched up early (sometimes the numbers and the beeps go down to the last second). The first frame of recorded program material should appear at the zero countdown.

When cueing a videotape for playback, you can stop the tape at a particular leader number, say, 4; or you can advance the tape right to the first video frame. When you stop the tape at leader number 4, you must preroll the tape exactly 4 seconds before the program material is to appear on the air. If, after starting the tape on leader number 4, there is a longer interval between the last leader number 3 (and two seconds of black and silence) and the first video frame, the TD will not know exactly when to punch up the videotape and almost certainly will miss the first second of the playback. If you start the tape on leader number 4, the first video frame should come up after exactly 4 seconds.

When all the footage is on a video server, you can use the leader numbers for cueing the segment right to the first frame. One of the big advantages of using a server over a VTR is that it has an instant start without any video or audio breakup or distortion.

Recording checks As the VR operator, you are responsible for seeing that the pictures and the sound are actually recorded. Here are some checkpoints that greatly reduce recording problems.

- Always do a brief test recording, then play back the recording to ensure that the whole system works properly. Just because you see a picture on the video-record monitor and see the VU meter fluctuate during the test recording, it does not mean that the video and audio signals are actually being recorded. But once the test recording plays back all right, you can trust that the subsequent video and audio feeds will be recorded.

- Reset the tape counter on the VTR or go back to your starting time code number before beginning the actual program recording. If you need to record time code during the actual production (instead of laying in time code after the video recording), make sure it is recorded with the picture.

- When using a VTR, wait until the machine has reached operating speed and has stabilized before starting to record. This lockup time may take anywhere from 1 to 4 seconds. The VTR has a control light that flashes during the lockup period and remains steady once the system is locked, that is, sufficiently stabilized for recording. As the VTR operator, you should watch the flashing light and, when you see a steady light, call out, "speed" or "in record." The director will then proceed with the actual recording.

- Watch the audio and video levels during the recording. If you do not have a separate audio setup but instead feed the mic directly to the video recorder, pay particular attention to the audio portion. You may find that some directors become so captivated by the beautiful camera shots that they don't even hear, for example, the talent giving an African country the wrong name, an airplane noise interrupting the medieval scene shot on-location, or the wireless mic cutting out briefly during an especially moving moment of a song.

- If you are using an external video recorder in the field, always monitor the audio by wearing headphones.

- When recording for postproduction, record enough of each segment so that the action overlaps the preceding and following scenes. At the end of each take, record a few seconds of black before stopping the recorder. This run-out signal acts as a pad and greatly facilitates editing.

- Ask the director whether you should video-record the camera rehearsals. Sometimes you get a better performance during rehearsal than during the actual take. The camera rehearsals (when run like a full dress rehearsal) can then be edited into the rest of the production.

- Again, be sure to slate every take, rehearsal or not. When you are in a hurry, audio-slate each take by having the audio operator use the console mic. When in the field, have the floor manager read the brief slate information into the talent's lavalier or fishpole shotgun mic: "President's Chat, take 7." Some directors like an additional brief verbal countdown, such as "five, four, three" with the last two seconds silent before the cue to the talent. Many field productions are slated more extensively only at the beginning of the video recording, with subsequent takes being only verbally slated.

- Do not waste time between takes. If you are properly prepared, you can keep the intervals to a minimum. Although the playback of each take may occasionally improve the subsequent performance by cast and crew, it often does not justify the time it takes away from the actual production. If you pay close attention during the video recording, you do not need to review each take. Long interruptions not only waste time but also lower the energy level of the production team and the talent. On the other hand, do not rush through recording sessions at a frenetic pace. If you feel that another take is warranted, do it right then and there. It is far less expensive and time-consuming to repeat a take immediately than to re-create a whole scene later simply because one take turned out to be unusable.

Time code If you need to record time code simultaneously with each take, verify that the time code is recorded on its designated address track or, if necessary, on a free audio track (but make sure that the audio track is not already used for recording actual audio information). Unless the camera or video recorder has a built-in time code generator, you need a separate time code generator for the address system. Time code can also be laid down later in postproduction (as explained in chapter 19). ZVL4 EDITING→ Postproduction guidelines→ time code

Recordkeeping Keeping accurate records of what you video-record and the proper labeling of the recording media may seem insignificant while in the middle of a production, but they are critical when you want to locate a particular scene or a specific tape or disk. You will be surprised at how quickly you can forget the "unforgettable" scene and especially the number and the sequence of takes.

Keeping accurate records during the video recording saves much time in postproduction editing. Although you will most likely log the various takes and scenes when reviewing the video recording after the production, you are still greatly aided by a rough record kept during the production, called a ***field log***. As a VR operator, you should keep a field log even when recording in the studio. A field log is especially useful in more complex field productions (hence its name) that involve a number of locations. Mark the good takes (usually with a circle) and identify especially those takes that seem unusable at that time. Label each recording media and its box, and mark the field log with the corresponding information. **SEE 13.22** ZVL5 EDITING→ Production guidelines→ field log

Specific aspects of disk-based video recording The preproduction and production elements discussed here apply equally whether you record with an analog or a digital VTR or some tapeless recording device. There are, of course, some different production requirements when you use the disk-based system for editing, which is the subject of chapter 20.

13.22 FIELD LOG
The field log is kept by the VR operator during the recording. It normally indicates the media (tape, SD card, hard drive) or reel number, scene and take numbers, approximately where the take is located on the recording media, and other information useful in postproduction editing.

PRODUCTION TITLE: "impressions" PRODUCER/DIRECTOR: Hamid Khani

TAPING DATE: 4/15 LOCATION: BECA Newsroom

MEDIA NUMBER	SCENE	TAKE	OK or NO GOOD	TIME CODE IN	TIME CODE OUT	EVENT / REMARKS
C-005	2	1	NG	01:57:25	02:07:24	student looks into Camera CU – Z-axis
		(2)	OK	02:09:04	02:14:27	monitor + L news anchor MS getting ready
		(3)	OK	02:14:28	02:34:22	Pan R to reveal anchor in news set
		4	NG	02:34:22	02:45:18	Rack focus from Floor Mgr to L anchor - OUT OF FOCU
		5	NG	02:48:05	02:55:12	Rack focus Both out of focus
		6	NG	02:58:13	03:05:11	Rack OK LOST Audio
		(7)	OK	03:12:02	03:46:24	Hurrah! Rack OK
	3	(1)	OK	04:16:03	04:28:11	MS Floor Mgr + camera op from behind
		2	NG	04:35:13	04:49:05	CU of R anchor Lost audio
		(3)	OK	05:50:00	06:01:24	CU of R anchor
		4	NG	06:03:10	06:30:17	CU of L anchor audio problem
		5	NG	06:40:07	07:04:08	LS of both anchors Floor Mgr walks through shot
		(6)	OK	07:07:15	07:28:05	Good! Floor Mgr silhouette against set
		(7)	OK	07:30:29	07:45:12	slow pullout
C-006	4	(1)	OK	49:48:28	51:12:08	MCU Marty talks to anchors
		2	NG	51:35:17	51:42:01	Lav comes off R anchor

MAIN POINTS

- Video recording is used primarily for building a whole show by assembling parts that have been recorded at different times and/or locations; for time delay; for program duplication and distribution; and for records for preservation, reference, and study.
- The production purpose should determine the type of video recorder (VR) used. Simple material destined for home consumption does not need a top-of-the-line video recorder.
- The important preproduction steps for video recording include scheduling, equipment checklists, and specific edit preparations. Take along enough recording media and be sure that it fits the camcorder or separate recording device. Check that the record-protect tab is in the closed (recording OK) position.
- The major production factors in video recording are the video leader (color bars, test tone, slate information, and leader numbers and beeps), recording checks, time code, recordkeeping, and specific aspects of disk-based video recording.
- Slate all takes, either visually and/or verbally.
- The field log is kept during the actual studio or field production. It lists all media, scene, and take numbers as well as comments about shots and audio.

ZETTL'S VIDEOLAB

For your reference or to track your work, the *Zettl's VideoLab* program cues in this chapter are listed here with their corresponding page numbers.

ZVL1 EDITING→ Postproduction guidelines→ tape basics 276

ZVL2 EDITING→ Postproduction guidelines→ leader 289

ZVL3 EDITING→ Production guidelines→ slate 290

ZVL4 EDITING→ Postproduction guidelines→ time code 291

ZVL5 EDITING→ Production guidelines→ field log 292

CHAPTER

14

Switching, or Instantaneous Editing

When watching a television director during a live multicamera show, such as a newscast or a basketball game, you might be surprised to find that the primary activity of the director is not telling the camerapersons what to do but rather selecting the most effective shots from the variety of video sources displayed on a row of preview monitors. In fact, the director is engaged in a sort of editing, except that it's the selection of shots *during* rather than after the production. Cutting from one video source to another or calling for other transitions, such as dissolves, wipes, and fades, while a show is in progress is known as ***switching***, or instantaneous editing.

Unlike postproduction editing, in which you have the time to deliberate exactly which shots and transitions to use, switching demands snap decisions. Although the aesthetic principles of switching are identical to those used in postproduction, the technology involved is quite different. Instead of linear and nonlinear editing systems, the major editing tool is the video switcher or a computer that performs the switcher functions.

Section 14.1, How Switchers Work, acquaints you with the basic functions, layout, and operation of a production switcher in a television control room. Section 14.2, Electronic Effects and Switcher Functions, looks at standard and digital video effects and additional switcher functions.

You should realize that you cannot really learn switching by reading about it—even from this fine chapter. Like learning to drive a car, you must practice switching to become proficient at it. Nevertheless, once you know the principles of how a switcher works, you will have less trouble operating one.

KEY TERMS

auto-transition An electronic device that functions like a fader bar.

bus A row of buttons on the switcher.

character generator (C.G.) A dedicated computer system that electronically produces letters, numbers, and simple graphic images for video display. Any desktop computer can become a C.G. with the appropriate software.

chroma keying An effect that uses color (usually blue or green) for the backdrop, which is replaced by the background image during a key.

delegation controls Buttons on the switcher that assign specific functions to a bus.

digital video effects (DVE) Visual effects generated by a computer or digital effects equipment in the switcher.

downstream keyer (DSK) A control that allows a title to be keyed (cut in) over the picture (line-out signal) as it leaves the switcher.

effects buses Program and preview buses on the switcher, assigned to perform effects transitions.

fader bar A lever on the switcher that activates preset transitions, such as dissolves, fades, and wipes, at different speeds. It is also used to create superimpositions. Also called *T-bar.*

genlock The synchronization of two or more video sources (such as cameras and video recorders) or origination sources (such as studio and remote) to prevent picture breakup during switching. A house sync will synchronize all video sources in a studio.

key An electronic effect. Keying means cutting with an electronic signal one image (usually lettering) into a different background image.

key bus A row of buttons on the switcher, used to select the video source to be inserted into a background image.

key-level control A switcher control that adjusts the key signal so that the title to be keyed appears sharp and clear. Also called *clip control* and *clipper.*

layering Combining two or more key effects for a more complex effect.

matte key A keyed (electronically cut-in) title whose letters are filled with shades of gray or a specific color.

M/E bus Short for *mix/effects bus.* A row of buttons on the switcher that can serve a mix or an effects function.

mix bus Rows of buttons on the switcher that permit the mixing of video sources, as in a dissolve or super.

preview/preset bus Rows of buttons on the switcher used to select the upcoming video (preset function) and route it to the preview monitor (preview function) independently of the line-out video. Also called *preset/background bus.*

program bus The bus on a switcher whose inputs are directly switched to the line-out. Allows cuts-only switching. Also called *direct bus* and *program/background bus.*

special-effects generator (SEG) An image generator built into the switcher that produces special-effects wipe patterns and keys.

super Short for *superimposition.* A double exposure of two images, with the top one letting the bottom one show through.

switching A change from one video source to another during a show or show segment with the aid of a switcher. Also called *instantaneous editing.*

wipe Transition in which a second image, framed in some geometric shape, gradually replaces all or part of the first image.

SECTION

14.1 How Switchers Work

When you look at a large production switcher with all the different-colored rows of buttons and the various levers, you may feel as intimidated as when looking into the cockpit of an airliner. But once you understand the basic principles and functions of a switcher, you can learn to operate it faster than running a new computer program. Even the most elaborate digital video-switching system performs the same basic functions as a simple production switcher, except that large switchers have more video inputs and can perform more visual tricks.

This section explores what a production switcher does and how it works.

▶ **BASIC SWITCHER FUNCTIONS**
Monitor display, selecting video sources, performing transitions between them, and creating special effects

▶ **SIMPLE SWITCHER LAYOUT**
Program bus, mix buses, preview bus, effects buses, multifunction switchers, additional switcher controls, and large production switchers

▶ **OTHER SWITCHERS**
Portable switchers, switching software, routing switchers, master control switchers, and genlock

BASIC SWITCHER FUNCTIONS

There are three basic functions of a production switcher: to select an appropriate video source from several inputs, to perform basic transitions between two video sources, and to create or access special effects. Some switchers can automatically switch the program audio with the video.

Each video input on a switcher has a corresponding button. If you have only two cameras and all you want to do is cut from one to the other, two buttons (one for camera 1 and the other for camera 2) are sufficient. By pressing the camera 1 button, you put camera 1 "on the air," that is, route its video to the line-out, which carries it to the video recorder or transmitter. Pressing the camera 2 button will put camera 2 on the air.

What if you wanted to expand your switching to include a video recorder (VR), a character generator (C.G.), and a remote feed? You would need three additional buttons—one for the VR, one for the C.G., and one for the remote feed. When you want the screen in black before switching to one of the video sources and then to go to black again at the end of the show, you need an additional *BLK* (black) button. The row of buttons, called a ***bus***, has increased to six inputs. Production switchers have not only many more buttons but several buses as well. Let's find out why. ZVL1 SWITCHING→ Switching functions→ select | connect

Monitor Display

If you are to select among various video sources, wouldn't you have to see all of them before making your choice? Yes, and this is why a switcher is usually located in front of a monitor bank or a divided monitor panel much like the screen on a laptop computer. SEE 14.1 As you can see, in this control room of a remote truck, there are preview monitors for all kinds of inputs, from multiple cameras to a variety of special effects. There are three larger monitors in the middle of the display: one for presetting effects (such as titles and split screens), another for previewing the upcoming shot, and the third—the line, or program, monitor—that shows the pictures going out on the air and/or to a video recorder. The upper-left corner of the switcher is visible in the lower-right corner of the photograph.

SIMPLE SWITCHER LAYOUT

It may be easier to understand the parts of a switcher by constructing one that fulfills the basic switcher functions: cuts, dissolves, supers, and fades. This switcher should also let you see the selected video inputs or effects before you punch them up on the air. While building a switcher, you will realize that even a simple switcher can get quite complicated and that we need to combine several functions to keep it manageable.[1]

1. See Stuart W. Hyde, *Television and Radio Announcing*, 4th ed. (Boston: Houghton Mifflin, 1983), pp. 226–35. He explains the workings of an audio console by building one. I am using his construction metaphor with his permission.

14.1 PHOTO BY EDWARD AIONA

14.1 MONITOR BANK
This monitor bank, located in a large remote truck, shows preview monitors of a great variety of video sources. Note the larger preview, preset, and program (line) monitors. The switcher is located below the monitor bank. In a production the director and the technical director sit side by side.

Program Bus

If all you wanted to do was cut (switch instantaneously) from one video source to another without previewing them, you could do it with a single row of buttons, each representing a different video input. **SEE 14.2** This row of buttons, which sends everything you punch up directly to the line-out (and from there to the VR or transmitter), is called the ***program bus***. Also called program/background bus, the program bus represents, in effect, a selector switch for the line-out. It is a direct input/output link and is therefore also called the direct bus. Note that there is an additional button at the beginning of the program bus, labeled *BLK* or *BLACK*. Instead of calling up a specific picture, the *BLK* button puts the screen to black.

Mix Buses

If you want the switcher to do dissolves (during which one image gradually replaces the other through a temporary double exposure), supers (a double exposure of two images, with the top one letting the bottom one show through),

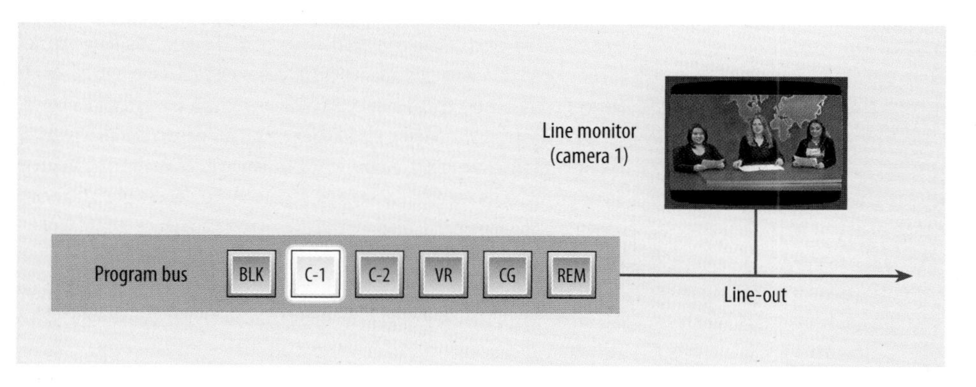

14.2 PHOTO BY EDWARD AIONA

14.2 PROGRAM BUS
This switcher has six video inputs: black, camera 1, camera 2, video recorder, character generator, and remote feed. Whatever source is punched up on the program bus (camera 1) goes directly to the line-out.

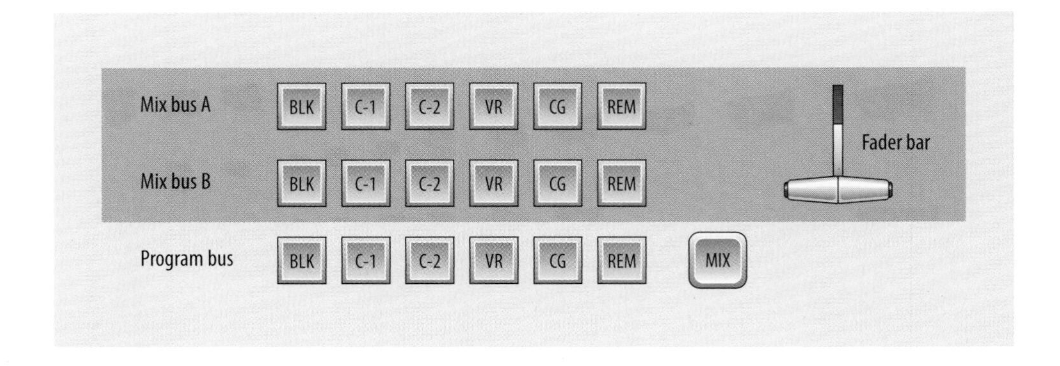

14.3 PROGRAM BUS WITH MIX BUSES AND FADER BAR
The mix buses A and B enable the mixing of two video sources.

and fades (the gradual appearance of an image from black or disappearance to black) in addition to simple cuts, you need two more buses—the ***mix buses***—and a lever, called the ***fader bar***, or T-bar, that controls the speed of the mix (dissolves and fades) and the nature of the super. **SEE 14.3**

When moving the fader bar to the full extent of travel, the picture of one bus is faded in while the picture of the other bus is faded out. The actual dissolve happens when the video images of the two buses temporarily mix. When you stop the fader bar somewhat in the middle, you arrest the dissolve and create a superimposition of the two video sources.

How does the program bus get this "mix" to the line-out? You must add still another button to the program bus that can transfer to the line-out the video generated by the mix buses. This *MIX* button is at the far right of the program bus.

Preview Bus

The preview bus is identical to the program bus in the number, type, and arrangement of buttons. The functions of the buttons are also similar, except that the "line-out" of the preview bus does not go on the air or to a recording device but simply to a preview (P/V) monitor. If, for example, you press the camera 2 *(C-2)* button on the preview bus, camera 2's picture appears on the preview monitor without affecting the output of the program bus, such as the *C-1* button in figure 14.2. If you don't like camera 2's picture and want to switch to what the video recorder is showing, you simply press the *VR* button on the preview bus. The preview bus is also called the preset bus if it also functions as a monitor that shows preset effects.

The preview and line monitors are usually side-by-side to show whether two succeeding shots will cut together well, that is, preserve vector continuity and mental map positions (see chapter 20).

As you can see, our simple switcher has grown to 26 buttons, arranged in four buses, and has a fader bar added. **SEE 14.4**

Effects Buses

If you now wanted your switcher to perform some transitions or effects, such as wipes (one image framed in a geometrical shape gradually replacing the other), ***keys*** (lettering or an image inserted into a different background picture), and other image manipulations (shape and/or color transformations), the basic design would have to include at least two or more ***effects buses*** and one additional fader bar. You would probably then want to expand the other video inputs to accommodate several more cameras, two or three VRs, electronic still store (ESS) and clip store functions, and remote feeds. In no time your switcher would have so many buttons and T-bars that operating them would require roller skates to get to all of them in a hurry.

Multifunction Switchers

To keep switchers manageable, manufacturers have designed buses that perform multiple functions. Rather than have separate program, mix, effects, and preview buses, you can assign a minimum of buses various mix/effects (M/E) functions. When you assign two such ***M/E buses*** (A and B) to the mix mode, you can dissolve from A to B or even do a super (by stopping the dissolve midway). By assigning them to the effects mode, you can achieve special effects, such as wipes from A to B. You can even assign the program and preview buses various M/E functions while still preserving their original functions.

The buttons with which you delegate what a bus is to do are, logically enough, called ***delegation controls***. The following discussion identifies the different buses and how they interact on a simple multifunction switcher. **SEE 14.5**

Preview bus BLK C-1 C-2 VR CG REM MIX
Mix bus A BLK C-1 C-2 VR CG REM
Mix bus B BLK C-1 C-2 VR CG REM
Fader bar
Program bus BLK C-1 C-2 VR CG REM MIX

14.4 BASIC SWITCHER LAYOUT
This basic production switcher has a preview bus, two mix buses, and a program bus. Note that the preview bus is identical to the program bus except that its output is routed to the preview monitor rather than to the line-out.

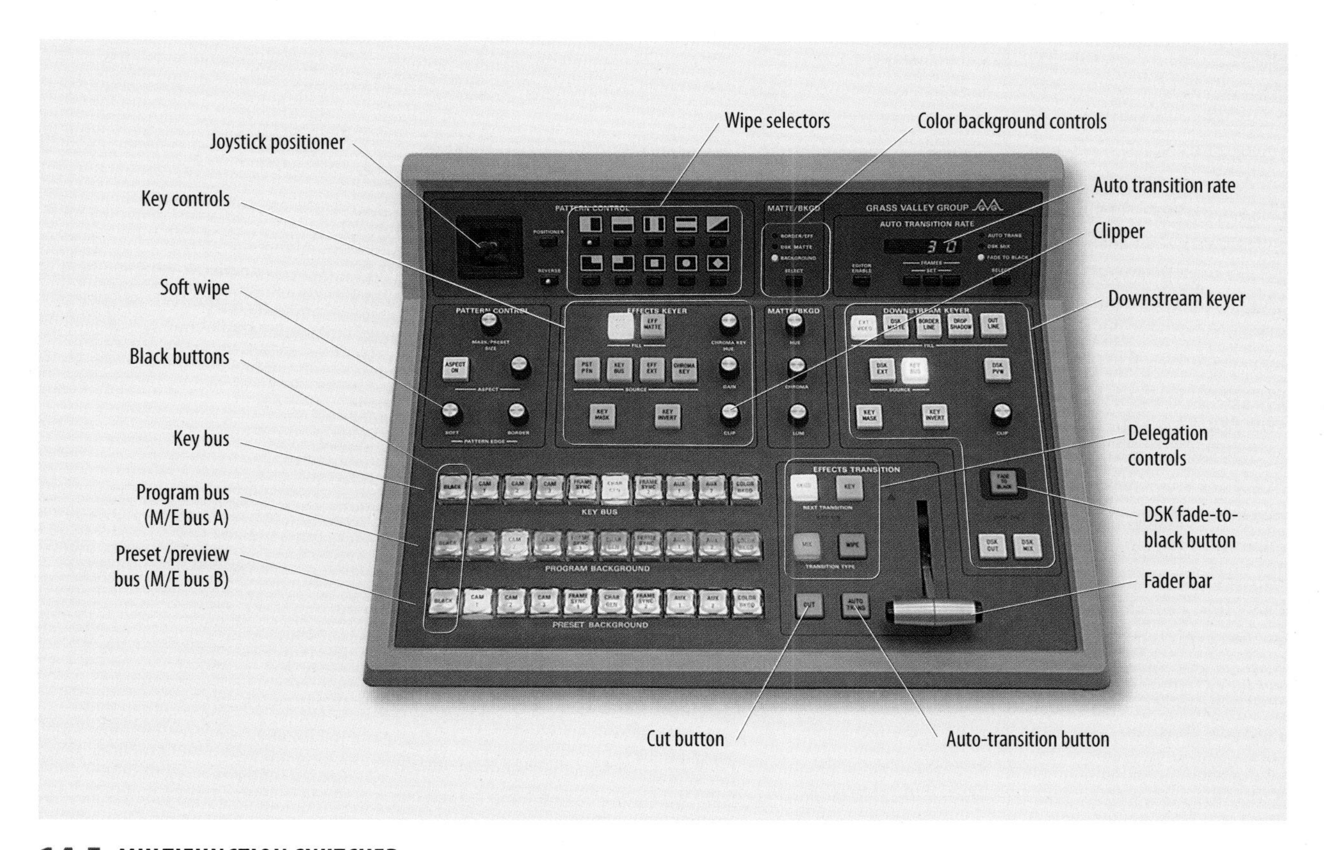

14.5 PHOTO BY EDWARD AIONA

14.5 MULTIFUNCTION SWITCHER
This multifunction switcher (Grass Valley 100) has only three buses: a preview/preset bus, a program bus, and a key bus. You can delegate M/E functions to the program and preview/preset buses.

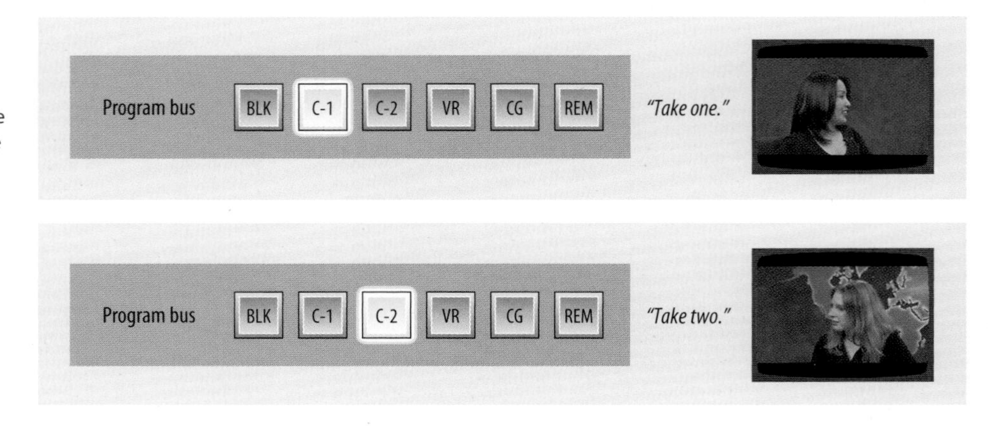

14.6 SWITCHING ON THE PROGRAM BUS
When switching on the program bus, the transitions will be cuts-only. With camera 1 on the air, you can cut to camera 2 by pressing the *C-2* button.

14.6 PHOTOS BY EDWARD AIONA

Although this analog switcher—a Grass Valley 100—was superseded years ago by larger and much more versatile digital HDTV switchers, it still serves as one of the best examples of a multifunction mix/effects architecture, which it pioneered.

Major buses As you can see, the switcher in figure 14.5 has only three buses: a preview/preset bus (lower row of buttons), a program bus (middle row); and a key bus (upper row). It also has a number of button groups that let you create certain effects.

Let's briefly review the functions of the buses. The program bus always directs its output to the line-out. If you don't need to preview the upcoming pictures and your switching is "cuts-only," you can do it all on the program bus. SEE 14.6 When assigned a mix or an effects function, however, it becomes one of the two M/E buses—in this case M/E bus A.

The ***preview/preset bus*** (also called preset/background bus) lets you preview the video source that you selected as your next shot. Whenever you press the corresponding button on the preview bus, the selected shot will automatically appear on the preview/preset monitor. But this preview bus can also function as the second M/E bus—on our switcher, M/E bus B. As soon as you activate a certain transition (cut, dissolve, or wipe), this preview picture will replace the on-the-air picture as shown on the line monitor. Despite its dual function, this bus is generally known as the preview bus.

Complicating the terminology a little more, both the program and the preview/preset buses are sometimes called "background" buses because they can serve as background for special effects, such as a title key.

The third (top) row of buttons is the ***key bus***. It lets you select the video source, such as lettering supplied by the C.G., to be inserted into the background image, supplied by the program bus. ZVL2 SWITCHING→ Architecture→ program bus | preview bus | mix buses | fader bar automatic transition | try it

Delegation controls These controls let you choose a transition or an effect. On this multifunction switcher, they are located to the immediate left of the fader bar. SEE 14.7

By pressing the background button *(BKGD)*, you activate the program and preview/preset (A and B) buses for cuts-only switching. Whatever you punch up on the program bus (A) will go on the air and, therefore, show

14.7 PHOTO BY EDWARD AIONA

14.7 DELEGATION CONTROLS
The delegation controls assign the function of the buses and the specific transition mode.

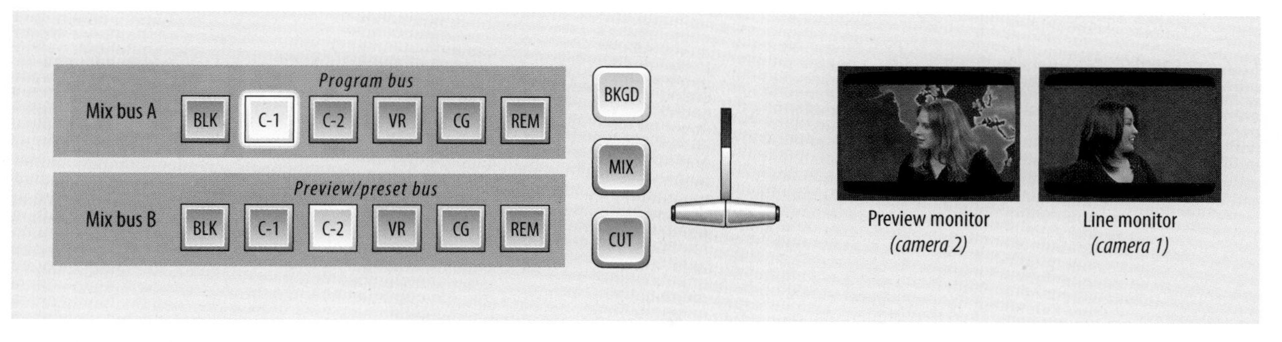

14.8 PHOTOS BY EDWARD AIONA

14.8 DUAL FUNCTION OF PROGRAM AND PRESET BUSES
When delegated a background and mix function, the program bus becomes M/E bus A and the preview/preset bus becomes M/E bus B. Here camera 1 is punched up on bus A and is on the air. Camera 2 is preset to replace camera 1 as soon as you press the *CUT* button.

14.9 PHOTOS BY EDWARD AIONA

14.9 IMAGE CHANGE AFTER CUT
When the cut is completed, the program bus shows camera 2 on the air, and the preview/preset bus switches automatically to camera 1.

up on the line monitor. Whatever you press on the preview/preset bus (B) will show up on the preview monitor, ready to replace—through a cut—the picture from bus A currently on the air. To perform the actual cut from the line-out picture (on the air) to the new picture selected on the preview bus, you need to press the red *CUT* button. If you press the *CUT* button repeatedly, it will toggle from preview to program and back. This is especially important when switching from one person to another in a fast dialogue. **SEE 14.8 AND 14.9**

By additionally pressing the red *MIX* button in the delegation controls section of the switcher, you expand the transitions from cuts-only to include dissolves as well. You can now cut from one video source to another or dissolve between them. **SEE 14.10** When you press the red *WIPE* button instead of the *MIX* button, the transition will be a wipe instead of a dissolve.

Fading from black or to black is actually dissolving from black to a picture or from a picture to black. Instead of choosing an actual image on the preview bus, you choose the *BLK* button. **SEE 14.11**

By pressing the *KEY* button, you activate the top (key) bus. On this bus you can select a proper key source, such as the C.G., that is to be inserted into the background picture currently activated on the program bus (A) and therefore on the air.

The advantage of a multifunction switcher is that you can achieve all these effects with only three buses. If you had continued the architecture—the electronic design logic—of the switcher we were building, you would have needed at least five buses, two fader bars, and several additional buttons to achieve the same key effect. ZVL3 SWITCHING→ Switching functions→ transitions | create effects

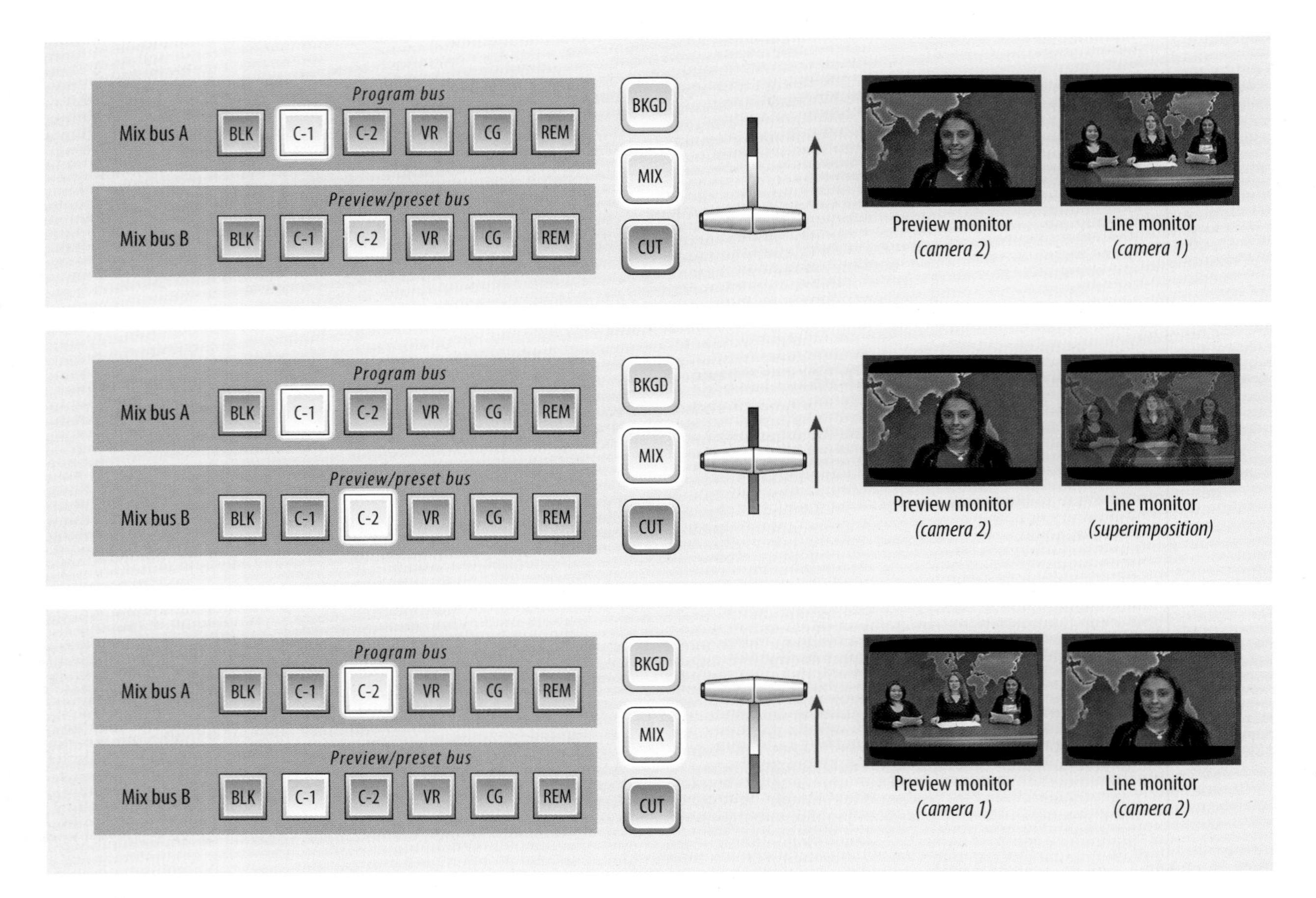

14.10 DISSOLVE
Once you have assigned the mix function through the mix delegation control, you can dissolve from camera 1 to camera 2. Assuming that camera 1 is on the air on bus A, you need to preset camera 2 on bus B. By moving the fader bar to the full extent of travel (in this case, up), you activate the dissolve from camera 1 to camera 2. Once the dissolve is completed, camera 2 will replace camera 1 on the program bus. Note that you can move the fader bar either up or down for the dissolve.

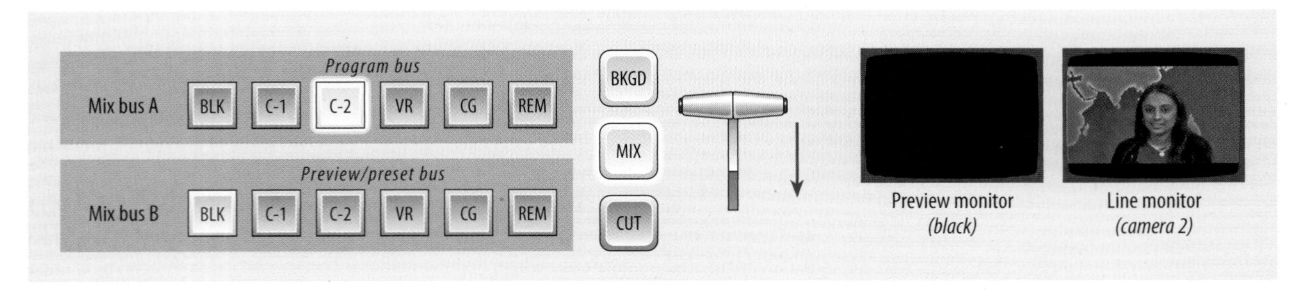

14.11 FADE TO BLACK
When fading to black from camera 2, you need to punch up the *BLK* button on bus B (preview/preset) and then dissolve into it by moving the fader bar to the full extent of travel (in this case, down).

Additional Switcher Controls

Now that you are familiar with the basic switcher controls, there are a few more you need to know: auto-transition, wipe controls and wipe patterns, key and key-level controls, the downstream keyer, and color background controls.

At this point don't worry about exactly how these controls are operated. Although all professional production switchers have these additional controls, they often require different means of operation. To become efficient in using a particular switcher, you need to study its operations manual and, above all, practice, as you would when learning to play a musical instrument.

Realize that these controls do not by themselves create the effect; rather, it is the special-effects generator (SEG) that performs this task. (We discuss special effects in section 14.2.) All production switchers have a built-in SEG. In fact, you will find that the manufacturers of most digital production switchers pride themselves not so much on operational ease but on the many visual tricks their SEGs can perform. The buzzword is ***layering***, which means combining several key effects into a more complex one. ZVL4 SWITCHING→ Transitions→ cut | mix/dissolve | wipe | fade | try it

Auto-transition The ***auto-transition*** is an automated fader bar. Instead of your moving the T-bar toward and away from you, the auto-transition control does it automatically. Such a device is especially useful if you must time each dissolve or wipe exactly the same. You can set the number of frames to establish the rate of the transition. With 30 frames per second, a 60-frame setting will give you a 2-second (rather slow) dissolve. On our switcher (figure 14.5), the auto-transition control is in the upper-right corner, set at 30, which means a 1-second transition. To activate the transition, you press the *AUTO TRANS* button next to the *CUT* button.

Wipe controls and wipe patterns When pressing the *WIPE* button in the delegation controls section in addition to the *BKGD* button, all transitions will be wipes. During a ***wipe*** the source video is gradually replaced by the second image that is framed in a geometrical shape. You can select the specific pattern in the group of buttons called wipe mode or pattern selectors. SEE 14.12

Common wipe patterns are expanding vertical wipes and horizontal wipes. SEE 14.13 AND 14.14 When you stop

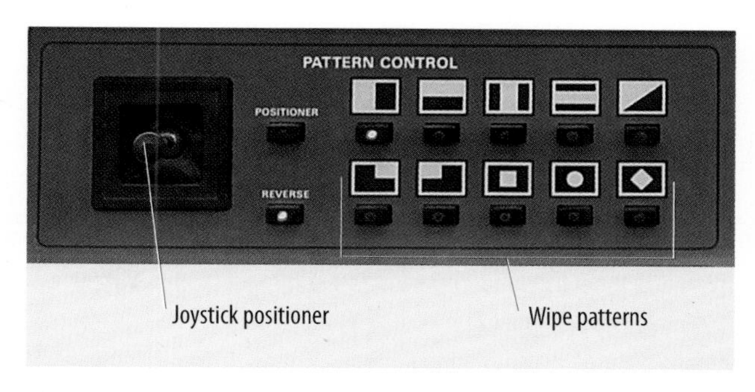

14.12 WIPE PATTERNS
The wipe selectors offer a choice of geometrical wipe patterns. The configurations can be placed in a specific screen position using the joystick positioner.

14.13 VERTICAL WIPE
In a vertical wipe, one picture is gradually replaced by another from the bottom up or from the top down.

14.14 HORIZONTAL WIPE
In a horizontal wipe, one picture is gradually replaced by another from the side.

14.12 PHOTO BY EDWARD AIONA
14.13 PHOTO BY EDWARD AIONA
14.14 PHOTO BY EDWARD AIONA

14.15 PHOTOS BY EDWARD AIONA

14.15 SPLIT SCREEN

A In this horizontal split-screen effect, camera 1 frames in the left side of the viewfinder the image designated to become the left half of the split screen.

B Camera 2 places its image in the right side of the viewfinder.

C In the completed split-screen wipe, the two images appear in the designated sides of the frame.

a horizontal wipe midway, you get a split screen. SEE 14.15 Other popular wipes are diamond wipes and an assortment of box wipes whereby the new image replaces the old image as an expanding box. SEE 14.16

On large switchers these controls can be extended to nearly 100 different patterns by inputting a code into the switcher. You can also control the direction of the wipe (whether a horizontal wipe, for example, starts from screen-left or screen-right during the transition). The joystick positioner lets you move patterns on different locations on-screen. Other controls give the wipes a soft or hard edge and give letters different borders and shadows.

14.16 PHOTO BY EDWARD AIONA

14.16 DIAMOND WIPE

In a diamond wipe, the second video source is gradually revealed in an expanding diamond-shaped cutout.

Key and key-level controls As mentioned, keying lets you insert lettering or other picture elements into the existing, or background, scene. The most common use of keys is to put lettering over people or scenes, or the familiar box over the newscaster's shoulder. With the key bus, you can select the video source to insert into the background image, such as the titles from the C.G. The ***key-level control***, also called the clip control or clipper, adjusts the key signal so that the letters appear sharp and clear during the key.

Regardless of the switcher model, you must always first select a background image (the main image into which you want to insert the title) and the key source (the title) and then work with the clip control so that the key has sharp and clear edges.

Downstream keyer The "downstream" in ***downstream keyer (DSK)*** refers to the manipulation of the signal at the line-out (downstream), rather than at the M/E (upstream) stage. With a downstream keyer, you can insert (key) a title or other graphic over the signal as it leaves the switcher. This last-minute maneuver, which is totally independent of any of the controls on the switcher buses, is done to keep as many M/E buses as possible available for the other switching and effects functions. Most switchers with a DSK have a master fader, which consists of an additional fader bar or, more commonly, a fade-to-black button, with which you can automatically fade to black the base picture together with the downstream key effect (see figure 14.5).

You may ask why this fade-to-black control is necessary when, as just demonstrated, you can fade to black by simply dissolving to black on the program bus. The reason for the extra fade control is that the effect produced by the DSK is independent of the rest of the (upstream) switcher controls. The *BLK* button on the program bus will eliminate the background but not the key itself. Only the button in the downstream keyer section (to the right of the T-bar) will fade the entire screen to black. ZVL5 SWITCHING→ Effects→ keys | key types | downstream keyer | special effects

As an example, let's set up a simple DSK effect at the end of a product demonstration of the latest computer model and then fade to black. The final scene shows a close-up (CU) of the computer as the background, with the name of the computer inserted by the DSK. Recall that one way to fade to black is to press the *BLK* button on the preset bus and then dissolve into it by moving the fader bar or pressing the *AUTO TRANS* button. But when you look at the line monitor, the background image (CU of the computer) has been replaced by black as it should be, but the name of the computer remains on the line monitor. You now know why. The downstream keyer is unaffected by what you do in the upstream part of the switcher—such as going to black on the M/E bus. Totally independent of the rest of the switcher controls, the DSK obeys only those controls in its own (downstream) territory, hence the need for its own black controls.

Color background controls Most switchers have controls with which you can provide color backgrounds to keys and even give the lettering of titles and other written information colors or colored outlines. Color generators built into the switcher consist of dials that you can use to adjust hue (the color itself), saturation (the color strength), and brightness or luminance (the relative darkness and lightness of the color) (see figure 14.5). On large production switchers, these controls are repeated on each M/E bus.

Large Production Switchers

Just to give you an idea of how large production switchers look, the following figure shows one with 32 video inputs, 16 separate outputs, and multiple M/E buses that permit complex, multilevel effects, such as running video clips into secondary screens or other graphic frames. Just in case you forgot to key some names over those screens, you can use any one or all four of the downstream keyers. Such complex wonder switchers nevertheless operate on the same basic principles you have learned from studying the architecture of the simple multifunction switcher. **SEE 14.17**

OTHER SWITCHERS

Besides studio production switchers, which are also used in large remote trucks for the coverage of sporting events, there are four other types that you may find in television stations or independent production houses: portable switchers, switching software, routing switchers, and master control switchers. An additional concern during switching is genlock.

Portable Switchers

Portable switchers are actually small control centers that contain a production switcher, preview monitors for the video sources, larger preview and line monitors, an audio console, a character generator, remote camera controls, and

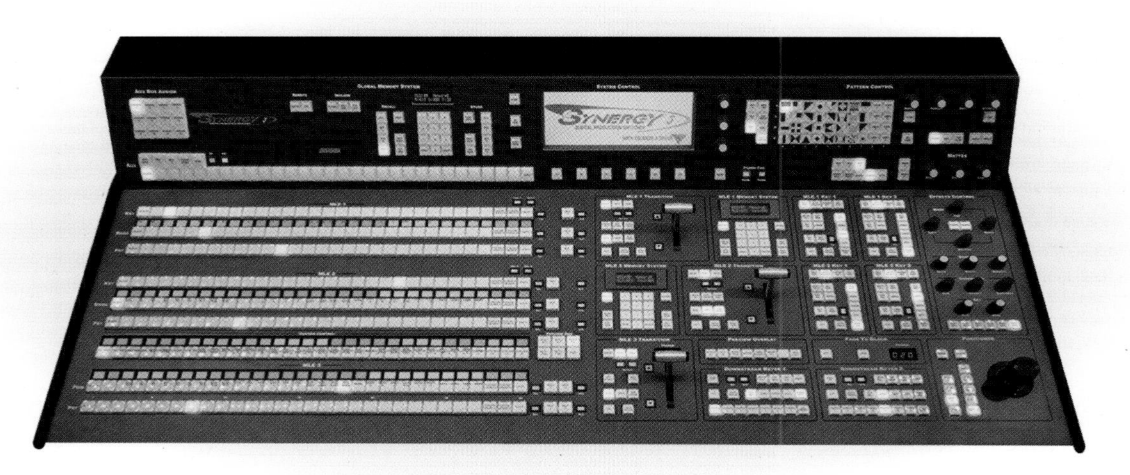

14.17 PHOTO COURTESY ROSS VIDEO LIMITED

14.17 LARGE PRODUCTION SWITCHER
This large production switcher has more of everything. It can handle all standard and high-definition video systems and import more than 30 video sources. You may select any one of 40 buttons for designated wipes and digital video effects, and you can operate four DSKs and an array of additional multilevel effects. This switcher operates on the same principles as a much simpler switcher.

14.18 PORTABLE SWITCHER
This portable switcher, packed in a small suitcase, represents a control room, with six video source monitors, larger preview and program (line) monitors, a fader bar, a six-channel stereo audio mixer, a C.G., and operating controls for remote cameras.

14.18 PHOTO COURTESY SONY ELECTRONICS, INC.

an intercom system—all packed in a small but expensive suitcase that weighs less than 18 pounds. **SEE 14.18** In effect, the suitcase holds all the key elements you would need for a moderately complex electronic field production (EFP). The portable switcher in figure 14.18 has the customary layout of preview and program buses, with six video inputs and a seventh one for ESS images. The video inputs are visible on a row of preview monitors, with larger monitors for preview and program sources. It has a special-effects generator, a C.G., and pan, tilt, zoom, and focus controls for remote cameras. It also has a six-channel audio mixer with individual faders and a master pot and volume-unit (VU) meter, an interface to connect to an external VR, and several additional pieces of production control equipment.

Switching Software

Some companies have developed software that allows you to use a desktop or even a notebook computer as a switcher. The computer interface displays the customary preview, program, and key buses; a fader bar; a monitor bank; and, among the more advanced systems, various video and audio control equipment. **SEE 14.19**

14.19 COMPUTER SWITCHER INTERFACE
This software program of the NewTek switcher VT[4] displays and activates all basic production and postproduction switcher functions. It has an amazing array of built-in test and video-quality equipment as well as a multitrack audio console. This switcher can be used for live switching or postproduction work.

14.19 PHOTO COURTESY NEWTEK

The switcher in *Zettl's VideoLab* is a simple version of a computer switcher.[2] ZVL6 SWITCHING→ Architecture→ try it

Routing Switchers

Routing switchers route video signals to specific destinations. For example, you would need a routing switcher to switch the line-out or preview video to specific monitors and then to the satellite video. Or you may assign the line-out signal to the video server instead of video recorder 2 because VR 2 is involved in editing. The buttons on a routing switcher are usually arranged in rows that look very much like the program bus on a production switcher or part of a computer-controlled system.

2. Although the switcher layout and functions were designed by the author, the software programs were specially developed by the highly creative computer people of the former THEmedia company in Vancouver, Canada. Some of these people are now part of the TBAdigital company, also in Vancouver.

Master Control Switchers

The master control switcher retrieves all the program material stored in the server according to the program log time line; it cues, rolls, and stops video recorders, and it switches automatically to remote feeds, such as a network program or live event. SEE 14.20

Genlock

When switching among various video sources in the studio or among remote sources, you normally need to provide the same synchronization information—a signal called house sync—to all video sources. This procedure guarantees that all video sources march in step to the same drummer—in this case, the same scanning clock—and that there will be no picture breakup during the switching. Locking the synchronization generators from different origination sources is called ***genlock***.

14.20 PHOTO BY EDWARD AIONA

14.20 MASTER CONTROL SWITCHER

The computerized master control switcher switches specific video and audio sources automatically. An operator can override the automation if necessary.

When you have to import a video for live switching without genlock (that is, without supplying house sync to all video sources), the framestore synchronizer will take over and help prevent picture breakup during switching.

MAIN POINTS

- Instantaneous editing is the switching from one video source to another, or the combining of two or more sources while the show, or show segment, is in progress.
- All switchers, simple and complex, perform the same basic functions: selecting an appropriate video source from several inputs, performing basic transitions between two video sources, and creating or accessing special effects.
- The switcher has a separate button for each video input. There is a button for each camera, video recorder (VR), character generator (C.G.), and other video source, such as a remote input. The buttons are arranged in rows, called buses.
- The basic multifunction switcher has a preset bus for selecting and previewing the upcoming shot, a program bus that sends its video input to the line-out, a key bus for selecting the video to be inserted over a background picture, a fader bar (or T-bar) to activate mix effects, a delegation control section, and various special-effects controls.
- The program bus is a direct input/output link and is therefore also called the direct bus. Whatever is punched up on the program bus goes directly to the line-out. It can also serve as a mix/effects (M/E) bus.
- The preview/preset bus is used to select the upcoming video (preset function) and route it to the preview monitor (preview function). It also serves as an M/E bus.
- The M/E buses (program and preview buses in a mix/effects mode) can serve a mix (dissolve, super, or fade) or an effects function.
- The key bus is used to select the video source to be inserted (keyed) into a background image.
- Delegation controls are used to assign the buses specific functions.
- The actual transition is activated by moving the fader bar from one limit of travel to the other, or by an auto-transition button that takes on the functions of the T-bar.
- Most switchers offer a variety of digital effects, such as wipe patterns, borders, background colors, and the possibility of effects layering.
- Other types of switchers include portable switchers, switching software, routing switchers, and master control switchers.
- The synchronization of two or more video sources or origination sources to prevent picture breakup during switching is called genlock.

SECTION

14.2

Electronic Effects and Switcher Functions

When watching television you are bombarded with an awesome array of effects. Even in a simple weathercast, the weathercaster stands in front of the weather map with the landscape of your area, zoomed in on the town in which you live. The map may show little suns moving over areas where it is sunny, or show rain and snow falling, depending on what is happening outside. The three-dimensional landscape tilts while the prevailing temperatures pop in and out. What used to be the province of MTV has become commonplace.

Even news presentations are so loaded with video special effects that they often rival, if not surpass, the latest video games. Titles dance across the screen, change color, and zoom in and out. News anchors, field reporters, and guests are squeezed into side-by-side boxes when talking to one another. The brief stories that introduce the latest troubles in the world often end in a freeze-frame and then peel off the screen and tumble out of sight to make room for the next batch of maladies.

The screen is often bursting with simultaneous information. While the anchor's comments about a tragic traffic accident are accompanied by graphic video footage, the stock market quotes crawl across the bottom of the screen, and the side panel reveals the latest sports scores and weather. Throughout it all, the station or network logo is solidly embedded in a corner.

Such electronic wizardry is so easy to use that it may tempt you to substitute effect for content. Do not fall into the trap of camouflaging insignificant content or poorly shot or edited pictures with electronic effects. As dazzling as the effects may be, they cannot replace a basically solid message. When applied judiciously, however, many effects can enhance production considerably and give the message added impact.

Whenever you intend to use a visual effect, ask yourself: *Is it really necessary? Does it help clarify and intensify my message? Is it appropriate?* If the answer to all three questions is yes, such effects are readily available in even modest switchers, and many more can be created with special-effects software.

Some effects can be created during a production, such as title keys and wipes; others need to be built with digital equipment in the pre- or postproduction phase and stored in the switcher or integrated during editing. This section discusses the two major types of visual effects—standard video effects and digital video effects—as well as additional functions of the switcher.

- **STANDARD VIDEO EFFECTS**
  Superimposition, key, and chroma key
- **DIGITAL VIDEO EFFECTS**
  Manipulation of image size, shape, light, and color; manipulation of motion; and manipulation of multi-images.
- **ADDITIONAL SWITCHER FUNCTIONS**
  Character generator, still store, and clip store

STANDARD VIDEO EFFECTS

The ***special-effects generator (SEG)*** is built into all production switchers. It can produce or recall a dazzling variety of special effects reliably and with ease. Many electronic effects have become so commonplace in television production that they have lost their specialty status and are simply considered part of the standard visual arsenal. These include the superimposition, the key, and the chroma key. You can accomplish all these effects "on the fly" without having to build and store them in the switcher ahead of time.

Superimposition

As you certainly know, a superimposition, or ***super*** for short, is a form of double exposure. The picture from one video source is electronically superimposed over the picture from another. As explained in section 14.1, the super is easily achieved by activating both mix buses with the fader bar (see figure 14.10). A distinct characteristic of a super is that you can see through the superimposed image to the one that lies beneath it. You can then vary the strength of either picture (signal) by moving the T-bar toward one mix bus or the other.

Supers are often used for creating the effects of inner events—thoughts, dreams, or processes of imagination.

The traditional (albeit overused) super of a dream sequence shows a close-up of a sleeping person, with images superimposed over his or her face. Sometimes supers are used to make an event more complex. For example, you may want to super a close-up of a dancer over a long shot of the same dancer. If the effect is done properly, we are given new insight into the dance. You are no longer photographing a dance but helping create it. ZVL7 SWITCHING→ Transitions→ mix/dissolve

Key

Keying means using an electronic signal to cut out parts of a television picture and fill them in with colors or another image. The basic purpose of a key is to add titles to a background picture or to cut another image (the weathercaster) into the background picture (the satellite weather map). (Because chroma keying works on a different principle, we discuss it separately later in this section.) The lettering for the title is generally supplied by a built-in or standalone character generator. **SEE 14.21** ZVL8 SWITCHING→ Effects→ keys

Matte key The most common keys are the ***matte keys***, in which one signal is doing the cutting and another is filling the hole with color. To achieve a clean key, in which the letters are cut into the base picture without any tearing or breakup, you may have to first use the key-level (or clip) control, which adjusts the brightness difference between foreground and background, and the gain control, which adjusts the key signal strength. Most switchers have rotary knobs for these key controls. You can preset the key effect and then watch the preview monitor to check that the key shows up and whether the letters are tearing or otherwise fuzzy. On many switchers with downstream keyers, you can push down the key-level control to display the total key effect on the preview monitor.

The most common modes for keying letters are the edge mode, the drop-shadow mode, and the outline mode. In the edge mode, each letter has a black outline around it. **SEE 14.22** In the drop-shadow mode, the letters have a black shadow contour that makes them appear three-dimensional. **SEE 14.23** In the outline mode, the letters themselves appear in outline form, with the base picture filling the inside. **SEE 14.24** These key modes are especially effective in preventing the lettering from getting lost in a busy background.

You may be somewhat bewildered hearing about keys, mattes, and matte keys—all seemingly referring to the same thing. It really doesn't matter what term you use, so long as you are consistent and all members of the production team know what you mean.

You can, of course, also key shapes of objects into the base picture so long as there is enough contrast relative to the base that the edges do not tear. The box over the news anchor's shoulder is an example of a widely used key effect.

14.21 KEY
When keying a title over a base picture, the key signal cuts a hole into the base picture in the shape of the letters supplied by the C.G., and another fills the cutout letters with color (here with white). The letters seems to be pasted on top of the background picture.

14.21 PHOTO BY EDWARD AIONA

Chroma Key

Chroma keying is a special effect that uses a specific color (chroma), usually blue or green, as the backdrop for a person or an object that is to appear in front of the background scene. During the key the blue (or green) backdrop is replaced by the background video source without affecting the foreground object. A typical example is the weathercaster in front of a weather map or satellite picture. During the chroma key, the computer-generated weather map or satellite image replaces all blue (or green) areas—but not the weathercaster. The key effect makes the weathercaster appear to be standing in front of the weather map or satellite image. **SEE 14.25**

Because the chroma key responds to the hue of the backdrop rather than to the brightness (luminance) contrast as in a regular key, be sure that the chroma-key area is uniformly painted (even blue or green with a fairly high saturation throughout the area) and especially evenly lighted. Uneven background lighting will prevent a full replacement of the backdrop by the background video or cause the foreground image to tear.

If the on-camera talent wears something similar to the backdrop color, such as a blue sweater, while standing in

14.22 MATTE KEY IN EDGE MODE
The edge mode matte key puts a black border around the letter to make it more readable than with the normal key.

14.23 MATTE KEY IN DROP-SHADOW MODE
The drop-shadow matte key adds a prominent cast shadow below the letter as though it were illuminated by a spotlight.

14.24 MATTE KEY IN OUTLINE MODE
The outline matte key makes the letter appear in outline form. It shows only the contour of the letter.

14.25 PHOTOS BY EDWARD AIONA

14.25 CHROMA-KEY EFFECT: WEATHERCAST

A In this chroma key, the weathercaster stands in front of a green backdrop.

B During the key the green backdrop is replaced by the computer-enhanced satellite photo.

C The weathercaster appears to be standing in front of the satellite photo.

front of the blue chroma-key area, the blue of the sweater will also be replaced by the background image during the key. Unless you want to amuse your audience with a special effect in which part of the weathercaster disappears, don't let him or her wear anything blue in front of the blue chroma-key set. If the talent likes to wear blue, use green as the backdrop color.

If the weathercaster stands too close to the chroma-key backdrop, the blue reflections on part of his or her clothing or hair may cause the key to tear. Such problems may also occur especially if the lighting on the weathercaster has extremely fast falloff. The deep shadows, which are apt to be seen as blue by the camera, may cause the image to tear during the key.

Recall that you can counteract this nuisance to some extent by using yellow or amber gels on the back lights (not the background lights) because the yellow or amber back light neutralizes the blue shadows, thus eliminating the tearing. When using green as the chroma-key color, you need to use a light (desaturated) magenta (bluish red) gel to counteract the green reflections. Most digital chroma keys produce clean edges, however, even without colored back lights.

Because an even backdrop color and even lighting are important to getting a clean chroma key, alternative chroma-key methods have been developed. One uses a small light ring that is attached to the front of the camera lens. It emits blue (or green) light that is reflected off a beaded gray cloth backdrop. But shouldn't this cloth be blue or green? No—the backdrop has millions of embedded glass beads that reflect enough of the blue (or green) light to enable the switcher to perform the chroma key. SEE 14.26

You may wonder why the person standing in front of the reflective backdrop doesn't look blue (or green). It's because the light is simply too weak to reflect much off skin or clothing. If the person stands too close to the camera

14.26 CHROMA-KEY LIGHT RING
This light ring, attached to the front of the lens, emits enough light to have the beaded gray backdrop reflect enough blue (or green) light for a functional chroma key.

14.26 PHOTO COURTESY RELFLECMEDIA, WWW.REFLECMEDIA.COM

and wears a highly reflective shirt, however, you may see a bluish or greenish tint on the person. This coloring process also tends to occur when objects are too close to the camera during the chroma key.

Studio chroma key Despite the availability of highly sophisticated digital video effects, the chroma-key process is used extensively in studio production. The previous discussion focused on one of the most popular uses of the chroma key—for weathercasts—but chroma keying can also be done to enhance the background of an office set, for example, or to place a person in a specific outdoor setting. SEE 14.27 AND 14.28 In both cases the key source (skyline or museum pillars) was supplied by the switcher's electronic still store system.

EFP chroma key Chroma keying is also useful during electronic field productions and big remotes, especially

14.27 CHROMA-KEY EFFECT: WINDOW

A In this chroma key, a suitable background view is selected from the ESS system.

B A studio camera focuses on the office set in front of a green chroma-key backdrop.

C Through chroma keying there seems to be a picture window behind the CEO sitting at her desk.

14.27A PHOTO BY ROBAIRE REAM
14.27B PHOTO BY EDWARD AIONA
14.27C PHOTOS BY ROBAIRE REAM AND EDWARD AIONA

14.28 CHROMA-KEY EFFECT: SIMULATED LOCATION

A The source for the background image is a video frame of the museum exterior from the ESS system.

B The studio camera focuses on the actor playing a tourist in front of a blue chroma-key backdrop.

C All blue areas are replaced by the background image; the actor appears to be a tourist in front of the museum.

14.28 PHOTOS BY HERBERT ZETTL

if the talent is not able to stand directly in front of the desired background scene, such as a stadium or government building. When using a chroma-key effect during a sports remote, for example, the announcer may even be in the studio, with the remote feed (long shot of the football stadium) serving as the chroma-key background.

If you do chroma keying on-location, with the talent standing outdoors, watch out for blue reflections from the sky. To avoid such problems, switch to green for the chroma-key color and put the talent in front of a green backdrop. Because green is usually less common than blue in skin tones, it is slowly replacing blue as the dominant chroma-key color.

14.29 SHRINKING
Through shrinking, also called a squeeze-zoom, you can reduce the total full-frame image to a smaller frame that contains the same picture information.

14.29 PHOTO COURTESY BROADCAST AND ELECTRONIC COMMUNICATION ARTS DEPARTMENT AT SAN FRANCISCO STATE UNIVERSITY

DIGITAL VIDEO EFFECTS

To make sense of the potential of ***digital video effects (DVE)***, we divide them into three areas: manipulation of image size, shape, light, and color; manipulation of motion; and creation and manipulation of multi-images.

Image Size, Shape, Light, and Color

A great variety of effects are available for manipulating the size, shape, light, and color of an image. Some of the more prominent are: shrinking and expanding, perspective, mosaic, and posterization and solarization. Many of these DVE change a realistic picture into a basically graphical image.

Shrinking and expanding Shrinking refers to decreasing a picture's size while keeping the entire picture and its aspect ratio intact. Unlike cropping, where you actually remove some of the picture information, shrinking simply renders the picture smaller. Because the visual effect is similar to a zoom-out (shrinking) or a zoom-in (expansion), this effect is also called a squeeze-zoom. **SEE 14.29**

14.30 PERSPECTIVE
Through DVE you can distort an image so that it seems to float in the three-dimensional screen space.

14.30 PHOTO BY EDWARD AIONA

Perspective You can distort an image in such a way that it looks as though it is floating in the three-dimensional video space. When combined with motion, such a 3D video space is greatly intensified. **SEE 14.30**

Mosaic In the mosaic effect, the video image (static or in motion) is distilled into many discrete, equally sized squares of limited brightness and color. The resulting screen image looks like an actual tile mosaic. Such an image appears to contain greatly enlarged pixels. **SEE 14.31** This technique is sometimes used to obscure parts of the

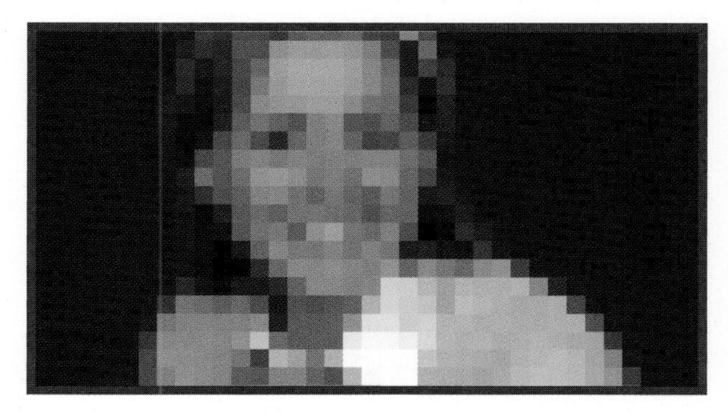

14.31 MOSAIC EFFECT
Here the image is reduced to equally sized squares resembling mosaic tiles. In the electronic mosaic, as in a traditional tile mosaic, you can change the size of the tiles.

14.31 PHOTO BY EDWARD AIONA

14.32 POSTERIZATION
In posterization the brightness values are severely reduced. The picture takes on a high-contrast look.

14.32 PHOTO BY EDWARD AIONA

14.33 SOLARIZATION
Solarization is a special effect that is produced by a partial polarity reversal of an image. In a color image, the reversal results in a combination of complementary hues.

14.33 PHOTO BY EDWARD AIONA

14.34 PEEL EFFECT
In a peel, or page-turn, effect, video A seems to curl and peel off a stack of images, revealing video B underneath.

14.34 PHOTO BY GARY PALMATIER

body or a guest's identity. The mosaiclike distortion shows the person's face but renders the features unrecognizable.

Posterization and solarization In posterization the brightness values (luminance) and the shades of the individual colors are collapsed so that the image is reduced to a few single colors and brightness steps. For example, the colors on a face show up as though they were painted by number with only a few paints. This image looks like a poster, hence the name of the effect. SEE 14.32

Solarization combines a positive and a negative image of the subject. Some solarization effects result in a complete polarity reversal, in which white becomes black, and colors appear as their complementary hue (yellow turns to blue, and red turns to green). When combined such effects often look like highly overexposed images. SEE 14.33

Motion

There are so many possibilities for making effects move that a sensible and common terminology has not yet been developed. Don't be surprised to hear the director in the control room or the editor in the editing room using the sound language of cartoons—"squeeze," "bounce," or "fly"—when calling for motion effects. Some terms have been coined by DVE manufacturers, others by imaginative production personnel. Let's look at a few of the more popular effects: peel effect, rotation and bounce effects, and cube spin.

Peel effect In the peel, or page-turn, effect, the top (existing) image curls up as though it were peeled off a pad of paper, revealing the next "page"—a new image that seems to lie underneath. SEE 14.34

Rotation and bounce effects With the rotation effect, you can spin any image on all three axes, individually or simultaneously: the x-axis (representing width), the y-axis (representing height), and the z-axis (representing depth). Although rotation terminology varies, normally a "tumble" refers to an x-axis rotation, a "flip" to a y-axis rotation, and a "spin" to a z-axis rotation. SEE 14.35

Cube spin The rotation can also be applied to three-dimensional effects. The well-known cube spin shows a rotating cube, with each of the three visible sides displaying a different static or moving image. SEE 14.36

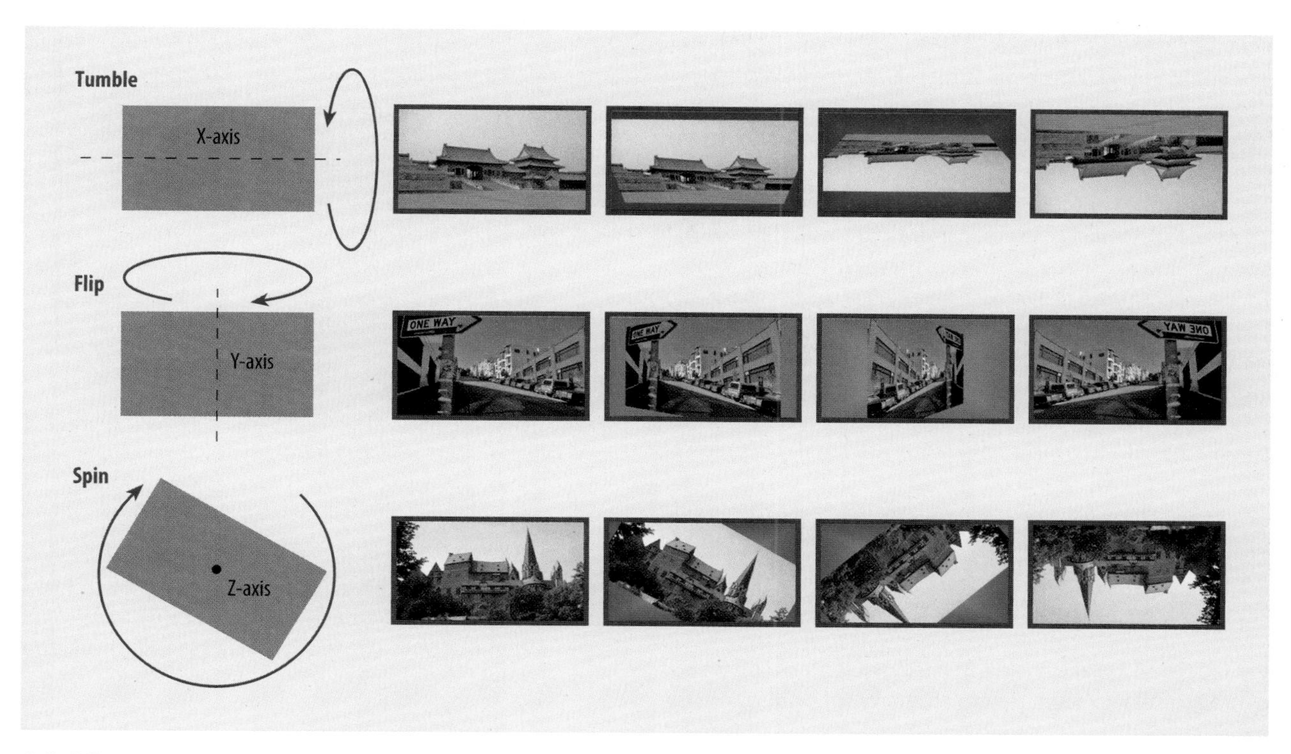

14.35 PHOTOS BY HERBERT ZETTL

14.35 ROTATION EFFECT
In a rotation effect, the image can be revolved around the x-axis (tumble), the y-axis (flip), and the z-axis (spin).

14.36 PHOTOS BY RENEE CHILD

14.36 CUBE SPIN
In a cube spin, a rotating cube displays a different static or moving image on each of the three visible sides.

Multi-images

The multi-image effects include the many possibilities of dividing the screen into sections or of having a specific image update itself. The former we call secondary-frame effects; the latter, echo effects.

Secondary frame The secondary-frame effect shows several images, each clearly contained within its own frame. A common use of such an effect is to show host and guest simultaneously in separate frames, talking to each other from different locations. To emphasize that they are speaking to each other, although both are actually looking into the camera (at the viewer), the frames are sometimes tilted toward each other through a digital perspective change. **SEE 14.37**

As you have seen in many show formats, even dramatic programs, the screen is sometimes divided into multiple frames, each showing a different event occurring at the same time.

Echo The echo effect is created when a static object is mirrored (similar to seeing yourself many times in opposing barbershop mirrors) or a moving object leaves a continuous trail of previous positions. **SEE 14.38 AND 14.39**

14.37 SECONDARY-FRAME EFFECT
The trapezoidal distortion of the frames makes us perceive two people talking to each other rather than to the viewer.

14.37 PHOTOS BY EDWARD AIONA

ADDITIONAL SWITCHER FUNCTIONS

In addition to the standard selection, mix, and effects functions, many switchers incorporate features that effectively make the switcher a small production center. The most common are the character generator, the electronic still store system, and clip storage. These features are primarily intended to help the technical director perform tasks that would otherwise require the services of additional personnel. As you can see, the job of a TD gets more and more complex.

Character generator For creating standard titles or a credit crawl (names that move up the screen), you can plug in a keyboard to activate the built-in ***character generator (C.G.)***. You can store these titles and call them up when needed.

Still store As you recall, the ESS (electronic still store) system lets you grab any video frame, store it in an ESS file, and retrieve it for a news story, background, or key source.

14.38 ECHO EFFECT: STATIC TITLE
In this echo effect, a static image is repeated many times, and the copies are placed in close proximity to one another.

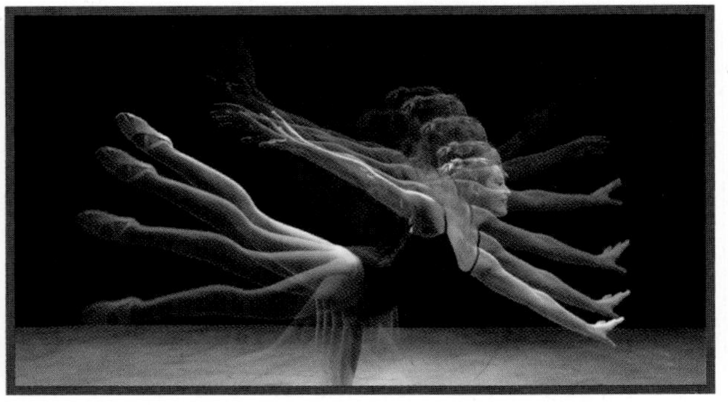

14.39 ECHO EFFECT: MOVING DANCER
In this echo effect, the moving dancer trails her previous movements.

14.39 PHOTO BY EDWARD AIONA

This is much like when you grab a single frame from a motion clip and store it in a still-photo file.

Clip store With this feature you can import, store, and retrieve a number of video clips without the need for an external server or video recorder. You can store a fair number of video clips (in some models up to two or more hours' worth) and retrieve them for bumpers (short videos announcing upcoming news stories), brief news stories, or animated titles. You can also use them to show events in secondary frames, similar to the multiple screens that show simultaneous events happening in different locations.

MAIN POINTS

- The special-effects generator (SEG) is built into all production switchers. It can produce or recall a dazzling variety of special effects reliably and with ease.
- The three standard video effects are the super (superimposition), key, and chroma key.
- The main purpose of a key is to add titles or objects to a base (background) picture.
- In a matte key, one video signal cuts a hole in the base picture, and the other signal fills the cutout with colors generated by the switcher. The standard matte key modes are edge, drop shadow, and outline.
- Chroma keying uses a blue or green backdrop, which, during the key, is replaced by the background image.
- For a standard chroma key, the blue or green backdrop must be evenly lighted.
- To neutralize blue (or green) reflections on the talent standing in front of the chroma-key backdrop, back lights fitted with complementary color gels can be used: light yellow or amber for a blue background, and light magenta for a green background.
- Digital video effects (DVE) show images whose size, shape, light, and motion are manipulated in a predetermined fashion.
- The more common DVE include shrinking and expanding; perspective changes; mosaic; posterization and solarization; peel, or page-turn, effect; rotation and bounce effects; and cube-spin effect.
- Multi-images are secondary frames that show simultaneous events happening in different locations.
- Additional switcher functions include a built-in character generator (C.G.), an electronic still store (ESS) system, and a clip store that holds and plays more than two hours of video clips.

ZETTL'S VIDEOLAB

For your reference or to track your work, the *Zettl's VideoLab* program cues in this chapter are listed here with their corresponding page numbers.

CHAPTER 15

Design

Although you are probably very conscious of design and style when buying clothes or an automobile, you may be unaware of specific design elements when watching an opening show title, a news set, or the living room set of a situation comedy. You may be dazzled by an animated title that does everything but pop out of the screen, but you're probably not motivated to analyze its aesthetic qualities. And you probably perceive the living room in the daytime drama as exactly that—a living room—not carefully placed scenery and properties. We all know, of course, that all such design elements are meticulously planned.

In fact, design, or the lack of it, permeates everything a television production company shows on the air and off. It sets the style of the video presentation if not of the production company as a whole. Design includes not only the colors and the letters of a show title and the look of a studio set but also the production company's stationery, office furniture, hallway artwork, and logo. The CNN logo, for example, suggests up-to-date, no-nonsense news. **SEE 15.1**

But a handsome logo does not automatically carry its design qualities over to the programming or the on-air graphics or scenery. It is important to develop a design consciousness for everything you do; a well-executed logo is merely the symbol for such awareness, not its sole cause.

15.1 CNN LOGO

15.1 LOGO COURTESY NEWSCOM

Section 15.1, Designing and Using Television Graphics, stresses the major design considerations of graphics for television. Section 15.2, Scenery and Props, looks at major aspects of television scenery and properties and discusses some elements of scene design.

KEY TERMS

character generator (C.G.) A dedicated computer system that electronically produces letters, numbers, and simple graphic images for video display. Any desktop computer can become a C.G. with the appropriate software.

color compatibility Color signals that can be perceived as black-and-white pictures on monochrome television sets. Generally used to mean that the color scheme has enough brightness contrast for monochrome reproduction with a good gray-scale contrast.

essential area The section of the television picture, centered within the scanning area, that is seen by the home viewer regardless of masking or slight misalignment of the receiver. Also called *safe title area* and *safe area.*

flat A piece of standing scenery used as a background or to simulate the walls of a room.

floor plan A diagram of scenery and properties drawn on a grid pattern. Can also refer to *floor plan pattern.*

floor plan pattern A plan of the studio floor, showing the walls, main doors, location of the control room, and the lighting grid or battens.

graphics generator Dedicated computer or software that allows a designer to draw, color, animate, store, and retrieve images electronically. Any desktop computer with a high-capacity RAM and a hard drive can become a graphics generator with the use of 2D and 3D software.

letterbox The aspect ratio that results from fitting the full width of a 16 × 9 image onto a 4 × 3 screen by blocking the top and bottom screen edges with stripes.

pillarbox The aspect ratio that results from fitting a 4 × 3 image onto a 16 × 9 screen by blocking the sides with stripes.

props Short for *properties.* Furniture and other objects used for set decoration and by actors or performers.

scanning area Picture area that is scanned by the imaging device; in general, the picture area usually seen in the camera viewfinder and the preview monitor.

virtual set A computer-generated environment, normally a studio set, which provides the background for a chroma-key action, such as a newscast.

windowbox A smaller picture that is centered in the actual display screen, with the leftover space of the 16 × 9 frame surrounding it.

SECTION

15.1

Designing and Using Television Graphics

When watching television you may be more captivated by the opening titles than the show that follows. Even when the program consists of a no-nonsense interview or a simple product demonstration, we seem obliged to have the title burst onto the scene, make its dancing letters change shape and color, and have at least three different backgrounds moving slowly underneath it. Such titles are usually supported by high-energy sound effects.

You may wonder whether we spend an inordinate amount of time and effort on the graphics compared with the program itself. Even if we don't, video graphics have become a major factor in television production. Because creating such titles requires highly specialized computer skills, rather than competence in television production, we limit our discussion to the following aspects of television graphics.

- **SPECIFICATIONS OF TELEVISION GRAPHICS**
  Scanning and essential areas, out-of-aspect-ratio graphics, and matching STV and HDTV aspect ratios
- **INFORMATION DENSITY AND READABILITY**
  Making screen lettering legible
- **COLOR**
  Color aesthetics: high- and low-energy colors
- **STYLE**
  Contemporary look, personal touch, and consistency
- **SYNTHETIC IMAGES**
  Computer-generated pictures

SPECIFICATIONS OF TELEVISION GRAPHICS

When comparing a television screen with a movie screen, you will see two obvious differences: the standard television (STV) screen is much smaller and much narrower than the movie screen. But even the large and wide high-definition television (HDTV) flat-panel displays are considerably smaller than even a small movie screen. These two factors have a profound influence on the design specifications of television graphics.

The relatively small size of the STV screen limits the amount of writing you can display and demands fonts (lettering) that can be clearly seen. The limited screen width in relation to its height means that the titles do not have as much room to play across the screen and therefore must be kept closer to the center. Other design requirements of television graphics are common to all graphic design and deal more with readability, color, and style.

This section takes a closer look at the following design requirements and specifications: scanning and essential areas, out-of-aspect-ratio graphics, and matching STV and HDTV aspect ratios.

Scanning and Essential Areas

Unlike the painter or still photographer who has full control over how much of the picture shows within the frame, we cannot be so sure about how much of the video-recorded or broadcast pictures are actually seen on the home screen. Even in digital television, there is often a picture loss or slight distortion during transmission, in the editing process, or by how a specific flat-panel display is programmed. Also, not all television receivers are equally adjusted to match the preview monitors in a control room or editing room. Even if you gave the proper headroom when framing a close-up shot (CU) in a studio interview, your shots may have lost some or all of the headroom by the time they reach the home receiver. The same is true for titles that are framed too close to the screen edge. Because the edge information is often lost, you may end up with incomplete titles or the first and last digits missing from a telephone number.

How can you ensure that the information you send is actually seen on the home screen? Is there a standard that will more or less guarantee that all essential picture information, such as a title or telephone number, will appear in its entirety? The answer is a qualified yes. Although not mathematically precise, guidelines are available that can help you keep picture information from getting lost during dubbing or transmission. Basically, these guidelines tell you to keep vital information away from the screen edges. This is especially important when framing shots in the wider 16 × 9 ratio that are then broadcast mainly in the STV (standard 4 × 3) format. Just how far away you should be from the edge when framing a shot is prescribed by the scanning and essential areas.

The ***scanning area*** includes the picture you see in the camera viewfinder and on preview monitors in the control room. It is the area actually scanned by the camera imaging device (the CCD). The ***essential area***, also called safe title area and, simply, safe area, is centered within the scanning area. It is the portion seen by the home viewer regardless of the masking of the set, transmission loss, or slight misalignment of the receiver. **SEE 15.2**

Obviously, information such as titles and telephone numbers should be contained within the essential area. Also, if two people are facing each other from one screen edge to the other (blocked along the x-axis), be sure to place them well within the left and right borders of the essential area. But just how large is the essential area? It is usually smaller than you think—about 80 percent of the total area.

Some ***character generators (C.G.s)*** automatically keep a title within the essential area. The better studio cameras and camcorders as well as major editing programs have a device that electronically generates a frame outline within the viewfinder, showing the safe area.

If your C.G. or camera does not have such a built-in safety net, you need to create your own. Most word-processing or drawing programs let you create a rectangle and then reduce it by a specific percentage. You could, for example, draw a 4 × 3 rectangle that simulates a 100 percent TV screen area and then reduce it to an 80 percent one (exclusive of the lettering, of course). These new borders would then outline the essential area for you. You could use the same method for creating the essential area for a 16 × 9 format.[1]

After some practice you will be able to compensate in the camera framing for the picture loss or place a title within the essential area without having to juggle percentages. The surest way to test a title is to project it on the preview monitor. If the letters come close to the edges of the preview monitor, the title extends beyond the essential area and will certainly be cut off when seen by the home viewer. **SEE 15.3**

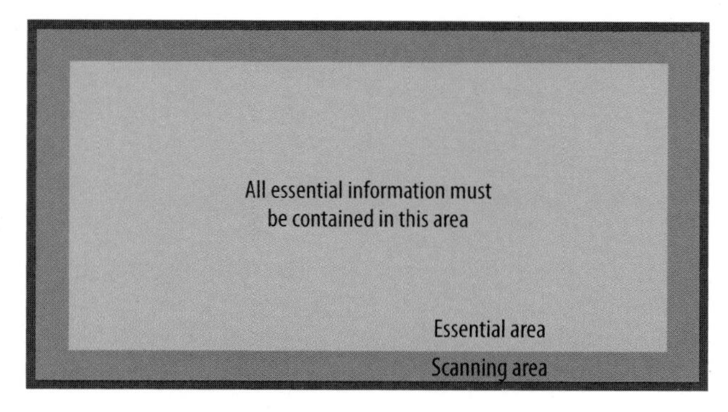

15.2 SCANNING AND ESSENTIAL AREAS
The scanning area is what the camera viewfinder and the preview monitor show. The essential, or safe title, area is what appears on the home television screen.

1. Adobe Photoshop software, for example, lets you specify the aspect ratio and the desired area percentage, and Apple's Final Cut Pro editing software offers two different frames—one for safe titles (essential area) and the other for safe action (scanning area).

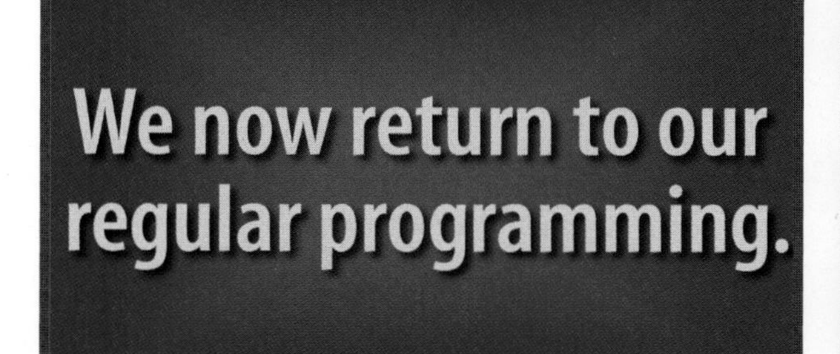

15.3 TITLE BEYOND ESSENTIAL AREA
A On the preview monitor, you can still see the complete title, although it comes close to the edges.

B When viewed on the home receiver, the information that lies outside the essential area is lost.

Out-of-aspect-ratio Graphics

Although you can change the aspect ratio of pictures within the television screen through various digital video effects (DVE), you can't change the dimensions of the screen itself. You can divide the screen into secondary frames of various aspect ratios, and you can block off areas of the screen and thus simulate different aspect ratios, but you are nevertheless confined to the set aspect ratio of the television display screen.[2]

You will inevitably run into situations in which the pictures to be shown do not fit the requirements of the television aspect ratio and essential area. Most often you encounter this problem when someone brings in an out-of-aspect-ratio chart or poster to pitch an upcoming event during a promotional interview or to illustrate a point in a sales meeting. More often than not, you must cover such a meeting live or record it without much chance for post-production doctoring. Many oversized graphics are vertical and do not adhere to the 4 × 3, much less the 16 × 9, aspect ratio. The problem with an out-of-aspect-ratio graphic is that, when shown in its entirety, the information on the graphic becomes so small that it is no longer readable. **SEE 15.4** By moving the camera close enough that the graphic fits the aspect ratio of the television screen, you inevitably cut out important information. **SEE 15.5**

If the lettering and other visual information are simple and bold enough, you can mount the entire out-of-aspect-ratio graphic on a larger card that is in aspect ratio. Then you simply pull back with the camera and frame up on the large card, keeping the out-of-aspect-ratio information as screen-center as possible. On a vertically oriented graphic without lettering, you could possibly tilt up and reveal the information bit by bit. If done smoothly, this gradual revelation adds drama. With lettering, however, such a tilt does not add drama but simply makes the graphic more difficult to read.

You encounter the same framing difficulty when trying to show a line of musical notation, a long mathematical formula or flow chart, or writing on a blackboard or whiteboard. If you zoom out all the way to show the entire whiteboard, the text is difficult to read. If you zoom in to a close-up, you can see only part of the writing. **SEE 15.6** The correct way of presenting whiteboard information is to divide the whiteboard into 4 × 3 or 16 × 9 aspect ratio fields and contain the writing within each of these fields. The camera can then get a close-up of the entire sentence. Even when working in the HDTV aspect ratio, you should write the information in blocks rather than across the width of the whiteboard. This is especially important if you work in a situation that uses remotely controlled cameras. It is much easier to zoom in on such a lettering block than to make the camera pan sideways. **SEE 15.7**

Matching STV and HDTV Aspect Ratios

You may wonder why digital television did not maintain the traditional 4 × 3 aspect ratio. The main reason for the horizontally stretched 16 × 9 aspect ratio is that it readily accommodates the wide-screen movie format. When showing a movie on a traditional television set, either both sides of the wide frame are crudely amputated or the images are displayed in the ***letterbox*** that shows the movie in its full width but necessitates the black stripes at the top and the bottom of the screen. **SEE 15.8**

2. See Herbert Zettl, *Sight Sound Motion: Applied Media Aesthetics*, 6th ed. (Boston: Wadsworth, 2011), pp. 87–93.

15.4 OUT-OF-ASPECT-RATIO GRAPHIC
When trying to frame this out-of-aspect-ratio graphic in its entirety, most of the information becomes difficult to read if not totally illegible.

15.5 INFORMATION LOSS IN CLOSE-UP
When you try to get a closer shot, all information outside the aspect ratio is lost.

15.6 PHOTOS BY EDWARD AIONA

15.6 ASPECT RATIO PROBLEM WITH A LONG LINE
Normal writing on a whiteboard can present a typical aspect ratio problem. The camera cannot show a close-up of a message that spans the full width of the whiteboard.

15.7 PHOTOS BY EDWARD AIONA

15.7 PROPER USE OF ASPECT RATIO BLOCKS
If the whiteboard is divided into proper aspect ratio fields, the camera can see the entire message even in a close-up.

Sometimes the 16 × 9 frame is digitally squeezed into the 4 × 3 frame, making everything look taller and skinnier than in the original shot. To avoid such picture distortions, some films are subjected to the pan-and-scan process whereby the more important portions of the wide-screen frame are selected to fit the 4 × 3 frame. But this process is quite time-consuming and costly and often violates the integrity of the original shot compositions.

When shown on the 16 × 9 HDTV screen, wide-screen movies suffer only slight picture loss, but what about standard 4 × 3 television programs? We can either stretch or enlarge the STV image so that it fills the full width of the HDTV screen. When we stretch the STV image to fill the width of the wide screen, everything looks fat, including the people. And when enlarging the STV image so that it fills the 16 × 9 screen, objects and people lose some of their headroom—and sometimes even their heads and feet!

15.8 LETTERBOX
Making the entire frame of a wide-screen movie fit into the 4 × 3 aspect ratio of STV results in empty (black) space at the top and the bottom of the screen. The resulting horizontal aspect ratio is called a letterbox.

15.9 PILLARBOX
When showing a standard 4 × 3 television frame on the 16 × 9 screen, there are empty dead zones, or side bars, on both sides of the screen. The resulting vertical aspect ratio is called a pillarbox.

15.10 WINDOWBOX
When a slightly reduced 4 × 3 picture is centered in a 16 × 9 screen, it's called a windowbox.

You can also place the full 4 × 3 frame in the center of the 16 × 9 screen so that the height of the 4 × 3 screen matches the height of the 16 × 9 screen. But then you have space—usually black stripes—left over on either side of the 4 × 3 screen insert. These stripes are called dead zones or side bars; the resulting aspect ratio is called a ***pillarbox***. **SEE 15.9**

If you simply shrink the 4 × 3 frame somewhat and center it in the 16 × 9 screen, you get what is called a ***windowbox***. Windowboxing is usually done to save bandwidth (reduce the file size) during transmission. **SEE 15.10**

Interestingly enough, some programs make a virtue out of this unavoidable handicap. You may have seen MTV presentations or commercials that are letterboxed, with black borders at the top and the bottom of the screen. This is to imply that they were originally shot for wide-screen movie presentation rather than television distribution, which supposedly lends more prestige to the program. Also, many producers are quite pleased to have this additional screen space. They consider the side bars anything but "dead" space and fill it with additional program information and advertisements. The side bars are also a time-saver: often the previous show's credits are shown on a side bar while the new program segment is already under way. We as viewers seem to accept quite readily the stretching or fattening effect of digital manipulation.

INFORMATION DENSITY AND READABILITY

Just as with overcrowded Web pages, there is a tendency to load the screen with a great amount of information. And in our quest to squeeze as much information as possible onto the relatively small television screen, the print used for on-the-air copy gets smaller and smaller. Unless you have a large, high-resolution monitor, such information is virtually unreadable.

Information Density

There is some justification for crowding the screen if the data simultaneously displayed are related and add relevant information. For example, if in a home-shopping program you show a close-up of an item and simultaneously display the retail price, the sale price, and the telephone number to call, you are providing the viewer with a valuable service. On the other hand, if you show a newscaster reading the news in one corner of the screen, display the weather report in another, run stock market numbers and sports scores across the top and the bottom, and show station logos and ads all at the same time, you run the risk of information overload in addition to excessive screen clutter. **SEE 15.11**

When the elements are properly arranged according to the principles of composition, however, such additional information can add significantly to the basic communication. With thoughtfully arranged secondary frames, we are less likely to be overwhelmed and can pick and choose among the information presented. **SEE 15.12**

Readability

In television graphics readability means that you should be able to read the words that appear on-screen quickly and without straining. As obvious as this statement is, it seems to have eluded many a graphic artist. Sometimes titles explode onto and disappear from the screen so quickly that only video game champs and people with superior perceptive abilities can actually see and make sense of them;

15.11 SCREEN CLUTTER
This screen has so much unrelated information that it is difficult to make sense of it amid the visual clutter.

15.12 PROPER STRUCTURE OF MULTIPLE SCREEN ELEMENTS
The arrangement of these secondary frames and information areas makes it relatively easy to seek out the desired information.

15.12 PHOTO BY EDWARD AIONA

or the letters are so small and detailed that you can't read them without a magnifying glass.

Such readability problems occur regularly when motion picture credits are shown in a traditional 4 × 3 aspect ratio. First, as already pointed out, the titles generally extend beyond the essential area, so you can see only parts of them. Second, the credit lines are so small that they are usually impossible to read even on a relatively large television screen. Third, the letters themselves are not bold enough to show up well on television, especially if the background is busy. These problems are greatly minimized on a large HDTV screen but are greatly magnified when watching a small cell-phone display. In consideration of the many viewers watching standard television and the ever-growing number of mobile media users, you need to aim for a high degree of readability. What, then, makes for optimal readability? Here are some recommendations.

- Keep all written information within the essential area.
- Choose fonts that have a bold, clean contour. Thin-lined fonts are hard to read, even on an HDTV screen. Also, the fine strokes and serifs of the letters are susceptible to breakup when keyed. Sometimes even bold, sans serif fonts can get lost in the background and need to be reinforced with a drop shadow or color outline.
- Limit the amount of information. The less information that appears on-screen, the easier it is to comprehend. Some television experts suggest a maximum of seven lines per page. It is more sensible to prepare a series of titles on several C.G. "pages," each displaying a small amount of text, than a single page with an overabundance of information. As you well know from text messaging, even brief messages gain by their own visual shorthand. When programming for the small screen of mobile media, keep the words short and the letters bold.
- Format all lettering into blocks for easily perceivable graphical units. **SEE 15.13** This block layout is often used

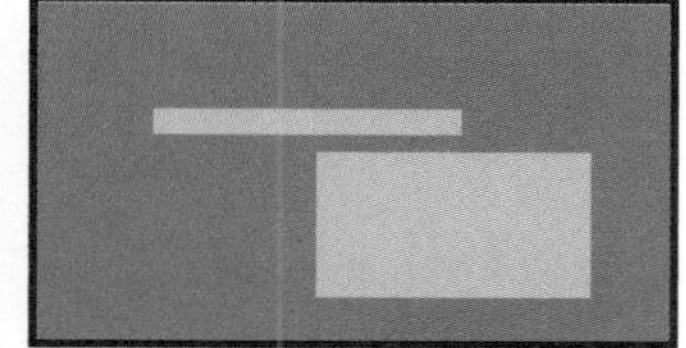

15.13 BLOCK ORGANIZATION OF TITLES
When titles are arranged in blocks, related information is graphically organized for easy perception.

15.14 SCATTERED TITLES
When titles are scattered, the information is difficult to read.

15.15 PHOTO BY HERBERT ZETTL

15.15 BOLD LETTERS OVER A BUSY BACKGROUND
This title reads well despite the busy background. The letters are bold and differ sufficiently in brightness from the background.

in well-designed Web pages. If the titles are scattered, they look unbalanced and are hard to read. **SEE 15.14** Scattered information is a typical characteristic of a poorly designed Web page, where bits of text, little secondary frames, and pop-ups randomly drop onto the screen.

- Avoid keying lettering into a busy background. If you must add lettering, select a simple, bold font. **SEE 15.15**

The same principles apply when you animate a title using special effects. In fact, if the title twists and tumbles around the screen, the letters must be even more legible than if they were used for a static, straightforward title.

Bear in mind that whenever you use printed material as on-air graphics, including reproductions of famous paintings, professional photographs, illustrated books, and similar matter, you must obtain copyright clearance (see section 2.2). If you have subscribed to a stock image service, ask whether it holds the copyright to those images. Your copyright limits normally depend on the amount of user fees you pay.

COLOR

You read in chapter 6 about the attributes of color and the technical aspects of additive color mixing. In general, the difference between HDTV and HDV (high-definition video) is not the relative sharpness of the picture—they look pretty much the same—but the fidelity of the color. Because color is an important design element, you need to learn more about its aesthetics—that is, how various colors influence design and how the television system reacts to them.

Color Aesthetics

The recognition and the application of color harmony cannot be explained in a short paragraph; they require experience, practice, sensitivity, and taste. Rather than try to dictate which colors go with what other colors, you can more easily divide the colors into "high-energy" and "low-energy" and then balance their energies.

The high-energy colors include bright, highly saturated hues, such as rich reds, yellows, and blues. The low-energy group contains more subtle hues with a low degree of saturation, such as pastels. Normally, you should keep the colors of the background low-energy and the colors in foreground high-energy. In a set (as in your home), the background (walls) is usually less colorful than the set pieces and dressings, such as rugs, sofas, pictures, and pillows. **SEE 15.16 AND 15.17** Titles work on the same principle: you will find that an easily readable title has high-energy lettering on a low-energy background.

Of course, the colors must also be appropriate for the event. For example, if the titles are intended to announce a high-energy show, such as a vivacious dance program, high-energy colors are certainly fitting. If, however, you use the same high-energy colors to introduce a discussion on the perils and the benefits of atomic energy, the choice is inappropriate even if the title has good readability.

Independent of aesthetics, only top-of-the-line television cameras can handle highly saturated reds. Unless there is an abundance of baselight, the video camera "sees red" when looking at red—at best, distorting the red color or, in some cases, making red areas in the shot vibrate (excessive video noise) or bleed into adjacent areas. This color bleeding is not unlike the bleeding of one sound track into another. When working with lesser-quality camcorders,

15.16 PHOTOS BY EDWARD AIONA (KITES), HERBERT ZETTL (FLOWER), AND STEVE RENICK (PLACE SETTING)

15.16 HIGH-ENERGY COLORS
The energy of a color is determined mainly by its saturation. High-energy colors are highly saturated hues, usually at the red and yellow end of the spectrum. They are especially effective when set against a low-energy background.

15.17 PHOTOS BY EDWARD AIONA

15.17 LOW-ENERGY COLORS
Low-energy colors are desaturated hues. Most pastels are low-energy.

suggest that the talent not wear highly saturated red clothing and that scene designers not paint large areas with saturated reds. This problem becomes especially noticeable in electronic field production (EFP), where you generally work in less-than-optimal lighting conditions.

The new HDTV systems not only produce superior resolution (picture sharpness) but their brightness range is considerably larger, one of the major factors in achieving the "film look" of HDTV. Like film, HDTV can display many more subtle steps of gray between TV white and TV black (see figure 6.24). Note that the middle value of the grayscale is not 30 percent but 18 percent. This means that you need more light to get from the very bright step 3 to an even brighter step 2 than when moving from the darker 8 to the slightly lighter 7. This principle is the same as using a flashlight at night and in daylight. At night a flashlight can illuminate a room; in bright sunlight you need to look at the lamp to tell whether it is even turned on.

Compatible color Technically, ***color compatibility*** means that a graphic is equally readable on a black-and-white television set as on a color one. In production it simply means that a color picture shows up well and with distinct grayscale (brightness) steps and contrast between the brightest and darkest part of the graphic. You may want to revisit the discussion of the grayscale in section 6.2.

For titles there should be a significant brightness contrast between the background and the foreground lettering. When you use exclusively high-energy colors for a title, such as red lettering on a green or blue background, the difference in hue is so obvious that you might be tempted to neglect the grayscale difference. As different as they seem on a color monitor, if they have the same brightness they are unreadable on a monochrome monitor. **SEE 15.18**

Even if the colors you use are not intended for reproduction on a black-and-white television set, good brightness contrast is also important for color rendition. It aids the picture's resolution and three-dimensionality and helps separate the colors. With a little experience, you will find that just by squinting your eyes while looking at the set, you can determine fairly well whether two colors have enough brightness contrast to ensure compatibility.

STYLE

Style, like language, is nonstatic. It changes according to the specific aesthetic demands of a given time and location. To ignore it means to communicate less effectively. You learn

15.18 TITLE WITH INCOMPATIBLE COLORS
Although these colors show up prominently on a color monitor, their brightness is almost the same, so they are illegible on a black-and-white monitor.

style not from a book but primarily through being sensitive to your environment—by experiencing life with open eyes and ears and, especially, an open heart. The way you dress now compared with the way you dressed 10 years ago is an example of a change in style. Some people not only sense the prevailing style but also manage to enhance it with a personal, distinctive flair.

Sometimes it is the development of television equipment that influences presentation styles more than personal creativity or social need. As emphasized in chapter 14, DVE equipment contributed not only to a new graphical awareness but also to an abuse of style. Often animated titles are generated not to reflect the prevailing aesthetic taste or to signal the nature of the upcoming show but simply because it is fun to see letters dance on-screen. Although flashy graphics in news may be tolerated because they express and intensify the urgency of the message, they are inappropriate for shows that explore a natural disaster or plays that delve into an intense relationship between two people.

You may have noticed that contemporary television graphics are imitating the colors and the layout of computer Web pages. Some television graphics even parrot the shortcomings of the computer image, such as the aliasing ("jaggies") of diagonals or curves in lines and letters, differently colored horizontal strips that contain lettering and small product icons, or the scattering of tiny secondary frames on the main television screen (see figure 15.11). One of the reasons for such emulation is to be hip and on the cutting edge. More often than not, however, such screen clutter works against a distinctive style and, unfortunately, against effective communication.

Regardless of whether you are a trendsetter, you should try to match the style of the artwork with that of the show. But do not go overboard and identify your guest from China with Chinese-like lettering or your news story about the devastating flood with titles that bob across the screen. Do not abandon good taste for effect. In a successful design, all images and objects interrelate and harmonize with one another—from the largest, such as the background scenery, to the smallest, such as the fruit bowl on the table. Good design displays a continuity and a coherence of style.

SYNTHETIC IMAGES

Synthetic images refer to pictures that are created entirely with the computer. Most desktop imaging software offers millions of different colors, thin and thick lines, shapes, and various brush strokes and textures for creating digital art. A television weathercast is a good example of the many capabilities of a large-scale ***graphics generator***. The lettering, basic geographic map, temperature zones and numbers, high- and low-pressure zones, symbols for sunshine and precipitation, and moving clouds and falling rain—all are generated and animated by the digital graphics generator. **SEE 15.19**

Depending on the storage capacity and the sophistication of the software, you can create and store complex graphical sequences, such as clips of animated three-dimensional titles that unfold within an animated 3D environment, or multilevel mattes that twist within a 3D video space.[3] **SEE 15.20 AND 15.21**

Some computer programs, based on complex mathematical formulas, allow you to paint irregular shapes, called fractals, which are used to create realistic and fantasy landscapes and countless abstract patterns. **SEE 15.22**

3. *Zettl's VideoLab* has many 2D and 3D animated renderings, which you can control interactively.

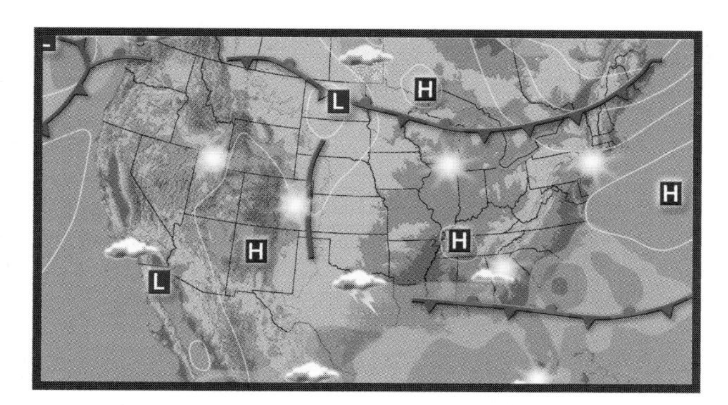

15.19 DIGITALLY GENERATED WEATHER MAP
This weather map is entirely computer-generated. Some of the multilevel effects are animated.

15.20 PHOTOS COURTESY VIZRT

15.20 3D RENDERING
Software specifically for graphics generators can create a variety of three-dimensional titles and moving images.

15.21 PHOTOS COURTESY VIZRT

15.21 DIGITAL RENDERING: PIAZZA SAN PIETRO
This animated fly-by sequence shows the Piazza San Pietro in Vatican City from various points of view.

15.22 FRACTAL LANDSCAPE
Most painting software allows you to "paint" irregular images using mathematical formulas.

MAIN POINTS

- Design is an overall concept that includes such elements as the fonts for titles, the station logo, the look of the news set, and even the office furniture.
- The major purposes of television graphics are to give you specific information, to tell you something about the nature of the event, and to grab your attention.
- The scanning area is what the camera viewfinder and the preview monitor show. The essential, or safe title, area is the portion seen by the viewer regardless of transmission loss or slight misalignment of the receiver.
- Out-of-aspect-ratio graphics need special consideration to make them fit the STV (standard television) or HDTV (high-definition television) screen.
- Making the entire frame of a wide-screen movie fit into the 4 × 3 aspect ratio of STV results in empty (black) space at the top and the bottom of the frame. The resulting horizontal aspect ratio is called a letterbox.
- To show a standard 4 × 3 television frame on the 16 × 9 screen, empty dead zones, or side bars, are added on both sides of the frame. The resulting vertical aspect ratio is called a pillarbox.
- Centering a slightly reduced 4 × 3 frame in a 16 × 9 screen, with empty space on all four sides of the 4 × 3 image, is called a windowbox.
- To avoid information overload when showing unrelated information simultaneously on a single screen, arrange the elements in easy-to-read secondary frames or text blocks.
- Good readability results when the written information is within the essential area, the letters are relatively large and have a clean contour, the background is not too busy, and there is good color and brightness contrast between the lettering and the background.
- Color compatibility means that the color image translates into distinct brightness values (grayscale steps) when seen on a monochrome receiver. Most STV systems reproduce at best nine separate brightness steps, ranging from TV white to TV black. HDTV delivers a much wider grayscale.
- Synthetic images are generated entirely by computer graphics. They can be still or animated.

SECTION

15.2

Scenery and Props

Although you may never be called upon to design or build scenery, you will likely set up scenery in the studio or rearrange an interior at a remote location. Setting up even a small interview set requires that you know what the pieces of scenery are called and how to read a floor plan. Your ability to see an existing on-location interior as a "set" will not only speed up camera placement and lighting but also help you determine if it needs redecorating for maximally effective camera shots. Knowing how to manage studio space through scenery and properties will also help you structure screen space in general.

- **TELEVISION SCENERY**
  Standard set units, hanging units, platforms and wagons, and set pieces
- **PROPERTIES AND SET DRESSINGS**
  Stage props, set dressings, hand props, and the prop list
- **ELEMENTS OF SCENE DESIGN**
  Floor plan, set backgrounds, and virtual sets

TELEVISION SCENERY

Because the television camera looks at a set both at close range and from a distance, scenery must be detailed enough to appear realistic yet plain enough to prevent cluttered pictures. Regardless of whether it's a simple interview set or a realistic living room, a set should allow for optimal camera angles and movement, appropriate lighting, microphone placement and occasionally boom movement, and maximum action by performers. Fulfilling these requirements are four types of scenery: standard set units, hanging units, platforms and wagons, and set pieces.

Standard Set Units

Standard set units consist of softwall and hardwall ***flats*** and a variety of set modules. Both are used to simulate interior or exterior walls. Although television stations and nonbroadcast production houses use hardwall scenery almost exclusively, softwall scenery is more practical for rehearsals as well as for high-school and college television operations.

Softwall flats The flats for standard softwall set units are constructed of a lightweight wood frame covered with muslin or canvas. They have a uniform height but various widths. The height is usually 10 feet (about 3 meters) or 8 feet (about 2½ meters) for small sets or studios with low ceilings. Width ranges from 1 to 5 feet (30 centimeters to 1½ meters). When two or three flats are hinged together, they are called twofolds (also called a book) or threefolds. Flats are supported by jacks—wood braces that are hinged or clamped to the flats and weighted down with sandbags or metal weights. SEE 15.23

Softwall scenery has numerous advantages: it is relatively inexpensive to construct and can usually be done in the scene shops of theater departments; it lends itself to a great variety of set backgrounds; it is easy to move and store; it is simple to set up, brace, and strike; and it is relatively easy to maintain and repair. The problems with softwall scenery are that it is difficult to hang pictures on the flats, and they often shake when someone closes a door or a window on the set or when something brushes against them.

Hardwall flats Hardwall flats are much sturdier than softwall flats and are preferred for more ambitious television productions. Hardwall scenery does have a few drawbacks: hardwall units do not always conform to the standard set dimensions of softwall scenery, and the flats are heavy and difficult to store. (In the interest of your—and the flats'—well-being, do not try to move hardwall scenery by yourself.) Hardwall flats also reflect sound more readily than do softwall flats, which can interfere with good audio pickup. For example, if a set design requires that two hardwall flats stand opposite and in close proximity to each other, the talent operating in this space will most likely sound as though they are speaking inside a barrel.

Most hardwall scenery is built for specific shows—such as newscasts, interview areas, crime labs, police stations, or soap operas—and remains set up for the length of the series. Carefully constructed hardwall scenery is a must

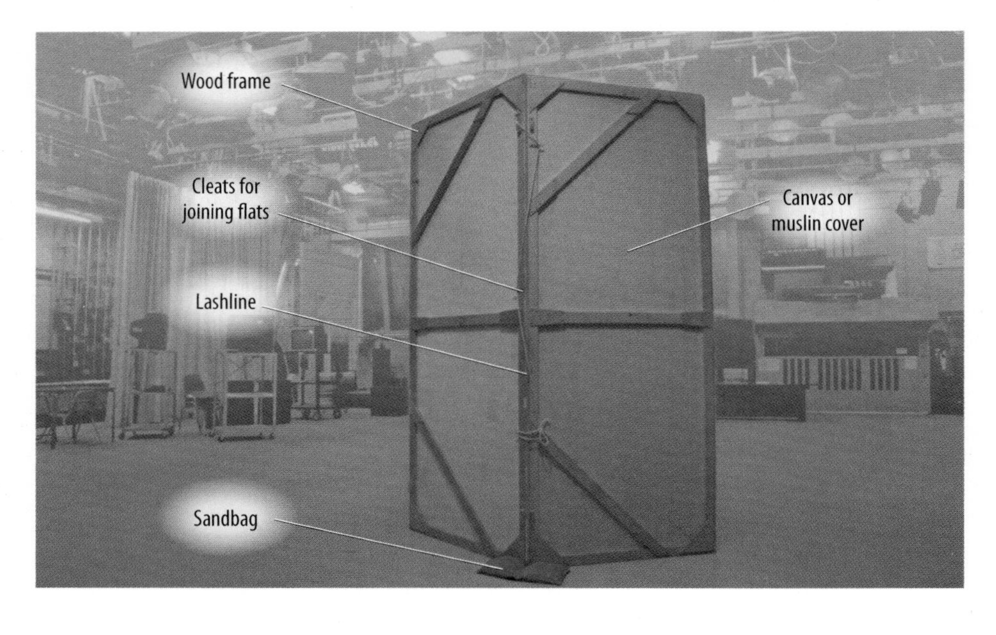

15.23 SOFTWALL FLATS
Softwall flats consist of a wood frame covered with muslin or canvas.

15.23 PHOTO BY HERBERT ZETTL

15.24 HARDWALL SET
This set was constructed with hardwall flats for a specific television drama. Note the specific set props that give the set its character.

15.24 PHOTO COURTESY BROADCAST AND ELECTRONIC COMMUNICATION ARTS DEPARTMENT AT SAN FRANCISCO STATE UNIVERSITY

for HDTV or any other form of digital television that has a higher picture resolution than STV. SEE 15.24

Set modules For small television stations or educational institutions, where you do not have the luxury of building new sets for every show, you may consider versatile set modules that can be used in a variety of configurations. A set module is a series of flats and three-dimensional set pieces whose dimensions match, whether they are used vertically (right side up), horizontally (on their sides), or in combinations. A wide variety of set modules are commercially available.

Hanging Units

Whereas flats stand on the studio floor, hanging units are supported from overhead tracks, the lighting grid, or lighting battens. They include the cyclorama, drops, and drapes and curtains.

15.25 MUSLIN CYCLORAMA
The muslin cyc runs on overhead tracks and normally covers three sides of the studio.

15.25 PHOTO BY EDWARD AIONA

Cyclorama The most versatile hanging background is a cyclorama, or cyc, a continuous piece of muslin or canvas stretched along two, three, and sometimes even all four studio walls. **SEE 15.25**

A fairly light color (light gray or beige) is more advantageous than a dark cyc. You can always make a light cyc dark by keeping the light off it, and you can colorize it easily using floodlights (scoops or softlights) with color gels attached. A dark cyc will let you do neither; at best it would take a great amount of light to make a dark cyc brighter, but this would still not allow you to colorize it. Some studios have hardwall cycs, which are not actually hanging units but are built solidly against the studio wall. **SEE 15.26**

Drops A chroma-key drop is a wide roll of chroma-key blue or green material that can be pulled down and even stretched over part of the studio floor for chroma keying.

You can make a simple and inexpensive drop by suspending a roll of seamless paper (9 feet wide by 36 feet long), which comes in different colors. Seamless paper hung from a row of flats provides a continuous cyclike background. Simply roll it sideways and staple the top edge to the flats. You can paint some design on it, or some stylized landscape or cityscape as background for a music group or dance. It also lends itself well for cookie projections. **SEE 15.27**

15.26 HARDWALL CYC
This cyc is made of hardwall material and is permanently installed on two sides of the studio.

15.26 PHOTO BY HERBERT ZETTL

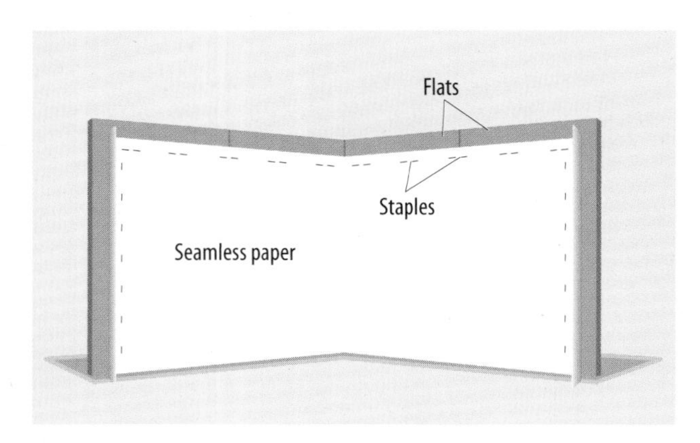

15.27 SEAMLESS PAPER DROP
A simple yet effective backdrop can be constructed by hanging a roll of seamless paper from a row of flats.

Drapes and curtains Stay away from detailed patterns and fine stripes when choosing drapes. Unless you shoot with high-end HDTV cameras, fine patterns tend to look smudgy, and contrasting stripes often cause moiré interference. Drapes are usually stapled to 1 × 3 battens and hung from the tops of the flats. Most curtains should be translucent enough to let the back light come through without revealing scenic pieces that may be behind the set.

Platforms and Wagons

Platforms are elevation devices, typically 6 or 12 inches (roughly 15 or 30 centimeters) high, that can be stacked. If you use a platform for interviews, you may want to cover it with carpeting. This cover not only will look good on-camera but will also absorb the hollow sounds of people moving on the platform. You can further dampen this sound by filling the platform interior with foam rubber.

Some 6-inch platforms have four casters so that they can be moved around. Such platforms are called wagons. **SEE 15.28** You can mount a portion of a set, or even a whole set, on a series of wagons and, if the doors are big enough, move these sections with relative ease in and out of the studio. Once in place, wagons should be secured with wood wedges or sandbags so that they do not move unexpectedly.

Larger platforms and hardwall scenery are often supported by a slotted-steel frame, which works like a big erector set.

Set Pieces

Set pieces are important scenic elements. They consist of freestanding three-dimensional objects, such as pillars, pylons (which look like thin, three-sided pillars), sweeps (curved pieces of scenery), folding screens, steps, and periaktoi (pronounced "pear-ee-*ack*-toy"; plural for *periaktos* but also used in the singular), a three-sided standing unit that looks like a large pylon. Most periaktoi move and swivel on casters and are painted differently on each side to allow for quick scene changes. **SEE 15.29**

There are numerous advantages to using set pieces: you can move them easily, they are self-supporting, and they quickly and easily establish three-dimensional space. Although set pieces are freestanding and self-supporting (which are, after all, their major advantages), always check

15.28 PLATFORMS AND WAGONS
Platforms are usually 6 or 12 inches high. When equipped with sturdy casters, they are called wagons.

15.28 PHOTOS BY HERBERT ZETTL

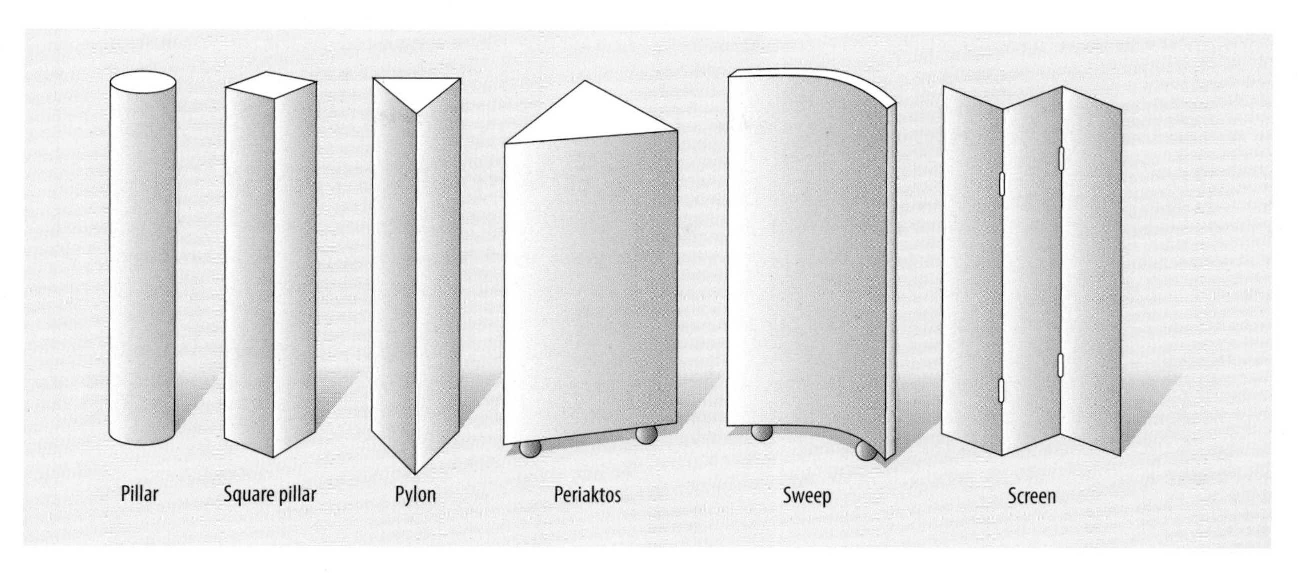

15.29 SET PIECES
Set pieces are freestanding scenic elements that roll on casters for quick and easy repositioning.

whether they need additional bracing. At a minimum they must be able to withstand bumps by people and cameras.

As a general rule, *it is always better to overbrace than to underbrace a set.* Do not forsake safety for convenience or speed.

PROPERTIES AND SET DRESSINGS

After having struggled with softwall and hardwall scenery, you will find that it is really the properties and the set dressings that give the environment a specific look and style. Much like decorating your room, it is primarily the furniture and what you hang on the walls that distinguish a particular environment, rather than the walls themselves. Because good television has more close-ups than medium and long shots, the three types of ***props***—stage props, set dressings, and hand props—must be realistic enough to withstand the close scrutiny of the camera.

Stage Props

Stage props include common furniture and items constructed for a specific purpose, such as news desks, panel tables, and chairs. You should also have enough furniture to create settings for a modern living room, a study, an office, a comfortable interview area, and perhaps some type of outdoor area with a patio table and chairs. For an interview set, relatively simple chairs are more useful than large, upholstered ones, even if you see such oversized chairs occasionally used for network interviews. You don't want the chairs to take on more prominence than the people sitting on them. Try to get chairs and couches that are not too low so that sitting and rising gracefully is not problematic, especially for tall people.

Set Dressings

Set dressings are a major factor in determining the style and the character of a set. Although the flats may remain the same from one show to another, the dressings help give each set individual character. They include such items as draperies, pictures, lamps and chandeliers, fireplaces, flowerpots, plants, candleholders, and artwork. Secondhand stores provide an unlimited source for these things. In an emergency you can always raid your own home or a friend's office.

Hand Props

Hand properties consist of all items that are actually handled by the talent during a show. They include dishes, silverware, telephones, radios, and desktop computers. In television the hand props must be realistic: use only real objects. A papier-mâché chalice may look regal and impressive on the stage, but on the television screen it looks dishonest if not ridiculous. Television is very dependent on human action. Think of hand props as extensions of gestures. If you want the actions to be sincere and genuine, the extension of them must be real as well. If an actor is supposed to carry a heavy suitcase, make sure the suitcase is actually heavy. Pretending that it is heavy does not go over well on television.

If you must use food, check carefully that it is fresh and that the dishes and the silverware are meticulously clean. Liquor is generally replaced with water (for clear spirits), tea (for whiskey), or fruit juice (for red wine). With all due respect for realism, such substitutions are perfectly appropriate.

As obvious as it sounds, see to it that hand props actually work and that they are on the set for the performers to use. A missing prop or a bottle that doesn't open at the right time may cause costly production delays.

Prop List

In small routine productions, the floor manager or a member of the floor crew normally takes care of the props. For elaborate productions, however, have a person assigned exclusively to the handling of props—the property manager. To procure the props and ensure that they are available at the camera rehearsal and recording sessions, you need to prepare a prop list. SEE 15.30

If you need to strike a set and set it up again for subsequent recording sessions, mark all the props and take digital photos of the set before putting the props away. This way you will have an instant record of what props were used and where they were placed on the set. A missing prop or one that is in a different location for the next recording session can create a serious continuity problem for the editor.

five outside bushes	low square cabinet
two rubber plants	6' blue couch
tall cactus	set of eight family photos
transparent curtains	sunflower painting
low 8' cabinet	Picasso print
square end table	magazines
round end table	newspaper
two lamps	books
two bookcases	stereo
chair (with armrests)	tea set
blue wing chair	Indian sculpture
coffee table	folding screen
fruit bowl	

15.30 PROP LIST
This prop list contains all set props, set dressings, and hand props shown in the set in figure 15.32.

ELEMENTS OF SCENE DESIGN

Even if you are not designing a set yourself, you must know enough about scene design to tell the art director what you want. This requires that you know what the show is all about.

You arrive at a concept for set design by defining the necessary spatial environment for optimal communication rather than by copying what you see on the air. For example, you may feel that the best way to inform viewers is not by having an authoritative newscaster read stories from a pulpitlike contraption but by moving the cameras into the newsroom itself or out into the street where events are happening. If the show is intended to be shot with a single camera for heavy postproduction editing, it may be easier to take the camera to the street corner rather than to re-create the street corner in the studio.

But even if the show is slated for studio production, you can often streamline the set design by taking some time to visualize the entire show in screen images and work from there. For example, even if the interview guest is a famous defense lawyer, you don't automatically have to set up a typical lawyer's office complete with antique desk, leather chairs, and law books in the background. Ask about the nature of the interview and its intended communication objective. This is where a clear process message proves immensely helpful. If, for example, the process message is to probe the conscience and the feelings of the defense lawyer rather than hear about future defense strategies, you could visualize the entire interview being conducted in intimate close-ups of the guest. Does this interview require an elaborate lawyer's set? Not at all. Considering the shooting style that includes a majority of tight close-ups, two comfortable chairs in front of a simple background will do just fine.

When the show is sketched out on a detailed storyboard, the set design is frequently predetermined. Nevertheless, speak to the producer and the director if you think you have a much better idea.

There are three major elements of scene design: the floor plan, set backgrounds, and virtual sets.

Floor Plan

A set design is drawn on the ***floor plan pattern,*** which is literally a plan of the studio floor. It shows the floor area, the main studio doors, the location of the control room, and the studio walls. The lighting grid or batten locations are normally drawn on the floor area to give a specific orientation pattern according to which the sets can be placed. In effect, the grid resembles the orientation squares of a city map. **SEE 15.31**

The completed ***floor plan*** should convey enough information that the floor manager and the crew can put up the set and dress it, even in the absence of the director or set designer. You may find that both the floor plan pattern and the finished floor plan that shows the scenic design are called "floor plan."

The scale of the floor plan pattern varies, but it is normally ¼ inch = 1 foot. All scenery and set properties are drawn on the floor plan pattern in the proper position relative to the studio walls and the lighting grid. **SEE 15.32** For simple setups you may not need to draw the flats and the set properties to scale; you can approximate their size and placement relative to the grid.

Floor plan functions The floor plan is an important tool for all production and engineering personnel. The director uses it to visualize the show and to block the major actions of performers, cameras, and microphone booms. It is essential to the floor crew, who must set up the scenery and place the major properties. The LD (lighting director) needs it for designing the general light plot. The audio technician can become familiar with specific microphone placement and possible audio problems.

Set positioning Whenever possible, *try to locate the set where the lights are.* Position it so that the back lights, key lights, and fill lights hang in their approximate positions. Sometimes an inexperienced designer will place a set in a studio corner, where most of the lighting instruments have to be rehung for proper illumination, whereas in another part of the studio the same set could have been lighted with the instruments already in place.

As you can see once again, you cannot afford to specialize in a single aspect of television production. Everything interrelates, and the more you know about the various production techniques and functions, the better your coordination of those elements will be.

Problem areas When drawing a floor plan, watch for potential problems. Many times a carelessly drawn floor plan will indicate scenery backing, such as the walls of a living room, that is not wide enough to provide adequate cover for the furniture or other items placed in front of it. One way to avoid such design mistakes is to draw the in-scale furniture on the floor plan first, then add the flats for the backing. Limit the set design to the actual space available.

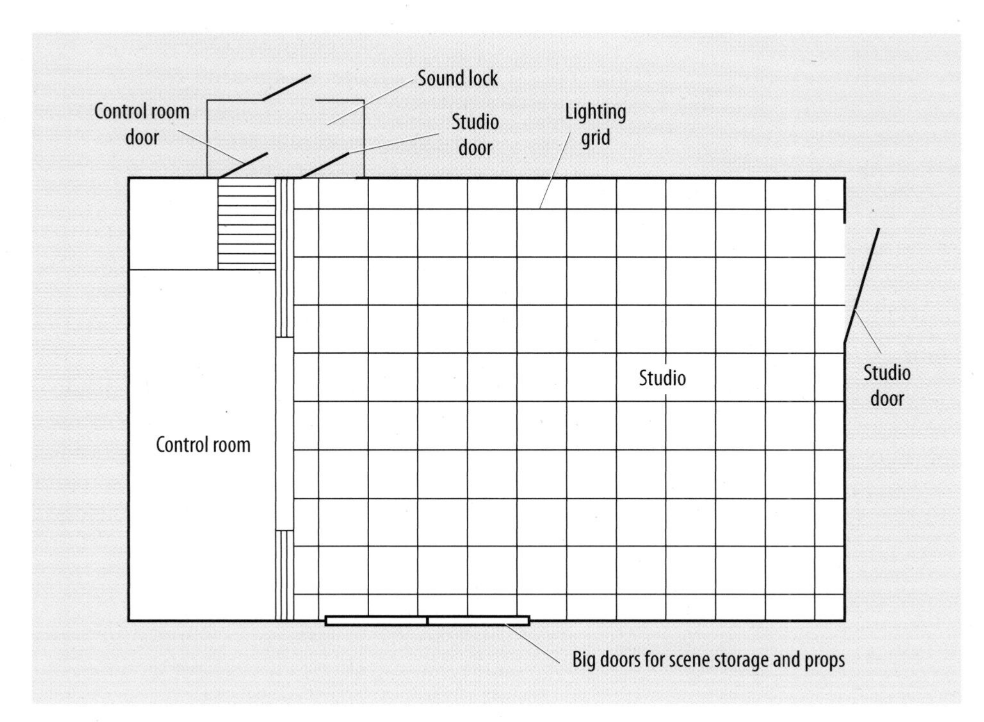

15.31 FLOOR PLAN PATTERN
The floor plan pattern shows the dimensions of the studio floor, which is further defined by the lighting grid or similar pattern. The set is drawn on this basic studio grid.

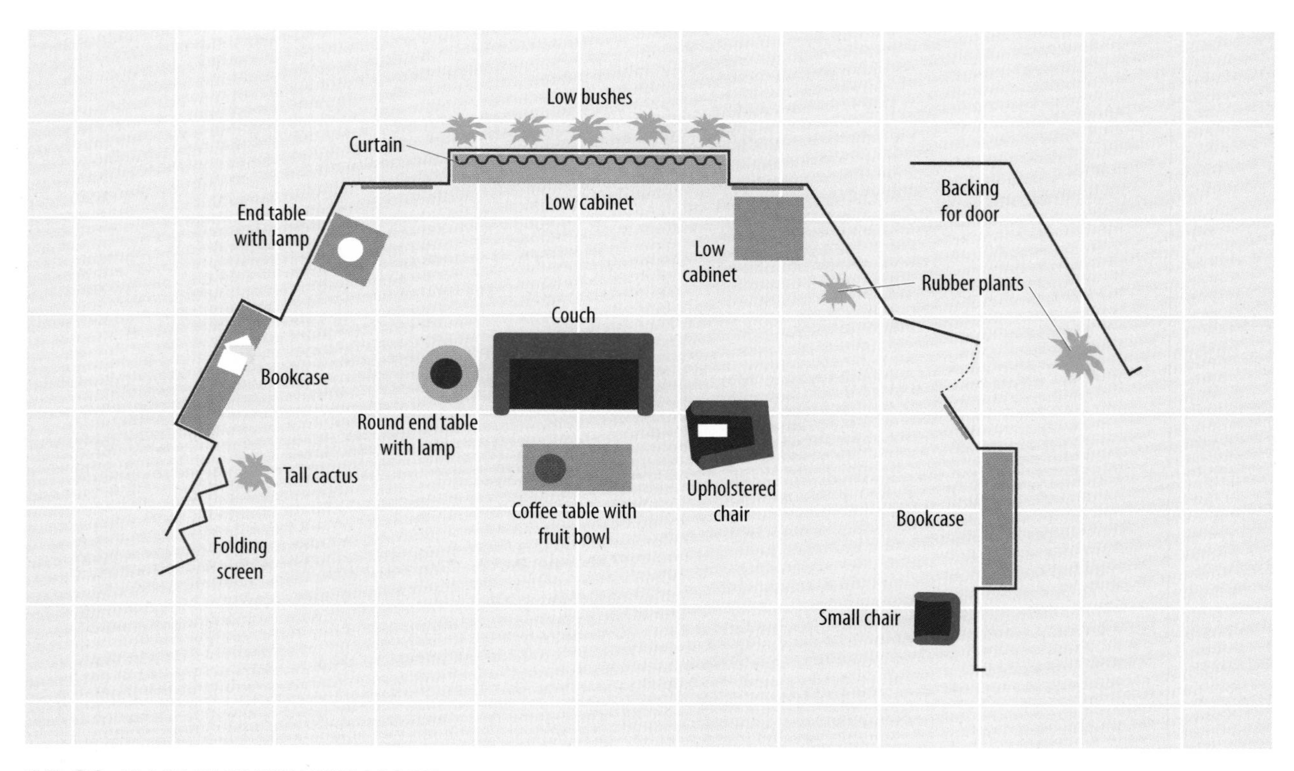

15.32 FLOOR PLAN FOR LIVING ROOM
This floor plan shows all the necessary scenery, set props, and dressings as well as the more prominent hand props. It is usually not drawn precisely to scale.

Always place active furniture (used by the performers) at least 6 feet (roughly 2 meters) from the set wall so that the back lights can be directed at the performance areas at not too steep an angle. Also, the director can use the space between wall and furniture for camera placement and talent movement.

Set Backgrounds

The background of a set helps unify a sequence of shots and places the action in a single continuous environment. It can also provide visual variety behind relatively static foreground action. Although set continuity is an important element in scene design, a plain background is not the most interesting scenic background. You need to "dress" the set by hanging artwork, posters, or other objects on the wall to break it up into smaller yet related areas. When you dress a plain background with pictures or other objects, place them so that they are in camera range. For example, if you hang a picture between two interview chairs, it will show only in the straight-on two-shot but not in the individual close-ups. If you want more background texture in the close-up shots, position pictures so that they are seen by the cameras during cross shooting. SEE 15.33

Virtual Sets

Virtual sets are computer-generated and keyed into a chroma-key area behind the talent, much like a weather map. Such a technique, however, rarely saves time and money. The virtual background must match the perspective of the actual foreground set pieces, and the lighting of the background must continue with the foreground shadow logic. If the key light of the foreground action is coming from the right, with the attached and cast shadows falling to the left, you can't have the key lights of the background coming from the left, causing the attached and cast shadows to fall to the right.

The major problem is actually for the talent, who have to work in an often disorienting blue or green environment. Parts of virtual sets, such as keying in ceilings and floors in postproduction, however, are an effective technique to skirt potentially difficult design problems.

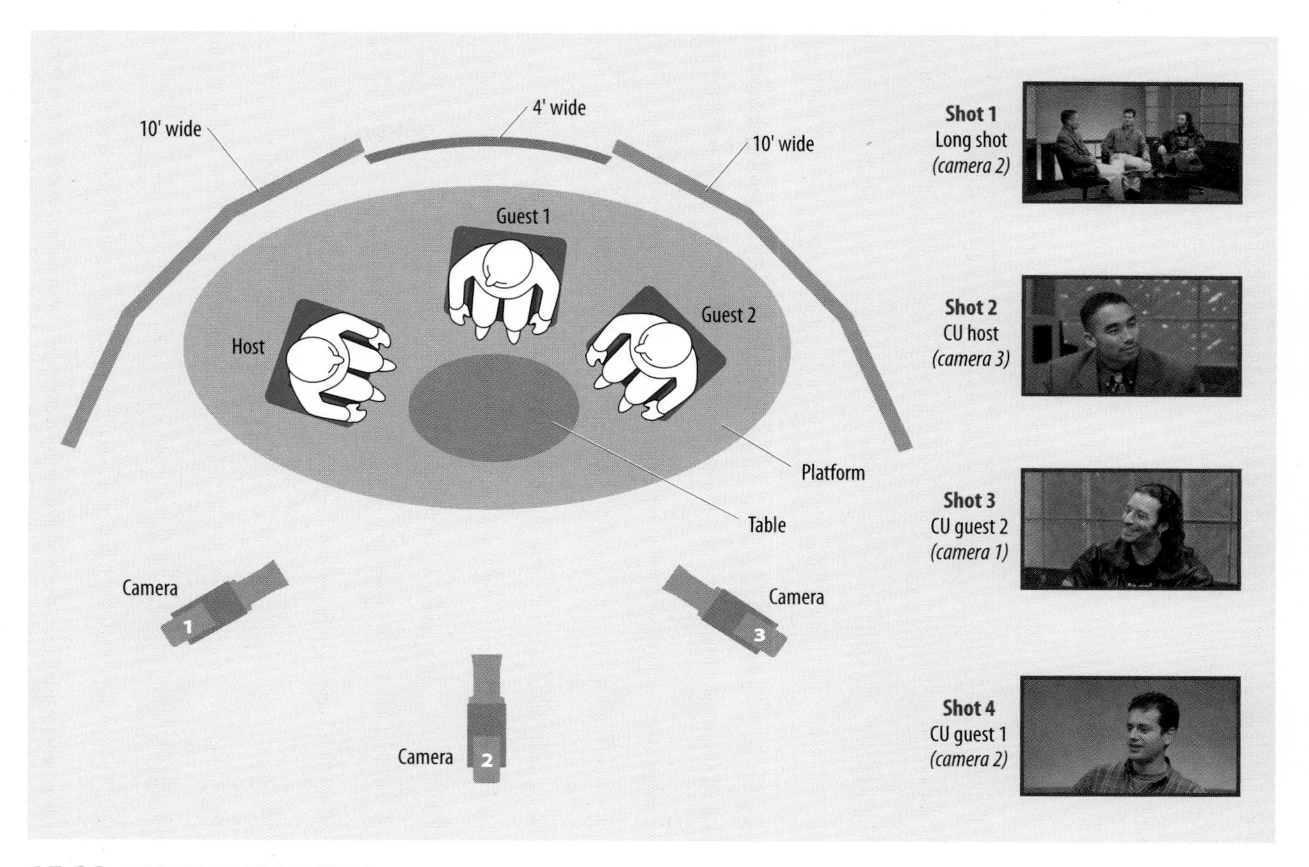

15.33 PHOTOS BY HERBERT ZETTL

15.33 BACKGROUND DRESSING
The establishing (long) shot of this interview set shows that the background flats provide some visual texture and interest for the host (camera-left) and guest 2 (far right), but not for guest 1 in the middle. The subsequent close-ups confirm this design problem.

MAIN POINTS

- Television scenery encompasses the three-dimensional aspects of design.
- There are four types of scenery: standard set units, that is, hardwall and softwall flats and set modules; hanging units, such as cycloramas (cycs), drops, and curtains; platforms and wagons; and set pieces, such as pillars, screens, and periaktoi.
- The three basic types of properties are stage props, such as furniture, news desks, and chairs; set dressings, such as pictures, draperies, and lamps; and hand props—items such as dishes, telephones, and computers that are actually handled by the talent.
- To procure the props and ensure that they are available at the camera rehearsal and recording sessions, you need to prepare a prop list.
- When a set must be struck and set up again for a subsequent video-recording session, take digital photos of all set details to ensure consistency of the setup.
- A floor plan is drawn on a floor plan pattern and shows the exact location of the scenery and the set properties relative to the lighting grid. The floor plan is essential for the director to prepare the preliminary blocking of talent, cameras, and microphone booms; for the floor crew to set up the scenery and place the major set properties; and for the lighting director to design the basic light plot.
- The background of a set helps unify a sequence of shots and places the action in a single continuous environment. It can also provide visual variety behind relatively static foreground action.
- Virtual sets are complicated to set up because they require that the foreground match the background perspective and lighting, including the logical placement of shadows. For the talent, they are difficult to move within. They are, however, quite successfully used to key in ceilings and floors.

CHAPTER

16

Television Talent

When you look at the people appearing regularly on television and talking to you—telling you what to buy, what is happening around the world, or what the weather is going to be like—you may feel that the job is not too difficult and that you could easily do it without too much sweat. After all, most of them are simply reading copy that appears on a teleprompter. But when you actually stand in front of the camera, you quickly learn that the job is not as easy as it looks. Appearing relaxed on-camera, and pretending that the camera lens or the teleprompter is a real person to whom you are talking, takes concentration and a good amount of talent and skill. This is why we call all people appearing regularly on television ***talent***. Although television talent may have varied communication objectives—some seek to entertain, educate, or inform; others seek to persuade, convince, or sell—all strive to communicate with the television audience as effectively as possible.

Section 16.1, Television Performers and Actors, concerns the major on-camera performance and acting techniques. Section 16.2, How to Do Makeup and What to Wear, briefly describes the makeup performers and actors use and what type of clothing looks good on-camera.

KEY TERMS

actor A person (male or female) who appears on-camera in dramatic roles. Actors always portray someone else.

blocking Carefully worked-out movement and actions for the talent and for all mobile television equipment.

cake A makeup base, or foundation makeup, usually water-soluble and applied with a sponge. Also called *pancake.*

cue card A large, hand-lettered card that contains copy, usually held next to the camera lens by floor personnel.

foundation A makeup base over which further makeup such as rouge and eye shadow is applied. For HDTV the foundation is usually sprayed on.

makeup Cosmetics used to enhance, correct, or change appearance.

pan-stick A foundation makeup with a grease base. Used to cover a beard shadow or prominent skin blemish.

performer A person who appears on-camera in nondramatic shows. Performers play themselves and do not assume someone else's character.

talent Collective name for all performers and actors who appear regularly on television.

teleprompter A prompting device that projects the moving (usually computer-generated) copy over the lens so that the talent can read it without losing eye contact with the viewer. Also called *auto-cue.*

SECTION

16.1

Television Performers and Actors

You can divide television talent into two categories: performers and actors. The difference between them is fairly clear-cut. Television ***performers*** are engaged basically in nondramatic activities: they play themselves and do not assume roles of other characters; they sell their own personalities to the audience. Television ***actors***, on the other hand, always portray someone else: they project a character's personality rather than their own, even if the character is modeled after their own experience or mannerisms. Their stories are always fictional.

Although there are distinct differences between television performers and television actors, the groups do share several functions. All talent communicate with the viewers through the television camera and must keep in mind the nuances of audio, movement, and timing. And all talent interact with other television personnel—the producer, the director, the floor manager, the camera operator, and the audio technician.

This section covers the basics of working in front of the camera.

▶ **PERFORMANCE TECHNIQUES**
Performer and camera, performer and audio, performer and timing, floor manager's cues, prompting devices, and maintaining continuity

▶ **ACTING TECHNIQUES**
Actor and audience, actor and blocking, memorizing lines, actor and timing, maintaining continuity, and the director/actor relationship

▶ **AUDITIONS**
Preparation, appearance, and creativity

PERFORMANCE TECHNIQUES

The performer speaks directly to the camera, plays host to television guests, and communicates with other performers or the studio audience; he or she is also fully aware of the presence of the television audience at home. This latter audience, however, is not the large, anonymous, and heterogeneous television audience that modern sociologists study. For the television performer, the audience is an individual or an intimate group who have gathered in front of a television set.

If you are a performer, try imagining your audience as a family of three, seated in their favorite room, about 10 feet away from you. With this picture in mind, you have no reason to scream at the "millions of viewers out there in videoland"; a more successful approach is to talk quietly and comfortably to the family who were kind enough to let you into their home.

When you assume the role of a television performer, the camera becomes your audience. You must adapt your performance techniques to its characteristics and to other production aspects such as audio and timing. In this section we discuss the performer and the camera, the performer and audio, the performer and timing, the floor manager's cues, prompting devices, and maintaining continuity.

Performer and Camera

The camera is not simply an inanimate piece of machinery; it sees everything you do or don't do. It sees how you look, move, sit, and stand—in short, how you behave in a variety of situations. At times it looks at you much more closely and with greater scrutiny than a polite person would ever dare to do. It reveals the nervous twitch of your mouth when you are ill at ease and the expression of mild panic when you have forgotten a name. The camera does not look away when you scratch your nose. It faithfully reflects your behavior in all pleasant and unpleasant details. As a television performer, you must carefully control your actions without letting the audience know that you are conscious of doing so.

Camera lens Because the camera represents your audience, you must look directly into the lens (or at the prompting device in front of it) whenever you intend to establish eye contact with the viewer. As a matter of fact, you should try to look *through* the lens, rather than at it, and keep eye contact much more than you would with an actual person. If you merely look at the lens instead of looking through it, or if you pretend that the camera operator is your audience and therefore glance away from the lens ever so slightly, you break the continuity and the intensity of the communication between you and the viewer; you break, however temporarily, television's magic.

Camera switching If two or more cameras are used, you must know which one is on the air so that you can remain in direct contact with the audience. When the director switches cameras, you must follow the floor manager's cue (or, when working with robotic cameras, the change of tally lights) quickly but smoothly. Do not jerk your head from one camera to the other. If you suddenly discover that you have been talking to the wrong one, look down as if to collect your thoughts and then casually look up and glance into the "hot" camera. Continue talking in that direction until you are again cued to the other camera. This method works especially well if you work from notes or a script, as in an interview or a newscast. You can always pretend to be looking at your notes when, in reality, you are changing your view from the wrong camera to the right one.

If the director has one camera on you in a medium shot (MS) and the other camera in a close-up (CU) of the object you are demonstrating, such as the guest's book during an interview, it is best to keep looking at the host (medium-shot) camera during the whole demonstration, even when the director switches to the close-up camera. You will not be caught looking the wrong way because only the medium-shot camera is focused on you. SEE 16.1 You will also find that it is easier to read the copy off a single teleprompter rather than switch from one to another in midsentence.

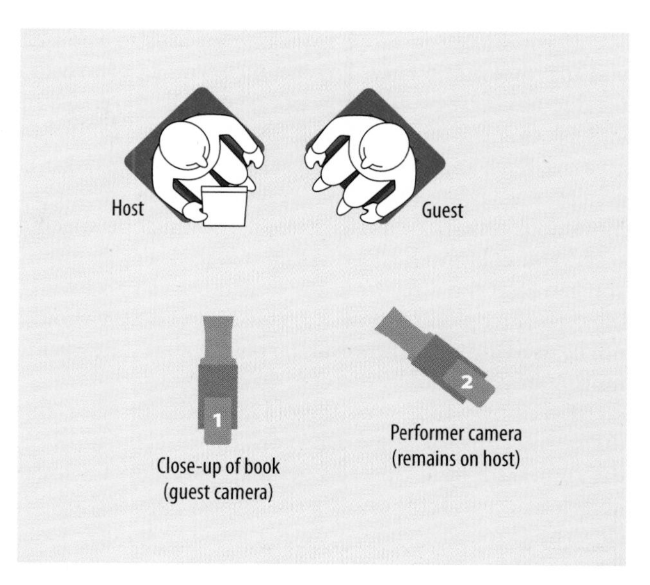

16.1 PERFORMER CAMERA
When one camera (camera 1) is on a close-up of the product (the book) and the other (camera 2) is on a medium shot of the host, the host should continue looking into camera 2 during the close-up.

Close-up techniques The tighter the shot, the harder it is for the camera to follow movement. If a camera is on a close-up, you must restrict your motions severely and move with great care. While you're performing a song, for example, the director may want to shoot very closely to intensify an especially emotional passage. Try to stand as still as possible, and do not wiggle your head. The close-up itself is intensification enough. All you have to do is sing well.

When demonstrating small objects on a close-up, hold them steady. If they are arranged on a table, do not pick them up. You can either point to them or tilt them a little to give the camera a better view. There is nothing more frustrating to the camera operator and the director than a performer who snatches the product off the table just when the camera has a good close-up of it. A quick look at the studio monitor usually tells you how to hold the object for maximum visibility on-screen. If two cameras are used, "cheat" (orient) the object somewhat toward the close-up camera, but do not turn it so much that it looks unnaturally distorted on the medium-shot camera.

Warning cues In most nondramatic shows—lectures, demonstrations, interviews, and the like—there is generally not enough time to work out a detailed blocking scheme. The director usually just walks the performers through some of the most important crossovers from one performance area to the other and through a few major actions, such as complicated demonstrations. During the on-the-air performance, you as a performer must therefore give the director and the studio crew visual and audible warnings of your unrehearsed actions. Before you stand up, for instance, first shift your weight and position your arms and legs; this signals the camera operator as well as the microphone boom operator to prepare for your move. If you pop up unexpectedly, the camera may stay in one position, focusing on the middle part of your body, which won't make for the most interesting shot to say the least, and the boom microphone will be clearly visible on-screen.

If you intend to move from one set area to another, you may use audio cues. For instance, you can warn the production crew by saying: "Let's go over to the children and ask them…" or "If you will follow me over to the lab area…." Such cues sound natural to the viewer, who is generally unaware of the fast reactions these seemingly innocuous remarks actually trigger. You must be specific when you cue unrehearsed visual material. For example, you can alert the director to the upcoming visuals by saying: "The first picture shows…" This cueing device should not be used too often, however. If you can alert the director more subtly yet equally directly, do so.

Do not try to convey the obvious. The director, not the talent, runs the show. Don't tell the director to bring the cameras a little closer to get a better view of a small object, especially if the director has already obtained a good close-up through a zoom-in. Also, avoid walking toward the camera to demonstrate an object. You may walk so close to the camera that it has to tilt up into the lights to keep your face in the shot or so close that the zoom lens can no longer focus. The zoom lens allows the camera to get to you much faster than you can get to the camera.

Performer and Audio

As a television performer, besides looking natural and relaxed, you must speak clearly and effectively; it rarely comes naturally. Don't be misled into believing that a resonant voice and affected pronunciation are the two prime requisites for a good announcer or other performer. On the contrary: first, you need to have something important to say; second, you need to say it with conviction and sincerity; third, you must speak clearly so that everyone can understand you.

Thorough training in television announcing is an important prerequisite for any performer. Most novices speak much too fast, as though they wanted to get through the on-camera torture as quickly as possible. Don't speed up or lower your volume too much when you come to the end of a sentence or paragraph. Professional performers do not drop the endings of sentences but maintain the same volume and energy from beginning to end. Take a deep breath and slow down. You will be amazed how much more relaxed you will be.

Microphone technique The following summarizes the main points about handling microphones and assisting the audio technician. (See chapter 9 for a detailed discussion of the basic microphone techniques.)

- Most often you will work with a lavalier microphone. Once it is properly fastened, you do not have to worry about it, especially if you are relatively stationary during the performance. If you have to move from one set area to another on-camera, watch that the mic cord does not get tangled up with the set or props. Gently pull the cable behind you to keep the tension off the mic itself. A wireless lavalier will enable you to move within the performance area without having to worry about a cable.

- When using a hand mic, check that you have enough cable for your planned actions. Speak across it, not into it. If you are interviewing someone in noisy surroundings, such as a downtown street, hold the microphone close to your mouth when you are talking, then point it toward the person as he or she responds to your questions.

- When working with a boom mic (including a handheld shotgun or one that is mounted on a fishpole), be aware of the boom movements without letting the audience know. Give the boom operator enough warning so that he or she can anticipate your movements. Move slowly so that the boom can follow. In particular, do not make fast turns because they involve a great amount of boom movement. If you have to turn fast, try not to speak until the boom has been repositioned.

- Do not move a desk mic unless it is pointing in the opposite direction. Even if the microphone is pointing more toward another performer than at you, it was probably placed that way by the audio technician to achieve better audio balance.

- In all cases, treat microphones gently. Mics are not intended to be hand props, to be tossed about or twirled by their cords like a lasso, even if you see such misuse occasionally in a rock performance.

Audio level A good audio technician will ask you for an audio level before you go on the air. Many performers have the bad habit of rapidly counting to 10 or mumbling and speaking softly while the level is being taken, then, when they go on the air, blasting their opening lines. If a level is taken, speak as loudly as you will in your opening remarks and for as long as required for the audio technician to adjust the volume to an optimal level.

Opening cue At the beginning of a show, all microphones are dead until the director gives the cue for audio. You must therefore wait until you receive the opening cue from the floor manager or through the I.F.B. (interruptible foldback, or feedback) system. If you speak beforehand, you will not be heard. Do not take your opening cue from the red tally lights on the cameras unless you are so instructed. When waiting for the opening cue, look into the camera that is coming up on you and not at the floor manager.

Performer and Timing

Live and live-recorded television operate on split-second timing. Although it is ultimately the director's responsibility to get the show on and off the air as scheduled, you as the performer have a great deal to do with successful timing. Aside from careful pacing throughout the show, you must learn how much program material you can cover after you have received a 3-minute, a 2-minute, a 1-minute, a 30-second, and a 15-second cue. You must, for example, still look comfortable and relaxed although you may have to cram a lot of material into the last minute while at the same time listening to the director's or producer's I.F.B. On the other hand, you must be prepared to fill an extra 30 seconds without appearing to grasp for words and things to do. This presence of mind, of course, is achieved through practical experience and cannot be learned solely from a textbook. It also requires that you are at least somewhat familiar with the news stories that you read off the teleprompter.

Floor Manager's Cues

Unless you are connected with the producer and the director via I.F.B., it is the floor manager who provides the link between the director and you, the performer.

With the use of robotic cameras, the floor manager may be the only other human being in the studio besides you. The floor manager can tell you whether your delivery is too slow or too fast, how much time you have left, and whether you are speaking loudly enough or holding an object correctly for a close-up.

Although stations and production houses may use slightly different cueing signals and procedures, they normally consist of time cues, directional cues, and audio cues. If you are working with an unfamiliar production crew, ask the floor manager to review the cues before you go on the air. **SEE 16.2**

React to each cue immediately, even if you think it is not appropriate at that particular time. The director would not give the cue if it were not necessary. A truly professional performer is not one who never needs cues but rather one who can react to all signals quickly and smoothly. Do not look nervously for the floor manager if you think you should have received a cue; he or she will find you and draw your attention to the signal. When you receive a cue, do not acknowledge it in any way; the floor manager will know whether you noticed it.

You will find that receiving and reacting to I.F.B. information during a performance is no easy task. We all know how difficult it can be to continue a telephone conversation when someone close by is trying to tell us what else to communicate to the other party. But when reporting news in the studio or in the field, such simultaneous communication is common. You must learn to listen carefully to the I.F.B. instructions of the director or producer without letting the audience know that you are listening to someone else while talking to them. Do not interrupt your communication with the audience when getting I.F.B. instructions, even if the transmission is less than perfect. If during a live remote you can't understand what is being said on the I.F.B. channel, however, you may have to stop your narration to tell the audience that you are getting some important information from your director. Listen carefully to the I.F.B. instructions, then go on with what you were saying. Try not to adjust your earpiece while on the air. If at all possible, wait until the camera cuts away from you to do an adjustment.

Prompting Devices

Prompting devices have become an essential production tool, especially for news and speeches. The audience has come to expect the newscaster to talk directly to them rather than read the news from a script, although we all know that the newscaster cannot possibly remember the entire news copy. We expect speakers to deliver copious and complicated information without having to think about what to say next. Prompting devices are also helpful to performers who fear they may suddenly forget their lines or who have had no time to memorize a script.

Prompting devices must be totally reliable, and the performer must be able to read the copy without appearing to lose eye contact with the viewer. Two devices have proved especially successful: cue cards and the teleprompter.

Cue cards Used for relatively short amounts of copy, ***cue cards*** are of many types, and the choice depends largely on what the performer is used to and what he or she prefers. Usually, they are fairly large posterboards on which the copy is hand-lettered with a felt-tipped marker. The size of the cards and the lettering depends on how well the performer can see and how far away the camera is.

16.2 FLOOR MANAGER'S CUES

The floor manager uses a set of standard hand signals to relay the director's commands to the on-the-air talent.

CUE	SIGNAL	MEANING	SIGNAL DESCRIPTION
TIME CUES			
Standby		Show about to start.	Extends hand above head.
Cue	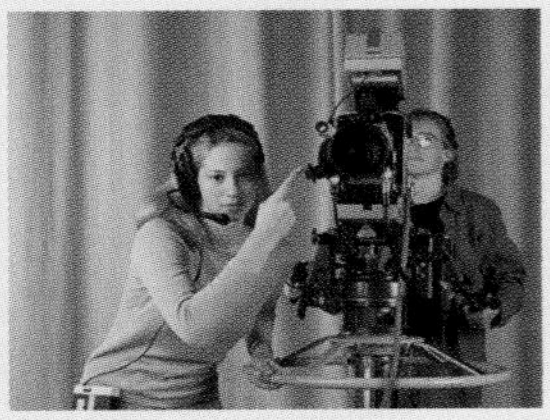	Show goes on the air.	Points to performer or live camera.
On time		Go ahead as planned (on the nose).	Touches nose with forefinger.
Speed up		Accelerate what you are doing. You are going too slowly.	Rotates hand clockwise with extended forefinger. Urgency of speed-up is indicated by fast or slow rotation.
Stretch	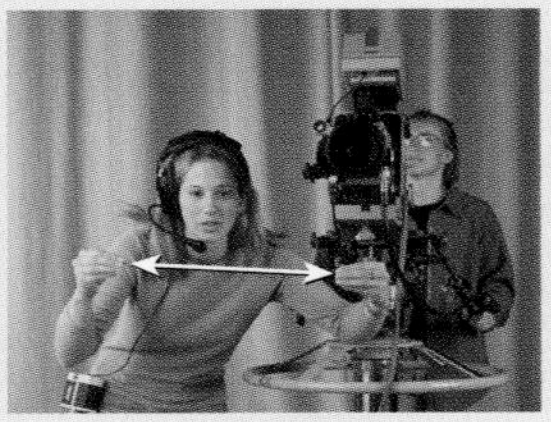	Slow down. Too much time left. Fill until emergency is over.	Stretches imaginary rubber band between hands.

16.2 PHOTOS BY EDWARD AIONA

16.2 FLOOR MANAGER'S CUES *(continued)*

CUE	SIGNAL	MEANING	SIGNAL DESCRIPTION
TIME CUES			
Wind up		Finish up what you are doing. Come to an end.	Similar motion to speed-up but usually with arm extended above head. Sometimes expressed with raised fist, good-bye wave, or hands rolling over each other as if wrapping a package.
Cut		Stop speech or action immediately.	Pulls index finger in knifelike motion across throat.
5 (4, 3, 2, 1) minute(s)		5 (4, 3, 2, 1) minute(s) left until end of show.	Holds up five (four, three, two, one) finger(s) or small card with number on it.
30 seconds (half minute)		30 seconds left in show.	Forms a cross with two index fingers or arms. Or holds card with number.
15 seconds		15 seconds left in show.	Shows fist (which can also mean wind up). Or holds card with number.
Roll VR (and countdown) 2, 1, take VR		Video recorder is rolling.	Holds extended left hand in front of face, moves right hand in cranking motion. Extends two, one finger(s); clenches fist or gives cut signal.

16.2 PHOTOS BY EDWARD AIONA

16.2 FLOOR MANAGER'S CUES *(continued)*

CUE	SIGNAL	MEANING	SIGNAL DESCRIPTION
DIRECTIONAL CUES			
Closer		Performer must come closer or bring object closer to camera.	Moves both hands toward self, palms in.
Back		Performer must step back or move object away from camera.	Uses both hands in pushing motion, palms out.
Walk		Performer must move to next performance area.	Makes a walking motion with index and middle fingers in direction of movement.
Stop		Stop right here. Do not move any more.	Extends both hands in front of body, palms out.
OK		Very well done. Stay right there. Do what you are doing.	Forms an O with thumb and forefinger, other fingers extended, motioning toward talent.

16.2 PHOTOS BY EDWARD AIONA

16.2 FLOOR MANAGER'S CUES *(continued)*

CUE	SIGNAL	MEANING	SIGNAL DESCRIPTION
AUDIO CUES			
Speak up		Performer is talking too softly for current conditions.	Cups both hands behind ears or moves hand upward, palm up.
Tone down		Performer is too loud or too enthusiastic for the occasion.	Moves both hands toward studio floor, palms down, or puts extended forefinger over mouth in shhh-like motion.
Closer to mic		Performer is too far away from mic for good audio pickup.	Moves hand toward face.
Keep talking		Keep on talking until further cues.	Extends thumb and forefinger horizontally, moving them like a bird's beak.

16.2 PHOTOS BY EDWARD AIONA

Even the handling of cue cards is easier said than done. A good floor person holds the cards as close to the lens as possible, the hands do not cover any of the copy, and he or she follows the performer's lines so that the changes from one card to the next are smooth. **SEE 16.3**

As a performer you must learn to read by peripheral vision so that you will not lose eye contact with the lens. Get together with the floor person handling the cards to double-check their correct order. If the floor person forgets to change the card at the appropriate moment, snap your fingers to attract his or her attention; in an emergency you may have to ad-lib until the system is functioning again. You should study the topic long before the show begins, enabling you to improvise sensibly at least for a short time.

16.3 HANDLING CUE CARDS

A This is the wrong way to hold a cue card: the card is too far away from the lens, and the hands cover part of the copy. The floor person cannot see the copy and does not know when to change the card.

B This is the correct way to hold a cue card: the floor person does not cover the copy, holds the card close to the lens, and reads along with the talent.

16.4 TELEPROMPTER DISPLAY OF COPY

The monitor or flat-panel display reflects the copy onto a glass plate directly over the lens. The lettering remains invisible to the camera, but the talent can read the copy while keeping eye contact with the audience.

If your performance is shot for postproduction, ask the director to stop the recording so that the cards can be put in the correct order.

Studio teleprompter The most effective prompting device is the ***teleprompter***, or auto-cue, which normally uses a flat-panel video display upon which the copy scrolls. The video display is then reflected onto a glass angled over the camera lens. You can read the copy, which now appears in front of the lens but which remains invisible to the camera. This way you do not have to glance to the side but can maintain eye contact with the viewer at all times. **SEE 16.4**

Most often the copy is typed into a computer that acts as a combination word processor and character generator. It can produce the text in several font sizes and scroll (often referred to as crawl) the copy up and down the screen at various speeds. The copy is then sent to the teleprompter flatbed screen mounted on the camera. All cameras used in the production display the same copy at the same speed.

In newscasts the anchorperson should have the text as it appears on the teleprompter also printed out as hard copy. This script serves as backup in case the prompting device fails. Such copy also gives the anchor a reason to glance down to indicate a story transition, to change cameras, or to see during a commercial break what is coming up next.

When using a teleprompter, the distance between you and the camera is no longer arbitrary. The camera must be close enough for you to read the copy without squinting but not so close that the home viewer can see your eyes moving back and forth. If the minimum camera distance is too far to see the teleprompter copy comfortably, have the font enlarged.

Field prompter Have you ever wondered how some correspondents can stand on a busy city street and report a well-written story without ever stumbling or searching for words? Although some certainly have that skill, others use some kind of prompting device. If the copy is brief, handheld cue cards or even some notes will do. Longer copy calls for a field prompter.

There are several models of field prompters, including a miniature version of the studio teleprompter. The flat-panel video displays are so lightweight that they can be attached to a tripod. Most high-end field prompters can be hooked up to a laptop computer with prompting software. You can adjust the size of the font and scroll the copy at different speeds. The Apple iPad makes the ideal field prompter. You can hold it and read the copy off it as you would hard-copy notes, or you can attach it to the camera as you would a laptop. The advantage of the iPad is, of course, its light weight. The camera-plus-iPad prompter can be handheld or shoulder-mounted—at least for a reasonable period of time. SEE 16.5

Low-end prompters use a paper roll mounted immediately below or to one side of the lens. Similar units can be used independently of the camera and held by a floor person or mounted on a tripod directly above or below the camera lens. Regardless of the quality of the teleprompter, you should always be familiar enough with the subject matter to be able to talk about it intelligently if the prompting device fails. If you don't have a prompter, read the package intro or story from a regular news script. If it is windy, use a clipboard or a piece of foam core as backing for the script. This is especially important when you need the other hand to hold a mic.

16.5 FIELD PROMPTER
This lightweight prompter uses an iPad to display the text. Like a studio prompter, it projects the text directly over the lens. The whole prompter mechanism is so light that it can be mounted even on small HDTV cameras and handheld by the camera operator.

Continuity

When you work on a brief commercial or announcement that presents a continuous event but that is shot film-style over a period of several days or even weeks for postproduction, you must look the same in all the video-recording sessions. Obviously, you must wear the same clothes. You must also wear the same jewelry, scarf, and tie from one recording session to the next. You cannot have your coat buttoned on one take and unbuttoned on the next. Makeup and hairstyle too must be identical for all sessions. Have digital photos taken of yourself from the front, sides, and back immediately after the first recording session for an easy and readily available reference.

Most importantly, you must maintain the same energy level throughout all recording sessions. For example, you cannot end one session full of energy and then be very low-key the next day when the video recording resumes, especially when the edited version does not suggest any passage of time between takes. On repeat takes, try to maintain identical energy levels (not always easy).

ACTING TECHNIQUES

In contrast to the television performer, the television actor assumes someone else's character and personality. (In this discussion the term *actor* refers to both male and female talent.) To become a good television actor, you obviously must first master the art of acting, a subject beyond the scope of this chapter. This discussion focuses on how to adapt your acting to the peculiarities of the television medium. Many excellent actors consider television the most difficult medium in which to work. Actors must function effectively within an environment crowded with confusing and impersonal technical gear, and they often get less attention from the director than do the camera operator and the sound technician.

Audience

The biggest difference between stage acting and screen acting is that you are not playing for a stationary live audience but for a constantly moving camera that never blinks or offers feedback on your performance. Worse, your performance is chopped up into short takes that rarely, if ever, allow you to work up to a memorable performance pitch. Each of the little performance segments must be on the mark.

In most takes the camera moves around you, looking at you at close range and from a distance as well as from above and below. It may look at your eyes, your feet, your hands, your back—whatever the director selects for the audience to see. And at all times you must look completely convincing and natural; the character you are portraying must appear on-screen as a believable human being. Keep in mind that you are playing to a virtual audience that is almost always standing right next to you, looking at you from very close range and from all angles. You need not (and should not) project your motions and emotions as you would when acting on-stage. The television camera does the projecting—the communicating—for you. Be aware of the camera or cameras, but don't ever acknowledge their presence.

Internalization of your role, as opposed to externalization, is a key factor in your performance. You must attempt to become as much as possible the person you are portraying, rather than act out the character. Because of the close scrutiny of the camera and the intimacy of the close-up, keep your gestures minimal. Also, your reactions become as important as your actions. You can often communicate feelings more readily by reacting to a situation than by being a part of it.

Blocking

You must be meticulous in following rehearsed ***blocking***—where you should move and what you should do in relation to the set, the other actors, and the television equipment. Sometimes inches are significant, especially if the show is shot primarily in close-ups. Precise television lighting and the limited microphone radius when booms are used also force you to adhere strictly to the established blocking.

Once the show is on the air or recorded live, you have an obligation to follow the rehearsed action. This is not the time to innovate just because you have a sudden inspiration. If the director has not been warned of your change, the new blocking will always be worse than the previously rehearsed one. The camera has a limited field of view; if you want to be seen, you must stay within it.

Some directors have the floor manager mark the exact spots for you to stand or the paths of movement. This is called spiking your position. Look for these tape or chalk marks and follow them without being too obvious. If such spike marks are not used, establish a mental blocking map by remembering where you stand for specific shots in relation to the set and the props. For example, for your scene with the office manager you stand to the left of the file cabinet; for the scene in the doctor's office, you walk counterclockwise around the desk and stop at the camera-right corner of the desk.

In over-the-shoulder and cross-shots, you need to see the camera lens if you are to be seen by the camera. *If you cannot see the lens, the camera cannot see you.* Even the lighting instruments can help you with blocking. For example, to be sure you're in the light when coming through a door, move forward until you feel the warmth of the lights on your forehead.

Sometimes the director will position you in a way that seems entirely wrong to you, especially in relation to the other actors. Don't try to correct this on your own by arbitrarily moving away from the designated spot. A certain camera angle and zoom lens position may very well warrant unusual blocking to achieve a certain effect. Do not second-guess the director.

When you are handling props, the camera is often on a close-up. This means that you must remember all the rehearsed actions and execute them in exactly the same way and with the same speed as they were initially rehearsed. Don't appear nervous when using props (unless the director calls for it); handle them routinely as extensions of your gestures. The way you handle props, such as taking off your glasses, cleaning them, and putting them on again, can often sharpen your character.

Memorizing Lines

As a television actor, you must be able to learn your lines quickly and accurately. If, as is the case in soap operas, you have only one evening to learn a great number of lines for the next day, you must indeed be a quick study. You cannot ad-lib during such performances simply because you have played the role for so long. Most of your lines are important not only from a dramatic point of view but also because they serve as video and audio cues for the whole production team. Your last line of a speech is often a trigger for several key actions in the control room: to switch to another camera, to start a video-recorded insert, or to call up a special effect.

For a single-camera EFP (electronic field production) or digital cinema studio production, each shot is set up and recorded separately. Such a production approach often gives you a chance to read over your lines for each take. Although this approach may make it easier to remember lines, it is harder to maintain continuity of action and emotion. Good screen actors do not rely on prompting devices; after all, you should live, not read, your role. Nevertheless, many good actors like to have all their lines backed up by cue cards, just in case. Most of the time, they never look at them. But even if the cue cards function only as a safety net, their contribution to a good performance more than justifies their use.

Timing

Like the television performer, as an actor you must have an acute sense of timing. Timing matters for pacing your performance, for building to a climax, for delivering a punch line, and for staying within a tightly prescribed clock time. Even if a play is video-recorded scene-by-scene or even shot-by-shot, you still need to carefully observe the stipulated running times for each take. You may have to stretch a scene without making it appear to drag, or you may need to gain a few seconds by speeding up a scene without destroying its solemn mood. You must be flexible without stepping out of character.

Always respond immediately to the floor manager's cues. Do not stop in the middle of a scene simply because you disagree with a specific cue; you're not privy to all the goings-on in the control room. Play the scene to the end and then speak up. Minor timing errors are usually corrected in postproduction.

Maintaining Continuity

As you know, digital cinema and television plays are video-recorded piecemeal, which means that you are not able to perform a play from beginning to end as in a theater production. You cannot be upbeat during the first part of the video recording and then, a few days later when the scene is continued, project a low-energy mood because you didn't sleep well the night before. Often scenes are shot out of sequence for production efficiency and, ultimately, to save money, so it is not possible to have a continuous and logical development of emotions, as is the case in a continuous live or live-recorded pickup. Scenes are inevitably repeated to make them better or to achieve various fields of view and camera angles. This means that, as an actor, you cannot psych yourself up for a single show-stopping performance but must maintain your energy and motivation for each take. Television unfailingly detects subtle nuances and energy levels and the accompanying acting continuity—or lack thereof.

One of the most important qualities to watch for when continuing a scene that was started some days before is the tempo of your performance. If you moved slowly in the first part of the scene, do not race through the second part unless the director wants such a change. It usually helps to watch a recording of your previous performance so that you can continue the scene with the same energy level and tempo.

Director/Actor Relationship

As a television actor, you cannot afford to be temperamental; the director has too many people and elements to coordinate. You may sometimes feel utterly neglected by the director, who seems to worry more about a camera shot than advising you on how to play a difficult scene. The fact is that often you will indeed be neglected by the director in favor of some technical detail. You need to realize that you are not the only significant element in the production; other production people are too—the camera operators, the technical director, the audio technician, the recording technician, and the lighting director, to name but a few.

Even if you have no intention of becoming a television actor, you should make an effort to learn as much about acting as possible. An able actor is generally an effective television performer; a television director with training in acting is generally better equipped to deal with talent than one who has no knowledge of the art.

AUDITIONS

All auditions are equally important, whether you are trying out for a one-line off-camera utterance or a principal role in a dramatic series. Whenever you audition give your best. You can prepare yourself even if you don't know beforehand what you will be reading. Wear something appropriate that looks good on-camera and be properly groomed. Keep your energy up even if you have to wait half a day before you are called to deliver your line.

When you get the script beforehand, study it carefully. For example, if you are doing a commercial for a soft drink, become as familiar as possible with the product, the company that makes it, and the advertising agency producing the commercial. Knowing about the product gives you a certain confidence that inevitably shows up in your delivery. Listen carefully to the instructions given to you before and during the audition. Remember that television is an intimate medium.

When instructed to demonstrate a product, practice before you are on-camera to make sure you know how, for example, to open the easy-to-open package. Ask the floor crew to help you prepare a product for easy handling. Also find out how close the majority of shots will be so that you can keep your actions within camera range.

As an actor be sure to understand thoroughly the character you are to portray. If the script does not tell you much about the character, ask the director or producer to explain how he or she perceives it. You should be able to sense the specifics of the character even when given only minimal cues. Decide on a behavior pattern and follow it, even if your interpretation may be somewhat off base. If the director's perceptions run counter to your interpretation, do not argue. Most importantly, do not ask the casting director to provide you with the "proper motivation" as you may have learned in acting school. At this point it is assumed that you can analyze the script and motivate yourself for the reading. Realize that you are auditioned primarily on how well and how quickly you perceive the script's image of the character and how close you can come to it in speech and sometimes also in looks and actions.

Be creative without overdoing it. When auditions were held for the male lead in a television play about a lonely woman and a rather crude and unscrupulous man who wanted to take advantage of her, one of the actors auditioning added a little of his own interpretation of the character that eventually got him the part. While reading an intimate scene in which he was supposed to persuade the leading lady to make love to him, he manicured his fingernails with slightly rusty fingernail clippers. In fact, this aggravating fingernail clipping was later written into the scene.

Finally, when auditioning—as when participating in athletics or any competitive activity—be aware, but not afraid, of the competition. Innate acting talent is not always the deciding factor in casting a part. Sometimes the director may have a particular image in mind of the physical appearance and the behavior of the actor—heavy and awkward, light and agile, or lean and muscular—that overrides acting skill. Sometimes a well-known actor who can guarantee a large audience may win out. As an actor you need to be prepared to take it repeatedly on the chin.

MAIN POINTS

- Television talent refers to all persons who perform regularly in front of the camera. They are classified into two large groups: performers and actors.
- Television performers are basically engaged in nondramatic shows, such as newscasts, interviews, and game shows. They portray themselves. Television actors portray someone else.
- The television performer must adapt his or her techniques to the characteristics of the camera and the other production elements, including audio, timing, the floor manager's cues, prompting devices, and continuity.
- Because the camera lens represents the audience, performers must look *through* the lens to establish and maintain eye contact with the viewer. If cameras are switched, performers must transfer their gaze to the hot camera smoothly and naturally.
- Timing is an important performance requirement. A good performer must respond quickly yet smoothly to the floor manager's time, directional, and audio cues.
- Prompting devices have become essential in television production. The two most frequently used devices are cue cards and the teleprompter.
- Television acting requires that the actor overcome the lack of an actual audience and internalize the role, restrict gestures and movements because of close-ups, follow exactly the rehearsed blocking, memorize lines quickly and accurately, have a good sense of timing, maintain continuity in physical appearance and energy level over a series of takes, and keep a positive attitude despite occasional neglect by the director.
- Performers and actors should prepare as much as possible for auditions, dress appropriately for the occasion (role), and sharpen the character through some prop or mannerism.

SECTION

16.2

How to Do Makeup and What to Wear

When you hear of makeup, you may think of movies in which actors are transformed into monsters or aliens or of how to fake a variety of wounds. But most television makeup is done not so much to transform appearance as to make someone look as good as possible on-camera. The same goes for clothing. Unless they act in a period play, most actors wear clothes that fit the role, and performers choose clothes that make them look attractive on-camera.

The aim of this section is to help you choose makeup, clothing, and costumes that not only fit but also add to the overall production values and communication intent.

▶ **MAKEUP**
Materials, application, and technical requirements

▶ **CLOTHING AND COSTUMING**
Line, texture and detail, and color

MAKEUP

All ***makeup*** is used for three basic reasons: to enhance appearance, to correct appearance, and to change appearance. Standard over-the-counter cosmetics are used daily by many people to accentuate and improve their features. Minor skin blemishes are covered up, and the eyes and lips are emphasized. Makeup can also be used to correct closely or widely spaced eyes, sagging flesh under the chin, a short or long nose, a slightly too prominent forehead, and many similar minor "flaws."

If a person is to portray a specific character in a play, a complete change of appearance may be necessary. Dramatic changes of age, ethnicity, and character can be accomplished through creative makeup techniques. Makeup artists working for crime shows or medical series have a field day. Their grisly renderings of all sorts of bodily harm or close-ups of bloody operations are often so realistic that they border on the repulsive; somehow we seem to have lost the respect for the miracle of life and go to any extreme to capture the audience's attention. Nevertheless, these effects are testimony to the high artistic skills of makeup artists.

The various purposes for applying cosmetics require different techniques, of course. Enhancing someone's appearance calls for the most straightforward procedure, correcting someone's appearance is slightly more complicated, and changing an actor's appearance may require involved and complex makeup techniques. Making a young actor look 80 years old is best left to the professional makeup artist. You need not learn all about corrective and character makeup methods, but even the normal enhancing makeup for high-definition television (HDTV) calls for new and slightly more elaborate procedures than brushing on some powder to keep the key light from reflecting off your forehead.

Even if you are not planning to be talent but to work as a floor manager, camera person, or director of an EFP crew, you should have some idea of the basic materials, techniques, and technical requirements of television makeup.

Materials

A great variety of excellent television makeup materials are available. Most makeup artists in the theater arts department of a college or university have up-to-date product lists. In fact, most large drugstores can supply you with the basic materials for enhancing a performer's appearance. Female performers are generally experienced in cosmetic materials and techniques; men may, at least initially, need some advice.

The most basic makeup item is a ***foundation*** that covers minor blemishes and cuts down light reflections on oily skin. You can use water-based ***cake*** makeup or a variety of creams for foundation. ***Pan-stick*** is a type of foundation with a grease base used to cover a beard shadow or prominent skin blemish. Any one of the Kryolan Paint Stick, Aqua Color Cake, Supracolor Cream Foundation, Maybelline EverFresh cake series, or the Max Factor cake and pan-stick or similar products will do the job. The colors range from a warm, light ivory to dark shades for dark-skinned performers.

Women can use their own lipsticks so long as the reds do not contain too much blue. For dark-skinned talent, a warm red, such as coral, is often more effective than a darker red that contains a great amount of blue.

Other materials, such as eyebrow pencil, mascara, and eye shadow, are generally a part of every female performer's makeup kit.

Materials such as hairpieces or even latex masks and prosthetics are the purview of the professional makeup artist and are of little use in most nondramatic productions.

Application

It is not always easy to persuade nonprofessional performers, especially men, to put on necessary makeup. You may do well to look at the guests on-camera before deciding whether they need any. If they do, you must be tactful in suggesting its application. Try to appeal not to the performer's vanity but to his or her desire to contribute to a good performance. Explain the necessity for makeup in technical terms, such as color and light balance.

All makeup rooms have large mirrors so that talent can watch the entire makeup procedure. Adequate, even illumination is critical. The color temperature of the light in which the makeup is applied must match, or at least closely approximate, that of the production lighting. Most makeup rooms have two illumination systems that can be switched from the indoor (3,200K) standard to the outdoor (5,600K) standard. Ideally, makeup should be applied under the lighting conditions in which the production is recorded. This is because each lighting setup has its own color temperature. Reddish light (low color temperature) may require cooler, more bluish makeup than would bluish light (high color temperature), in which you need to use warmer, more reddish makeup. (For a review of color temperature, see chapter 12.) ZVL1 LIGHTS→ Color temperature→ light sources

When makeup is applied in the studio, have a small mirror on hand. Most female performers are glad to apply the more complicated makeup themselves—lipstick and mascara, for instance. In fact, most professional television talent prefer to apply their own makeup; they usually know what kind they need for a specific television show.

When using a water-based cake makeup, apply it evenly with a wet sponge over the face and adjacent exposed skin areas. Get the base right up into the hairline, and have a towel ready to wipe off the excess. If close-ups of hands are shown, apply cake base to them and the arms as well. This is especially important for performers who demonstrate small objects on-camera. If an uneven suntan is exposed (especially when female performers wear backless dresses or different kinds of bathing suits), all bare skin areas must be covered with cake makeup. Bald men need a generous amount of cake foundation to tone down inevitable light reflections and to cover up perspiration.

HDTV has spawned not only new makeup materials but also new application techniques. Instead of a sponge or brush being used to apply cream foundation, it is customarily sprayed on in liquid form with an airbrush. Here are the basic steps for applying foundation makeup for HDTV.

- Clean the face using a cleanser and a cotton pad.
- Use a concealer (cream-based pan-stick) to cover up any blemishes.
- Set the concealer with a very light dusting of powder.
- Using the airbrush, spray on the liquid foundation. Keep the nozzle about 6 to 8 inches away from the face.
- After allowing it to dry for about 30 seconds, use a large, soft brush to distribute a thin layer of light powder over the foundation.

To prevent it from clogging, the airbrush should be cleaned between uses.

Be careful not to give male performers a baby-face complexion through too much makeup. It is sometimes desirable to have a little beard area show. A very thin application of yellow- or orange-tinged translucent makeup satisfactorily counteracts the blue of a heavy five-o'clock shadow (much like the backlight for the blue chroma-key area). If additional foundation is necessary, pan-stick around the beard area should be applied first and then set with powder. You may find that brushing on a light layer of powder is all that is needed, not only to improve your looks, but especially to keep light reflection off your forehead, nose, and cheekbones.

Because your face is the most expressive communication agent, try to keep your hair out of your face as much as possible.

Technical Requirements

Like so many other production elements, makeup must yield to the demands of the television camera. These limitations include color distortion, color balance, and close-ups.

Color distortion As mentioned, skin tones are the only real reference the viewer has for color adjustment on a home receiver. Their accurate rendering is therefore of the utmost importance. Because cool colors (hues with a blue tint) have a tendency to overemphasize bluishness, especially in high-color-temperature lighting, warm colors (reds, oranges, browns, and tans) are preferred for television makeup. They usually provide more sparkle, especially when used on a dark-skinned face.

The color of the foundation makeup should match the natural skin tones as closely as possible, regardless of whether the face is naturally light or naturally dark. Again, to avoid bluish shadows, warm rather than cool foundation colors are preferred. Be careful, however, that light-colored skin does not turn pink. As much as you should guard against too much blue in a dark face, you must watch for too much pink in a light face.

The skin reflectance of a dark face can produce unflattering highlights. These should be toned down with cake foundation or translucent powder. Otherwise the video operator will have to compensate for the highlights through shading, making the dark picture areas unnaturally dense.

Color balance Generally, the art director, scene designer, makeup artist, and costume designer coordinate in production meetings all the colors in a scene. In nonbroadcast productions, where freelance people are usually hired for scene design and makeup, such coordination is not always easy. In any case try to communicate the color requirements to these people as best you can. Some attention beforehand to the coordination of the colors used in the scenery, costumes, and makeup certainly facilitates the production process.

Sometimes the surrounding colors reflect on the performer's clothing or face, which the camera shows as noticeable color distortions. One way of avoiding such reflections is to have the talent step far enough away from the reflecting surfaces. When such a move is not possible, apply an adequate amount of foundation makeup and additional powder to the discolored skin areas. The viewer will tolerate to some extent the color distortion on clothing but not on skin unless the colors are purposely distorted for dramatic effect.

Close-ups Makeup must be smooth and subtle enough that the talent's face looks natural even in an extreme close-up with an HDTV camera. The skin should have a normal sheen, neither too oily (high reflectance) nor too dull (low reflectance but no brilliance—the skin looks lifeless). The subtlety of television makeup goes directly against theater makeup techniques, in which features and colors are greatly exaggerated for the benefit of the spectators in the back row. Good television makeup remains largely invisible, so a close-up of a person's face under actual production lighting conditions is the best criterion for judging the necessity for and the quality of makeup. If the performer or actor looks good on-camera without makeup, none is needed. A light dusting of translucent powder, however, will prevent undesirable light reflections, especially when the talent begins to perspire. If a performer needs makeup and the close-up of his or her finished face looks normal, the makeup is acceptable. If it shows, the makeup must be toned down.

CLOTHING AND COSTUMING

In small-station operations and most nonbroadcast productions, you are mainly concerned with clothing the performer rather than costuming the actor. The performer's clothes should be attractive and stylish but not too conspicuous or showy. Television viewers expect a performer to be well dressed but not overdressed. After all, a television performer is a guest in the viewer's home, not a nightclub entertainer.

Clothing

The type of clothing you wear as a performer depends largely on your personal taste. It also depends on the type of program or the setting and the occasion. Obviously, you dress differently when reporting live in the field during a snowstorm than when taking part in a panel discussion on the effectiveness of schools in your city. Whatever the occasion, some types of clothing look better on television than others. Because the camera may look at you both from a distance and at close range, the lines, texture, and details are as important as the overall color scheme.

Line Television has a tendency to add a few pounds to the performer, even if they are not digitally stretched to make a 3 × 4 picture fit the 16 × 9 screen. Clothing cut to a slim silhouette helps combat this problem. Slim dresses and closely tailored suits look more attractive than do heavy, horizontally striped fabrics and baggy styles. The overall silhouette of the clothing should look pleasing from a variety of angles and should appear slim-fitting yet comfortable.

Texture and detail Whereas line is especially important in long shots, the texture and the detail of clothing become important on close-ups. Textured material often looks better than plain, but avoid patterns that have too much contrast or are too busy. Closely spaced geometric patterns such as herringbone weaves and checks cause a moiré

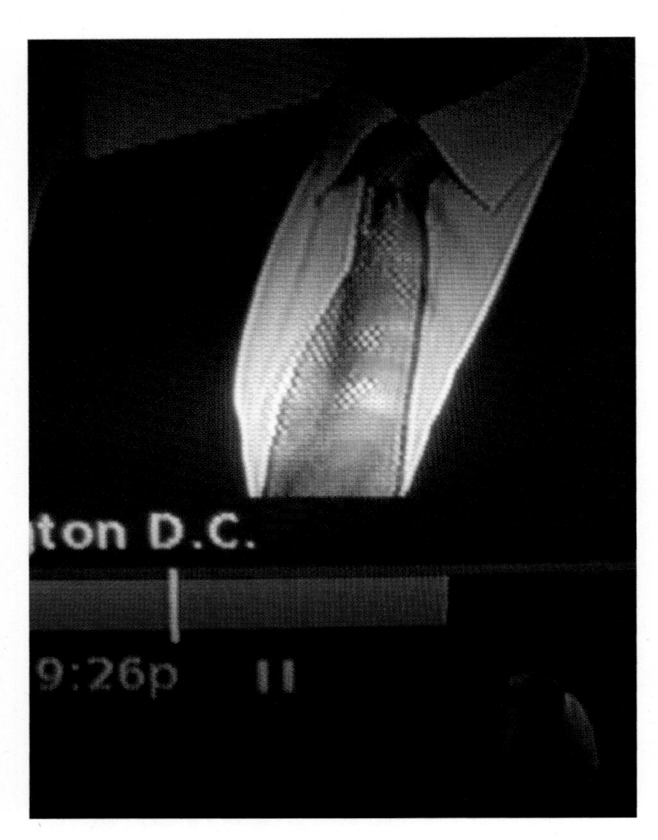

16.6 MOIRÉ PATTERN
Clothing with thinly striped patterns can cause an undesirable artifact, called moiré pattern, which usually vibrates and sometimes produces rainbow colors, as in this shiny gray spotted tie.

16.6 PHOTO BY ELIZABETH VON RADICS

effect, which looks like superimposed vibrating rainbow colors. SEE 16.6

Such moiré patterns are caused when the scanning lines of the camera literally bump into the lines of the optical pattern. HDTV does not help and in fact frequently makes this problem worse. One proven technique is to defocus the camera somewhat, which will render the optical lines less precise. Note that most high-quality studio monitors have moiré suppression circuits built-in, but most home receivers do not. You may not always be aware of the moiré problem a herringbone jacket or checked tie causes when watching yourself in the mirror or on a high-end monitor. If you suspect possible moiré problems, view the attire on a television set that does not contain such preventive circuits.

Stripes may also cause problems. They sometimes extend beyond the garment and bleed through surrounding objects and set. Unless you have a high-quality HDTV camera, extremely fine detail in a pattern will either look too busy or appear smudgy.

Make your clothing more interesting on-camera not by choosing a detailed cloth texture but by adding decorative accessories, such as scarves and jewelry. Although jewelry style depends, of course, on your taste, in general you should limit it to one or two distinctive pieces. The sparkle of rhinestones can become an exciting visual accent when dressing for a special occasion, such as the televised fund-raising dinner or a concert by the community symphony, but they are obviously out of place when interviewing a crime victim.

Color The most important consideration when choosing clothing colors is that they harmonize with the set. If the set is lemon yellow, do not wear a lemon yellow dress. As mentioned before, avoid saturated red, unless you are working with very good studio cameras. If you are taking part in blue chroma keying (such as in weathercasting), don't wear a blue unless you want to become transparent during the chroma key. Even a blue scarf or tie may give you trouble. The same goes for wearing green if you work in front of a green chroma-key backdrop.

You can wear black or a very dark color, or white or a very light color, so long as the material is not glossy and highly reflective. But avoid wearing a combination of the two, even if the high-end digital camera can handle such a contrast. If the set is very dark, avoid a starched white shirt. If the set colors are extremely light, do not wear black. As desirable as a dramatic color contrast is, extreme brightness variations cause difficulties for even the best cameras and video operators.

Stark white, glossy clothes can turn exposed skin areas dark on the television screen or distort the more subtle colors, especially when the cameras are on auto-iris. Dark-skinned performers should avoid highly reflecting white or light-yellow clothes. If you wear a dark suit, reduce the brightness contrast by wearing a pastel shirt. Light blue, pink, light green, tan, or gray—all show up well on television. As always, when in doubt as to how well a certain color combination photographs, preview it on-camera on the set and under the actual lighting conditions.

Costuming

For most normal productions in nonbroadcast, or non-network, operations, you do not need costumes. If you do a play or a commercial that involves costumed actors, you can always rent the necessary articles from a costume company or borrow them from the theater arts department of a local high school or college. Theater departments usually have a well-stocked costume room from which you can draw most standard uniforms and period costumes. If you use stock costumes on television, they must look convincing even in a tight close-up. The general construction and, especially, the detail of theater accessories are often too coarse for the HDTV camera.

The color and pattern restrictions for clothing also apply to costumes. The total color design—the overall balance of colors among scenery, costumes, and makeup—is important in some television plays, particularly in musicals and variety shows, where long shots often reveal the total scene, including actors, dancers, scenery, and props. As mentioned, rather than try to balance all the hues, it is easier to balance the colors by their relative aesthetic energy. You can accomplish this balance by keeping the set relatively low-energy (colors with low saturation) and the set accessories and costumes high-energy (high-saturation colors).

MAIN POINTS

- Makeup and clothing are important aspects of the talent's preparation for on-camera work.
- Makeup is used for three basic reasons: to enhance, to correct, and to change appearance.
- Warm colors generally look better than cool colors because the camera tends to emphasize the bluishness of cool colors; but avoid wearing red unless you work with high-end cameras.
- Makeup must be smooth and subtle to appear natural in the actual production lighting and on extreme close-ups. The most basic makeup item is a foundation that covers minor blemishes. Water-based cake foundations, which come in a variety of skin tones, are often used for standard television makeup.
- The techniques of television makeup do not differ drastically from applying everyday makeup, especially if the purpose is to enhance or correct appearance. Ordinary makeup techniques may not be smooth enough for the high-definition television (HDTV) camera, however. Foundation makeup for HDTV is therefore sprayed on with an airbrush.
- These factors are important when choosing clothing: line, whereby a slim cut is preferred; texture and detail, which must not make the clothing appear too busy; and color, which should harmonize yet contrast with the dominant color of the set.
- Tightly striped or checkered patterns and herringbone weaves, as well as highly saturated reds and a combination of black-and-white fabrics, should be avoided.

ZETTL'S VIDEOLAB

For your reference or to track your work, the *Zettl's VideoLab* program cue in this chapter is listed here with its corresponding page number.

ZVL1 LIGHTS→ Color temperature→ light sources 356

CHAPTER

17

The Director in Production

Now you are ready to put on your director's cap again, sit in the director's high chair, and do some directing. But instead of your notebook in which you recorded all the preproduction steps, you now need a megaphone to call for "lights, camera, action!" You will quickly find, however, that calling for action is only a small part of the director's role. You must be skilled in multitasking but also able to pay full attention to every single production element. Not an easy job by any means!

All the meticulous preparation you made in the preproduction phase means little if you cannot direct or coordinate the various elements during the production phase. Section 17.1, Multicamera Control Room Directing, gives an overview of what is required when directing various multicamera studio and remote productions. In section 17.2, Single-camera and Digital Cinema Directing, you will read about so-called film-style shooting: using a high-definition television (HDTV), or digital cinema, camera for the acquisition of the video scenes.

KEY TERMS

camera rehearsal Full rehearsal with cameras and other pieces of production equipment. Often identical to the dress rehearsal.

dress rehearsal Full rehearsal with all equipment operating and with talent in full dress. The dress rehearsal is often video recorded. Often called *camera rehearsal* except that the camera rehearsal does not require full dress for talent.

dry run Rehearsal without equipment, during which the basic actions of the talent are worked out. Also called *blocking rehearsal.*

intercom Short for *intercommunication system.* Used by all production and technical personnel. The most widely used system has telephone headsets to facilitate voice communication on several wired or wireless channels. Includes other systems, such as I.F.B. and cell phones.

multicamera directing Simultaneous coordination of two or more cameras for instantaneous editing (switching). Also called *control room directing.*

running time The duration of a program or program segment.

schedule time The time at which a program starts and ends.

single-camera directing Directing method for a single camera. For digital cinema it may mean moving from an establishing long shot to medium shots, then to close-ups of the same action. Also called *film-style shooting.*

walk-through Orientation session with the production crew *(technical walk-through)* and the talent *(talent walk-through)* wherein the director walks through the set and explains the key actions.

SECTION

17.1

Multicamera Control Room Directing

As a television director, you are expected to be an artist who can translate ideas into effective pictures and sounds, a psychologist who can encourage people to give their best, a technical adviser who can solve problems the crew would rather give up on, and a coordinator and a stickler for detail who leaves nothing unchecked. This section explains the director's roles and lingo and takes you through the major steps of multicamera directing and the use of multiple cameras for digital cinema.

- **THE DIRECTOR'S ROLES**
  Artist, psychologist, technical adviser, and coordinator
- **THE DIRECTOR'S TERMINOLOGY**
  Terms and cues for visualization, sequencing, special effects, audio, video recording, and the floor manager
- **MULTICAMERA DIRECTING PROCEDURES**
  Directing from the control room, and control room intercom systems
- **DIRECTING REHEARSALS**
  Script reading, dry run (blocking rehearsal), walk-through, camera and dress rehearsals, and walk-through/camera rehearsal combination
- **DIRECTING THE SHOW**
  Standby procedures and on-the-air procedures
- **CONTROLLING CLOCK TIME**
  Schedule time and running time, back-timing and front-timing, and converting frames into clock time

THE DIRECTOR'S ROLES

The various roles you must assume as a director are not as clear-cut as you will see them described in this section. They frequently overlap, and you may have to switch from one to another several times just in the first five minutes of rehearsal. Even when pressed for time and pressured by people with a variety of problems, always pay full attention to the task at hand before moving on to the next one.

Director as Artist

In the role of an artist, a director is expected to produce pictures and sound that not only convey the intended message clearly and effectively but do so with flair. You need to know how to look at an event or a script, quickly recognize its essential quality, and select and order those elements that help interpret it for a specific audience. Flair and style enter in when you do all these things with a personal touch—when, for example, you shoot a certain scene very tightly to heighten its energy or when you select unusual background music that helps convey a specific mood. But unlike the painter, who can wait for inspiration and can retouch the painting until it is finally right, the television director is expected to be creative by a specific clock time and to make the right decisions the first time around.

Director as Psychologist

Because you deal with a variety of people who approach television production from different perspectives, you need to also assume the role of psychologist. For example, in a single production you may have to communicate with a producer who worries about the budget, technicians who are primarily concerned with the technical quality of pictures and sound, temperamental talent, a designer who has strong ideas about the set, and the mother of a child actor, who thinks your close-ups of her daughter are not tight enough.

Not only must you get everyone to perform at a consistently high level, you also have to get them to work as a team. Although there is no formula for directing a team of such diverse individuals, the following are some basic guidelines to help you exercise the necessary leadership.

- Be well prepared and know what you want to accomplish. You cannot possibly get people to work for a common goal if you do not know what that goal is.

- Know the specific functions of each team member. Explain to all the individuals what you want them to do before holding them accountable for their work.

- Be precise about what you want the talent to do. Do not be vague with your instructions or intimidated by a celebrity. The more professional the talent, the more readily they will follow your direction.

- Project a secure attitude. Be firm but not harsh when giving instructions. Listen to recommendations from other production staff but do not yield your decision-making to them.

- Do not ridicule someone for making a mistake. Point out the problems and suggest solutions. Keep the overall goal in mind.

- Treat the talent and all members of the production team with respect and compassion. A good director knows no hierarchy of importance among the team members. All are equally responsible for getting the production done.

Director as Technical Adviser

Although you do not have to be an expert in operating the technical equipment, as a director you should still be able to give the crew helpful instructions on how to use it to achieve your communication goal. In the role of technical adviser, you are acting much like a conductor of a symphony orchestra. The conductor may not be able to play all the instruments in the orchestra, but he certainly knows the sounds that the various instruments can generate and how they ought to be played to produce the music he wants to hear.

Director as Coordinator

In addition to your artistic, psychological, and technical skills, you must be able to coordinate a great many production details and processes. The role of coordinator goes beyond directing in the traditional sense, which generally means blocking the talent and helping them give peak performances.

Especially when directing nondramatic shows, you must expend most of your effort on cueing members of the production team (both technical and nontechnical) to initiate certain video and audio functions, such as getting appropriate camera shots, rolling the video recorder (VR), watching audio levels, switching among cameras and special effects, retrieving electronically generated graphics, and switching to remote feeds. You still need to pay attention to the talent, who sometimes (and rightly so) feel that they play second fiddle to the television machine. You also need to coordinate productions within a rigid time frame in which every second has a hefty price tag attached. Such coordination requires practice, and you should not expect to be a competent director immediately even after reading this book.

THE DIRECTOR'S TERMINOLOGY

As does any other human activity in which many people work together at a common task, television directing demands a precise and specific language. Your first task in becoming a director is, of course, to learn to speak this lingo with clarity and confidence. Only then can you fulfill your difficult task as master juggler of schedules, equipment, people, and artistic vision. This jargon, which must be understood by all members of the team, is generally called the director's language or, more specifically, the director's terminology. It is essential for efficient, error-free communication between the director and the production team.

By the time you learn television directing, you will probably have mastered most production jargon in general and perhaps even the greater part of the director's specific lingo. Like any language the director's terminology is subject to habit and change. Although the basic vocabulary is fairly standard, you will hear some variations among directors. And as new technology emerges, the director's language changes accordingly. For example, we still don't have a standard term for an entirely tapeless operation. Some directors still call for "SOT" (sound on tape) even if the clips are delivered by a server; or they refer to "live-on-tape" when they record a show on a hard drive during a live telecast.

The terminology listed here reflects primarily multicamera directing from the studio or remote truck control room—the type of directing that requires the most precise terminology. A single inaccurate call can cause a number of serious mistakes. You can also use most of these terms in single-camera directing, regardless of whether the production happens in the studio or in the field.

Whatever terminology you use, it must be consistent and understood by everyone on the production team; there is little time during a show to explain. The shorter and less ambiguous the signals, the better the communication. The following tables list the director's terminology for visualization, sequencing, special effects, audio, video recording, and cues to the floor manager. SEE 17.1–17.6

17.1 DIRECTOR'S VISUALIZATION CUES

The visualization cues are directions for the camera to achieve optimal shots. Some of these visualizations can be achieved in postproduction (such as an electronic zoom through digital magnification), but they are much more easily done with proper camera handling.

17.1 PHOTOS BY EDWARD AIONA

17.1 DIRECTOR'S VISUALIZATION CUES *(continued)*

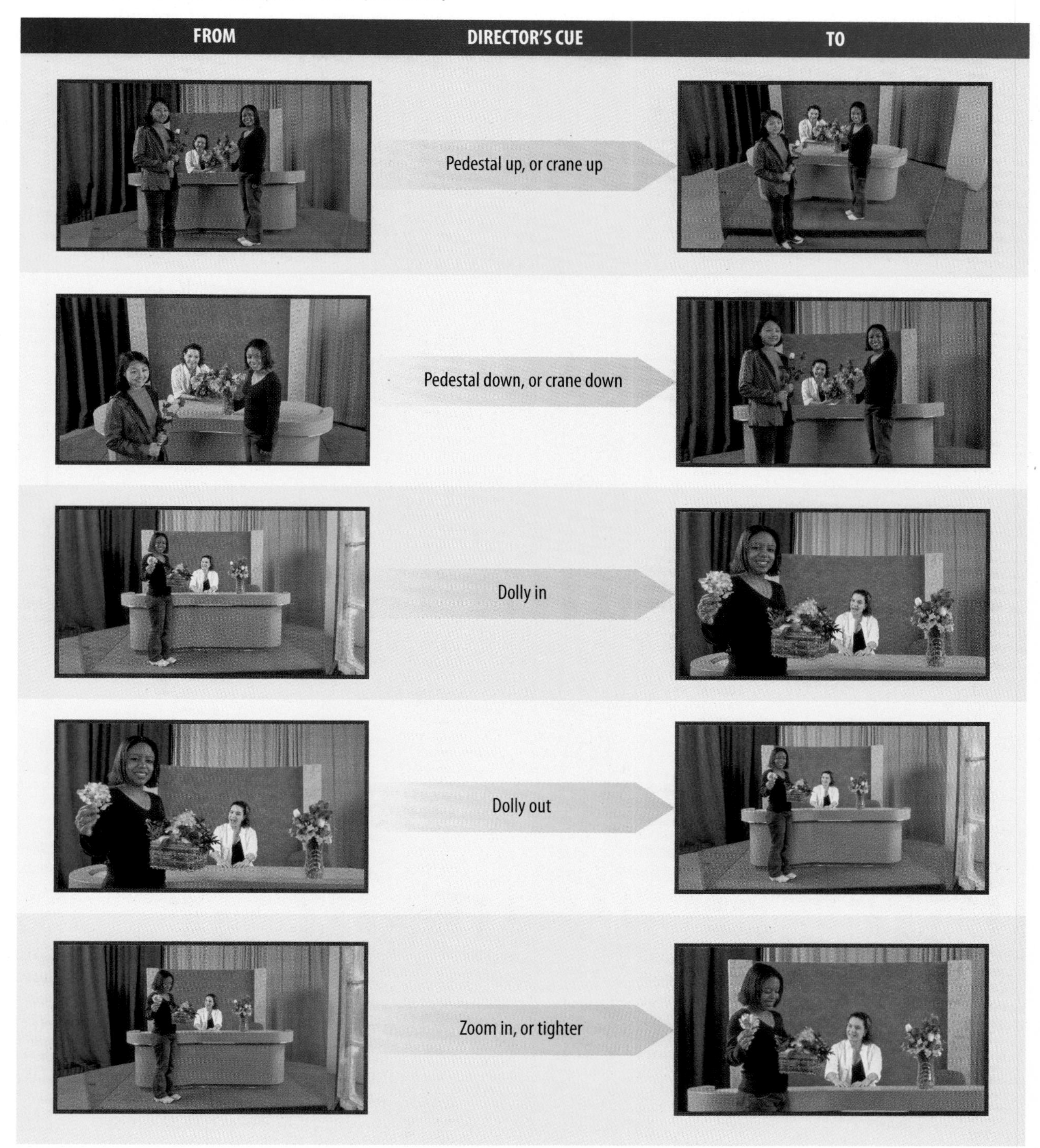

17.1 PHOTOS BY EDWARD AIONA

17.1 DIRECTOR'S VISUALIZATION CUES *(continued)*

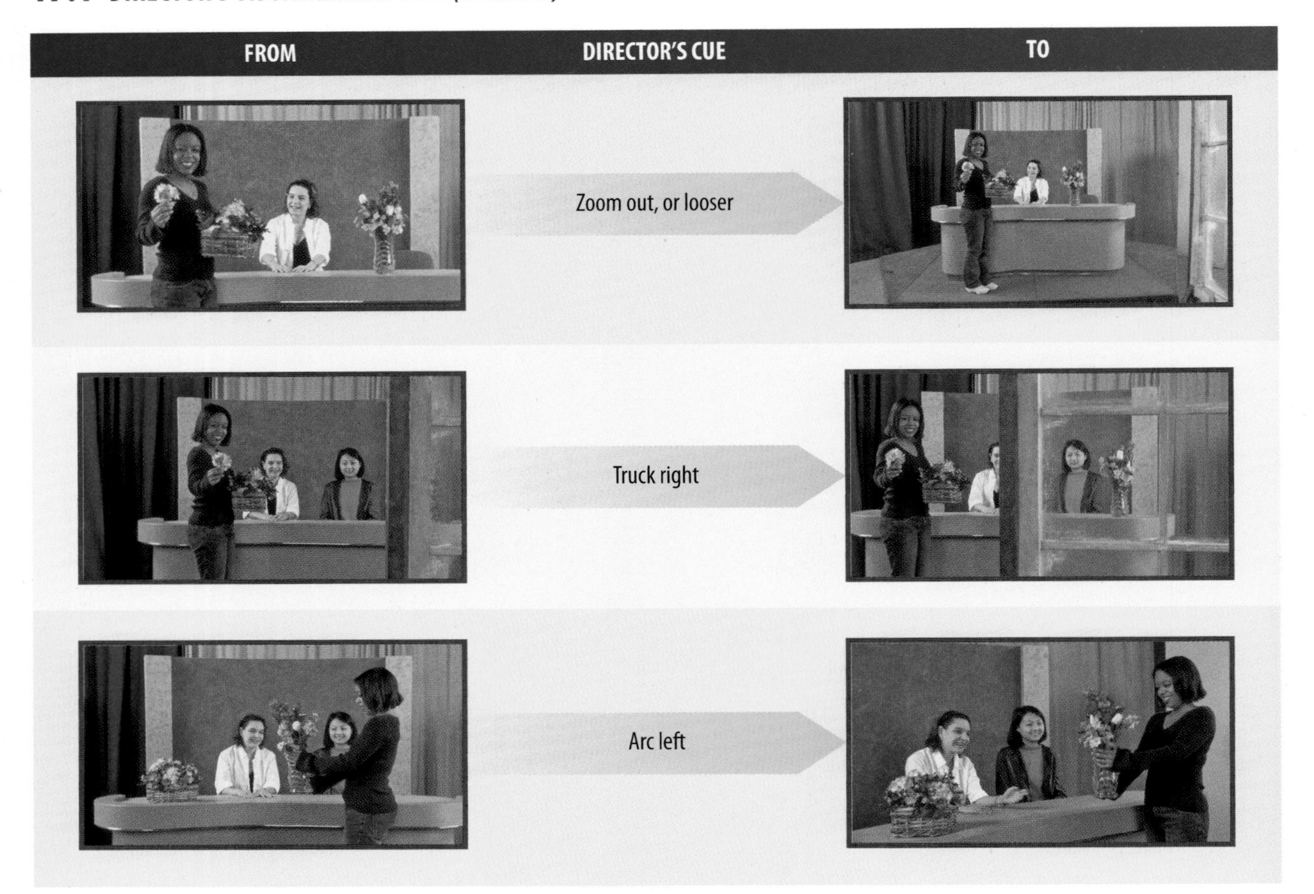

17.1 PHOTOS BY EDWARD AIONA

MULTICAMERA DIRECTING PROCEDURES

When ***multicamera directing*** you direct and coordinate various production elements simultaneously from a television control room in the studio or the remote truck (see chapter 18). You generally try to create as finished a product as possible, which may or may not need some postproduction editing. When doing a live telecast, you have no chance to fix anything in postproduction; your directing is the final cut. Multicamera directing involves the coordination of many technical operations as well as the actions of the talent. You will find that, at first, managing the complex machinery—cameras, audio, graphics, video recorder, remote feeds, and the clock—provides the greatest challenge. But once you have mastered the machines to some extent, the most difficult job will be dealing with people—those in front of the camera (talent) as well as those behind it (production people and crew). ZVL1 PROCESS→ Methods→ multicamera

Directing from the Control Room

In multicamera directing you need to be concerned with not only the visualization of each shot but also the immediate sequencing of the various shots. It includes the directing of live shows, live-recorded productions, and longer show segments that are later assembled but not otherwise altered in relatively simple postproduction. Multicamera directing always involves the use of a control room, regardless of whether it is attached to the studio, inside a remote truck, or temporarily assembled in the field for digital cinema. Because the control room is designed specifically for multicamera productions and for the smooth coordination of all other video, audio, and recording facilities and

17.2 DIRECTOR'S SEQUENCING CUES

The sequencing cues help get from one shot to the next. They include the major transitions.

ACTION	DIRECTOR'S CUE
Cut from camera 1 to camera 2	**Ready two — take two.**
Dissolve from camera 3 to camera 1	**Ready one for dissolve — dissolve.**
Horizontal wipe from camera 1 to camera 3	**Ready three for horizontal wipe** [over 1] — **wipe.** *or:* **Ready effects number** ***x*** [the number being specified by the switcher program] — **effects.**
Fade in camera 1 from black	**Ready fade in one — fade in one.** *or:* **Ready up on one — up on one.**
Fade out camera 2 to black	**Ready black — go to black.**
Short fade to black between cameras 1 and 2	**Ready cross-fade to two — cross-fade.**
Cut between camera 1 and clip 2 from server	**Take clip two.** [Sometimes you simply call the server number. If, for example, the server is labeled 6 on the switcher, you say: **Ready six — take six.**]
Cut between VR and C.G.	**Ready C.G. — take C.G.** *or:* **Ready effects on C.G. — take effects.**
Cut between C.G. titles	**Ready change page — change page.**

people, multicamera directing is often called control room directing.

Your lifeline during camera rehearsals and on-the-air directing is a reliable and flawlessly working intercommunication system that connects you with the rest of the control room personnel, the studio crew, and, if necessary, the talent.

Control Room Intercom Systems

The control room ***intercom*** systems provide immediate voice communication among all production and technical personnel. The most common are the P.L., the I.F.B., and the S.A. systems.

P.L. intercom Most small stations and independent production studios use the telephone intercommunication, or P.L. (private line or phone line), system. All production and technical personnel who need to be in voice contact with one another wear standard telephone headsets with an earphone and a small microphone for talkback. Although most studio P.L. systems are wireless, you should still have a wired system for backup.

17.3 DIRECTOR'S SPECIAL-EFFECTS CUES

Special-effects cues are not always uniform, and, depending on the complexity of the effect, directors may invent their own verbal shorthand. Whatever cues are used, they need to be standardized among the production team members.

ACTION	DIRECTOR'S CUE
To super camera 1 over 2	**Ready super one over two — super.**
To return to camera 2	**Ready to lose super — lose super.** *or:* **Ready to take out one — take out one.**
To go to camera 1 from the super	**Ready to go through to one — through to one.**
To key the C.G. over the base picture on camera 1	**Ready key C.G.** [over 1] **— key.**
To key the studio card title on camera 1 over the base picture on camera 2	**Ready key one over two — key.**
To fill the keyed-out title from the studio card on camera 1 with a yellow hue over the base picture on camera 2	**Ready matte key one, yellow, over two — matte key.**
To have the title from the C.G. appear in drop-shadow outline over the base picture on camera 1	**Ready C.G. drop shadow over one — key C.G.** [Sometimes the director may use the name of the C.G. manufacturer, such as Chyron. Thus you would say: **Ready Chyron over one — key Chyron.** Because the C.G. information is almost always keyed, the "key" is usually omitted in the ready cue.] *or:* **Ready effects, drop shadow — take effects.** [Some directors simply call for an "insert," which refers to the downstream keyer. Usually the lettering mode (drop shadow or outline) is already programmed into the C.G., so you just say: **Ready insert seven — take insert.**]
To have a wipe pattern appear over a picture, such as a scene on camera 2, and replace a scene on camera 1 through a circle wipe	**Ready circle wipe two over one — wipe.** [Any other wipe is called for in the same way except that the specific wipe pattern is substituted for "circle wipe." If you need a soft wipe, simply call for "Ready soft wipe" instead of "Ready wipe."]
To have an insert (video B) grow in size in a zoomlike motion, replacing the base picture (video A)	**Ready squeeze out — squeeze.** *or:* **Ready effect sixteen — squeeze out.**
To achieve the reverse squeeze (video B getting smaller)	**Ready squeeze in — squeeze.**
To achieve a great many transitions through wipes	**Ready wipe effect twenty-one — wipe.**

Many of the complicated effects are preset and stored in the computer program. The retrieval goes by numbers. All you do to activate a whole effects sequence is call for the number: **Ready effects eighty-seven — take effects.**

17.4 DIRECTOR'S AUDIO CUES

Audio cues involve cues for microphones, starting and stopping audio sources such as CD players, and integrating or mixing those sources.

ACTION	DIRECTOR'S CUE
To activate a microphone in the studio	**Ready to cue talent.** [Or something more specific, like **Mary — cue her.** The audio engineer will automatically open her mic.] *or:* **Ready to cue Mary — open mic, cue her.**
To start music	**Ready music — music.**
To bring the music under for an announcer	**Ready to fade music under — music under, cue announcer.**
To take the music out	**Ready music out — music out.** *or:* **Fade music out.**
To close the microphone in the studio (announcer's mic) and switch over to the sound on source, in this case, a clip from the server	**Ready SOS** [sound on source] **— close mic, track up.** *or:* **Ready SOS — SOS.**
To roll an audio recording (such as a clip or CD)	**Ready audio recorder, clip two** [or **CD two**]. **Roll audio.** *or:* **Ready audio clip two — play.**
To fade one sound source under and out while simultaneously fading another in (similar to a dissolve)	**Ready cross-fade from *[source]* to *[other source]* — cross-fade.**
To go from one sound source to another without interruption (usually two pieces of music)	**Ready segue from *[source]* to *[other source]* — segue.**
To increase program speaker volume for the director	**Monitor up, please.**
To play a sound effect from a CD	**Ready CD cut *x* — play.** *or:* **Ready sound effect *x* — play.**
To put slate information on the recording media (either open floor manager's mic or talkback patched to the VR)	**Ready to read slate — read slate.**

17.5 DIRECTOR'S VIDEO-RECORDING CUES

The video-recording cues are used to start and stop the video recorder, to slate a video recording, and to switch to the VR output.

ACTION	DIRECTOR'S CUE
To start recording a program	**Ready to roll VR one — roll VR one.** [Now you have to wait for the "in-record" or "speed" confirmation from the VR operator if the recorder is a VTR.]
To "slate" the program after the VR is in the record mode. The slate is on camera 2 or on the C.G.; the opening scene is on camera 1. (We are assuming that the color bars and the reference level audio tone are already on the recording media.)	**Ready two** [or **C.G.**], **ready to read slate — take two** [or **C.G.**], **read slate.**
To put the opening 10-second beeper on the audio track and fade in on camera 1. (Do not forget to start your stopwatch as soon as camera 1 fades in.)	**Ready black, ready beeper — black, beeper.** **Ten — nine — eight — seven — six — five — four — three — two — cue Mary — up on one.** [Start your stopwatch.]
To stop the recording on a freeze-frame	**Ready freeze — freeze.**
To roll the recording out of a freeze-frame	**Ready to roll VR three — roll VR three.**
To roll the recording for a slow-motion effect	**Ready VR four slo-mo — roll VR four.** *or:* **Ready VR four slo-mo — slo-mo four.**
To roll a VR as a program insert while you are on camera 2; sound is on source. Assuming a 2-second roll.	**Ready to roll VR three, SOS — roll VR three.** **Two — one, take VR three, SOS.** If you do not use a countdown because of instant start, simply say: **Ready VR three, roll and take VR three.** [Start your stopwatch for timing the VR insert.]
To return from VR to camera and Mary on camera 1. (Stop your watch and reset it for the next insert.)	**Ten seconds to one, five seconds to one.** **Ready one, ready cue Mary — cue Mary, take one.**

Most production studios have several intercom outlets for plugging in the headsets for the floor manager, floor crew, and microphone boom operator. In addition, each camera has at least two intercom outlets: one for the camera operator and the other for a member of the floor crew. It is best, however, to have the floor crew connect their headsets to the wall outlets rather than the extra camera outlet. A tie to the camera not only limits their operation radius but also interferes with the camera's flexibility, and there is always the danger of the floor manager's yanking the camera while trying to catch the talent's attention.

Some shows require a simultaneous feed of program sound and control room signals to other production personnel, such as the microphone boom operator or studio musicians (usually the band or orchestra leader), who have to coordinate their actions with both the program sound and the director's cues. In such cases you can use a double headset wherein one of the earphones carries the intercom signals and the other carries the program sound.

Sometimes when you work in noisy surroundings or near a high-volume sound source, such as a rock band, you may need a double-muff headset, which filters out the

17.6 DIRECTOR'S CUES TO THE FLOOR MANAGER

The directional cues are always given from the camera's point of view, not the talent's. "Left" means camera-left; "right" means camera-right.

17.6 PHOTOS BY EDWARD AIONA

high-volume sounds at least to some degree. The mic in such headsets does not transmit the ambient sound and is activated only when you speak into it.

In most television operations, production and technical crews use the same intercom channel, which means that everyone can be heard by everyone else. Most of these systems also have provisions for separating the lines for different functions. For example, while the TD (technical director) confers with the video operator on one channel, the director may simultaneously give instructions to the floor crew. Larger studios and remote trucks provide a dozen or more separate intercom channels.

I.F.B. system You use an I.F.B. (interruptible foldback or interruptible feedback) system in shows with highly flexible formats or when program changes are likely to occur while on the air. The I.F.B. system connects the control room (director and producer) directly with the performers, bypassing the floor manager. The performer wears an earpiece that carries the total program sound (including his or her own voice) unless the director, the producer, or any other member of the control room team connected with the system interrupts the program sound with instructions.

For example, an on-camera field reporter in Washington who is describing the arrival of foreign dignitaries can hear herself until the director cuts in and says, "Throw it back to New York"—that is, for the talent to tell the viewers that the program is returning to the origination center in New York. But while the director is giving these instructions, the viewer still hears the field reporter's continuous description of the event. Relaying such messages through an off-camera floor manager would be much too slow and inaccurate in as tight a show as a newscast or a live telecast of a special event.

The producer uses I.F.B. to supply the host with follow-up questions while interviewing a guest. Needless to say, such a system works only with experienced talent and producer (or director). There are countless occasions when the I.F.B. system unfortunately acts as a performer-interrupt device because the inexperienced performer cannot maintain effective commentary while listening simultaneously to the producer's instructions.

S.A. system The S.A. (studio address) system is used by the control room personnel, primarily the director, to give instructions to people in the studio not connected by the P.L. system. Also called studio talkback, the S.A. system uses a studio loudspeaker similar to a public address system, helping communicate directly with everyone in the studio. For example, you may use it to give some general instructions to everybody, especially at the beginning of a rehearsal, or to inform talent and production personnel of a temporary delay. Also, if most personnel happen to be off the P.L. headsets, as is frequently the case during a short break, you can use the talkback system to call them back to work.

Considering the importance of the intercom systems, you should include them in routine facilities checks. If you discover faulty headsets or a noisy intercom line, report it to the maintenance crew and have it fixed right away. A malfunctioning intercom can be more detrimental to a multicamera production than a defective camera.

DIRECTING REHEARSALS

Unless you are directing routine shows or doing a remote pickup of a live event, you need to rehearse as much as possible. Rehearsals not only give you and the rest of the production team practice in what to do during the recording session but they also readily reveal any major and minor flaws or omissions in your production preparation.

Ideally, you should be able to rehearse everything that goes on the video recorder or on the air. Unfortunately, in practice this is hardly the case. Because the amount of scheduled rehearsal time always seems insufficient, the prerehearsal preparations (discussed in chapter 4) become extremely important. To make optimal use of the available time during scheduled rehearsals, you might try the following methods: script reading; dry run, or blocking rehearsal; walk-through; camera and dress rehearsals; and walk-through/camera rehearsal combination.

Note, however, that you rarely go through all these steps. Many nondramatic shows are rehearsed, if at all, simply by walking the talent through certain actions, such as moving to a display table and holding items properly for close-ups, or crossing the performance area to greet the guest.

Script Reading

Under ideal conditions every major production should begin with a script-reading session. Even for a relatively simple show, you should meet at least once with the talent, the producer, the PA (production assistant), and the key production personnel—AD (assistant director), TD, and floor manager—to discuss and read the script. Bring the floor plan along; it will help everyone visualize just where the action takes place and may reveal some potential production problems.

In a script-reading session, which normally doubles as a production meeting, you should address these points:

- Explain the process message, including the purpose of the show and its intended audience.
- Outline the major actions of the performers, the number and the use of hand props, and the major crossovers (walking from one performance area to another while on-camera).
- Discuss the performer's relationship to the guests, if any. In an interview, for example, discuss with the host the key questions and what he or she should know about the guest. Normally, such talent preparation is done by the producer before rehearsal.

Script-reading sessions are particularly important if you are rehearsing TV specials or a television drama. You will find that the time you invest in thorough script interpretation is regained during subsequent rehearsals.

In script-reading sessions for onetime plays, you should discuss the process message, the structure of the play (theme, plot, and environment), and the substance of each character. An extremely detailed analysis of the characters is probably the most important aspect of the dramatic script-reading session. The actor who really understands his or her character, role, and relationship to the event has mastered the principal part of his or her performance. After this analysis the actors tend to block themselves (under your careful guidance, of course) and move and "act" naturally. You no longer need to explain the motivation for each move. More than any other, the television actor must understand a character so well that he or she is no longer acting out but rather living the role. The internalization of the character, which can be readily achieved through extensive script-reading sessions, will almost always enhance the actor's performance.

When you direct a daily daytime serial or a weekly situation comedy with an ensemble cast, such intense and repeated character explorations are obviously superfluous. By the second or third episode, the actors will have a firm grip on their roles and how to relate to the rest of the cast.

Dry Run, or Blocking Rehearsal

In the ***dry run***, also called blocking rehearsal, the basic actions of the talent are worked out. By that time you must have a very good idea of where the cameras should be in relation to the set and where the actors should be in relation to the cameras and to one another.

The dry run presupposes a detailed floor plan and thorough preparation by the director. It also presumes that the actors have pretty much internalized their characters and roles. Tell them where the action should take place (the approximate location in the imagined set area—the actual set is rarely available at this point) and let them block as naturally as possible. Watch their actions as screen images, not from the point of view (POV) of a live audience. Adjust their blocking and your imagined camera positions so that you are reasonably assured that you will achieve the visualized screen image in the actual camera rehearsal. Do not fuss about specially framed shots at this time; you can always make such adjustments during the camera rehearsal. Be ready to give precise directions to the actor who is asking what to do next. Rather than always knowing what to do without the director's help, a good actor asks what to do and then does it convincingly and with precision.

The following are some general conventions to observe during a dry run.

- Hold the dry run in the studio or a rehearsal hall. In an emergency any room will do. Use tables, chairs, and chalk marks on the floor in place of sets and furniture.
- Work out the blocking problems. Use a director's viewfinder or a consumer camcorder, bearing in mind that a studio camera is not as flexible. Have the PA take notes of the major blocking maneuvers. Allow time later for a "notes" session to revisit any problems that need correcting.
- Try to block according to the actors' most natural movements, but keep in mind the camera and microphone positions and moves. Some directors walk right to the spot where the active camera will be and watch the proceedings from the camera's POV. If you block nondramatic action, observe first what the performers would do without the presence of a camera. As much as possible, try to place the cameras to suit the action rather than the other way around.
- If it will help, call out the major cues, such as "cue Lisa," "ready two, take two," and so forth.
- Run through the scenes in the order in which they are to be recorded. If you do the show live or live-recorded, try to go through the whole script at least once. If you cannot rehearse the whole script, pick the most complicated parts for rehearsal. In a nondramatic show, rehearse the opening as much as time allows. Inexperienced talent often stumble over the opening lines, with the show going downhill from there.

- Time each segment and the overall show. Allow time for long camera movements, music bridges, announcer's intro and closing, opening and closing credits, and so forth.

- Reconfirm the times and the dates of subsequent rehearsals.

Walk-through

The ***walk-through*** is an orientation session that helps the production crew and the talent understand the necessary medium and performance requirements. You can have both a technical walk-through and a talent walk-through. When pressed for time, or when doing a smaller production, combine the two.

The walk-throughs as well as the camera rehearsals occur shortly before the actual on-the-air performance or recording session. Walk-throughs are especially important when you are shooting on-location. The talent will get a feel for the environment, and the crew will discover possible obstacles to camera and microphone moves.

Technical walk-through Once the set is in place, gather the production crew—AD, floor manager, floor personnel, TD, LD (lighting director), camera operators, audio engineer, and boom or fishpole operator—for the technical walk-through. Explain the process message and your basic concept of the show. Then walk them through the set and explain these key factors: basic blocking and actions of talent, camera locations and traffic, specific shots and framings, mic locations and moves, basic cueing, scene and prop changes (if any), and major lighting effects.

The technical walk-through is especially important for electronic field production (EFP) and big remotes. Have the AD or PA take notes of all your major decisions, then have them read back and discussed at a scheduled "notes" session so that the technical crew can take care of the problems (reset).

Talent walk-through While the production people go about their tasks, take the performers on a talent walk-through—a brief excursion through the set or location; reiterate their major actions, positions, and crossovers. Tell them where the cameras will be in relation to their actions and whether they are to address the camera directly.

Here are some of the more important aspects of the talent walk-through:

- Point out to each performer his or her major positions and walks. If the performer is to look directly into the camera, indicate which camera and where it will be positioned.

- Explain briefly where and how the talent should work with specific props. For example, tell the actor that the coffeepot will be here and how he should walk with the coffee cup to the couch—in front of the table, not behind it. Explain your blocking to the talent from the POV of the camera. Have the performer go through part of the demonstration, and watch this simulation from the camera's POV. Watch that the performer does not block close-ups of the product he or she is demonstrating.

- Have all the talent go through their opening lines, then have them skip to the individual cue lines (often at the end of their dialogue). If the script calls for ad-lib commentary, ask the talent to ad-lib so that everyone will get an idea of what it sounds like.

- Allow enough time for makeup and dressing before the camera rehearsal. During the talent walk-through, try to stay out of the crew's way as much as possible. Again, have the AD or PA take notes. Finish the walk-through early enough so that you can have the "notes" session before everyone takes a break before the camera rehearsal.

Try to rehearse by yourself the opening and closing of a show prior to camera rehearsal. Sit in a quiet corner with the script and start calling out the opening shots: "Roll VR. Ready slate—take slate. Ready black, ready beeper. Black, beeper. Ready to cue Lynne. Ready to fade up on two. Open mic. Cue Lynne, up on two," and so on. By the time you enter the control room, you will practically have memorized the opening and closing of the show and will be able to pay full attention to the control room monitors and the audio.

Camera and Dress Rehearsals

The following discussion of camera rehearsals is primarily for studio productions and multicamera big remotes that are directed from a control room. (Camera rehearsals for EFPs are discussed in section 17.2.)

Essentially, the ***camera rehearsal*** is a full rehearsal that includes all cameras and other production equipment. In minor productions the camera rehearsal and the final ***dress rehearsal***, or dress, are almost always the same, the only difference being that the talent is already properly dressed and made up for the final recording.

Frequently, the camera rehearsal time is cut short by technical problems, such as lighting or mic adjustments. Do not get nervous when you see most of the technical crew working frantically on the intercom system or audio console five minutes before airtime. Be patient and stay calm. Realize that you are working with a highly skilled group who know just as well as you do how much depends on a successful performance. Like all other systems, the television system sometimes breaks down. Be ready to suggest alternatives should the problem prevail.

The two basic methods of conducting a camera rehearsal for a live or live-recorded production are the stop-start method and the uninterrupted run-through. A stop-start rehearsal is usually conducted from the control room, but it can also be done, at least partially, from the studio floor. An uninterrupted run-through is always conducted from the control room.

In both types of rehearsals, you should call for a "cut" (stop all action) only when a grave mistake has been made—one that cannot be corrected later. All minor mistakes and fumbles are corrected after the run-through. Dictate notes of all minor problems to the AD or PA and allow enough time for a reset.

In larger productions camera rehearsals and the dress rehearsal are conducted separately. Whereas in camera rehearsals the actors are not yet dressed and you may stop occasionally to correct some blocking or technical problem, dress rehearsals are normally done in full costume and are run straight through. You stop only when major production problems arise. Many times, as in the video recording of a situation comedy before a live audience, the recording of the dress rehearsal is combined in postproduction with that of the "on-the-air" performance to make the final edit master recording that is then broadcast.

Walk-through/Camera Rehearsal Combination

As necessary as the preceding rehearsal procedures seem, they are rarely possible in smaller operations. First, most directing chores in nonbroadcast or nonnetwork productions are of a nondramatic nature, demanding less rehearsal effort than dramatic shows. Second, because of time and space limitations, you are lucky to get rehearsal time equal to or slightly more than the running time of the entire show; 45 or even 30 minutes of rehearsal time for a half-hour show is not uncommon. In some cases you have to jump from a cursory script reading to a camera rehearsal immediately preceding the on-the-air performance or recording session. In these situations you have to resort to a walk-through/camera rehearsal combination.

Because you cannot rehearse the entire show, you simply rehearse as well as possible the most important parts. Usually, these are the transitions rather than the parts between them. *Always direct this rehearsal from the studio floor.* If you try to conduct it from the control room, you will waste valuable time explaining shots and blocking over the intercom system.

The following are some of the major points for conducting a walk-through/camera rehearsal combination.

- Get all production people into their respective positions: all camera operators at their cameras (with the cameras uncapped and ready to go); the fishpole mic ready to follow the sound source; the floor manager ready for cueing; and the TD, the audio console operator, and, if appropriate, the LD ready for action in the control room.

- Have the TD feed the studio monitor with a tripart or quad-split screen, with each of the three or four mini-screens showing a respective camera feed. This split-screen display will serve as your preview monitors. If you don't have a wireless intercom system, use a wireless or wired lavalier to relay your directing calls from the studio floor to the control room. If you cannot do such a split-screen display for your studio monitor, have the TD execute all your switching calls and feed the line-out pictures to the studio monitor. This way everybody can see the shots and the sequence you selected.

- Walk the talent through all the major parts of the show. Rehearse only the critical transitions, crossovers, and specific shots. Watch the action on the studio monitor.

- Give all cues for music, sound effects, lighting, VR rolls, slating procedures, and so forth to the TD via the open studio mic, but do not have them executed (except for the music, which can be easily reset).

- Even if you are on the floor yourself, have the floor manager cue the talent and mark the crucial spots on the studio floor with chalk or masking tape.

If everything goes well, you are ready to go to the control room. Do not allow yourself or the crew to get hung up on some insignificant detail.

Once you are in the control room, contact the cameras by number and verify that the operators can communicate with you. Then briefly rehearse once more from the control room the most important parts of the show: the opening, closing, major talent actions, and camera movements.

When in the control room, the only way you can see the floor action is via the preview monitors. You should therefore develop a mental map of the major talent and camera movements and of where the cameras are in relation to the primary performance areas. To help you construct and maintain this mental map, always try to position the cameras counterclockwise, with camera 1 on the left and the last camera on the far right.

As pressed for time as you may be, try to remain cool and courteous to everyone. This is not the time to make drastic changes; there will always be other ways in which the show might be directed and even improved, but the camera rehearsal is not the time to try them out. Reserve sudden creative inspirations for your next show. Stick as closely as possible to the time line. Do not rehearse right up to video recording or airtime. Give the talent and the crew a brief break before the actual performance. As mentioned earlier, don't just say, "Take five"—tell them the exact time to be back in the studio and ready to go.

DIRECTING THE SHOW

Directing the on-the-air performance or the final live recording session is, of course, the most important part of your job as a director. After all, the viewers do not sit in on the rehearsals—all they see and hear is what you finally put on the air. This section gives some pointers about standby procedures and on-the-air directing. Again, we assume that you are doing a live or live-recorded multicamera show or at least the video recording of fairly long, uninterrupted show segments that require a minimum of postproduction editing. You will notice that you can transfer multicamera-directing skills much more readily to single-camera direction than the other way around.

Standby Procedures

Here are some of the most important standby procedures to observe immediately preceding the recording session or on-the-air telecast.

- Call on the intercom every member of the production team who needs to react to your cues—TD, camera operators, boom operator, floor manager and other floor personnel, VR operator, LD or lighting technician, audio technician, and character generator (C.G.) operator. Ask them if they are ready.

- Check with the floor manager to make sure everyone is in the studio and ready for action. Tell the floor manager who gets the opening cue and which camera will be on first. From this point the floor manager is an essential link between you and the studio.

- Announce the time remaining until the telecast or recording. If you are directing a live recording or show segments, have the TD, C.G. operator, and audio engineer ready for the opening slate identification. You can save time by having the AD or TD direct the recording of the video leader (bars and tone) before airtime. Check that the slate shows the correct information. Verify the spelling of names that you will use as key inserts and for the credit crawl.

- Again, alert everyone to the first cues.

On-the-air Procedures

Assuming you direct a largely uninterrupted video recording, such as the interview with the singers described in chapter 4, you must first go through the usual VR rolling procedures (see figure 17.5).

Directing from the control room Once the video recorder is rolling and active, and the scene and the take are slated, you can begin the actual recording. You are now on the air. Imagine the following opening sequence:

> Ready to come up on three [CU of Lynne, the interview host]. Ready to cue Lynne. Open mic, cue Lynne, up on [or "fade in"] three [Lynne addresses camera 3 with opening sentence]. Ready C.G. opening titles—take C.G. Cue announcer. Change page. Change page. Ready three [which is still on Lynne]. Cue Lynne—take three [introduces guests]. One, two-shot of singers. Two, cover [wide shot of all three]. Ready one—take one. Ready two, open mics [guest mics in the interview set]—take two. Ready three—take three [Lynne is asking her first question]. One, on Ron [CU of one of the singers]. Ready one—take one. Two, on Marissa [the other singer]. Ready two—take two.

By now you are well into the show. Listen carefully to what is being said so that you can anticipate the upcoming shots. Have the floor manager stand by to give Lynne time cues to the crossover. The end of the interview marks the talent's crossover to the performance area. Before you stop the recording for the crossover, wait until the singers have stepped out of the frame. This way you can logically cut from a close-up of the singers to an establishing shot of the performance area. Stop down again for the singers to return to the interview area. When the singers are back on the interview set, watch the time carefully and give closing time cues to Lynne, who is thanking the singers for their great performance.

After the one-minute cue, you must prepare for the closing. Are the closing credits ready? Again, watch the time.

> Thirty seconds. Wind her up. Wind her up [or "give her a wrap-up"]. Fifteen [seconds]. C.G. closing credits. Two, zoom in on Ron's guitar. Ready two, ready C.G. Cut Lynne. Cut mics. Roll credits. Take two. Key C.G.

[over camera 2]. Lose key. Ready black—fade to black. Hold. Stop VR.

OK, all clear. Good job, everyone.

Unfortunately, not every show goes this smoothly. You can contribute to a smooth performance, however, by paying attention to the following on-the-air directing procedures.

- Give all signals clearly and precisely. Be relaxed but alert. If you are too relaxed, everyone will become somewhat lethargic, thinking that you don't really take the show seriously.

- Cue talent before you come up on him or her with the camera. By the time he or she speaks, the TD will have faded in the picture.

- Indicate talent by name. Do not tell the floor manager to cue just "him" or "her," especially if the talent consists of several "hims" or "hers" anticipating a cue sooner or later.

- Do not give a ready cue too far in advance or the TD or camera operator may have forgotten it by the time your take cue finally arrives. Repeating the same ready cue may trigger a take by the TD.

- Do not pause between the take and the number of the camera. Do not say, "Take [pause] two." Some TDs may punch up the camera before you say the number.

- Keep in mind the number of the camera already on the air, and do not call for a take or dissolve to that camera. Watch the preview monitors as much as possible. Do not bury your head in your script or fact sheet.

- Do not ready one camera and then call for a take to another. In other words, do not say, "Ready one—take two." If you change your mind, nullify the ready cue—"No" or "Change that"—then give another.

- Talk to the cameras by number, not by the name of the operator.

- Call the camera first before you give instructions. For example: "One, stay on the guitar. Two, give me a close-up of Ron. Three, CU of Marissa. One, zoom in on the guitar."

- After you have put one camera on the air, immediately tell the other camera what to do next. Do not wait until the last second; for example, say, "Take two. One, stay on this medium shot. Three, tight on Ron's guitar." If you reposition a camera, give the operator time to recalibrate the zoom lens; otherwise the camera will not stay in focus during subsequent zooms.

- If you make a mistake, correct it as well as you can and go on with the show. Do not meditate on how you could have avoided it while neglecting the rest of the show. Pay full attention to what is going on.

- If you record each take separately, spot-check the video recording after each take to make sure that the take is actually recorded and technically acceptable. Check the audio. Then go on to the next one. It is always easier to repeat a take immediately than to go back at the end of a strenuous recording session and try to recapture the mood and the energy level of the original take.

- If there is a technical problem that you must solve from the control room, tell the floor manager about it on the intercom or use the S.A. system to inform the whole studio about the delay. The talent then know that there is a technical delay and that it was not caused by them. The people on the floor can use this time to relax, however busy it may be for those in the control room.

- During the show speak only when necessary. If you talk too much, people will stop listening and may miss important instructions. Worse, the crew will follow your example and start chatting on the intercom.

- When you have the line in black (your final fade to black), call for a VR stop and give the all-clear signal. Thank the crew and the talent for their efforts. If something went wrong, do not storm into the studio to complain. Take a few minutes to catch your breath, then talk calmly to the people responsible for the problem. Be constructive in your criticism to help them avoid the mistake in the future. Just telling them that they made a mistake helps little at this point.

CONTROLLING CLOCK TIME

In commercial television, time is indeed money: each second of broadcast time has a monetary value. Salespeople sell time to their clients as a tangible commodity. One second of airtime may cost much more than another, depending on the potential audience an event may command.

As director, you are ultimately responsible for getting a program on and off the air on time and for the proper timing of segments, such as news packages. Precise timing is also important because as a successful director you will inevitably be called upon to connect with international broadcasting services. The Internet has local-time clocks for most major cities in the world. If you were asked to switch during your morning show to Cape Town for a live interview with a winemaker, you should know that South Africa is nine hours ahead. If the interview is scheduled for 6:30 a.m. PDT (Pacific daylight time), the floor manager in Cape Town would have to cue the winemaker at 3:30 p.m.

Schedule Time and Running Time

Schedule time is the clock time at which a program starts and ends. When doing a live show, you need to start and stop exactly according to the start and end schedule times as indicated in the daily program log. When video-recording a show, you don't have to worry about its schedule time, but you are still responsible for the accurate ***running time***—the actual duration of a program or program segment—so that it can fit the prescribed time slot in the day's programming.

When directing a live show such as a newscast, you use the control room clock for meeting the schedule times (the switch to network news) and the stopwatch for measuring the running times of the program inserts (the individual video-recorded stories).

Back-timing and Front-timing

Although the master control computer calculates almost all the start and end times of programs and program inserts, and pocket calculators help you add and subtract clock times, you should nevertheless know how to do time calculations even in the absence of electronic devices. For example, a performer may request at the last minute specific time cues, which you then have to figure by hand. Note that you need to convert the seconds and the minutes on a scale of 60 rather than 100.

Back-timing One of the most common time controls involves cues to the talent so that he or she can end the program as indicated by the schedule time. In a 30-minute program, the talent normally expects a 5-minute cue and subsequent cues with 3 minutes, 2 minutes, 1 minute, 30 seconds, and 15 seconds remaining in the show. To figure out such time cues quickly, you simply back-time from the scheduled end time or the start time of the new program segment (which is the same thing). For example, if the program log shows that your live *What's Your Opinion?* show is followed by a Salvation Army public service announcement (PSA) at 4:29:30, at what clock times do you give the talent the standard time cues, assuming that the recorded close takes 30 seconds?

You should start with the end time of the panel discussion, which is 4:29:00, and subtract the various time segments. (You do not back-time from the end of the program at 4:29:30 because the standard video-recorded close will take up 30 seconds.) When, for example, should the moderator get her 3-minute cue or the 15-second wind-up cue?

Let's proceed with back-timing this program:

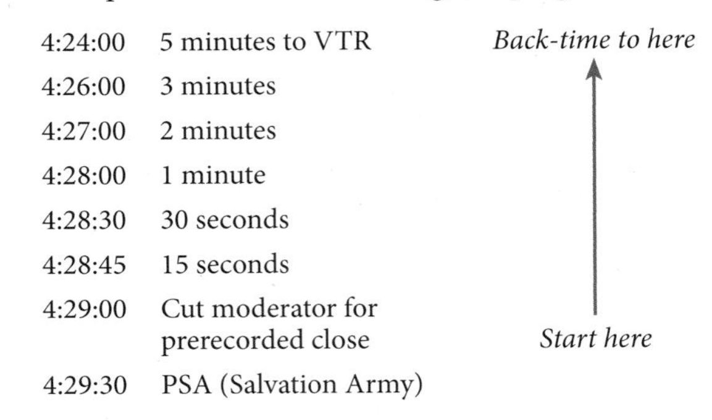

4:24:00	5 minutes to VTR	*Back-time to here* ↑
4:26:00	3 minutes	
4:27:00	2 minutes	
4:28:00	1 minute	
4:28:30	30 seconds	
4:28:45	15 seconds	
4:29:00	Cut moderator for prerecorded close	*Start here*
4:29:30	PSA (Salvation Army)	

Front-timing To keep a show—such as a live newscast with many recorded inserts—on time, you need to know more than the start and end times of the program and the running times of the inserts. You also need to know when (using clock time) the inserts are to be run; otherwise you cannot figure out whether you are ahead or behind with the total show.

To figure out the additional clock times for each break or insert, simply add the running times to the initial schedule time as shown on the program log or show format.

Converting Frames into Clock Time

Because in the NTSC system we figure that 30 frames make up 1 second, the frames roll over after 29, with the thirtieth frame starting the new second. But seconds and minutes roll over after 59. You must therefore convert frames into seconds, or seconds into frames, when front- or back-timing time code numbers. Again, you need to compute the frames, seconds, minutes, and hours individually, then convert the frames on the 30 scale and the seconds and the minutes on the 60 scale.

For example:

```
  00:01:58:29
+ 00:00:03:17
  00:01:61:46 ——> 00:01:62:16 ——> 00:02:02:16
```

Note that you simply added the frames, then subtracted 30 for the additional second.

Fortunately, the time code will do this for you automatically. There are also computer programs and handheld calculators available that take care of the rollovers of clock time as well as frame time.

MAIN POINTS

- As a television director, you must be an artist who can translate a script or an event into effective television pictures and sound, a psychologist who can work with people of different temperaments and skills, a technical adviser who knows the potentials and the limitations of the equipment, and a coordinator who can initiate and keep track of myriad production processes.
- A precise directing terminology is essential for the proper coordination of talent and crew and the instantaneous editing during a show.
- Multicamera directing involves the simultaneous use of two or more cameras and instantaneous editing with a switcher. It is done from the control room.
- Directing from the control room involves not only the proper visualization of shots but also their immediate sequencing.
- Well functioning intercom systems—P.L. (private line or phone line), I.F.B. (interruptible foldback or interruptible feedback), and S.A. (studio address)—are essential for successful control room directing.
- The different types of rehearsals include script reading; dry run, or blocking rehearsal; walk-through; camera and dress rehearsals; and walk-through/camera rehearsal combination.
- The two important clock times are schedule time (start and end of a program) and running time (program length).
- Back-timing means figuring specific clock times (usually for cues) by subtracting the running time from the schedule time at which the program ends.
- Front-timing means starting at the schedule time that marks the beginning of a program and then adding specific running times.
- When converting frames into clock time, you must have the frames roll over to the next second after 29, but seconds and minutes after 59.

SECTION

17.2

Single-camera and Digital Cinema Directing

In *single-camera directing* you are concerned primarily with directing individual takes for later assembly in postproduction. The big difference between directing multicamera and single-camera productions is that multicamera productions are continuous and single-camera productions are discontinuous. *Continuous* in this context means that you do not stop after each shot but rather sequence a series of shots through instantaneous editing (switching) with a minimum of interruption or none at all. In single-camera studio productions, the video recording is discontinuous: you no longer intend to video-record a finished product that needs little or no postproduction for broadcast but rather to produce effective source material that can be shaped into a continuous program through extensive postproduction editing.

Digital cinema is included because it is mostly single-camera but can also be done with a multicamera setup.

- **SINGLE-CAMERA DIRECTING PROCEDURES**
  Visualization, script breakdown, rehearsals, and video recording

- **DIRECTING MULTICAMERA DIGITAL CINEMA**
  Directing from the studio floor and directing on-location

SINGLE-CAMERA DIRECTING PROCEDURES

This section focuses on the following major aspects of single-camera studio directing: visualization, script breakdown, rehearsals, and video recording. ZVL2 PROCESS→ Methods→ single-camera

Visualization

Even if you are video-recording a production discontinuously—shot-by-shot—your basic visualization is not much different from what it would be when continuous-shooting with multiple cameras and instantaneous editing. As described in chapter 4, the first reading of a script may conjure up some locking-in points—key visualizations that set the style for the entire production. This process is intuitive and depends a great deal on your own perception of the environment and the situation. It also depends on how you perceive the people in a documentary or the development of characters in a scripted drama.

Once you have established locking-in points that determine your general shooting style, you must go back to the script and break it down for discontinuous video recording. The order in which you record the shots is no longer guided by the script context, the narrative, or even aesthetic continuity but strictly by convenience and efficiency. For example, you may want to video-record all the scenes in the hospital corridor, then the waiting-room scenes, then all the operating-room scenes, then all the scenes in the patient's room, and so forth, irrespective of when they actually occur in the story.

To give you an idea of how script preparation for multicamera shooting differs from the single-camera approach, take another look at figure 4.13, showing the director's markings of a brief multicamera drama script. How would you break down the very same script segment for a single-camera shoot? Write down a series of shots that show Yolanda meeting Carrie in the hospital hallway, then compare it with the breakdown in figure 17.7. SEE 17.7

Script Breakdown

As you can see, the breakdown is more detailed and not necessarily in the order of the action. Note that this script breakdown is just one of many possibilities. Some breakdowns list the corresponding page numbers to the master script and some continuity information, such as the time of day. For example, the motion picture industry (including digital cinema) has established its own script breakdown conventions.[1] You may find, however, that such detailed markings would make most of the single-camera video productions unnecessarily complicated.

If you are shooting several scenes in the same location, you may want to video-record the scenes involving specific characters first, and then move on to the next set of characters. This way they can be scheduled at different times of the day. Shooting a scene piecemeal requires that the actors repeat their lines and actions several times

1. See Michael Rabiger, *Directing: Film Techniques and Aesthetics*, 3rd ed. (Boston: Focal Press, 2003), p. 113.

RECEPTION ROOM AND HALLWAY

1. Yolanda in the reception room
2. Hallway: Yolanda pacing up and down the hallway in the vicinity of the emergency room
3. Hallway: Typical hospital traffic--nurses, a gurney, a wheelchair, visitors with flowers, a doctor and a nurse, a physical therapist protecting a person on crutches
4. Hallway: Doctor pushes Carrie in wheelchair
5. POV Carrie: Yolanda
6. CU Yolanda
7. POV Yolanda: Doctor and Carrie

YOLANDA RUSHING TOWARD DOCTOR AND CARRIE

1. Hallway: Yolanda rushes toward Doctor and Carrie
2. Reverse-angle shot (POV Carrie): Yolanda
3. Same shots with gurney traffic interfering with Yolanda's approach (Steadicam)

CARRIE AND YOLANDA

1. CU Carrie: "Hi, Mom!"
2. CU Yolanda: "Carrie--are you all right? What happened?"
3. CU swish pan from Carrie to Yolanda: "Carrie--are you all right? What happened?"
4. CUs and ECUs of Carrie
5. CUs and ECUs of Yolanda
6. CUs and ECUs of Doctor

17.7 SINGLE-CAMERA SCRIPT BREAKDOWN
Takes are grouped for convenience and efficiency, not for narrative order.

identically; you must watch carefully that the individual shots will eventually cut together into a seamless scene. This means that, besides vector continuity, you must also connect the various visualization points so that the scene and the sequences have narrative (story) continuity and continuity of aesthetic energy.

Continuity All the shots in a sequence must connect seamlessly so that they are no longer recognized by the audience as isolated shots but as a single scene. As explained in chapter 4, a detailed storyboard will aid you greatly in seeing individual shots as a sequence. Even if you don't have the time or resources to design storyboards for each sequence, you must try to visualize how well the shots cut together and watch for continuity errors during the video recording. If, for example, Yolanda kisses her daughter on the left cheek in the medium shot, do not let her switch to the right cheek during the close-ups of the same scene.

Close-ups and cutaways Whatever breakdown you have, be sure to get some close-ups and extreme close-ups of all the people in the scene for intensification and possible cutaways. You can assign the AD or second-unit director to take care of such pickups. How you start and end a specific take can make the postproduction editor's job a delight or a nightmare. As a director you are responsible for providing the editor with shots that can eventually be assembled into a continuous and sensible sequence.

Rehearsals

As mentioned earlier, in single-camera directing you rehearse each take immediately before video-recording it. Walk the talent and the camera and microphone operators through each take, explaining what they should and should not do. Have the single camera connected to a monitor so that you can watch the action on-screen and, if necessary, make necessary corrections before the video recording. When on-location, you can set up battery-powered monitors on a simple card table and relay your directions to the camera operators via a P.L. system.

Video Recording

Slate each take. When on-location have the floor manager use a simple handheld slate or clapstick. If in a hurry, ask the floor manager to audio-slate the takes by calling the next take number and title into the hot mic. Have the VR operator or the PA keep an accurate field log. Always watch for continuity mistakes but be careful not to wear out talent and crew with too many retakes; there is a point where retakes become counterproductive because of talent and crew fatigue. Finally, have the VR operator or the PA label all recording media and boxes and verify that the labels correspond with the field log.

DIRECTING MULTICAMERA DIGITAL CINEMA

When using a multicamera setup for digital cinema or a complex EFP, and assuming that you have a competent AD, you can direct some fully scripted scenes from the studio floor, very much like doing a walk-through/camera rehearsal combination, or from a temporary control room that is set up on-location.

Directing from the Studio Floor

In this case, all cameras function as iso cameras that also feed the switcher. You rehearse each take from the floor and then watch the performance on a studio monitor while the AD calls the shots from the control room. One camera is usually on an establishing shot, another with a Steadicam mount gets the CUs and more fluid shots, and the rest of the cameras are assigned to different-angle shots. Much like in a walk-through/camera rehearsal combination, you can have the studio monitor show a quad-split screen that serves as four separate preview monitors of the various camera views.

Directing On-location

When on-location you need to set up a control center on the set unless you use a remote truck (see chapter 18). This "control room" usually consists of a large flat-panel display with preview images (smaller screens for each camera preview, and larger preview and line monitor displays), a portable switcher, and intercom systems that connect the control center with the cameras and the floor manager.

Much like in the studio setup, each HDTV camcorder is placed in an iso position but also feeds into the switcher. You can then either call for the various shots on the intercom or have the AD conduct the instantaneous editing while you watch the performance. Such a setup is especially helpful when covering onetime-only events, such as complicated stunts, explosions, and the like. If all goes well, the customary repeats for close-ups and different-angle shots are no longer necessary.

The advantage of such a multicamera setup is that you end up with a director's cut through instantaneous editing in addition to the recordings of the individual cameras that can be used for postproduction editing. You can then watch the flow of the scene or scenes and, if necessary, do some retakes right then and there.

MAIN POINTS

- Like multicamera directing, single-camera directing starts with the visualization of key shots.
- The script breakdown in single-camera directing is guided more by production convenience and efficiency than by visualization and sequencing. The production sequence is dictated not by the event sequence but by such production factors as location and getting various points of view or close-ups of the same action.
- Each take is normally rehearsed immediately before it is recorded.
- When video-recording the separate takes, you need to slate each with a handheld slate or clapstick, or at least have the floor manager do a verbal audio slate. Label the boxes and the recording media.
- When using a multicamera setup for digital cinema or a complex EFP, you can direct some fully scripted scenes from the studio floor or from a temporary control room that is set up on-location.

ZETTL'S VIDEOLAB

For your reference or to track your work, the *Zettl's VideoLab* program cues in this chapter are listed here with their corresponding page numbers.

ZVL1 PROCESS→ Methods→ multicamera 366

ZVL2 PROCESS→ Methods→ single-camera 380

CHAPTER

18

Field Production and Big Remotes

When you see one of those big television trailer rigs pull up and production crews start unloading cameras, microwave and satellite uplink dishes, miles of cable, and other pieces of television equipment, you know that a big remote is in the offing. Why undergo such an effort when you could simply grab a few small camcorders to shoot the same field event? This chapter provides some answers.

Section 18.1, Field Production, takes a closer look at each of the three field production methods: ENG (electronic news gathering), EFP (electronic field production), and big remotes. Section 18.2, Covering Major Events, offers further information about standard television setups of sports remotes and other field events, interpreting location sketches, field communication systems, and signal transport.

KEY TERMS

big remote A production outside of the studio to televise live and/or live-record a large scheduled event that has not been staged specifically for television. Examples include sporting events, parades, political gatherings, and studio shows that are taken on the road. Also called *remote.*

broadband A high-bandwidth standard for sending information (voice, data, video, and audio) simultaneously over cables.

direct broadcast satellite (DBS) Satellite with a relatively high-powered transponder (transmitter/receiver) that broadcasts from the satellite to small, individual downlink dishes; operates on the Ku-band.

downlink The antenna (dish) and equipment that receive the signals from a satellite.

field production All productions that happen outside of the studio; generally refers to electronic field production.

instant replay Repeating for the viewer, by playing back a key play or an important event immediately after its live occurrence. In drama it is used as a visualized recall of a prior event.

iso camera Stands for *isolated camera.* Feeds into the switcher and into its own separate video recorder.

Ku-band A high-frequency band used by certain satellites for signal transport and distribution. The Ku-band signals can be influenced by heavy rain or snow.

live recording The uninterrupted video recording of a live show for later unedited playback. Formerly called live-on-tape, live recording includes all recording devices.

location sketch A rough map of the locale of a remote shoot. For an indoor remote, it shows the room dimensions and the furniture and window locations. For an outdoor remote, it indicates the location of buildings, the remote truck, power sources, and the sun during the time of the telecast.

microwave relay A transmission method from the remote location to the station and/or transmitter involving the use of several microwave units.

mini-link Several microwave setups that are linked together to transport the video and audio signals past obstacles to their destination (usually the television station and/or transmitter).

remote survey A preproduction investigation of the location premises and event circumstances. Also called *site survey.*

remote truck The vehicle that carries the program control, the audio control, the video-recording and instant-replay control, the technical control, and the transmission equipment.

uplink Earth station transmitter used to send video and audio signals to a satellite.

uplink truck A truck that sends video and audio signals to a satellite.

VJ Stands for *video journalist.* A television news reporter who shoots, edits, and writes his or her own material.

SECTION

18.1 Field Production

When a television production happens outside the studio, it is known as a ***field production***. We normally distinguish among electronic news gathering that covers daily news events, electronic field production that deals with smaller scheduled events, and big remotes that are done for major events, such as sports, parades, and political conventions.

There are advantages to taking a production out of the studio and into the field:

- You can place or observe an event in its real setting or select a specific setting for a fictional event.
- You can choose from an unlimited number and a variety of highly realistic settings.
- You can use available light and background sounds so long as they accomplish your technical and aesthetic production requirements.
- You can save on production people and equipment because many EFP productions require less equipment and crew than similar studio productions (unless you do a complex EFP or a big remote).
- You avoid considerable rental costs for studio use and, if you work for a station, studio scheduling problems.

There are disadvantages as well:

- You do not have the production control the studio affords. Good lighting is often difficult to achieve in the field, in both indoor and outdoor locations, as is high-quality audio.
- On outdoor shoots the weather always presents a hazard. For example, rain or snow can cause serious delays simply because it is too wet or too cold to shoot outside. A few clouds may give you considerable continuity problems when the preceding takes showed clear skies.
- You are always location-dependent, which means that some locations require the close cooperation of nonproduction people. For example, if you shoot on a busy downtown street, you will need the help of the police to control traffic and onlookers.
- When shooting on city, county, or federal property, you may need a permit from these agencies plus additional insurance stipulated by them.
- Field productions also normally require crew travel and lodging as well as equipment transportation.

As a television professional, you need to cope with these disadvantages. After all, you can't squeeze a football field into a studio. With ENG and relatively simple EFP, the production efficiency of shooting in the field usually outweighs the lack of production control. Although ENG and EFP have been discussed throughout this book, the focus here is on their specific field production requirements.

▶ **ELECTRONIC NEWS GATHERING**
ENG production features and satellite uplink

▶ **ELECTRONIC FIELD PRODUCTION**
EFP preproduction; production with equipment check, setup, rehearsals, video recording, and strike and equipment check; and postproduction

▶ **BIG REMOTES**
Preproduction remote survey, equipment setup and operation, and production procedures for the director, the floor manager, and the talent

ELECTRONIC NEWS GATHERING

Electronic news gathering is the most flexible remote operation. As pointed out in previous chapters, one person with a camcorder can handle a complete ENG assignment. Even if the signal must be relayed back to the station or transmitter, ENG requires only a fraction of the equipment and the people of a big remote. Sometimes the shooter, or videographer (news camera operator), or even the ***VJ*** (video journalist) will also take care of the signal feed from the news vehicle to the station. With WiFi access, the video clips can be uploaded and sent to the station much like e-mail.

ENG PRODUCTION FEATURES

The major production features of ENG are the readiness with which you can respond to an event, the mobility possible in the coverage of an event, and the flexibility of ENG equipment and people. Because ENG equipment is compact and self-contained, you can get to an event and video-record or broadcast it faster than with any other type of television equipment. An important operational difference between ENG and EFP or big remotes is that ENG requires no standard preproduction. ENG systems are specifically designed for immediate response to a breaking story. In ENG you exercise no control over the event but merely observe it with your camcorder and microphone as best you can.

Even when working under extreme conditions and time restrictions, experienced VJs and shooters can quickly analyze an event, select its most important parts, and video-record pictures that edit together well. For important events the ENG team normally consists of two people—the videographer and the field reporter; but many ENG stories are covered by a single VJ or shooter and narrated later by the anchor during the newscast. ENG equipment can go wherever you go. It can operate in a car, an elevator, a helicopter, or a small kitchen. Your shoulder or arms usually substitute for a heavy tripod.

With ENG equipment you can either video-record an event or transmit it live. Note that video recording includes a variety of recording methods, such as videotape, hard drives, optical discs, and memory cards (flash drives). The transmission equipment has become so compact and flexible that a single camera operator can accomplish even a live transmission. Most ENG vehicles (usually vans) are equipped with a microwave transmitter, which, when extended, can establish a transmission link from the remote location to the station. **SEE 18.1** Some ENG camcorders have a transmitter attached for signal transmission over relatively short distances.

18.1 ENG VAN WITH EXTENDED MICROWAVE TRANSMITTER
This regular-sized van houses the extendible microwave transmission device and a variety of recording and intercommunications equipment.

When doing a live transmission, you connect the camera cable to the microwave transmitter. You can also use such a microwave link to quickly transmit to the station the unedited video recording directly from the camcorder or the video recorder (VR) in the ENG van.

Satellite Uplink

If you cannot establish a signal connection between your ENG location and the station, you need to request a satellite uplink van. This van looks like a small remote truck and contains computer-assisted uplink equipment and, when used for news, one or two video recorders as well as editing equipment (often laptops). You can also play back the captured footage directly from the camcorder. When time permits, you should transfer the video from the camcorder to the VR in the van. **SEE 18.2**

Newspeople prefer uplinking "hot" video recordings (those made moments before the transmission) to live transmission because it permits repeated transmission in case the satellite feed is temporarily interrupted or lost altogether. To further safeguard against signal loss, two video recorders are sometimes used for the recording and

18.2 SATELLITE UPLINK VAN
The satellite news vehicle is a portable earth station. It sends television signals to the Ku-band satellites.

18.2 PHOTO BY GARY PALMATIER

the playback of the same news story. If something goes wrong with one VR, you can quickly switch over to the next for the same material.

Such uplinks are pressed into service whenever big and especially newsworthy events are scheduled, such as a presidential election, a summit of heads of state, a high-profile criminal trial, or the world soccer finals. But the uplink truck is also used locally for the distribution of news stories, national and international teleconferencing, and whenever a signal cannot be sent readily by microwave or cable.

ELECTRONIC FIELD PRODUCTION

As you already know, EFP uses both ENG and studio techniques. From ENG it borrows its mobility and flexibility; from the studio it borrows its production care and quality control. The following discussion of some fundamental steps of preproduction, production, and postproduction in the field assumes that you are still functioning as the director, so you have to deal with production detail that is important for each member of the EFP team regardless of the specific jobs assigned. ZVL1 PROCESS→ Methods→ location | studio

Preproduction

Compared with ENG, in which you simply respond to a situation, EFP requires careful planning. In general, the preproduction steps are almost the same for EFP as for a studio production. As a first step, you need to formulate a workable process message, which you then need to translate into the most effective and efficient production method—whether to shoot it indoors or outdoors, single- or multi-camera, in the normal sequence of events or shot-by-shot.

Production Preparation

The preparation for the actual production starts with deciding on the medium requirements—the equipment and the people. Assuming that you have practiced this translation of process message into production requirements (see chapter 1), we jump to the production phase and the initial production steps: location survey, initial production meeting, and field production time line.

Location survey To get to know the environment in which the production will take place, make an accurate ***location sketch***—a map of the locale of a remote telecast. For an indoor remote, the sketch shows the room dimensions and the furniture and window locations. For an

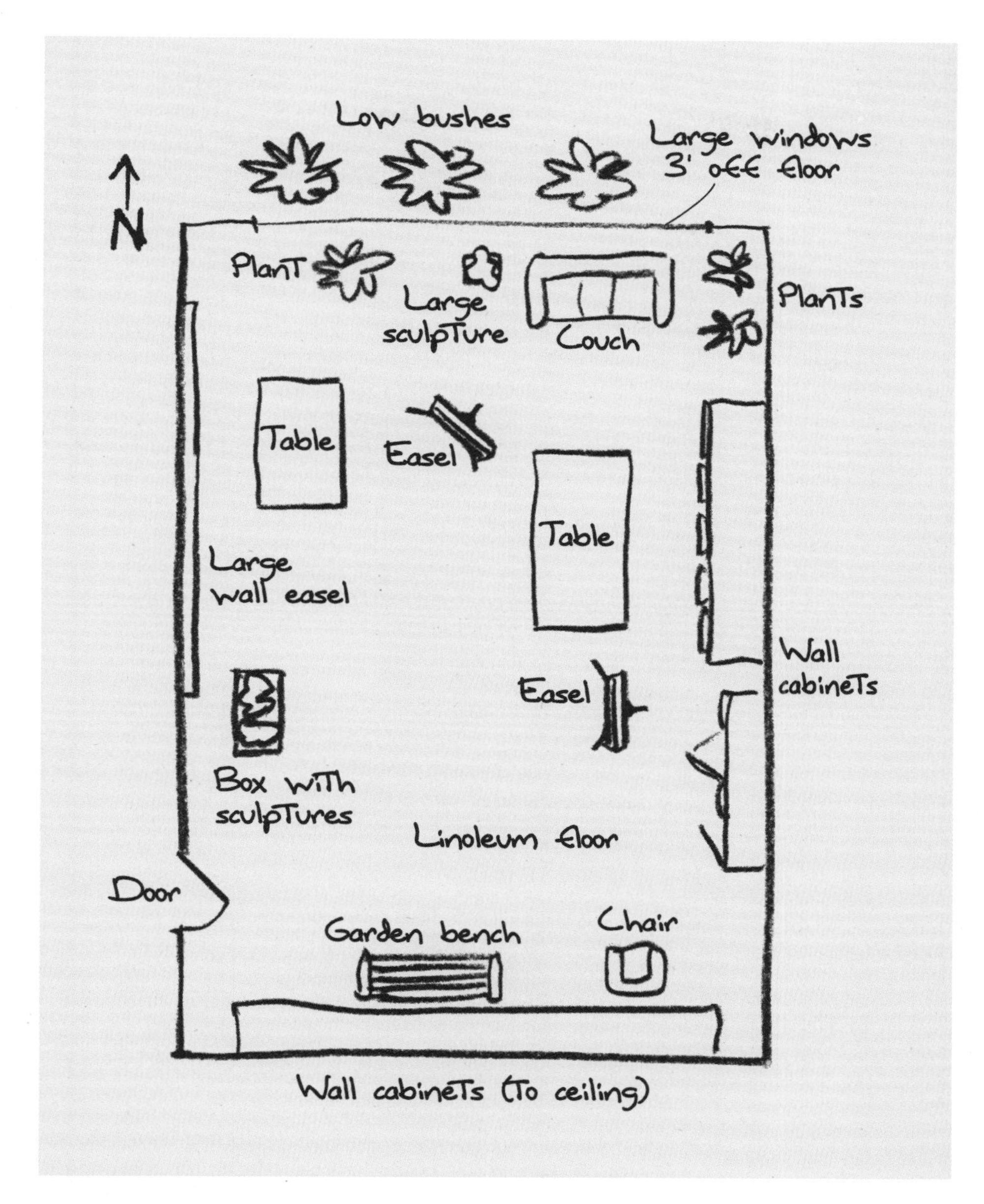

18.3 INDOOR LOCATION SKETCH: ARTIST'S STUDIO
This location sketch of an artist's studio shows the basic dimensions and the doors, windows, and furnishings.

outdoor remote, it indicates the location of buildings, the EFP vehicle, power sources, and the sun during the time of the telecast.

Take a look at the location sketch of an artist's studio. **SEE 18.3** This sketch gives important information about lighting and audio requirements, camera positions, and shooting sequences. Although technical preparations may not be your immediate concern, check on the availability of power (wall outlets), the acoustics (small room, reflective walls and floor, and traffic noise from nearby freeway), and potential lighting problems (large windows).

If the production is literally in a field, are the most basic conveniences available? (Location surveys are discussed further in the context of big remotes later in this chapter.)

Initial production meeting The initial production meeting ideally involves all key personnel, including the

PA (production assistant), the floor manager, and the DP (director of photography) or camera operator (who will also take care of the lighting). At a minimum you should meet with the PA (who may double as the audio/VR operator) and the camera operator. Explain the process message and what you hope to accomplish. Distribute the location sketch and discuss the major production steps.

It is critical that everyone knows the exact location of the production and how to get there. Can everyone fit into the EFP van? Who is riding with whom? Who needs to come first to the station or production facility for equipment check-out, and who will go directly to the location? Who will drive the van? Hand out the time line and ask the PA to distribute it (via fax and/or e-mail) to all other crewmembers who may not be at the meeting. As you can see, transportation to and from the location is an essential scheduling issue. If the field production is outdoors, what do you do in case of rain or snow? Always have an alternate time line ready.

Field production time line A shooting schedule for a fairly elaborate field production may look like this:

TIME LINE: FIELD PRODUCTION

7:30–8:15 a.m.	Equipment check-out
8:15 a.m.	Departure
9:15 a.m.	Estimated arrival time
9:30–10:00 a.m.	Production meeting with talent and crew
10:00–11:00 a.m.	Technical setup
11:00–11:30 a.m.	Lunch
11:30 a.m.–12:00 p.m.	Technical and talent walk-through
12:00–12:20 p.m.	Notes and reset
12:20–12:30 p.m.	Break
12:30–1:00 p.m.	Segment 1 video recording
1:00–1:15 p.m.	Notes and reset for segment 2
1:15–1:45 p.m.	Segment 2 video recording
1:45–1:55 p.m.	Break
1:55–2:10 p.m.	Notes and reset for segment 3
2:10–2:40 p.m.	Segment 3 video recording
2:40–3:00 p.m.	Spill
3:00–3:30 p.m.	Strike
3:30 p.m.	Departure
4:30 p.m.	Estimated arrival time at station
4:30–4:45 p.m.	Equipment check-in

Production: Equipment Check

Again, apprise the crew of the schedule and the aim of the production. Go over the time line and the fact sheet of the major locations and recording sessions. Be extracareful when loading the equipment. Unlike studio productions, where all the equipment is close at hand, in field productions you need to bring everything to the location. Even if you and the crew are involved in shooting an outdoor series, which requires for each EFP similar or identical equipment, use an equipment checklist before dashing off to the location. A wrong cable or adapter can cause undue delays or even the cancellation of the entire production.

Before loading equipment onto a vehicle, check each item to see that it works. At a minimum do a test recording of picture and sound before leaving for the location.

Equipment checklist The following equipment checklist is intended as a general guide and may not include all the items you need to take along. Depending on the relative complexity of the EFP, you may need considerably less or more than the items listed.

- *Cameras.* How many? Have they been checked out? Do you have the appropriate lens attachments (usually filters), if any? What camera mounts do you need: tripods, tripod dollies, special mounts? Do you have enough batteries? Are they fully charged? Do they fit the specific camcorders you use in the EFP?

- *Video recorders and recording media.* If you use field cameras instead of camcorders, do they connect with the video recorders? Do you have the proper recording media for the VRs? Do you have enough? Check that the box actually contains the necessary media (specific tape, memory cards, optical discs). Do all audio cables connect properly from camera to VR?

- *Intercom and technical equipment.* If the EFP requires the simultaneous use of multiple cameras, do you have provisions for intercom and tally lights for the cameras? Will the camera and audio cables fit the jacks of the portable switcher (see figure 14.18)?

In case you do not have access to AC (alternating current) power, do you have batteries to drive all the equipment used? In critical (film-style) field productions for which you use a single high-quality camera, check with the technical supervisor about such additional equipment as an RCU (remote control unit) and test equipment to enable you to adjust the camera for optimal performance; a waveform monitor (oscilloscope) to help you adjust the brightness (keeping the white and black levels within tolerable limits); and a vector scope to help adjust the camera so that it produces true colors.

■ *Audio.* If you have not checked out the acoustics of the location, take several types of mics. Check the wireless lavaliers. Do you have enough batteries for the wireless mic transmitters? Do the portable mics fit the channel frequency of the receiver? All remote mics, including the lavaliers, should have windscreens. Shotgun mics need additional wind jammers. Choose the most appropriate mounting equipment, such as clamps, stands, and fishpoles. Do you need a small field mixer? Does it work? If you use a separate audio recorder, check it out before taking it on-location. Don't forget headsets for the fishpole operator and the audio-recording technician.

■ *Power supply.* Do you have the right batteries for the monitors, camcorders or field cameras, and audio equipment? Are they fully charged? If using AC power, do you have the right AC/DC adapters? Do you have enough AC extension cords to reach the outlets? Unless they are battery-driven, you also need AC power and extension cords for the monitors. Take a few power strips along, but be careful not to overload the circuits.

■ *Cables and connectors.* Do you have enough camera cables, especially if there is a long run between the camera and the RCU or portable switcher? Are there enough coax and AC cables for monitor feeds? Always take a sufficient number of mic cables along, even if you plan to use wireless mics. The mic cables may save a whole production day if the wireless system breaks down or is unusable on-location. Do you have the right audio connectors for the mic cables and jacks (usually XLR connectors, but sometimes RCA phono)? Bring some adapters for video and audio cables (BNC to RCA phono, and XLR to RCA phono and vice versa). Although you should avoid adapters as much as possible, bring some along that fit the cables and a variety of input jacks, just for insurance.

■ *Lighting.* You can light most interiors with portable lighting instruments. Bring several lighting kits. Check that the kits actually contain the standard complement of lights, stands, and accessories. Do the lights work? Always pack a few spare lamps. Do the lamps actually fit the lighting instruments? Do they burn with the desired color temperature (3,200K or 5,600K)? Do you have enough reflectors (white foam core), umbrella reflectors, diffusion material (scrims, screens, diffusion gels), and color gels for regulating color temperature? The color gels most often needed are the amber or light-orange ones for lowering the color temperature and the light-blue ones for raising it.

Other important items to take along are: a light meter, light stands and clamps, sandbags to secure the portable light stands, some pieces of 1 × 3 lumber to construct a light bridge for back lights, a roll of aluminum foil for heat shields, extra barn doors, flags, and a dozen or so wood clothespins to attach scrims or color gels to barn doors.

■ *Intercom.* If the single-camera EFP is taking place in a confined area, you don't need elaborate intercom systems; you can call your shots right from the production area. But if the event covers a large outdoor area, you need a small power megaphone and walkie-talkies and cell phones to communicate with the widely dispersed crew and the station, if necessary. If you use the portable switcher for a live or live-recorded pickup, such as for the high-school state championship basketball game, you need headsets and intercom (often regular audio) cables.

■ *Miscellaneous.* There are a few more items that are often needed for a field production: extra scripts and time lines; field log sheets; slate and dry-erase markers; regular rain umbrellas and "raincoats" (plastic covers) for cameras; white cards for white-balancing; a teleprompter (flat-panel display and laptop), if applicable; blank cue cards or large newsprint pad and marker; an easel; and several rolls of gaffer's tape and masking tape. You also need white chalk, extra sandbags, wood clothespins, rope, makeup kit and bottled water, towels, flashlights, trash bags, and a first-aid kit.

Production: Setup

Once everyone knows what is supposed to happen, the setup will be relatively smooth and free of confusion. Although as director you may not be responsible for the technical setup, you should watch carefully that the equipment is put in the right places.

For example, when shooting indoors, will the lights be out of camera range? Are they far enough away from combustible materials (especially curtains) and properly insulated (with aluminum foil, for example)? Are the back lights high enough so that they will be out of the shot? Is there a window in the background that might cause lighting problems? (The window in the artist's studio in figure 18.3 would certainly present a problem if you had to shoot the large sculpture.) Does the room look too cluttered? Too clean? Are there any audio problems you can foresee? If the talent wears a wired lavalier mic, does the mic cord restrict talent mobility? If you use a shotgun mic, can the fishpole operator get close enough to the talent and, especially, move with the talent without stumbling over furniture? Are pictures hung where the camera can see them? Look behind the talent to see whether the background will cause any problems (such as lamps or plants seeming to extend from the talent's head).

When outdoors check for obvious obstacles that may be in the way of the camera, fishpole operators, and talent. Look past the shooting location to see whether the background fits the scene. Are there bushes, trees, or telephone poles that may, again, appear to extend from the talent's head? Large billboards are a constant background hazard. What are the potential audio hazards? Although the country road may be quiet now, will there be traffic at certain times? Are there any factory whistles that may go off in the middle of your scene? Is there any air or train traffic near your location?

Production: Rehearsals

Most often rehearsals are limited to a quick walk-through. But you may need more rehearsals if the EFP requires the interaction of more than one or two persons.

Walk-through Before you start with the actual rehearsal and video recording, you should have a brief walk-through first with the crew and then with the talent to explain the major production points, such as camera positions, specific shots, and principal actions. In relatively simple productions, you can combine the technical and talent walk-throughs. Always follow the walk-through with the "notes" session, then have the crew take care of the remaining problems. Don't forget to give the talent and the crew a short break before starting with the rehearsal and recording sessions.

Rehearsal As pointed out earlier, single-camera field directing has its own rehearsal technique. Basically, you rehearse each take immediately before recording it. You walk the talent and the camera through the take, explaining what they should and should not do. If you use a fishpole for the audio pickup, include the fishpole operator in the walk-through. Record some of the critical scenes and watch and listen to the playback. You may want to adjust or change the mic or the mic position for a better pickup.

Production: Video Recording

Just before the actual recording, ask the DP and/or the camera operator whether the camera is properly white-balanced for the scene location. Sometimes clouds or fog move in between the rehearsal and the actual recording, changing the color temperature of the light. Slate all takes and have the PA or the VR operator record them on the field log.

Watch the background action as well as the main foreground action. For example, curious onlookers may suddenly appear out of nowhere and get in your shot, or the talent may stop her action exactly in line with a distant power pole that then appears to grow out of her head. Listen carefully to the various foreground and background sounds during the take. Do not interrupt the recording because there was a faint airplane noise. Most likely, such noise will get buried by the main dialogue or the additional sounds added in postproduction (such as music). But the noise of a police siren that interrupts a Civil War scene definitely calls for a retake.

At the end of each take, have the talent stand quietly and record a few seconds of additional material. This pad will be of great help to the editor in postproduction. Always record some usable cutaways and location sounds and room ambience for each location. The recorded "silence" will help bridge possible audio drops at edit points in postproduction.

When you feel that you have a series of good takes, play them back on the field monitor to see whether they are indeed acceptable for postproduction. If you detect gross problems, you can still do some retakes before moving on to the next scene or location. But watch the time line.

Production: Strike and Equipment Check

Have the location reset (furniture, curtains, and the like) the way you found it and the place cleaned before you leave. Pick up all scripts, shot sheets, and log sheets. Do not leave pieces of gaffer's tape stuck on floors, doors, or walls, and take away your trash. When loading the EFP vehicle, the floor manager, crew chief, or PA should run down the equipment checklist to see that everything is back in the vehicle before leaving or changing locations. Check that the source media are all properly labeled and—most importantly—loaded onto the vehicle. Some directors insist on carrying the source media personally.

Postproduction

EFP postproduction activities are, for all practical purposes, identical to those of single-camera studio productions: capturing the various takes on the hard drive of the editing computer, logging all takes on the source media, doing a rough-cut, and finally doing an on-line edit that is transferred to the edit master tape or DVD.

BIG REMOTES

A ***big remote***, or simply remote, is done to televise the live or ***live recording*** of large scheduled events that have not been staged specifically for television. Examples are sporting events, parades, and political gatherings. All big remotes use high-quality field cameras (studio cameras with high-zoom-ratio lenses) in key positions, a number of ENG/EFP cameras—as many as 20 or more for big events— and an extensive audio setup.

The cameras and the various audio elements are coordinated from a mobile control center—the ***remote truck***. Remote trucks are usually powered by a portable generator, with a second one standing by in case the first one fails. If there is enough power available at the remote site, the truck is connected to the available power, with a single generator serving as backup.

The remote truck represents a compact studio control room and equipment room. It contains the following control centers:

- Program control, also called production, with preview and line monitors, a switcher with special effects, a character generator (C.G.), and various intercom systems: P.L., P.A., and elaborate I.F.B. systems
- Audio control with a fairly large audio console, digital recording media, monitor speakers, and intercom systems
- Video-recording control with several high-quality VRs that can do regular recordings, do instant replays, and play in slow-motion and freeze-frame modes
- Technical control with CCUs (camera control units), line monitors, patchboards, a generator, and signal transmission equipment SEE 18.4 AND 18.5

In very big remotes, one or more additional trailers may be used for supplemental production and control equipment.

Because the telecast happens away from the studio, some production procedures are different from those for studio productions. We therefore examine the following production aspects specific to the field: remote survey,

18.4 REMOTE TRUCK
The remote truck is a complete control center on wheels. It contains program, audio, video, and technical control centers as well as C.G. and recording facilities.

18.4 PHOTO BY HERBERT ZETTL

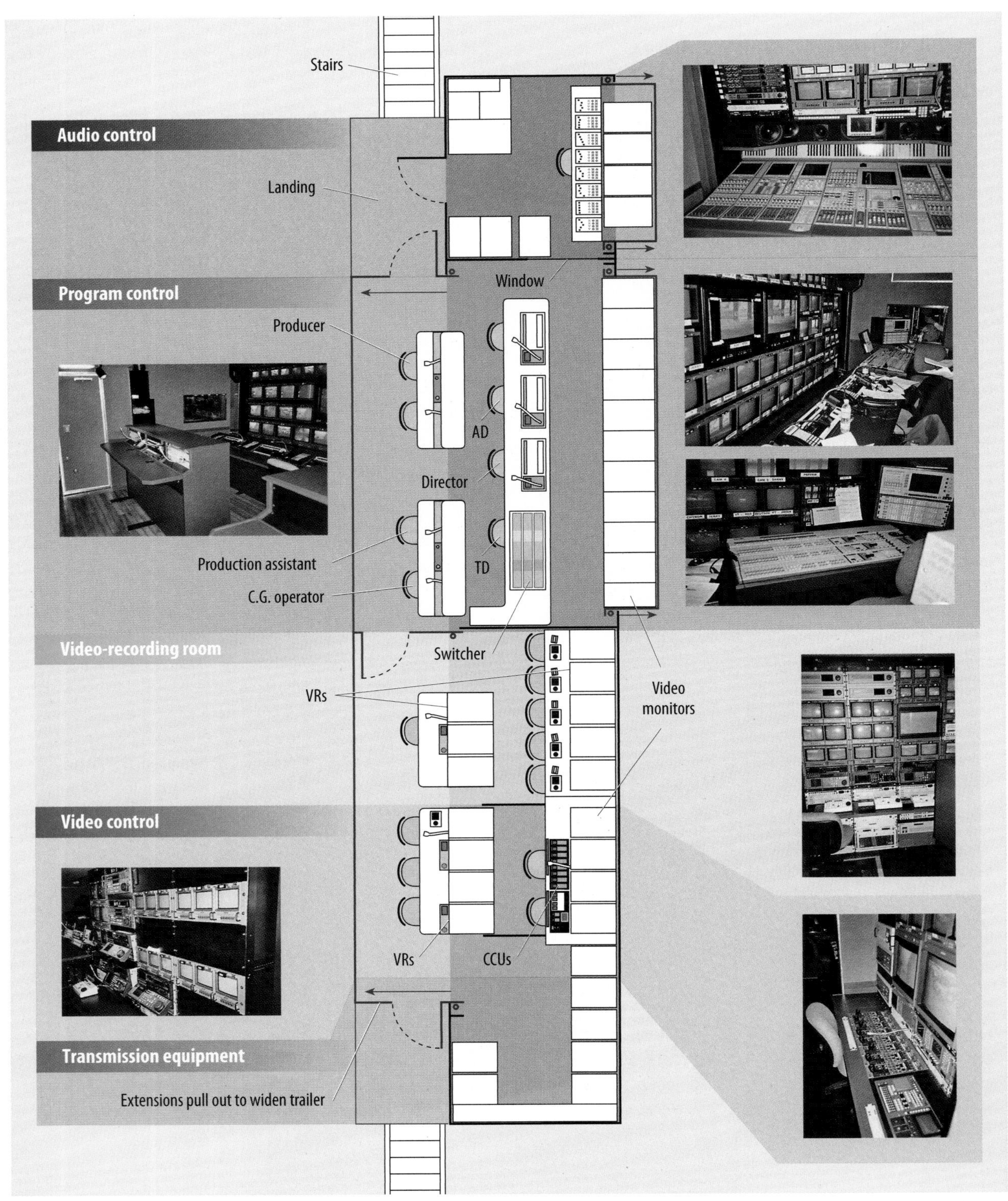

18.5 PHOTOS BY HERBERT ZETTL

18.5 CONTROL CENTERS IN REMOTE TRUCK
The remote truck carries the audio control, the program control, the video-recording and instant-replay control, the video control, and the transmission equipment.

equipment setup and operation, and director, floor manager, and talent procedures.

Remote Survey

Like any other scheduled production, a big remote requires thorough preparation—only more so. One problem with preparing for big remotes is that the event you cover is normally a onetime happening that you cannot rehearse. It would be ridiculous to ask two national hockey teams to repeat the whole game for you, or to ask political leaders to restate their lively debate verbatim just so you can have your rehearsal. A remote of an award ceremony, however, allows some limited rehearsals; you can rehearse with stand-ins filling in for the master of ceremonies and the award recipients. Most sporting events (such as football, baseball, basketball, ice hockey, tennis, and so forth) have standard setups that don't change much, or at all, from one game to the next. Still, you have no control over the event itself but must follow it as best you can. Your production preparations must take these and several other such considerations into account. Yet another problem is that you can truck only the control room and the technical facilities to the site—not the studio itself. Cameras, microphones, and often lighting need to be brought to the remote location and made operational.

One of the key preparations is the remote survey. Many of the survey items for big remotes are equally applicable to other EFPs, such as an MTV segment or a visit to a manufacturing plant. As the name implies, a ***remote survey***, or site survey, is an investigation of the location premises and the event circumstances. It should provide you with answers to some key questions about the nature of the event and the technical facilities necessary to televise it. Such surveys are obviously not necessary if you repeatedly do the same remote in the same location.

Contact person Your first concern is to talk to someone who knows about the event. This individual, called the contact person, or simply contact, may be the public relations officer of an institution or someone in a supervisory capacity. Call the contact to find out what he or she knows about the event and whether he or she can refer you to others who might answer your questions. In any case, get the contact's full name and position, business and e-mail addresses, and business, home, fax, cell, and pager numbers. Then make an appointment for the actual remote survey.

The survey should ideally be conducted at the same time of day as the scheduled remote telecast because the location of the sun is extremely important for outdoor remotes as well as for indoor remotes where windows will be in the shots. Arrange to have the contact person with you during the production. Establish an alternate contact and copy him or her on information you send to the primary contact. Sometimes the contact person comes down with a cold on your production day.

Survey party The remote survey itself is concerned with production and technical considerations. The survey party therefore includes people from production and engineering. The minimum party usually comprises the remote's producer, director, and technical director (TD) or technical supervisor. Additional production and technical supervisory personnel, such as the production manager and the chief engineer, may join the survey party, especially if the remote covers an important event and includes such elements as complex microwave or satellite links.

In general, the production requirements are first determined, then the technical people try to make the planned production procedures technically possible. Depending on the complexity of the telecast, extensive compromises must often be made by production people as well as by technical personnel.

As director you can make such compromises only if you know what the particular technical setup and pickup problems are and what changes in procedures will help overcome them. You should therefore familiarize yourself with the production as well as the engineering requirements of television remotes. Although many production and technical survey questions overlap, for better clarification we consider them separately here.

Production survey The table on the following pages lists the key questions you should ask during the production survey. **SEE 18.6** A good location sketch can help you prepare for the production and anticipate major problems (see figures 18.15 through 18.18).

Technical survey The technical survey on page 398 lists only those items that directly influence the production procedures and, ultimately, your portion of the remote survey. **SEE 18.7** Technical points that have already been mentioned in the production survey, such as cameras and microphones, are not listed again.

Equipment Setup and Operation

There is no clear-cut formula for setting up equipment for a remote telecast. As with a studio production, the number of cameras, the type and the number of microphones,

18.6 REMOTE SURVEY: PRODUCTION

These are the key questions you should ask during the production survey of a special event.

SURVEY ITEM	KEY QUESTIONS
Contact	Who are your principal and alternate contacts? Title; business and e-mail addresses; business, home, and cell phone numbers; and fax and pager numbers.
Place	Where is the exact location of the telecast? Street address, telephone number. If you need traffic control, have the police been notified?
Time	When is the remote telecast? Date, time. What is the arrival time of the truck? Who is meeting you at the site for positioning the truck?
Nature of event	What is the exact nature of the event? Where does the action take place? What type of action do you expect? The contact person should be able to supply the necessary information.
Cameras (stationary)	How many cameras do you need? Use as few as possible. Where will you place the cameras? Do not place cameras on opposite sides of the action. In general, the closer together they are, the easier and less confusing the cutting will be. Shoot with the sun, not against it. Try to keep it behind or to the side of the cameras for the entire telecast. The press boxes of larger stadiums are generally located on the shadow side. If possible, survey the remote location during the exact time of the scheduled telecast. If it is not a sunny day, determine the position of the sun as closely as possible. Are there any large objects blocking the camera view, such as trees, telephone poles, or billboards? Will you have the same field of view during the actual telecast? A stadium crowd, for instance, may block the camera's field of view, although the view is unobstructed during the survey. Can you avoid large billboards in the background of shots, especially if the advertising competes with your sponsor's product? Do you need camera platforms? Where? How high? Can the platforms be erected at a particular point? Can you use the remote truck as a platform? If competing stations are also covering the event, have you obtained exclusive rights for your camera positions? Where do you want iso cameras positioned? Do you need umbrellas to shield viewfinders from the sun?
Cameras (mobile)	Do you need to move certain cameras? What kind of floor is there? Can the camera be moved on a field dolly, or do you need remote dollies (usually with large, inflatable rubber tires)? Will the dolly with the camera fit through narrow hallways and doors? Can you use ENG/EFP cameras instead of large studio/field cameras? What is their action radius? Can you connect them to a remote truck by cable (less chance of signal interference or signal loss), or do you have to send the signal back to the remote truck via microwave? If there are robotic cameras, where are they located?
Lighting	Do you need additional lighting? Where and what kind? Can the instruments be hung conveniently, or do you need light stands? Do you need to make arrangements for back lights? Will the lights be high enough so that they are out of camera range? Do you have to shoot against windows? If so, can they be covered or filtered to block out undesirable daylight? Can you use reflectors?

18.6 REMOTE SURVEY: PRODUCTION *(continued)*

SURVEY ITEM	KEY QUESTIONS
Audio	What type of audio pickup do you need? Where do you need to place the mics? What is the exact action radius so far as audio is concerned? How long must the mic cables be? Which are stationary mics and which are handled by the talent? Do you need long-distance mics, such as shotgun or parabolic mics? Do you need wireless mics? Can their signals reach the wireless receivers? Do you need such audio arrangements as audio foldback or a speaker system that carries the program audio to the location? Can you tie into the "house" public address system? Do you need long-distance mics for sound pickups over a long range?
Intercommunications	What type of intercom system do you need? Do you have to string intercom lines? How many I.F.B. channels and/or stations do you need and where do they go? Is there a need for a P.A. talkback system? In case cell phones do not work in this area, are there enough telephone lines available?
Miscellaneous production items	Do you need a clock? Where? Do you need line monitors, especially for the announcer? How many? Where should they be located? Will the announcer need a preview monitor to follow iso playbacks? Do you have a camera slate in case the C.G. cannot be used?
Permits and clearances	Have you (or the producer) secured clearances for the telecast from the police and fire departments? Do you have written clearances from the originators of the event? Do you have parking permits and assistance from security people for the remote truck and other production vehicles? Do you have passes and parking for all technical and production personnel, especially when the event requires entrance fees or has some kind of admission restrictions?
Other production aids	Does everyone have a fact sheet of the approximate order of events? These sheets are essential for the director, floor manager, and announcer and are extremely helpful to the camera operators, audio engineer, and additional floor personnel. Does the director have a spotter who can identify the major action and the people involved? In sports remotes, spotters are essential.

the lighting, and so forth depend entirely on the event or, more precisely, on the process message as defined in the preproduction meetings. Employing a great number of cameras, microphones, and other types of technical equipment does not necessarily guarantee a better telecast than when using less equipment. In fact, one or two camcorders are often more flexible and effective than a cumbersome remote truck with the fanciest VRs and audio-recording and switching gear. Even for multicamera EFPs, a portable switcher is usually easier to set up and operate than using large trucks. For such big-remote operations as the live coverage of major sporting events, however, which may use many cameras in various positions, the remote truck provides essential equipment and production control.

Once set up, many of the production routines of big remotes do not differ significantly from studio productions. There are nevertheless some procedures in big-remote operations, such as instant replays, that you will not find in normal studio productions. For the following discussion of a onetime event, let's assume that you are functioning first as a director of a big remote, then as a floor manager, and finally as talent.

Directing the setup Because the actual on-the-air telecast of a big remote is usually live, the directing procedures bear little resemblance to the other field production methods; they more closely resemble live or live-recorded studio productions.

18.7 REMOTE SURVEY: TECHNICAL

The technical survey lists only those items that directly influence the production procedures.

SURVEY ITEM	KEY QUESTIONS
Power	Assuming you do not work from a battery pack or your own generator, is enough electricity available on-site? Where? Does the contact person have access to the power outlets? If not, who does? Make sure the contact is available during the remote setup and the actual production. Do you need extensions for the power cables? If you use a generator, do you have a second one for backup?
Location of remote truck and equipment	Where should the remote truck be located? Its proximity to the available power is critical if you do not have a power generator. Are you then close enough to the event location? Keep in mind that there is a maximum length for camera cables beyond which you will experience video loss. Watch for possible sources of video and audio signal interference, such as nearby X-ray machines, radar, or any other high-frequency electronic equipment. Does the remote truck block normal traffic? Does it interfere with the event itself? Reserve parking for the truck. Have you asked the police for assistance?
Recording devices	If the program is recorded, do you have the necessary video recorders in the truck? Do you need additional VRs for instant replays? If you have to feed the audio and video signals back to the station separately, are the necessary phone lines cleared for the audio feed? Do you have enough media to record the full event? Have you made provisions for switching media without losing part of the event? Are the iso cameras properly patched into the switcher and into separate VRs?
Signal transmission	If the event is fed back to the station for video recording or directly to the transmitter for live broadcasting, do you have a good microwave or satellite uplink location? Do you need microwave mini-links? Double-check on the requirements for feeding the satellite uplink.
Cable routing	How many camera cables do you need? Where do they have to go? How many audio cables do you need? Where do they have to go? How many intercom lines do you need? Where do they have to go? How many AC power lines do you need? Where do they go? Route the cables in the shortest possible distance from remote truck to pickup point, but do not block heavily traveled hallways, doors, walkways, and so on. Do the cables have to cover a great span? If so, string a rope and tie the cable to it to relieve the tension.
Lighting	Are there enough AC outlets for all lighting instruments? Are the outlets fused for the lamps? Do not overload ordinary household circuits (usually 15 amps). Do you have enough extension cords and power strips (or simple multiple wall plugs) to accommodate all lighting instruments and the power supply for monitors and electric clocks?
Communication systems	What are the specific communication requirements? P.L.s? I.F.B. channels? Telephone lines? Cell phones? P.A. systems? Long-range walkie-talkies? Two-way radios?

■ As soon as the remote truck is in position, conduct a thorough technical walk-through. Tell the technical staff where you want the stationary cameras located and what field of view you require (how close or wide a shot you need to get with each camera). Get the cameras as close to the action as possible to avoid overly narrow-angle zoom lens positions. Apprise the crew of the approximate moves and ranges of mobile cameras and what audio needs you have. Unless they will be in a booth, specify where the announcers are going to be so that their monitors, mics, and intercom can be properly routed.

■ While the technical crew is setting up, hold a production meeting with the contact person, the producer, the AD (associate director), the floor manager, the PA, the talent, and, if not directly involved in the setup, the TD or technical supervisor. Have the contact describe the anticipated event. Explain how you intend to cover it. Although it is the producer's job to alert the talent to the prominent features of the event, such as a prize-winning float in the parade, be prepared to take over in case the producer is sidetracked by some other problem. Delegate the setup supervision to the AD, floor manager, and TD. Do not try to do everything yourself.

■ Pay attention to all communication systems, especially the intercom. During the telecast you will have no chance to run in and out of the remote truck to the actual site; all your instructions will come via voice communication from the truck. Discuss the coverage of the event in detail with the floor manager, who holds one of the most critical production positions during a remote.

■ Usually, you as director have no control over the event itself; you merely try to observe it as faithfully as possible. Once again, check with the contact person and the announcer on the accuracy of the fact sheet and the specific information concerning the event. Ask the talent to double-check on the pronunciation of the names of participants and places.

■ Walk through the site again and visualize the event from the cameras' perspectives. Are they in the optimal shooting positions? Are they all on one side of the principal vector line so that you will not reverse the action on-screen when cutting from one to another? If shooting outdoors, will any of the cameras be blinded by the sun? Will the sun be directly behind the cameras (which, in effect, will wash out the viewfinder images)? Experienced camera operators will use an umbrella or flags attached to the camera to prevent the sun from washing out their viewfinders.

■ Keep in mind that you are a guest while covering a remote event. Unless television is an integral part of the event, such as in most sports, try to work as quickly and as unobtrusively as possible. Do not make a big spectacle of your production. Realize that you are basically intruding on the event and that the people involved are usually under some stress.

Directing the on-the-air telecast Once you are on the air, try to keep on top of the event as well as possible. If you have a good spotter (the contact person, event expert, or the AD), you will be able to anticipate certain happenings and be ready for them with the cameras. The following are some general points to remember.

■ Speak loudly and clearly. Usually, the site is noisy and the camera operators and the floor crew may not hear you very well. Put your headset mic close to your mouth. Yell if you have to, but do not get frantic. Tell the crewmembers to switch off their headset talkbacks to prevent the ambient sound from entering the intercom system.

■ Listen to the floor manager and the camera operators. They may help spot event details and report them to you as they occur.

■ Watch the monitors carefully. Often the off-air cameras get especially interesting shots, but do not be tempted by cute yet meaningless or even event-distorting shots. If, for example, the great majority of an audience listens attentively to the orchestra, do not single out the one person who is conducting in the back row, as colorful a shot as this may be.

■ Listen to the audio. A good announcer will give you clues as to the development of the event and sometimes direct your attention to a significant event detail.

■ If things go wrong, stay calm. For example, if a spectator blocks the key camera or if the camera operator swish-pans to another scene because he thinks his camera is off the air, don't scream at the camera operator that he is still "hot" or at the floor manager to "get this jerk out of the way." Simply cut to another camera.

■ Exercise propriety and good taste in what you show the audience. Avoid capitalizing on accidents (especially during sporting events) or situations that are potentially embarrassing to the person on-camera, even if such situations might appear hilarious to you and the crew at the moment.

Instant replay In an ***instant replay***, a key play or event segment is repeated for the viewer. Instant-replay operations usually use ***iso cameras***, which feed into the switcher as well as into their own separate video recorders. Some large sports remotes employ a second, separate switcher that is dedicated exclusively to inserting instant replays.

During the replay digital video effects (DVE) are often used to explain a particular play. The screen may be divided into several squeezed boxes or corner wipes, each displaying a different aspect of the play; or it may function as an electronic blackboard that shows simple line drawings over the freeze-frame of an instant replay, very much like the sketches on a traditional blackboard. Game and player statistics are displayed using the C.G. Some of the information is preprogrammed and stored on disk, but up-to-date statistics are continuously entered by a C.G. operator. The whole instant-replay and C.G. operation is usually guided by the producer or the AD. The director is generally much too occupied with the real-time coverage to worry about instant replays and special effects.

When watching an instant replay of a key action, you may notice that the replay either duplicates exactly the sequence you have just seen or, more frequently, shows the action from a slightly different perspective. In the first case, the picture sequence of the regular game coverage—that is, the line output—has been recorded and played back; in the second case, it is the pickup of an iso camera that has been recorded and played back.

In sports the principal function of the iso cameras is to follow key plays or players for instant replay. Iso cameras are also used in remote productions such as rock concerts or orchestra performances with a multicamera setup. When covering an orchestra, you may want to have an iso camera on the conductor at all times, which provides a convenient cutaway in postproduction.

When a remote production is done for postproduction rather than live, all cameras may be used in iso positions, with each camera's output recorded by a separate VR. The output of all iso cameras is then used as source material for extensive postproduction editing.

Postshow activities The remote is not finished until all the equipment is struck and the site is restored to its original condition. As a director of big remotes, you should pay particular attention to the following postshow procedures.

- If something went wrong, do not storm out of the remote truck, accusing everyone, except yourself, of making mistakes. Cool off first.

- Thank the crew and the talent for their efforts. Nobody ever wants a remote to look bad. Especially thank the contact person and others responsible for making the event and the remote telecast possible. Leave as good an impression as you can of you and your team. Remember that when you are at a remote location, you are representing your company and, in a way, the whole of "the media."

- Thank the police for their cooperation in reserving parking spaces for the remote vehicles, controlling the spectators, and so forth. Remember that you will need them again for your next remote telecast. See to it that the floor manager returns all the production equipment to the station.

Production: Floor Manager and Talent Procedures

As the floor manager, you play a key role in big remotes. The "floor" activities you have to manage have increased considerably in size and complexity.

Floor manager's procedures As a floor manager (also called stage manager or unit manager on big remotes), you have, next to the director and the TD, the major responsibility for the success of the remote telecast. Because you are close to the scene, you often have a better overview of the event than does the director, who is isolated in the remote truck. The following points will help you make the big-remote production a successful one.

- Familiarize yourself with the event ahead of time. Find out where it is taking place, how it will develop, and where the cameras and the microphones are positioned relative to the remote truck. Make a sketch of the major event developments and the equipment setup (see section 18.2).

- Triple-check all intercom systems. Find out whether you can hear the instructions from the remote truck and if you can be heard there. Check that the intercom is working properly for the other floor personnel. Check all wireless headsets, I.F.B. channels, walkie-talkies, cell-phone connections, and any other field communication devices.

- Be aware of the traffic in the production area. Try to keep onlookers away from the equipment and the action areas. Be polite but firm. Work around the crews from other stations. Be especially aware of reporters from other media. It would not be the first time that a news photographer snapping pictures just happens to stand right in front of your key camera. Appeal to the photographer's sense of responsibility: say that you too have a job to do in trying to inform the public.

- If the telecast is to be video-recorded, have the slate ready, unless the C.G. is used for slating.

- Check that all cables are properly secured to minimize potential hazards to the people in the production area. If it has not been done by the technical crew, tape cables to the floor or sidewalk and put a mat over the cables at major pedestrian traffic areas.

- Introduce yourself to the police officers assigned to the remote and fill them in on the major event details. Introduce them to the talent. The police are generally more cooperative and helpful when they get to meet on-the-air personalities and feel that they are part of the remote operation.

- Help the camera operators with spotting key event details. Discuss with the cable pullers (floor personnel) the action radius of the portable cameras. These people are often the key to getting good shots with a portable EFP camera or a camcorder that is also attached to an RCU.

- Relay all director's cues immediately and precisely. Position yourself so that the talent sees the cues without having to look for you. (Most of the time, talent are hooked up to the I.F.B. via earphones, so the director can cue them without the floor manager as an intermediary.)

- Have several 3 × 5 cards handy so you can write cues and pass them to the talent, just in case you lose the I.F.B. channel.

- When the talent are temporarily off the air, keep them informed about what is going on. Help keep their appearance intact for the next on-the-air performance and offer encouragement and positive suggestions.

- After the telecast pick up all the production equipment for which you are directly responsible—easels, platforms, sandbags, slates, and headsets. Double-check whether you have forgotten anything before you leave the remote site. Make use of the director's or TD's equipment checklist.

Talent procedures The general talent procedures (as discussed in chapter 16) also apply to remote operations, but there are some points that are especially pertinent for you as talent.

- Familiarize yourself thoroughly with the event and your specific assignment. Know the process message and do your part to effect it. Review the event with the producer, the director, and the contact person.

- Test your microphone and your intercommunication system. If you work with an I.F.B. system, check it out with the director or producer.

- Verify that your monitor is working. Ask the floor manager to have the TD punch up the line-out picture as soon as the cameras are uncapped. Ask for at least color bars to be put on-line.

- If you have the help of a contact person or a spotter to identify the players in a particular event, discuss again the major aspects of the event and the communication system between the two of you once you're on the air. For example, how is the spotter going to tell you what is going on while the microphone is hot?

- Verify the pronunciation of names and places. Little is more embarrassing for all involved than when names of locally well-known people are mispronounced by the television commentators.

- When you're on the air, tell the audience what they cannot see for themselves. Do not report the obvious. For example, if you see the celebrity stepping out of the airplane and shaking hands with the people on the tarmac, do not say, "The celebrity is shaking hands with some people"; tell who is shaking hands with whom. If a football player lies on the field and cannot get up, do not tell the audience that the player apparently got hurt—they can see that for themselves; tell them who the player is and what might have caused the injury. Also, follow up this announcement periodically with more detailed information about the injury and how the player is doing.

- Do not get so involved in the event that you lose your objectivity. On the other hand, do not remain so detached that you appear to be uninterested or to have no feelings whatsoever.

- If you make a mistake in identifying someone or something, admit it and correct it as soon as possible.

- Do not identify event details solely by color, as colors are often distorted on home receivers. For instance, refer to the runner not only as the one in the red trunks but also as the one on the left side of the screen.

- As much as possible, let the event itself do the talking. Keep quiet during extremely tense moments. For example, do not talk during the intense pause between the starter's "Get set" command and the firing of the starter's pistol in the 100-meter track finals.[1]

1. For a more detailed description of sports reporting, see Stuart W. Hyde, *Television and Radio Announcing,* 11th ed. (Boston: Allyn and Bacon, 2009), pp. 285–308.

MAIN POINTS

- The three types of remotes are ENG (electronic news gathering), EFP (electronic field production), and big remotes.
- ENG is the most flexible remote operation. It offers speed in responding to an event, maximum mobility while on-location, and flexibility in transmitting the event live or in live-recording it with camcorders for immediate transmission by the station or for postproduction editing.
- Unlike ENG, which has little or no preparation time for covering a breaking story, EFP must be carefully planned. In this respect it is similar to big remotes.
- EFP is normally done with an event that can be interrupted and restaged for video-recording various event segments. It is most often done with a single camera or sometimes with multiple iso (isolated) cameras that shoot an event simultaneously.
- A big remote televises live, or live-records, a large scheduled event that has not been staged specifically for television, such as a sporting event, parade, political gathering, or congressional hearing.
- All big remotes use high-quality cameras in key positions and ENG/EFP cameras for mobile coverage. Big remotes usually require extensive audio setups.
- Big remotes are coordinated from the remote truck, which contains a program control, an audio control, video-recording and instant-replay control, and a technical control with CCUs (camera control units) and transmission equipment.
- Big remotes require extensive production and technical surveys as part of the production activities.
- In sports remotes instant replay is one of the more complicated production procedures. It is normally handled by an instant-replay producer or an AD (associate director).

SECTION

18.2

Covering Major Events

As you know by now, EFP and especially big remotes require meticulous planning. Such careful preparation is particularly important for onetime happenings, such as sporting events that are not routinely televised, a rock concert, or a political rally. No two remotes are exactly the same, and there are always unique circumstances that require adjustments and compromises. This section includes some simple setups for sports remotes, how to read location sketches and some examples of typical indoor and outdoor remote setups, and an overview of field communication systems and signal transport.

▶ **SPORTS REMOTES**
Pickup requirements for baseball, football, soccer, basketball, tennis, boxing or wrestling, and swimming

▶ **LOCATION SKETCH AND REMOTE SETUPS**
Reading location sketches, indoor remotes, and outdoor remotes

▶ **COMMUNICATION SYSTEMS**
ENG, EFP, and big-remote communication systems

▶ **SIGNAL TRANSPORT**
Microwave transmission, cable distribution, and communications satellites

SPORTS REMOTES

Many big remotes are devoted to the coverage of sporting events. The number of cameras used and their functions depend almost entirely on who is doing the remote. Networks use a great amount of equipment and many personnel for the average sports remote. For especially important games, such as the Super Bowl or World Cup soccer, a crew of 100 or so people set up and operate 30 or more field cameras; additional robotic cameras, some of which run on cables stretched along the field; countless mics; multiple monitors; and intercom and signal-distribution systems. There may be several large trailers that house the control room and the production equipment, and two or three satellite uplink trucks. For the coverage of a local high-school game, however, you must get by with far less equipment. You may have only two cameras and three mics to do the pickup. Local stations or smaller production companies usually supply only the key production and technical personnel (producer, director, AD, PA, floor manager, TD, engineering supervisor, and audio technician) and hire a remote service that includes a remote truck, all equipment, and, on request, extra technical personnel.

The following figures illustrate the minimum video and audio pickup requirements for baseball, football, soccer, basketball, tennis, boxing or wrestling, and swimming. **SEE 18.8–18.14** Sometimes ENG/EFP cameras are used in place of the larger high-quality studio/field cameras or are added to the minimal setups described here.[2]

LOCATION SKETCH AND REMOTE SETUPS

To simplify preproduction you as the director, or your AD, should prepare a location sketch. Like the studio floor plan, the location sketch shows the principal features of the environment in which the event takes place (stadium and playing field, street and major buildings, or hallways and rooms). This location sketch will help you decide on the placement of cameras, it will help the audio technician decide on the type and the possible location of microphones, and it will give the TD some idea about the location of the remote truck and the cable runs. Finally, if indoors, it will help the LD determine the type and the placement of lighting instruments.

2. For detailed miking setups of multiple sporting events, see Stanley R. Alten, *Audio in Media,* 9th ed. (Boston: Wadsworth, 2011), pp. 257–75.

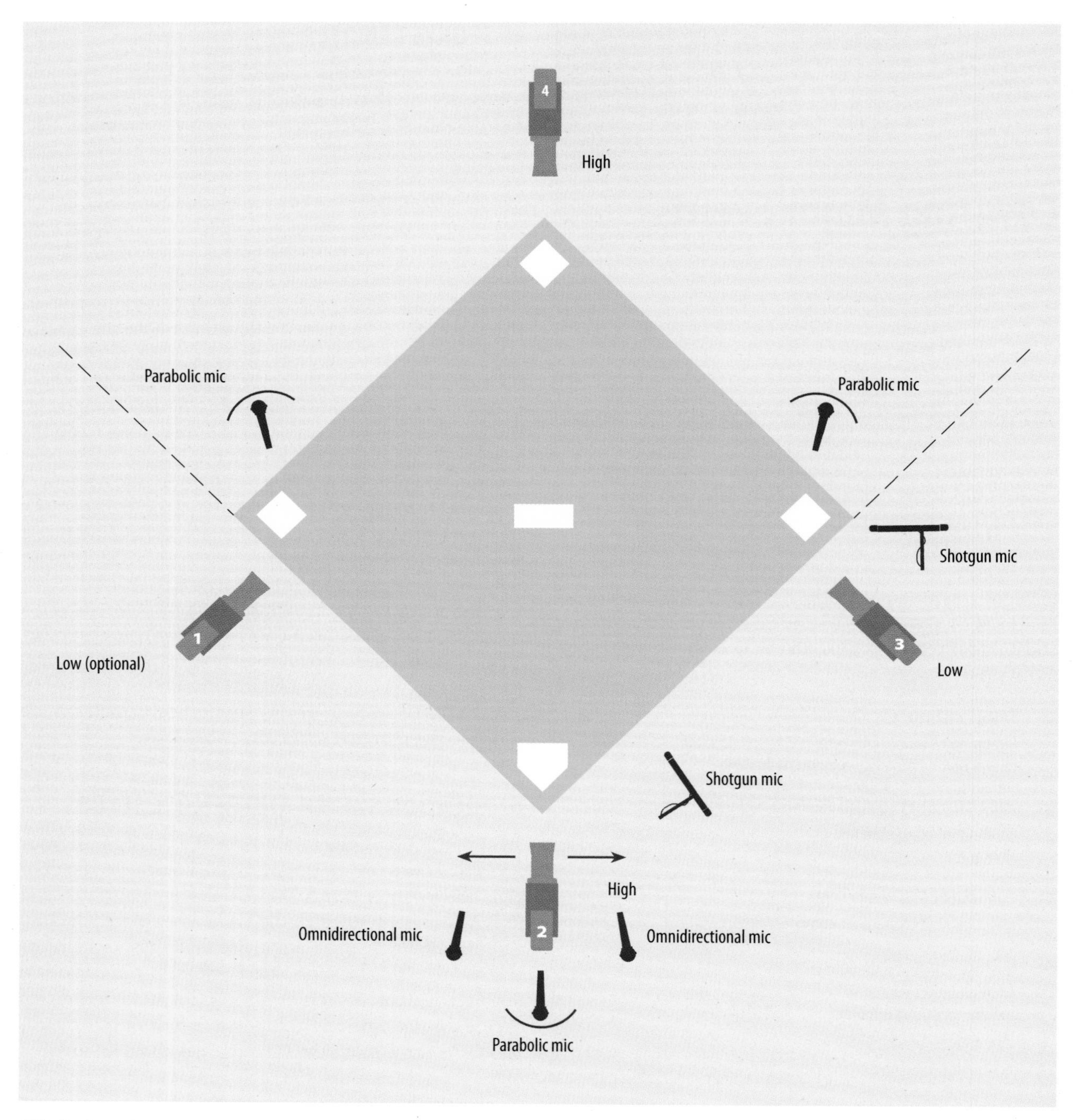

18.8 BASEBALL SETUP

Number of cameras: 3 or 4

C1: Near third base; low, optional

C2: Behind home plate; high

C3: Near first base; low; watch for action reversal when intercutting with C1

C4: Opposite C2 in center field; high; watch for action reversal

Number of mics: 6 or 7

2 omnidirectional mics high in stands for audience

2 shotgun mics or a parabolic (mobile) mic behind home plate for game sounds

2 or 3 parabolic mics for field and audience sounds

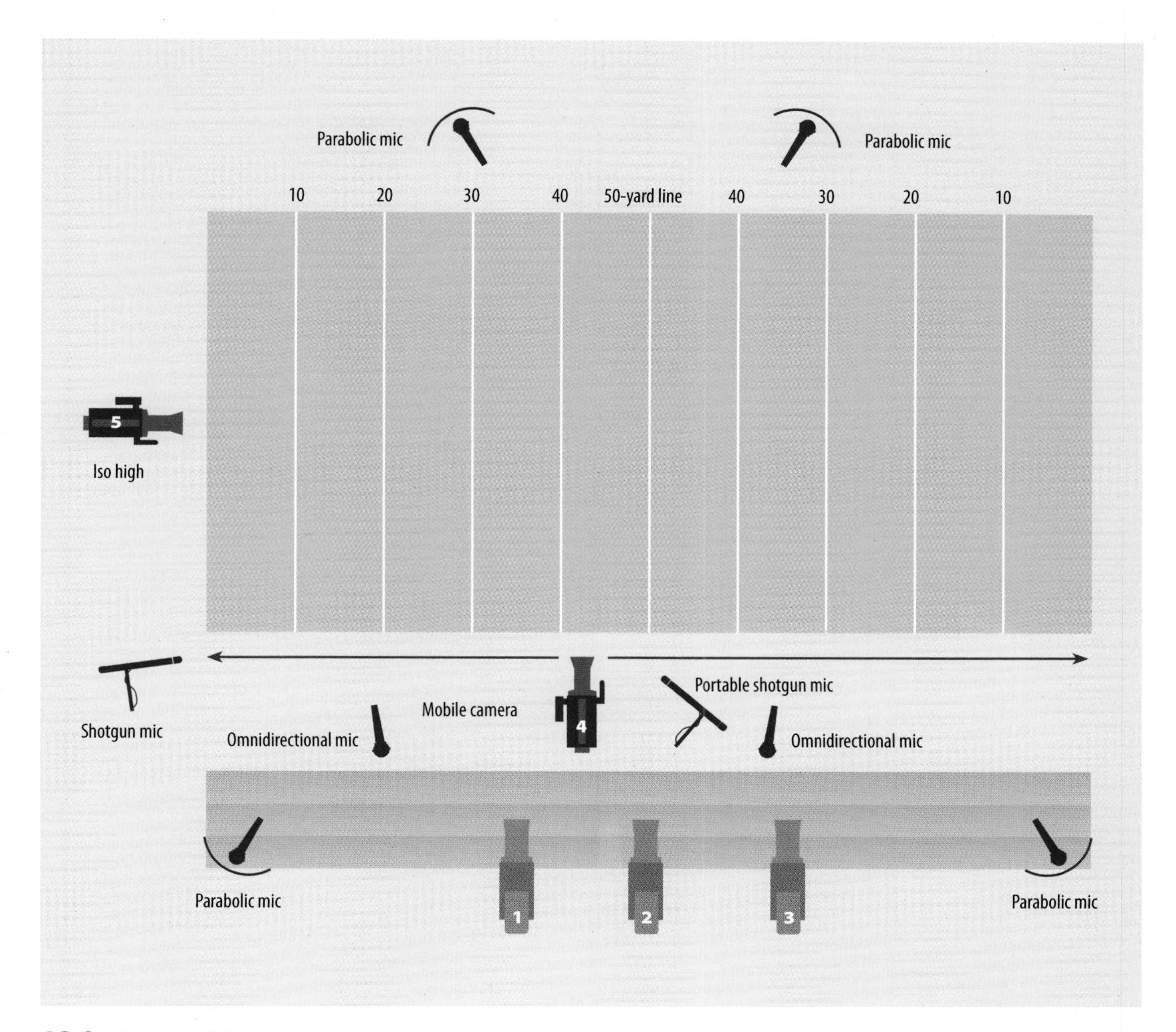

18.9 FOOTBALL SETUP

Number of cameras: 4 or 5

C1, C2, C3: High in the stands, near the 35-, 50-, and 35-yard lines (press box, shadow side)

C4: Portable or on dolly in field

C5: Optional iso camera behind goal (portable ENG/EFP or big camera)

Number of mics: 8

2 omnidirectional mics in stands for audience

2 shotgun or parabolic mics on field

2 parabolic reflector mics in stands and 2 on opposite side of field

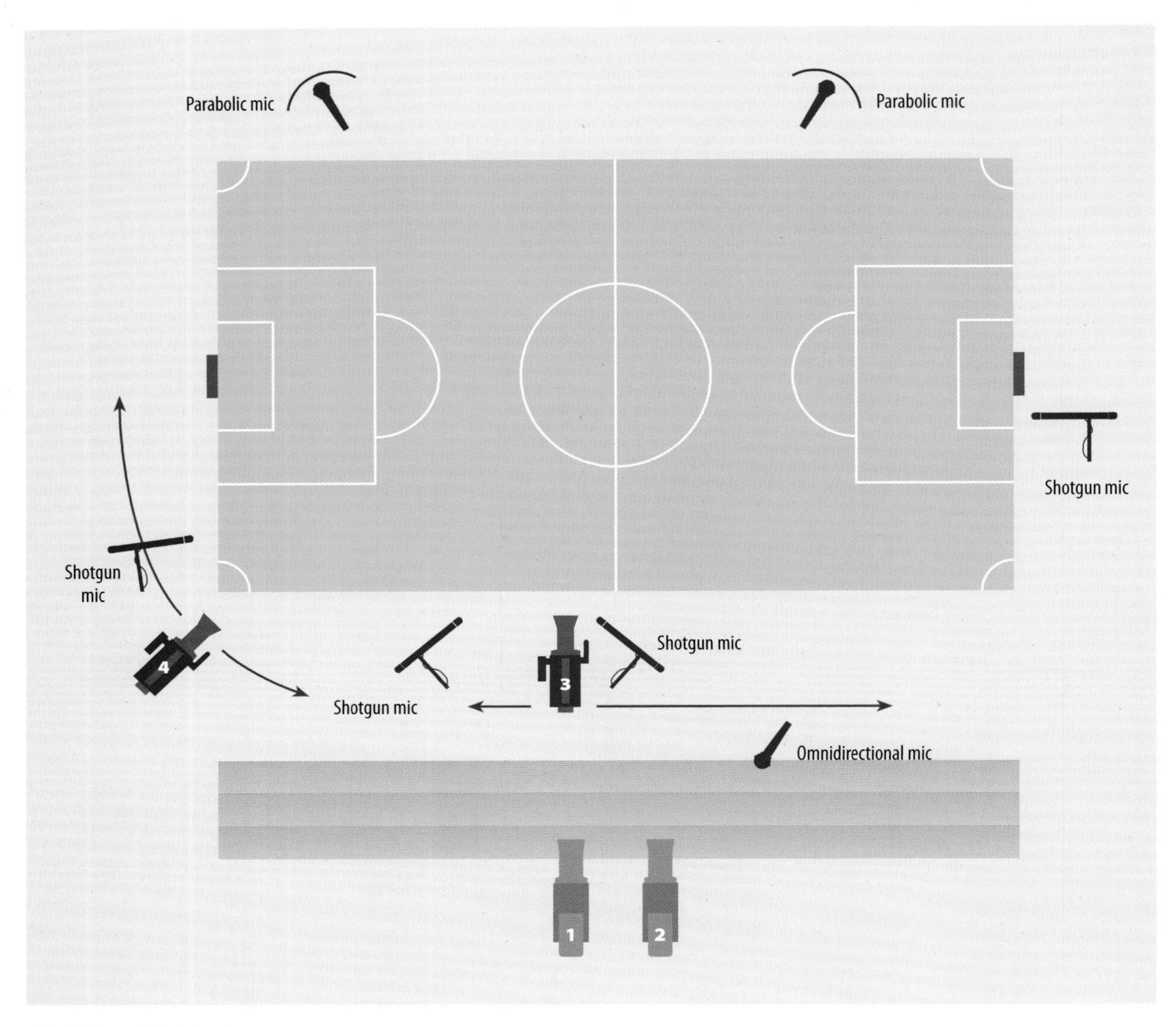

18.10 SOCCER SETUP

Number of cameras: 3 or 4

C1: Left of centerline (high)

C2: Right of centerline (high)

C3: Mobile on field

C4: Optional, left corner; may be used as iso camera and mobile camera on field

All four cameras are on shadow side of field.

Number of mics: 7

1 omnidirectional mic in stands for audience

4 shotgun mics on field

2 parabolic mics on opposite side of field

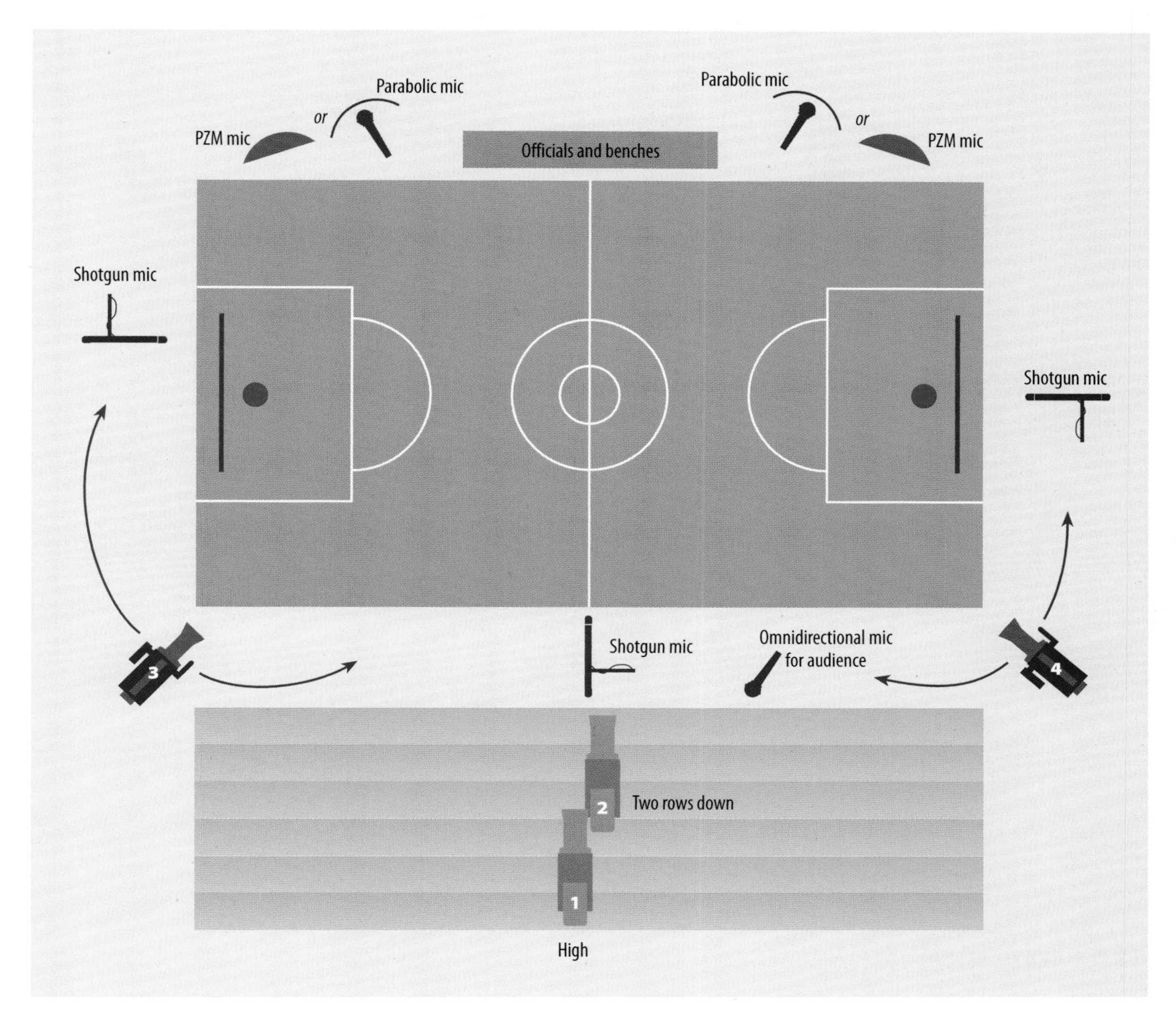

18.11 BASKETBALL SETUP

Number of cameras: 4

C1: High in stands, left of centerline—follows game

C2: Lower (2 rows down) in stands, right of centerline (fairly close to C1)—gets close-ups

C3: In left corner (mobile)

C4: In right corner (mobile)

Number of mics: 6

1 omnidirectional mic in stands for audience

2 PZM or parabolic mics in stands for audience

2 shotgun mics behind each basket for game sounds

1 shotgun mic at center court

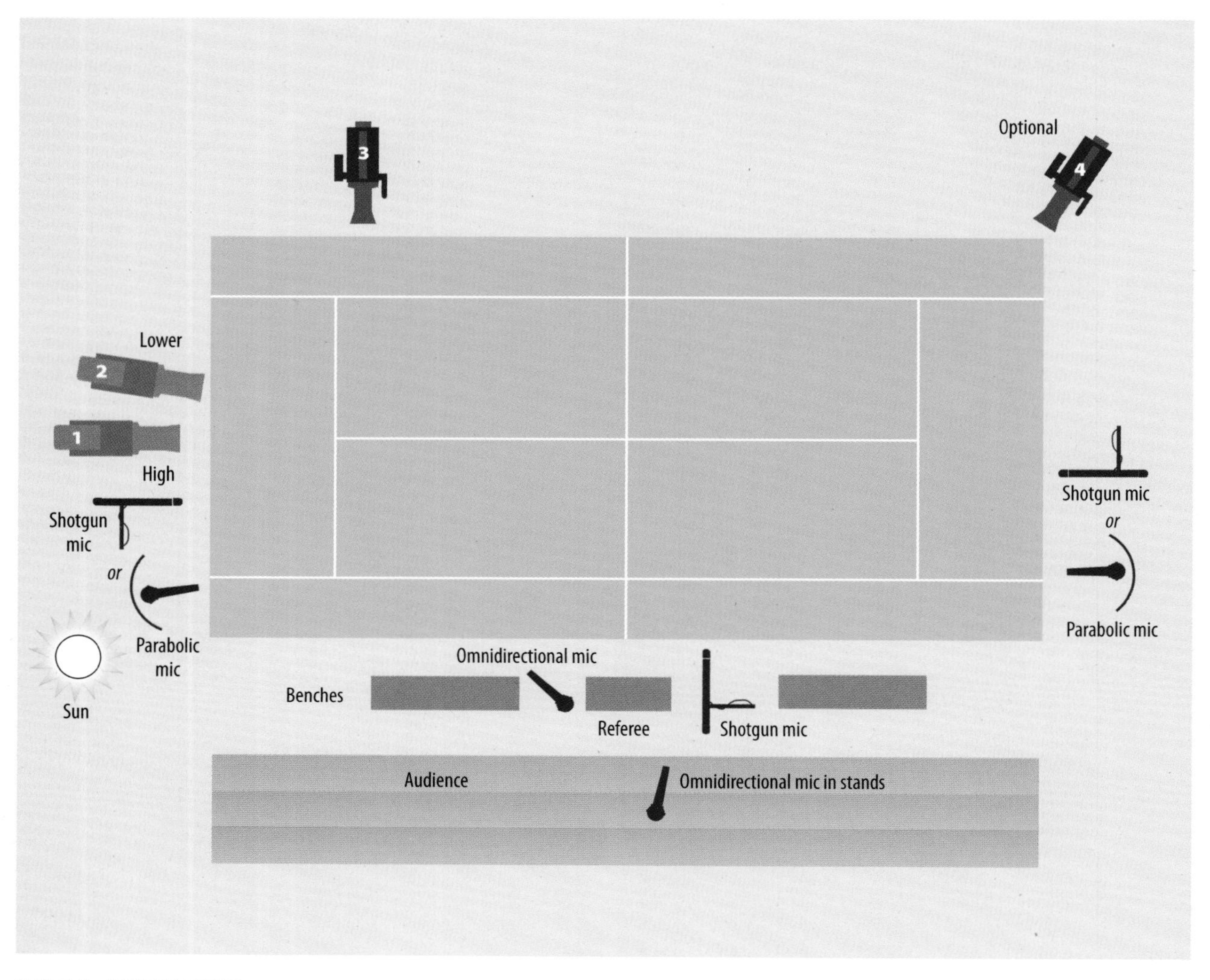

18.12 TENNIS SETUP

Number of cameras: 3 or 4

C1: At end of court, high enough so that it can cover total court, shooting with the sun

C2: Next to C1, but lower

C3: At side of court, opposite officials or where players rest between sets (mobile); also shoots CUs of players

C4: If four cameras are used, C3 shoots CUs of left player, C4 of right player

Number of mics: 5

1 omnidirectional mic in stands for audience

1 omnidirectional mic for referee's calls

3 shotgun mics at center court and on each end of court for game sounds, or 2 parabolic mics on the ends of the court

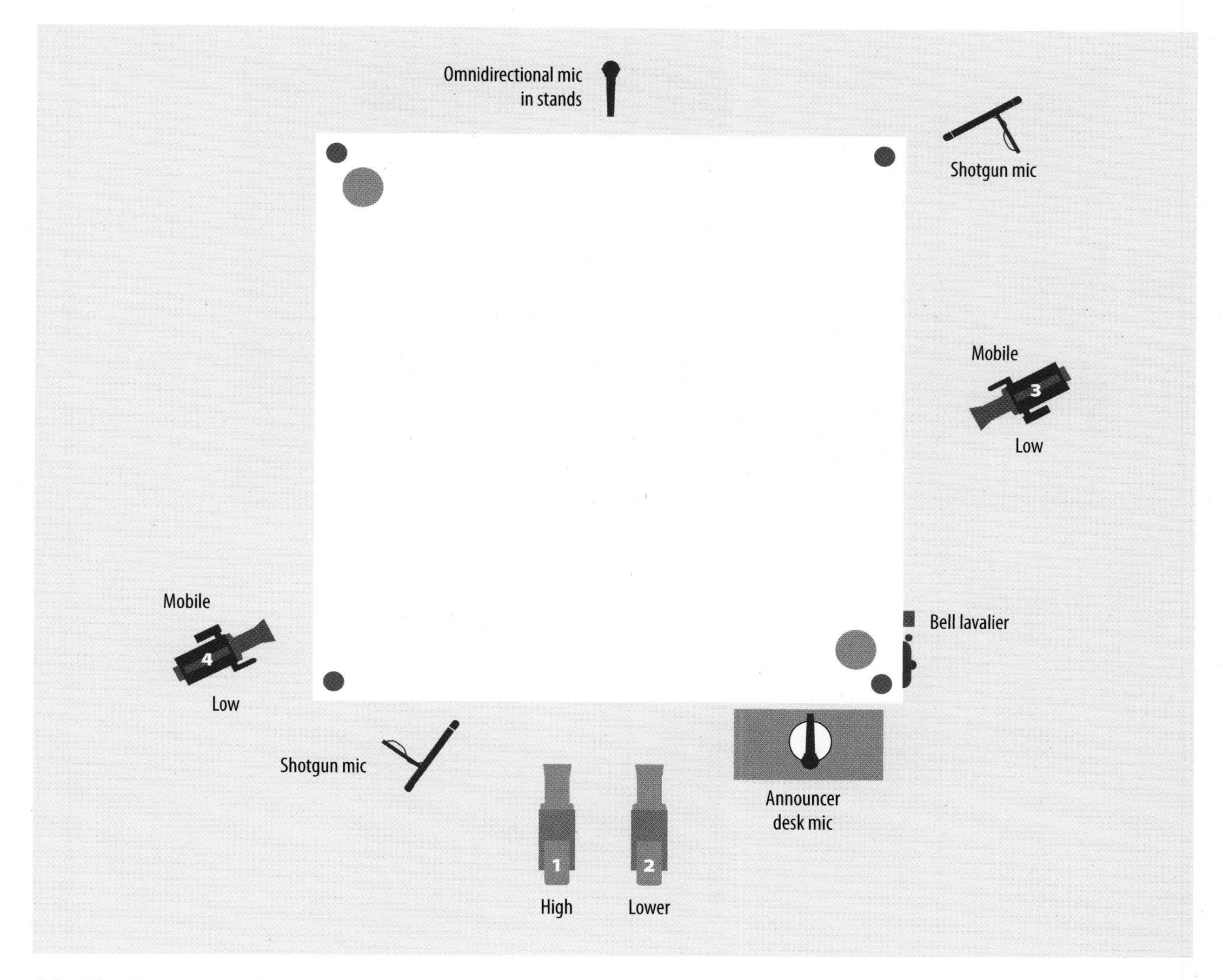

18.13 BOXING OR WRESTLING SETUP

Number of cameras: 3 or 4

C1: High enough to overlook the entire ring

C2: About 10 feet to the side of C1; high, slightly above ropes; used for replays

C3: ENG/EFP mobile camera carried on floor, looking through the ropes

C4: ENG/EFP mobile camera carried on floor, looking through the ropes

All mobile cameras have their own camera shotgun mics.

Number of mics: 5

1 omnidirectional mic for audience

2 shotgun mics for boxing sounds and referee

1 lavalier for bell

1 desk mic for announcer

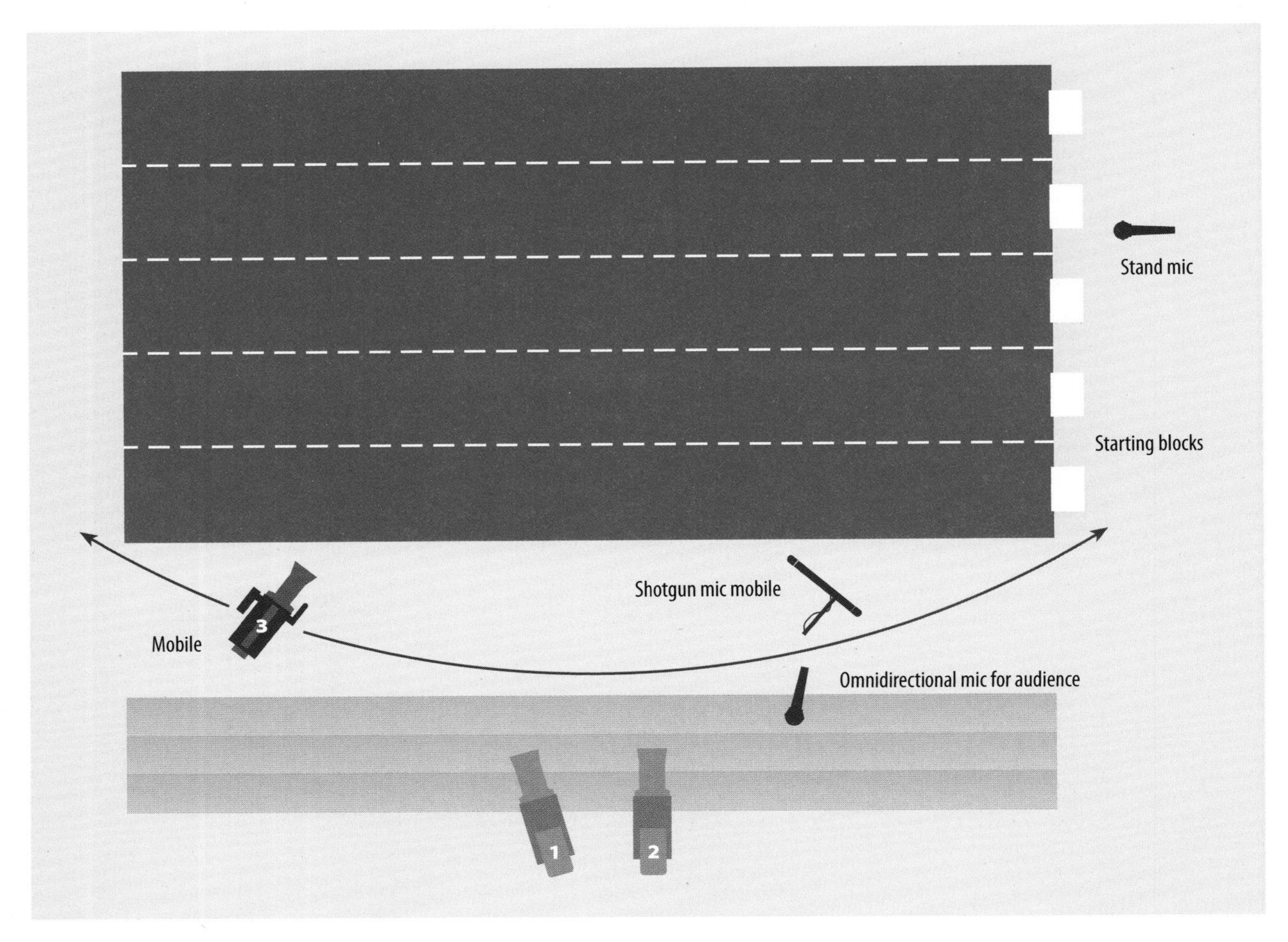

18.14 SWIMMING SETUP

Number of cameras: 2 or 3

C1: High in stands, about at center of pool—follows swimmer

C2: Next to C1—gets close-ups

C3: Optional ENG/EFP mobile camera on side and ends of pool

Number of mics: 3

1 omnidirectional mic in stands for audience

1 shotgun mic at pool level for swimmers

1 omnidirectional mic on stand

Reading Location Sketches

The location sketch for indoor events should show the general dimensions of the room or hallway and the location of windows, doors, and furniture. Ideally, it should also indicate the principal action (where people are seated or where they will be walking) and such details as power outlets; actual width of especially narrow hallways, doors, and stairs; direction the doors open; and prominent thresholds, rugs, and other items that may present problems for the movement of cameras mounted on tripod dollies. The following two examples show a location sketch for a public hearing room and an outdoor location sketch for a parade.

Public hearing The occasion is a newsworthy public hearing at city hall. SEE 18.15 Assuming you are the director of the remote, what can you tell from this sketch? How much preparation can you do? What key questions does the sketch generate? Limiting the questions to the setup within this hearing room, what are the camera, lighting, audio, and intercom requirements?

Parade The outdoor remote is intended for a Sunday afternoon live multicamera telecast. The estimated time of the telecast is from 3:30 to 5:30 p.m. The location sketch in figure 18.16 shows the action area as well as the major facilities. What setup and production clues can you devise from this location sketch? SEE 18.16

Before continuing, study the indoor location sketch (figure 18.15) and the outdoor location sketch (figure 18.16) and list as many production requirements as you can determine, then pencil in the type and the placement of cameras and microphones. In the following section, you can compare your lists and equipment placement with the suggested setups in figures 18.17 and 18.18.

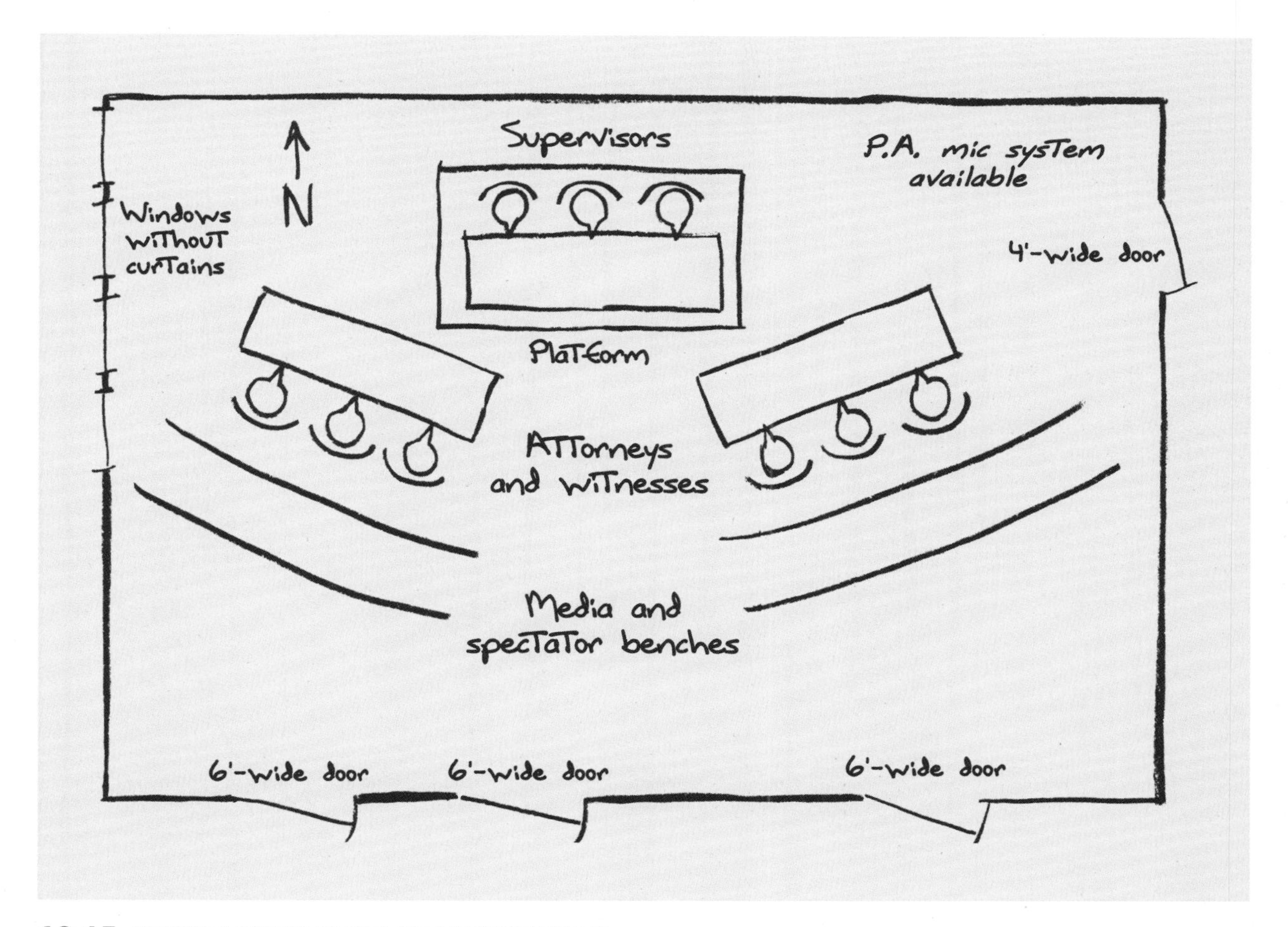

18.15 LOCATION SKETCH OF CITY HALL HEARING ROOM

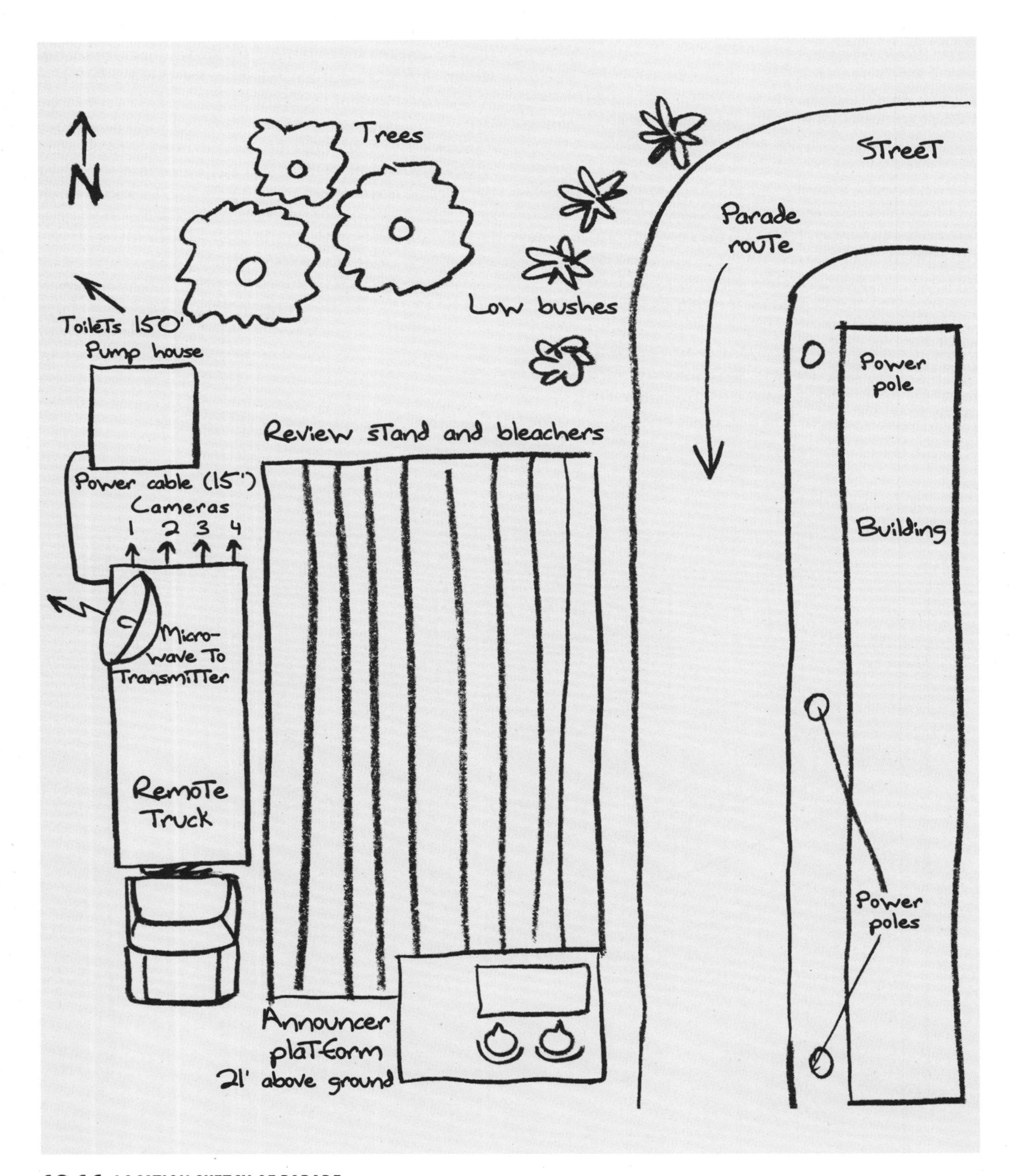

18.16 LOCATION SKETCH OF PARADE

Production Requirements for Public Hearing (Indoor Remote)

Figure 18.17 shows the possible camera placement, mic setup, and lighting solutions for the public hearing room. **SEE 18.17**

■ *Cameras.* This setup requires two cameras on tripod dollies connected by cable to the remote truck. C1 will cover the supervisors; C2 will cover the attorneys, witnesses, and spectators.

■ *Lighting.* The hearing is scheduled for 3 p.m. The large window presents a definite lighting problem. There are two solutions: (1) cover it with drapes and add floodlights or (2) have camera 2 truck closer to the supervisors' table and try to avoid shooting the window when covering the attorneys at the window-side table. Now the window can act as a large key light. In this case, all additional floodlighting must have blue gels to raise the color temperature to the outdoor (5,600K) standard. If the ceiling is high enough, place some back lights. Are there enough AC outlets for the lights? Are they on different circuits? There may be some access problems if the mic and lighting cables are strung across the doorways.

■ *Audio.* Because the hearing room is already equipped with a P.A. system, tie into the existing mics. If the system is not operational, desk mics are the most logical solution. One additional mic should be placed on each of the three tables (supervisors' table and two witness tables) just in case the existing audio system malfunctions.

■ *Intercommunications.* Because there is little or no cueing involved (usually for the start and the end of the recording

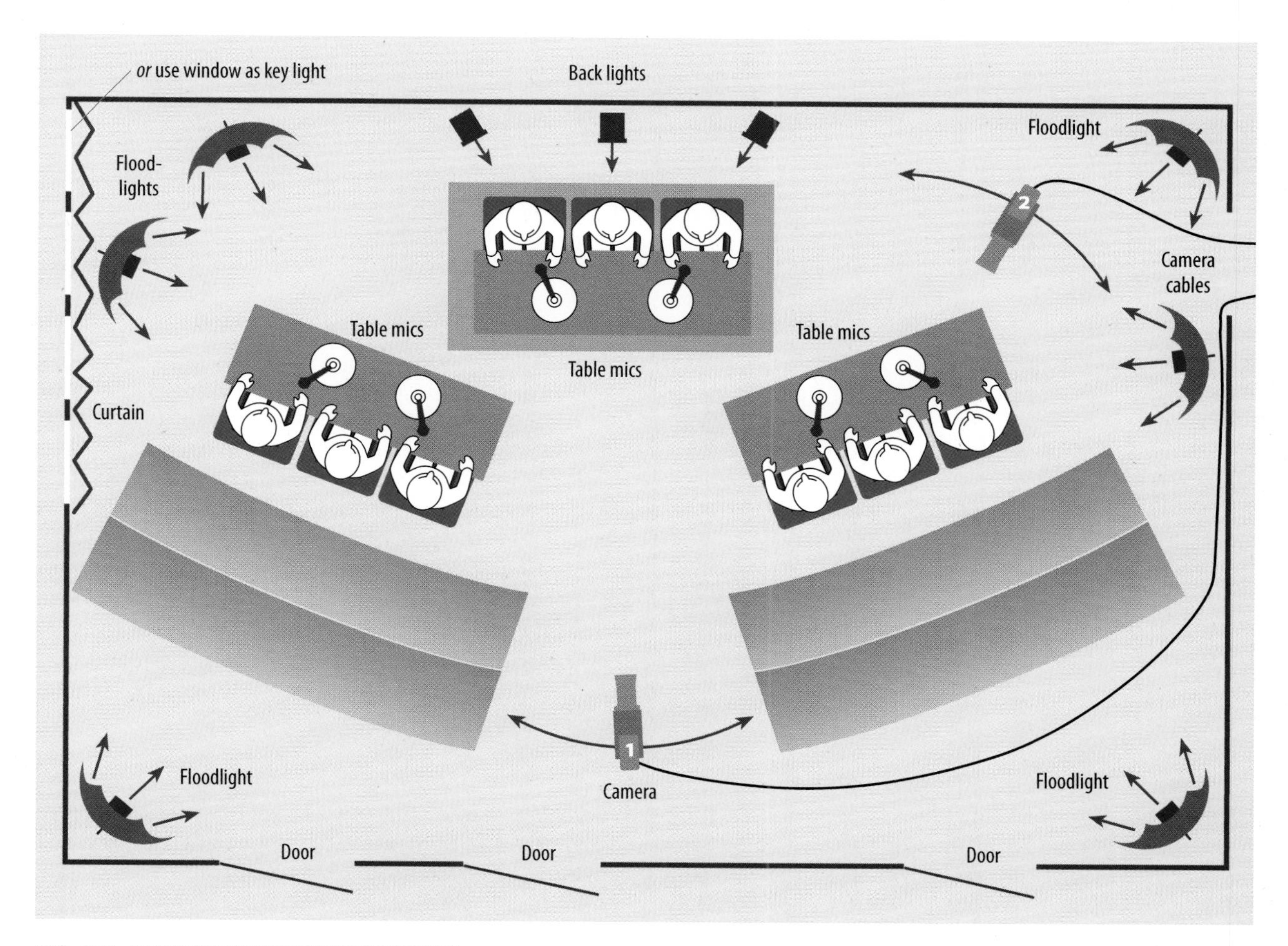

18.17 HEARING ROOM WITH FACILITIES

only), the floor manager can plug the headsets into one of the cameras. If ENG/EFP cameras are used, separate intercom cables may have to be strung for the floor manager and each camera operator.

■ *Other considerations.* Camera cables can be routed through the side door. If the room has a hardwood floor, the cameras could dolly into position for optimal shots. Because there is much traffic in the room, all cables must be taped to the floor and covered by rubber mats. Camera 1 will be in heavy traffic because of the public access doors.

Production Requirements for Parade (Outdoor Remote)

Now compare your list and sketch for the parade with the setup shown in the next figure. SEE 18.18

■ *Location of remote truck.* The truck is in a good location. It's fairly close to a power source (pump house) and the camera positions, minimizing cable runs.

■ *Cameras.* You'll need a minimum of four cameras: C1 and C2 (studio/field cameras) on top of the bleachers, and C3 and C4 (ENG/EFP) on the street. C2 can also cover talent. If possible, C4 should be mounted on a field jib at street level.

■ *Lighting.* Because the videotaping is scheduled for 3:30 to 5:30 p.m., there is sufficient light throughout the telecast. Because the sun is mostly behind the cameras, you may need a device to prevent the sun from washing out the viewfinders of cameras 1 and 2.

■ *Audio.* There are three types of audio pickup: (1) the voice pickup of the two announcers, (2) the bands in the parade, and (3) the sounds of the spectators. Use lavalier mics with windscreens or headset mics for the talent. Use three shotgun or parabolic mics (one high in the stands, the others just above ground level) for the bands. Use two omnidirectional mics on the platform for the crowd noise. All mics need windscreens, and the shotgun mics need wind jammers. One of the shotgun mics could be aimed at the street turn to increase the audio pickup zone. This will enable you to hear the band when you use a tight zoom shot of it coming around the bend.

■ *Intercommunications.* The camera operators are connected to the normal P.L. lines of the camera cables. You'll need a separate intercom line for the floor manager's headset. Use I.F.B. for the talent. Unless you rely on your cell phones, you will need at least two telephone lines for intercommunication from the truck: a direct line to the station and the transmitter and another line for general voice communication.

■ *Signal transmission.* You'll use a direct microwave link to the transmission tower (and from there to the station). Audio is sent via telephone lines (independent of microwave). This separation ensures audio continuity if the microwave link fails.

■ *Other considerations.* Cameras 1 and 2 need a field lens to catch close-ups of the action around the bend (40×). You will need a large monitor for the talent and a second monitor for backup. Shade the monitors from the sun. Route the cables underneath the platform to reduce the potential tripping hazard. The ENG/EFP cameras (C3 and C4) need cable pullers in addition to the camera operators. Toilet facilities are fairly close to the pump house. Raincoats and umbrellas may be needed for crew, talent, and cameras just in case the weather report predicting a beautiful day is wrong.

COMMUNICATION SYSTEMS

Well-functioning communication systems are especially important for production people in the field, regardless of whether the "field" is the street corner across from the station or one in London. These systems must be highly reliable and must enable the people at home base to talk with the field personnel, and the field personnel to talk with one another. When doing ENG you must be able to receive messages from the newsroom as well as the police and fire departments. As a producer or director, you need to reach the talent directly with specific information, even while the talent is on the air.

We have come to expect the relatively flawless transporting of television pictures and sound, regardless of whether they originate from the mayor's downtown office or the orbiting space station. Although communication systems and signal distribution are the province of the technical crew, you should still be familiar with them so that you will know what is available to you. This section provides a brief overview of ENG, EFP, and big-remote communication systems.

ENG Communication Systems

Electronic news gathering has such a high degree of readiness not only because of the mobile and self-contained camera/recorder/audio unit but also because of elaborate communication devices. Most ENG vehicles are equipped with cell phones, scanners that continuously monitor the

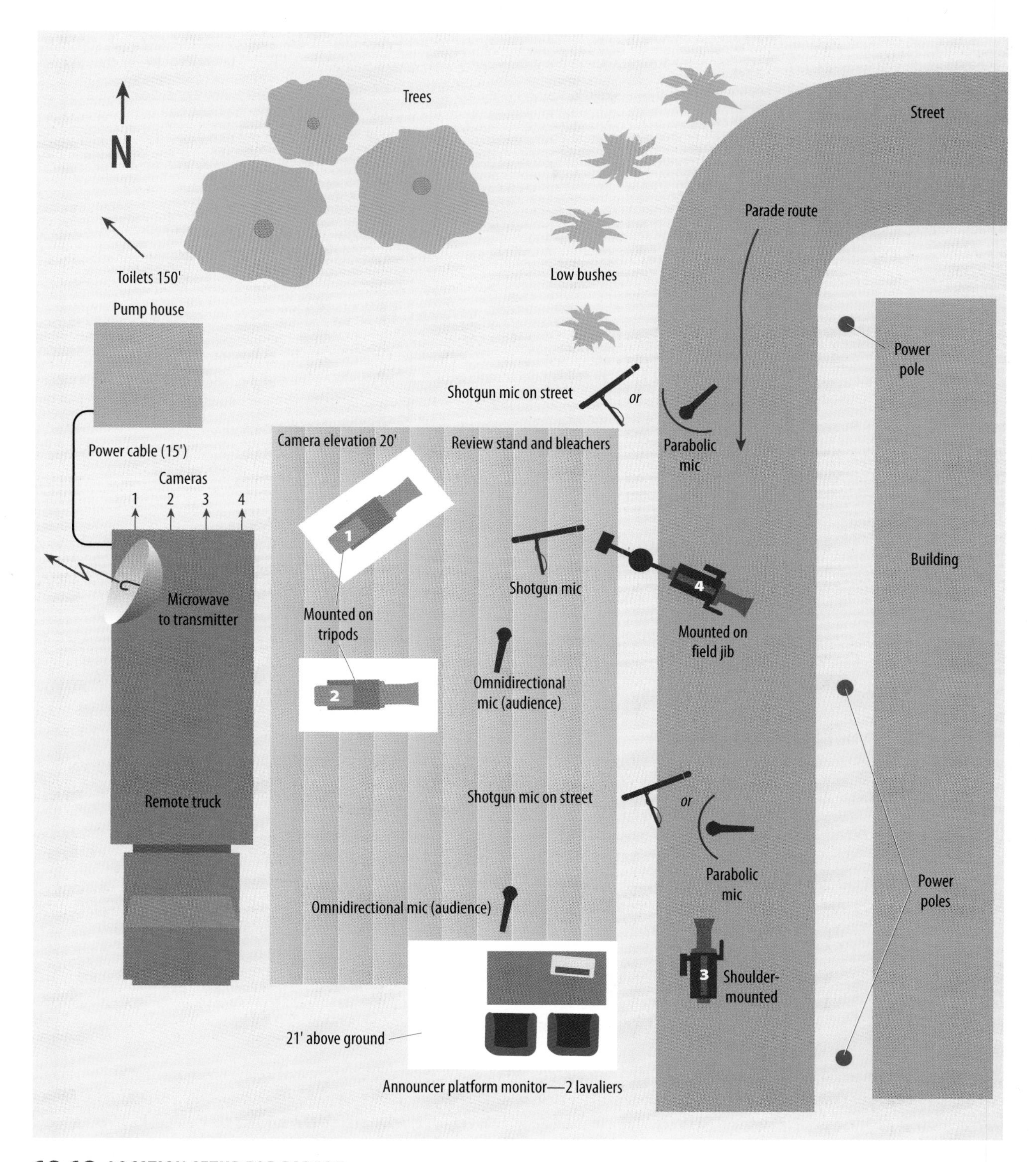

18.18 LOCATION SETUP FOR PARADE

frequencies used by police and fire departments, a paging system, and two-way radios. Scanners lock in on a certain frequency as soon as they detect a signal and let you hear the conversation on that frequency.

These communication systems also make it possible for your station's newsroom to contact you while you are on the road and give you a chance to respond immediately to police and fire calls. Sometimes news departments use codes to communicate with their "cruising" field reporters to prevent the competition from getting wind of a breaking story.

EFP Communication Systems

A single-camera EFP needs the least sophisticated communication system. Because the director is in direct contact with the crew and the talent at the event location, no intercom system is needed. Generally, widely dispersed crewmembers keep in touch with one another via walkie-talkies or cell phones. As pointed out earlier, a small power megaphone might save your voice when giving directions collectively to talent and crew.

The EFP van is normally equipped with phone jacks for regular phone connections, but cell phones will usually suffice. If the EFP uses multiple cameras that are coordinated from a central location, a headset intercom system must be set up for the communication among director, TD, and crew. When doing a live telecast from the field, an I.F.B. system is added.

Big-remote Communication Systems

Big remotes need communication systems between the remote truck (or any other remote control room) and the production people and crew, between the truck and the station, and between the truck and the talent. The truck and the production crew communicate through a regular P.L. (private line or phone line) system, which uses the P.L. channels in the camera cable, separately wired P.L. lines, or wireless P.L.s. During a complicated setup in which the crew is widely scattered (such as when covering a downhill ski race), walkie-talkies are also used. If necessary, the P.L. communication can be carried by telephone lines from truck to station. For the setup and signal transmission details, the engineers use their own phone line.

The I.F.B. (interruptible foldback or feedback) system is one of the most important lines of communication between the producer or director and the talent during a big remote. If several reporters or commentators are involved in the same event, you can switch among several I.F.B. channels so that, if necessary, you can address the field reporters and the commentators individually. If needed, your I.F.B. instructions to the talent can be transmitted via satellite over great distances. Realize, however, that there is inevitably a slight delay before the talent receives your instructions. In sports remotes the talent occasionally wear headsets through which they receive instructions from the truck.

The remote truck is, of course, equipped with several wired telephone lines, two-way radios, cell phones, paging systems, and walkie-talkies.

SIGNAL TRANSPORT

Signal transport refers to the systems available to you when transmitting the video and audio signals from their origin (microphone and camera) to the recording device or transmitter, and from the point of origin to various reception points. Signal transport includes microwave transmission, cable distribution, and communications satellites.

Microwave Transmission

If you need to maintain optimal camera mobility during a live pickup, such as when shooting interviews from a convention floor, you cannot use a camera cable but must send the signal to the remote truck via microwave.

From camera to remote truck There are small, portable, battery-powered transmitters that can be mounted on the camera. If the distance from camera to receiving station in the truck is not too great, you can use this system to relay the camera video and audio signals to the remote truck without too much difficulty. To minimize interference from other stations covering the same event, you can transmit on several frequencies, a practice called frequency agility.

If you need a more powerful microwave transmitter, you can mount a small microwave dish on a tripod and place it close to the camera action radius. That way you can work at a considerable distance from the remote truck while using only a relatively short cable run from camera to dish. This type of link is especially useful if a cable run would create potential hazards, such as when strung from a building across high-tension wires or across busy city streets. SEE 18.19

An ever-present problem with camera-to-truck microwave links is interference, especially if several television crews are covering the same event. Even if you use a system with relatively great frequency agility, your competition may be similarly agile and overpower you with a stronger signal.

18.19 PHOTO BY HERBERT ZETTL

18.19 TRIPOD-MOUNTED MICROWAVE TRANSMITTER
This small tripod-mounted microwave transmitter can relay camera signals over a considerable distance.

From remote van to station or transmitter The longer, and usually much more complex, signal link is from the remote van to the television station. (Although sometimes the signal is sent directly to the transmitter, we will call the end point of this last link before the actual broadcast the "station.") You can send the signals from the remote van directly to the station only if you have a clear, unobstructed line of sight. SEE 18.20

Because the microwave signal travels in a straight line, tall buildings, bridges, or mountains that are in the line of sight between the remote van and the station may block the signal transmission. In such cases several microwave links, called ***mini-links***, have to be established to carry the signal around the obstacles. SEE 18.21

Television stations in metropolitan areas have permanent ***microwave relays*** installed in strategic locations so that remote vans can send their signals back from practically any point in their coverage area. If these permanent installations do not suffice, helicopters are used as microwave relay stations (see figure 18.21). Permanent microwave relays are also used for transmitting the video of permanently installed cameras that monitor the weather and traffic.

There are several other ways to transmit signals from the remote location to the station. Although this is not your responsibility unless you are a broadcast engineer, you should nevertheless have some idea of what is available.

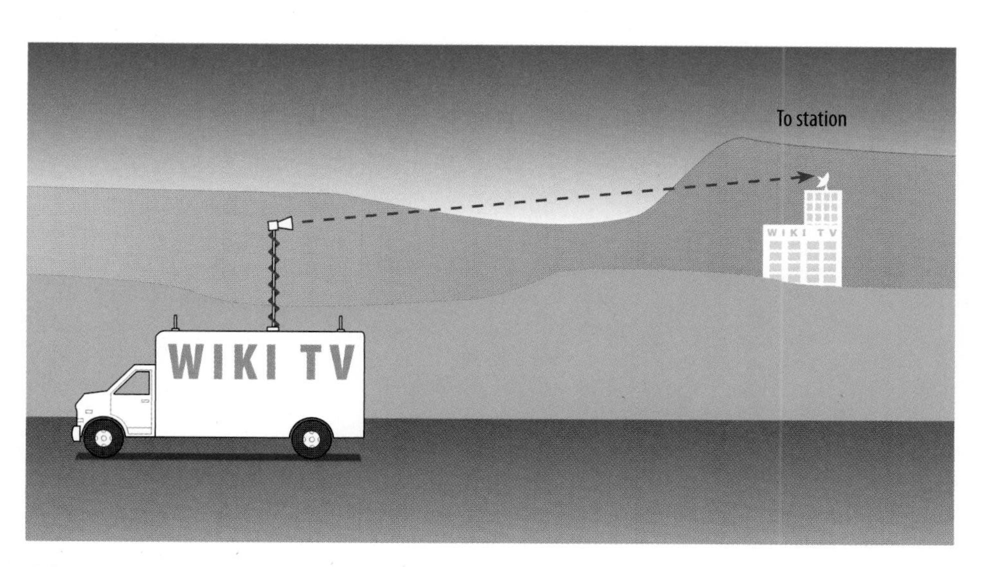

18.20 DIRECT MICROWAVE LINK
You can transmit a signal via microwave from the remote van back to the station only if there is a clear, unobstructed line of sight.

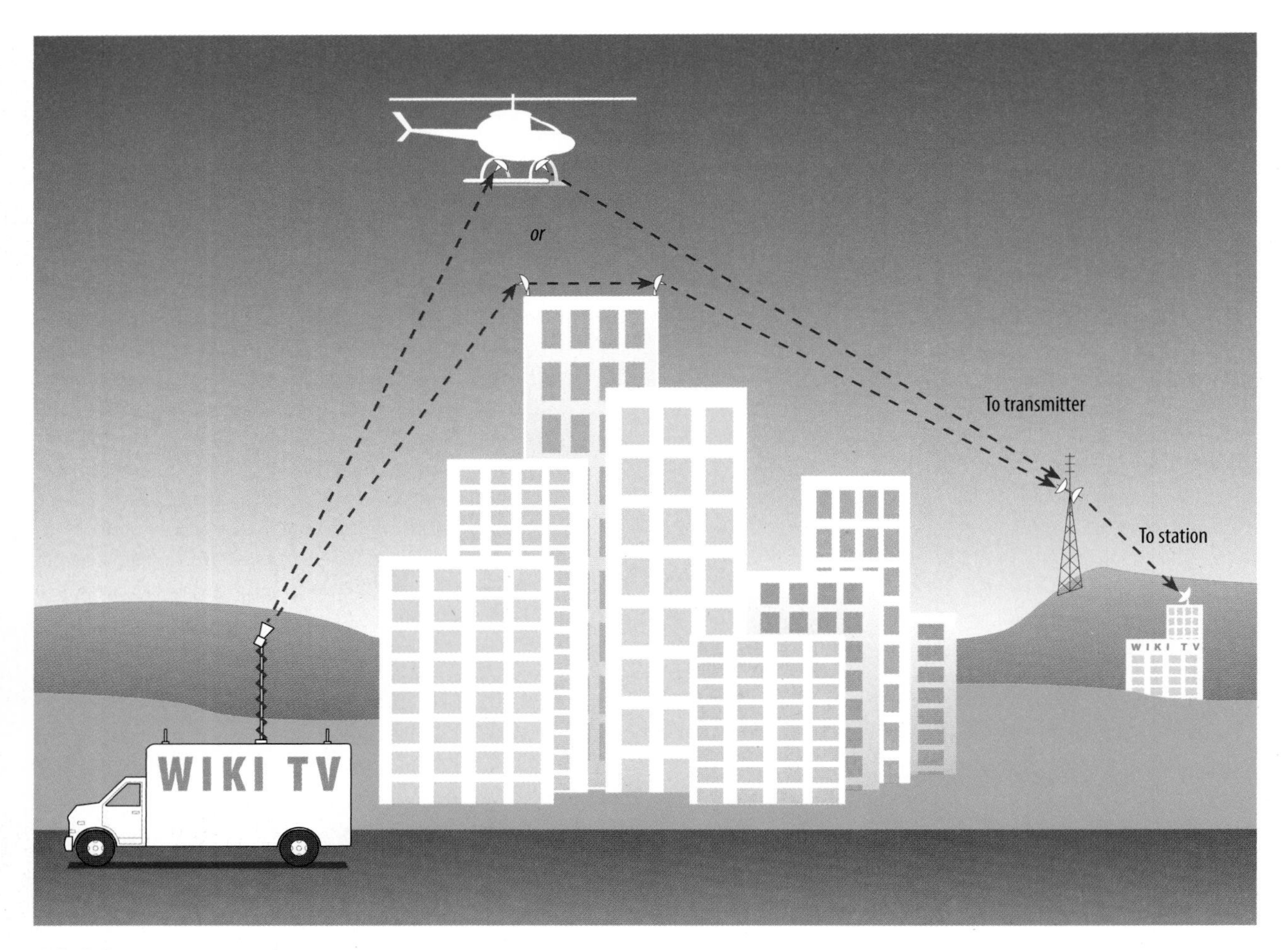

18.21 MINI-LINK FROM REMOTE VAN TO STATION
If there is no clear line of sight between the remote van and the station, the microwave signal must be transported via mini-links.

Cable Distribution

Telephone companies can provide broadband high-definition digital cable service (coax or fiber-optic) that transports all video and audio signals, including HDTV 1080i signals. ***Broadband*** is a high-bandwidth standard for sending information (voice, data, video, and audio) simultaneously over cables.

Communications satellites

The communications satellites used for broadcast are positioned in a geosynchronous orbit 22,300 miles above the earth. In this orbit the satellite moves synchronously with the earth, thereby remaining in the same position relative to it.

Communications satellites operate on two frequency bands: the lower-frequency C-band and the higher-frequency ***Ku-band*** (pronounced "kay-*you*-band"). Some satellites have transponders for C-band as well as Ku-band transmission and can convert internally from one to the other.

A ***direct broadcast satellite (DBS)*** has a relatively high-powered transponder (transmitter/receiver) that broadcasts from the satellite to small individual downlink dishes (which you can buy in larger electronics stores and install yourself). A DBS operates on the Ku-band.

The C-band is a highly reliable system that is relatively immune to weather interference. Because the C-band works with microwave frequencies, it may interfere with ground-based microwave transmission. To avoid such interference, the C-band operates with relatively low power; because of the low power, the ground stations need large dishes, which range anywhere from 15 to 30 feet. Such large dishes are obviously not suitable for mobile uplink trucks. To use the C-band, the television signals must be transported to and from permanent ground stations.

The Ku-band, on the other hand, operates with more power and smaller dishes (3 feet or less) that can be mounted and readily operated on mobile trucks or on a rooftop. The Ku-band is also less crowded than the C-band and allows immediate, virtually unscheduled access to uplinks. One of the major problems with the Ku-band is that it is susceptible to weather; because of its ultrashort wavelength, rain and snow can interfere with transmission.

Uplinks and downlinks Television signals are sent to a satellite through an ***uplink*** (earth station transmitter), received, amplified by the satellite, and beamed back at a different frequency (actually rebroadcast) by the satellite's own transmitter to one or several receiving earth stations, called ***downlinks***. The receiver-transmitter unit in the satellite is called a transponder, a combination of *trans*mitter and res*ponder* (receiver). Many satellites used for international television transmission have built-in translators that automatically convert one electronic signal standard, such as the NTSC system, into another, such as the European PAL system.

Because the satellite transmission covers a large area, simple receiving stations (downlinks) can be set up in many widely dispersed parts of the world. SEE 18.22 These

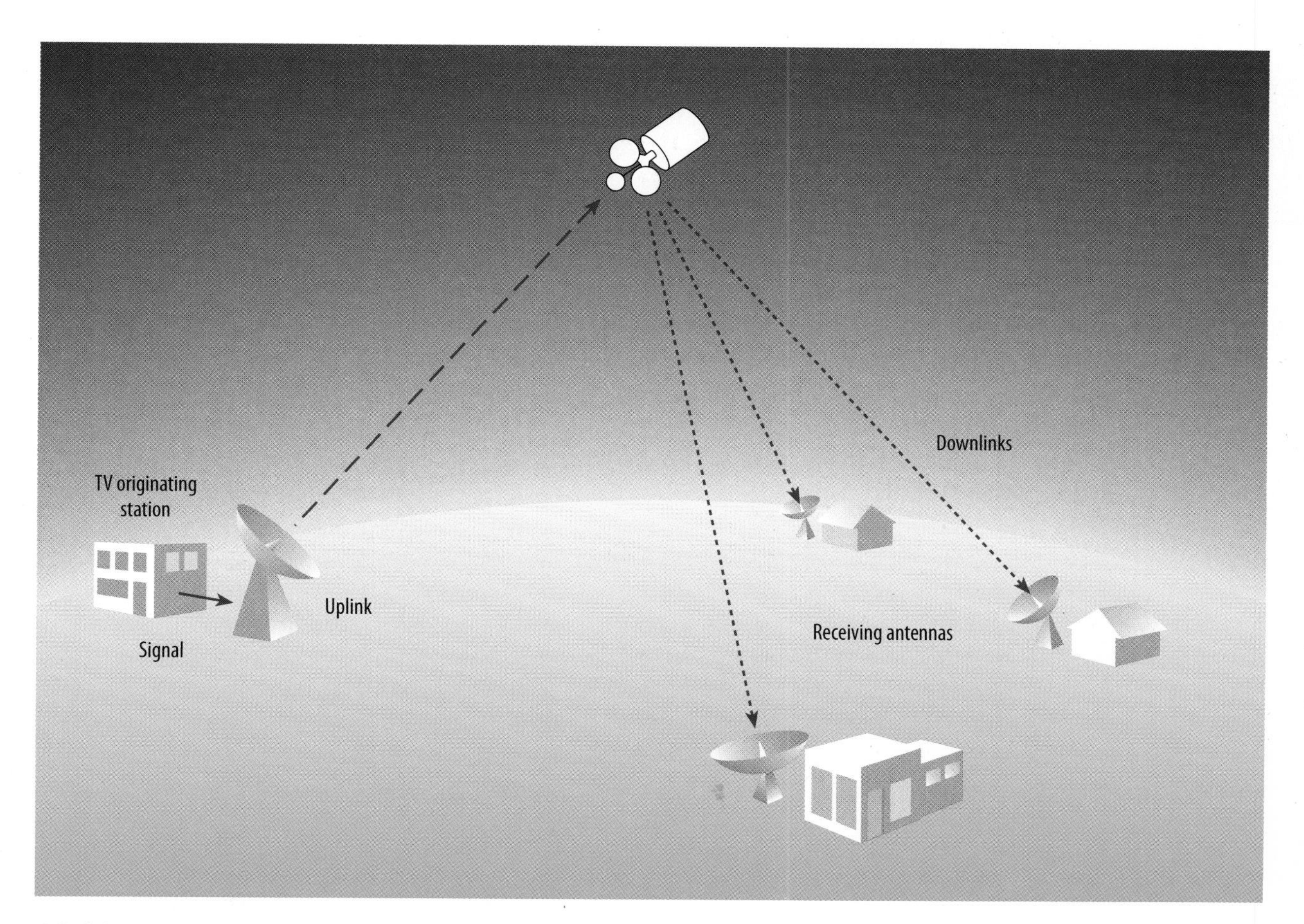

18.22 SATELLITE UPLINK AND DOWNLINKS
The uplink sends television signals to the satellite. The downlinks receive television signals from the satellite.

strategically placed satellites can spread their footprint (coverage area) over the whole earth.

Specialized vans can provide mobile uplinks for the transport of television signals. These ***uplink trucks*** operate on the very same principle as a microwave van except that they send the television signals to a satellite rather than to a receiving microwave dish. As noted in section 18.1, satellite uplink vans usually contain additional equipment, such as video recorders and editing computers.

MAIN POINTS

- Many big remotes are devoted to the coverage of sporting events. Networks typically use a great amount of equipment and many personnel for sports remotes, but good coverage is also possible with less equipment.
- There are standard setups for most sporting events, which can be augmented with more cameras and audio equipment.
- Location sketches are a valuable preproduction aid for big remotes. For an indoor remote, they may show the general dimensions of a room or hallway; the locations of windows, doors, and furniture; and the principal action areas. Outdoor location sketches may show buildings, remote truck location, power source, steep inclines or steps, the path of the sun, and the location and/or direction of the main event.
- A good location sketch can aid the director and the technical supervisor in deciding on camera locations, focal lengths of zoom lenses, lighting and audio setups, and intercommunication systems.
- Big-remote operations depend heavily on reliable intercommunication systems, including the P.L. system, walkie-talkies, pagers, cell phones, and multichannel I.F.B. systems. The I.F.B. information can be transmitted via telephone lines and satellite to widely scattered talent in remote locations.
- Remote signals are usually transported via microwave, cable, or satellite.
- Communications satellites used for broadcast operate in the lower-frequency C-band and the higher-frequency Ku-band.

ZETTL'S VIDEOLAB

For your reference or to track your work, the *Zettl's VideoLab* program cue in this chapter is listed here with its corresponding page number.

PART IV

Postproduction

CHAPTER 19

Postproduction Editing: How It Works

Almost all programs you see on television have been edited in some way, either during or after the actual production. When editing is done after (*post* in Latin), it is known as postproduction editing. Its processes differ considerably from switching, the instantaneous editing done during production. In postproduction you have more time to deliberate just which shot to include in your masterpiece and which to discard, but you also have the responsibility to select the one that tells the story most effectively.

Today most postproduction editing is done with a computer and editing software rather than with videotape recorders (VTRs), whereby you copy the selected shots from one videotape to another. When using a computer for editing, you are engaged in nonlinear editing. When transferring shots in a particular order from one VTR to another, you are doing linear editing.

Regardless of the system you are using, your aesthetic choices are the most important aspect of the postproduction phase. You must now apply everything you have learned about television production to identify the most impressive shots and sequence them so that they will tell your story with clarity and impact.

Section 19.1, How Nonlinear Editing Works, examines the basic components and procedures of the nonlinear editing system and describes some pre-editing tasks. Section 19.2, How Linear Editing Works, gives an overview of the basic linear editing process.

KEY TERMS

AB-roll editing Creating an edit master recording by combining the A-roll, which contains a set of shots (such as long and medium shots), and the B-roll, which contains different but related shots (such as cutaways or close-ups of the same scene).

assemble editing Adding shots in linear editing on videotape in consecutive order without first recording a control track on the edit master tape.

automated dialogue replacement (ADR) The synchronization of speech with the lip movements of the speaker in postproduction.

batch capture Having the computer use the logging information to capture all the selected clips for editing.

capture Transferring video and audio information to a computer hard drive for nonlinear editing. Also called *importing.*

clip A shot or brief series of shots as captured on the hard drive and identified by a file name.

control track system Linear editing system that counts the control track sync pulses and translates that count into elapsed time and frame numbers. It is not frame-accurate. Also called *pulse-count system.*

edit controller Machine that assists in linear editing functions, such as marking edit-in and edit-out points, rolling source and record videotape recorders, and activating effects equipment. Often a desktop computer with specialized software. Also called *editing control unit.*

edit decision list (EDL) Consists of edit-in and edit-out points, expressed in time code numbers, and the nature of transitions between shots.

edit master recording The recording of the final edit on videotape or other media. Used for broadcast or duplication.

insert editing Requires for linear editing the prior laying of a control track on the edit master tape. The shots are edited in sequence or inserted into an already existing recording. Necessary mode for editing audio and video tracks separately.

linear editing Analog or digital editing that uses tape-based systems. The selection of shots is nonrandom.

nonlinear editing (NLE) Allows instant random access to shots and sequences and easy rearrangement. The video and audio information is stored in digital form on computer hard drives or other digital recording media.

off-line editing In linear editing it produces an edit decision list or a videotape not intended for broadcast. In nonlinear editing the selected shots are captured in low resolution to save computer storage space.

on-line editing In linear editing it produces the final high-quality edit master recording for broadcast or program duplication. In nonlinear editing it can mean recapturing the selected shots at a higher resolution.

record VTR The videotape recorder that edits the program segments as supplied by the source VTR(s) into the final edit master tape. Also called *edit VTR.*

shot The smallest convenient operational unit in video and film, usually the interval between two transitions. In cinema it may refer to a specific camera setup.

slate (1) Visual and/or verbal identification of each video-recorded segment. (2) A small blackboard or whiteboard upon which essential production information is written. It is recorded at the beginning of each take.

SMPTE/EBU time code Stands for *Society of Motion Picture and Television Engineers/European Broadcasting Union time code.* Electronic signal recorded on the cue or address track of a videotape or a track of a multitrack audiotape to give each frame a specific address. The time code reader translates this signal into a specific number (hours, minutes, seconds, and frames) for each frame.

source media The media (videotape, hard disk, optical disc, or memory card) that holds the recorded camera footage.

source tape The videotape with the original camera footage.

source VTR The videotape recorder that supplies the program segments to be assembled by the record VTR. Also called *play VTR.*

split edit Technically, the audio of a shot is substituted with related sounds or narration. In common practice the audio precedes the shot or bleeds into the next one. Viewers hear the audio of the next shot before seeing it, or they still hear the audio of the previous shot at the beginning of the new one.

take Any one of similar repeated shots taken during video recording or filming. Usually assigned a number. A good take is the successful recording of a shot. A bad take is an unsuccessful recording, requiring another take.

time code Gives each television frame a specific address (number that shows hours, minutes, seconds, and frames of elapsed tape).

vector Refers to a perceivable force with a direction and a magnitude. Vector types include graphic vectors, index vectors, and motion vectors.

video recorder (VR) All devices that record video and audio. Includes videotape, hard disks, read/write optical discs, and memory cards.

VR log A list of all takes—both good (acceptable) and no good (unacceptable)—in consecutive order by scene number and time code address. Often done with computerized logging programs. A vector column facilitates shot selection.

window dub A "bumped-down" copy of all source media that has the time code keyed over each frame.

SECTION

19.1

How Nonlinear Editing Works

The operational principle of ***nonlinear editing (NLE)*** is selecting video and audio data files and making the computer play them back as a specific sequence. All nonlinear editing is done by selecting and sequencing shots that have been transferred from a camcorder or other video storage device to the computer hard drive of an editing system. Once you use editing software to transfer the source footage to the hard drive of the editing computer, you are engaged in nonlinear editing. ZVL1 EDITING→ Editing introduction

This section explains the nonlinear editing system and its use in postproduction editing.

▶ **NONLINEAR EDITING**
Difference between linear and nonlinear editing

▶ **NONLINEAR EDITING SYSTEM**
Computer hardware and software, source media, audio/video capture, and exporting the final edit

▶ **PRE-EDIT PHASE**
Thinking about shot continuity, recordkeeping, and reviewing and ordering the source footage

▶ **PREPARATION PHASE**
Time code, logging, capture, and audio transcription

▶ **EDITING PHASE: VIDEO**
Editing video to audio, editing audio to video, and transitions and effects

▶ **EDITING PHASE: AUDIO**
Linear audio editing, nonlinear audio editing, condensing, correcting, mixing, controlling quality, and automated dialogue replacement

NONLINEAR EDITING

As indicated in the introduction to this chapter, the fundamental difference between linear and nonlinear editing systems is that linear systems copy selected portions from the source tapes to another videotape—the edit master (see section 19.2). Nonlinear editing, on the other hand, does not copy anything. All that the computer really does is mark the files that contain specific ***clips*** and play the clips back in a particular order. Instead of editing one shot next to another, with NLE you are engaged in file management.

Nonlinear editing allows you to try, compare, and keep as many editing versions as you like, without being committed to any one. All you do is create ***edit decision lists (EDLs)***, which the computer remembers and applies when recalling the clips. Once you decide on a particular arrangement, you can output that version to the edit master recorder, which may be a DVD burner, a server, a hard drive, a videotape, or some other digital storage media.

Difference Between Linear and Nonlinear Editing

Why are these editing modes called linear and nonlinear? Consider a scenario in which you are editing film or videotape: you need to start with shot 1, then add shot 2, then shot 3, and so forth until you have added the last shot. The whole process is a linear affair that starts at the beginning title and ends with the closing credits. Even when looking for a particular frame or shot on the source tapes, you cannot jump to it directly but must patiently roll through the footage until you find it. This procedure is especially trying if the shot happens to be buried in the middle of the tape. Once the final cut is done and you change your mind, or if the director or client wants you to replace a shot or scene in the middle of the final cut, you may have to re-edit everything that follows the insert.

On the other hand, the nonlinear editing system allows random access. There is no need to roll through a videotape until the desired shot finally shows up. Once the source material is stored on the hard drive of the editing system, you can access any clip randomly and almost instantly. You can also insert or delete a shot anywhere in the edited sequence without affecting the footage preceding or following the change. There is no linearity in storing the source material, in retrieving it, or even in assembling the selected clips. Because even the final cut is nonlinear, you can accommodate the director's or client's request for a change in the middle of the production as easily as changing the opening title or last credit line. Learning this type of file management is just a bit more difficult than a simple word-processing program.

Because there are so many nonlinear systems on the market, we concentrate on how a basic nonlinear system works rather than go into detail about a particular software program. If you want to learn a specific NLE system, you need to study the often-voluminous instructions and tutorials, and, most importantly, do some actual editing with the new system.

NONLINEAR EDITING SYSTEM

A nonlinear editing system consists of a playback facility for the source material; a high-speed computer and software for the capture, storage, and manipulation of video and audio clips; and a ***video recorder (VR)*** that produces the final ***edit master recording*** for broadcast or distribution. **SEE 19.1** ZVL2 EDITING→ Nonlinear editing→ system

Let's take a closer look at each element and the major processes of the NLE system: computer hardware and software, source media, audio/video capture, and exporting the final edit.

Computer Hardware and Software

All nonlinear editing systems are basically computers that store digital video and audio information on high-capacity hard drives and compatible software that allows the selection and the sequencing of video and audio clips. An important factor in the choice of an editing computer is that it has a large storage capacity and a fast enough processor to ensure smooth playback even if the video and audio files are only slightly compressed. Some NLE systems have their own computer configuration, some have keyboards whose keys are marked with edit commands, and most editing software supplies stickers with symbols that you can apply to a standard keyboard.

Although you may have a relatively large flat-panel monitor that can be divided into two screens, you will find that a second, separate monitor greatly facilitates editing. You can use the large one for the editing display and the second one as a playback monitor of the edited segments. Sometimes you may want to mix sound before importing it into the computer. In this case you will need an audio mixer that feeds into the computer. **SEE 19.2**

Thanks to readily available and sophisticated editing software, even a laptop computer can be a powerful NLE

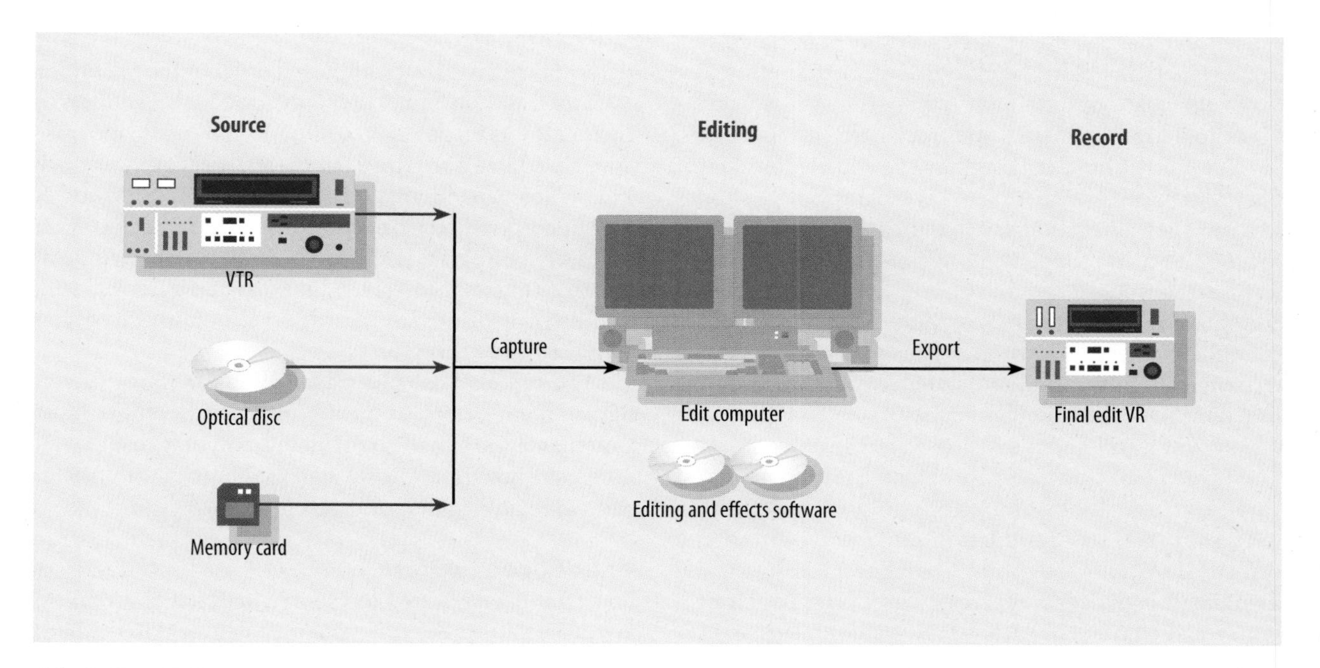

19.1 NONLINEAR EDITING SYSTEM
The components of a nonlinear editing system are: playback facility for the source footage, computer with software, and video recorder that produces the final edit master recording.

19.2 BASIC NONLINEAR EDITING SETUP
The nonlinear setup normally has a high-speed, high-capacity computer; a keyboard; a display for the editing interface and one for viewing the edited sequences; a small audio mixer; and speakers for stereo sound.

19.2 PHOTOS BY HAMID KHANI (NLE SYSTEM) AND ROBAIRE REAM (MONITOR IMAGES)

system. In fact, some of the simpler software programs, such as Apple's iMovie, offer more than you need for the average editing job. When watching the final edit, nobody can tell whether you used iMovie software or the latest version of Final Cut Pro for a cut from one shot to another. The major drawback of the simple editing software is its relatively limited audio-mixing and special-effects capabilities.

It would be futile to describe a specific nonlinear editing program. Although most NLE programs have similar features, they differ enough to require study and especially practice. Even the simpler editing programs offer so many features that you will definitely need to consult the manual and the tutorials. Nevertheless, there are features that are common to most NLE software. SEE 19.3 As you can see, the terminology differs from one system to another as do the functions. Once you are familiar with one NLE program, however, you will find it easier to learn the others.

If you plan to do extensive color correction or a great variety of transitions and effects in a high-definition television (HDTV) production, high-end editing and special-effects software is a must. Nonlinear editing lets you manipulate audio clips and create a great variety of transitions and digital video effects (DVE)—all with relative ease. For example, with Adobe's After Effects you can build even complex animated effects in a short rendering time.

A final caveat: The ease with which you can create DVE might tempt you to pay more attention to the special effects than to the story. Use such knock-your-socks-off effects only when they are appropriate to the content of the story and energize the message.[1]

Source Media

All the footage you shot with a camcorder or received from a camera team constitutes the source material, regardless of whether it was recorded on a mini- or full-sized videocassette, a portable hard drive, an optical disc, or a memory card. When using digital ***source media***, you can connect the camcorder directly to the editing computer via USB 3.0, eSATA, FireWire (IEEE 1394), or any such peripheral interface (high-speed cable). Some systems let you plug flash drives (such as a P2 card) directly into the editing computer for easy capture.

You can also use older analog tapes so long as you digitize them before or during the capture phase. Most computers have a built-in digitizing card that enables you to convert analog to digital signals. You can also use an external digitizing unit or bridge.

Audio/Video Capture

Once you have selected the source material you want to have available for the edit, you must deliver it to the hard

1. A good example is the comparison between the films *Avatar* (2009) and *The Hurt Locker* (2008). The former had a trite story that had to live off its effects; the latter used the effects as an integral part of an incredibly intense story. The 2010 Oscar for Best Motion Picture was justifiably awarded to *The Hurt Locker* rather than *Avatar*.

19.3 PHOTOS BY GARY PALMATIER

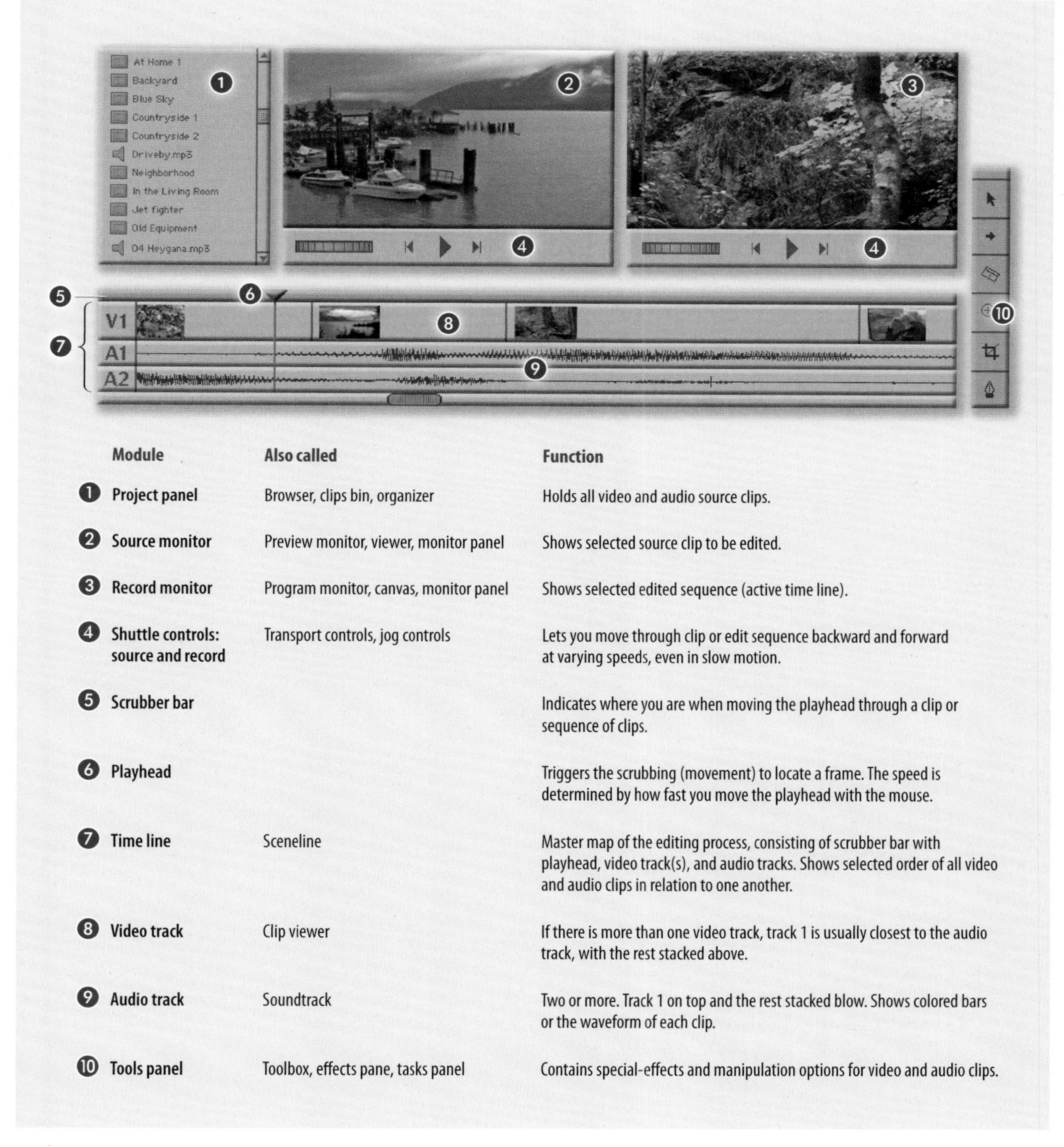

	Module	Also called	Function
1	Project panel	Browser, clips bin, organizer	Holds all video and audio source clips.
2	Source monitor	Preview monitor, viewer, monitor panel	Shows selected source clip to be edited.
3	Record monitor	Program monitor, canvas, monitor panel	Shows selected edited sequence (active time line).
4	Shuttle controls: source and record	Transport controls, jog controls	Lets you move through clip or edit sequence backward and forward at varying speeds, even in slow motion.
5	Scrubber bar		Indicates where you are when moving the playhead through a clip or sequence of clips.
6	Playhead		Triggers the scrubbing (movement) to locate a frame. The speed is determined by how fast you move the playhead with the mouse.
7	Time line	Sceneline	Master map of the editing process, consisting of scrubber bar with playhead, video track(s), and audio tracks. Shows selected order of all video and audio clips in relation to one another.
8	Video track	Clip viewer	If there is more than one video track, track 1 is usually closest to the audio track, with the rest stacked above.
9	Audio track	Soundtrack	Two or more. Track 1 on top and the rest stacked blow. Shows colored bars or the waveform of each clip.
10	Tools panel	Toolbox, effects pane, tasks panel	Contains special-effects and manipulation options for video and audio clips.

19.3 NONLINEAR EDITING INTERFACE
This generic nonlinear editing interface shows the major components of an NLE system. Note that it does not show or list the items accessible through the menus.

drive of the computer you are using for the NLE system. This process, called ***capture***, or importing, is usually done by connecting the source recorder (usually the camcorder) to the editing computer via a high-speed cable (peripheral interface). Bridges facilitate this process, enabling various audio and video inputs, including standard definition (SD), high-definition (HD), and even analog video, to be converted to the capture standards of the NLE computer. If you use such a bridge, you obviously don't need the external digitizing unit for converting analog footage. SEE 19.4

When importing the source material, you can usually select a specific codec that will compress the video to a certain degree. Recall from chapter 5 that the higher the image compression, the less space the files occupy on the hard disk and the faster the import will be. The downside is that the pictures look worse (lower definition and color

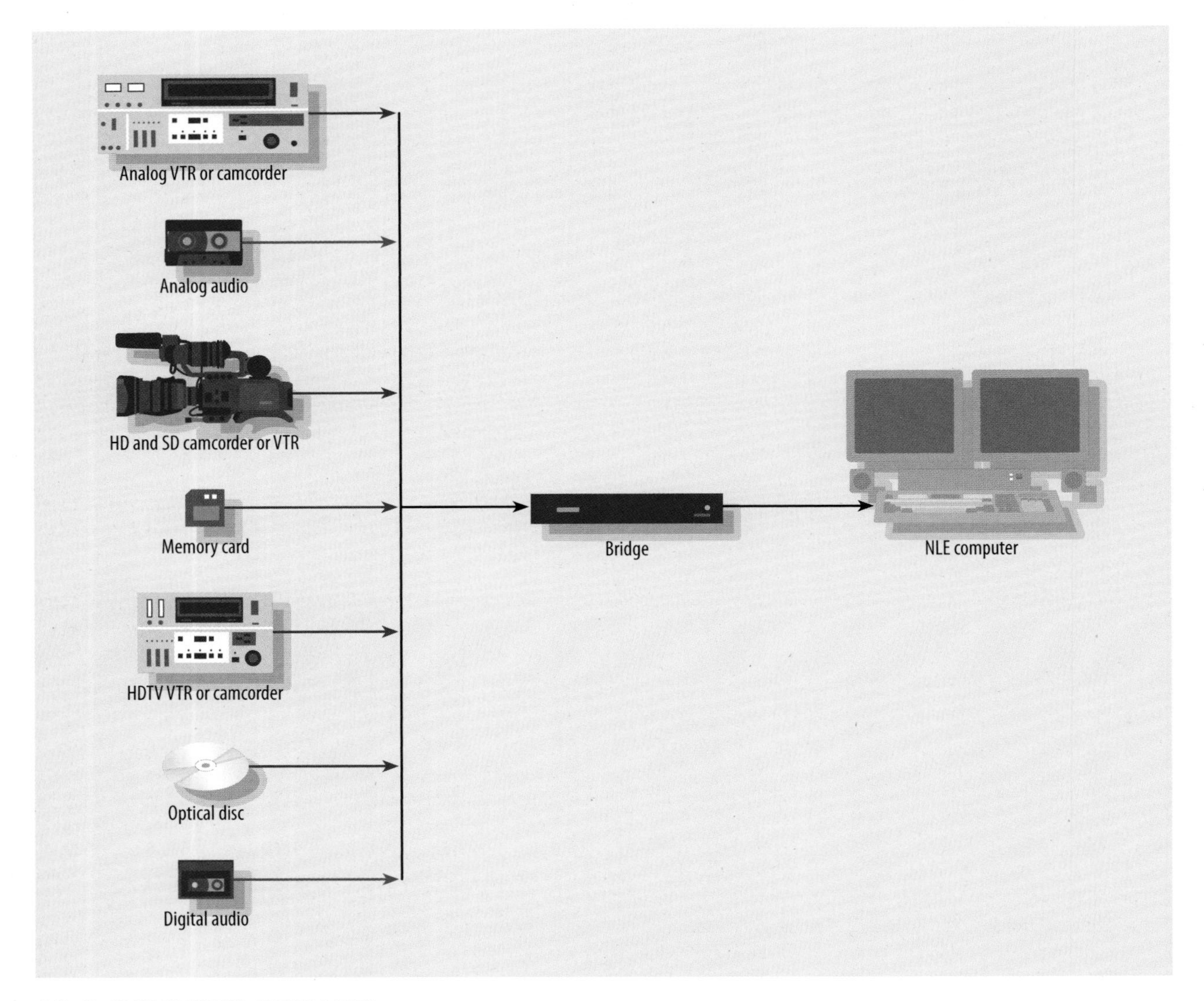

19.4 SOURCE MEDIA CONNECTION
The source footage is usually delivered via a high-speed cable (a peripheral interface, such as USB 3.0, FireWire, or eSATA) to the NLE system. A bridge connects a variety of video and audio inputs to the computer, enabling the signals to be converted to the capture standards of the NLE system.

fidelity) with higher compression. Make sure that the NLE software can decode the codec used in the recording system of the camcorder.

Unless you have already logged the source material, you can do it during the capture phase. You will read more about logging later in this chapter.

Exporting the Final Edit

When you are finished with the actual editing, you need to export the final version and record it on a DVD, videotape, or other media.

PRE-EDIT PHASE

Editing starts with turning on the camcorder for the shoot. Although this may seem strange, it means that experienced videographers are already thinking about postproduction editing when video-recording during the production phase. The actual editing will also be greatly facilitated if you spend some time looking over the recorded footage and arranging the source media according to the intended event sequence. Specifically, the pre-edit phase includes thinking about shot continuity, keeping accurate records, and reviewing and ordering the source footage.

Thinking About Shot Continuity

Much of the editing is predetermined by the way the material is shot. Some novice directors or camera operators stop one shot or scene and begin the next without any pads (overlapping action) or consideration for continuity. Others have the ability and the foresight to visualize transitions between shots and scenes and to provide images that cut together well in postproduction. The key is to visualize not just individual shots but shot sequences. Imagining a shot sequence will help you compose shots that join together well to form seamless transitions. Here are a few pointers.

- When recording an event, such as a standup field report, start recording a few seconds both before the cue and after the reporter has finished speaking. These pauses will give you a video pad, also called trim handles, to cut to the next shot at just the right moment. A change in the angle and the field of view (moving the camera nearer or farther away from the event) between shots will also help make the cuts look organic and smooth. ZVL3 EDITING→ Production guidelines→ pickup

- Always get some cutaway shots. A cutaway is a brief shot that establishes continuity between two other shots; provides the necessary video pad when editing according to sound bites (portion of video-recorded interview in which we see and hear the person talk); and, in more ambitious productions, helps connect jumps in time and/or location. The cutaway may or may not be part of the principal action, but it must be thematically related to the event. Good cutaways are relatively static and neutral as to screen direction. Examples are straight-on shots of onlookers, reporters with still cameras or camcorders, buildings that show the location, house numbers, or objects that are part of the story (such as the book the guest has written). ZVL4 EDITING→ Production guidelines→ cutaways

- When on an ENG assignment, try to get some cutaways that identify the event location. For example, after covering the downtown fire, get a shot of the street signs of the nearest intersection, the traffic that has backed up, and some close-ups of onlookers and exhausted firefighters. For good measure also get several wide shots of the event location. You will then have cutaways that not only facilitate transitions but also show exactly where the fire occurred.

- Always record the ambient (background) sound with the cutaways. The sound is often as important as the pictures for smooth transitions. The continuity of ambient sound can also help immensely in preserving shot continuity, even if the visuals do not cut together too well.

- Record a minute or two of "room tone" or any other kind of environmental sound, even if the camera has nothing interesting to shoot.

- Whenever possible during ENG and EFP, verbally ***slate*** (identify) the individual takes or at least the shot series. You can do this by simply calling out the name of the event and the take number, such as: "Market Street police station, take 2." You can do the slating on the camera mic or the portable (reporter's) mic. After saying the take number, count backward from five or three to one. This counting is similar to the beeper after the slate in studio productions. Although not essential, it helps locate the take and cue it up during editing, especially if no address code is used.

Recordkeeping

Entering data in a field log or labeling the recording media seems like an unnecessary interruption of your creative fervor while in the middle of a production, and this is true for most ENG assignments. In productions that are shot for extensive postproduction, however, a field log will greatly facilitate editing. During the recording session, you may readily recollect certain scenes, especially when they didn't go well, but you might have a hard time remembering the cutaways and just where in the various source media specific scenes are buried. Accurate records kept during the production will be an invaluable memory aid during postproduction.

Field log As you recall from chapter 13, you need to slate and log all takes unless you are engaged in ENG. You should write on the field log the number of the recording media; the scene and take number; and in and out time code numbers for all takes. If the time code numbers are not available, you should nevertheless write down the scene title and number, the take numbers, and whether the takes are OK or NG (no good).

In a pinch you can audio-slate each take. If you are operating the camera, quickly call out "scene 3, take 5" and record it on the camera-mic track. The field log will save you a considerable amount of time when editing short scenes directly from the recorded material or when constructing the VR log in postproduction.

Labeling the recording media Even if it seems like a waste of time and effort when you video-record some especially difficult scenes for a documentary, you must nevertheless take the time to label each recording media as you extract it from the camcorder. Be sure to label both the box and the media it contains. Cross-reference the information with your field log entry. A good system is to consecutively number (starting with 1) the boxes and the media they contain and note the title of the production.

Reviewing and Ordering the Source Footage

Basically, this means looking at what you've got. If you were actively involved in the field shoot or studio production, you start with source media 1 through however many recordings you made. This is to check that all takes were properly recorded, regardless of the quality of the individual takes. This process is much like looking at the series of still shots you took on your vacation, to get some idea of which ones you may want to keep.

If you are handed a stack of videocassettes or other recording media from a news videographer or a producer of a documentary or drama series, your first-impression review is even more important than if you were part of the production team. Not only will you get an overall impression of the quality of the shots but it's also your first chance to get a feel for the story line.

Getting to know the story If you are editing a fully or partially scripted show, the script is your primary editing guide. Otherwise, the first quick review will usually give you some idea of what the story or process message is all about.

When you work in corporate television, where many of the productions have specific instructional objectives, the process message may not reveal itself even after several viewings of the source material. In this case, ask someone who knows. After all, the story and the communication objective will greatly influence your selection of shots or scenes and their sequencing. Read the script and discuss the communication objectives with the writer or producer/director. Discussions about overall story, mood, and style are especially important when editing documentaries and plays. When reviewing the source footage, look at the content but also for vector continuity. Will the shots cut together relatively well, or will you need to insert cutaways?

When editing ENG footage, you rarely get a chance to learn enough about the total event. Worse, you have to keep to a rigid time frame ("Be sure to keep this story to 20 seconds!") and work with limited footage ("Sorry, I just couldn't get close enough to get good shots" or "The mic was working when I checked it"). Also, you have precious little time to get the job done ("Aren't you finished yet? We go on the air in 45 minutes!"). Very much like an ENG videographer or an emergency-room doctor, the ENG editor has to work quickly yet accurately and with little preproduction preparation. While shot continuity is still important, it clearly plays second fiddle to telling the story.

Unless you are the VJ (video journalist) who did the videography and is now doing the editing, you must get as much information as you can about the story before you start editing. Ask the reporter, the camera operator, or the producer to fill you in. After some practice you will be able to sense the story and edit it accordingly.

In all editing, you will also find that it is often the audio that gives you quicker clues to the story than looking at the pictures.

Ordering the recording media Once you are done with this cursory review of the source material, you can refer to the field logs while you stack the cassettes, discs, or memory cards in the order in which you will tell the story. This order may be different from the numbers the media were assigned during the production. Is this necessary if the NLE system gives you random access anyway? Yes. The main reason is for convenience when you start logging the footage. Not only will you save time with such a basic ordering of shots but it will also help with searching for a specific shot and keeping track of the overall story continuity.

PREPARATION PHASE

Before you can engage in the actual editing, you still need to take care of a few items and operational procedures: time code, logging, capture, and audio transcription.

Time Code

A ***time code*** marks each frame of a video recording with a specific numeric address, which represents elapsed recording time: hours, minutes, seconds, and frames. For example, if you see a time code read-out of 00:45:16:29, this means the recording contains 45 minutes, 16 seconds, and 29 frames of footage. If you had recorded one more frame, you would end up with a time code reading of 00:45:17:00. Because there are roughly 30 frames per second in the NTSC system, the time code counter rolls over to the next second after 29 frames, to the next minute after 59 seconds, and to the next hour after 59 minutes.[2]

SEE 19.5 ZVL5 EDITING→ Postproduction guidelines→ time code

2. In countries that use the 25-frame-per-second PAL system, the rollover occurs, of course, after the twenty-fourth frame instead of the twenty-ninth as in the 30-frame NTSC system.

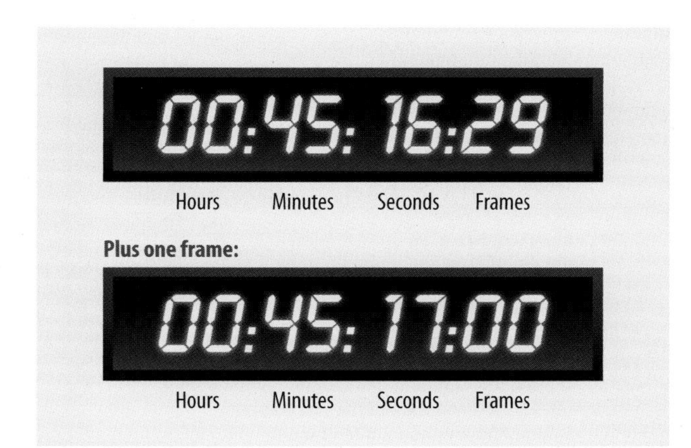

19.5 TIME CODE
The time code gives each television frame a specific address—a number that shows hours, minutes, seconds, and frames of elapsed video and audio.

But, true to all things in television production, even time code gets a bit more complicated. One of the most widely used time codes, called ***SMPTE/EBU time code*** (for Society of Motion Picture and Television Engineers/European Broadcasting Union), lets you select from two modes: non–drop frame and drop frame.

Non–drop frame mode The non–drop frame time code dutifully stamps 30 frames every second, although the NTSC system can display only 29.97 frames per second. This means that the non-drop time code reads out an inflated number of frames from what you actually see on the screen in 1 second. This is of little consequence so long as your program is relatively short. But for a longer program of, say, 30 minutes, the difference between the time code display and the actual elapsed time is quite noticeable. When the time code says it has marked every frame to make up 30 minutes, 0 seconds, and 0 frames of your program, you will find that it is actually 54 frames (or 1 second and 24 frames) shorter than the actual running (stopwatch) time of your program.

Drop frame mode To synchronize exactly the time code read-out with actual time, you need to switch the camcorder to the drop frame mode. This causes the counter to skip a few frames at regular intervals so that the time code window shows the same time as your stopwatch. Note that the drop frame mode does not drop any actual frames but simply skips a few during counting.

The DV code, which is normally used instead of the SMPTE time code in DV camcorders, is automatically in the drop frame mode; it starts with 00:00:00:00 for the first frame with all MiniDV cassettes.

All of this means that unless you need to identify every single frame, you should use the drop frame mode. That way you will know how long the clips actually are no matter how long they run.[3]

Time code use Time code is routinely recorded with each take. The time code helps you mark the precise spots where a clip starts and ends. This is important for logging the clips and also for locating them. Time code is especially useful for matching audio with video. Because the audio portion of the source material shows up as clips on the editing

3. Because the refresh rate in PAL is a true 25 fps, there is no need for the drop-frame or non–drop frame mode.

screen much like the video portion, the audio and video time codes can be synchronized for a perfect marriage.

Synchronized time code A synchronized time code is especially helpful when you shoot a scene with several cameras simultaneously. To synchronize the time code, you need to start the cameras at approximately the same time and have them all point at a clapboard, somebody clapping with outstretched arms, or—best of all—at a still camera triggering a flash. The first frame that shows the clapboard or hands closed, or that is overexposed, will serve as the synchronization point for the time code even if the cameras were not all started at the exact same moment.

Logging

Haven't we done a log already? Yes. Can't we use the field log for editing? No. The field log is only a rough guide to where the shots are on specific source media, but it is usually not accurate enough for editing. The ***VR log*** takes the field log as a rough guide but then lists all takes—both good (OK) and no good (NG)—in consecutive order by scene number and time code address. **SEE 19.6**

As you can see in figure 19.6, this log is similar to the field log except that it contains the scene numbers and the in and out time code numbers of each take as actually recorded on the source media. It also contains additional information that is especially useful in editing. ZVL6 EDITING→ Postproduction guidelines→ VTR log

Let's take a closer look at each of the VR log columns.

- *Source media numbers.* These numbers refer to the ones you or somebody else wrote on each source media and box, such as S1 (source 1), S2 (source 2), et cetera.
- *Scene/shot and take numbers.* If you properly slated the scenes, shots, and takes, copy the numbers from the

PRODUCTION TITLE: Traffic Safety PRODUCTION NO: 114 DATE: 07/15

PRODUCER: Hamid Khani DIRECTOR: Elan Frank

MEDIA NO.	SCENE/ SHOT	TAKE NO.	IN	OUT	OK/ NG	SOUND	REMARKS	VECTORS
4	2	1	04 44 21 14	04 44 23 12	NG		mic problem	m ←
		(2)	04 44 42 06	04 47 41 29	OK	car sound	car A moving Through sTop sign	m ←
		(3)	04 48 01 29	04 50 49 17	OK	brakes	car B puTTing on brakes (Toward camera)	⊙m
		(4)	04 51 02 13	04 51 42 08	OK	reacTion	pedesTrian reacTion	→ i
5	5	1	05 03 49 18	05 04 02 07	NG	car brakes ped. yelling	ball noT in fronT of car	⊙m ← m ball
		2	05 05 02 29	05 06 51 11	NG	"	Again, ball problem	⊙m ← m ball
		(3)	05 07 40 02	05 09 12 13	OK	car brakes ped. yelling	car swerves To avoid ball	⊙m ← m ball
	6	(1)	05 12 03 28	05 14 12 01	OK	ped. yelling	Kid running inTo sTreeT	→ i ← m child
		(2)	05 17 08 16	05 21 11 19	OK	car	cuTaways car moving	⊙ · m
		3	05 22 15 03	05 26 28 00	NG	sTreeT		↔ g

19.6 VR LOG

The VR log contains all the necessary information about the video and audio recorded on the source media. Notice the notations in the Vectors column: *g, i,* and *m* refer to graphic, index, and motion vectors. The arrows show the principal direction of the index and motion vectors. Z-axis index and motion vectors are labeled with ⊙ (toward the camera) or ● (away from the camera).

slates. Otherwise, simply list all shots as they appear on the source media in ascending order. A ***shot*** is the smallest convenient operational unit in video and film, usually the interval between two transitions. In cinema it may mean a specific camera setup. A ***take*** is any one of similar repeated shots taken during video-recording or filming. It is usually assigned a number. A good take is the successful completion of a shot. A bad take is an unsuccessful recording, requiring another take.

■ *Time code.* Enter the time code number of the first frame of the shot in the In column and the last frame of the shot in the Out column regardless of whether the shot is OK or no good.

■ *OK or no good.* Mark the acceptable shots by circling the shot number and/or by writing "OK" or "NG" (no good) in the appropriate column. Unless you have already eliminated all shots that were labeled "NG" on the field log, you can now see if you agree with previous assessments of whether a take was OK.

When evaluating shots look for obvious mistakes—but also look for whether the shot is suitable in the context of the defined process message and the overall story. An out-of-focus shot may be unusable in one context but quite appropriate if you try to demonstrate impaired vision. Look behind the principal action: Is the background appropriate? Too busy or cluttered? It is often the background rather than the foreground that provides the necessary visual continuity. Will the backgrounds facilitate continuity when the shots are edited together?

■ *Sound.* Here you note in- and out-cues for dialogue and sound effects that need attention during editing. Listen carefully not only to the foreground sounds but also to the background sounds. Is there too much ambience? Not enough? Note any obvious audio problems, such as trucks going by, somebody hitting the microphone or kicking the table, intercom chatter of the crew, or talent flubs in an otherwise good take. Write down the nature of the sound problem and its time code address.

■ *Remarks.* Use this column to indicate what the shot is all about, such as "CU of watch," and to record the audio cues (unless you have a designated audio column).

■ *Vectors.* ***Vectors*** indicate the major directions of lines or motions within a shot. Noting such directional vectors will help you locate specific shots that continue or purposely oppose a principal direction.

There are three types of vectors: graphic, index, and motion. A graphic vector is created by stationary elements that guide our eyes in a specific direction, such as a line formed by the window frame or the edge of a book. An index vector is created by something that points unquestionably in a specific direction, such as an arrow or a person's gaze. A motion vector is brought about by something moving. Take another look at the Vectors column in figure 19.6. The *g, i,* and *m* refer to the vector type (graphic, index, or motion); the arrows indicate the principal direction. The circled-dot symbol indicates movement or pointing toward the camera; the dot alone indicates movement or pointing away from the camera. ZVL7 EDITING→ Continuity→ vectors

Logging method You can do the VR log by hand or with the editing software.

When logging by hand, you review each source media as ordered by the scene number and fill in the pertinent information. Do not use the camcorder for this logging process. Except for memory cards, which don't have moving parts, the drive mechanisms of camcorders are not designed for the stop-and-go torture of constant shuttling. It is better to dub the source material onto a sturdier recording device such as a VTR, hard drive, or server before doing any logging or editing.

You can now play each take through the monitor you normally use for your video display and enter by hand on the logging sheet the media number, scene number, in and out time code numbers, and additional remarks for each desired clip (see figure 19.6). A better way is to enter the information on a basic spreadsheet (without using any editing software), which you can then save and later use as a guide for the capture phase, when you transfer the selected clips onto the hard drive of the editing system.

Most high-end editing software allows you to log the source material before the actual capture. When using the logging function, you can import the footage from the source device (camcorder hard drive, server, VTR, disc, or memory card) and log it by media and scene number, the in and out time code numbers, and remarks. The only category missing is the vector information, but you can devise your own symbols for the three types of vectors and enter them in the log notes section.

Capture

Once you have logged the source material, you can proceed with the video and audio capture, where you import the selected clips into the actual editing file on the NLE system hard drive. Again, there are different means of capture. One way is to use the VR log and import one clip after another from the source media or from the copy of the footage you used for logging. Another is to use your log information (clip names and time code numbers per the spreadsheet or computer log) and let the computer perform a ***batch capture***, whereby it will do the work for you: reading your log and importing the clips as specified by name and time code. This requires that you have everything set up just right for such an operation. For smaller projects, you will find that importing each clip is often faster than initiating a batch capture. ZVL8 EDITING→ Nonlinear editing→ capture

Off-line and on-line capture Some editors who are engaged in large editing projects prefer to perform the original capture in low resolution to save disk space, do the editing in low resolution, and then recapture the clips in high resolution for the final cut. The low-resolution import is called off-line, and the high-resolution import, on-line. These designations are a holdover from linear editing, where off-line editing was done with less expensive equipment, and, when everybody was satisfied with the off-line version, the edit was repeated with top-of-the-line equipment. This linear editing version, which produces the edit master tape, is called on-line. In nonlinear editing, the off-line editing usually results in the EDL (edit decision list) that specifies which clips need to be imported again at a higher resolution. ZVL9 EDITING→ Nonlinear editing→ off-line and on-line

Audio Transcription

There's one more chore to do before indulging in sorting out pictures and sound. Unless you are editing sports, fully scripted plays, or routine daily news, you still need to transcribe the audio. A transcription is an essential aid when editing interviews and documentaries. Transcribing audio involves typing every word on the recorded clips. You will find that reading words will let you scan the content much more quickly than listening to them. A printed page is not linear like an audio recording and enables you to skip over sections without having to listen to the recording over and over again.

If, for example, you need to cut down a 64-minute interview to 30 seconds, it would take you hours to go back and forth in the recording to find the most significant moments, even if you have a good memory. When looking at the page, you can grasp the important quotes more quickly and simply mark them with a highlighter.

EDITING PHASE: VIDEO

Just how you edit depends entirely on the NLE software you are using. Despite some major operational differences, the basic editing principles and the approach to the actual editing phase as described here are common to all NLE programs. Just keep in mind that there is no need for the latest complex editing system if all you do is order your shots and join them mostly with cuts and a few dissolves. If, however, you include a great number of special effects, perform extensive color correction, and engage in elaborate audio mixing and audio/video matching, you need a high-end editing system that will provide the tools to go beyond the cut-and-paste approach.

Whatever system you use, don't expect to become a master editor overnight. Good editing takes patience and especially practice, much like learning to play a musical instrument. Once you are adept at using the software, you will realize that the real editing skill lies in making the right decisions in selecting and sequencing the most effective shots. Just how to do this is explored in chapter 20.

Editing Video to Audio

Whenever you deal with speech, you must match the video to the audio. This is where the audio transcription can save you hours of tedious listening. The transcribed audio lets you quickly locate a specific part of an interview without having to run the audio track over and over again. If you have provided the audio clips with time code, you can locate the audio clips as easily as the video clips.

The audio—the spoken text—is your primary editing guide. Although you ordinarily cut at the end of a sentence, whoever is uttering it, in a ***split edit*** you can cut in mid-sentence to the interviewer listening, or to the interviewee listening to the questions, to capture the all-important reaction shots.

If somebody gives a speech, you can save time by replacing the speaker's audio with a summary voice-over (V/O) narration by the news anchor while showing the speaker for a while longer. SEE 19.7

If you do a feature story that is primarily advanced by voice-over (off-camera) narration, or even brief news features, it is best to first lay down the audio track of the narration and match it with the appropriate video. In effect, the audio becomes the A-roll and the video becomes the B-roll. In this way you establish a guide for the length of the video clips and the sequence rhythm much more easily

19.7 PHOTO BY EDWARD AIONA

19.7 VOICE-OVER NARRATION
The voice-over narration replaces from time to time the speaker's comments and summarizes what she said.

than when trying to match the narration to the video sequence. SEE 19.8

Obviously, when editing video to music, the sound track becomes the primary editing guide regardless of whether you are matching video with a classical piece or a hard rock song. Although you normally cut on the beat, you may decide to place the video edits just a little before or after the beat. When not overdone, such cutting "around the beat" can greatly intensify the scene.[4]

Editing Audio to Video

When the video drives the narrative, such as in a recording of a football game, the video becomes the A-roll, and the audio is the B-roll. This means that the editing of the game's highlights is guided by what happened on the field, not by what the play-by-play announcer said. The editing of a fully scripted drama is guided primarily by its dramatic development, the story's inherent conflict, and the director's visualization rather than by the dialogue or background music. The music is *added* to the edited play rather than used as the principal guide for the video edit.

Transitions and Effects

All that remains to do is select the transitions between shots and the special effects, if any. NLE software offers myriad transitions and effects, and you can always add to the effects arsenal by using additional effects software, but recall the warning from chapter 14: any effect that does not intensify the scene or is in the wrong place will most certainly degrade your shot if not your whole effort.

A cut is still the least obtrusive transition if the preceding and following shots have good continuity. All other transitions occupy screen time and therefore slow down the edited sequence. They also introduce a new element which, if not in line with the feel of the show, will look out of place or, worse, ridiculous and amateurish.

But the ease of generating transitions and effects in nonlinear editing is also a blessing: you can try any number of them and their relative speed without affecting the overall edit. Just click on the transition or effects menu and preview the choices. If you don't like what you see, try something else or go back to what you had in the first place. In the old days of film editing, such experimentation was not possible. Each transition or special effect had to be

4. For a lucid treatment of editing video to audio, see Nancy Graham Holm, *Fascination: Viewer Friendly TV Journalism* (Århus, Denmark: Ajour Danish Media Books, 2007), pp. 43–46. See also Herbert Zettl, *Sight Sound Motion: Applied Media Aesthetics*, 6th ed. (Boston: Wadsworth, 2011), pp. 337–50.

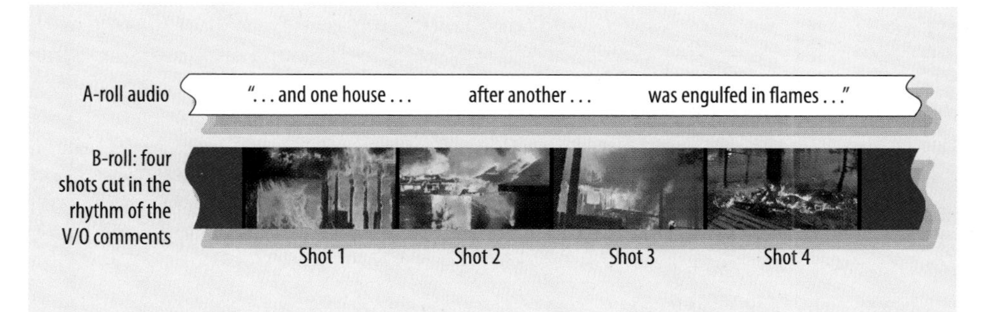

19.8 PHOTOS BY HERBERT ZETTL

19.8 EDITED VOICE-OVER A-ROLL AND B-ROLL VISUALS
In this sequence the audio track represents the A-roll (edited voice-over), and the video track represents the B-roll (video clip sequence as guided by the audio A-roll). The video edit is guided by the voice-over track.

rendered in the processing lab, which added considerably to production time and cost. Even if you didn't like the effect, you were pretty much stuck with it.

EDITING PHASE: AUDIO

Think back to chapter 10 and the brief description of the audio postproduction room and equipment. Much like video postproduction systems, audio postproduction systems can be relatively simple but can also get quite complex.

Even if you had the money, a large audio postproduction room is unnecessary if all you need to do is select some key phrases from an interview. If, however, your audio editing involves the mixing of dialogue and sound effects, you obviously need a more sophisticated setup.

Generally, audio production tasks involve condensing, correcting, mixing, and controlling sound quality. As in video there are considerable differences between linear and nonlinear audio editing. Although you may be editing almost exclusively with NLE systems, you should nevertheless have some idea of how to approach linear sound editing.

Linear Audio Editing

When editing the audio track of a videotape, you need to select the video and audio portions from the source VTR that contains the original footage, then copy the video and the audio (or the audio only) onto the edit master tape of the record VTR. You can adjust the record VTR so that it reads the audio track independently of the video track. To accomplish this split, the video-editing system must be in the insert mode. If you want to add audio that is not on the source tape, you need to feed the new audio to the record VTR via a mixer.

Nonlinear Audio Editing

If you work with a nonlinear video-editing system, all the audio files—much like the video files—are stored on the computer hard drive. Like video editing, audio editing resembles cutting and pasting words and sentences with a word-processing program. The great advantage of nonlinear audio editing is that you can not only hear the sounds but also see them as a waveform graphic on-screen. Such a visual representation of sound enables you to cut the sound at exactly the right frame. Another advantage is that you can synchronize specific sounds with the selected video or move them from place to place with relative ease.[5] SEE 19.9

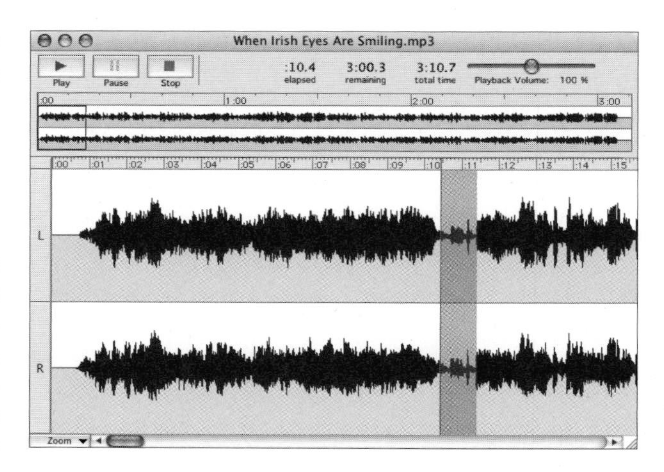

19.9 AUDIO WAVEFORM
All nonlinear editing systems show a visual representation—a waveform—of the sound track.

Condensing

You will find that the most common postproduction audio assignment is making sense of what somebody says, even though that particular person's comments are drastically cut. You may see in a newscast a political candidate speaking behind a lectern but not hear her actual words. Instead the news anchor summarizes what she has said so far before we hear her actual sound bite about lowering taxes and improving education (synchronized video and audio of her statement). Or, we start with the sound bite and then, while the camera still shows her behind the lectern, switch the audio to the anchor's summary of her numerous other promises (see figure 19.7).

To dramatize the danger of children playing with matches, you may sneak in the sound of a fire engine while the kids, playing in the attic, laugh and strike the first match, and then cut to the actual fire engine racing to the resulting fire. You could also show the fire engine racing to the fire while we still hear the children talking about the fun of striking one match after another. Both examples are versions of a split edit. The most common split edit is

5. For a detailed discussion of digital audio editing, see Stanley R. Alten, *Audio in Media*, 9th ed. (Boston: Wadsworth, 2011), pp. 420–47.

when the audio of a shot precedes its pictures or bleeds into the next one. This means that viewers hear the audio of the next shot before seeing it, or they still hear the audio of the previous shot at the beginning of the new one.

As simple as it looks when watching television, such audio-editing tasks can get complicated and require practice. Whenever you have to condense a speech to a few sound bites, you will be thankful for the audio transcription. Instead of playing the audio clips over and over again to search for the appropriate sound bites, you can identify them quickly by skimming the written pages and highlighting the more memorable utterances.

Correcting

Fixing a seemingly simple mistake, such as the talent's mispronouncing a word or giving the wrong address, can become a formidable if not impossible postproduction task. When a politician says "I am not a *cook*" instead of "crook" in the middle of video-recording his defense, "fixing it in post" can be very labor-intensive. It is much easier to correct the problem right away and have the politician repeat his comments a few sentences before he made the mistake. Such problems become almost impossible to fix if you have to try to lip-sync the new word or words in postproduction. ZVL10 EDITING→ Functions→ correct

Filtering out the low rumble of wind during an outdoor shoot or the hum of a lighting instrument in the studio is possible with sophisticated equipment but is nevertheless difficult and time-consuming. Even experienced audio production people labor long hours correcting what may seem like a relatively simple problem. The more care you take during the audio acquisition, the more time you save in the postproduction phase.

Mixing

Postproduction mixing is not much different from live mixing except that you remix separately recorded sound tracks instead of live inputs. Because you mix recorded sound tracks, you can be much more discriminating in how to combine the sounds for optimal quality. In digital cinema, sound designers and engineers can spend weeks if not months on audio postproduction. But don't worry—nobody will ask you to do complicated audio postproduction for video right away; however, if audio is your main interest in television production, try to participate in as many mixing sessions as possible, even if you can only watch and listen.

Controlling Quality

The management of sound quality is probably the most difficult aspect of audio control. You must be thoroughly familiar with signal-processing equipment (such as equalizers, reverberation controls, and filters), and, most importantly, you must have a trained ear. As with the volume control in live mixing, you must be careful how you use these quality controls. If there is an obvious hum or hiss that you can filter out, by all means do, but do not try to adjust the quality of each input before you have done at least a preliminary mix.

For example, you may decide that the sound effect of a police siren sounds much too thin, but when mixed with the traffic sounds, the thin and piercing siren may be perfect for communicating mounting tension. Before making any final quality judgments, listen to the audio track in relation to the video. An audio mix that sounds warm and rich by itself may lose those qualities when juxtaposed with a high-impact video scene. As in all other aspects of television production, the communication goal and your aesthetic sensitivity, not the availability and the production capacity of the equipment, should determine your quality control. No volume meter or equalizer in the world can substitute for aesthetic judgment.

Automated Dialogue Replacement

Some large postproduction houses have a room specifically for ***automated dialogue replacement (ADR)***. Technically, ADR refers to the postdubbing of dialogue, but it sometimes refers to the synchronization of sound effects as well. This audio-dubbing process is borrowed directly from motion pictures. Many sounds, including dialogue, that are recorded simultaneously with pictures do not always live up to the expected sound quality, so they are replaced by painstaking re-creations and mixing of dialogue, sound effects, and ambient sounds.

Elaborate ADR has the actors repeat their lines while watching footage of themselves on a large-screen projection. Recording sound effects is usually done with the Foley stage, which consists of a variety of props and equipment that are set up in a recording studio to produce common sound effects, such as footsteps, opening and closing of doors, and so forth. The Foley stage uses sound-effects equipment much like that of traditional radio and film productions. Foley offers this equipment in efficiently packaged boxes so that it can be transported by truck, sound-effects artists included.

MAIN POINTS

- Nonlinear editing (NLE) allows random access to all source material, multiple editing versions, and any number of transitions and special effects.
- The nonlinear editing system consists of a playback facility for the source media; a high-speed computer with software for the capture, storage, and manipulation of audio and video clips; and a video recorder for the final edit master recording.
- Nonlinear editing enables you to create edit decision lists (EDLs), which the computer remembers and applies when recalling the clips.
- The pre-edit phase includes thinking about shot continuity during the production, labeling the source media and keeping accurate records by entering each take into the field log, reviewing the source footage, and ordering the source media (videocassettes, hard disks, optical discs, memory cards) according to the sequence of the story.
- The preparation phase includes using and synchronizing the time code, creating a VR log, capturing the source footage, and transcribing the audio track.
- Time code shows the elapsed time of a recording in hours, minutes, seconds, and frames. The SMPTE/EBU time code has two modes: drop frame and non–drop frame. The drop-frame mode shows the correct running time of the recording.
- The VR log lists the source media numbers, the scene/shot numbers, and all takes by the time code in- and out-numbers. It usually includes additional information such as shot evaluation (OK or no good), audio instructions, vector directions, and general remarks.
- All footage recorded by the camcorder must be imported in digital form by the editing computer, a process called capture.
- During video capture, the recorded event is transported from the camcorder via high-speed cable (called a peripheral interface) to the hard drive of the editing system. When memory cards are used as the recording media, they can usually be directly inserted into the editing computer for capture.
- Whenever time allows, all speech on the audio track should be transcribed into readable form, which facilitates editing.
- Editing can be done by editing the audio track first and then matching the video to it, or the other way around.
- Transitions and special effects are added as the last editing task, before the final edit is exported to the edit master video recorder.
- The audio-editing phase includes linear and nonlinear editing, condensing, correcting, mixing, controlling quality, and automated dialogue replacement (ADR).

SECTION

19.2

How Linear Editing Works

Why bother with linear editing when almost all video editing is nonlinear? Because, as a video production professional, you will undoubtedly be asked to do some analog videotape editing without the benefit of digital equipment and techniques. Besides, linear editing—like traditional film editing—gives you a good insight of what editing is all about. This section outlines the principal elements and practices of linear editing.

▶ **BASIC AND EXPANDED LINEAR EDITING SYSTEMS**
Single-source system, edit controller, multiple-source system, AB-roll editing, and time code and window dub

▶ **CONTROL TRACK AND TIME CODE EDITING**
Finding a clip or frame

▶ **ASSEMBLE AND INSERT EDITING**
The two principal methods of linear editing

▶ **LINEAR OFF- AND ON-LINE EDITING**
Creating a rough cut and the final edit master tape

BASIC AND EXPANDED LINEAR EDITING SYSTEMS

The basic principle of ***linear editing*** is selecting shots from one videotape and copying them in a specific order onto another tape. It is called linear because once the footage is recorded on tape, you can no longer retrieve it randomly.

To locate shot 25 on a videotape, for example, you need to roll through the previous 24 shots; you cannot simply jump to shot 25. All tape-based editing systems are linear, regardless of whether the tapes contain analog or digital video and audio information.

However complex tape-based linear editing systems may be, they all work on the same basic principle: one or several VTRs play back portions of the tape with the original source footage, and another VTR records on its own tape the selected material from the source tape. The different tape-based systems fall into two categories: the single-source system and the multiple-source system. ZVL11 EDITING→ Linear editing→ system

Single-source System

A basic system that has only one VTR supplying the material to be edited is called a single-source, or cuts-only, editing system. The machine that plays back the tape with the original footage is called the ***source VTR*** or play VTR. The machine that copies the selected material is called the ***record VTR*** or edit VTR. In the same manner, the videotape with the original footage is the ***source tape***, and the one onto which the selected portions are recorded in a specific editing sequence is the edit master tape. To see what is on both the source and edit master tapes, you need monitors for both VTRs. **SEE 19.10**

When doing the actual editing, you use the source VTR to find the exact in- and out-points of the source footage you want to copy to the edit master tape. The record VTR does the actual copying of the material supplied by

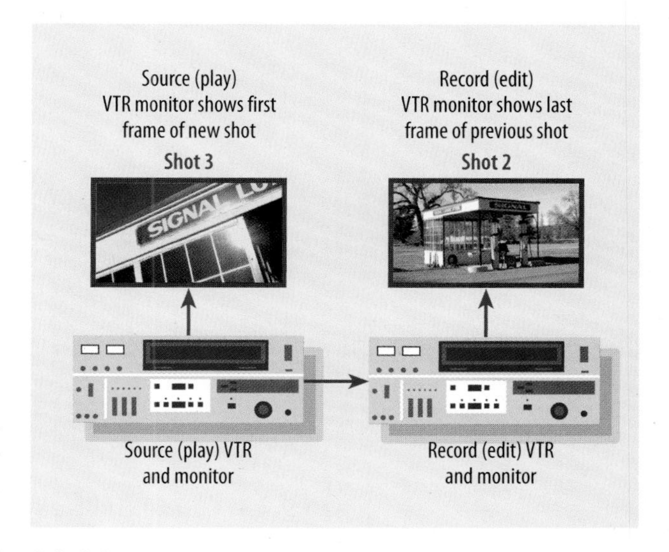

19.10 BASIC SINGLE-SOURCE SYSTEM
The source VTR supplies specific sections of the source tape (displayed on the source VTR monitor). The record VTR copies in a particular sequence and adds each new shot to the previously recorded shot (displayed on the record VTR monitor).

the source VTR and joins the frames at predetermined points—the edit points. You have to tell the record VTR when to start recording (copying) the source material and when to stop recording. An in-cue (or entrance cue) tells the record VTR when to start recording the source material; an out-cue (or exit cue) tells it when to stop.

Edit controller Assisting you in this task is a piece of equipment called the ***edit controller***, or editing control unit, which automates editing to a certain extent. It memorizes your commands and executes them with precision and reliability. **SEE 19.11**

A basic edit controller performs the following functions:

- Controls VTR search modes (variable forward and reverse speeds) separately for the source and record VTRs to locate scenes
- Reads and displays elapsed time and frame numbers or time code (frame address) for accurate cueing of the source and edit master tapes
- Marks and remembers precise edit-in and -out points (cues)
- Backs up, or "backspaces," both VTRs to precisely the same preroll point (on some edit controllers, a switch gives you several preroll choices, such as a 2-second or a 5-second preroll; preroll helps the VTRs achieve optimal speed for jitter-free recording)
- Simultaneously starts both machines and synchronizes their tape speeds
- Makes the record VTR perform in either the assemble- or the insert-edit mode (discussed later in this section)

Most single-source system edit controllers can do additional tasks, such as letting you do a trial run before performing the actual edit, performing separate edits for video and audio tracks without one affecting the other, and producing intelligible sounds at fast-forward tape speeds.

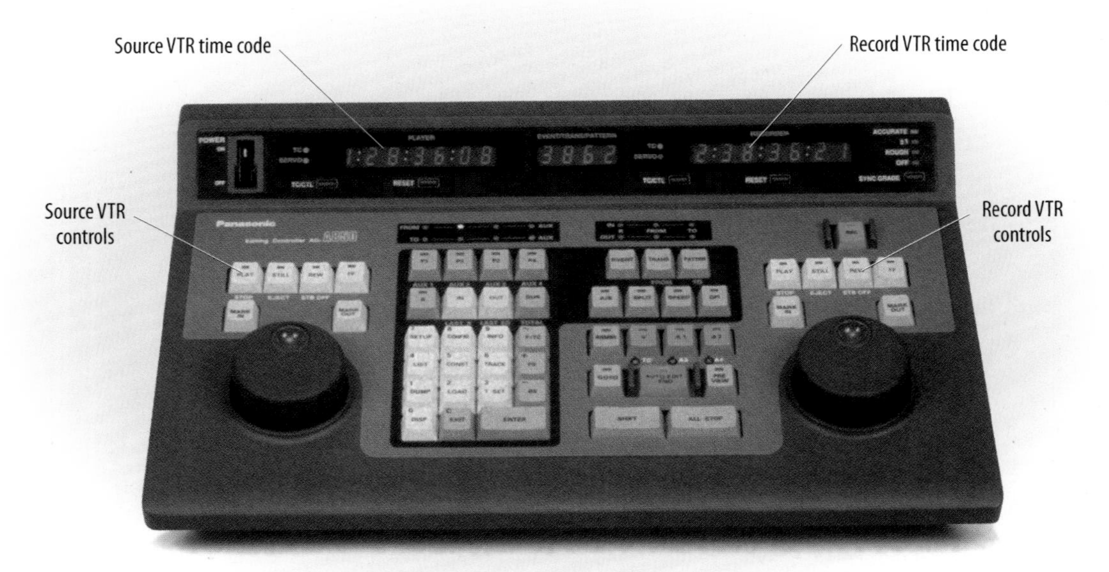

19.11 PHOTO COURTESY PANASONIC

19.11 EDIT CONTROLLER
The edit controller is an interface between the source and record VTRs. It displays elapsed tape time and frames, controls source and record VTR rolls, stores edit-in and edit-out points and tells the VTRs to locate them on the tape, and offers previewing before the edit and reviewing after the edit.

Multiple-source System

The tape-based multiple-source editing system consists of two or more source VTRs (generally labeled with letters *A*, *B*, *C*, etc.), a single record VTR, and a computer-assisted edit controller. Multisource systems can, and usually do, include an audio mixer, a switcher, and special-effects equipment. The computerized edit controller directs the functions of the source A and B VTRs, the character generator (C.G.) or effects generator (unless part of the software program), the audio mixer, and the edit and record functions of the record VTR. SEE 19.12

The multiple-source editing system allows you to synchronously run two or more source VTRs and combine the shots and audio tracks from any of them quickly and effectively through a variety of transitions or other special effects. The big advantage of this system is that it facilitates a great variety of transitions (such as cuts, dissolves, and wipes) and allows the mixing of audio tracks from two or more source tapes. Another advantage is that you can arrange all even-numbered shots on the A-roll (the tape used for the source A VTR) and all odd-numbered shots on the B-roll (the tape for the source B VTR). By switching

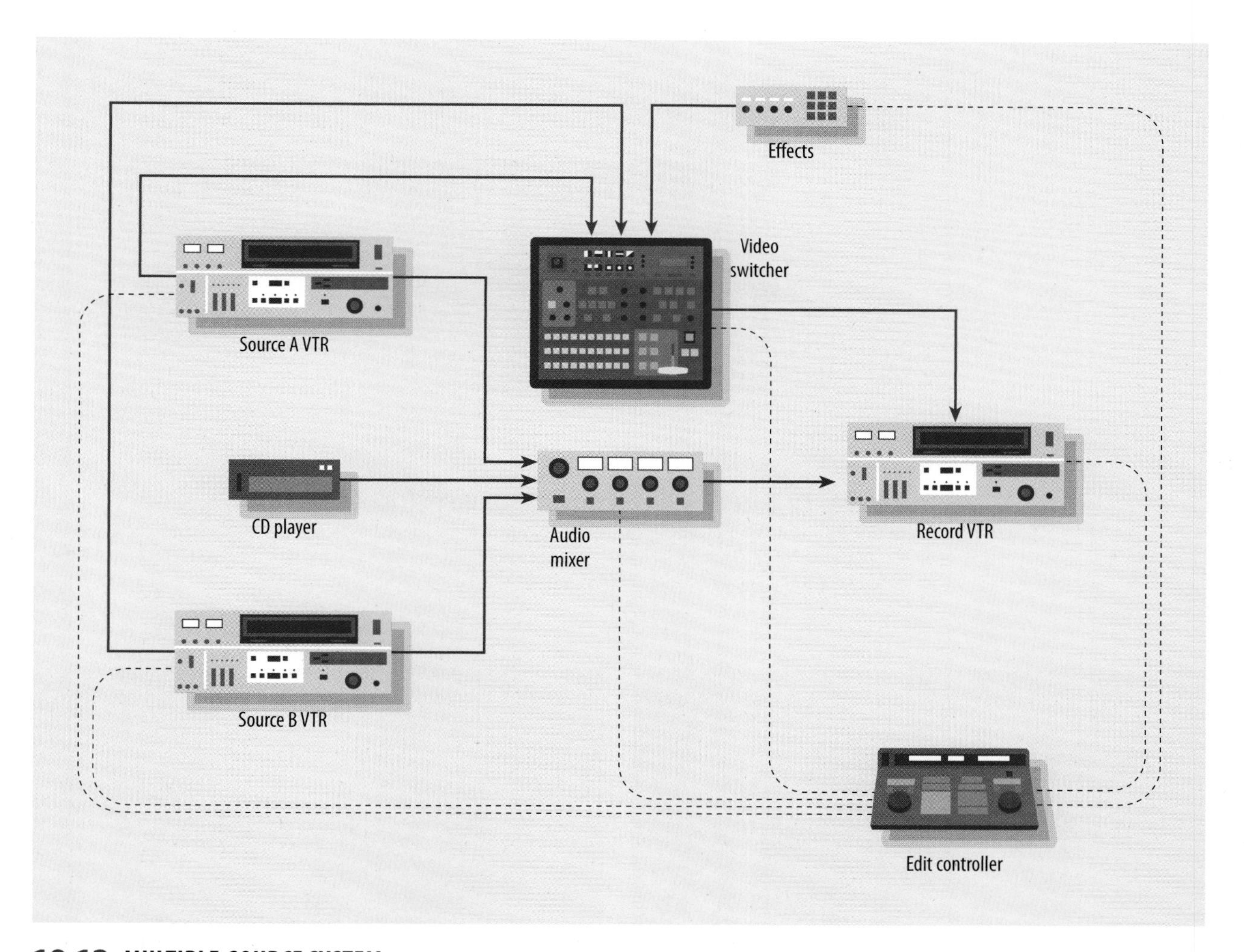

19.12 MULTIPLE-SOURCE SYSTEM

The multiple-source linear editing system has two or more source VTRs (A and B) and interfaces special effects, audio, and switcher equipment or functions.

from the A-roll to the B-roll during editing, you can quickly assemble the "pre-edited" shots. As you can see, transitions and effects were considerably more difficult in analog television than with digital software.

AB-roll Editing

Here is an example of an editing assignment that lends itself well to ***AB-roll editing:*** Assume that one of the source tapes—the A-roll—contains primarily long and medium shots of a rock band. The second source tape—the B-roll—has close-ups (CUs) of the band members. Assuming that the source tapes have a common time code and, in this case, a common audio track (the band playing), you can tell the edit controller when to switch from the A-roll long and medium shots to the B-roll CUs at certain edit points. SEE 19.13

Time Code and Window Dub

Once the videotape is striped with time code, for convenient logging you can make a window dub and log the time code of the A and B rolls, showing when to switch from one to the other. A ***window dub*** is a "bumped-down" (lower-quality, such as VHS) copy of all source tapes, which has the time code keyed over each frame. The window dub also helps when establishing a preliminary EDL for linear editing. SEE 19.14

19.13 AB-ROLL EDITING TO COMMON SOUND TRACK
In this AB-roll editing example, the A-roll consists of long and medium shots of the band; the B-roll contains CUs of the individual musicians. Because both source VTRs are synchronized, you can use the A-roll sound track to insert the B-roll video.

19.13 PHOTOS COURTESY BROADCAST AND ELECTRONIC COMMUNICATION ARTS DEPARTMENT AT SAN FRANCISCO STATE UNIVERSITY

19.14 TIME CODE DISPLAY IN WINDOW DUB
The window dub shows the unique time code number keyed over each frame.

19.14 PHOTO BY EDWARD AIONA

CONTROL TRACK AND TIME CODE EDITING

All linear editing systems are guided by the control track or the time code. Recall from chapter 13 that the control track is the area of the videotape used for recording the synchronization information (sync pulse), providing a reference for the running speed of the VTR and for counting the number of frames. The ***control track system*** counts the sync pulses and translates this count into elapsed time and frames, similar to the time code. Unlike time code, however, no frame is given a unique address.

Control Track Editing

The control track on a videotape marks each frame of recorded material. It therefore takes 30 control track pulses to mark each second of tape play. SEE 19.15

Any one of the individual sync pulses of the control track can become an actual edit-in or edit-out point (frame). By counting the number of pulses, you can locate specific edit-in and edit-out points with greater accuracy than by simply looking at the video pictures. Control track editing is also called pulse-count editing because the edit controller counts the number of control track pulses. ZVL12 EDITING→ Postproduction guidelines→ tape basics

The edit controller counts the pulses of both the source and the edit master tapes from the beginning and displays—just like the time code—the count as elapsed time: hours, minutes, seconds, and frames.

Finding the right address Although the control track system can identify a specific frame with a pulse-count number, it is not frame-accurate. This drawback means that you may get different frames when advancing repeatedly to the same pulse-count number. Realize that when you advance the tape by only 2 minutes, the edit controller must count 3,600 pulses. If you were to back up the tape to the beginning and again run it for 2 minutes, you would probably end up at a different frame, mainly because during high-speed shuttles or repeated threading and unthreading, the tape may stretch or slip, or the unit may simply skip some pulses when counting thousands of them.

Time Code Editing

As you recall, whenever precise editing is required, such as when editing video to the beat of music or when synchronizing dialogue or sound effects to the video track, you need to edit with time code. Because each frame has its own time code address, you can locate a specific frame relatively quickly and reliably, even if it is buried in hours of recorded program material or despite occasional tape slippage during repeated high-speed shuttles. Once the edit controller is told which frame to use as an edit point, it will find it again no matter how many times you shuttle the tape back and forth, and it will not initiate an edit until the right address is located.

Most studio VTRs, portable professional VTRs, and camcorders have a built-in time code generator for writing time code during the production and reading it during playback. Otherwise you need a time code generator that can be connected to the camcorder, and a time code reader when playing back the tapes. Although you may encounter various types of time codes, they all work on the same principle.

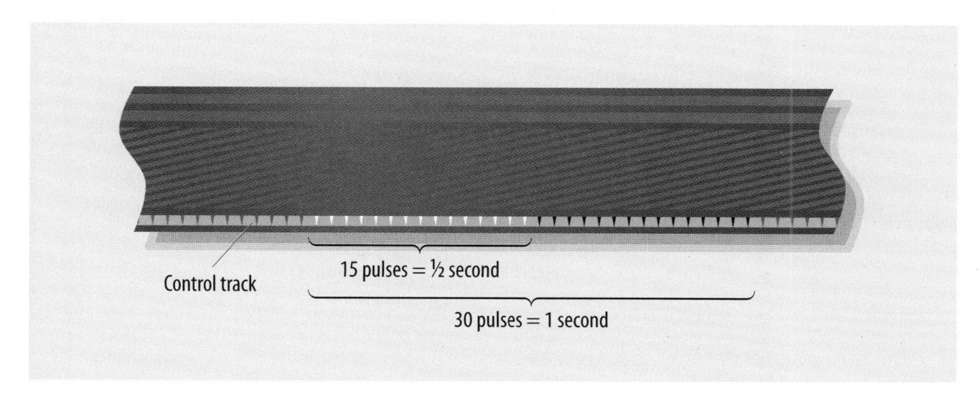

19.15 CONTROL TRACK PULSES

The control track, or pulse-count, system counts the sync pulses to mark a specific spot on the videotape. Every 30 pulses mark one second of elapsed tape time.

ASSEMBLE AND INSERT EDITING

Most professional VTRs let you switch between two editing modes: assemble editing and insert editing.

Assemble Editing

When in the assemble mode, the record VTR erases everything on its tape (video, audio, control, and address tracks) just ahead of copying the material supplied by the source VTR. When you use a tape in an analog or digital camcorder, the camcorder will, in effect, use ***assemble editing*** every time you shoot a new scene: it will simply erase what was there before and replace it with the new video and audio.

The same thing happens in a more sophisticated editing system. Even if the edit master tape has a previous recording on it (not recommended), the assemble mode will clear the portion of the tape that is needed for the new shot. When editing shot 2 onto shot 1, the record VTR will erase everything on the edit master tape following shot 1 to make room for copying the new video and audio information. The record VTR will then supply a new control track that is modeled exactly after the control track information contained in shot 2 of the source tape. This procedure is repeated in subsequent shots. SEE 19.16

If you can simply select and connect some shots from the source tapes in the order in which they were recorded, assemble editing is quite fast and efficient. The problem with assemble editing, however, is that the control track on the edit master tape, as reconstructed by the record VTR from the bits and pieces of the source tapes, is not always smooth and evenly spaced. A slight mismatch of sync pulses will cause some edits to "tear," causing a sync roll, which, as you recall from chapter 13, causes the picture to break or roll momentarily at the edit point (see figure 13.8).

The primary advantage of assemble editing is that it is fast. You do not have to prepare the edit master tape by first recording a continuous control track as you do when editing in the insert mode.

Insert Editing

For ***insert editing*** you need first to record a continuous control track on the edit master tape. The simplest way to do this is to record "black," with the video and audio inputs in the *off* position. Some editors prefer to record

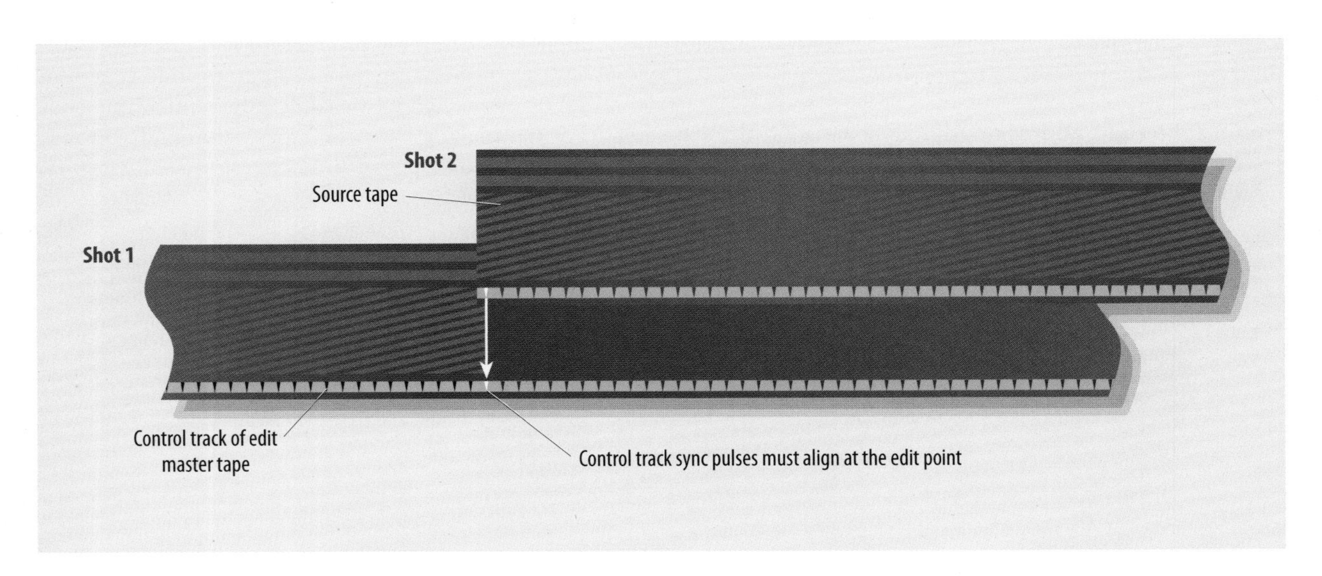

19.16 ASSEMBLE EDITING

In assemble editing, the record VTR produces the control track in bits and pieces. The record VTR copies from the source VTR all video and audio information of shot 2. The control track for shot 2 stays with the source VTR but is regenerated and attached to the shot 1 control track by the record VTR.

color bars as a continuous color reference. As though it were recording an important event, the VTR faithfully lays down a control track in the process. The "blackened" tape has now become an empty edit master, ready to receive the momentous scenes from your source tapes. Note that you do not have to blacken any of the source tapes you use in the camcorder but only the tape onto which the selected shots are copied. SEE 19.17

The recording of black or color bars (and thereby laying a control track) happens in real time, which means you cannot speed up the process but must take 30 minutes to lay a 30-minute control track. Though this may seem like wasted time, it has the following advantages:

- All insert edits are roll-free and tear-free.
- You can easily insert new video and audio material anywhere on the tape without affecting anything preceding or following the insert (hence the name).
- You can edit the video without affecting the sound track, or you can edit the audio without affecting the pictures. This is especially important when you want to edit the sound track first and then insert the pictures to match the sound. The process is the same as AB-roll editing, discussed earlier in this chapter.

Most professional videotape editing is, therefore, done in the insert mode.

LINEAR OFF- AND ON-LINE EDITING

Off- and on-line editing mean quite different things in linear and nonlinear editing environments. ***Off-line editing*** with a nonlinear system refers to a low-resolution import of clips, whereas in linear editing it means a preliminary edit with a lower-quality editing system to produce an EDL or a rough-cut—an edit not intended for broadcast or duplication. ***On-line editing*** in a nonlinear environment refers to the recapture of high-resolution images for the final cut; in linear editing it means to use a high-end editing system to produce the final edit master tape.

Off-line Edit

The linear off-line edit will give you a rough idea of how the intended shot sequence looks and feels. It is a sketch, not the final painting. It gives you a chance to check the

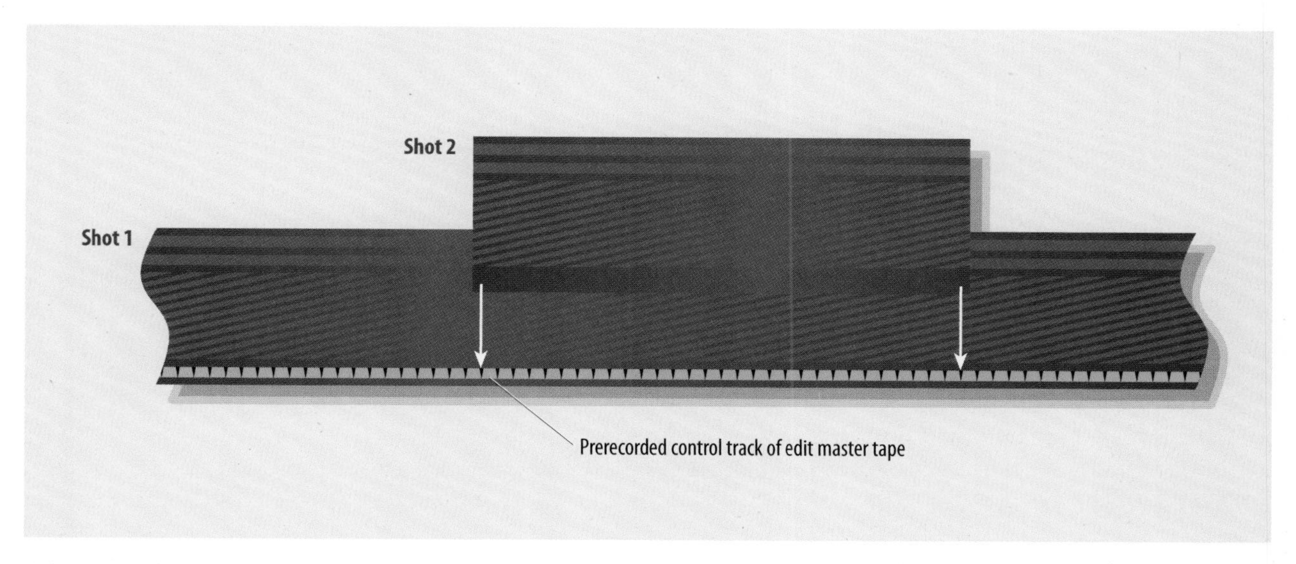

19.17 INSERT EDITING
In insert editing, the source material is transferred without its control track and placed according to the prerecorded continuous control track on the edit master tape.

rhythm of the shot sequence, decide on various transitions and effects, and get some idea of the audio requirements. You could even use two VHS recorders for an off-line rough-cut: one feeds the source tapes, and the other records the selected shots in the desired sequence. Never mind the sloppy transitions or audio—all you want to see is whether the sequences make sense, that is, tell the intended story. The most valuable by-product of off-line editing is a final edit decision list that you can then use for on-line editing.

Paper-and-pencil edit When you edit a longer and more complex production, such as a documentary or drama, your main concern is choosing the shots that most effectively fulfill the story and contribute to a smooth shot sequence. You can save a great amount of actual editing time by simply watching the window dubs (in linear editing) or the clips (in nonlinear editing) and making a list of the edit-in and edit-out points for each selected shot. This list will be your preliminary EDL. Because this list is usually written by hand, this decision-making activity is called paper-and-pencil editing, or paper editing for short. **SEE 19.18** ZVL13 EDITING→ Linear editing→ paper edit

On-line Edit

Much like in nonlinear editing, the on-line edit in linear editing results in the final edit master tape. It is actually a re-edit according to the EDL created in the off-line process. It contains all the transitions and effects as well as a clean audio track. You obviously use the best equipment available to produce the on-line version.

PRODUCTION TITLE: Traffic Safety PRODUCTION NO: 114 DATE: 07/15

PRODUCER: Hamid Khani DIRECTOR: Elan Frank

TAPE NO.	SCENE/ SHOT	TAKE NO.	IN	OUT	TRANSITION	SOUND	REMARKS
1	2	2	01 46 13 14	01 46 15 02	CUT	car	
		3	01 51 10 29	01 51 11 21	CUT	car	
	3	4	02 05 55 17	02 05 56 02	CUT	ped. yelling—brakes	
		5	02 07 43 17	02 08 46 01	CUT	brakes	
		6	02 51 40 02	02 51 41 07	CUT	ped. yelling—brakes	

19.18 HANDWRITTEN EDIT DECISION LIST

Paper-and-pencil off-line editing normally yields a handwritten EDL containing information similar to that generated by a computerized system.

MAIN POINTS

- All linear systems are tape-based and do not allow random access of information.
- The basic single-source editing system consists of a source (or play) VTR and a record (or edit) VTR. The basic principle of linear editing is selecting shots from the source VTR and copying them onto the record VTR in a specific order.
- The edit controller, or editing control unit, facilitates marking the in- and out-points of the selected shots on the source VTR and copying the shots in a specific order to the record VTR.
- The multiple-source editing system consists of two or more source VTRs, a single record VTR, and a computer-assisted edit controller. Multisource systems can also include an audio mixer, a switcher, and special-effects equipment, enabling a variety of transitions between shots.
- The control track, or pulse-count, linear editing system uses sync pulses for locating specific edit-in and edit-out points. It does not supply a unique frame address, however, and is not frame-accurate.
- Time code editing uses a specific code that gives each frame a unique address. It is frame-accurate.
- In assemble editing, all video, audio, control, and address tracks on the edit master tape are erased to make room for the shot to be copied over from the source tape (containing its own video, audio, and control track information for the record VTR). If the newly assembled control track is not perfectly aligned, the edits will cause brief video breakups—or sync rolls—at the edit points.
- In insert editing, the entire control track is prerecorded continuously on the edit master tape before any editing takes place. It prevents breakups at the edit points and allows separate video and audio editing.
- In linear editing, off-line means to use lower-quality equipment for the rough cut. It can be used to produce an edit decision list (EDL). On-line means to use high-end equipment to prepare the final edit master tape.

ZETTL'S VIDEOLAB

For your reference or to track your work, the *Zettl's VideoLab* program cues in this chapter are listed here with their corresponding page numbers.

ZVL1 EDITING→ Editing introduction 424

ZVL2 EDITING→ Nonlinear editing→ system 425

ZVL3 EDITING→ Production guidelines→ pickup 429

ZVL4 EDITING→ Production guidelines→ cutaways 429

ZVL5 EDITING→ Postproduction guidelines→ time code 431

ZVL6 EDITING→ Postproduction guidelines→ VTR log 432

ZVL7 EDITING→ Continuity→ vectors 433

ZVL8 EDITING→ Nonlinear editing→ capture 434

ZVL9 EDITING→ Nonlinear editing→ off-line and on-line 434

ZVL10 EDITING→ Functions→ correct 437

ZVL11 EDITING→ Linear editing→ system 439

ZVL12 EDITING→ Postproduction guidelines→ tape basics 443

ZVL13 EDITING→ Linear editing→ paper edit 446

CHAPTER

20

Editing Functions and Principles

You have practiced enough with the editing software to help your friend edit his short documentary on water conservation. He likes your postproduction job and is especially impressed by the special effects you used to "liven things up a little." But when you show your masterpiece to a family friend, who happens to be a professional video editor, she shows much less enthusiasm. All she initially says is: "Well, nice—but..." When you press her for a frank evaluation, she tells you that she likes your seamless cuts in the first part but dislikes the occasional jump cuts. She thinks you should do away with the many dissolves that destroy the rhythm of the otherwise pretty good montage in the second part. She then adds that there is a general problem with index vector continuity that works against forming a mental map. Finally, she warns about using so many effects transitions that they become too obvious and draw attention away from the story. Besides, they make the whole edit look amateurish.

What in the world is she talking about? You will find out in this last chapter. In fact, you will soon realize that the hardest part of editing is not learning a complex editing program but making the right aesthetic decisions—which shot to include and which to leave out to tell an effective story. This is what the editor meant when she said that editing, like most other television production activities, requires a mix of technical skills and creative aesthetics.

Section 20.1, Continuity Editing, addresses the major editing functions and the basic principles of editing for continuity—how to make it look seamless. Section 20.2, Complexity Editing, explores how to use editing to intensify a story. ZVL1 EDITING→ Editing introduction

KEY TERMS

complexity editing The juxtaposition of shots that primarily, though not exclusively, helps intensify a screen event. Editing conventions as advocated in continuity editing are often purposely violated.

continuity editing The establishment of a visual flow from shot to shot—to make the series appear seamless.

cutaway A shot that is thematically connected with the overall event. When inserted between jump-producing shots, it will camouflage the position shift. Used to facilitate continuity.

graphic vector A vector created by lines or by stationary elements in such a way as to suggest a line.

index vector A vector created by someone looking or something pointing unquestionably in a specific direction.

jump cut Cutting between shots that are identical in subject yet slightly different in screen location, or any abrupt transition between shots that violates the established continuity.

mental map Tells viewers where things are or are supposed to be in on- and off-screen space.

montage The juxtaposition of two or more separate event details that combine into a larger and more intense whole—a new gestalt.

motion vector A vector created by an object actually moving or perceived as moving on-screen.

vector Refers to a perceivable force with a direction and a magnitude. Vector types include graphic vectors, index vectors, and motion vectors.

vector line A dominant direction established by two people facing each other or through a prominent movement in a specific direction. Also called *the line, the line of conversation and action,* and *the hundredeighty.*

SECTION

20.1 Continuity Editing

As an editor, you have to work with the footage you shot or, more often, the footage that is handed to you. In some rare cases, such as in digital cinema production, you may be able to gather some missing material even in the postproduction stage, but usually you need to work with the footage you've got. As indicated in the introduction to this chapter, the shots you choose must not only create story continuity but also sequence continuity. This section is designed to help you in this task.

▶ **EDITING FUNCTIONS**
Combine, shorten, correct, and build

▶ **EDITING FOR CONTINUITY**
Story continuity, subject continuity, vectors and mental map, screen position continuity, motion continuity, light and color continuity, and sound continuity

EDITING FUNCTIONS

Editing is done for different reasons. Sometimes you need to arrange shots so that they tell a story. Other times you may have to eliminate extraneous material to make a story fit a given time slot, or you may want to cut out the shot where the talent stumbled over a word, or substitute a close-up for an uninteresting medium shot. These different reasons are all examples of the four basic editing functions: combine, shorten, correct, and build. ZVL2 EDITING→ Functions→ select

Combine

The simplest editing is combining program portions by hooking the video-recorded pieces together in the chosen sequence. The more care that was taken during the production, the less work you have to do in postproduction. For example, most soap operas are shot in long, complete scenes or in even longer sequences with a multicamera studio setup; the sequences are then combined in postproduction. Or, you may select various shots taken at a friend's wedding and simply combine them in the order in which they occurred.

Shorten

Many editing assignments involve cutting the available footage to make the final edit fit a given time slot or to eliminate extraneous material. As an ENG (electronic news gathering) editor, you will find that you often have to tell a complete story in an unreasonably short amount of time and that you have to pare down the available material to its bare minimum. For example, the producer may give you only 20 seconds to tell the story of a rescue scene of a car stuck on a flooded bridge, although the ENG team had proudly returned with 30 minutes of exciting footage.

Paradoxically, when editing ENG footage, you will discover that although you have an abundance of similar material, you may lack certain shots to tell the story coherently. For example, when screening the flood footage you may find that there are many beautiful shots of the flood waters creeping higher and higher to window-level but no pictures of the child who was pulled to safety on the opposite side. ZVL3 EDITING→ Functions→ condense

Correct

Much editing time is spent on correcting mistakes, either by eliminating unacceptable portions of a scene or by replacing them with better ones. This type of editing can be simple—merely cutting out the part during which the talent coughed and replacing it with a retake. But it can also be challenging, especially if the retake does not match the rest of the recording. You may find, for example, that some of the corrected scenes differ noticeably from the others in color temperature, sound quality, or field of view (shot too closely or too loosely in relation to the rest of the footage). In such cases the relatively simple editing job becomes a formidable postproduction challenge and, in some cases, a nightmare.

Although most nonlinear editing (NLE) software includes remarkably powerful color correction features, applying them is often a tedious and highly time-consuming affair. This is one reason why you should pay particular attention to color matching (white-balancing) during the production phase. Even if you can fix many mistakes in postproduction, editing should never be regarded as a convenient magic bullet for sloppy production practices. ZVL4 EDITING→ Functions→ correct

Build

The most difficult, but also most satisfying, editing assignments are those on which you can build a show from a great many takes. Postproduction is no longer ancillary to production but constitutes a crucial phase in the production process. For example, when a play or documentary is shot film-style, you need to select those shots that reveal not only the basic process message but also the major action. Film-style refers to the established motion picture technique whereby you repeat a brief scene several times and shoot it from a variety of angles and fields of view, irrespective of the scripted event sequence. When editing these shots, you cannot simply select some and combine them in the sequence in which they were recorded; rather, you have to go back to the script and rearrange the shots to fit the story line. The story is literally built shot-by-shot. ZVL5 EDITING→ Functions→ combine | try it

EDITING FOR CONTINUITY

Continuity editing refers to the achievement of story continuity despite the fact that great chunks of the story are actually missing, and to assemble the shots in such a way that viewers are largely unaware of the edits. These are the principal aspects of continuity to consider: story, subject, vectors and mental map, screen position, motion, light and color, and sound.[1]

Story Continuity

The overriding factor in making editing decisions is telling a story. Most of the time, your editing will probably not involve a major motion picture but rather commercials, documentaries, news features, or your personal footage. You will then find that some of the most difficult assignments are not the longer projects but the ones in which you have to tell your story in 30 seconds or less.

For example, if the story for the five out-of-order clips in figure 20.1 is a young woman getting into her car to drive home after work, how would you sequence the shots? **SEE 20.1** Without peeking ahead, renumber the shots in the order you would sequence them to tell the story. Now look at the three edited versions. **SEE 20.2–20.4** Identify the one you think is the best combination of shots.

1. For a more detailed discussion of continuity editing, see Herbert Zettl, *Sight Sound Motion: Applied Media Aesthetics,* 6th ed. (Boston: Wadsworth, 2011), pp. 355–75.

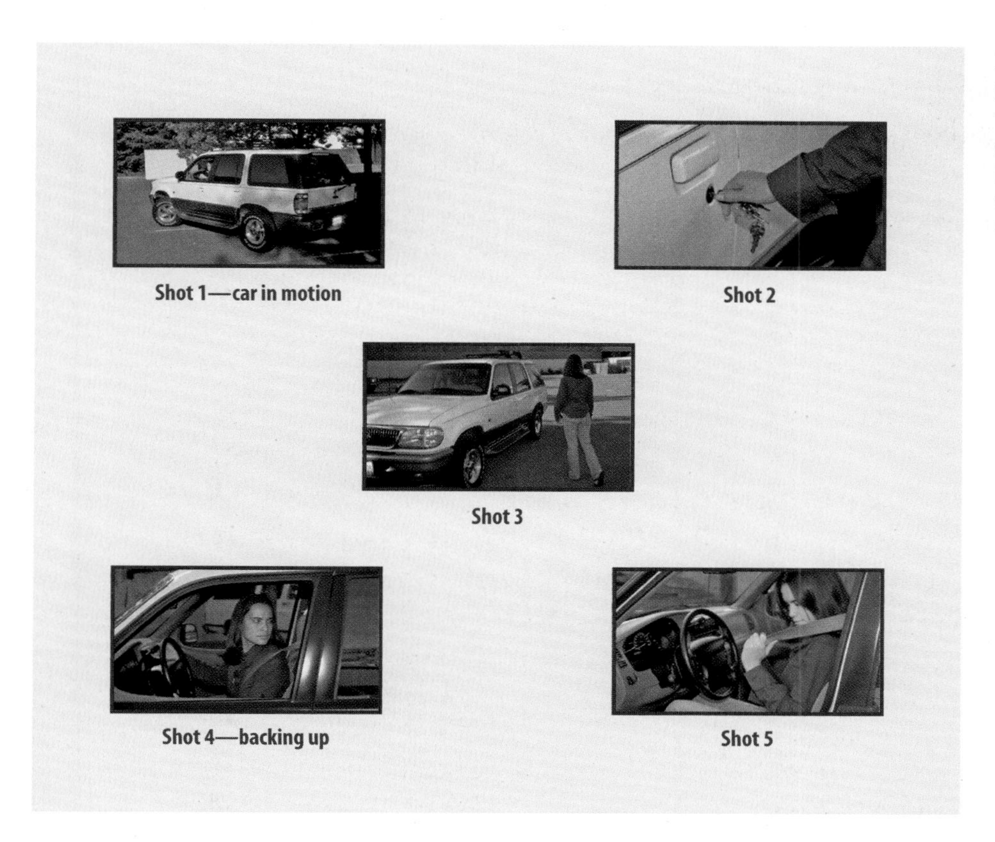

20.1 SOURCE RECORDING SHOT SEQUENCE
The story is about a woman getting into her car and driving home after work. This is the sequence in which the event was restaged and recorded.

20.1 PHOTOS BY EDWARD AIONA

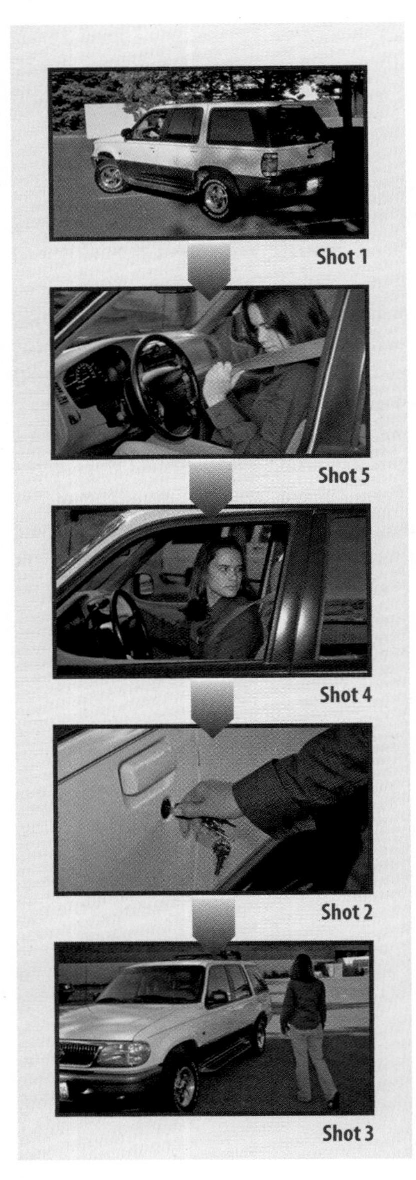

20.2 EDITING SEQUENCE 1
Evaluate this sequence to determine whether the shots are ordered to provide story continuity.

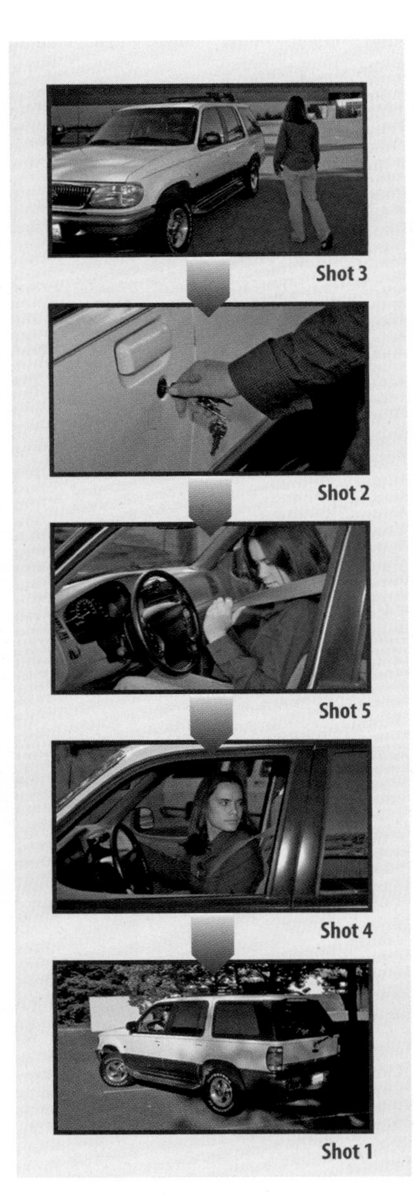

20.3 EDITING SEQUENCE 2
Evaluate this sequence to determine whether the shots are ordered to provide story continuity.

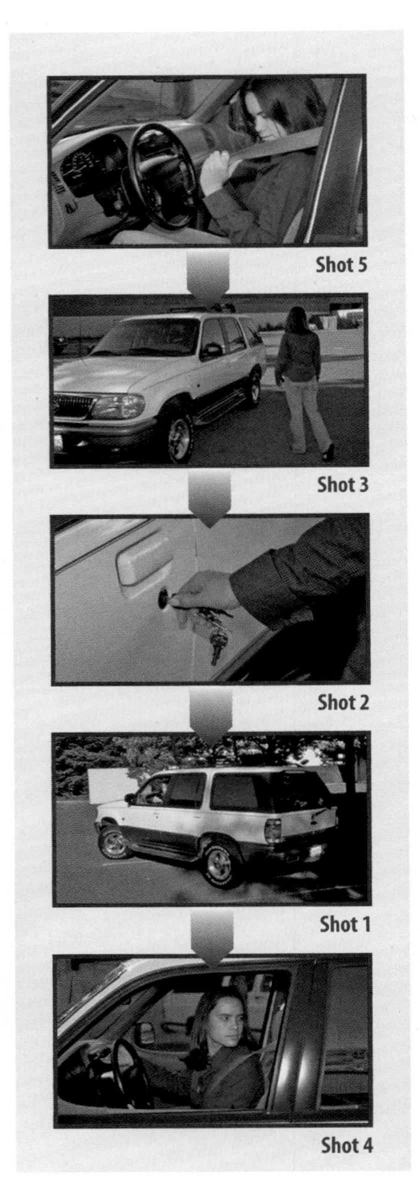

20.4 EDITING SEQUENCE 3
Evaluate this sequence to determine whether the shots are ordered to provide story continuity.

20.2 PHOTOS BY EDWARD AIONA
20.3 PHOTOS BY EDWARD AIONA
20.4 PHOTOS BY EDWARD AIONA

If you selected sequence 2, you made the right choice. Here is why: Obviously, the driver needs to walk to the car (shot 3) and unlock the door (shot 2) before fastening the seat belt (shot 5). The next action is to back the car out (shot 4). Finally, we see the car driving off (shot 1).

If you now had to reduce the edit to three clips, which ones would you select? You can certainly cut her walking to the car (shot 3). Then you could use either shot 2 (key) or shot 5 (seat belt) to qualify the story requirement of "getting into the car." The better sequence would probably be to drop shot 4, keeping this sequence: shot 2 (key), shot 5 (seat belt), and shot 1 (long shot of car driving off). SEE 20.5 ZVL6 EDITING→ Continuity→ try it

Subject Continuity

The viewer should be able to recognize a subject or an object from one shot to the next. Therefore avoid editing between shots of extreme changes in distance. SEE 20.6 If it is available, you can intercut a medium shot (MS) from a different angle or use a slow zoom-in. ZVL7 EDITING→ Continuity→ subject ID

Vectors and Mental Map

Learning about vectors and the mental map will help you greatly not only in continuity editing but also in the pre-editing production phase when visualizing the proper framing of shots or how they will edit together as well as the blocking of talent.

Vectors Think of ***vectors*** as graphic forces with a direction and a magnitude (strength). In video and film, we work with three types of vectors: graphic, index, and motion.

A ***graphic vector*** is created by lines or by stationary elements that are positioned in such a way as to suggest a line. Look around you and you see nothing but graphic vectors: the doorframe, the lines of the desk, the computer screen. The red cones that are lined up at a construction site also form a graphic vector. SEE 20.7

20.5 PHOTOS BY EDWARD AIONA

20.5 EDITING SEQUENCE 4
There are several possibilities, one of which is the following sequence: shot 2 (unlocking the car), shot 5 (getting ready to drive), and shot 1 (driving off).

20.6 PHOTOS BY EDWARD AIONA

20.6 EXTREME CHANGES IN DISTANCE
When you cut from an extreme long shot to a tight close-up, viewers may not recognize exactly whose close-up it is.

20.7 GRAPHIC VECTORS
The lines of the ceiling lights, the curved edge and the slanted lines of the baggage carousel, the vertical lines of the pillars, the lines of the vertical baggage carousel, and the lines on the floor—all are strong graphic vectors

20.8 INDEX VECTOR
An index vector points in a specific direction. The camera lens and the microphone point unquestionably screen-right.

20.7 PHOTO BY HERBERT ZETTL
20.8 PHOTO BY HERBERT ZETTL

Shot 1

Shot 2

20.9 GRAPHIC VECTOR CONTINUITY
The prominent graphic vectors of a horizon line must match from shot to shot.

20.9 PHOTOS BY HERBERT ZETTL

An ***index vector*** is formed by someone looking or something pointing unquestionably in a specific direction. SEE 20.8

A ***motion vector*** is generated by something actually moving or perceived to be moving on-screen. You can illustrate this type of vector with a video camera but not in a book. ZVL8 EDITING→ Continuity→ vectors

Mental map Like when we drive from our house to our favorite restaurant, when watching television we establish automatically a ***mental map*** that shows where things are and where they should be. Because even the large flat-panel television screen is relatively small compared with a movie screen, we normally see little of a total scene in the on-screen space. Rather, the many close-ups (CUs) suggest, or should suggest, that the event continues in the off-screen space. What you show in the on-screen space defines the off-screen space as well. All vector types play a significant role in achieving and preserving shot continuity and keeping intact the viewer's mental map of on- and off-screen space.

For example, if you show a horizon line in subsequent shots rather than by panning, the strong graphic vector of the horizon must continue at approximately the same screen height. SEE 20.9 If the horizon jumps from shot to shot, the graphic vector is no longer seen as continuing but rather suggests a new scene.

Proper handling of index vectors during editing plays a major role in preserving the mental map. If, for example, the screen shows person A looking screen-right

20.10 PHOTOS BY EDWARD AIONA

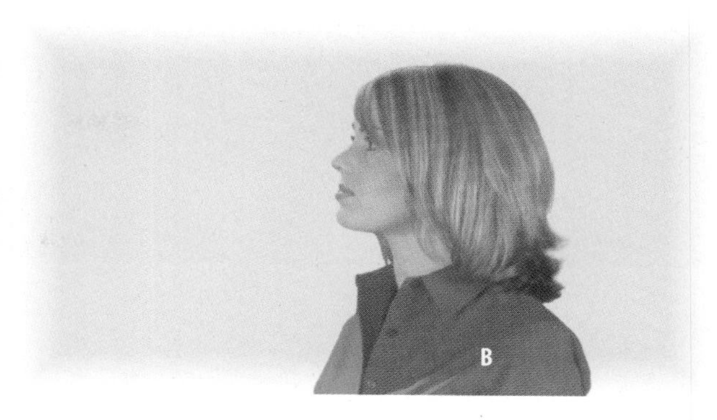

20.10 MENTAL MAP SHOT 1
Here person A's screen-right gaze (his index vector) suggests that person B must be located in the off-screen space to the right.

20.11 PHOTOS BY EDWARD AIONA

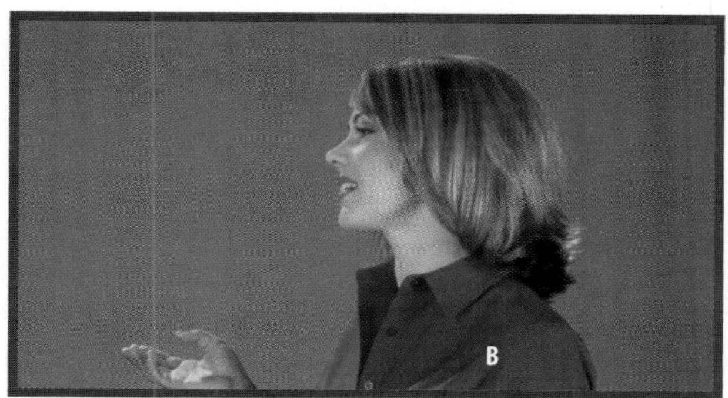

20.11 MENTAL MAP SHOT 2
When we now see person B in a close-up looking screen-left, we assume person A to be in the left off-screen space.

in a close-up, obviously talking to an off-screen person (B), the viewer would expect person B to look screen-left on a subsequent close-up. **SEE 20.10 AND 20.11**

Once the mental map is established, the viewer expects the subsequent screen positions to adhere to it. The mental map is so strong that if the subsequent shot showed person B also looking screen-right, the viewer would think that both person A and B person were talking to a third party. **SEE 20.12** ZVL9 EDITING→ Continuity→ mental map

As you can see, the on-screen index vectors have a considerable influence on how we perceive the off-screen vectors. If you were to apply the index vectors to the example in figures 20.10 and 20.11 of on-screen person A talking to off-screen person B, the screen-right index vector of A needs to be edited to the screen-left index vector of B. Although the index vectors of the two persons are converging in off-screen space, they indicate that A and B are talking with each other rather than away from each other. Obviously, if you were to cut to a two-shot of the two people, their index vectors would have to converge much like the ones on the individual close-ups.

Screen Position Continuity

According to the mental map we established in figures 20.10 and 20.11 of A and B talking to each other, we expect A to remain on screen-left and B on screen-right in the follow-up on-screen two-shot. A switch of A to screen-right and B to screen-left would not make A and B unrecognizable but would certainly disturb the viewer's mental map and, with it, inhibit rather than facilitate communication.

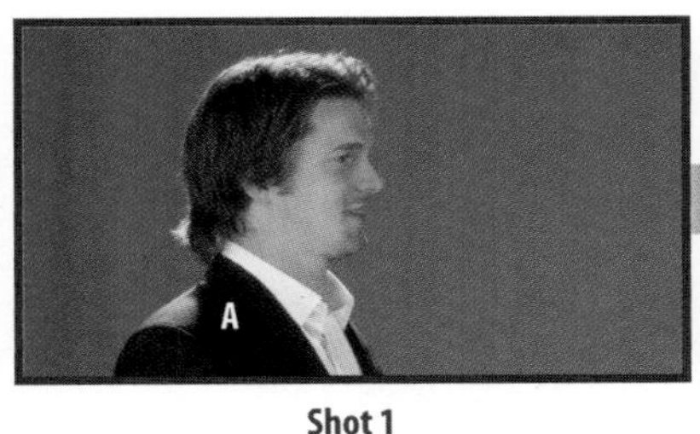

Shot 1

Shot 2

Imagined third person

20.12 PHOTOS BY EDWARD AIONA

20.12 MENTAL MAP SHOT 3
When both A and B look screen-right in successive shots, we conclude that they both are looking at somebody in the right off-screen space.

20.13 PHOTOS BY EDWARD AIONA

20.13 MAINTAINING SCREEN POSITIONS IN REVERSE-ANGLE SHOOTING
In this over-the-shoulder reverse-angle shot sequence, the interviewer and the interviewee maintain their basic screen positions.

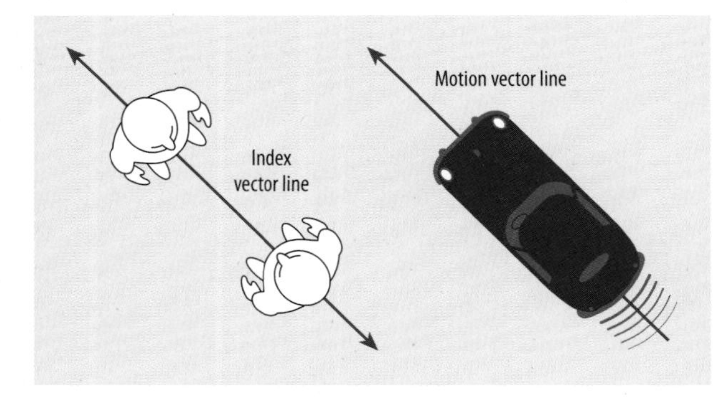

20.14 VECTOR LINE
The vector line is formed by extending converging index vectors or a motion vector.

Maintaining screen positions is especially important in over-the-shoulder shots. If, for example, you show a reporter interviewing somebody in an over-the-shoulder two-shot, the viewer expects the two people to remain in their relative screen positions and not switch places during a reverse-angle shot. **SEE 20.13**

Vector line One important aid in maintaining the viewer's mental map and keeping the subjects in their expected screen locations in reverse-angle shooting is the vector line. The ***vector line*** (also called the line, the line of conversation and action, and the hundredeighty) is an extension of converging index vectors or of a motion vector in the direction of object travel. **SEE 20.14**

When doing reverse-angle switching from camera 1 to camera 2, you need to position the cameras on the same side of the vector line. **SEE 20.15** Crossing the line with one of the two cameras will switch the subjects' screen positions and make them appear to be playing musical chairs, thus upsetting the mental map. **SEE 20.16** ZVL10 EDITING→ Continuity→ mental map | try it

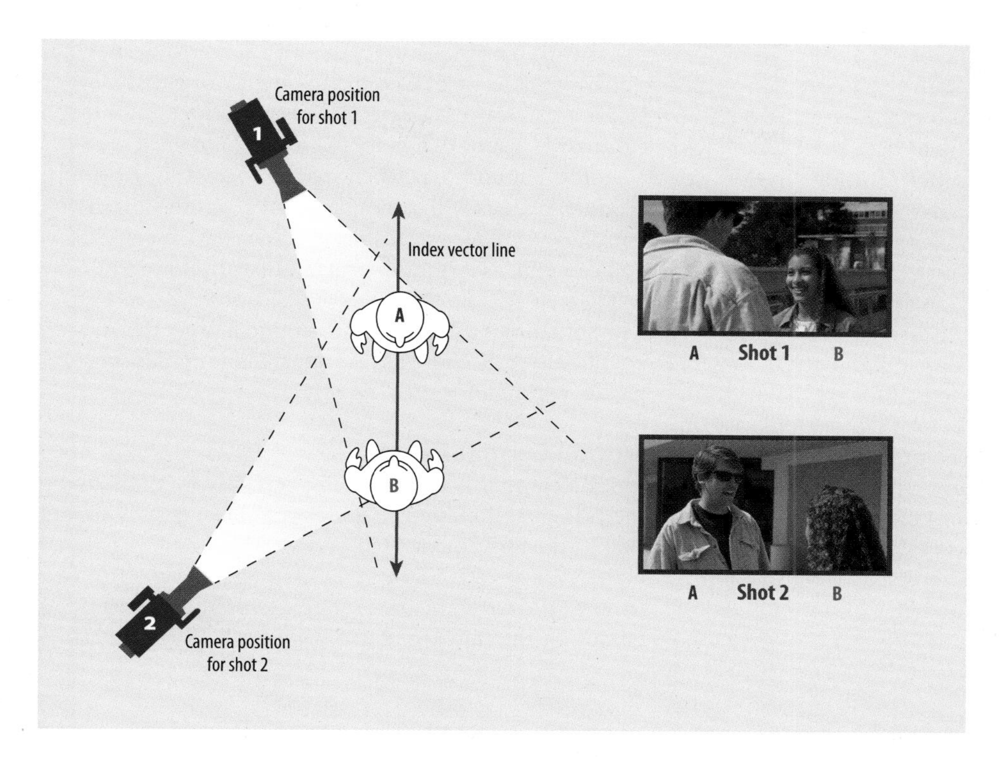

20.15 VECTOR LINE AND PROPER CAMERA POSITIONS
To maintain the screen positions of persons A and B in over-the-shoulder shooting, the cameras must be on the same side of the vector line.

20.15 PHOTOS BY HERBERT ZETTL

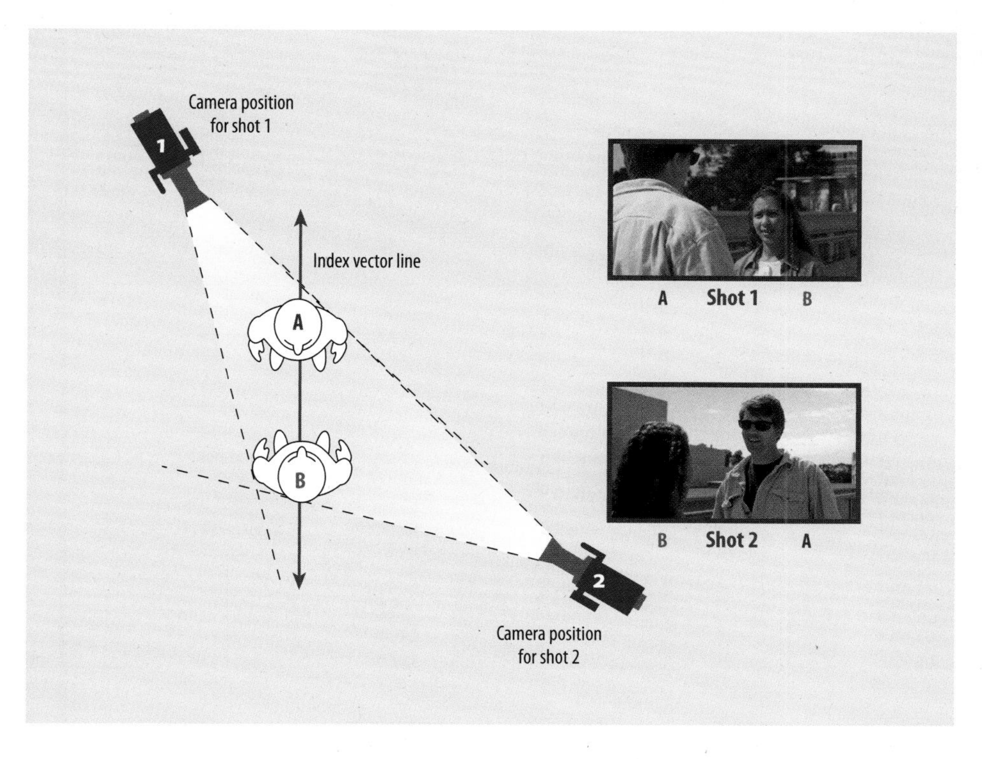

20.16 CROSSING THE VECTOR LINE
When one of the cameras crosses the vector line, persons A and B will switch positions every time you cut between the two cameras.

20.16 PHOTOS BY HERBERT ZETTL

Jump cut A small position change of an object from one shot to the next can cause an even bigger problem—the ***jump cut***. If the director stops a dance routine because of a lighting problem and then tries to have the dancer pick it up exactly where she left off, her positions in the two succeeding shots will rarely match exactly. **SEE 20.17** When cutting the two shots together, you will see her arm suddenly jerk up as if yanked by an unseen force. Even though the position change is relatively small, the cut produces an obvious position jump—hence its name.

You will also get an undesirable jump cut if you are shooting on a set or location where props are moved from one shooting day to the next. If, for example, the vase on the background dresser had been moved sideways even a few inches, you may see the vase jump left or right even when the person in the foreground is framed in the same way as on the previous production day.

Generally, anything that is caused by a cut to bounce from one screen position to another without motivation is considered a jump cut.

To avoid a jump cut, try to find a succeeding shot that shows the object from a different angle or field of view or insert a cutaway. **SEE 20.18** Recall from chapter 19 that a ***cutaway*** is a shot that is thematically connected with the overall event and that when inserted between the jump-producing shots will camouflage the position shift. If you don't have a cutaway available, you don't have much choice. You can either let it jump or use a dissolve to muddle the problem. ZVL11 EDITING→ Production guidelines→ cutaways

Sometimes a jump cut is intentional to get the viewer's attention. We discuss this technique in the context of complexity editing in section 20.2.

Motion Continuity

Motion continuity is determined not only by maintaining the principal direction in which an object is moving but also by the movement itself.

Direction Crossing the motion vector line with cameras (shooting the moving object from both sides or placing

20.17 JUMP CUT
If the size, screen position, or shooting angle of an object is only slightly different in two succeeding shots, the object seems to jump within the screen.

20.17 PHOTOS BY EDWARD AIONA

20.18 CUTAWAY
You can avoid a jump cut by changing image size and/or angle of view or by separating the two shots with a cutaway, as shown here.

20.18 PHOTOS BY EDWARD AIONA

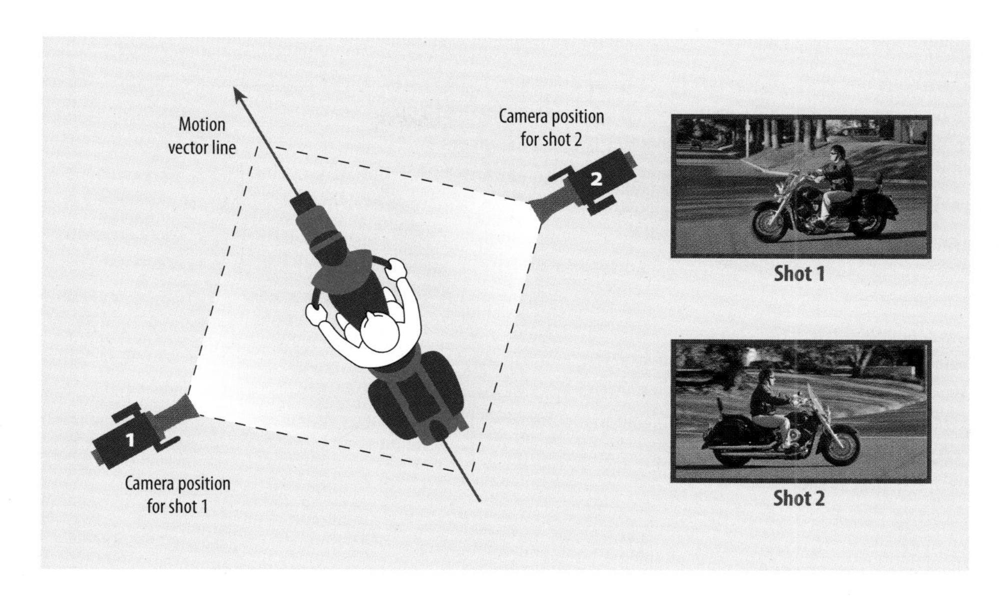

20.19 CROSSING THE MOTION VECTOR LINE
When crossing the motion vector line with cameras, the object motion will be reversed in each shot.

20.19 PHOTOS BY EDWARD AIONA

cameras on both sides of the line) will reverse the direction of object motion every time you cut. SEE 20.19 In a multicamera setup, you would also see the opposite camera in the background. To continue a screen-left or screen-right object motion, you must keep both cameras on the same side of the motion vector line.

Movement To preserve motion continuity, cut during the motion of the subject, not before or after it. For example, if you see in a long shot a man walking to a park bench and sitting down, cut to the medium shot of the man as he is still in the act of sitting. This way the motion appears quite seamless. If you wait until he is seated, the medium shot draws attention to his next actions rather than connecting the two shots. If you cut too early, you will be repeating some of the sitting action. ZVL12 EDITING→ Continuity→ cutting on motion

To trim the shots for a seamless cut, select the in-point of the second (closer) shot so that it matches exactly the out-point of the first (wider) shot. ZVL13 EDITING→ Continuity→ cutting on motion | try it

If one shot contains a moving object, do not follow it with a shot that shows the object stationary. Similarly, if you follow a moving object in one shot with a camera pan, do not cut to a stationary camera in the next shot. Equally jarring is a cut from a stationary object to a moving one. You need to have the subject or camera move in both the preceding and the subsequent shots.

When working with footage in which the action has been shot from both sides of the motion vector line (resulting in a reversal of screen direction), you must separate the two shots with a cutaway or a head-on shot so that the reversed screen direction can be perceived as continuing. SEE 20.20

20.20 PHOTOS BY GARY PALMATIER

20.20 CUTAWAY FOR MOTION VECTOR REVERSAL
If you want to suggest a continuing motion vector of two diverging shots (the Jeep going in opposite directions), you need to insert a cutaway that has some connection to the event (in this case, the Jeep's side mirror).

Light and Color Continuity

When editing indoor events, you must obviously watch that day doesn't change to night and back again in subsequent shots, especially when the action shows nothing but a brief conversation in the living room. A more subtle change may occur during an outdoor field production. The director may have shot both the greeting and good-bye scenes at the front entrance of a house during morning hours and the rest of the show in different locations under a progressively denser cloud cover. So long as the gradual lighting change from sunlight to overcast sky is coinciding with the time progression of the story, it poses no editing problem. But when the final good-byes at the doorstep happen in bright sunlight again, you have a big continuity problem. If too obvious, the good-bye scene needs to be reshot.

Despite careful white-balancing, you will find that the colors don't always match from shot to shot. Although viewers are generally tolerant of minor color shifts, seeing a bride's white wedding dress turn blue when she steps from the church into the sunlight is simply unacceptable, and you must resort to color correction using high-end editing software. Keep in mind, however, that color correction requires skill that goes way beyond the routine editing described in this text.

Sound Continuity

When editing dialogue or commentary, take extra care to preserve the general rhythm of the speech. The pauses between shots of a continuing conversation should be neither much shorter nor much longer than those in the unedited version. In an interview the cut (edit or switcher-activated) usually occurs at the end of a question or an answer. Reaction shots, however, are often smoother when they occur during rather than at the end of phrases or sentences. But note that action is generally a stronger motivation for a cut than dialogue. If somebody moves during the conversation, you must cut on the move, even if the other person is still in the middle of a statement. ZVL14 EDITING→ Continuity→ sound

As discussed in chapter 10, ambient (background) sounds are very important in maintaining editing continuity. If the background noise acts as environmental sounds, which give clues to where the event takes place, you need to maintain those sounds throughout the scene, even if it was built from shots actually taken from different angles and at different times. You may have to supply this continuity by mixing in additional sounds in the postproduction sweetening sessions.

When editing video to music, try to cut with the beat. Cuts determine the beat of the visual sequence and keep the action rhythmically tight, much as the bars measure divisions in music. If the general rhythm of the music is casual or flowing, dissolves are usually more appropriate than hard cuts, but do not be a slave to this convention. Cutting "around the beat" (slightly earlier or later than the beat) on occasion can make the cutting rhythm less mechanical and intensify the scene.

MAIN POINTS

- The four basic editing functions are: to combine—hook various video-recorded pieces together pretty much in the sequence in which they were recorded; to shorten—make the program fit a given time slot and eliminate extraneous material; to correct—cut out bad portions of a scene and replace them with good ones; and to build—select and sequence shots that will advance a specific story.
- Editing for continuity involves keeping in mind the continuity aspects of story, subject, vectors and the mental map, screen position, motion, light and color, and sound.
- Graphic, index, and motion vectors are pictorial forces that play an important part in establishing and maintaining continuity from shot to shot.
- Paying attention to vector continuity facilitates the viewer's mental map of where things are or move in on- and off-screen space.
- A small position change of an object from one shot to the next is known as a jump cut.
- Crossing the vector line with cameras will reverse object positions and the direction of object motion every time you cut; you must keep both cameras on the same side of the motion vector line.
- To preserve motion continuity, cut during the motion of the subject, not before or after it.
- Watch for light, color, and sound continuity of a scene that was shot piecemeal at various times of the day but is supposed to represent a continuous progression of the story.

SECTION

20.2 Complexity Editing

Complexity editing goes beyond the seamless sequencing of shots and is more concerned with the intensification of the screen event. Your selection and sequencing of shots is guided no longer by the need to maintain visual and aural continuity but by ways of getting and keeping viewers' attention and increasing their emotional involvement. To increase the complexity and the intensity of a scene, such editing often makes a deliberate break with the conventions discussed in section 20.1. Although by no means exhaustive, this discussion should give you some idea of what complexity editing involves.[2]

- **TRANSITIONS IN COMPLEXITY EDITING**
  Cut, dissolve, fade, wipe, and animated transitions
- **CROSSING THE VECTOR LINE**
  Background shift, position shift, and motion vector reversal
- **SPECIAL COMPLEXITY EFFECTS**
  Flashback and flashforward, instant replays, and multiple screens
- **MONTAGE**
  Filmic shorthand, complexity, meaning, and structure
- **ETHICS**
  Objectivity and honesty in editing

TRANSITIONS IN COMPLEXITY EDITING

The wide choice of readily available transitions may be as bewildering as it is tempting. Your choice of transitions will be less arbitrary if you know their functions and the importance of context in complexity editing. We'll first look at the cut, dissolve, wipe, and fade; then we'll focus on how animated transitions are perceived in various contexts.

Recall the discussion in chapter 14 of the more common digital video effects (DVE). We'll revisit some of these DVE to see how they might be perceived by viewers when used as transitions.

Cut

The most common transition, the cut, is invisible. It does not occupy any screen time because it does not actually exist. It gets its name from film editors, who literally cut the film apart at certain frames, sorted the pieces, and glued them back together again in a specific order. How "visible" a cut is, that is, how aware we are of the change from one shot to another, depends entirely on the pictures, sounds, and general vector structures of the preceding and following shots. Much like the bars in music, the cut is primarily responsible for the basic rhythm of a shot sequence. ZVL15 SWITCHING→ Transitions→ cut | try it

Dissolve

The dissolve, or lap dissolve, is a gradual transition from shot to shot, with the two images overlapping temporarily. Whereas the cut itself cannot be seen on-screen, the dissolve is a clearly visible transition. Dissolves are often used to provide a smooth bridge for action or to indicate the passage of time. Depending on the overall rhythm of an event, you can use slow or fast dissolves. A very fast one functions almost like a cut but makes the transition a little softer. It is therefore called a soft cut. For an interesting and smooth transition from a wide shot of a dancer to a close-up, for instance, simply dissolve from one camera to the other. When you hold the dissolve in the middle, you will create a superimposition, or super.

Because dissolves are so readily available in NLE software, you may be tempted to use them more often than is necessary or even desirable. A dissolve will inevitably slow down the transition and, with it, the scene. If dissolves are overused, the presentation will lack precision and accent and will bore the viewer. ZVL16 SWITCHING→ Transitions→ mix/dissolve | try it

Fade

In a fade the picture either goes gradually to black (fade-out) or appears gradually on-screen from black (fade-in). You use the fade to signal a definite beginning (fade-in) or end (fade-out) of a scene. Like the curtain in a theater, it defines the beginning or end of a portion of a screen event. As such the fade is technically not a true

2. The visual syntax of complexity editing is explained in Zettl, *Sight Sound Motion*, pp. 379–93.

transition, although some directors and editors use the term *cross-fade,* or *dip-to-black,* for a quick fade to black followed immediately by a fade-in to the next image to separate one story from the next. Here the fade acts as a transition device, decisively distinguishing the preceding and following images from each other. But do not go to black too often—the program continuity will be interrupted too many times by fades that all suggest final endings. ZVL17 SWITCHING→ Transitions→ fade | try it

Wipe

There are a great variety of wipes available, the simplest of which is when the base picture is replaced by another one that moves conspicuously from one screen edge to the other. The wipe is such an unabashed transition device that it is normally classified as a special effect. Wipes are pushy and show up especially prominently on the large 16 × 9 HDTV screen.

Vertical wipe Both horizontal and vertical wipes bully us into entering a new scene. This is almost like having someone change channels on us while watching television. The wipe signals a new theme. If it happens within a story, it disrupts continuity rather than establishes it.

Circle wipe Both the expanding and shrinking circle wipes are, for some reason, funny. Don't use them in a serious documentary. When we fade in with an expanding circle wipe, you certainly don't expect to see a scene of an accident. A shrinking circle wipe signals "The End" even if you have never seen it used in silent movies. ZVL18 SWITCHING→ Transitions→ wipe | try it

Sawtooth wipe Any kind of sawtooth wipe has a cutting effect. It has no place in the context of subtle and intimate scenes. Using it to spice up a series of mom-and-baby shots is not good idea; the teeth of the wipe seem to cut right into the tranquil scene.

Animated Transitions

Beware of animated transitions and use them only in the proper context. For example, the computer rendering of a photo album in which the various clips appear may be fitting for family videos, but it is certainly inappropriate for showing a series of flag-draped coffins.

Peels In a peel transition, the first image curls up as though it were torn off a pad of paper images, revealing the new one underneath. Such a transition may accentuate the different locations you visited when showing how kids play soccer around the world, but it is certainly out of place when showing children in various big-city slums.

Flips These are extended peel effects that imitate a stack of still pictures or clips that slide off the pile and then flip and tumble through space. You may have seen them used to introduce upcoming news stories. Unless the newscast consists entirely of flippant reports, don't use such animated transitions. They may catch our attention, but they also suggest, however subconsciously, that even the most grievous events shown are merely pictures. Because the images flip through the air as though they were disposable, the effect tends to distance us psychologically from the events shown; we will watch them but not empathize with them in any way.

Even a simple NLE program offers a wide array of such transition effects. Before using any of them, ask yourself whether the transition is appropriate in the context of the show. The advantage of nonlinear editing is that you can try a variety of effects before realizing that, perhaps, the "nonexistent" cut is the most fitting after all.

CROSSING THE VECTOR LINE

In section 20.1 you learned how to use the vector line in continuity editing; you will now be teased with breaking the rules of the line for event intensification. The idea is to cross the vector line on purpose to create a break in background continuity, cause a position change, and disrupt motion vector continuity.

Background Shift

Some directors like to indicate a person's confused state of mind by shooting him or her from both sides of the vector line and by canting the camera. Crossing the line causes a noticeable and disrupting background shift. Tilting the horizon line makes the event less stable. When edited into a sequence of brief shots, the shifts of POV (points of view) and the tilting horizon line emphasize the person's mental instability. SEE 20.21

Position Shift

Although not the result of crossing the line, the most jarring position shift—the jump cut—has been exonerated from the accusation of being an inexcusable editing error and has become a popular intensification device. You have probably seen commercials where the featured product or

20.21 PHOTOS BY EDWARD AIONA

20.21 INSTABILITY THROUGH COMPLEXITY EDITING
Here the shooting from both sides of the vector line creates a disturbing flip-flop of index vectors and backgrounds, conveying the subject's confusion.

20.22 PHOTOS BY EDWARD AIONA

20.22 INTENSIFICATION BY CROSSING THE VECTOR LINE
Converging motion vectors of a prominent moving object will intensify its power. The head-on shot acts as a connecting cutaway.

endorsing talent is suddenly yanked a little screen-left or screen-right. Didn't the editor, director, and client see that this was a jump cut? Of course they did. In fact, they placed it there to draw attention to the product or person and to shake the viewers out of their perceptual complacency.

MTV productions use jump cuts routinely, sometimes out of ignorance but mostly by design. Although hardly necessary, the jarring position shift and the discontinuity of shots intensify the high energy of the music and its performers. Although this is often an effective intensification technique for music videos, I wouldn't use it for a college president's monthly campus video address. In such a context, the jump cut would certainly be perceived as an editing mistake.

Motion Vector Reversal

As you well know, shooting a moving object from one side of the vector line and then from the opposite side will cause the object to reverse its direction. But this vector reversal is sometimes used to intensify the power of the moving object. If you had to magnify the raw power of a truck, for example, you can shoot the moving truck from both sides of the line and then show the sequence with converging motion vectors—the truck moving screen-right in shot 1 and screen-left in shot 2. Using a cutaway of a head-on shot of the truck between the converging motion vectors will make it easier to perceive the reversed screen directions (converging motion vectors) as continuing without detriment to the intensification factor. SEE 20.22

SPECIAL COMPLEXITY EFFECTS

By watching just one night's television programming, you can come across a great selection of complexity effects. Among the more popular ones are the flashback and the flashforward, instant replays, and multiple screens.

Flashback and Flashforward

You are certainly familiar with crime shows that have the victim recall parts of a dramatic experience through flashbacks. The screen flashes, and we see a re-creation of the scene the victim is describing to the authorities. The flashes are usually accompanied by *whooosh* sounds just to make sure we get the point that this is what the victim remembers. A similar technique is employed in flashforward scenes, where a character foresees an upcoming event. Such complexity effects are generally color-distorted to distinguish them from the "real" event in progress.

Instant Replays

In sports, instant replays are usually done to reshow an especially exciting moment in the live coverage of a game or to check on whether a player has violated a rule. In complexity editing, the instant replay is similar to the flashback except that the recall is almost immediately after the actual event. For example, it can show how two people perceive the same statement or situation in different ways. You must be careful, however, that such replays do not turn into unplanned comedy routines. Of course, it works great for Woody Allen–type bits of introspection.

Multiple Screens

You can use multiple screens to show the complexity of a moment. A well-known technique is to show on multiple screens related events that occur in different locations simultaneously. You can also use multiple screens to show an event from different points of view or how we perceive the identical event over time. You should realize, however, that when you show a secondary frame on the actual television display, the image in the second-order space will more likely be perceived as a picture rather than the primary event happening in the first-order space of the television display.[3]

MONTAGE

Montage in its original French means "assembly." It can refer to assembling the parts of an automobile, but in a media context it usually means a deliberate assemblage of shots. In fact, ***montage***—the juxtaposition of two or more separate event details that combine into a larger and more intense whole—is considered the basic building block of complexity editing.

Montage was essential to express specific ideas in the days of silent movies but has now become a filmic shorthand to lead up to an event or describe the complexity of a moment, similar to multiple screens. Montage can also be used as a structural device to give a series of shots rhythmic unity. Many commercials are built on the montage principle.

Filmic Shorthand

If, for example, you need to edit the footage for a public service announcement in which a drunk driver hits a bicyclist, you can use a montage to tell the story in a very short time: man sitting at bar having still another drink, having trouble inserting ignition key, can't see the road well in the rain, wet bicyclist struggling uphill on a curve, extremely quick CU cuts of driver-bicycle-driver-bicycle-driver, police and ambulance next to a twisted bicycle. What happened to the accident?

In this type of montage, the main event is often left to the viewer's imagination.

Complexity

To describe the complexity and the tension of the moment when the judge reads the jury's decision, you can use the montage principle to look around the courtroom, using quick CU cuts of the principals involved. You can intensify the tension by making the shots progressively shorter up to the guilty verdict. Such a montage is often used in sports coverage just ahead of the decisive action.

Meaning

You can also engender new meaning by juxtaposing two different shots. For example, you can intensify the anguish of a veteran in a wheelchair by showing a young girl skipping happily past him. Such collision montages were quite popular and effective in the days of silent movies, but today they are often so blunt that we tend to be alienated rather than moved by them. Also beware of unintentional juxtapositions. To follow a commercial for laxatives with one for toilet paper is not a good idea.

Structure

A montage is often determined not so much by its content as by its length and rhythm. All montages are relatively brief and the shots are cut to equal lengths or, as in the accident and courtroom examples, grow shorter toward the climax.

You should be able to clap rhythmically to the edits in a montage.

ETHICS

Because as an editor you have even more power than the cameraperson over what and what not to show and the ability to construct different meanings of the basic event footage, this section ends with a brief discussion of ethics, or principles of right conduct.

The willful distortion of an event through editing is not a case of poor aesthetic judgment but a question of ethics. The most important principle for the editor, as for all production people working with the presentation of nonfictional events (news and documentaries rather than drama), is to remain as true to the actual event as possible.

For example, if you were to add applause simply because your favorite political candidate said something you happen to support, although in reality there was

3. For a more thorough discussion of multiple screens and first- and second-order space, see Zettl, *Sight Sound Motion*, pp. 145–51, 193–94, 383–84.

dead silence, you would definitely be acting unethically. It would be equally wrong to edit out all the statements that go against your convictions and leave only those with which you agree. If someone gives pro and con arguments, be sure to present the most representative of each. Do not edit out all of one side or the other to meet the prescribed length of the segment.

Be especially careful when juxtaposing two shots that may generate by implication a third idea not contained in either of the two shots. To follow a politician's plea for increased armaments with the explosion of an atomic bomb may unfairly imply that this politician favors nuclear war. These types of montage shots are as powerful as they are dangerous.

Montage effects between video and audio information are especially effective; they may be subtler than the video-only montages but are no less potent. For example, adding the penetrating and aggravating sounds of police sirens to the footage of "for sale" signs of several houses in a wealthy neighborhood would probably suggest that the neighborhood is changing for the worse. The implied message is to not buy any houses in this "crime-ridden" neighborhood.

Do not stage events just to get exciting pictures. For example, if a police officer made a successful rescue but the only footage you got was the rescued person on a stretcher, do not ask the officer to return to the scene of the accident to simulate the daring feat. Although reenactments of this sort have become routine for some ENG teams, stay away from them. There is enough drama in all events if you look closely enough and take effective pictures. You do not have to stage anything.

Finally, you are ultimately responsible to the viewers for your choices as an editor. Do not violate the trust they put in you. As you can see, there is a fine line between intensifying an event through careful editing practices and distorting an event through careless or unethical ones. The only safeguard the viewers have against irresponsible persuasion and manipulation is your responsibility as a professional communicator and your basic respect for your audience.

MAIN POINTS

- Complexity editing is a deliberate break with editing conventions to reveal the stratification of a scene and increase its intensity.
- Any one of the many available transitions—cut, dissolve, fade, wipe, and animated—can enhance a complexity edit so long as the transition fits the story content and context.
- In complexity editing, the vector line is sometimes deliberately crossed. The resulting background and position shifts as well as motion vector reversals can intensify the scene.
- Special complexity effects such as the flashback and flash-forward, instant replays, and multiple screens have become widely used techniques in complexity editing.
- The montage is a filmic shorthand that uses a series of rhythmic cuts to advance a story, show various points of view, create new meaning through juxtapositions, and help establish a beat.
- Because as editor you have power over what the audience will see in the final edit, you must make your decisions in not only an aesthetic context but an ethical one as well.

ZETTL'S VIDEOLAB

For your reference or to track your work, the *Zettl's VideoLab* program cues in this chapter are listed here with their corresponding page numbers.

Epilogue

You are now in command of one of the most powerful means of communication and persuasion. Use it wisely and responsibly.

Treat your audience with respect and compassion. Whatever role you play in the production process—pulling cables or directing a network show—you influence many people. Because they cannot communicate back to you very readily, they must—and do—trust your professional skills and judgment.

Do not betray that trust.

Glossary

480i The scanning system for standard analog television. Each complete television frame consists of 525 lines, of which only 480 are visible on the screen. The *i* stands for *interlaced scanning.*

480p The lowest-resolution scanning system of digital television. The *p* stands for *progressive,* which means that each complete television frame consists of 480 visible, or active, lines that are scanned one after the other (out of 525 total scanning lines). Used for standard digital television transmission.

720p A progressive scanning system of high-definition television. Each frame consists of 720 visible, or active, lines (out of 750 total scanning lines).

1080i An interlaced scanning system of high-definition television. The *i* stands for *interlaced,* which means that a complete frame is formed from two fields, each carrying half of the picture information. Each field consists of 540 visible, or active, lines (out of 1,125 total scanning lines). Just like the standard NTSC television system, the digital HDTV 1080i system produces 60 fields, or 30 complete frames, per second.

1080p A progressive scanning system. All visible lines (1,080 of 1,125) are scanned for each frame. Although the acquisition rate for digital cinema may be 24 frames per second, to avoid flicker during playback the refresh rate is generally boosted to 60 fps.

AB-roll editing Creating an edit master recording by combining the A-roll, which contains a set of shots (such as long and medium shots), and the B-roll, which contains different but related shots (such as cutaways or close-ups of the same scene).

AC Stands for *alternating current.* Electric energy as supplied by normal wall outlets.

actor A person (male or female) who appears on-camera in dramatic roles. Actors always portray someone else.

AD Stands for *associate director* or *assistant director.* Assists the director in all production phases.

additive primary colors Red, green, and blue. Ordinary white light (sunlight) can be separated into the three primary light colors. When these three colored lights are combined in various proportions, all other colors can be reproduced. The process is called *additive color mixing.*

address code An electronic signal that marks each frame with a specific address. *See* **SMPTE/EBU time code.**

ad-lib Speech or action that has not been scripted or specially rehearsed.

ADR *See* **automated dialogue replacement**

advanced television *See* **digital television (DTV)**

AFTRA Stands for *American Federation of Television and Radio Artists.* A broadcasting talent union.

AGC *See* **automatic gain control**

ambience Background environmental sounds.

analog A signal that fluctuates exactly like the original stimulus.

analog recording systems Record the continually fluctuating video and audio signals generated by the video and/or audio source.

aperture Iris opening of a lens, usually measured in *f*-stops.

arc To move the camera in a slightly curved dolly or truck.

aspect ratio The width-to-height proportions of the standard television screen and therefore of all standard television pictures: 4 units wide by 3 units high. For HDTV the aspect ratio is 16 × 9. The small mobile media (cell-phone) displays have various aspect ratios.

assemble editing Adding shots in linear editing on videotape in consecutive order without first recording a control track on the edit master tape.

ATR *See* **audiotape recorder**

ATSC Stands for *Advanced Television Systems Committee.* Sets the standards for digital television, which largely replaces the analog NTSC standard in the United States.

ATSC signals Video signals transported as packets, each packet holding only part of a television picture and the accompanying sound plus instructions for the receiver on how to reassemble the parts into complete video pictures and sound. This system is not compatible with NTSC.

ATV Stands for *advanced television. See* **digital television (DTV)**.

audio The sound portion of television and its production. Technically, the electronic reproduction of audible sound.

audio control booth Houses the audio, or mixing, console; analog and digital playback machines; a turntable; a patchbay; a computer display; speakers; intercom systems; a clock; and a line monitor.

audio monitor *See* **program speaker**

audio postproduction room For postproduction activities such as sweetening; composing music tracks; adding music, sound effects, or laugh tracks; and assembling music bridges and announcements.

audiotape recorder (ATR) A reel-to-reel audiotape recorder.

audio track The area of the videotape used for recording the sound information.

auto-cue *See* **teleprompter**

auto-focus Automated feature whereby the camera focuses on what it senses to be the target object.

auto-iris Automatic control of the lens diaphragm.

automated dialogue replacement (ADR) The synchronization of speech with the lip movements of the speaker in postproduction.

automatic gain control (AGC) Can regulate sound volume or video brightness and contrast levels.

auto-transition An electronic device that functions like a fader bar.

A/V format Another name for the two-column AV (audio/video) script.

background light Illumination of the set, set pieces, and backdrops. Also called *set light.*

back light Illumination from behind the subject and opposite the camera.

back-timing The process of figuring additional clock times by subtracting running times from the schedule time at which the program ends.

balanced mic or line Professional audio wiring with three connectors or wires: two that carry substantially the same audio signal out of phase and one that is a ground shield. Relatively immune to hum and other electronic interference.

barn doors Metal flaps mounted in front of a lighting instrument that control the spread of the light beam.

base *See* **baselight**

baselight Even, nondirectional (diffused) light necessary for the camera to operate optimally. Normal baselight levels are 150 to 200 foot-candles (1,500 to 2,000 lux) at *f*/8 to *f*/16. Also called *base.*

baselight level *See* **operating light level**

batch capture Having the computer use the logging information to capture all the selected clips for editing.

batten A horizontal metal pipe that supports lighting instruments in a studio.

beam splitter Compact internal optical system of prisms within a television camera that separates white light into the three primary colors: red, green, and blue. Also called *prism block.*

big boom *See* **perambulator boom**

big remote A production outside of the studio to televise live and/or live-record a large scheduled event that has not been staged specifically for television. Examples include sporting events, parades, political gatherings, and studio shows that are taken on the road. Also called *remote.*

binary A number system with the base of 2.

binary digit (bit) The smallest amount of information a computer can hold and process. A charge is either present, represented by a 1, or absent, represented by a 0. One bit can describe two levels, such as on/off or black/white.

bit *See* **binary digit**

black Darkest part of the grayscale, with a reflectance of approximately 3 percent; called *TV black.* "To black" means to fade the television picture to black.

blocking Carefully worked-out movement and actions for the talent and for all mobile television equipment.

blocking rehearsal *See* **dry run**

BNC Standard coaxial cable connector for professional video equipment.

book Two flats hinged together. Also called *twofold.*

boom (1) *Audio:* microphone support. (2) *Video:* part of a camera crane. (3) To move the camera via the boom of the camera crane; also called *crane.*

boundary microphone Microphone mounted or put on a reflecting surface to build up a pressure zone in which all the sound waves reach the microphone at the same time. Ideal for group discussions and audience reaction. Also called *pressure zone microphone (PZM).*

brightness The color attribute that determines how dark or light a color appears on the monochrome television screen or how much light the color reflects. Also called *lightness* and *luminance.*

broad A floodlight with a broadside, panlike reflector.

broadband A high-bandwidth standard for sending information (voice, data, video, and audio) simultaneously over cables.

bump-down Dubbing (copying) picture and sound information from a higher-quality VTR format to a lower-quality one.

bump-up Dubbing (copying) picture and sound information from a lower-quality videotape format to a higher-quality one.

burn-in A permanent trace of an image in a video display.

bus (1) A row of buttons on the switcher. (2) A common central circuit that receives electrical signals from several sources and that feeds them to a common or several separate destinations.

bust shot Framing of a person from the upper torso to the top of the head.

byte Eight bits. Can define 256 discrete levels (28 bits), such as shades of gray between black and white. *See also* **binary digit (bit)**.

cake A makeup base, or foundation makeup, usually water-soluble and applied with a sponge. Also called *pancake.*

calibrate (1) *Audio:* To make all VU meters (usually of the audio console and the video recorder) respond in the same way to a specific audio signal. (2) *Video:* to preset a zoom lens to remain in focus throughout the zoom.

camcorder A portable camera with the videotape recorder or some other recording device built into it to form a single unit.

cameo lighting Foreground figures are lighted with highly directional light, with the background remaining dark.

camera The general name for the camera head, which consists of the lens (or lenses), the main camera with the imaging device and the internal optical system, electronic accessories, and the viewfinder.

camera chain The television camera (head) and associated electronic equipment, including the camera control unit, sync generator, and power supply.

camera control unit (CCU) Equipment, separate from the camera head, that contains various video controls, including color fidelity, color balance, contrast, and brightness. The CCU enables the video operator to adjust the camera picture during a show.

camera head The actual television camera, which is at the head of a chain of essential electronic accessories. It comprises the imaging device, lens, and viewfinder. In ENG/EFP cameras, the camera head contains all the elements of the camera chain.

camera light Small spotlight mounted on the front of the camera, used as an additional fill light. (Frequently confused with tally light.) Also called *eye light* and *inky-dinky.*

camera mounting head A camera mounting device that permits smooth tilts and pans. Also called *pan-and-tilt head.*

camera rehearsal Full rehearsal with cameras and other pieces of production equipment. Often identical to the dress rehearsal.

camera stabilizing system Camera mount whose mechanism holds the camera steady while the operator moves.

cant To tilt the shoulder-mounted or handheld camera sideways.

capacitor microphone *See* **condenser microphone**

capture Transferring video and audio information to a computer hard drive for nonlinear editing. Also called *importing.*

cardioid Heart-shaped pickup pattern of a unidirectional microphone.

cassette A video- or audiotape recording or playback device that uses tape cassettes. A cassette is a plastic case containing two reels—a supply reel and a takeup reel.

C-band A frequency band for certain satellites. It is relatively immune to weather interference. *See also* **Ku-band**.

CCD *See* **charge-coupled device**

C-clamp A metal clamp with which lighting instruments are attached to the lighting battens.

C channel *See* **chrominance channel**

CCU *See* **camera control unit**

CD *See* **compact disc**

C.G. *See* **character generator**

character generator (C.G.) A dedicated computer system that electronically produces letters, numbers, and simple graphic images for video display. Any desktop computer can become a C.G. with the appropriate software.

charge-coupled device (CCD) The imaging sensor in a television camera. It consists of horizontal and vertical rows of tiny image-sensing elements, called *pixels,* that translate the optical (light) image into an electric charge that eventually becomes the video signal.

chip A common name for the camera's imaging device. *See* **imaging device**.

chroma-key backdrop A well-saturated blue or green canvas backdrop that can be pulled down from the lighting grid to the studio floor as a background for chroma keying.

chroma keying An effect that uses color (usually blue or green) for the backdrop, which is replaced by the background image during a key.

chrominance channel Consists of the three color (chroma) signals in a video system. The chrominance channel is responsible for each of the basic color signals: red, green, and blue (RGB). Also called *C channel.*

classical dramaturgy The technique of dramatic composition.

clip (1) To compress the white and/or black picture information or prevent the video signal from interfering with the sync signals. (2) A shot or brief series of shots as captured on the hard drive and identified by a file name.

clip control *See* **key-level control**

clip light Small internal reflector spotlight that is clipped to scenery or furniture with a gator clip. Also called *PAR (parabolic aluminized reflector) lamp.*

clipper *See* **key-level control**

close-up (CU) Object or any part of it seen at close range and framed tightly. The close-up can be extreme (extreme or big close-up—ECU) or rather loose (medium close-up—MCU).

closure Short for *psychological closure.* Mentally filling in spaces of an incomplete picture. *See also* **mental map**.

CMOS A camera imaging sensor similar to a CCD but which operates on a different technology. It translates light into an electronic video charge that eventually becomes the video signal.

codec Short for *compression/decompression.* A specific method of compressing and decompressing digital data.

coding To change the quantized values into a binary code, represented by 0's and 1's. Also called *encoding.*

color bars A color standard used by the television industry for the alignment of cameras and videotape recordings. Color bars can be generated by most professional portable cameras.

color compatibility Color signals that can be perceived as black-and-white pictures on monochrome television sets. Generally used to mean that the color scheme has enough brightness contrast for monochrome reproduction with a good grayscale contrast.

color media *See* **gel**

color temperature The standard by which we measure the relative reddishness or bluishness of white light. It is measured on the Kelvin (K) scale. The standard color temperature for indoor light is 3,200K; for outdoor light it is 5,600K. Technically, the numbers express Kelvin degrees.

compact disc (CD) A small, shiny disc that contains information (usually sound signals) in digital form. A CD player reads the encoded digital information using a laser beam.

complexity editing The juxtaposition of shots that primarily, though not exclusively, helps intensify a screen event. Editing conventions as advocated in continuity editing are often purposely violated.

component system A process in which the luminance (Y, or black-and-white) signals and the color (C) signals, or all three color signals (RGB), are kept separate throughout the recording and storage process. *See also* **Y/C component system**, **Y/color difference component system**, and **RGB component system**.

composite system A process in which the luminance (Y, or black-and-white) signal and the chrominance (C, or color) signal as well as sync information are encoded into a single video signal and transported via a single wire. Also called *NTSC signal.*

compression (1) *Data:* reducing the amount of data to be stored or transmitted by using coding schemes (codecs) that pack all original data into less space (lossless compression) or by throwing away some of the least important data (lossy compression). (2) *Optical:* the crowding effect achieved by a narrow-angle (telephoto) lens wherein object proportions and relative distances seem shallower.

condenser microphone A mic whose diaphragm consists of a condenser plate that vibrates with the sound pressure against another fixed condenser plate, called the *backplate.* Also called *electret microphone* and *capacitor microphone.*

contact A person, usually a public relations officer, who knows about an event and can assist the production team during a remote telecast.

continuity editing The establishment of a visual flow from shot to shot—to make the series appear seamless.

continuous-action lighting Overlapping triangle lighting for all major performance areas. Also called *zone lighting.*

contrast ratio The difference between the brightest and the darkest portions in the picture (often measured by reflected light in foot-candles). The contrast ratio for low-end cameras and camcorders is normally 50:1, which means that the brightest spot in the picture should be no more than 50 times brighter than the darkest portion without causing loss of detail in the dark or light areas. High-end digital cameras can exceed this ratio and can tolerate a contrast ratio of 1,000:1 or more.

control room A room adjacent to the studio in which the director, the technical director, the audio engineer, and sometimes the lighting director perform their various production functions.

control room directing *See* **multicamera directing**

control track The area of the videotape used for recording the synchronization information (sync pulse). Provides reference for the running speed of the videotape recorder, for the reading of the video tracks, and for counting the number of frames.

control track system Linear editing system that counts the control track sync pulses and translates this count into elapsed time and frame numbers. It is not frame-accurate. Also called *pulse-count system.*

cookie A popularization of the original term *cucoloris.* Any pattern cut out of thin metal that, when placed inside or in front of an ellipsoidal spotlight (pattern projector), produces a shadow pattern. Also called *gobo.*

crab Sideways motion of the camera crane dolly base.

crane (1) Motion picture camera support that resembles an actual crane in both appearance and operation. The crane can lift the camera from close to the studio floor to more than 10 feet above it. (2) To move the boom of the camera crane up or down. Also called *boom.*

crawl The horizontal movement of electronically generated copy (the vertical movement is called a *roll*). Can also refer to the program that activates such a movement.

cross-fade (1) Audio: transition method whereby the preceding sound is faded out and the following sound is faded in simultaneously; the sounds overlap temporarily. (2) Video: transition method whereby the preceding picture is faded to black and the following picture is faded in from black.

cross-keying The crossing of key lights for two people facing each other.

cross-shot (X/S) Similar to the over-the-shoulder shot except that the camera-near person is completely out of the frame.

CU *See* **close-up**

cucoloris *See* **cookie**

cue card A large, hand-lettered card that contains copy, usually held next to the camera lens by floor personnel.

cue-send *See* **foldback**

cue track The area of the videotape used for such information as in-house identification or SMPTE address code. Can also be used for an additional audio track.

cut (1) The instantaneous change from one shot (image) to another. (2) Director's signal to interrupt action.

cutaway A shot that is thematically connected with the overall event. When inserted between jump-producing shots, it will camouflage the position shift. Used to facilitate continuity.

cuts-only editing system *See* **single-source editing system**

cyc *See* **cyclorama**

cyc light *See* **strip light**

cyclorama A U-shaped continuous piece of canvas or muslin for backing of scenery and action. Also called *cyc.*

DAT *See* **digital audiotape**

DBS *See* **direct broadcast satellite**

DC Stands for *direct current.*

decoding The reconstruction of a video or audio signal from a digital code.

definition How sharp an image appears. In television, the number and size of pixels that make up the screen image. *See also* **resolution**.

delegation controls Buttons on the switcher that assign specific functions to a bus.

demographics Audience research factors concerned with such data as age, gender, marital status, and income.

depth of field The area in which all objects, located at different distances from the camera, appear in focus. Depth of field depends on the focal length of the lens, its *f*-stop, and the distance between the object and the camera.

depth staging Arrangement of objects on the television screen so that foreground, middleground, and background are each clearly defined.

diaphragm (1) Audio: the vibrating element inside a microphone that moves with the air pressure from the sound. (2) Video: *see* **iris**.

dichroic filter A mirrorlike color filter that singles out from the white light the red light (red dichroic filter) and the blue light (blue dichroic filter), with the green light left over. Also called *dichroic mirror.*

dichroic mirror *See* **dichroic filter**

diffused light Light that illuminates a relatively large area with an indistinct beam. Diffused light, created by floodlights, produces soft shadows.

diffusion filter Filter that attaches to the front of the lens, which gives a scene a soft, slightly out-of-focus look.

digital Usually means the binary system—the representation of data in the form of binary digits (on/off pulses).

digital audiotape (DAT) Recording system in which the sound signals are encoded on audiotape in digital form. Includes digital recorders as well as digital recording processes.

digital cart system Digital audio system that uses built-in hard drives, removable disks, or read/write optical discs to store and access audio information almost instantaneously. It is normally used for the playback of brief announcements and music bridges.

digital cinema camera A high-definition television camera with sensors that can produce extremely high-resolution pictures exceeding 4,000 (4K) pixels per line. It records on videotape or memory cards, with a variable frame rate for normal, slow, and accelerated motion capture.

digital recording systems Sample the analog video and audio signals and convert them into discrete on/off pulses. These digits are recorded as 0's and 1's.

digital television (DTV) Digital television systems that generally have a higher image resolution than standard television (STV). Also called *advanced television (ATV).*

digital versatile disc (DVD) The standard DVD is a read-only, high-capacity (4.7 gigabytes or more) storage device of digital audio and video information.

digital video effects (DVE) Visual effects generated by a computer or digital effects equipment in the switcher.

digital zooming Simulated zoom that crops the center portion of an image and electronically enlarges the cropped portion. Digital zooms lose picture resolution.

digital zoom lens A lens that can be programmed through a built-in computer to repeat zoom positions and their corresponding focus settings.

digitize To convert analog signals into digital (binary) form or to transfer information in a digital code.

dimmer A device that controls the intensity of light by throttling the electric current flowing to the lamp.

direct broadcast satellite (DBS) Satellite with a relatively high-powered transponder (transmitter/receiver) that broadcasts from the satellite to small, individual downlink dishes; operates on the Ku-band.

direct bus *See* **program bus**

direct input *See* **direct insertion**

direct insertion Recording technique wherein the sound signals of electric instruments are fed into an impedance-matching box and from there into the mixing console without the use of a speaker and a microphone. Also called *direct input.*

directional light Light that illuminates a relatively small area with a distinct beam. Directional light, produced by spotlights, creates harsh, clearly defined shadows.

director of photography *See* **DP**

disk-based video recorder All digital video recorders that record or store information on a hard disk or read/write optical disc. All disk-based systems are nonlinear.

dissolve The gradual replacement of one image by another through a temporary double exposure. Also called *lap dissolve.*

distortion Unnatural alteration or deterioration of sound.

diversity reception Setup for a single wireless microphone wherein more than one receiving station is established, so one can take over when the signal from the other gets weak.

documentary format *See* two-column A/V script

dolly (1) Camera support that enables the camera to move in all horizontal directions. (2) To move the camera toward (dolly in) or away from (dolly out or back) the object.

double headset A telephone headset (earphones) that carries program sound in one earphone and the P.L. information in the other. Also called *split intercom.*

downlink The antenna (dish) and equipment that receive the signals from a satellite.

downloading The transfer of files that are sent in data packets. Because these packets are often transferred out of order, the file cannot be seen or heard until the downloading process is complete. *See also* **streaming**.

downstream keyer (DSK) A control that allows a title to be keyed (cut in) over the picture (line-out signal) as it leaves the switcher.

DP Stands for *director of photography.* In major motion picture production, the DP is responsible for the lighting (similar to the LD in television). In smaller motion picture productions and in EFP, the DP will operate the camera. In television it refers to the camera operator, or shooter.

drag Degree of friction needed in the camera mounting head to allow smooth panning and tilting.

dramaturgy The technique of dramatic composition.

dress (1) What people wear on-camera. (2) Decorating a set with set properties. (3) Dress rehearsal.

dress rehearsal Full rehearsal with all equipment operating and with talent in full dress. The dress rehearsal is often video recorded. Often called *camera rehearsal* except that the camera rehearsal does not require full dress for talent.

drop Large, painted piece of canvas used for scenery backing.

drop frame A video-recording mode in which single frames are periodically overlooked (dropped) by the SMPTE time code to make it match the actual elapsed clock time.

dropout Loss of part of the video signal, which shows up on-screen as white or colored glitches. Caused by uneven videotape iron-oxide coating (bad tape quality or overuse) or dirt.

dry run Rehearsal without equipment, during which the basic actions of the talent are worked out. Also called *blocking rehearsal.*

DSK *See* **downstream keyer**

DTV *See* **digital television**

dual-redundancy The use of two identical microphones for the pickup of a sound source, whereby only one of them is turned on at any given time. A safety device that permits switching over to the second microphone in case the active one becomes defective.

dub The duplication of an electronic recording. Dubs can be made from tape to tape, or from disc to tape and vice versa. The dub is always one generation away from the recording used for dubbing. In analog systems each dub shows increased deterioration. Digital dubbing produces copies almost identical in quality to the original.

DVCAM Digital videotape-recording system developed by Sony.

DVCPRO Digital videotape-recording system developed by Panasonic.

DVD *See* **digital versatile disc**

DVE *See* **digital video effects**

dynamic microphone A mic whose sound pickup device consists of a diaphragm that is attached to a movable coil. As the diaphragm vibrates with the air pressure from the sound, the coil moves within a magnetic field, generating an electric current. Also called *moving-coil microphone.*

echo A sound that is reflected from a single surface and perceived as consecutive, rapidly fading, and repetitious. *See also* **reverberation**.

ECU *See* **extreme close-up**

edit controller Machine that assists in linear editing functions, such as marking edit-in and edit-out points, rolling source and record videotape recorders, and activating effects equipment. Often a desktop computer with specialized software. Also called *editing control unit.*

edit decision list (EDL) Consists of edit-in and edit-out points, expressed in time code numbers, and the nature of transitions between shots.

editing The selection and assembly of shots in a logical sequence.

editing control unit *See* **edit controller**

edit master recording The recording of the final edit on videotape or other media. Used for broadcast or duplication.

edit master tape The videotape on which the selected portions of the source tapes are edited. Used with the record VTR.

edit VTR *See* **record VTR**

EDL *See* **edit decision list**

effects buses Program and preview buses on the switcher, assigned to perform effects transitions.

effect-to-cause model Moving from idea to desired effect on the viewer, then backing up to the specific medium requirements to produce such an effect.

EFP Stands for *electronic field production.* Television production outside of the studio that is normally shot for postproduction (not live). Part of field production.

EFP camera High-quality portable, shoulder-mounted field production camera that must be connected to an external video recorder.

electret microphone *See* **condenser microphone**

electron gun Produces the electron (scanning) beam in a television receiver.

electronic field production *See* **EFP**

electronic still store (ESS) system An electronic device that can grab a single frame from any video source and store it in digital form. It can retrieve the frame randomly in a fraction of a second.

ellipsoidal spotlight Spotlight that produces a very defined beam, which can be shaped further by metal shutters.

ELS *See* **extreme long shot**

electronic news gathering See **ENG**

encoding *See* **coding**

ENG Stands for *electronic news gathering.* The use of portable camcorders or cameras with separate portable video recorders, lights, and sound equipment for the production of daily news stories. ENG is usually not planned in advance and is often transmitted live or immediately after postproduction editing.

ENG/EFP camcorder High-quality portable field production camera with the recording device built-in.

environment General ambience of a setting.

equalization Controlling the quality of sound by emphasizing certain frequencies while de-emphasizing others.

eSATA A high-speed interface (cable) to transport digital information.

essential area The section of the television picture, centered within the scanning area, that is seen by the home viewer regardless of masking or slight misalignment of the receiver. Also called *safe title area* and *safe area.*

ESS system *See* **electronic still store (ESS) system**

establishing shot *See* **extreme long shot (ELS)** and **long shot (LS)**

event order The way event details are sequenced.

extender *See* **range extender**

external key The cutout portion of the base picture is filled by the signal from an external source, such as a second camera.

extreme close-up (ECU) Shows the object with very tight framing.

extreme long shot (ELS) Shows the object from a great distance. Also called *establishing shot.*

eye light *See* **camera light**

facilities request A list that contains all technical facilities needed for a specific production.

fact sheet Lists the items to be shown on-camera and their main features. May contain suggestions of what to say about the product. Also called *rundown sheet.*

fade The gradual appearance of a picture from black (fade-in) or its disappearance to black (fade-out).

fader A sound-volume control that works by means of a button sliding horizontally along a specific scale. Identical in function to a pot. Also called *slide fader.*

fader bar A lever on the switcher that activates preset transitions, such as dissolves, fades, and wipes, at different speeds. It is also used to create superimpositions. Also called *T-bar.*

falloff (1) The speed with which light intensity decays. (2) The speed (degree) with which a light picture portion turns into shadow area. Fast falloff means that the light areas turn abruptly into shadow areas and there is a great brightness difference between light and shadow areas. Slow falloff indicates a very gradual change from light to dark and a minimal brightness difference between light and shadow areas.

fast lens A lens that permits a relatively great amount of light to pass through at its maximum aperture (relatively low *f*-stop number at its lowest setting). Can be used in low-light conditions.

fc *See* **foot-candle**

feed Signal transmission from one program source to another, such as a network feed or a remote feed.

feedback (1) *Audio:* piercing squeal from the loudspeaker, caused by the accidental reentry of the loudspeaker sound into the microphone and subsequent overamplification of sound. (2) *Communications:* reaction of the receiver of a communication back to the communication source. (3) *Video:* wild streaks and flashes on the monitor screen caused by the reentry of a video signal into the switcher and subsequent overamplification.

fiber-optic cable Thin, transparent fibers of glass or plastic used to transfer light from one point to another. When used in broadcast signal transmission, the electrical video and audio signals use optical frequencies (light) as the carrier wave to be modulated.

field (1) A location away from the studio. (2) One-half of a complete scanning cycle, with two fields necessary for one television picture frame. There are 60 fields, or 30 frames, per second in standard NTSC television.

field log A record of each take during the video recording. *See also* **VR log**.

field of view The portion of a scene visible through a particular lens; its vista. Expressed in abbreviations, such as *CU* for close-up.

field production All productions that happen outside of the studio; generally refers to electronic field production.

figure/ground (1) *Audio:* emphasizing the most important sound source over the general background sounds. (2) *Video:* objects seen in front of a background; the ground is perceived to be more stable than the figure.

fill light Additional light on the opposite side of the camera from the key light to illuminate shadow areas and thereby reduce falloff. Usually done with floodlights.

film-style shooting *See* **single-camera directing**

fishpole A suspension device for a microphone; the mic is attached to a pole and held over the scene for brief periods.

fixed-focal-length lens A lens whose focal length cannot be changed (contrary to a zoom lens that has a variable focal length). Also called *prime lens.*

flag A thin, rectangular sheet of metal, plastic, or cloth used to block light from falling on specific areas. Also called *gobo.*

flash drive *See* **memory card**

flash memory card *See* **memory card**

flat (1) *Lighting:* even illumination with minimal shadows (slow falloff). (2) *Scenery:* a piece of standing scenery used as a background or to simulate the walls of a room.

flat response Measure of a microphone's ability to hear equally well over its entire frequency range. Is also used as a measure for devices that record and play back a specific frequency range.

floodlight Lighting instrument that produces diffused light with a relatively undefined beam edge.

floor plan A diagram of scenery and properties drawn on a grid pattern. Can also refer to *floor plan pattern. See also* **floor plan pattern**.

floor plan pattern A plan of the studio floor, showing the walls, main doors, location of the control room, and the lighting grid or battens. *See also* **floor plan**.

floor stand Heavy stand mounted on a three-caster dolly, designed specifically to support a variety of lighting instruments. An extension pipe lets you adjust the vertical position of the lighting instrument to a certain degree.

fluorescent Lamps that generate light by activating a gas-filled tube to give off ultraviolet radiation, which lights up the phosphorous coating inside the tube.

focal length The distance from the optical center of the lens to the front surface of the camera's imaging device at which the image appears in focus with the lens set at infinity. Focal lengths are measured in millimeters or inches. Short-focal-length lenses have a wide angle of view (wide vista); long-focal-length (telephoto) lenses have a narrow angle of view (close-up). In a variable-focal-length (zoom) lens, the focal length can be changed continuously from wide-angle (zoomed out) to narrow-angle (zoomed in) and vice versa. A fixed-focal-length (prime) lens has a single designated focal length.

focus A picture is in focus when it appears sharp and clear on-screen (technically, the point at which the light rays refracted by the lens converge).

foldback The return of the total or partial audio mix to the talent through headsets or I.F.B. channels. Also called *cue-send.*

Foley stage A variety of props and equipment set up in a recording studio to produce common sound effects, such as footsteps, doors opening and closing, and glass breaking.

follow focus Maintaining the focus of the lens in a shallow depth of field so that the image of an object is continuously kept sharp and clear even when the camera or object moves.

follow spot Powerful special-effects spotlight used primarily to simulate theater stage effects. It generally follows action, such as dancers, ice skaters, or single performers moving in front of a stage curtain.

foot-candle (fc) The amount of light that falls on an object. One foot-candle is the amount of light from a single candle that falls on a 1-square-foot area located 1 foot away from the light source. *See also* **lux**.

format Type of television script indicating the major programming steps; generally contains a fully scripted show opening and closing.

foundation A makeup base over which further makeup such as rouge and eye shadow is applied. For HDTV the foundation is usually sprayed on.

fps Stands for *frames per second. See* **frame rate**.

frame (1) The smallest picture unit in film, a single picture. (2) A complete scan from top to bottom of all picture lines by the electron beam, or one single frame of a motion series. *See* **interlaced scanning** and **progressive scanning**.

frame rate The number of complete video frames the video system is producing each second. Also expressed as fps. The NTSC standard of traditional American television is 30 fps. The 480p and 720p scanning systems normally have a frame rate of 60 fps. Some HD digital cinema cameras have a frame rate of 24 fps and/or variable frame rates. The standard 1080i HDTV system has a frame rate of 30 fps.

framestore synchronizer Image stabilization and synchronization system that stores and reads out one complete video frame at a time. Used to synchronize signals from a variety of video sources that are not genlocked.

freeze-frame Continuous replaying of a single frame, which is perceived as a still shot.

frequency Cycles per second, measured in hertz (Hz).

frequency response Measure of the range of frequencies a microphone can hear and reproduce.

Fresnel spotlight One of the most common spotlights, named after the inventor of its lens. Its lens has steplike concentric rings.

front-timing The process of figuring out clock times by adding given running times to the clock time at which the program starts.

***f*-stop** The calibration on the lens indicating the aperture, or iris opening (and therefore the amount of light transmitted through the lens). The larger the *f*-stop number, the smaller the aperture; the smaller the *f*-stop number, the larger the aperture.

full shot *See* **long shot**

fully scripted format A complete script that contains all dialogue or narration and major visualization cues.

gain (1) *Audio:* level of amplification for audio signals. "Riding gain" means keeping the sound volume at a proper level. (2) *Video:* electronic amplification of the video signal, boosting primarily picture brightness.

gel Generic term for color filters put in front of spotlights or floodlights to give the light beam a specific hue. *Gel* comes from *gelatin,* the filter material used before the invention of more durable plastics. Also called *color media.*

generating element The primary part of a microphone. It converts sound waves into electric energy.

generation The number of dubs away from the original recording. A first-generation dub is struck directly from the source tape. A second-generation tape is a dub of the first-generation dub (two steps away from the original tape), and so forth. In analog recordings, the greater the number of nondigital generations, the greater the quality loss. Digital recordings remain virtually the same through many generations.

genlock The synchronization of two or more video sources (such as cameras and video recorders) or origination sources (such as studio and remote) to prevent picture breakup during switching. A house sync will synchronize all video sources in a studio.

gigabyte 1,073,741,824 bytes (2^{30} bytes); usually figured as roughly 1 billion bytes.

goal-directed information Program content intended to be learned by the viewer.

gobo In television, a scenic foreground piece through which the camera can shoot, thus integrating the decorative foreground with the background action. In film a gobo is an opaque shield used for partially blocking a light, or the metal cutout that projects a pattern on a flat surface. *See also* **cookie** and **flag**.

graphics generator Dedicated computer or software that allows a designer to draw, color, animate, store, and retrieve images electronically. Any desktop computer with a high-capacity RAM and a hard drive can become a graphics generator with the use of 2D and 3D software.

graphic vector A vector created by lines or by stationary elements in such a way as to suggest a line.

grayscale A scale indicating intermediate steps from TV white to TV black. Usually measured with a nine-step scale for standard television. HDTV and digital cinema cameras deliver many more steps.

hand props Objects, called *properties,* that are handled by the performer.

hard drive A high-capacity computer storage disk. Often called *hard disk.*

HDMI Stands for *High-Definition Multimedia Interface.* A high-speed cable of great bandwidth that can carry a great number of digital video and audio signals, and connect with a variety of audio/video equipment.

HDV *See* **high-definition video**

HDTV *See* **high-definition television**

head assembly (1) Audio: small electromagnets that erase the signal from the tape (erase head), put the signals on the tape (record head), and read (induce) them off the tape (playback head). (2) *Video:* small electromagnets that put electrical signals on the videotape or read (induce) the signals off the tape. Video heads, as well as the tape, are in motion.

headroom (1) *Video:* the space left between the top of the head and the upper screen edge. (2) *Audio:* the allowable amount of upper amplification before the sound becomes distorted.

headset microphone Small but good-quality omni- or unidirectional mic attached to padded earphones; similar to a telephone headset but with a higher-quality mic.

helical scan The diagonally slanted path of the video signal when recorded on the videotape. Also called *helical VTR* or *slant-track.*

high-definition television (HDTV) Has at least twice the picture detail of standard (NTSC) television. The 720p uses 720 visible, or active, lines that are normally scanned progressively each $1/60$ second. The 1080i standard uses 60 fields per second, each field consisting of 540 visible, or active, lines. In interlaced scanning, a complete frame consists of two scanning fields of 540 visible lines. In the 1080p standard, all 1,080 lines are scanned for each frame. The refresh rate (complete scanning cycle) for HDTV systems is usually 60 fps but can vary.

high-definition television (HDTV) camera Video camera that delivers pictures of superior resolution, color fidelity, and light-and-dark contrast; uses high-quality imaging sensors and lenses.

high-definition video (HDV) A recording system that produces images of the same resolution as high-definition television (720p and 1080i) with equipment that is similar to standard digital video camcorders. The video signals are much more compressed than those of HDTV, however, which results in lower overall video quality.

high-key Light background and ample light on the scene. Has nothing to do with the vertical positioning of the key light.

high-Z High impedance. *See also* **impedance**.

HMI light Stands for *hydragyrum medium arc-length iodide light.* Uses a high-intensity lamp that produces light by passing electricity through a specific type of gas. Needs a separate ballast.

horizontal blanking The temporary starvation of the electron beam when it returns to write another scanning line.

hot (1) A current- or signal-carrying wire. (2) A piece of equipment that is turned on, such as a hot camera or a hot microphone.

hot spot Undesirable concentration of light in one spot.

house number The in-house system of identification for each piece of recorded program material. Called the *house number* because the code numbers differ from station to station (house to house).

hue One of the three basic color attributes; hue is the color itself—red, green, yellow, and so on.

hundredeighty *See* **vector line**

HUT Stands for *households using television.* Used in calculating share, the HUT figure represents 100 percent of all households using television. *See also* **share**.

Hz Hertz, which measures cycles per second.

IATSE Stands for *International Alliance of Theatrical Stage Employees, Moving Picture Technicians, Artists and Allied Crafts of the United States, Its Territories and Canada.* Trade union.

IBEW Stands for *International Brotherhood of Electrical Workers.* Trade union for studio and master control engineers; may include floor personnel.

I.F.B. *See* **interruptible foldback or interruptible feedback**

I-F lens *See* **internal focus (I-F) lens**

imaging device The imaging element in a television camera. Its sensor (CCD or CMOS) transduces light into electric energy that becomes the video signal. Also called *chip* and *sensor.*

impedance Type of resistance to electric current. The lower the impedance, the better the signal flow.

impedance transformer Device allowing a high-impedance mic to feed a low-impedance recorder or vice versa.

importing *See* **capture**

incandescent The light produced by the hot tungsten filament of ordinary glass-globe or quartz-iodine lamps (in contrast to fluorescent light).

incident light Light that strikes the object directly from its source. An incident-light reading is the measure of light in foot-candles (or lux) from the object to the light source. The light meter is pointed directly into the light source or toward the camera.

index vector A vector created by someone looking or something pointing unquestionably in a specific direction.

inky-dinky *See* **camera light**

input overload distortion Distortion caused by a microphone when subjected to an exceptionally high-volume sound. Condenser microphones are especially prone to input overload distortion.

insert editing Requires for linear editing the prior laying of a control track on the edit master tape. The shots are edited in sequence or inserted into an already existing recording. Necessary mode for editing audio and video tracks separately.

instant replay Repeating for the viewer, by playing back a key play or an important event immediately after its live occurrence. In drama it is used as a visualized recall of a prior event.

instantaneous editing *See* **switching**

intercom Short for *intercommunication system.* Used by all production and technical personnel. The most widely used system has telephone headsets to facilitate voice communication on several wired or wireless channels. Includes other systems, such as I.F.B. and cell phones.

interframe compression A compression technique that borrows recurring pixels from previous frames, thus reducing the number of pixels.

interlaced scanning In this system the electron beam skips every other line during its first scan, reading only the odd-numbered lines. After the beam has scanned half of the last odd-numbered line, it jumps back to the top of the screen and finishes the unscanned half of the top line and continues to scan all even-numbered lines. Each such even- or odd-numbered scan produces a *field*. Two fields produce a complete *frame*. Standard NTSC television operates with 60 fields per second, which translates to 30 frames per second.

internal focus (I-F) lens A mechanism of an ENG/EFP lens that allows focusing without having the front part of the lens barrel extend and turn.

interruptible foldback or interruptible feedback (I.F.B.) Communication system that allows communication with the talent while on the air. A small earpiece worn by on-the-air talent carries program sound or instructions from the producer or director.

in-the-can A term borrowed from film, which referred to when the finished film was literally in the can. It now refers to a finished television recording; the show is preserved and can be broadcast at any time.

intraframe compression A compression method that looks for and eliminates redundant pixels in each frame.

inverse square law The intensity of light falls off as $1/d^2$ from the source, where d is distance from the source. It means that light intensity decreases as distance from the source increases. Valid only for light sources that radiate light isotropically (uniformly in all directions) but not for light whose beam is partially collimated (focused), such as from a Fresnel or an ellipsoidal spot.

ips Stands for *inches per second*. An indication of tape speed.

iris Adjustable lens-opening that controls the amount of light passing through the lens. Also called *diaphragm* and *lens diaphragm*.

iso camera Stands for *isolated camera*. Feeds into the switcher and into its own separate video recorder.

jack (1) A socket or phone-plug receptacle. (2) A brace for scenery.

jib arm Similar to a camera crane. Permits the jib arm operator to raise, lower, and tongue (move sideways) the jib arm while tilting and panning the camera.

jogging Frame-by-frame advancement of videotape with a VTR. *See also* **stop-motion**.

JPEG A video compression method mostly for still pictures, developed by the Joint Photographic Experts Group.

jump cut Cutting between shots that are identical in subject yet slightly different in screen location, or any abrupt transition between shots that violates the established continuity.

Kelvin (K) Refers to the Kelvin temperature scale. In lighting it is the specific measure of color temperature—the relative reddishness or bluishness of white light. The higher the K number, the more bluish the white light. The lower the K number, the more reddish the white light.

key An electronic effect. *Keying* means cutting with an electronic signal one image (usually lettering) into a different background image.

key bus A row of buttons on the switcher, used to select the video source to be inserted into a background image.

key-level control A switcher control that adjusts the key signal so that the title to be keyed appears sharp and clear. Also called *clip control* and *clipper*.

key light Principal source of illumination.

kicker light Usually directional light that is positioned low and from the side and the back of the subject.

kilobyte 1,024 bytes (2^{10} bytes); usually figured as roughly 1,000 bytes.

kilowatt (kW) 1,000 watts.

knee shot Framing of a person from approximately the knees up.

Ku-band A high-frequency band used by certain satellites for signal transport and distribution. The Ku-band signals can be influenced by heavy rain or snow. *See also* **C-band**.

kW *See* **kilowatt**

lag Smear that follows a moving object or motion of the camera across a stationary object under low light levels.

lap dissolve *See* **dissolve**

lav *See* lavalier microphone

lavalier microphone A small mic that can be clipped onto clothing. Also called *lav.*

layering Combining two or more key effects for a more complex effect.

LCD Stands for *liquid crystal display.* A flat television display screen that uses an electric charge to activate liquid crystals. These in turn filter colors from white light for each pixel.

LD Stands for *lighting director.*

LED light Stands for *light emitting diode light.* Its light source is an array of semiconductors (a solid-state electronic device) that emits light when electricity passes through. Can produce different-colored light.

leader numbers Numerals used for the accurate cueing of the videotape and film during playback. The numbers from 10 to 3 flash at 1-second intervals and are sometimes synchronized with short audio beeps.

leadroom The space left in front of a person or an object moving toward the edge of the screen. *See also* **noseroom**.

lens Optical lens, essential for projecting an optical (light) image of a scene onto the film or the front surface of the imaging device. Lenses come in various fixed focal lengths or in variable focal lengths (zoom lenses) and with various maximum apertures (iris openings).

lens diaphragm *See* **iris**

lens prism A prism that, when attached to the camera lens, produces special effects, such as the tilting of the horizon line or the creation of multiple images.

letterbox The aspect ratio that results from fitting the full width of a 16 × 9 image onto a 4 × 3 screen by blocking the top and bottom screen edges with stripes. *See also* **pillarbox**.

level (1) *Audio:* sound volume. (2) *Video:* signal strength (amplitude).

libel Written or televised defamation.

lighting The manipulation of light and shadows: to provide the camera with adequate illumination for technically acceptable pictures; to tell us what the objects onscreen actually look like; and to establish the general mood of the event.

lighting triangle The triangular arrangement of key, back, and fill lights. Also called *triangle lighting. See also* **photographic lighting principle**.

light level Light intensity measured in lux or foot-candles. *See also* **foot-candle (fc)** and **lux**.

lightness *See* **brightness**

light plot A plan, similar to a floor plan, that shows the type, size (wattage), and location of the lighting instruments relative to the scene to be illuminated and the general direction of the beams.

light ratio The relative intensities of key, back, and fill. A 1:1 ratio between key and back lights means that both light sources burn with equal intensities. A 1:½ ratio between key and fill lights means that the fill light burns with half the intensity of the key light. Because light ratios depend on many production variables, they cannot be fixed. A key:back:fill ratio of 1:1:½ is often used for normal triangle lighting.

limbo Any set area that has a plain, light background.

line *See* **vector line**

linear editing Analog or digital editing that uses tape-based systems. The selection of shots is nonrandom.

line monitor The monitor that shows only the line-out pictures that go on the air or on videotape. Also called *master monitor* and *program monitor.*

line of conversation and action *See* **vector line**

line-out The line that carries the final video or audio output for broadcast.

line producer Supervises daily production activities on the set.

lip-sync Synchronization of sound and lip movement.

live recording The uninterrupted video recording of a live show for later unedited playback. Formerly called *live-on-tape,* live recording includes all recording devices.

location sketch A rough map of the locale of a remote shoot. For an indoor remote, it shows the room dimensions and the furniture and window locations. For an outdoor remote, it indicates the location of buildings, the remote truck, power sources, and the sun during the time of the telecast.

location survey Written assessment, usually in the form of a checklist, of the production requirements for a remote.

locking-in An especially vivid mental image—visual or aural—during script analysis that determines the subsequent visualizations and sequencing.

lockup time The time required by a videotape recorder for the picture and sound to stabilize once the tape has been started.

log The major operational document: a second-by-second list of every program aired on a particular day. It carries such information as program source or origin, scheduled program time, program duration, video and audio information, code identification (house number, for example), program title, program type, and additional pertinent information.

long-focal-length lens *See* **narrow-angle lens**

long shot (LS) Object seen from far away or framed loosely. Also called *establishing shot* and *full shot.*

lossless compression Rearranging but not eliminating pixels during storage and transport. *See also* **compression**.

lossy compression Throwing away redundant pixels during compression. Most compression methods are of the lossy kind. *See also* **compression**.

low-angle dolly Dolly used with high hat to make a camera mount for particularly low shots.

low-key Dark background and illumination of selected areas. Has nothing to do with the vertical positioning of the key light.

low-Z Low impedance. *See also* **impedance**.

LS *See* **long shot**

lumen The light intensity power of one candle (light source radiating isotropically, i.e., in all directions).

luminaire Technical term for a lighting instrument.

luminance The measured brightness (black-and-white) information of a video signal (reproduces the grayscale). Also called *Y signal.*

luminance channel A separate channel within color cameras that deals with brightness variations and allows them to produce a signal receivable on a black-and-white television. The luminance signal is usually electronically derived from the chrominance signals. Also called *Y channel.*

luminant Lamp that produces the light; the light source.

lux European standard unit for measuring light intensity: 1 lux is the amount of 1 lumen (one candlepower of light) that falls on a surface of 1 square meter located 1 meter away from the light source; 10.75 lux = 1 fc; usually roughly translated as 10 lux = 1 fc. *See also* **foot-candle (fc)**.

macro position A lens setting that allows it to be focused at very close distances from an object. Used for close-ups of small objects.

makeup Cosmetics used to enhance, correct, or change appearance.

master control Nerve center for all telecasts. Controls the program input, storage, and retrieval for on-the-air telecasts. Also oversees technical quality of all program material.

master monitor *See* **line monitor**

matte key A keyed (electronically cut-in) title whose letters are filled with shades of gray or a specific color.

MCU Medium close-up.

MD *See* **mini disc**

M/E bus Short for *mix/effects bus.* A row of buttons on the switcher that can serve a mix or an effects function.

medium requirements All content elements, production elements, and people needed to generate the defined process message.

medium shot (MS) Object seen from a medium distance. Covers any framing between a long shot and a close-up. Also called *waist shot.*

megabyte 1,048,576 bytes (2^{20} bytes); usually figured roughly as 1 million bytes.

megapixel A CCD or digital image containing about 1 million pixels. The higher the number of pixels, the higher the picture resolution. Generally used to indicate the relative quality of digital still cameras.

memory card A read/write solid-state storage device for large amounts of digital video and audio data. Also called *flash drive* and *flash memory card.*

mental map Tells viewer where things are or are supposed to be in on- and off-screen space. *See also* **closure**.

mic Stands for *microphone.*

microwave relay A transmission method from the remote location to the station and/or transmitter involving the use of several microwave units.

MIDI *See* **musical instrument digital interface**

mini-cassette A small (2½ × 1⅝ inch or 65mm × 47mm) ¼-inch tape cassette used in digital cameras. It allows one hour of recording with standard recording speed.

mini disc (MD) Optical 2½-inch-wide disc that can store one hour of CD-quality audio.

mini-link Several microwave setups that are linked together to transport the video and audio signals past obstacles to their destination (usually the television station and/or transmitter).

minimum object distance (MOD) How close the camera can get to an object and still focus on it.

mix bus (1) *Audio:* a mixing channel for audio signals. The mix bus combines sounds from several sources to produce a mixed sound signal. (2) *Video:* rows of buttons on the switcher that permit the mixing of video sources, as in a dissolve or super.

mixdown Final combination of sound tracks on a single or stereo track of an audio- or videotape.

mix/effects bus *See* **M/E bus**

mixing (1) *Audio:* combining two or more sounds in specific proportions (volume variations) as determined by the event (show) context. (2) *Video:* creating a dissolve or superimposition via the switcher.

mix-minus Type of multiple audio feed missing the part that is being recorded, such as an orchestra feed with the solo instrument being recorded. Also refers to a program sound feed without the portion supplied by the source that is receiving the feed.

mm Millimeter, one-thousandth of a meter: 25.4mm = 1 inch.

MOD *See* **minimum object distance**

moiré effect Color vibrations that occur when narrow, contrasting stripes of a design interfere with the scanning lines of the television system.

monitor (1) *Audio:* speaker that carries the program sound independently of the line-out. (2) *Video:* high-quality television set used in the television studio and control rooms. Cannot receive broadcast signals.

monochrome One color. In television it refers to a camera or monitor that reads only various degrees of brightness and produces a black-and-white picture.

monopod A single pole onto which you can mount a camera.

montage The juxtaposition of two or more separate event details that combine into a larger and more intense whole—a new gestalt.

mounting head A device that connects the camera to the tripod or studio pedestal to facilitate smooth pans and tilts. Also called *pan-and-tilt head.*

motion vector A vector created by an object actually moving or perceived as moving on-screen.

moving-coil microphone *See* **dynamic microphone**

MP3 A widely used lossy compression system for digital audio. Most Internet-distributed audio is compressed in the MP3 format.

MPEG A compression technique for moving pictures, developed by the Moving Picture Experts Group.

MPEG-2 A compression standard for motion video.

MPEG-4 A compression standard for motion video. It has many codec variations for use in video compression and Internet streaming.

MS *See* **medium shot**

multicamera directing Simultaneous coordination of two or more cameras for instantaneous editing (switching). Also called *control room directing.*

multiple-microphone interference The canceling out of certain sound frequencies when two identical microphones close together are used to record the same sound source on the same tape.

multiple-source editing system Editing system having two or more source VTRs.

musical instrument digital interface (MIDI) A standardized protocol that allows the connection and interaction of various digital audio equipment and computers.

NAB Stands for *National Association of Broadcasters.*

NABET Stands for *National Association of Broadcast Employees and Technicians.* Trade union for studio and master control engineers; may include floor personnel.

narrow-angle lens Gives a close-up view of an event relatively far away from the camera. Also called *long-focal-length lens* and *telephoto lens.*

ND filter *See* **neutral density filter**

neutral density (ND) filter Filter that reduces the incoming light without distorting the color of the scene.

news production personnel People assigned exclusively to the production of news, documentaries, and special events.

noise (1) Audio: unwanted sounds that interfere with the intentional sounds, or unwanted hisses or hums inevitably generated by the electronics of the audio equipment. (2) Video: electronic interference that shows up as "snow."

NLE *See* **nonlinear editing**

non–drop frame mode A video-recording mode in which the slight discrepancy between actual frame count and elapsed clock time is ignored by the SMPTE time code.

nonlinear editing (NLE) Allows instant random access to shots and sequences and easy rearrangement. The video and audio information is stored in digital form on computer hard drives or other digital recording media.

nontechnical production personnel People concerned primarily with nontechnical production matters that lead from the basic idea to the final screen image.

normal lens A lens or zoom lens position with a focal length that approximates the spatial relationships of normal vision.

noseroom The space left in front of a person looking or pointing toward the edge of the screen. *See also* **leadroom**.

NTSC Stands for *National Television System Committee.* Normally designates the composite television signal, consisting of the combined chroma information (red, green, and blue signals) and the luminance information (black-and-white signal). *See also* **composite system**.

NTSC signal *See* **composite system**

off-line editing In linear editing it produces an edit decision list or a videotape not intended for broadcast. In nonlinear editing the selected shots are captured in low resolution to save computer storage space.

omnidirectional Pickup pattern in which the microphone can pick up sounds equally well from all directions.

on-line editing In linear editing it produces the final high-quality edit master recording for broadcast or program duplication. In nonlinear editing it can mean recapturing the selected shots at a higher resolution.

operating light level Amount of light needed by the camera to produce a video signal. Most high-end cameras still need from 100 to 250 foot-candles of illumination for optimal performance at a particular f-stop, such as $f/8$. Also called *baselight level.*

optical disc A digital storage device whose information is recorded and read by laser beam.

O/S *See* **over-the-shoulder shot**

oscilloscope *See* **waveform monitor**

over-the-shoulder shot (O/S) Camera looks over a person's shoulder (shoulder and back of head included in shot) at another person.

PA Stands for *production assistant.*

P.A. Stands for *public address.* Loudspeaker system. Also called *studio talkback* and *S.A. (studio address) system.*

pace Perceived duration of the show or show segment. Part of subjective time.

pan To turn the camera horizontally.

pan-and-tilt head *See* **camera mounting head**

pancake *See* **cake**

pan-stick A foundation makeup with a grease base. Used to cover a beard shadow or prominent skin blemish.

pantograph Expandable hanging device for lighting instruments.

paper-and-pencil editing The process of examining window-dubbed, low-quality (VHS) source tapes and creating a preliminary edit decision list by writing down edit-in and edit-out numbers for each selected shot. Also called *paper editing.*

PAR (parabolic aluminized reflector) lamp *See* **clip light**

parabolic reflector microphone A small parabolic dish whose focal center contains a microphone. Used for the pickup of faraway sounds.

partial two-column A/V script Describes a show for which the dialogue is indicated but not completely written out. *See also* **two-column A/V script**.

patchbay *See* **patchboard**

patchboard A device that connects various inputs with specific outputs. Also called *patchbay.*

pattern projector An ellipsoidal spotlight with a cookie (cucoloris) insert, which projects the cookie's pattern as a cast shadow.

peak program meter (PPM) Meter in an audio console that measures loudness. Especially sensitive to volume peaks, it indicates overmodulation.

pedestal (1) Heavy camera dolly that permits raising and lowering the camera while on the air. (2) To move the camera up and down via a studio pedestal. (3) The black level of a television picture; can be adjusted against a standard on the waveform monitor.

pencil mic A short condenser shotgun mic normally used on a fishpole boom. It has exchangeable heads for omnidirectional or cardioid pickup patterns.

perambulator boom Mount for a studio microphone. An extension device, or boom, is mounted on a dolly, called a *perambulator,* that permits rapid and quiet relocation anywhere in the studio. Also called *big boom.*

performer A person who appears on-camera in nondramatic shows. Performers play themselves and do not assume someone else's character.

periaktos A triangular piece of scenery that can be turned on a swivel base.

phantom power The power for preamplification in a condenser microphone, supplied by the audio console rather than a battery.

phone plug A ¼-inch plug most commonly used at both ends of audio patch cords. These plugs are also used to route sound signals over relatively short distances from various musical instruments, such as electric guitars or keyboards.

photographic lighting principle The triangular arrangement of key, back, and fill lights, with the back light opposite the camera and directly behind the object, and the key and fill lights on opposite sides of the camera and to the front and the side of the object. Also called *triangle lighting.*

pickup (1) Sound reception by a microphone. (2) Reshooting parts of a scene for postproduction editing.

pickup pattern The territory around the microphone within which the mic can "hear equally well," that is, has optimal sound pickup.

picture element *See* **pixel**

pillarbox The aspect ratio that results from fitting a 4 × 3 image onto a 16 × 9 screen by blocking the sides with stripes. *See also* **letterbox**.

pipe grid Heavy steel pipes mounted above the studio floor to support lighting instruments.

pixel Short for *picture element.* (1) A single imaging element (like the single dot in a newspaper picture) that can be identified by a computer. The more pixels per picture area, the higher the picture quality. (2) The light-sensitive elements on a CCD that contain a charge.

P.L. Stands for *private line* or *phone line.* Major intercommunication system in television production.

plasma display A flat-panel television screen whose RGB pixels are activated by specific gases.

playhead In nonlinear editing, can be moved with the mouse to various locations on the scrubber bar. Helps to locate a frame within a clip or sequence.

play VTR *See* **source VTR**

plot How a story develops from one event to the next.

point of view (POV) As seen from a specific character's perspective. Gives the director a clue to camera position.

polar pattern The two-dimensional representation of a microphone pickup pattern.

polarity reversal The reversal of the grayscale; the white areas in the picture become black, and the black areas become white. Color polarity reversal in colors results in the complementary color.

pop filter A bulblike attachment (either permanent or detachable) on the front of a microphone, which filters out sudden air blasts, such as plosive consonants (*p, t,* and *k*) delivered directly into the mic.

ports (1) Slots in the microphone that help achieve a specific pickup pattern and frequency response. (2) Jacks on the computer for plugging in peripheral hardware.

posterization Visual effect that reduces the various brightness values to only a few (usually three or four) and gives the image a flat, posterlike look.

postproduction Any production activity that occurs after the production. Usually refers to either video editing or audio sweetening (a variety of quality adjustments of recorded sound).

postproduction editing The assembly of recorded material after the actual production.

pot Short for *potentiometer.* A sound-volume control.

POV *See* **point of view**

PPM *See* **peak program meter**

preamp Short for *preamplifier.* Strengthens weak electrical signals produced by a microphone or imaging device before they can be further processed (manipulated) and amplified to normal signal strength.

preproduction The preparation of all production details.

preroll To start a videotape and let it roll for a few seconds before it is put in the playback or record mode so that the electronic system has time to stabilize.

preset/background bus *See* **preview/preset bus**

preset monitor (PST) Allows previewing of a shot or an effect before it is switched on the air. Its feed can be activated by the *CUT* button. Similar to or the same as the preview monitor.

pressure zone microphone (PZM) *See* **boundary microphone**

preview/preset bus Rows of buttons on the switcher used to select the upcoming video (preset function) and route it to the preview monitor (preview function) independently of the line-out video. Also called *preset/background bus.*

preview (P/V) monitor (1) Any monitor that shows a video source, except for the line (master) and off-the-air monitors. (2) A color monitor that shows the director the picture to be used for the next shot.

prime lens *See* **fixed-focal-length lens**

prism block *See* **beam splitter**

process message The message actually perceived by the viewer in the process of watching a television program. The program objective is the defined process message.

producer Creator and organizer of television shows.

production The actual activities in which an event is recorded and/or televised.

production schedule The calendar that shows the preproduction, production, and postproduction dates and who is doing what, when, and where.

production switcher Switcher located in the studio control room or remote truck, designed for instantaneous editing.

program (1) A specific television show. (2) A sequence of instructions, encoded in a specific computer language, to perform predetermined tasks.

program/background bus *See* **program bus**

program bus The bus on a switcher whose inputs are directly switched to the line-out. Allows cuts-only switching. Also called *direct bus* and *program/background bus.*

program monitor *See* **line monitor**

program proposal Written document that outlines the process message and the major aspects of a television presentation.

program speaker A loudspeaker in the control room that carries the program sound. Its volume can be controlled without affecting the actual line-out program feed. Also called *audio monitor.*

progressive scanning In this system the electron beam starts with line 1, then scans line 2, then line 3, and so forth, until all lines are scanned, at which point the beam jumps back to its starting position to repeat the scan of all lines.

project panel A folder in nonlinear editing that holds all video and audio source clips. Also called *browser* and *clips bin.*

props Short for *properties.* Furniture and other objects used for set decoration and by actors or performers.

PSA Public service announcement.

PST *See* **preset monitor**

psychographics Audience research factors concerned with such data as consumer buying habits, values, and lifestyles.

psychological closure *See* **closure**

pulse-count system *See* **control track system**

P/V *See* **preview monitor**

pylon Triangular set piece, similar to a pillar.

PZM Stands for *pressure zone microphone. See* **boundary microphone**.

quad-split Switcher mechanism that makes it possible to divide the screen into four quadrants and fill each one with a different image.

quantization *See* **quantizing**

quantizing A step in the digitization of an analog signal. It changes the sampling points into discrete values. Also called *quantization.*

quartz A high-intensity incandescent light whose lamp consists of a quartz or silica housing (instead of the customary glass) that contains halogen gas and a tungsten filament. Produces a very bright light of stable color temperature (3,200K). Also *called TH (tungsten-halogen) lamp. See* also **tungsten-halogen (TH)**.

quick-release plate A mounting plate used to attach camcorders and ENG/EFP cameras to the mounting head.

rack focus To change focus from one object or person closer to the camera to one farther away or vice versa.

radio frequency (RF) Broadcast frequency divided into various channels. In an RF distribution, the video and audio signals are superimposed on the radio frequency carrier wave. Usually called *RF.*

radio mic *See* **wireless microphone**

range extender An optical attachment to the zoom lens that extends its focal length. Also called *extender.*

rating Percentage of television households tuned to a specific station in relation to the total number of television households. *See also* **share**.

RCA phono Video and audio connector for consumer equipment.

RCU *See* **remote control unit**

rear projection (R.P.) Translucent screen onto which images are projected from the rear and photographed from the front.

record VTR The videotape recorder that edits the program segments as supplied by the source VTR(s) into the final edit master tape. Also called *edit VTR.*

reel-to-reel A tape recorder that transports the tape past the heads from one reel (the supply reel) to the other reel (the takeup reel).

reference black The darkest element in a set, used as a reference for the black level (beam) adjustment of the camera picture.

reference white The brightest element in a set, used as a reference for the white level (beam) adjustment of the camera picture.

reflected light Light that is bounced off of the illuminated object. A reflected-light reading is done with a light meter held close to the illuminated object.

refresh rate The number of complete digital scanning cycles (frames) per second. *See* also **frame**.

remote *See* **big remote**

remote control unit (RCU) (1) The CCU control separate from the CCU itself. (2) A small, portable CCU that is taken into the field with the EFP camera. *See also* **camera control unit (CCU)**.

remote survey A preproduction investigation of the location premises and event circumstances. Also called *site survey.*

remote truck The vehicle that carries the program control, the audio control, the video-recording and instant-replay control, the technical control, and the transmission equipment.

resolution The measurement of picture detail, expressed in the number of pixels per scanning line and the number of visible scanning lines. Resolution is influenced by the imaging device, the lens, and the television set that shows the camera picture. Often used synonymously with *definition.*

reverberation Reflections of a sound from multiple surfaces after the sound source has ceased vibrating. Generally used to liven sounds recorded in an acoustically "dead" studio. *See also* **echo**.

RF *See* **radio frequency**

RF mic *See* **wireless microphone**

RGB Stands for *red, green, and blue*—the basic colors of television.

RGB component system Analog video-recording system wherein the red, green, and blue (RGB) signals are kept separate throughout the entire recording and storage process and are transported via three separate wires.

ribbon microphone A mic whose sound pickup device consists of a ribbon that vibrates with the sound pressures within a magnetic field. Also called *velocity mic.*

riser (1) Small platform. (2) The vertical frame that supports the horizontal top of the platform.

robotic *See* **robotic pedestal**

robotic pedestal Remotely controlled studio pedestal and mounting head. It is guided by a computerized system that can store and execute a great number of camera moves. Also called *robotic.*

roll (1) Graphics (usually credit copy) that move slowly up the screen; *see also* **crawl**. (2) Command to start the videotape recorder.

rough-cut The first tentative arrangement of shots and shot sequences in the approximate order and length. Done in off-line editing.

R.P. *See* **rear projection**

rundown sheet *See* **fact sheet**

running time The duration of a program or program segment.

runout signal The recording of a few seconds of black at the end of each videotape recording to keep the screen in black for the video changeover or editing.

run-through Rehearsal.

S.A. Stands for *studio address.* Loudspeaker system. *See also* **studio talkback**.

safe area *See* **essential area**

safe title area *See* **essential area**

sampling The process of reading (selecting and recording) from an analog electronic signal a great many equally spaced, tiny portions (values) for conversion into a digital code.

saturation The color attribute that describes a color's richness or strength.

scanning The movement of the electron beam from left to right and from top to bottom on the television screen.

scanning area Picture area that is scanned by the imaging device; in general, the picture area usually seen in the camera viewfinder and the preview monitor.

scene Event details that form an organic unit, usually in a single place and time. A series of organically related shots that depict these event details.

scenery Background flats and other pieces (windows, doors, pillars) that simulate a specific environment.

schedule time The time at which a program starts and ends.

scoop A scooplike television floodlight.

scrim (1) *Lighting:* a spun-glass material that is put in front of a lighting instrument as an additional light diffuser or intensity reducer. (2) *Scenery:* loosely woven curtain hanging in front of a cyclorama to diffuse light, producing a soft, uniform background.

script Written document that tells what the program is about, who says what, what is supposed to happen, and what and how the audience should see and hear the event.

script marking A director's written symbols on a script to indicate major cues.

scrubber bar Guide for the playhead when moving through a clip or sequence of clips to locate a frame.

SDTV Standard digital television.

secondary-frame effect Visual effect in which the screen shows several images, each of which is clearly contained in its own frame.

SEG (1) Stands for *Screen Extras Guild.* Trade union. (2) *See* **special-effects generator**.

selective focus Emphasizing an object in a shallow depth of field through focus while keeping its foreground and/or background out of focus.

sensor The CCD or CMOS imaging device in a video camera.

sequencing The control and the structuring of a shot series during editing.

servo zoom control Zoom control that activates motor-driven mechanisms.

set Arrangement of scenery and properties to indicate the locale and/or mood of a show.

set light *See* **background light**

set module Series of flats and three-dimensional set pieces whose dimensions match, whether they are used vertically, horizontally, or in various combinations.

shader *See* **video operator (VO)**

shading Adjusting picture contrast to the optimal contrast range; controlling the color and the white and black levels.

share Percentage of television households tuned to a specific station in relation to all households using television (HUT); that is, all households with their sets turned on. *See also* **rating**.

shooter An ENG/EFP camera operator. Sometimes called *DP (director of photography)* in EFP.

shot The smallest convenient operational unit in video and film, usually the interval between two transitions. In cinema it may refer to a specific camera setup.

shot list *See* **shot sheet**

shot sheet A list of every shot a particular camera has to get. It is attached to the camera to help the camera operator remember the shot sequence. Also called *shot list.*

shotgun microphone A highly directional mic for picking up sounds from a relatively great distance.

show format Lists the show segments in order of appearance. Used in routine shows, such as daily game or interview shows.

shuttle Fast-forward and fast-rewind movement of videotape to locate a particular address (shot) on the tape.

side light Usually directional light coming from the side of an object. Acts as additional fill light or a second key light and provides contour.

signal processing The various electronic adjustments or corrections of the video signal to ensure a stable and/or color-enhanced picture. Usually done with digital equipment.

signal-to-noise (S/N) ratio The relation of the strength of the desired signal to the accompanying electronic interference (the noise). A high S/N ratio is desirable (strong video or audio signal relative to weak noise).

silhouette lighting Unlighted objects or people in front of a brightly illuminated background.

single-camera directing Directing method for a single camera. For digital cinema it may mean moving from an establishing long shot to medium shots, then to close-ups of the same action. Also called *film-style shooting.*

single-column drama script Traditional script format for television and motion picture plays. All dialogue and action cues are written in a single column.

single-source editing system Basic editing system that has only one source VTR. Also called *cuts-only editing system.*

site survey *See* **remote survey**

slander Oral defamation.

slant-track *See* **helical scan**

slate (1) Visual and/or verbal identification of each video-recorded segment. (2) A small blackboard or whiteboard upon which essential production information is written. It is recorded at the beginning of each take.

slide fader *See* **fader**

sliding rod Steel pipe that supports a lighting instrument and can be moved into various vertical positions. It is attached to the lighting batten by a modified C-clamp.

slow lens A lens that permits a relatively small amount of light to pass through at its maximum aperture (relatively high *f*-stop number at its lowest setting). Can be used only in well-lighted areas.

SMPTE/EBU time code Stands for *Society of Motion Picture and Television Engineers/European Broadcasting Union time code.* Electronic signal recorded on the cue or address track of a videotape or a track of a multitrack audiotape to give each frame a specific address. The time code reader translates this signal into a specific number (hours, minutes, seconds, and frames) for each frame.

S/N *See* **signal-to-noise ratio**

snow Electronic picture interference; looks like snow on the television screen.

softlight Television floodlight that produces extremely diffused light.

soft wipe Transition in which the demarcation line between two images is softened so the images blend into each other.

solarization A special effect produced by a partial polarity reversal of an image. In a monochrome image, thin black lines are sometimes formed where the positive and negative image areas meet. In a color image, the reversal results in a combination of complementary hues.

SOS Stands for *sound on source.* The video is played back with pictures and sound. Formerly called *SOT.*

SOT Stands for *sound on tape.* The videotape is played back with pictures and sound. *See also* **SOS**.

sound bite Brief portion of someone's on-camera statement.

sound perspective Distant sound must go with a long shot, close sound with a close-up.

source media The media (videotape, hard disk, optical disc, memory card) that holds the recorded camera footage.

source tape The videotape with the original camera footage.

source VTR The videotape recorder that supplies the program segments to be assembled by the record VTR. Also called *play VTR.*

special-effects controls Buttons on a switcher that regulate special effects. They include buttons for specific wipe patterns, the joystick positioner, DVE, color, and chroma-key controls.

special-effects generator (SEG) An image generator built into the switcher that produces special-effects wipe patterns and keys.

spiking To mark on the studio floor with chalk or gaffer's tape critical positions of talent, cameras, or scenery.

split edit Technically, the audio of a shot is substituted with related sounds or narration. In common practice the audio precedes the shot or bleeds into the next one. Viewers hear the audio of the next shot before seeing it, or they still hear the audio of the previous shot at the beginning of the new one.

split intercom *See* **double headset**

split screen Multi-image effect caused by stopping a directional wipe before its completion, each screen portion therefore showing a different image within its own frame.

spotlight A lighting instrument that produces directional, relatively undiffused light with a relatively well-defined beam edge.

spreader A triangular base mount that provides stability and locks the tripod tips in place to prevent the legs from spreading.

standard television (STV) A system based on the NTSC scanning system of 525 (480 visible) interlaced lines.

stand-by A warning cue for any kind of action in television production.

star filter Filter that attaches to the front of the lens; changes prominent light sources into starlike light beams.

Steadicam Camera mount whose built-in springs hold the camera steady while the operator moves.

stock shot An image of a common occurrence—clouds, storm, traffic, crowds—that can be repeated in a variety of contexts because its qualities are typical. There are stock-shot libraries from which any number of such shots can be obtained.

stop-motion A slow-motion effect in which one frame jumps to the next, showing the object in a different position. *See also* **jogging**.

storyboard A series of sketches of the key visualization points of an event, with the corresponding audio information.

streaming A way of delivering and receiving digital audio and/or video as a continuous data flow that can be listened to or watched while the delivery is in progress. *See also* **downloading**.

strike To remove certain objects; to break down scenery and remove equipment from the studio floor after the show.

striped filter Extremely narrow, vertical stripes of red, green, and blue filters attached to the front surface of the single pickup device (single chip). They divide the incoming white light into the three light primaries without the aid of a beam splitter. More efficient filters use a mosaic-like pattern instead of stripes to generate the light primaries.

strip light Several self-contained lamps arranged in a strip; used mostly for illumination of the cyclorama or chroma-key area. Also called *cyc light.*

studio camera High-quality camera with a large zoom lens that cannot be maneuvered properly without the aid of a studio pedestal or other camera mount.

studio monitor A video monitor located in the studio, which carries assigned video sources, usually the video of the line-out.

studio pan-and-tilt head A camera mounting head for heavy cameras that permits extremely smooth pans and tilts.

studio talkback A public address loudspeaker system from the control room to the studio. Also called *S.A. (studio address)* or P.*A. (public address) system.*

STV *See* **standard television**

subtractive primary colors Magenta (bluish red), cyan (greenish blue), and yellow. When mixed these colors act as filters, subtracting certain colors. When all three are mixed, they filter out one another and produce black.

super Short for *superimposition.* A double exposure of two images, with the top one letting the bottom one show through.

supply reel Reel that holds film or tape, which it feeds to the takeup reel.

surround sound Sound that produces a soundfield in front of, to the sides of, and behind the listener by positioning loudspeakers either to the front and the rear or to the front, sides, and rear of the listener.

S-video *See* **Y/C component system**

sweep Curved piece of scenery, similar to a large pillar cut in half.

sweetening A variety of quality adjustments of recorded sound in postproduction.

switcher (1) Technical crewmember doing the video switching (usually the technical director). (2) A panel with rows of buttons that allow the selection and the assembly of various video sources through a variety of transition devices, and the creation of electronic special effects.

switching A change from one video source to another during a show or show segment with the aid of a switcher. Also called *instantaneous editing.*

sync generator Part of the camera chain; produces an electronic synchronization signal, which keeps all scanning in step.

sync pulses Electronic pulses that synchronize the scanning in the various video origination sources (studio cameras and/or remote cameras) and various recording, processing, and reproduction sources (videotape, monitors, and television receivers). *See also* **control track**.

sync roll Vertical rolling of a picture caused by switching among video sources whose scanning is out of step. Also noticeable on a bad edit in which the control tracks of the edited shots do not match.

system microphone Mic consisting of a base upon which several heads can be attached that change its sound pickup characteristic.

take (1) Signal for a cut from one video source to another. (2) Any one of similar repeated shots taken during video recording or filming. Usually assigned a number. A *good take* is the successful recording of a shot. A *bad take* is an unsuccessful recording, requiring another take.

takeup reel Reel that receives (takes up) film or tape from the supply reel. Must be the same size as the supply reel to maintain proper tension.

talent Collective name for all performers and actors who appear regularly on television.

tally light Red light on the camera and/or inside the viewfinder, indicating when the camera is on the air.

tape-based video recorder All video recorders (analog and digital) that record or store information on videotape. All tape-based systems are linear.

tape cassette *See* **cassette**

tapeless system Refers to the recording, storage, and playback of audio and video information via digital electronic recording devices other than videotape.

tapeless video recorder All digital video recorders that record or store digital information on media other than videotape.

target audience The audience selected or desired to receive a specific message.

T-bar *See* fader bar

TBC *See* **time base corrector**

TD Stands for *technical director.*

technical personnel People who operate and maintain the technical equipment.

telephoto lens *See* **narrow-angle lens**

teleprompter A prompting device that projects the moving (usually computer-generated) copy over the lens so that the talent can read it without losing eye contact with the viewer. Also called *auto-cue.*

television system Equipment and people who operate the equipment for the production of specific programs. The basic television system consists of a television camera and a microphone, which convert pictures and sound into electrical signals, and a television set and a loudspeaker, which convert the signals back into pictures and sound.

test tone A tone generated by the audio console to indicate a 0 VU volume level. The 0 VU test tone is recorded with the color bars to give a standard for the recording level.

theme (1) What the story is all about; its essential idea. (2) The opening and closing music in a show.

TH (tungsten-halogen) lamp *See* **quartz**

threefold Three flats hinged together.

three-shot Framing of three people.

tilt To point the camera up or down.

time base corrector (TBC) Electronic accessory to a video recorder that helps make playbacks or transfers electronically stable.

time code Gives each television frame a specific address (number that shows hours, minutes, seconds, and frames of elapsed tape). *See also* **SMPTE/EBU time code**.

time cues Cues to the talent about the time remaining in the show.

time line (1) *Production:* a breakdown of time blocks for various activities on the actual production day, such as crew call, setup, and camera rehearsal. (2) *Nonlinear editing:* shows all video and audio tracks of a sequence and the clips they contain. Each track has individual controls for displaying and manipulating the clips.

tongue To move the boom or jib arm with the camera from left to right or right to left.

track Another name for *truck* (lateral camera movement).

tracking (1) An electronic adjustment of the video heads so that in the playback phase they match the recording phase of the tape. It prevents picture breakup and misalignment, especially in tapes that have been recorded on a machine other than the one used for playback. (2) Another name for *truck* (lateral camera movement).

transponder A satellite's own receiver and transmitter.

treatment A brief narrative description of a television program.

triangle lighting *See* **photographic lighting principle**

triaxial cable Thin camera cable in which one central wire is surrounded by two concentric shields.

trim (1) *Audio:* to adjust the signal strength of mic or line inputs. (2) *Video:* to lengthen or shorten a shot by a few frames during editing; also to shorten a videotaped story.

tripod A three-legged camera mount. Can be connected to a dolly for easy maneuverability.

truck To move the camera laterally by means of a mobile camera mount. Also called *track*.

tungsten-halogen (TH) The kind of lamp filament used in quartz lights. The tungsten is the filament itself; the halogen is a gaslike substance surrounding the filament enclosed in a quartz housing. *See also* **quartz**.

two-column A/V script Traditional script format with video information on page-left and audio information on page-right for a variety of television scripts, such as for documentaries and commercials. Also called *A/V format* and *documentary format.*

twofold Two flats hinged together. Also called *book*.

two-shot Framing of two people.

unbalanced mic or line Microphone or other audio-producing source that has as output two wires: one that carries the audio signal and the other acting as ground. Susceptible to hum and electronic interference.

unidirectional Pickup pattern in which the microphone can pick up sounds better from the front than from the sides or back.

uplink Earth station transmitter used to send video and audio signals to a satellite.

uplink truck A truck that sends video and audio signals to a satellite.

USB 3.0 A high-speed interface (cable) for transporting digital information.

variable-focal-length lens *See* **zoom lens**

vector Refers to a perceivable force with a direction and a magnitude. Vector types include graphic vectors, index vectors, and motion vectors. *See also* **graphic vector**, **index vector**, and **motion vector**.

vector line A dominant direction established by two people facing each other or through a prominent movement in a specific direction. Also called *the line, the line of conversation and action,* and *the hundredeighty.*

vector scope A test instrument for adjusting color in television cameras.

velocity mic *See* **ribbon microphone**

vertical blanking The return of the electron beam to the top of the screen after each cycle of the basic scanning process.

vertical key light position The relative distance of the key light from the studio floor, specifically with respect to whether it is above or below the eye level of the performer. Not to be confused with *high-* and *low-key lighting,* which refer to the relative brightness and contrast of the overall scene.

VHS Stands for *video home system.* A consumer-oriented ½-inch VTR system. Used extensively in all phases of television production for previewing and off-line editing.

video Picture portion of a television program.

videocassette A plastic case in which a videotape moves from a supply reel to a takeup reel, recording and playing back program segments through a videotape recorder.

video leader Visual material and a control tone recorded before the program material. Serves as a technical guide for playback.

video operator (VO) In charge of initial camera setup (white-balancing the camera and keeping the brightness contrast within tolerable limits) and for picture control during the production. Also called *shader.*

video recorder (VR) All devices that record video and audio. Includes videotape, hard disks, read/write optical discs, and memory cards.

videotape recorder (VTR) Electronic recording device that records video and audio signals on videotape for later playback or postproduction editing.

videotape tracks Most videotape systems have a video track, two or more audio tracks, a control track, and sometimes a separate time code track.

video track The area of the videotape used for recording the picture information.

viewfinder Generally means electronic viewfinder (in contrast to the optical viewfinder in a film or still camera); a small television set that displays the picture as generated by the camera.

virtual set A computer-generated environment, normally a studio set, which provides the background for a chroma-key action, such as a newscast.

visualization Mentally converting a scene into a number of key video images and sounds, not necessarily in sequence. The mental image of a shot.

VJ Stands for *video journalist.* A television news reporter who shoots, edits, and writes his or her own material.

VO *See* **video operator**

V/O *See* **voice-over**

voice-over Off-camera narration over a live event or recorded video.

volume The relative intensity of the sound; its relative loudness.

volume-unit (VU) meter Measures volume units, the relative loudness of amplified sound.

VR *See* **video recorder**

VR log A list of all takes—both good (acceptable) and no good (unacceptable)—in consecutive order by scene number and time code address. Often done with computerized logging programs. A vector column facilitates shot selection. *See also* **field log**.

VTR *See* **videotape recorder**

VU meter *See* **volume-unit meter**

W Watt.

wagon A platform with casters, which can be moved about the studio.

waist shot *See* **medium shot**

walk-through Orientation session with the production crew *(technical walkthrough)* and the talent *(talent walk-through)* wherein the director walks through the set and explains the key actions.

waveform monitor Electronic measuring device showing a graph of an electrical signal on a small CRT (cathode-ray tube) screen. Also called *oscilloscope.*

wedge mount Wedge-shaped plate attached to the bottom of a studio camera; used to attach the heavier cameras to the mounting head.

white balance The adjustments of the color circuits in the camera to produce a white color in lighting of various color temperatures (relative reddishness or bluishness of white light).

wide-angle lens A short-focal-length lens that provides a broad vista of a scene.

windowbox A smaller picture that is centered in the actual display screen, with the leftover space of the 16 × 9 frame surrounding it.

window dub A "bumped-down" copy of all source media that has the time code keyed over each frame.

windscreen Material (usually foam rubber) that covers the microphone head or the entire microphone to reduce wind noise.

wipe Transition in which a second image, framed in some geometric shape, gradually replaces all or part of the first image.

wireless microphone A system that transmits audio signals over the air rather than through mic cables. The mic is attached to a transmitter, and the signals are received by a receiver connected to the audio console or recording device. Also called *RF (radio frequency) mic* and *radio mic.*

workprint (1) A dub of the original videotape recording for viewing or off-line editing. (2) In film a dub of the original footage for doing a rough-cut.

wow Sound distortions caused by a slow start or variations in speed of an audiotape.

XLR connector Three-wire audio connector used for all balanced audio cables.

X/S *See* **cross-shot**

Y channel *See* **luminance channel**

Y/C component system Analog video-recording system wherein the luminance (Y) signals and the chrominance (C) signals are kept separate during signal encoding and transport but are combined and occupy the same track when actually laid down on videotape. The Y/C component signal is transported via two wires. Also called *S-video.*

Y/color difference component system Video-recording system in which three signals—the luminance (Y) signal, the red signal minus its luminance (R–Y) signal, and the blue signal minus its luminance (B–Y)—are kept separate throughout the recording and storage process.

Y signal *See* **luminance**

z-axis An imaginary line representing an extension of the lens from the camera to the horizon—the depth dimension.

zone lighting *See* **continuous-action lighting**

zoom To change the lens gradually to a narrow-angle position (zoom-in) or to a wide-angle position (zoom-out) while the camera remains stationary.

zoom lens A variable-focal-length lens. It can gradually change from a wide shot to a close-up and vice versa in one continuous move.

zoom range The degree to which the focal length can be changed from a wide shot to a close-up during a zoom. The zoom range is often stated as a ratio; a 20:1 zoom ratio means that the zoom lens can increase its shortest focal length 20 times.

Index

Television Production Workbook

Television Production Workbook

ELEVENTH EDITION

Herbert Zettl

San Francisco State University

Australia • Brazil • Japan • Korea • Mexico • Singapore • Spain • United Kingdom • United States

To all the students using this workbook,
which is meant not to reprimand you for what you don't know
but to help you identify what you need to learn

Preface to Workbook

This preface is divided into two parts—one for the student and another for the instructor—to facilitate giving each party specific and relevant information.

FOR THE STUDENT

The *Television Production Workbook* is designed to help you learn the rather complicated subject of television production. To move from being a gifted amateur to a true professional who understands the finer points of video and is able to produce high-quality work consistently, you must go beyond talent and innate skills. The *Workbook* will help you accomplish this task in two ways: by reinforcing the information that you have read in the *Television Production Handbook* and by revealing what you still need to learn.

If you could correctly fill in all the bubbles in the *Workbook* without using any outside aids, you would be eligible for an advanced class. Assuming that you are not quite there yet, this workbook is laid out for ease of use and optimal learning. Here are some of its main features:

- Each chapter begins with a review of key terms that tests your understanding of the chapter's basic terminology.
- The middle section of each chapter offers a variety of objective questions and production diagrams to analyze. The aim is to help you recognize and apply the production principles discussed in the *Handbook.*
- A true/false review quiz tests whether you can recall the basic terminology and specific production principles.
- The problem-solving applications at the end of each chapter are intended to enable you to apply the material you have learned to typical production situations. Your instructor will probably come up with many more examples—all in the interest of making the translation from classroom learning to field application maximally efficient and effective.

You will discover that the problems in each chapter differ considerably in degree of difficulty. Some are designed simply for quick recall, others for a more careful weighing of several possible options. Don't get overconfident when you can correctly answer the more obvious questions; you may miss the answers to the more demanding ones. Be aware that some of the problems require the filling in of two or more bubbles. In such a case, the problems that need multiple bubbles are clearly indicated.

Dealing with a workbook is similar to being involved in extensive preproduction activities: both at first seem somewhat irrelevant or at best time-consuming busywork.

This is especially true if you think that learning television production consists primarily of mastering the equipment. But later, when facing production challenges as a professional, you will undoubtedly draw on your rigorous academic problem-solving practice, whether or not you are cognizant of your classroom training.

FOR THE INSTRUCTOR

Because the *Workbook* is primarily an instrument for testing, it is not always embraced by students who may feel that their time is better spent running around capturing exciting video. Underlying such an attitude is often a justified test-anxiety. To minimize or even eliminate such a mindset, you may try using the *Workbook* as a diagnostic tool, choosing not to grade the results but simply make students aware of what they have yet to learn. The initial resistance to the *Workbook* quickly dissipates when even the more experienced students realize that they still have some brushing up to do and that the course offers the opportunity to overcome their deficiencies. You could also have students complete part of the *Workbook* problems at the beginning of each classroom period. In any case, you should encourage students to solve the assigned *Workbook* problems at least initially without the aid of the *Handbook*.

The chapters in this edition of the *Workbook* correspond to those of the *Television Production Handbook*, Eleventh Edition, without necessarily being tied to them. You can assign them in the chapter order you use for the *Handbook* or any other order if more convenient or effective.

To expedite scoring the assignments in the *Workbook*, filling in bubbles substitutes for handwritten answers. Although this binary method may confine some students' urge to express themselves creatively, it greatly facilitates evaluation and, more importantly, enables you to compare the standardized scores. The *Instructor's Manual with Answer Key to Workbook*, available both in print and online, suggests ways of managing the bubble answers most efficiently.

The problem-solving applications are intended for classroom discussion, but you can also assign them as homework.

If you are using *Zettl's VideoLab* DVD-ROM, its various modules dovetail smoothly with the *Workbook* exercises. For example, you can use the disc to demonstrate some motion concepts that are impossible to properly show in the main text or the *Workbook*, such as the reversal of motion vectors when crossing the line.

Both releases of this Windows- and Mac-compatible DVD-ROM—3.0 and the new 4.0—give students virtual hands-on practice and a proven shortcut from reading about production techniques to actually applying them in the studio and in the field. The in-text ZVL cues in the *Handbook* work with both *Zettl's VideoLab 3.0* and *4.0*. The new 4.0 release incorporates additional advanced exercises.

Finally, you may want to remind your students that they have chosen your class precisely because they want to go beyond simple equipment skills—to move from gifted amateur to creative and responsible professional, from somebody who conveniently submits to the industry routines to one who innovates new and more effective ways of communicating significant ideas to media consumers.

To access additional course materials and companion resources, please visit *www.cengagebrain.com*. At the CengageBrain.com home page, search for the ISBN of the *Television Production Handbook* (from the back cover of the book), using the search box at the top of the page. This will take you to the product page, where free companion resources can be found.

ACKNOWLEDGMENTS

The people at Wadsworth Cengage Learning deserve much praise for managing the publication of the *Television Production Workbook,* Eleventh Edition: Michael Rosenberg, publisher; Megan Garvey, development editor; and Erin Pass, assistant editor.

I was privileged to work, once again, with my proven A-team of experts: Gary Palmatier of Ideas to Images, art director and project manager; Elizabeth von Radics, copy editor; and Ed Aiona, photographer. Not only are they exceptionally gifted professionals who can create a book from manuscript fragments and scribbles but they also draw on their competency in video and the photographic arts. Best of all, they are fun to work with.

As with previous editions, I am especially indebted to my former colleagues: Hamid Khani, Marty Gonzales, Winston Tharp, all of San Francisco State University, and Paul Rose of Utah University. They were always ready to help in a variety of ways. Thanks also to my students who contributed significantly, however unknowingly, to my formulating the various problems. The many students who doubled as on-camera talent deserve special praise.

Finally, I owe a big thank-you to my wife, Erika, who as a longtime classroom teacher, administrator, and educational consultant helped with translating the more complicated problems into a binary format that enables students to give the answers by filling in bubbles.

Introduction: Process and System

Course No. ______ Date ______ Name ______

The Television Production Process

REVIEW OF KEY TERMS

Match each term with its appropriate definition by filling in the corresponding bubble.

1. **medium requirements**
2. **EFP**
3. **nonlinear editing**
4. **preproduction**
5. **effect-to-cause model**
6. **television system**
7. **technical personnel**
8. **production**
9. **linear editing**
10. **ENG**
11. **process message**
12. **postproduction**

A. The people, content, and production elements needed to generate the desired viewer effect

A ○1 ○2 ○3 ○4 ○5 ○6 ○7 ○8 ○9 ○10 ○11 ○12

B. Television production that covers daily events and is usually transmitted live or after immediate postproduction

B ○1 ○2 ○3 ○4 ○5 ○6 ○7 ○8 ○9 ○10 ○11 ○12

C. Video and audio editing phase

C ○1 ○2 ○3 ○4 ○5 ○6 ○7 ○8 ○9 ○10 ○11 ○12

PAGE TOTAL ______

1. medium requirements	**5. effect-to-cause model**	**9. linear editing**
2. EFP	**6. television system**	**10. ENG**
3. nonlinear editing	**7. technical personnel**	**11. process message**
4. preproduction	**8. production**	**12. postproduction**

D. People who primarily operate television equipment

D ○ 1 ○ 2 ○ 3 ○ 4 ○ 5 ○ 6 ○ 7 ○ 8 ○ 9 ○ 10 ○ 11 ○ 12

E. Analog or digital editing that uses tape-based systems

E ○ 1 ○ 2 ○ 3 ○ 4 ○ 5 ○ 6 ○ 7 ○ 8 ○ 9 ○ 10 ○ 11 ○ 12

F. Allows random access to, and flexible sequencing of, recorded video and audio material

F ○ 1 ○ 2 ○ 3 ○ 4 ○ 5 ○ 6 ○ 7 ○ 8 ○ 9 ○ 10 ○ 11 ○ 12

G. Preparation of all production details

G ○ 1 ○ 2 ○ 3 ○ 4 ○ 5 ○ 6 ○ 7 ○ 8 ○ 9 ○ 10 ○ 11 ○ 12

H. The information that the viewer actually receives

H ○ 1 ○ 2 ○ 3 ○ 4 ○ 5 ○ 6 ○ 7 ○ 8 ○ 9 ○ 10 ○ 11 ○ 12

PAGE TOTAL

Course No. ______________ Date ______________ Name ______________________________

I. All activities during the recording or televising of an event

J. The basic equipment necessary to produce video and audio signals and reconvert them into pictures and sound

K. Moving from the idea to the program objective, then backing up to the specific medium requirements to produce this objective

L. A relatively uncomplicated field production shot for postproduction

I	○ 1	○ 2	○ 3	○ 4
	○ 5	○ 6	○ 7	○ 8
	○ 9	○ 10	○ 11	○ 12
J	○ 1	○ 2	○ 3	○ 4
	○ 5	○ 6	○ 7	○ 8
	○ 9	○ 10	○ 11	○ 12
K	○ 1	○ 2	○ 3	○ 4
	○ 5	○ 6	○ 7	○ 8
	○ 9	○ 10	○ 11	○ 12
L	○ 1	○ 2	○ 3	○ 4
	○ 5	○ 6	○ 7	○ 8
	○ 9	○ 10	○ 11	○ 12

PAGE TOTAL ______

SECTION TOTAL ______

REVIEW OF EFFECT-TO-CAUSE MODEL

Select the correct answers and fill in the bubbles with the corresponding numbers.

1. Identify each part of the effect-to-cause diagram below and fill in the bubbles with the corresponding numbers.

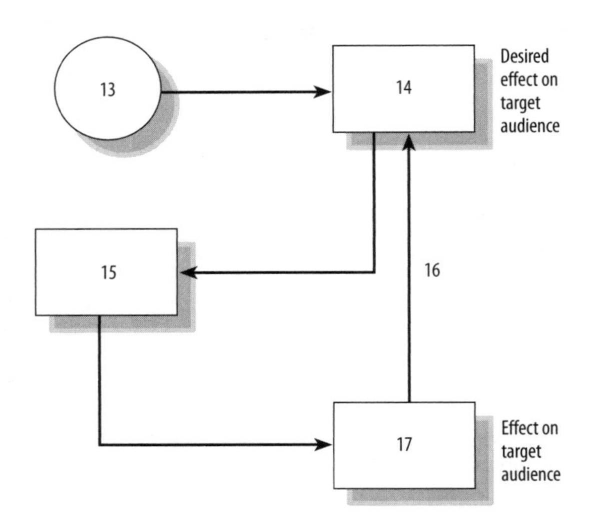

a. program content, people, and production elements

b. defined process message

c. initial idea

d. feedback

e. actual process message

1a ○ 13 ○ 14 ○ 15 ○ 16 ○ 17

1b ○ 13 ○ 14 ○ 15 ○ 16 ○ 17

1c ○ 13 ○ 14 ○ 15 ○ 16 ○ 17

1d ○ 13 ○ 14 ○ 15 ○ 16 ○ 17

1e ○ 13 ○ 14 ○ 15 ○ 16 ○ 17

PAGE TOTAL

Course No. ______________ Date ______________ Name ______________

2. The effect-to-cause model is especially helpful in (18) *preproduction* (19) *production* (20) *postproduction.*

3. The effect-to-cause model is especially useful in the production of (21) *breaking news stories* (22) *documentaries* (23) *dramas.*

4. The most important initial step in the effect-to-cause approach is (24) *determining the medium requirements* (25) *determining the available production equipment* (26) *defining the process message.*

5. Feedback helps determine (27) *whether the production was efficient* (28) *how close the actual effect came to the process message* (29) *how close the actual effect came to the original idea.*

6. The medium requirements include (30) *equipment and people* (31) *equipment but not people.*

7. The angle will define the (32) *process message* (33) *basic production approach* (34) *position of the camera.*

8. A defined process message must contain at least the (35) *target audience* (36) *medium requirements* (37) *desired effect on the viewer.*

9. Medium requirements are basically determined by the (38) *chief engineer* (39) *available equipment* (40) *process message.*

10. In the effect-to-cause model, we move from (41) *idea to medium requirements to production* (42) *idea to production to process message* (43) *idea to process message to medium requirements.*

2	○ 18	○ 19	○ 20
3	○ 21	○ 22	○ 23
4	○ 24	○ 25	○ 26
5	○ 27	○ 28	○ 29
6	○ 30	○ 31	
7	○ 32	○ 33	○ 34
8	○ 35	○ 36	○ 37
9	○ 38	○ 39	○ 40
10	○ 41	○ 42	○ 43

PAGE TOTAL

SECTION TOTAL

REVIEW OF PRODUCTION PERSONNEL

1. Match each job title with the most appropriate function by filling in the corresponding bubble.

(44) director
(45) PA
(46) AD
(47) VJ
(48) VR operator
(49) LD
(50) producer
(51) floor manager
(52) DP
(53) TD

a. In charge of all production activities on the production day

1a ○ 44 ○ 45 ○ 46 ○ 47 ○ 48 ○ 49 ○ 50 ○ 51 ○ 52 ○ 53

b. In EFP, works the camera; in cinema, is in charge of the lighting and film exposure

1b ○ 44 ○ 45 ○ 46 ○ 47 ○ 48 ○ 49 ○ 50 ○ 51 ○ 52 ○ 53

c. In charge of lighting

1c ○ 44 ○ 45 ○ 46 ○ 47 ○ 48 ○ 49 ○ 50 ○ 51 ○ 52 ○ 53

d. Supports the director in directing activities

1d ○ 44 ○ 45 ○ 46 ○ 47 ○ 48 ○ 49 ○ 50 ○ 51 ○ 52 ○ 53

e. In charge of video recording

1e ○ 44 ○ 45 ○ 46 ○ 47 ○ 48 ○ 49 ○ 50 ○ 51 ○ 52 ○ 53

f. In charge of all preproduction activities

1f ○ 44 ○ 45 ○ 46 ○ 47 ○ 48 ○ 49 ○ 50 ○ 51 ○ 52 ○ 53

g. Relays the director's messages to talent

1g ○ 44 ○ 45 ○ 46 ○ 47 ○ 48 ○ 49 ○ 50 ○ 51 ○ 52 ○ 53

PAGE TOTAL

Course No. ____________ Date ____________ Name ____________

h. Shoots and edits own news footage

1h ○ 44 ○ 45 ○ 46 ○ 47 ○ 48 ○ 49 ○ 50 ○ 51 ○ 52 ○ 53

i. In charge of a crew; usually does the switching

1i ○ 44 ○ 45 ○ 46 ○ 47 ○ 48 ○ 49 ○ 50 ○ 51 ○ 52 ○ 53

j. Assists producer and director in all production phases

1j ○ 44 ○ 45 ○ 46 ○ 47 ○ 48 ○ 49 ○ 50 ○ 51 ○ 52 ○ 53

2. Match each title of news personnel with its appropriate definition by filling in the corresponding bubble.

(54) writer
(55) sportscaster
(56) news producer
(57) news director
(58) reporter
(59) assignment editor
(60) videographer/shooter
(61) anchor
(62) VJ

a. Prepares on-the-air copy for the anchorpersons

2a ○ 54 ○ 55 ○ 56 ○ 57 ○ 58 ○ 59 ○ 60 ○ 61 ○ 62

b. Responsible for all the news operations

2b ○ 54 ○ 55 ○ 56 ○ 57 ○ 58 ○ 59 ○ 60 ○ 61 ○ 62

c. Combines in a single person the camera operator, the editor, the writer, and sometimes even the on-camera talent in news gathering.

2c ○ 54 ○ 55 ○ 56 ○ 57 ○ 58 ○ 59 ○ 60 ○ 61 ○ 62

d. Sends reporters and videographers to specific events

2d ○ 54 ○ 55 ○ 56 ○ 57 ○ 58 ○ 59 ○ 60 ○ 61 ○ 62

e. Principal presenter of newscast, normally from a studio set

2e ○ 54 ○ 55 ○ 56 ○ 57 ○ 58 ○ 59 ○ 60 ○ 61 ○ 62

PAGE TOTAL ____

(54) writer	(57) news director	(60) videographer/shooter
(55) sportscaster	(58) reporter	(61) anchor
(56) news producer	(59) assignment editor	(62) VJ

f. Operates camcorder and, in the absence of a reporter, decides what part of the event to cover

2f ○ 54 ○ 55 ○ 56 ○ 57 ○ 58 ○ 59 ○ 60 ○ 61 ○ 62

g. Gathers the news stories and often reports on-camera from the field

2g ○ 54 ○ 55 ○ 56 ○ 57 ○ 58 ○ 59 ○ 60 ○ 61 ○ 62

h. Responsible for individual news stories for on-the-air use

2h ○ 54 ○ 55 ○ 56 ○ 57 ○ 58 ○ 59 ○ 60 ○ 61 ○ 62

i. On-camera talent, giving sports content

2i ○ 54 ○ 55 ○ 56 ○ 57 ○ 58 ○ 59 ○ 60 ○ 61 ○ 62

PAGE TOTAL

SECTION TOTAL

Course No. ____________ Date ____________ Name ____________

REVIEW OF TECHNICAL SYSTEMS

1. Identify each of the major elements of the basic television system by filling in the corresponding bubble.

a. audio signal

b. microphone

c. video signal

d. TV receiver sound

e. TV receiver image

f. VR

g. TV camera

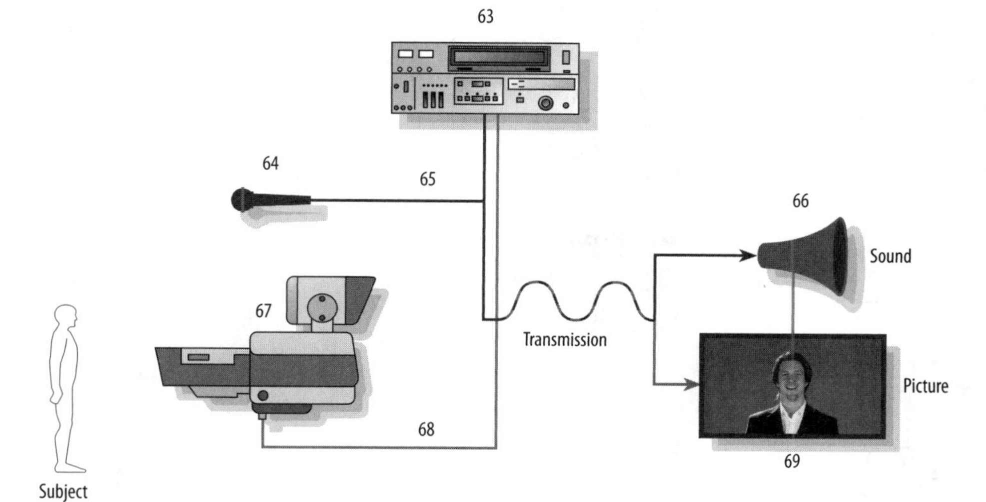

1a ○ 63 ○ 64 ○ 65 ○ 66 ○ 67 ○ 68 ○ 69

1b ○ 63 ○ 64 ○ 65 ○ 66 ○ 67 ○ 68 ○ 69

1c ○ 63 ○ 64 ○ 65 ○ 66 ○ 67 ○ 68 ○ 69

1d ○ 63 ○ 64 ○ 65 ○ 66 ○ 67 ○ 68 ○ 69

1e ○ 63 ○ 64 ○ 65 ○ 66 ○ 67 ○ 68 ○ 69

1f ○ 63 ○ 64 ○ 65 ○ 66 ○ 67 ○ 68 ○ 69

1g ○ 63 ○ 64 ○ 65 ○ 66 ○ 67 ○ 68 ○ 69

PAGE TOTAL

2. Identify each major component of the expanded television system by filling in the corresponding bubble.

a. cameras 1 and 2
b. transmitter
c. preview monitors
d. video switcher
e. home TV receiver
f. audio console
g. video recorder
h. line monitor
i. CCUs 1 and 2
j. audio monitor speaker

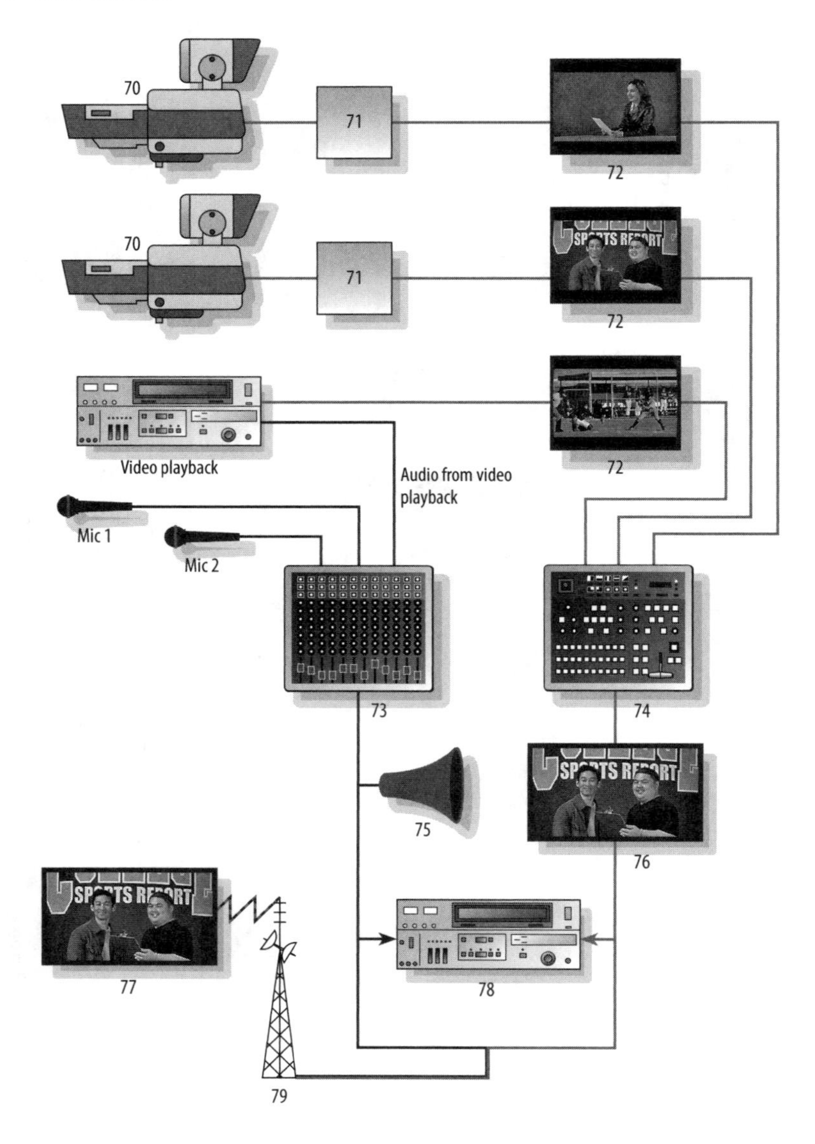

2a	○ 70	○ 71	○ 72	○ 73	○ 74
	○ 75	○ 76	○ 77	○ 78	○ 79
2b	○ 70	○ 71	○ 72	○ 73	○ 74
	○ 75	○ 76	○ 77	○ 78	○ 79
2c	○ 70	○ 71	○ 72	○ 73	○ 74
	○ 75	○ 76	○ 77	○ 78	○ 79
2d	○ 70	○ 71	○ 72	○ 73	○ 74
	○ 75	○ 76	○ 77	○ 78	○ 79
2e	○ 70	○ 71	○ 72	○ 73	○ 74
	○ 75	○ 76	○ 77	○ 78	○ 79
2f	○ 70	○ 71	○ 72	○ 73	○ 74
	○ 75	○ 76	○ 77	○ 78	○ 79
2g	○ 70	○ 71	○ 72	○ 73	○ 74
	○ 75	○ 76	○ 77	○ 78	○ 79
2h	○ 70	○ 71	○ 72	○ 73	○ 74
	○ 75	○ 76	○ 77	○ 78	○ 79
2i	○ 70	○ 71	○ 72	○ 73	○ 74
	○ 75	○ 76	○ 77	○ 78	○ 79
2j	○ 70	○ 71	○ 72	○ 73	○ 74
	○ 75	○ 76	○ 77	○ 78	○ 79

PAGE TOTAL

PHOTOS: EDWARD AIONA

Course No. ____________ Date ______________ Name ______________________________

3. Match each system element with its appropriate function by filling in the corresponding bubble.

(80) cameras
(81) preview monitors
(82) TV receiver
(83) audio monitor speaker
(84) line monitor
(85) microphones
(86) switcher
(87) audio console
(88) CCUs
(89) VR

a. To record video and audio signals on recording media

b. To control the picture quality of the television cameras

c. To translate the broadcast signals into pictures and sound

d. To convert what we hear into electrical signals

e. To control the audio quality of the various audio inputs

f. To convert what the lens sees into electrical signals

g. To select video inputs

3a ○ 80 ○ 81 ○ 82 ○ 83 ○ 84 ○ 85 ○ 86 ○ 87 ○ 88 ○ 89

3b ○ 80 ○ 81 ○ 82 ○ 83 ○ 84 ○ 85 ○ 86 ○ 87 ○ 88 ○ 89

3c ○ 80 ○ 81 ○ 82 ○ 83 ○ 84 ○ 85 ○ 86 ○ 87 ○ 88 ○ 89

3d ○ 80 ○ 81 ○ 82 ○ 83 ○ 84 ○ 85 ○ 86 ○ 87 ○ 88 ○ 89

3e ○ 80 ○ 81 ○ 82 ○ 83 ○ 84 ○ 85 ○ 86 ○ 87 ○ 88 ○ 89

3f ○ 80 ○ 81 ○ 82 ○ 83 ○ 84 ○ 85 ○ 86 ○ 87 ○ 88 ○ 89

3g ○ 80 ○ 81 ○ 82 ○ 83 ○ 84 ○ 85 ○ 86 ○ 87 ○ 88 ○ 89

PAGE TOTAL

(80) cameras	(85) microphones
(81) preview monitors	(86) switcher
(82) TV receiver	(87) audio console
(83) audio monitor speaker	(88) CCUs
(84) line monitor	(89) VR

h. To display the pictures supplied by the various video sources

3h ○ 80 ○ 81 ○ 82 ○ 83 ○ 84 ○ 85 ○ 86 ○ 87 ○ 88 ○ 89

i. To display the line-out pictures

3i ○ 80 ○ 81 ○ 82 ○ 83 ○ 84 ○ 85 ○ 86 ○ 87 ○ 88 ○ 89

j. To reproduce the line-out sound

3j ○ 80 ○ 81 ○ 82 ○ 83 ○ 84 ○ 85 ○ 86 ○ 87 ○ 88 ○ 89

PAGE TOTAL

SECTION TOTAL

Course No. ____________ Date ________________ Name ______________________________

REVIEW QUIZ

*Mark the following statements as true or false by filling in the bubbles in the **T** (for true) or **F** (for false) column.*

1. The switcher does not allow for instantaneous editing.
2. In the television studio, we use spotlights and floodlights.
3. Digital memory devices (memory cards) can be used to record video segments.
4. The primary function of the C.G. is to enhance picture quality.
5. A microphone converts sound into digital video.
6. Linear editing involves copying shots onto another videotape in a specific order.
7. At least two VRs are needed for linear postproduction editing.
8. With nonlinear editing you can edit directly from the source tapes to the edit master tape.
9. All audio consoles can select the signals from multiple incoming audio sources and control sound volume.
10. A digital camcorder can use videotape as its recording media.

	T	F
1	○ 90	○ 91
2	○ 92	○ 93
3	○ 94	○ 95
4	○ 96	○ 97
5	○ 98	○ 99
6	○ 100	○ 101
7	○ 102	○ 103
8	○ 104	○ 105
9	○ 106	○ 107
10	○ 108	○ 109

SECTION TOTAL ______

PROBLEM-SOLVING APPLICATIONS

Think through each production problem and consider the various options. Then pick the most effective solution and justify your choice.

1. List in any order the major components (equipment) of the expanded television system that will allow you to produce and select optimal pictures from three studio cameras, produce optimal sound from four microphones, and video-record and simultaneously transmit the signals to a television receiver. Now order these components and connect them with lines that show the basic signal flow from cameras, microphones, and the various video and audio selections to the video recorder and the home television receiver.

2. List the components of a linear videotape-editing system that permits a dissolve; then draw a diagram that shows the basic signal flow for these components.

3. What system elements are incorporated into a single camcorder? What are some of the advantages and the disadvantages of the camcorder system compared with those of the expanded television system?

4. What exactly distinguishes ENG from EFP?

5. Apply the effect-to-cause model to a variety of goal-directed programs. Pay particular attention to a precise process message.

6. How does a clearly stated process message help with the medium requirements?

7. List two tapeless recording media and describe the advantages and the disadvantages of each.

Preproduction

Course No. ______ Date ______ Name ______

The Producer in Preproduction

REVIEW OF KEY TERMS

Match each term with its appropriate definition by filling in the corresponding bubble.

1. **target audience**
2. **program proposal**
3. **time line**
4. **demographics**
5. **production schedule**
6. **psychographics**
7. **share**
8. **rating**
9. **treatment**

A. A breakdown of time blocks for various activities on the actual production day

A ○ 1 ○ 2 ○ 3 ○ 4 ○ 5 ○ 6 ○ 7 ○ 8 ○ 9

B. Audience factors concerned with such data as consumer buying habits, values, and lifestyles

B ○ 1 ○ 2 ○ 3 ○ 4 ○ 5 ○ 6 ○ 7 ○ 8 ○ 9

C. Percentage of television households tuned to a specific station in relation to all HUT

C ○ 1 ○ 2 ○ 3 ○ 4 ○ 5 ○ 6 ○ 7 ○ 8 ○ 9

D. Percentage of television households tuned to a specific station in relation to the total number of television households

D ○ 1 ○ 2 ○ 3 ○ 4 ○ 5 ○ 6 ○ 7 ○ 8 ○ 9

PAGE TOTAL ______

1. target audience	**4. demographics**	**7. share**
2. program proposal	**5. production schedule**	**8. rating**
3. time line	**6. psychographics**	**9. treatment**

E. The calendar dates for preproduction, production, and postproduction activities

E ○1 ○2 ○3 ○4 ○5 ○6 ○7 ○8 ○9

F. Written document that outlines the process message and the major aspects of a television presentation

F ○1 ○2 ○3 ○4 ○5 ○6 ○7 ○8 ○9

G. Audience factors concerned with such data as age, gender, marital status, and income

G ○1 ○2 ○3 ○4 ○5 ○6 ○7 ○8 ○9

H. Narrative description of a television program

H ○1 ○2 ○3 ○4 ○5 ○6 ○7 ○8 ○9

I. Viewers identified to receive a specific message

I ○1 ○2 ○3 ○4 ○5 ○6 ○7 ○8 ○9

PAGE TOTAL

SECTION TOTAL

Course No. ____________ Date ________________ Name ______________________________

REVIEW OF PREPRODUCTION PLANNING: GENERATING IDEAS

1. Expand three of the following four clusters according to the key word. Develop a precise process message for each. Choose the key word for the fourth cluster and develop its process message accordingly.

   a. cluster 1

Water conservation

Process message: ____

____

____

____

____

____

b. cluster 2

Peace

Process message: ____

____

____

____

____

____

Course No. ______ Date ______ Name ______

c. cluster 3

Alternative energy sources

Process message: ______

d. cluster 4 (on a subject of your choice)

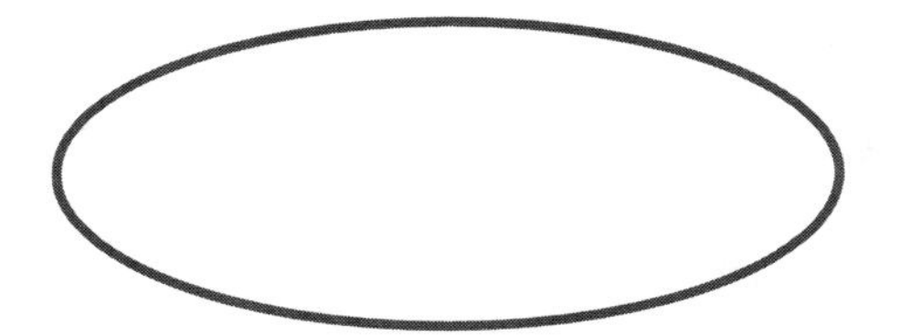

Process message: ____

____

____

____

____

____

Course No. ______ Date ______ Name ______

REVIEW OF EVALUATING IDEAS

Select the correct answers and fill in the bubbles with the corresponding numbers.

1. In the preproduction flowchart below, match the unmarked steps with the corresponding numbers.

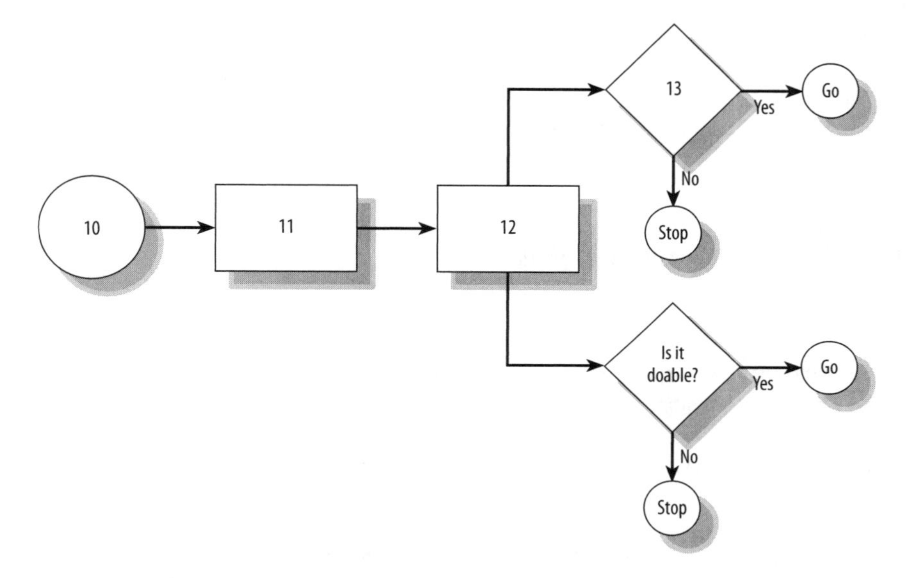

a. angle

b. idea

c. worth doing?

d. process message

1a ○ 10 ○ 11 ○ 12 ○ 13

1b ○ 10 ○ 11 ○ 12 ○ 13

1c ○ 10 ○ 11 ○ 12 ○ 13

1d ○ 10 ○ 11 ○ 12 ○ 13

SECTION TOTAL ______

REVIEW OF PROGRAM PROPOSAL

Select the correct answers and fill in the bubbles with the corresponding numbers.

1. The standard program proposal usually contains a (14) *program objective* (15) *list of studio or remote equipment* (16) *detailed storyboard.*

2. A show treatment usually contains a (17) *one-page sample of the dialogue and the video and audio cues* (18) *script sample with major visualization cues* (19) *brief narrative description of what we see and hear.*

3. A well-stated process message should (20) *state the steps of moving from idea to finished show* (21) *include the specific objective of the show* (22) *describe the process of moving from idea to detailed script.*

4. A good description of the target audience should include (23) *demographic and psychographic indicators* (24) *the number of potential viewers* (25) *only demographic indicators.*

5. When preparing a budget for an outside client, you can skip the cost for (26) *the writer* (27) *the equipment* (28) *neither the writer nor the equipment.*

6. In a large production, the daily activities are supervised by the (29) *PA* (30) *executive producer* (31) *line producer.*

7. Preproduction is necessary for (32) *every production except ENG* (33) *studio productions only* (34) *EFPs only.*

8. The individual who is normally absent during the beginning stage of preproduction is the (35) *producer* (36) *chief engineer* (37) *director.*

9. In the production schedule on the facing page, identify potential problems for each EFP shoot. From the list below, select the items that best describe the problems and fill in the corresponding bubbles. Most shoots have more than one problem. ***(Multiple answers are possible.)***

(38) different talent; break in continuity

(39) different director and crew; potential break in style and continuity

(40) need for remote truck questionable relative to the production scope

(41) shooting time too late; will cause lighting and continuity problems in postproduction

(42) should be done in conjunction with similar previous activity or opening

(43) facilities request very late

(44) facilities request too late

(45) too little time allotted

(46) too much time allotted

1	○ 14	○ 15	○ 16
2	○ 17	○ 18	○ 19
3	○ 20	○ 21	○ 22
4	○ 23	○ 24	○ 25
5	○ 26	○ 27	○ 28
6	○ 29	○ 30	○ 31
7	○ 32	○ 33	○ 34
8	○ 35	○ 36	○ 37

PAGE TOTAL

Course No. ______ Date ______ Name ______

Show/Scene/Subject	Date/Time	Location	Facilities	Talent/Personnel
Leisure City SHOOT 1 OPENING	Aug. 8 8:30 am 4:30 pm	In front of completed model home— Simple opening remarks. 1:00 min.	Normal EFP as per fac. req. Aug. 8	Talent: LYNNE Director: B.R. Crew A scheduled
Leisure City SHOOT 2	Aug. 9 12:30 pm 1:00 pm	Homes under construction. Show homes being constructed.	Special remote Truck. See equipment fac. req. Aug. 8	Talent: LYNNE Director: B.R. Crew A scheduled
Leisure City SHOOT 3	Aug. 10 8:30 am 9:00 am	Interior of model home. Shows how Typical home looks and works inside.	Normal EFP as per fac. req. Aug. 7	Talent: LYNNE Director: B.R. Crew A scheduled
Leisure City SHOOT 4	Aug. 11 7:30 pm 10:30 pm	Homes under construction.	Normal EFP as per fac. req. Aug. 11	Talent: SUSAN Director: JOHN HEWITT Crew B scheduled
Leisure City SHOOT 5	Aug. 12 8:00 pm 8:30 pm	In front of completed model home— Simple closing remarks. 1:30 min.	Special remote Truck. See equipment fac. req. Aug. 8	Talent: SUSAN Director: B.R. Crew A scheduled

a. shoot 1

9a ○ 38 ○ 39 ○ 40 ○ 41 ○ 42 ○ 43 ○ 44 ○ 45 ○ 46

b. shoot 2

9b ○ 38 ○ 39 ○ 40 ○ 41 ○ 42 ○ 43 ○ 44 ○ 45 ○ 46

c. shoot 3

9c ○ 38 ○ 39 ○ 40 ○ 41 ○ 42 ○ 43 ○ 44 ○ 45 ○ 46

d. shoot 4

9d ○ 38 ○ 39 ○ 40 ○ 41 ○ 42 ○ 43 ○ 44 ○ 45 ○ 46

e. shoot 5

9e ○ 38 ○ 39 ○ 40 ○ 41 ○ 42 ○ 43 ○ 44 ○ 45 ○ 46

PAGE TOTAL

SECTION TOTAL

REVIEW OF COORDINATION

Select the correct answers and fill in the bubbles with the corresponding numbers.

1. The person principally responsible for all preproduction communication is the (47) *director* (48) *producer* (49) *executive producer.*

2. When coordinating an EFP on water conservation, you don't have to include in your preproduction memos the (50) *technical personnel* (51) *sales manager* (52) *PA.*

3. In a typical television station, facilities requests are necessary (53) *only if you are planning a new production* (54) *for every production except ENG* (55) *for the preproduction conference.*

4. When sending your memos via e-mail, you must insist on a written response from (56) *production people only* (57) *technical personnel only* (58) *every person on your mailing list.*

5. The timeline is usually drawn up by the (59) *executive producer* (60) *producer* (61) *director.*

6. The production schedule is usually drawn up by the (62) *executive producer* (63) *producer* (64) *assistant director.*

7. A schedule change is normally communicated to the production team by the (65) *executive producer* (66) *producer* (67) *director.*

8. Informing the crew of the process message is the responsibility of the (68) *director* (69) *executive producer* (70) *chief engineer.*

1	○ 47	○ 48	○ 49
2	○ 50	○ 51	○ 52
3	○ 53	○ 54	○ 55
4	○ 56	○ 57	○ 58
5	○ 59	○ 60	○ 61
6	○ 62	○ 63	○ 64
7	○ 65	○ 66	○ 67
8	○ 68	○ 69	○ 70

SECTION TOTAL

Course No. ______ Date ______ Name ______

REVIEW OF UNIONS AND LEGAL MATTERS

Select the correct answers and fill in the bubbles with the corresponding numbers.

1. The theater department of the local high school would like to play its video production of Arthur Miller's *Death of a Salesman* on a local TV station. The student actors (71) *will* (72) *will not* need AFTRA clearance.

2. A potential sponsor would like a treatment of one of the proposed segments of your humanities series. Sending the script instead is (73) *acceptable* (74) *not acceptable.*

3. For each of the trade unions listed, mark whether it is a (75) *technical* or a (76) *nontechnical* union by filling in the appropriate bubble.

   a. NABET

   b. IATSE

   c. SEG

   d. WGA

   e. DGA

   f. SAG

   g. AFTRA

   h. IBEW

   i. AFM

1	○ 71	○ 72
2	○ 73	○ 74
3a	○ 75	○ 76
3b	○ 75	○ 76
3c	○ 75	○ 76
3d	○ 75	○ 76
3e	○ 75	○ 76
3f	○ 75	○ 76
3g	○ 75	○ 76
3h	○ 75	○ 76
3i	○ 75	○ 76

PAGE TOTAL ______

4. Determine whether each of the production cases below (77) *requires* or (78) *does not require* copyright clearance and fill in the appropriate bubble.

a. using a recent CD recording of Bach's *Toccata and Fugue in F Major* as the theme for a show on architecture

4a ○ 77 ○ 78

b. using a Beatles song as the theme for a documentary on the history of rock music

4b ○ 77 ○ 78

c. taking close-ups of paintings in your news coverage of the local outdoor art festival

4c ○ 77 ○ 78

d. using a sixteenth-century book to make a digital scan of a church floor plan for your Renaissance show

4d ○ 77 ○ 78

e. using a recently published art book to make a digital scan of a church floor plan for your show on Baroque art

4e ○ 77 ○ 78

f. using a record album cover as the background for your opening and closing titles on a music series

4f ○ 77 ○ 78

g. using three different scenes of published plays as the basis for your series about acting for the video camera

4g ○ 77 ○ 78

h. having your pianist friend play and record her own composition for use as a theme on your weekly music series

4h ○ 77 ○ 78

PAGE TOTAL

SECTION TOTAL

Course No. ________ Date ________ Name ________

REVIEW OF RATINGS

Select the correct answers and fill in the bubbles with the corresponding numbers.

1. HUT is a factor in figuring (79) *shares* (80) *ratings.*

2. A rating of 13 indicates that (81) *13 of 2,000* (82) *800 of 6,000* (83) *13 of 1,300* (84) *total television households* (85) *of all households using television* are tuned to your station. ***(Fill in two bubbles.)***

3. A share of 22 means that (86) *175 of 2,200* (87) *22 of 2,200* (88) *175 of 800* (89) *total television households* (90) *of all households using television* are tuned to your station. ***(Fill in two bubbles.)***

4. All rating services use (91) *audience samples* (92) *total populations* as a basis for their figures.

5. Share figures are usually (93) *higher* (94) *lower* than rating figures.

1	○ 79	○ 80	
2	○ 81	○ 82	○ 83
	○ 84	○ 85	
3	○ 86	○ 87	○ 88
	○ 89	○ 90	
4	○ 91	○ 92	
5	○ 93	○ 94	

SECTION TOTAL

REVIEW QUIZ

*Mark the following statements as true or false by filling in the bubbles in the **T** (for true) or **F** (for false) column.*

1. Normally, it is the writer who establishes the initial production process.
2. The facilities request for a specific production should contain equipment and technical facilities.
3. The two major criteria for evaluating program ideas are *Is it doable?* and *How much does it cost?*
4. A show treatment is necessary only for television documentaries.
5. CDs sold in record stores are in the public domain, so you can use them for television productions without securing copyright clearance.
6. Budgets must include expenses for preproduction, production, and all postproduction activities as well as personnel.
7. Whereas the budget is essential for a program proposal, a description of the target audience is not.
8. Broadcast unions include technical personnel only.
9. Demographic descriptors help define the target audience.
10. Once you have generated a worthwhile message, the producer can leave the day-to-day production details to the PA.
11. The producer works only with nontechnical personnel.
12. Because production is primarily a creative activity, any type of production system would prove counterproductive.
13. The line producer is responsible primarily for budgets.
14. A time line and a production schedule are the same thing.
15. Two of the important items in an effective program proposal are a treatment and a description of the target audience.

	T	F
1	○ 95	○ 96
2	○ 97	○ 98
3	○ 99	○ 100
4	○ 101	○ 102
5	○ 103	○ 104
6	○ 105	○ 106
7	○ 107	○ 108
8	○ 109	○ 110
9	○ 111	○ 112
10	○ 113	○ 114
11	○ 115	○ 116
12	○ 117	○ 118
13	○ 119	○ 120
14	○ 121	○ 122
15	○ 123	○ 124

SECTION TOTAL

Course No. ______________ Date ______________ Name ______________

PROBLEM-SOLVING APPLICATIONS

1. The art director asks you, the producer, whether her floor plan will allow optimal camera traffic. Are you the right person to answer this question? If so, why? If not, who would be the appropriate person to answer this question?

2. Your new comedy series is shot multicamera-style in the studio. You intend to video-record the dress rehearsal and the uninterrupted live-recorded show for later on-air scheduling. The production manager suggests that you prepare a budget that includes a generous amount of money for postproduction editing. Do you agree with the production manager? If so, why? If not, why not?

3. Write an effective program proposal for one or more of the following ideas. The proposal should include these points: (1) program title, (2) target audience, (3) process message (objective), (4) show treatment, (5) ideal program time and broadcast or other distribution channel, and (6) tentative budget.

    a. a series of shows about the effects of television on children

    b. a weekly fashion show

    c. a 10-week series about how to preserve water

    d. a three-show series about your favorite sport

    e. a five-show series for seventh- and eighth-graders about the dangers of drugs

    f. a 10-part series about human dignity and happiness

    g. a 5-part mini-documentary series about road rage and safe driving

    h. a 10-part series about the life and the work of a classical composer or a contemporary rock musician

    i. a 10-part series about the life and the work of your favorite sports figure

4. The local high-school video club has produced a music video, using magazine pictures that are synchronized with the latest recording of a rock band. The students plead with you to persuade the local cable company to put it on the air. What concerns, if any, do you have about airing this video recording? What can you do to accommodate the group's request?

5. The art director claims that she never received the director's e-mail about a set for the upcoming studio show of a dance recital. What simple steps would you suggest to remedy such problems?

Course No. ____________ Date ____________ Name ____________

The Script

REVIEW OF KEY TERMS

Match each term with its appropriate definition by filling in the corresponding bubble.

1. **partial two-column A/V script**
2. **event order**
3. **goal-directed information**
4. **two-column A/V script**
5. **classical dramaturgy**
6. **show format**
7. **single-column drama script**
8. **fact sheet**

A. Lists the items that have to be shown on-camera and their main features

A ○ 1 ○ 2 ○ 3 ○ 4 ○ 5 ○ 6 ○ 7 ○ 8

B. Used to describe a show for which the dialogue is indicated but not completely written out

B ○ 1 ○ 2 ○ 3 ○ 4 ○ 5 ○ 6 ○ 7 ○ 8

C. Program content intended to be learned by viewers

C ○ 1 ○ 2 ○ 3 ○ 4 ○ 5 ○ 6 ○ 7 ○ 8

D. A list of routine show segments

D ○ 1 ○ 2 ○ 3 ○ 4 ○ 5 ○ 6 ○ 7 ○ 8

PAGE TOTAL ____

1. partial two-column A/V script	**3. goal-directed information**	**6. show format**
2. event order	**4. two-column A/V script**	**7. single-column drama script**
	5. classical dramaturgy	**8. fact sheet**

E. The lineup of event details

E ○1 ○2 ○3 ○4 ○5 ○6 ○7 ○8

F. Traditional format for dramatic television and motion pictures scripts

F ○1 ○2 ○3 ○4 ○5 ○6 ○7 ○8

G. Traditional script with audio information in the right column and video information in the left

G ○1 ○2 ○3 ○4 ○5 ○6 ○7 ○8

H. Traditional composition of a play

H ○1 ○2 ○3 ○4 ○5 ○6 ○7 ○8

PAGE TOTAL

SECTION TOTAL

Course No. ______________ Date ________________ Name ______________________________

REVIEW OF BASIC SCRIPT FORMATS

Select the correct answers and fill in the bubbles with the corresponding numbers.

1. Each of the following three figures (**a** through **c**) shows a script segment that contains some format errors. For each figure identify the specific format errors: (9) *unnecessary talent instructions* (10) *video or audio instructions in wrong column* (11) *incomplete dialogue* (12) *unnecessary camera instructions* (13) *nonessential and confusing information.* ***(Multiple answers are possible.)***

   a. fully scripted serial drama (excerpt only)

1a ○ 9 ○ 10 ○ 11 ○ 12 ○ 13

GARY'S OFFICE: DAY

GARY is working intensely at his computer and ignores two telephone calls, when KIM bursts cheerfully into his office.

CUT TO CAMERA 3 WHEN KIM ENTERS

KIM

Let's go for coffee.

GARY (not looking up)

Don't have time.

KIM

Oh, shucks, make time.

GARY

You seem to be in a good mood today.

KIM

I'm always in a good mood…

GARY

[Says something about having to finish the report]

KIM

[Tries to persuade GARY to pay more attention to her]

CUE GARY TO STAND UP AND CUT TO CAMERA 1 WHEN HE GETS UP

PAGE TOTAL

b. standard A/V script of brief feature story on the value of books (excerpt only)

1b ○ ○ ○ ○ ○
9 10 11 12 13

Agency	Hot Stuff	**Writer**	Mary Smart
Client	Papermill Creek Publishing	**Producer**	Maurice Smart
Project	Book Promotion	**Director**	Chul Heo
Title	Books Are Alive!	**Art Director**	Buzz Palmer
Subject		**Medium**	EFP HDTV
Job #	011	**Contact**	Smart
Code #	HZWB11	**Draft**	2

VIDEO	AUDIO
CU Becky Take camera 3	BECKY: No, books are certainly not dead. On the contrary, 20 percent more books were printed worldwide last year than in any previous year.
Fade in sound of printing presses	Cue Becky to go to bookcase and look at some books.
CU of Peter. Must look annoyed.	PETER: [Says something about digital storage being so much better than clumsy books.]
Sound of books being dropped	BECKY: CU of her dropping books. Well, I think you are an ignorant nerd. How many books do you own?
	PETER: None!
Sound of Becky laughing	

PAGE TOTAL

Course No. ________ Date ________ Name ________

c. fact sheet

1c ○ ○ ○ ○ ○
9 10 11 12 13

SHOW: Tech News
DATE: Aug. 20
HOST: Larry W.
PROPS: New Extreme Printer (operational)

1. New super color laser printer.

2. Medium shot of open printer. Zoom in on ink cartridges. Prints are permanent. Will outlast Grandma's chemical photos.

3. Pan right to operating panel. Easy, intuitive operation. Just follow the instructions.

4. CU of operation manual.

5. CU of Larry:

LARRY: Let's do some printing right now and see how easy it is.

[Larry plugs printer into his laptop.]

Larry [looks enthusiastic]: As you can see, all I needed to do is access the picture on my laptop and click the Print command.

6. And it is quiet. [Larry cups his ears.]

7. Extreme close-up of Larry.

LARRY: Be sure to take advantage of our introductory offer. But hurry! This once-in-a-lifetime opportunity expires on Thursday.

GO TO BLACK

PAGE TOTAL

SECTION TOTAL

REVIEW OF STORY STRUCTURE, CONFLICT, AND DRAMATURGY

Select the correct answers and fill in the bubbles with the corresponding numbers.

1. Select the most common four elements of the basic dramatic story structure: (14) *theme, plot, story, characters* (15) *theme, plot, characters, environment* (16) *plot, characters, action, environment.*

2. Read the following very brief treatment excerpts and indicate whether the conflicts are (17) *plot-based* or (18) *character-based.*

   a. A drunk driver goes through a stoplight and hits a car in the intersection. By chance both drivers end up in the same emergency room. They begin discussing the physical danger of driving under the influence. The discussion turns into an argument. One of the drivers crawls out of bed and tries to hit the other one. The nurse stops the fight just in time.

   b. A young doctor, who has been fascinated with Africa since elementary school, decides to switch from being a successful family care provider to an AIDS researcher at a San Francisco clinic. She finally can't bear the slow progress in the lab any longer and decides to make her first Africa trip to help stem the AIDS epidemic in Zwamumbu [fictitious name]. With the help of an international relief organization, she opens a clinic and, within a short time, has gained the respect and the love of hundreds of adults and children for the "miracles" she performs as a doctor. But during a political uprising of a neighboring warlord, the clinic is invaded and she is accidentally shot in the crossfire.

   c. A young doctor is summoned by a world health organization to deliver AIDS medicine to a Zwamumbu international health clinic. This is a high-risk mission because the president of Zwamumbu has prohibited any use of AIDS medication. His justification is one of denial. In his words: "We do not have AIDS in our country." After several scary moments, such as the search at airport customs, the doctor not only manages to deliver the medicine but also helps administer it to the children who are already HIV infected. The president's secret service people learn about her activity and get permission to assassinate her. Despite tight security at the clinic, the order is carried out successfully. Posing as an AIDS patient, the assassin confronts the young doctor and, repeating the president's statement, shoots her.

1	○ 14	○ 15	○ 16
2a	○ 17	○ 18	
2b	○ 17	○ 18	
2c	○ 17	○ 18	

PAGE TOTAL

Course No. ______________ Date ______________ Name ______________________________

3. Fill in the bubbles whose numbers correspond with the numbers in the diagram below, identifying the various principal developmental steps of a classical dramaturgy.

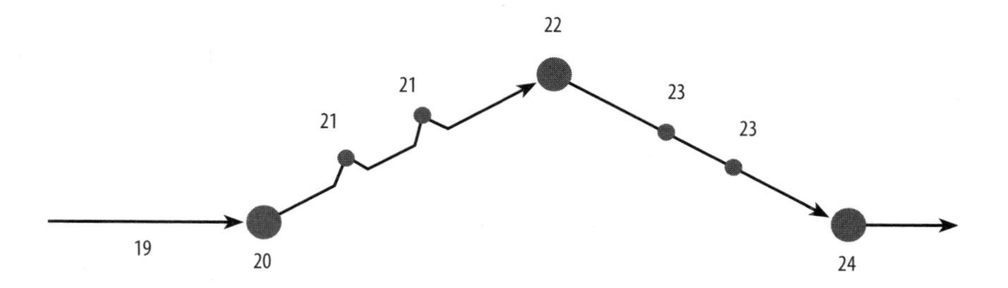

a. point of attack

3a ○ 19 ○ 20 ○ 21 ○ 22 ○ 23 ○ 24

b. climax

3b ○ 19 ○ 20 ○ 21 ○ 22 ○ 23 ○ 24

c. resolution

3c ○ 19 ○ 20 ○ 21 ○ 22 ○ 23 ○ 24

d. falling action and consequences of major crisis

3d ○ 19 ○ 20 ○ 21 ○ 22 ○ 23 ○ 24

e. exposition

3e ○ 19 ○ 20 ○ 21 ○ 22 ○ 23 ○ 24

f. rising action and additional conflicts

3f ○ 19 ○ 20 ○ 21 ○ 22 ○ 23 ○ 24

PAGE TOTAL

SECTION TOTAL

REVIEW QUIZ

*Mark the following statements as true or false by filling in the bubbles in the **T** (for true) or **F** (for false) column.*

1. In the resolution phase, the hero is always condemned.
2. The fact sheet and the show format have identical script formats.
3. The process message is especially useful for programs that contain primarily goal-directed information.
4. In a goal-directed program, the basic idea should lead to a precise process message.
5. The difference between a standard two-column A/V script format and a partial two-column A/V script is that the latter contains no video information.
6. In a classical dramaturgy, the point of attack marks the first major crisis.
7. The from-outside-in plots are primarily character-based.
8. A plot can develop from outside-in or from inside-out.
9. The rising as well as the falling action can include a number of crises.
10. The script is an effective communication device for all three production phases.
11. The standard two-column A/V script shows the video information on the left side and the audio information on the right.

	T	F
1	○ 25	○ 26
2	○ 27	○ 28
3	○ 29	○ 30
4	○ 31	○ 32
5	○ 33	○ 34
6	○ 35	○ 36
7	○ 37	○ 38
8	○ 39	○ 40
9	○ 41	○ 42
10	○ 43	○ 44
11	○ 45	○ 46

SECTION TOTAL

Course No. ______________ Date ______________ Name ______________________________

PROBLEM-SOLVING APPLICATIONS

1. Analyze three or four programs of a dramatic crime series and see whether the conflicts are primarily character-based, plot-based, or both. Justify your analyses.

2. How can you use plot to develop character?

3. Write a treatment of a one-hour special that includes all six elements of the classical dramaturgy.

4. Analyze a single program of a dramatic or comedy series and list all situations (verbal or action) that create an obvious conflict.

Course No. ______ Date ______ Name ______

The Director in Preproduction

REVIEW OF KEY TERMS

Match each term with its appropriate definition by filling in the corresponding bubble.

1. **time line**
2. **process message**
3. **facilities request**
4. **locking-in**
5. **sequencing**
6. **medium requirements**
7. **visualization**
8. **storyboard**
9. **production schedule**

A. A series of sketches of the key shots

A ○ 1 ○ 2 ○ 3 ○ 4 ○ 5 ○ 6 ○ 7 ○ 8 ○ 9

B. All content elements, production elements, and people needed to generate the defined process message

B ○ 1 ○ 2 ○ 3 ○ 4 ○ 5 ○ 6 ○ 7 ○ 8 ○ 9

C. The control and the structuring of a series of shots during editing

C ○ 1 ○ 2 ○ 3 ○ 4 ○ 5 ○ 6 ○ 7 ○ 8 ○ 9

D. An especially vivid mental image—visual or aural—during script analysis that determines the subsequent visualizations and sequencing

D ○ 1 ○ 2 ○ 3 ○ 4 ○ 5 ○ 6 ○ 7 ○ 8 ○ 9

PAGE TOTAL ______

1. **time line**
2. **process message**
3. **facilities request**
4. **locking-in**
5. **sequencing**
6. **medium requirements**
7. **visualization**
8. **storyboard**
9. **production schedule**

E. The mental image of a shot or several key images of a sequence

E ○1 ○2 ○3 ○4 ○5 ○6 ○7 ○8 ○9

F. The calendar that shows the preproduction, production, and postproduction dates and who is doing what, when, and where

F ○1 ○2 ○3 ○4 ○5 ○6 ○7 ○8 ○9

G. The message actually received by the viewer while watching a television program

G ○1 ○2 ○3 ○4 ○5 ○6 ○7 ○8 ○9

H. A list that contains all technical facilities needed for a specific production

H ○1 ○2 ○3 ○4 ○5 ○6 ○7 ○8 ○9

I. A breakdown of time blocks for various activities on the actual production day

I ○1 ○2 ○3 ○4 ○5 ○6 ○7 ○8 ○9

PAGE TOTAL

SECTION TOTAL

Course No. ____________ Date ______________ Name ______________________________

REVIEW OF PROCESS MESSAGE AND PRODUCTION METHOD

1. Evaluate to what extent the following four process messages will (10) *help* (11) *not help* you visualize key show elements and provide (12) *clear* (13) *only very few or no* clues to the various medium requirements. ***(Fill in two bubbles.)***

   a. The program should show racecar drivers.

   b. The program should make non–sports viewers admire, if not feel, the ballet-like skills of basketball players.

   c. The program should make people drive better.

   d. The program should show five different ways a family can save water during their morning shower and grooming.

   e. The program should demonstrate to the target audience (daily commuters) the benefits of turn signals and the consequences of ignoring them in rush-hour traffic.

   f. The program should help people save water.

   g. This program is a series of comedy shows.

   h. The program should help children learn about safety when walking to school.

1a	○ 10	○ 11
	○ 12	○ 13
1b	○ 10	○ 11
	○ 12	○ 13
1c	○ 10	○ 11
	○ 12	○ 13
1d	○ 10	○ 11
	○ 12	○ 13
1e	○ 10	○ 11
	○ 12	○ 13
1f	○ 10	○ 11
	○ 12	○ 13
1g	○ 10	○ 11
	○ 12	○ 13
1h	○ 10	○ 11
	○ 12	○ 13

PAGE TOTAL ______

2. A valuable process message should include (14) *a specific audience* (15) *specific production equipment* (16) *the intended effect on the audience.* **(Multiple answers are possible.)**

3. The translation of process message to video images is greatly aided by (17) *location scouting* (18) *a storyboard* (19) *a floor plan.*

4. Process message 1e suggests (20) *a series of location EFPs for extensive postproduction* (21) *a studio show* (22) *a single half-hour field pickup during which a videographer is riding with a highway patrol officer in rush-hour traffic.*

5. Look at process message 1b. It suggests (23) *a recorded live pickup of a professional basketball game* (24) *an EFP with staged plays on the basketball court* (25) *a live game restaged in the studio.*

2	○ 14	○ 15	○ 16
3	○ 17	○ 18	○ 19
4	○ 20	○ 21	○ 22
5	○ 23	○ 24	○ 25

PAGE TOTAL

SECTION TOTAL

Course No. ________ Date ________ Name ________

REVIEW OF SCRIPT MARKING

1. Match each field-of-view designation with its appropriate full term by filling in the bubble with the corresponding number.

(26) cross-shot	(29) extreme close-up	(32) close-up
(27) over-the-shoulder shot	(30) medium shot	(33) medium close-up
(28) long shot	(31) extreme long shot	(34) two-shot

a. 2-S

b. LS

c. O/S

d. ELS

e. MCU

f. X/S

g. CU

h. ECU

i. MS

1a ○ 26 ○ 27 ○ 28 ○ 29 ○ 30 ○ 31 ○ 32 ○ 33 ○ 34

1b ○ 26 ○ 27 ○ 28 ○ 29 ○ 30 ○ 31 ○ 32 ○ 33 ○ 34

1c ○ 26 ○ 27 ○ 28 ○ 29 ○ 30 ○ 31 ○ 32 ○ 33 ○ 34

1d ○ 26 ○ 27 ○ 28 ○ 29 ○ 30 ○ 31 ○ 32 ○ 33 ○ 34

1e ○ 26 ○ 27 ○ 28 ○ 29 ○ 30 ○ 31 ○ 32 ○ 33 ○ 34

1f ○ 26 ○ 27 ○ 28 ○ 29 ○ 30 ○ 31 ○ 32 ○ 33 ○ 34

1g ○ 26 ○ 27 ○ 28 ○ 29 ○ 30 ○ 31 ○ 32 ○ 33 ○ 34

1h ○ 26 ○ 27 ○ 28 ○ 29 ○ 30 ○ 31 ○ 32 ○ 33 ○ 34

1i ○ 26 ○ 27 ○ 28 ○ 29 ○ 30 ○ 31 ○ 32 ○ 33 ○ 34

PAGE TOTAL ________

Select the correct answers and fill in the bubbles with the corresponding numbers.

2. The shot sheet for (35) *camera 1* (36) *camera 2* is incorrect because it has (37) *consecutive shot numbers* (38) *discontinuous shot numbers* (39) *an unworkable shot sequence* (40) *insufficient information for the director.* **(Multiple answers are possible.)**

C1	
Shot #	
1	CU of Kim
2	Follow her
3	Zoom To ECU
4	Truck left
5	Zoom in To CU of Kim
6	Dolly out

C2	
Shot #	
9	CU of Gary
12	O/S Gary/Kim
16	Follow Kim
20	MS door; pick up Kim leaving
21	Follow Frank coming in

2 ○ 35 ○ 36

○ 37 ○ 38 ○ 39 ○ 40

PAGE TOTAL

3. Mark the following brief scene for a three-camera live-recorded studio production. Add any additional video cues you deem necessary. The scene takes place in the small office of a busy advertising executive. Draw a floor plan and prepare a shot sheet.

```
                         KIM
              (Bursts into Gary's office)

Let's go for coffee.

                         GARY

I don't have time.

                         KIM

Oh, shucks, make time.

                         GARY

You seem to be in a good mood today.

                         KIM

I'm always in a good mood...

                         GARY

Especially when I'm around.

                         KIM

I'm not so sure about that...but, yes, let's go.

                         GARY

I really don't...

                         KIM
(Walks behind Gary's desk and starts kissing his neck.)

Don't what?

                         GARY

Forget it. Let's go.

(The telephone rings. Gary turns to answer it, but then lets it
ring. He puts his arm around her. They both leave the office.)
```

4. Mark the following show opening of a series on basic video production. Memorize the cues so that you can devote your attention primarily to the preview monitors rather than to the script.

VIDEO BASICS SERIES
SHOW NO. 7
RECORDING DATE: July 15
AIR DATE: August 15

VIDEO	AUDIO
Opening Server 2 :08 sec	Music SOS (sound on source)
CU of Phil	PHIL Hi, I'm Phil Kipper. Welcome to the Broadcast and Electronic Communication Arts Series, "Video Basics." As promised last week, we will take you to a special room where magic takes place: the editing suite.
Pull out to reveal editing suite. Phil introduces Hamid. CU of Hamid.	PHIL Let me introduce to you the magician in charge, Hamid Khani, whose official title is senior postproduction editor.
2-shot	(SAYS HELLO TO HAMID AND HAS HAMID SAY HELLO TO THE AUDIENCE)

Course No. ______________ Date ______________ Name ______________________________

Select the correct answers and fill in the bubbles with the corresponding numbers.

5. The script markings in the following figure are (41) *acceptable* (42) *unacceptable* because they (43) *are too small* (44) *have unnecessary or redundant cues* (45) *are in the wrong place* (46) *show large, essential cues.* ***(Fill in two bubbles.)***

5 ○ 41 ○ 42
○ 43 ○ 44 ○ 45 ○ 46

JOHN

What's the matter?

Ready camera 1
Ready to cue Tammy

TAMMY

Nothing.

Cue Tammy and take camera 1

JOHN

What do you mean, "nothing"? I can feel something is wrong.

TAMMY

Ready to cue John
Ready to take camera 2

Well, I am glad you have some feeling left.

JOHN

Cue John and take camera 2

What's that supposed to mean?

TAMMY

Please, let's not start that again.

JOHN

Start what again?

TAMMY

Ready to take camera 3 for a two-shot

Well, I guess it's time to talk.

Take camera 3

JOHN

What do you think we have been doing all this time?

PAGE TOTAL

SECTION TOTAL

REVIEW OF INTERPRETING STORYBOARDS

1. Each of the following four storyboards shows one or several major problems. Fill in the bubbles whose numbers correspond with one or more of these major problems: (47) *poor continuity and disturbance of the mental map* (48) *wrong field-of-view designation* (49) *wrong above- or below-eye-level camera position.* ***(Note: Storyboards may exhibit more than one problem.)***

Storyboard a

1a ◯ 47 ◯ 48 ◯ 49

Storyboard b

1b ◯ 47 ◯ 48 ◯ 49

Storyboard c

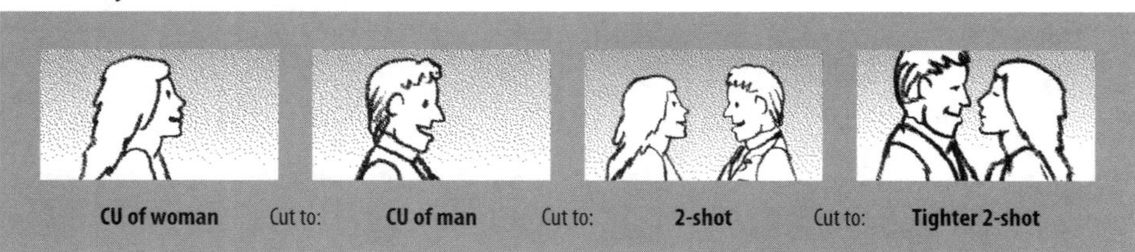

1c ◯ 47 ◯ 48 ◯ 49

Storyboard d

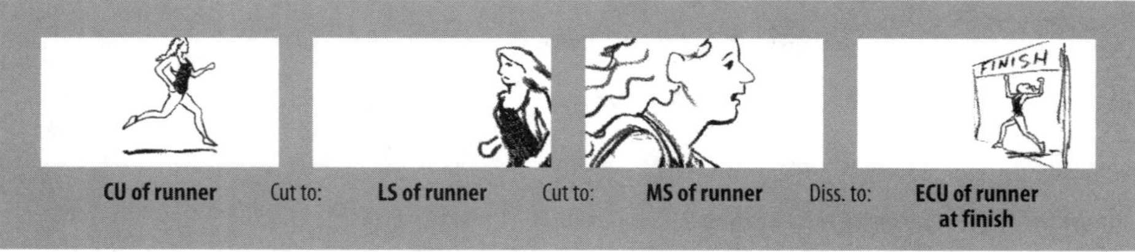

1d ◯ 47 ◯ 48 ◯ 49

SECTION TOTAL

Course No. ______ Date ______ Name ______

REVIEW OF SUPPORT STAFF

Identify the person mainly responsible for the following production activities and fill in the corresponding bubbles.

1. In elaborate multicamera productions or digital cinema, some of the scenes are sometimes directed by the (50) *floor manager* (51) *AD* (52) *PA.*
2. To put up and dress the new lawyer's set is the responsibility of the (53) *floor manager* (54) *art director* (55) *PA.*
3. In the absence of a property manager, props are usually handled by the (56) *floor manager* (57) *AD* (58) *art director.*
4. The novice news anchor would like to have the basic cues demonstrated. The cues should be demonstrated on the studio floor by the (59) *PA* (60) *floor manager* (61) *director.*
5. As the director, you want somebody to write down all major and minor problems that show up during rehearsal. For this job you would most likely ask the (62) *floor manager* (63) *producer* (64) *PA.*

1	○ 50	○ 51	○ 52
2	○ 53	○ 54	○ 55
3	○ 56	○ 57	○ 58
4	○ 59	○ 60	○ 61
5	○ 62	○ 63	○ 64

SECTION TOTAL ______

REVIEW OF TIME LINE

Select the correct answers and fill in the bubbles with the corresponding numbers.

1. Each of these time-line listings is (65) *acceptable* (66) *unacceptable,* because (67) *the time allotted for this production activity is appropriate* (68) *it allots too much time for the specific production activity* (69) *it allots too little time for the specific production activity.* ***(Fill in two bubbles.)***

   a. **Time Line May 25:** Live-recorded 20-minute interview with college president on a standard interview set. She brings a model of the new library building, which needs to be set up in an adjacent area.

(1)	8:15 a.m.	Crew call	**1a** (1) ○ 65 ○ 66 ○ 67 ○ 68 ○ 69
(2)	8:30–9:00 a.m.	Tech meeting	**1a** (2) ○ 65 ○ 66 ○ 67 ○ 68 ○ 69
(3)	9:00–11:00 a.m.	Setup and lighting	**1a** (3) ○ 65 ○ 66 ○ 67 ○ 68 ○ 69
(4)	11:00–11:30 a.m.	Lunch	**1a** (4) ○ 65 ○ 66 ○ 67 ○ 68 ○ 69
(5)	11:30–11:45 a.m.	Notes and reset	**1a** (5) ○ 65 ○ 66 ○ 67 ○ 68 ○ 69
(6)	11:45 a.m.–12:00 p.m.	Briefing of president (Green Room)	**1a** (6) ○ 65 ○ 66 ○ 67 ○ 68 ○ 69
(7)	12:00–12:30 p.m.	Run-through and camera rehearsal	**1a** (7) ○ 65 ○ 66 ○ 67 ○ 68 ○ 69

PAGE TOTAL

Course No. ____________ Date ____________ Name ____________

(8)	12:30–12:45 p.m.	Notes	**1a** (8) ○ 65 ○ 66 ○ 67 ○ 68 ○ 69
(9)	12:45–1:00 p.m.	Reset	**1a** (9) ○ 65 ○ 66 ○ 67 ○ 68 ○ 69
(10)	1:00–1:10 p.m.	Break	**1a** (10) ○ 65 ○ 66 ○ 67 ○ 68 ○ 69
(11)	1:10–1:45 p.m.	Record	**1a** (11) ○ 65 ○ 66 ○ 67 ○ 68 ○ 69
(12)	1:45–1:55 p.m.	Spill	**1a** (12) ○ 65 ○ 66 ○ 67 ○ 68 ○ 69
(13)	1:55–2:10 p.m.	Strike	**1a** (13) ○ 65 ○ 66 ○ 67 ○ 68 ○ 69

b. Time Line June 2: Multicamera shoot for postproduction of two songs by a local rock group.

(1)	6:00 a.m.	Crew call	**1b** (1) ○ 65 ○ 66 ○ 67 ○ 68 ○ 69
(2)	6:15–6:35 a.m.	Tech meeting	**1b** (2) ○ 65 ○ 66 ○ 67 ○ 68 ○ 69
(3)	6:35–7:00 a.m.	Setup and lighting	**1b** (3) ○ 65 ○ 66 ○ 67 ○ 68 ○ 69

PAGE TOTAL ____

Each of these time-line listings is (65) *acceptable* (66) *unacceptable,* because (67) *the time allotted for this production activity is appropriate* (68) *it allots too much time for the specific production activity* (69) *it allots too little time for the specific production activity.* ***(Fill in two bubbles.)***

(4) 7:00–9:30 a.m. Production meeting

1b (4) ◯ 65 ◯ 66 ◯ 67 ◯ 68 ◯ 69

(5) 9:30 a.m.–12:30 p.m. First run-through with cameras

1b (5) ◯ 65 ◯ 66 ◯ 67 ◯ 68 ◯ 69

(6) 12:30–2:30 p.m. Lunch

1b (6) ◯ 65 ◯ 66 ◯ 67 ◯ 68 ◯ 69

(7) 2:30–2:45 p.m. Record first song

1b (7) ◯ 65 ◯ 66 ◯ 67 ◯ 68 ◯ 69

(8) 2:45–3:30 p.m. Notes and reset

1b (8) ◯ 65 ◯ 66 ◯ 67 ◯ 68 ◯ 69

(9) 3:30–5:00 p.m. Record second song

1b (9) ◯ 65 ◯ 66 ◯ 67 ◯ 68 ◯ 69

(10) 5:00–6:00 p.m. Strike

1b (10) ◯ 65 ◯ 66 ◯ 67 ◯ 68 ◯ 69

PAGE TOTAL

SECTION TOTAL

Course No. ______ Date ______ Name ______

REVIEW QUIZ

*Mark the following statements as true or false by filling in the bubbles in the **T** (for true) or **F** (for false) column.*

#	Statement	T	F
1.	The time line is the responsibility of the PA.	○ 70	○ 71
2.	The drama script format requires the full dialogue of all actors but only a minimum of visualization cues.	○ 72	○ 73
3.	A good floor plan will greatly facilitate camera and talent blocking.	○ 74	○ 75
4.	Proper visualization is essential for correct sequencing.	○ 76	○ 77
5.	Experienced floor managers will cue on their own if they think the director has missed a cue.	○ 78	○ 79
6.	Script marking is important when directing from a dramatic script and when directing from a two-column A/V script.	○ 80	○ 81
7.	Dramas are always fully scripted.	○ 82	○ 83
8.	If the script marking simply indicates "(2)" for one shot and "(3)" for the next, it implies that no "Ready" cues need be given.	○ 84	○ 85
9.	In a complex multicamera show, the AD will normally give stand-by cues.	○ 86	○ 87
10.	A storyboard shows the key visualization points of an event.	○ 88	○ 89
11.	When preparing camera shot sheets, the shots for each camera are listed in the order they appear in the script.	○ 90	○ 91
12.	Although the process message is important to the director in the production phase, it is relatively unimportant in preproduction.	○ 92	○ 93
13.	A good storyboard helps the director visualize a shot and determine camera positions.	○ 94	○ 95
14.	In a properly scripted documentary, all audio information is on page-left and all video information is on page-right.	○ 96	○ 97
15.	Because the director is engaged in artistic activities, knowledge of technical production aspects is relatively unimportant.	○ 98	○ 99
16.	The locking-in point means that you conjure up a vivid visual or aural image while analyzing the script.	○ 100	○ 101

SECTION TOTAL ______

PROBLEM-SOLVING APPLICATIONS

1. The director of a weekly live sports show (consisting of a host and a prominent guest) tells the producer that she does not need a detailed script but that a show format will do just fine. What is your reaction?

2. When asked to direct an on-location television adaptation of a hit stage play, you are advised that the director of the play will determine the number and the positions of the cameras because he, after all, knows the stage blocking better than you do. What is your reaction? What would you suggest?

3. The director of a live-recorded segment of a new situation comedy tells you, the producer, that she has great difficulty deciding on optimal camera positions and marking the script because the art director has not yet finished the floor plan. What is your reaction? What would you suggest?

4. While you, the director, are preparing an EFP of a documentary segment on the lumber industry, the producer tells you not to worry too much about shot continuity because he intends to put the show together in extensive postproduction editing. Do you agree with the producer? If so, why? If not, why not?

5. The novice director proudly shows you, the producer, his marked show format for a live-recorded studio interview. He wrote out in longhand all the ready and take cues as well as all the cues for special effects. His writing takes up more space than the information of the show format. What is your reaction? Why?

6. If you were to describe to the producer what "locking-in" means when reading a dramatic script, what would you tell him?

7. The director of a number of successful digital movies tells you that "hearing" a shot can sometimes help the visualization process more than trying to "see" it. What does he mean by that?

8. Mark two or three scenes of a dramatic script, first for a three-camera live-recorded studio production and then for a single-camera studio production. Note the differences.

9. Prepare time lines for two studio productions and two EFPs.

10. Observe the scene while riding on a bus or train, waiting in line at an airport, eating lunch in a cafeteria, or sitting in a classroom listening to a lecture. How would you re-create and intensify one or all of these scenes for a multicamera or single-camera production?

Production

Course No. ______________ Date ______________ Name ______________

Analog and Digital Television

REVIEW OF KEY TERMS

Match each term with its appropriate definition by filling in the corresponding bubble.

1. **refresh rate**
2. **sampling**
3. **interlaced scanning**
4. **720p**
5. **compression**
6. **downloading**
7. **analog**
8. **frame**
9. **field**
10. **codec**
11. **HDTV**
12. **1080i**
13. **RGB**
14. **progressive scanning**
15. **streaming**
16. **aspect ratio**

A. A television standard with at least twice the picture detail of standard television

B. Delivering and receiving digital video as continuous data

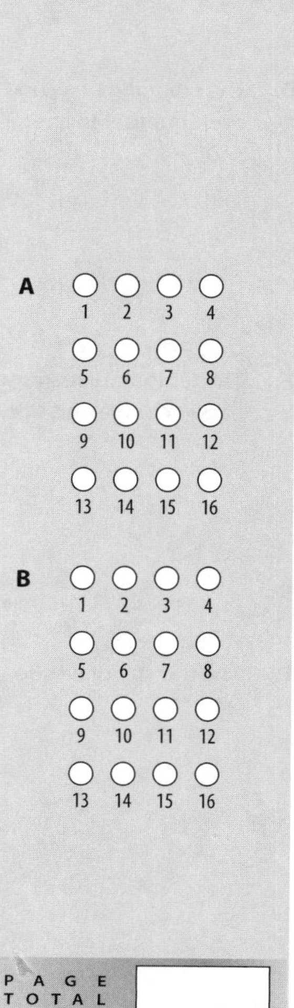

PAGE TOTAL 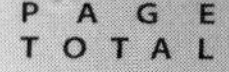

1. **refresh rate**
2. **sampling**
3. **interlaced scanning**
4. **720p**
5. **compression**
6. **downloading**
7. **analog**
8. **frame**
9. **field**
10. **codec**
11. **HDTV**
12. **1080i**
13. **RGB**
14. **progressive scanning**
15. **streaming**
16. **aspect ratio**

C. A system in which the electron beam starts scanning line 1, then line 2, then line 3, and so forth until all lines are scanned

C ○ 1 ○ 2 ○ 3 ○ 4 ○ 5 ○ 6 ○ 7 ○ 8 ○ 9 ○ 10 ○ 11 ○ 12 ○ 13 ○ 14 ○ 15 ○ 16

D. The scanning of all odd-numbered scanning lines and the subsequent scanning of all even-numbered lines

D ○ 1 ○ 2 ○ 3 ○ 4 ○ 5 ○ 6 ○ 7 ○ 8 ○ 9 ○ 10 ○ 11 ○ 12 ○ 13 ○ 14 ○ 15 ○ 16

E. The temporary rearrangement or elimination of redundant picture information for easier storage and signal transport

E ○ 1 ○ 2 ○ 3 ○ 4 ○ 5 ○ 6 ○ 7 ○ 8 ○ 9 ○ 10 ○ 11 ○ 12 ○ 13 ○ 14 ○ 15 ○ 16

F. The transfer of files sent in data packets

F ○ 1 ○ 2 ○ 3 ○ 4 ○ 5 ○ 6 ○ 7 ○ 8 ○ 9 ○ 10 ○ 11 ○ 12 ○ 13 ○ 14 ○ 15 ○ 16

PAGE TOTAL

Course No. ______ Date ______ Name ______

G. One-half of a complete scanning cycle

H. A complete interlaced scanning cycle

I. The basic colors of television

J. Progressive HDTV scanning system

K. A two-field interlaced scanning system for HDTV

G
1 2 3 4
5 6 7 8
9 10 11 12
13 14 15 16

H
1 2 3 4
5 6 7 8
9 10 11 12
13 14 15 16

I
1 2 3 4
5 6 7 8
9 10 11 12
13 14 15 16

J
1 2 3 4
5 6 7 8
9 10 11 12
13 14 15 16

K
1 2 3 4
5 6 7 8
9 10 11 12
13 14 15 16

PAGE TOTAL

1. refresh rate	**7. analog**	**12. 1080i**
2. sampling	**8. frame**	**13. RGB**
3. interlaced scanning	**9. field**	**14. progressive scanning**
4. 720p	**10. codec**	**15. streaming**
5. compression	**11. HDTV**	**16. aspect ratio**
6. downloading		

L. Selecting from an analog system a great many equally spaced signal values

L ○1 ○2 ○3 ○4 ○5 ○6 ○7 ○8 ○9 ○10 ○11 ○12 ○13 ○14 ○15 ○16

M. The width-to-height proportion of a television screen

M ○1 ○2 ○3 ○4 ○5 ○6 ○7 ○8 ○9 ○10 ○11 ○12 ○13 ○14 ○15 ○16

N. Stands for compressing and decompressing digital data

N ○1 ○2 ○3 ○4 ○5 ○6 ○7 ○8 ○9 ○10 ○11 ○12 ○13 ○14 ○15 ○16

O. The number of complete frames per second

O ○1 ○2 ○3 ○4 ○5 ○6 ○7 ○8 ○9 ○10 ○11 ○12 ○13 ○14 ○15 ○16

PAGE TOTAL

Course No. ______ Date ______ Name ______

P. Signal that fluctuates like the original stimulus

P			
◯ 1	◯ 2	◯ 3	◯ 4
◯ 5	◯ 6	◯ 7	◯ 8
◯ 9	◯ 10	◯ 11	◯ 12
◯ 13	◯ 14	◯ 15	◯ 16

PAGE TOTAL

SECTION TOTAL

REVIEW OF ANALOG AND DIGITAL TELEVISION

Select the correct answers and fill in the bubbles with the corresponding numbers.

1. Digital television (17) *must use progressive scanning* (18) *must use interlaced scanning* (19) *can use either of the two systems.*

2. The four major steps of digitizing an analog signal are (20) *compression* (21) *analyzing* (22) *aliasing* (23) *anti-aliasing* (24) *quantizing* (25) *sampling* (26) *coding* (27) *digital-to-analog conversion* (28) *scanning.* ***(Fill in four bubbles.)***

3. The aspect ratio for wide-screen HDTV is (29) *4 × 3* (30) *16 × 9* (31) *16 × 4.*

4. The standard television (STV) format is (32) *4 × 3* (33) *9 × 3* (34) *16 × 9.*

5. Standard NTSC television uses a (35) *progressive* (36) *interlaced* (37) *compressed* scanning system.

6. One of the major advantages of digital television is that (38) *it always uses interlaced scanning* (39) *its picture does not deteriorate over numerous generations* (40) *it still uses analog signals.*

7. The two operational DTV systems are (41) *1080i* (42) *1280p* (43) *720p.* ***(Fill in two bubbles.)***

8. Which of the diagrams below represents most appropriately a digital signal?

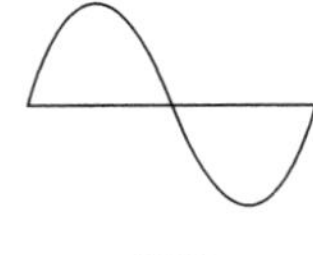

(44)

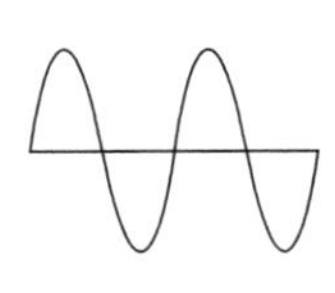

(45)

(46)

1	○ 17	○ 18	○ 19		
2	○ 20	○ 21	○ 22	○ 23	○ 24
	○ 25	○ 26	○ 27	○ 28	
3	○ 29	○ 30	○ 31		
4	○ 32	○ 33	○ 34		
5	○ 35	○ 36	○ 37		
6	○ 38	○ 39	○ 40		
7	○ 41	○ 42	○ 43		
8	○ 44	○ 45	○ 46		

SECTION TOTAL

Course No. ______ Date ______ Name ______

REVIEW OF BASIC IMAGE CREATION AND THE COLORS OF THE VIDEO DISPLAY

Select the correct answers and fill in the bubbles with the corresponding numbers.

1. The basic television image is created by activating pixels that are (47) *arranged in a stack of vertical and horizontal lines* (48) *arranged in vertical columns only* (49) *randomly distributed.*

2. All the colors that you see on the television screen are a mixture of (50) *red, green, and blue* (51) *red, green, and yellow* (52) *red, blue, and yellow.*

3. The scanning can be (53) *only interlaced* (54) *only progressive* (55) *either interlaced or progressive.*

4. In progressive scanning, each scanning cycle produces (56) *a field* (57) *a frame* (58) *an interlaced field.*

5. The scanning of a complete NTSC frame takes (59) $\frac{1}{30}$ (60) $\frac{1}{60}$ (61) $\frac{1}{20}$ second.

6. HDTV can use (62) *interlaced or progressive scanning* (63) *only interlaced scanning* (64) *only progressive scanning.*

7. The 4K in digital cinema means that (65) *each horizontal line consists of 4,000 or more pixels* (66) *there are 4,000 potential colors* (67) *there are 4,000 frames per second.*

8. The refresh rate in progressive scanning (68) *is fixed at 30 fps* (69) *is fixed at 60 fps* (70) *can be variable.*

9. All other factors being equal, the highest-quality HDTV pictures are delivered by the (71) *480p system* (72) *720p system* (73) *1080i system.*

10. In progressive scanning (74) *only the even-numbered lines are scanned* (75) *only the odd-numbered lines are scanned* (76) *each line is scanned in a top-to-bottom sequence.*

1	○ 47	○ 48	○ 49
2	○ 50	○ 51	○ 52
3	○ 53	○ 54	○ 55
4	○ 56	○ 57	○ 58
5	○ 59	○ 60	○ 61
6	○ 62	○ 63	○ 64
7	○ 65	○ 66	○ 67
8	○ 68	○ 69	○ 70
9	○ 71	○ 72	○ 73
10	○ 74	○ 75	○ 76

SECTION TOTAL ______

REVIEW QUIZ

*Mark the following statements as true or false by filling in the bubbles in the **T** (for true) or **F** (for false) column.*

1. Analog recordings can tolerate more tape generations without noticeable loss than digital recordings.
2. All flat-panel displays use LCD technology.
3. RGB are the basic primary colors of analog as well as digital television.
4. All digital video signals must be compressed before they can be recorded.
5. The 16 × 9 aspect ratio is especially advantageous for showing an ECU of a face.
6. Assigning 0's and 1's to the sampled signal is part of anti-aliasing.
7. Digital signals are more robust but less complete than analog signals.
8. A codec is a specific compression system.
9. Downloading allows you to view continuous images and sound while the process is ongoing.
10. In the digital process, sampling must precede quantizing.

	T	F
1	○ 77	○ 78
2	○ 79	○ 80
3	○ 81	○ 82
4	○ 83	○ 84
5	○ 85	○ 86
6	○ 87	○ 88
7	○ 89	○ 90
8	○ 91	○ 92
9	○ 93	○ 94
10	○ 95	○ 96

SECTION TOTAL

Course No. ____________ Date ______________ Name ______________________________

PROBLEM-SOLVING APPLICATIONS

1. Your editor tells you that, contrary to sampling, which is an essential step in the digitization process, compression is not. Do you agree with the editor? If so, why? If not, why not?

2. The same editor insists on all-digital equipment with as high a sampling ratio and as little compression as possible because your projects require extensive postproduction with a great number of complex effects. What is your reaction? Why?

3. Your friend, an ardent movie fan, is extremely happy about the new 16 × 9 aspect ratio because, according to him, it is especially well suited to playing back old movies from the 1920s. Do you share his enthusiasm? If so, why? If not, why not?

4. Your organization intends to deliver video content via the Internet. Some members of the organization want the content streamed; others think that downloading is a better choice. List a few justifications for each argument.

5. The salesperson in a television store tells you that a digital television receiver can change the scanning standard, regardless of how the signals were originally sent. Is she correct? If not, why not?

6. You have been asked to select one of the HDTV systems (720p or 1080i) for your television studio. Describe each and justify why you would choose one over the other.

Course No. ______ Date ______ Name ______

The Television Camera

REVIEW OF KEY TERMS

Match each term with its appropriate definition by filling in the corresponding bubble.

1. **camera control unit**
2. **camera chain**
3. **HDTV camera**
4. **resolution**
5. **pixel**
6. **sync generator**
7. **white balance**
8. **imaging device**
9. **beam splitter**
10. **standard television**
11. **high-definition video**
12. **hue**
13. **saturation**
14. **brightness**

A. The television system based on NTSC scanning

A ○ 1 ○ 2 ○ 3 ○ 4 ○ 5 ○ 6 ○ 7 ○ 8 ○ 9 ○ 10 ○ 11 ○ 12 ○ 13 ○ 14

B. How much light a color reflects; how dark or light a color appears on a black-and-white television screen

B ○ 1 ○ 2 ○ 3 ○ 4 ○ 5 ○ 6 ○ 7 ○ 8 ○ 9 ○ 10 ○ 11 ○ 12 ○ 13 ○ 14

C. Prism within a camera that separates white light into the three primary colors

C ○ 1 ○ 2 ○ 3 ○ 4 ○ 5 ○ 6 ○ 7 ○ 8 ○ 9 ○ 10 ○ 11 ○ 12 ○ 13 ○ 14

PAGE TOTAL

1. camera control unit	**6. sync generator**	**11. high-definition video**
2. camera chain	**7. white balance**	**12. hue**
3. HDTV camera	**8. imaging device**	**13. saturation**
4. resolution	**9. beam splitter**	**14. brightness**
5. pixel	**10. standard television**	

D. The relative sharpness of the picture as measured by number of pixels

D ○1 ○2 ○3 ○4 ○5 ○6 ○7 ○8 ○9 ○10 ○11 ○12 ○13 ○14

E. The camera connected with the CCU, power supply, and sync generator

E ○1 ○2 ○3 ○4 ○5 ○6 ○7 ○8 ○9 ○10 ○11 ○12 ○13 ○14

F. The color itself (such as red or yellow)

F ○1 ○2 ○3 ○4 ○5 ○6 ○7 ○8 ○9 ○10 ○11 ○12 ○13 ○14

G. Part of the camera chain; produces electronic synchronization signal

G ○1 ○2 ○3 ○4 ○5 ○6 ○7 ○8 ○9 ○10 ○11 ○12 ○13 ○14

H. A unit, separate from the camera, that is used to process signals coming from and going to the camera to ensure optimal television pictures

H ○1 ○2 ○3 ○4 ○5 ○6 ○7 ○8 ○9 ○10 ○11 ○12 ○13 ○14

PAGE TOTAL

Course No. ______ Date ______ Name ______

I. Color strength or richness

I ○1 ○2 ○3 ○4 ○5 ○6 ○7 ○8 ○9 ○10 ○11 ○12 ○13 ○14

J. Adjusting color circuits in a camera to produce a white color in lighting of various color temperatures

J ○1 ○2 ○3 ○4 ○5 ○6 ○7 ○8 ○9 ○10 ○11 ○12 ○13 ○14

K. A video camera that delivers superior resolution, color, and contrast

K ○1 ○2 ○3 ○4 ○5 ○6 ○7 ○8 ○9 ○10 ○11 ○12 ○13 ○14

L. The smallest single imaging element

L ○1 ○2 ○3 ○4 ○5 ○6 ○7 ○8 ○9 ○10 ○11 ○12 ○13 ○14

M. The television system that produces high-resolution pictures that are highly compressed

M ○1 ○2 ○3 ○4 ○5 ○6 ○7 ○8 ○9 ○10 ○11 ○12 ○13 ○14

N. The sensor mechanism in a camera that changes light into electrical energy

N ○1 ○2 ○3 ○4 ○5 ○6 ○7 ○8 ○9 ○10 ○11 ○12 ○13 ○14

PAGE TOTAL ______

SECTION TOTAL ______

REVIEW OF BASIC CAMERA ELEMENTS AND FUNCTIONS

Select the correct answers and fill in the bubbles with the corresponding numbers.

1. The three basic parts of the camera are (15) *tally light* (16) *viewfinder* (17) *VR* (18) *imaging device, or sensor* (19) *lens* (20) *pedestal.* ***(Fill in three bubbles.)***

2. Fill in the bubbles whose numbers correspond with the camera elements shown in the following figure.

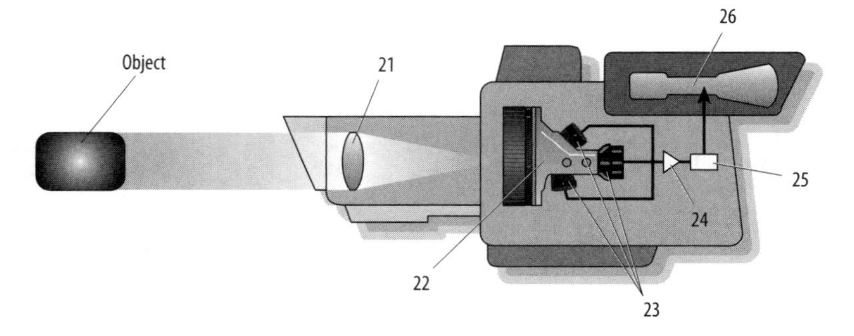

   a. gathers and transmits the light

   b. splits the white light into red, green, and blue light beams

   c. processes video signal

   d. converts signals back into visible screen images

   e. amplifies video signals

   f. transforms light into electric energy or video signals

1	○ 15	○ 16	○ 17
	○ 18	○ 19	○ 20
2a	○ 21	○ 22	○ 23
	○ 24	○ 25	○ 26
2b	○ 21	○ 22	○ 23
	○ 24	○ 25	○ 26
2c	○ 21	○ 22	○ 23
	○ 24	○ 25	○ 26
2d	○ 21	○ 22	○ 23
	○ 24	○ 25	○ 26
2e	○ 21	○ 22	○ 23
	○ 24	○ 25	○ 26
2f	○ 21	○ 22	○ 23
	○ 24	○ 25	○ 26

PAGE TOTAL

Course No. ____________ Date ____________ Name ____________

3. The camera imaging device is also called the (27) *ND filter* (28) *SD card* (29) *sensor.*

4. One of the following names describes a specific sensor: (30) *CCU* (31) *CMOS* (32) *chip.*

5. In some cameras the prism block is replaced by a (33) *CCD* (34) *lens* (35) *striped or mosaic filter array.*

6. The three basic parts of the camera chain are the (36) *power supply* (37) *prism block* (38) *lens* (39) *sync generator* (40) *camera head* (41) *viewfinder.* ***(Fill in three bubbles.)***

7. To adjust a studio camera to produce optimal pictures, the VO must operate the (42) *CCU* (43) *CCD* (44) *HDV.*

8. An 8K digital cinema camera refers to (45) *a specific color temperature* (46) *the number of pixels on each horizontal scanning line* (47) *the number of vertical scanning lines.*

3	○ 27	○ 28	○ 29
4	○ 30	○ 31	○ 32
5	○ 33	○ 34	○ 35
6	○ 36	○ 37	○ 38
	○ 39	○ 40	○ 41
7	○ 42	○ 43	○ 44
8	○ 45	○ 46	○ 47

PAGE TOTAL

SECTION TOTAL

REVIEW OF ELECTRONIC OPERATIONAL FEATURES

Select the correct answers and fill in the bubbles with the corresponding numbers.

1. Most HD cameras let you switch between the aspect ratios of (48) *4 × 9 and 16 × 9* (49) *4 × 9 and 16 × 3* (50) *4 × 3 and 16 × 9.*

2. Most cameras allow you to switch between these audio standards: (51) *16 bit and 32 kHz* (52) *32 kHz and 48 kHz* (53) *16 bit and 48 kHz.*

3. Long-distance camera cables commonly used for transporting digital signals are (54) *IEEE 1394 and HDMI* (55) *FireWire and USB3* (56) *triax and fiber-optic.*

4. XLR connectors are used for (57) *digital video cables only* (58) *audio and video cables* (59) *audio cables only.*

5. A focus-assist feature is (60) *especially important for focusing small STV camcorders* (61) *especially important for focusing HDTV cameras* (62) *identical to auto-focus.*

1	○ 48	○ 49	○ 50
2	○ 51	○ 52	○ 53
3	○ 54	○ 55	○ 56
4	○ 57	○ 58	○ 59
5	○ 60	○ 61	○ 62

SECTION TOTAL

Course No. ____________ Date ____________ Name ____________

REVIEW OF RESOLUTION, CONTRAST, AND COLOR

Select the correct answers and fill in the bubbles with the corresponding numbers.

1. Spatial resolution is determined by the number of (63) *scanning lines and pixels per line* (64) *frames per second* (65) *frame rate.*

2. Temporal resolution is determined by the (66) *number of frames per second* (67) *relative speed of a moving object* (68) *number of horizontal lines.*

3. In video "shading" means (69) *creating falloff electronically* (70) *adjusting the lighting for more shadows* (71) *controlling contrast.*

4. Contrast is a function of (72) *hue* (73) *saturation* (74) *brightness.*

5. The primary additive colors are (75) *LED* (76) *RGB* (77) *CMY.*

6. Television uses (78) *additive color mixing* (79) *color filter mixing* (80) *subtractive color mixing.*

7. The television color signal is processed as (81) *chrominance and color channels* (82) *luminance and chrominance* (83) *luminance and brightness.*

8. White-balancing adjusts the (84) *C channel* (85) *Y channel* (86) *z-axis.*

9. The C signal is responsible for the (87) *brightness* (88) *color* (89) *saturation* information.

10. The Y signal is responsible for the (90) *brightness* (91) *color* (92) *saturation* information.

1	○ 63	○ 64	○ 65
2	○ 66	○ 67	○ 68
3	○ 69	○ 70	○ 71
4	○ 72	○ 73	○ 74
5	○ 75	○ 76	○ 77
6	○ 78	○ 79	○ 80
7	○ 81	○ 82	○ 83
8	○ 84	○ 85	○ 86
9	○ 87	○ 88	○ 89
10	○ 90	○ 91	○ 92

SECTION TOTAL

REVIEW QUIZ

*Mark the following statements as true or false by filling in the bubbles in the **T** (for true) or **F** (for false) column.*

1. White-balancing adjusts the Y signals.
2. Digital data can be transferred by FireWire or with an HDMI cable.
3. An XLR plug is an audio connector.
4. Generally, studio cameras have higher-quality lenses than ENG camcorders.
5. A 720p scan always has a higher temporal resolution than a 480p scan.
6. You can use a neutral density filter to reduce the intensity of bright light.
7. RCA phono connectors can be used for digital video as well as audio signals.
8. A CMOS chip is similar in function to a CCD.
9. The camera sensor transduces (changes) electrical energy into light.
10. An S-video cable does not carry audio signals.
11. The sensor of a digital cinema camera normally is higher density than that of an HDTV camera.

	T	F
1	○ 93	○ 94
2	○ 95	○ 96
3	○ 97	○ 98
4	○ 99	○ 100
5	○ 101	○ 102
6	○ 103	○ 104
7	○ 105	○ 106
8	○ 107	○ 108
9	○ 109	○ 110
10	○ 111	○ 112
11	○ 113	○ 114

SECTION TOTAL

Course No. ______ Date ______ Name ______

PROBLEM-SOLVING APPLICATIONS

1. List and describe four major features that distinguish a professional camcorder from a consumer camcorder.

2. When on an ENG assignment, you are forced to shoot in an extremely dark environment. There is no time to turn on any auxiliary lights, and your camcorder is not equipped with a camera light. What, if anything, can you do to produce visible images however noisy they may be?

3. When moving from indoor studio lighting to midday outdoor light, the field reporter tells you not to worry about white-balancing the camera again because the outside light of the foggy day seems to match the studio lighting anyway. What is your response? Why?

4. When watching a rehearsal of a dance company, the TD expresses concern because the dancers wear white leotards while performing a number in front of a black background. Is the TD's concern justified? If so, why? If not, why not? What are your recommendations?

5. The TD tells the camera operator that a high shutter speed needs a considerable amount of light. What does the TD mean by *shutter speed?* When do you need a high shutter speed? How, if at all, is it related to light levels?

6. Draw and describe the parts of the camera chain and their primary functions.

7. The TD assures the director that he can use a USB 3.0 SuperSpeed cable for connecting older equipment built for standard USB 2.0 cables. Is the TD's advice correct? If so, why? If not, why not?

Course No. ______________ Date ______________ Name ______________________________

Lenses

REVIEW OF KEY TERMS

Match each term with its appropriate definition by filling in the corresponding bubble.

1. calibrate
2. depth of field
3. *f*-stop
4. wide-angle lens
5. fast lens
6. slow lens
7. zoom range
8. field of view
9. focal length
10. zoom lens
11. aperture
12. normal lens

A. The distance from the optical center of the lens to the front surface of the camera imaging device

A ○ 1 ○ 2 ○ 3 ○ 4 ○ 5 ○ 6 ○ 7 ○ 8 ○ 9 ○ 10 ○ 11 ○ 12

B. The area in which all objects, located at different distances from the camera, appear sharp and clear

B ○ 1 ○ 2 ○ 3 ○ 4 ○ 5 ○ 6 ○ 7 ○ 8 ○ 9 ○ 10 ○ 11 ○ 12

C. Lens opening measured in *f*-stops

C ○ 1 ○ 2 ○ 3 ○ 4 ○ 5 ○ 6 ○ 7 ○ 8 ○ 9 ○ 10 ○ 11 ○ 12

PAGE TOTAL ______

1. calibrate	**5. fast lens**	**9. focal length**
2. depth of field	**6. slow lens**	**10. zoom lens**
3. *f*-stop	**7. zoom range**	**11. aperture**
4. wide-angle lens	**8. field of view**	**12. normal lens**

D. The general lens focal length that approximates the spatial relationships of normal vision

D ○1 ○2 ○3 ○4 ○5 ○6 ○7 ○8 ○9 ○10 ○11 ○12

E. Variable-focal-length lens, which can change from a wide shot to a close-up and vice versa in one continuous movement

E ○1 ○2 ○3 ○4 ○5 ○6 ○7 ○8 ○9 ○10 ○11 ○12

F. The extent of a scene that is visible through a particular lens

F ○1 ○2 ○3 ○4 ○5 ○6 ○7 ○8 ○9 ○10 ○11 ○12

G. To make a lens keep focus throughout the zoom

G ○1 ○2 ○3 ○4 ○5 ○6 ○7 ○8 ○9 ○10 ○11 ○12

H. A lens that at its maximum aperture permits a relatively small amount of light to enter and pass through

H ○1 ○2 ○3 ○4 ○5 ○6 ○7 ○8 ○9 ○10 ○11 ○12

PAGE TOTAL

Course No. ______________ Date ______________ Name ______________

I. Same as short-focal-length lens, which gives a broad view of a scene

I			
○ 1	○ 2	○ 3	○ 4
○ 5	○ 6	○ 7	○ 8
○ 9	○ 10	○ 11	○ 12

J. The calibration on the lens indicating the diaphragm opening—and therefore the amount of light passing through the lens

J			
○ 1	○ 2	○ 3	○ 4
○ 5	○ 6	○ 7	○ 8
○ 9	○ 10	○ 11	○ 12

K. Shown in a focal-length ratio, such as 20:1

K			
○ 1	○ 2	○ 3	○ 4
○ 5	○ 6	○ 7	○ 8
○ 9	○ 10	○ 11	○ 12

L. A lens that at its maximum aperture permits a relatively great amount of light to enter and pass through

L			
○ 1	○ 2	○ 3	○ 4
○ 5	○ 6	○ 7	○ 8
○ 9	○ 10	○ 11	○ 12

PAGE TOTAL ______

SECTION TOTAL ______

REVIEW OF OPTICAL CHARACTERISTICS OF LENSES

Select the correct answers and fill in the bubbles with the corresponding numbers.

1. Telephoto prime lenses, or zoom lenses in a narrow-angle position, have a relatively (13) *great* (14) *narrow* (15) *shallow* depth of field.

2. In a digital zoom, the focal length is adjusted by (16) *moving the camera closer to or farther away from the object* (17) *shifting certain lens elements* (18) *cropping the image while magnifying it.*

3. The focal length of zoom lenses that are built into small camcorders is generally (19) *not long enough when zoomed in* (20) *not short enough when zoomed out* (21) *fixed while zooming.*

4. Zoom lenses used for sports coverage need (22) *a lower zoom ratio than* (23) *a higher zoom ratio than* (24) *the same zoom ratio as* those used for studio work.

5. A wide-angle lens has a (25) *short* (26) *normal* (27) *long* focal length.

6. A 15x zoom lens means that you can increase the focal length (28) *1.5 times* (29) *150 times* (30) *15 times* in one continuous zoom.

7. When presetting (calibrating) the zoom lens, you (31) *zoom out all the way to a long shot, focus on the target object, and zoom back in again* (32) *zoom in all the way, focus on the target object, and zoom back.*

8. Large apertures (iris openings) contribute to a (33) *great* (34) *shallow* depth of field.

9. Select the three variables that influence depth of field: (35) *focal length of lens* (36) *zoom speed* (37) *focus* (38) *camera-to-object distance* (39) *lens aperture* (40) *focus mechanism.* ***(Fill in three bubbles.)***

10. Given a fixed camera-to-object distance, short-focal-length lenses, or zoom lenses in the wide-angle position, have a relatively (41) *shallow* (42) *wide* (43) *great* depth of field.

11. The area in which all objects, although located at different distances from the camera, are in focus is called (44) *depth of focus* (45) *field of view* (46) *depth of field.*

12. In an optical zoom, the focal length is adjusted by (47) *shifting certain lens elements* (48) *shifting the pixels* (49) *moving the camera closer to or farther away from the object.*

1	○ 13	○ 14	○ 15
2	○ 16	○ 17	○ 18
3	○ 19	○ 20	○ 21
4	○ 22	○ 23	○ 24
5	○ 25	○ 26	○ 27
6	○ 28	○ 29	○ 30
7	○ 31	○ 32	
8	○ 33	○ 34	
9	○ 35	○ 36	○ 37
	○ 38	○ 39	○ 40
10	○ 41	○ 42	○ 43
11	○ 44	○ 45	○ 46
12	○ 47	○ 48	○ 49

PAGE TOTAL

Course No. ______________ Date ______________ Name ______________________________

13. Assuming maximum aperture, a slow lens (50) *transmits an image faster* (51) *transmits an image more slowly* (52) *permits more light to enter* (53) *permits less light to enter* than does a fast lens.

14. Assuming maximum aperture, a fast lens (54) *transmits an image faster* (55) *transmits an image more slowly* (56) *permits more light to enter* (57) *permits less light to enter* than does a slow lens.

15. In the diagram below, select the most appropriate *ƒ*-stop number for each of the four apertures (**a** through **d**) and fill in the bubbles with the corresponding number.

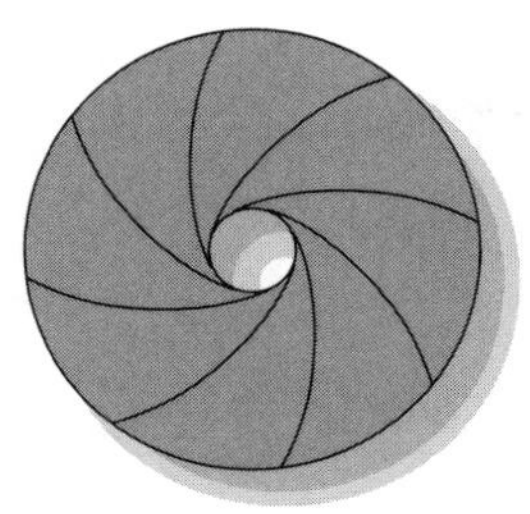

a. (58) *ƒ*/ 5.6 (59) *ƒ*/ 1.4 (60) *ƒ*/22

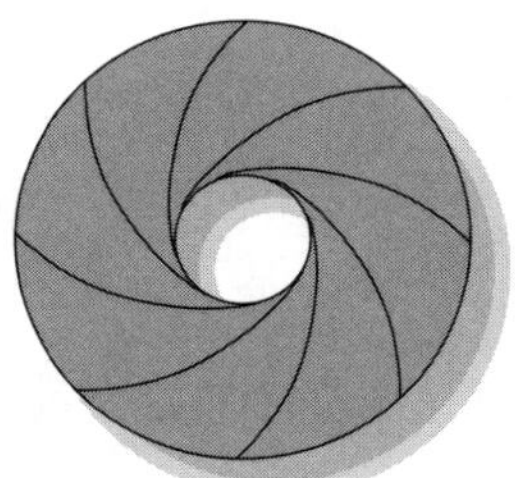

b. (61) *ƒ*/16 (62) *ƒ*/2.8 (63) *ƒ*/1.4

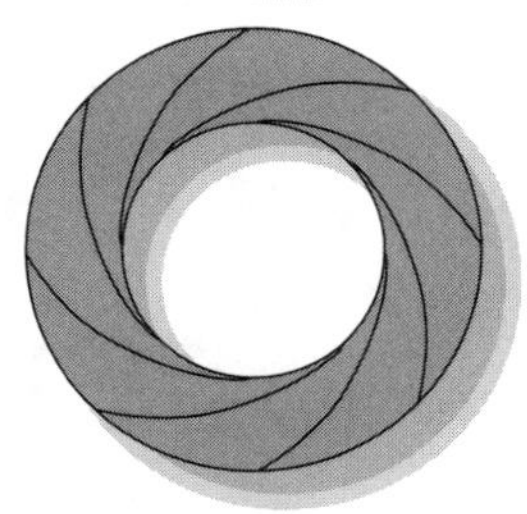

c. (64) *ƒ*/1.4 (65) *ƒ*/4 (66) *ƒ*/16

d. (67) *ƒ*/ 22 (68) *ƒ*/8 (69) *ƒ*/1.4

13 ○ 50 ○ 51 ○ 52 ○ 53

14 ○ 54 ○ 55 ○ 56 ○ 57

15a ○ 58 ○ 59 ○ 60

15b ○ 61 ○ 62 ○ 63

15c ○ 64 ○ 65 ○ 66

15d ○ 67 ○ 68 ○ 69

PAGE TOTAL

SECTION TOTAL

REVIEW OF HOW LENSES SEE

Select the correct answers and fill in the bubbles with the corresponding numbers.

1. A wide-angle lens (70) *increases* (71) *decreases* the illusion of depth and (72) *increases* (73) *decreases* the speed of an object moving toward or away from the camera. ***(Fill in two bubbles.)***

2. To apply selective focus, we need (74) *a great* (75) *a shallow* depth of field.

3. A narrow-angle lens makes objects positioned at different distances from the camera look (76) *more* (77) *less* crowded than they really are and (78) *increases* (79) *decreases* the speed of an object moving toward or away from the camera. ***(Fill in two bubbles.)***

4. To make a small room look larger, we use (80) *a wide-angle* (81) *a narrow-angle* lens.

5. The figure below shows the camera zoomed in all the way for a telephoto view and focused on object A. Object B will probably be (82) *in focus* (83) *out of focus*. The depth of field is therefore (84) *great* (85) *shallow. **(Fill in two bubbles.)***

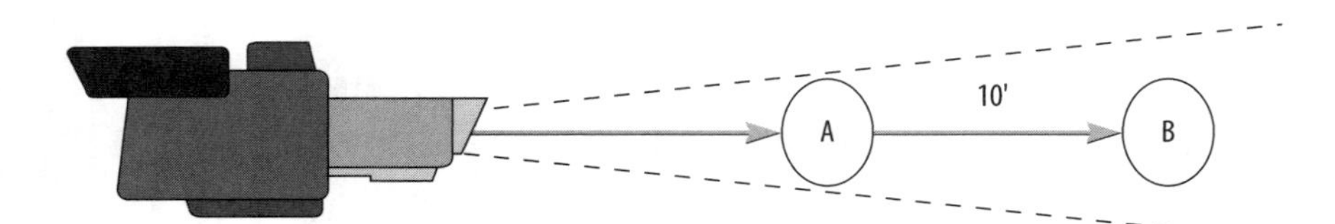

6. The figure below shows the camera zoomed out all the way for a wide-angle view and focused on object A. Object B will probably be (86) *in focus* (87) *out of focus*. The depth of field is therefore (88) *great* (89) *shallow. **(Fill in two bubbles.)***

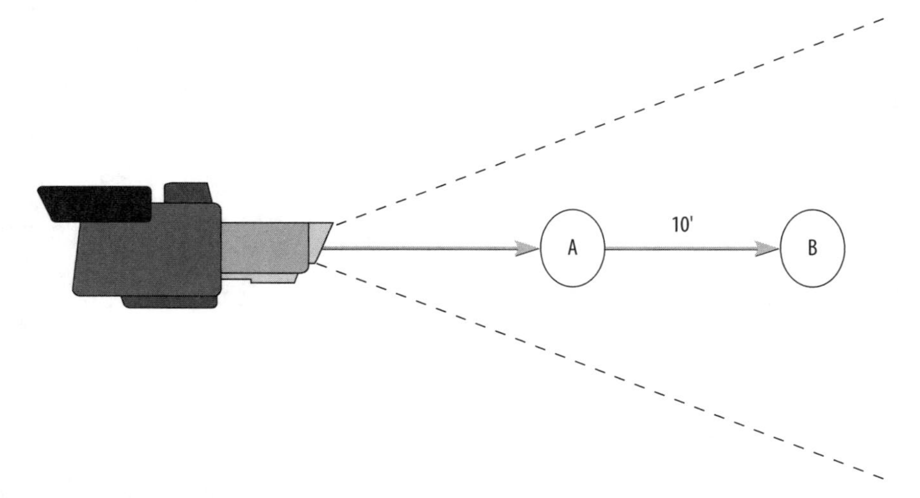

1	○ 70	○ 71
	○ 72	○ 73
2	○ 74	○ 75
3	○ 76	○ 77
	○ 78	○ 79
4	○ 80	○ 81
5	○ 82	○ 83
	○ 84	○ 85
6	○ 86	○ 87
	○ 88	○ 89

PAGE TOTAL

Course No. ________ Date ________ Name ________

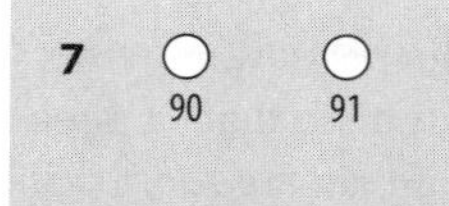

7. The screen image below displays a (90) *shallow* (91) *great* depth of field.

7 ◯ 90 ◯ 91

PHOTO: HERBERT ZETTL

8. The screen image below shows that the camera's zoom lens was in a (92) *narrow-angle* position (93) *wide-angle.*

8 ◯ 92 ◯ 93

PHOTO: HERBERT ZETTL

9. The opening shot of a documentary on city politics shows the city hall through a piece of sculpture. The camera operator used (94) *a wide-angle* (95) *a narrow-angle* position.

9 ◯ 94 ◯ 95

PHOTO: CHIRS ROZALES

PAGE TOTAL

10. Your preview monitors for cameras 1, 2, and 3 display the following images. Assuming that all three cameras are positioned right next to one another, which is the approximate zoom position for each? Choose among (96) *wide angle* (97) *normal* and (98) *narrow angle.*

Camera 1

Camera 2

Camera 3

PHOTOS: HERBERT ZETTL

11. The figure below simulates (99) *a digital zoom* (100) *an optical zoom.*

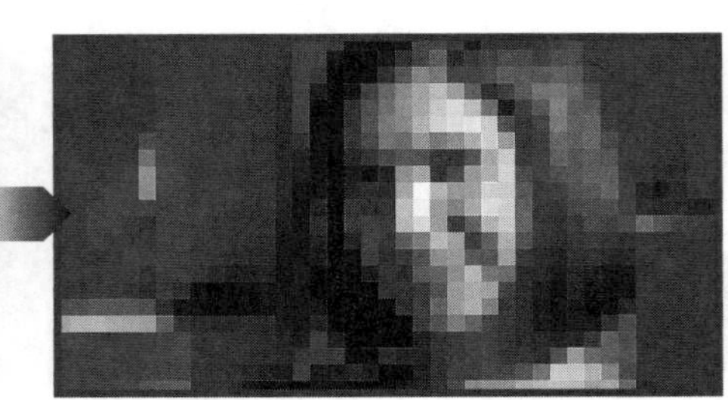

PHOTOS: HERBERT ZETTL

12. When zoomed in on somebody approaching the camera, the person seems to move (101) *slower than* (102) *about the same speed as* (103) *faster than* they actually do.

13. The closer the camera is to the object, the (104) *shallower* (105) *greater* (106) *wider* the depth of field becomes.

14. Assuming that your image stabilizer is disengaged, you can avoid handheld camera wobbles by (107) *zooming all the way in* (108) *zooming all the way out* (109) *keeping the zoom lens in the narrow-angle position.*

15. In the screen image below, the zoom lens was in the (110) *normal* (111) *narrow-angle* (112) *wide-angle* zoom position.

PHOTO: HERBERT ZETTL

10 C1 ○ 96 ○ 97 ○ 98
C2 ○ 96 ○ 97 ○ 98
C3 ○ 96 ○ 97 ○ 98

11 ○ 99 ○ 100

12 ○ 101 ○ 102 ○ 103

13 ○ 104 ○ 105 ○ 106

14 ○ 107 ○ 108 ○ 109

15 ○ 110 ○ 111 ○ 112

PAGE TOTAL

SECTION TOTAL

Course No. ______________ Date ______________ Name ______________________________

REVIEW QUIZ

*Mark the following statements as true or false by filling in the bubbles in the **T** (for true) or **F** (for false) column.*

1. Auto-focus and focus-assist features are the same.
2. A zoom can be simulated by gradually enlarging the image's center portion.
3. Depth of field is influenced only by the focal length of the lens.
4. Depth of field increases as focal length decreases.
5. When covering a news story with an ENG/EFP camera, you are best off with a shallow depth of field because there are few, if any, focusing problems.
6. A slow lens is one with a very low *f*-stop number, such as *f*/1.4.
7. After the initial calibration of a zoom lens, you need to preset it again each time the distance from object to camera changes substantially.
8. When calibrating a zoom lens, you must first zoom in on the target object and focus and then zoom out.
9. An object moving toward the camera looks faster than normal when shot with a wide-angle lens.
10. Because a 1080i HDTV image has so many scanning lines, the lens quality is relatively unimportant.
11. Digital and optical zooms work on the same principle.
12. Each time you stop a zoom from one extreme position to the next, you get a different focal length.
13. Compared with an optical zoom, a digital zoom permits a higher zoom ratio without picture deterioration.
14. We can dolly most easily when the lens is in the extreme wide-angle position.
15. Digital stabilizers can absorb all picture wobble when you are shooting a narrow-angle picture.
16. One of the major differences between a studio HDTV camera and a digital cinema camera is that the latter has a much higher-density sensor.

	T	F
1	○ 113	○ 114
2	○ 115	○ 116
3	○ 117	○ 118
4	○ 119	○ 120
5	○ 121	○ 122
6	○ 123	○ 124
7	○ 125	○ 126
8	○ 127	○ 128
9	○ 129	○ 130
10	○ 131	○ 132
11	○ 133	○ 134
12	○ 135	○ 136
13	○ 137	○ 138
14	○ 139	○ 140
15	○ 141	○ 142
16	○ 143	○ 144

SECTION TOTAL

PROBLEM-SOLVING APPLICATIONS

Let us now put the theory to work. You can observe the optical and performance characteristics of lenses easily by using a camcorder or a still camera that can accept various lenses. Think through each production problem and consider the various options; then pick the most effective solution and justify your choice.

1. Zoom all the way out with the camcorder, or attach a wide-angle lens (28mm or less focal length) to the single-lens reflex (SLR) still camera, and focus on an object 4 to 6 feet away from you. Look at the background objects (20 or so feet away from you). Are they visible? Do they appear in fairly sharp focus? Or are they blurred? Now do the same observations by zooming all the way in or by attaching a telephoto lens (with a focal length of 200mm) to the still camera. Explain depth-of-field characteristics.

2. When watching television or a movie, try to figure out what lenses were used for some of the shots. For example, when you see someone running toward the camera yet seemingly not getting closer, what lens was used? Or when you see the happy couple approach the dinner table through the flowers and the candles in the foreground, what lens was probably used, assuming that the couple, as well as the candles and the flowers, are in focus? Such observations will certainly help you become more aware of focal lengths and their effects.

3. You are the AD of a live telecast of a modern dance program, which is performed on a dimly lighted stage. The operators of the two key cameras express some concern about their lenses. The new lens of the handheld ENG/EFP camera 1 has a 25× zoom range and a maximum aperture of *f*/5.6. Although the lens was used successfully during the past three football games, the operator feels that it might be too slow for this type of application. Camera 2 has a 10× lens with a 2× range extender. Its maximum aperture is also *f*/5.6. Are the operators' concerns justified?

4. The novice director asks you, the operator of camera 3, to get the opening shot by zooming back slowly from an extreme close-up of the title of a book to an extreme wide shot that shows a large part of the studio (the other cameras, the floor manager, the overhead lighting) as the background for the opening titles. The director sets up the wide shot first to make sure that it shows enough of the studio. When your extreme-narrow-angle zoom lens position does not produce the desired close-up of the book title, he asks you to "simply pop in a range extender before we punch up your camera." What are the potential problems, if any? Be specific.

5. While shooting a dramatic program in the studio, you, the camera 1 operator, are told by the director that your camera shows the scratched background walls in sharp focus. The director asks you to make the walls look slightly out of focus without impeding the sharp focus on the foreground talent. What would you do?

Course No. ______________ Date ______________ Name ______________________________

Camera Operation and Picture Composition

REVIEW OF KEY TERMS

Match each term with its appropriate definition by filling in the corresponding bubble.

1. close-up
2. closure
3. cross-shot
4. field of view
5. headroom
6. noseroom
7. leadroom
8. over-the-shoulder shot
9. z-axis
10. monopod
11. arc
12. dolly
13. cant
14. truck
15. camera stabilizing system
16. robotic pedestal
17. tilt
18. quick-release plate
19. pan
20. studio pan-and-tilt head

A. Object or any part of it seen at close range

B. To move the camera laterally by means of a mobile camera mount

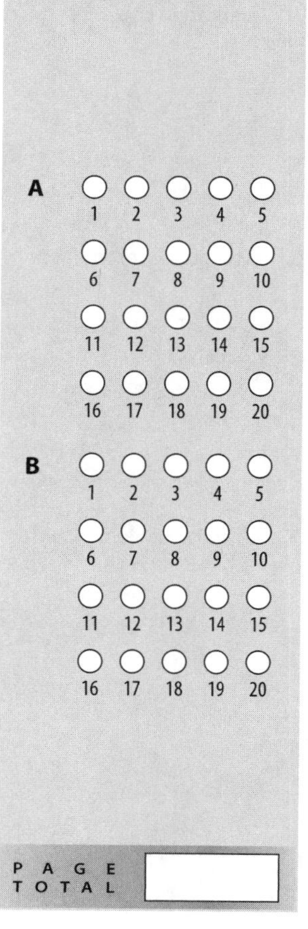

1. close-up	8. over-the-shoulder shot	15. camera stabilizing system
2. closure	9. z-axis	16. robotic pedestal
3. cross-shot	10. monopod	17. tilt
4. field of view	11. arc	18. quick-release plate
5. headroom	12. dolly	19. pan
6. noseroom	13. cant	20. studio pan-and-tilt head
7. leadroom	14. truck	

C. The space left between the top of the head and the upper screen edge

C ○1 ○2 ○3 ○4 ○5 ○6 ○7 ○8 ○9 ○10 ○11 ○12 ○13 ○14 ○15 ○16 ○17 ○18 ○19 ○20

D. Camera looks at the camera-far person with the back and shoulder of the camera-near person in the shot

D ○1 ○2 ○3 ○4 ○5 ○6 ○7 ○8 ○9 ○10 ○11 ○12 ○13 ○14 ○15 ○16 ○17 ○18 ○19 ○20

E. The space left in front of an object or a person moving toward the edge of the screen

E ○1 ○2 ○3 ○4 ○5 ○6 ○7 ○8 ○9 ○10 ○11 ○12 ○13 ○14 ○15 ○16 ○17 ○18 ○19 ○20

F. A single pole onto which you can mount a camera

F ○1 ○2 ○3 ○4 ○5 ○6 ○7 ○8 ○9 ○10 ○11 ○12 ○13 ○14 ○15 ○16 ○17 ○18 ○19 ○20

PAGE TOTAL

Course No. ______ Date ______ Name ______

G. To point the camera up or down

H. To move the camera in a slightly curved dolly or truck

I. To tilt a handheld camera sideways

J. Similar to the over-the-shoulder shot except that the camera-near person is completely out of the shot

K. Portion of a scene visible through a particular lens; its vista

G	1	2	3	4	5
	6	7	8	9	10
	11	12	13	14	15
	16	17	18	19	20
H	1	2	3	4	5
	6	7	8	9	10
	11	12	13	14	15
	16	17	18	19	20
I	1	2	3	4	5
	6	7	8	9	10
	11	12	13	14	15
	16	17	18	19	20
J	1	2	3	4	5
	6	7	8	9	10
	11	12	13	14	15
	16	17	18	19	20
K	1	2	3	4	5
	6	7	8	9	10
	11	12	13	14	15
	16	17	18	19	20

PAGE TOTAL ______

1. close-up	8. over-the-shoulder shot	15. camera stabilizing system
2. closure	9. z-axis	16. robotic pedestal
3. cross-shot	10. monopod	17. tilt
4. field of view	11. arc	18. quick-release plate
5. headroom	12. dolly	19. pan
6. noseroom	13. cant	20. studio pan-and-tilt head
7. leadroom	14. truck	

L. Imaginary line extending from the lens to the horizon

L ○1 ○2 ○3 ○4 ○5 ○6 ○7 ○8 ○9 ○10 ○11 ○12 ○13 ○14 ○15 ○16 ○17 ○18 ○19 ○20

M. Device to attach an ENG/EFP camera to the mounting head

M ○1 ○2 ○3 ○4 ○5 ○6 ○7 ○8 ○9 ○10 ○11 ○12 ○13 ○14 ○15 ○16 ○17 ○18 ○19 ○20

N. Mentally filling in spaces of an incomplete picture

N ○1 ○2 ○3 ○4 ○5 ○6 ○7 ○8 ○9 ○10 ○11 ○12 ○13 ○14 ○15 ○16 ○17 ○18 ○19 ○20

O. The space left in front of a person looking toward the screen edge

O ○1 ○2 ○3 ○4 ○5 ○6 ○7 ○8 ○9 ○10 ○11 ○12 ○13 ○14 ○15 ○16 ○17 ○18 ○19 ○20

PAGE TOTAL

Course No. ______________ Date ________________ Name ______________________________

P. To move the camera toward or away from an object

Q. Horizontal turning of the camera

R. Mounting head for heavy cameras that permits extremely smooth movements

S. Computer-controlled camera mount

T. Camera mount that helps produce jitter-free pictures even when the camera operator runs with it

P	1	2	3	4	5
	6	7	8	9	10
	11	12	13	14	15
	16	17	18	19	20
Q	1	2	3	4	5
	6	7	8	9	10
	11	12	13	14	15
	16	17	18	19	20
R	1	2	3	4	5
	6	7	8	9	10
	11	12	13	14	15
	16	17	18	19	20
S	1	2	3	4	5
	6	7	8	9	10
	11	12	13	14	15
	16	17	18	19	20
T	1	2	3	4	5
	6	7	8	9	10
	11	12	13	14	15
	16	17	18	19	20

PAGE TOTAL

SECTION TOTAL

REVIEW OF CAMERA MOVEMENTS AND CAMERA SUPPORTS

Select the correct answers and fill in the bubbles with the corresponding numbers.

1. Fill in the bubbles whose numbers correspond with the camera movements indicated in the following figure.

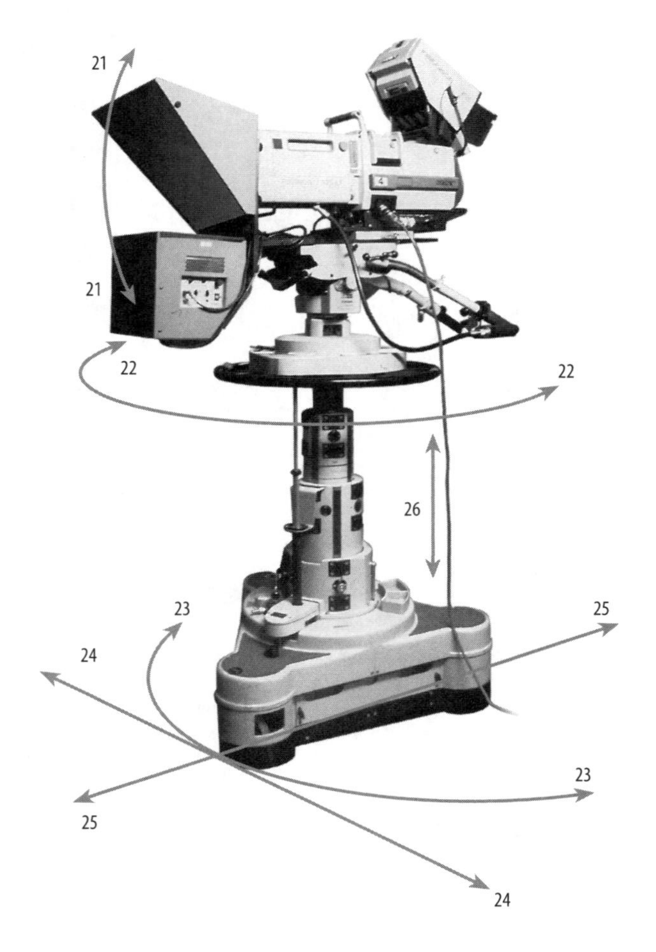

a. pedestal

b. truck

c. pan

d. tilt

e. dolly

f. arc

1a ○ 21 ○ 22 ○ 23 ○ 24 ○ 25 ○ 26

1b ○ 21 ○ 22 ○ 23 ○ 24 ○ 25 ○ 26

1c ○ 21 ○ 22 ○ 23 ○ 24 ○ 25 ○ 26

1d ○ 21 ○ 22 ○ 23 ○ 24 ○ 25 ○ 26

1e ○ 21 ○ 22 ○ 23 ○ 24 ○ 25 ○ 26

1f ○ 21 ○ 22 ○ 23 ○ 24 ○ 25 ○ 26

PAGE TOTAL

PHOTO: EDWARD AIONA

Course No. ______________ Date ______________ Name ______________________________

2. The spreader (27) *keeps the tripod legs from spreading too far* (28) *must always be fully extended* (29) *helps spread the tripod legs as much as possible.*

3. The (30) *leveling bowl* (31) *monopod* (32) *jib arm* helps maintain a camera in a horizontal position.

4. Mounting heads facilitate (33) *dollies and trucks* (34) *smooth tilts and pans* (35) *arcs and zooms.*

5. A (36) *wedge mount* (37) *robotic pedestal* (38) *quick-release plate* makes it easy to detach an ENG/EFP camera from the tripod and reattach it again.

6. Compared with a tripod, a studio pedestal allows these additional camera moves: (39) *canting left and right* (40) *raising and lowering the camera while on the air* (41) *booming up and down.*

7. To simultaneously boom, tongue, pan, and tilt the camera, you need a (42) *robotic pedestal* (43) *jib arm* (44) *camera stabilizing system.*

8. The camera support that allows the camera operator to run with the camera while keeping the picture steady is a (45) *robotic arm* (46) *Steadicam* (47) *jib arm.*

9. The camera support that allows the operator of a small handheld camera to walk or run without any picture wobbles is a (48) *Steadicam* (49) *monopod* (50) *handheld stabilizer.*

2	○ 27	○ 28	○ 29
3	○ 30	○ 31	○ 32
4	○ 33	○ 34	○ 35
5	○ 36	○ 37	○ 38
6	○ 39	○ 40	○ 41
7	○ 42	○ 43	○ 44
8	○ 45	○ 46	○ 47
9	○ 48	○ 49	○ 50

PAGE TOTAL ______

SECTION TOTAL ______

REVIEW OF HOW TO WORK A CAMERA

Select the correct answers and fill in the bubbles with the corresponding numbers.

1. When dollying with a studio camera or walking with an EFP camera, the depth of field should be as (51) *shallow* (52) *great* (53) *narrow* as possible.

2. To minimize camera wobbles when dollying with a studio camera or walking with an EFP camera, the zoom lens should be in a (54) *narrow-angle* (55) *wide-angle* (56) *telephoto* position.

3. After having calibrated the zoom lens, you need to preset it again (57) *only when the camera moves* (58) *only when the object moves relative to the camera* (59) *whenever camera or object moves relative to the other.*

4. When operating a camcorder in the field, you should always have the camera mic (60) *on* (61) *off* (62) *replaced by a shotgun mic.*

5. When panning with a shoulder-mounted ENG/EFP camera, you should point your knees toward (63) *the starting point of the pan* (64) *the end point of the pan* (65) *either direction.*

6. During a test recording with your ENG/EFP camcorder, you should (66) *leave the lens cap on but check the audio* (67) *make sure all camera features are working* (68) *ask the reporter to count to 10.*

7. When loading a VTR cassette or memory card for recording, the safety tab (69) *does not matter because it is primarily meant for playback protection* (70) *should be in the open position* (71) *should be in place or in the closed position.*

8. When on an ENG assignment, you should record ambient sound (72) *only if somebody is talking* (73) *only if there is no background noise* (74) *always.*

9. To calibrate a zoom lens (75) *zoom in, focus, zoom out* (76) *zoom out, focus, zoom in* (77) *adjust focus continuously while zooming.*

10. When calibrating a zoom lens, the tally light should be (78) *on* (79) *off* (80) *ignored.*

11. To achieve critical focus when operating an HDTV studio camera, you should (81) *adjust the viewfinder's sharpness* (82) *engage the auto-focus feature* (83) *engage the focus-assist feature.*

12. You should lock the camera mounting head (84) *every time you leave it* (85) *only when temporarily leaving the camera* (86) *at the end of the shoot.*

1	○ 51	○ 52	○ 53
2	○ 54	○ 55	○ 56
3	○ 57	○ 58	○ 59
4	○ 60	○ 61	○ 62
5	○ 63	○ 64	○ 65
6	○ 66	○ 67	○ 68
7	○ 69	○ 70	○ 71
8	○ 72	○ 73	○ 74
9	○ 75	○ 76	○ 77
10	○ 78	○ 79	○ 80
11	○ 81	○ 82	○ 83
12	○ 84	○ 85	○ 86

PAGE TOTAL

Course No. ________ Date ________ Name ________

13. When leaving a studio camera temporarily unattended, you should (87) *tighten the lock mechanism* (88) *tighten the drag control* (89) *point the camera toward the floor rather than into the lights.*

14. When trying to preserve battery power, you should switch off (90) *only the foldout monitor* (91) *only the image stabilizer* (92) *both the foldout monitor and the image stabilizer.*

15. When operating the camcorder in EFP, you should engage the audio AGC (93) *always* (94) *never* (95) *only when necessary.*

16. When operating a studio camera, the electronic adjustments for optimal picture quality are made for you by the (96) *VO* (97) *TD* (98) *AD.*

13	○ 87	○ 88	○ 89
14	○ 90	○ 91	○ 92
15	○ 93	○ 94	○ 95
16	○ 96	○ 97	○ 98

PAGE TOTAL

SECTION TOTAL

REVIEW OF FRAMING EFFECTIVE SHOTS

1. Using the set of numbered images below, fill in the bubbles for each of the following fields of view or shot designations.

99

100

101

102

103

104

105

106

107

PHOTOS: EDWARD AIONA

a. ELS (extreme long shot)

b. LS (long shot)

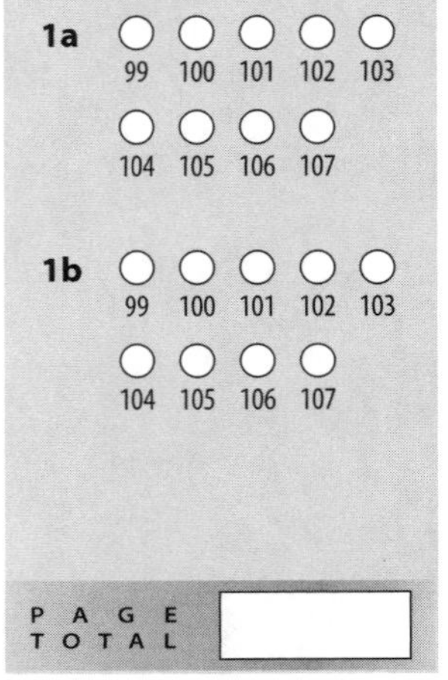

Course No. ______________ Date ________________ Name ______________________________

c. MS (medium shot)

d. bust shot

e. knee shot

f. three-shot

g. ECU (extreme close-up)

h. CU (close-up)

i. over-the-shoulder shot

1c	◯ 99	◯ 100	◯ 101	◯ 102	◯ 103
	◯ 104	◯ 105	◯ 106	◯ 107	
1d	◯ 99	◯ 100	◯ 101	◯ 102	◯ 103
	◯ 104	◯ 105	◯ 106	◯ 107	
1e	◯ 99	◯ 100	◯ 101	◯ 102	◯ 103
	◯ 104	◯ 105	◯ 106	◯ 107	
1f	◯ 99	◯ 100	◯ 101	◯ 102	◯ 103
	◯ 104	◯ 105	◯ 106	◯ 107	
1g	◯ 99	◯ 100	◯ 101	◯ 102	◯ 103
	◯ 104	◯ 105	◯ 106	◯ 107	
1h	◯ 99	◯ 100	◯ 101	◯ 102	◯ 103
	◯ 104	◯ 105	◯ 106	◯ 107	
1i	◯ 99	◯ 100	◯ 101	◯ 102	◯ 103
	◯ 104	◯ 105	◯ 106	◯ 107	

PAGE TOTAL ______

2. Evaluate the framing of shots in the next five figures by filling in the bubbles with the corresponding numbers. ***(Note: There may be more than one correct answer for some parts of a problem.)***

PHOTO: EDWARD AIONA

a. This CU is (108) *acceptable* (109) *unacceptable* because it has (110) *no headroom* (111) *too much headroom* (112) *adequate headroom* (113) *adequate noseroom.* If unacceptable, you should (114) *tilt up* (115) *tilt down* (116) *pan left.*

2a ◯ 108 ◯ 109
◯ 110 ◯ 111 ◯ 112 ◯ 113
◯ 114 ◯ 115 ◯ 116

PHOTO: EDWARD AIONA

b. This shot is (117) *acceptable* (118) *unacceptable* because it has (119) *sufficient noseroom* (120) *insufficient noseroom* (121) *sufficient headroom* (122) *insufficient leadroom.* If unacceptable, you should (123) *pan left* (124) *pan right* (125) *tilt up* (126) *tilt down* (127) *pedestal up* (128) *pedestal down.*

2b ◯ 117 ◯ 118
◯ 119 ◯ 120 ◯ 121 ◯ 122
◯ 123 ◯ 124 ◯ 125
◯ 126 ◯ 127 ◯ 128

PAGE TOTAL

PHOTO: HERBERT ZETTL

c. This shot is intended to emphasize the car's speed and risky driving. Its framing is therefore (129) *acceptable* (130) *unacceptable*. If unacceptable, you should (131) *level the horizon line* (132) *zoom out*.

2c ○ 129 ○ 130 ○ 131 ○ 132

PHOTO: EDWARD AIONA

d. This shot is (133) *acceptable* (134) *unacceptable* because it has (135) *no headroom* (136) *too much headroom* (137) *no noseroom* (138) *no leadroom* (139) *insufficient clues for closure in off-screen space*. If unacceptable, you should (140) *tilt up* (141) *tilt down*.

2d ○ 133 ○ 134 ○ 135 ○ 136 ○ 137 ○ 138 ○ 139 ○ 140 ○ 141

PHOTO: EDWARD AIONA

e. This over-the-shoulder shot is (142) *acceptable* (143) *unacceptable*. If unacceptable, you should (144) *zoom out* (145) *pedestal up* (146) *arc left* (147) *arc right*.

2e ○ 142 ○ 143 ○ 144 ○ 145 ○ 146 ○ 147

PAGE TOTAL ________

3. This shot makes (148) *good* (149) *poor* use of screen depth because (150) *it lacks foreground objects* (151) *the horizon is too high.*

PHOTO: HERBERT ZETTL

3	○ 148	○ 149
	○ 150	○ 151

4. This framing is (152) *acceptable* (153) *unacceptable* in terms of closure.

PHOTO: EDWARD AIONA

4	○ 152	○ 153

PAGE TOTAL

SECTION TOTAL

Course No. ________ Date ________ Name ________

REVIEW QUIZ

*Mark the following statements as true or false by filling in the bubbles in the **T** (for true) or **F** (for false) column.*

		T	F
1.	Once balanced, the wedge mount ensures that the camera is mounted in an optimally balanced position for each subsequent use.	○ 154	○ 155
2.	Psychological closure always ensures good composition.	○ 156	○ 157
3.	The studio pedestal permits very-low-angle shots.	○ 158	○ 159
4.	You should unlock the pan-and-tilt mechanism at the beginning of the show and lock it again every time you leave the camera unattended.	○ 160	○ 161
5.	HDTV cameras are easier to focus than standard cameras.	○ 162	○ 163
6.	To boom up means to raise the camera pedestal.	○ 164	○ 165
7.	The drag controls on a mounting head are used to lock down the camera.	○ 166	○ 167
8.	If your camcorder has an LCD foldout monitor, you should use it to focus whenever possible because it is bigger than the viewfinder display.	○ 168	○ 169
9.	Dolly and truck movements show up as similar movements on-screen.	○ 170	○ 171
10.	A robotic pedestal is motor-driven and remotely controlled.	○ 172	○ 173
11.	The medium shot is also called a bust shot.	○ 174	○ 175
12.	Because an ENG camera can be shoulder-mounted, the operator has no need for a tripod.	○ 176	○ 177
13.	The higher the zoom ratio, the more effective the lens is for studio work.	○ 178	○ 179
14.	The handheld stabilizer is especially effective for full-sized EFP cameras.	○ 180	○ 181
15.	The jib arm and the camera crane can make the camera move in similar ways.	○ 182	○ 183
16.	Field of view refers to how far or close the object appears relative to the camera.	○ 184	○ 185
17.	Before the studio or remote production, you should check the tightest and widest field of view of the zoom lens from the principal camera position.	○ 186	○ 187
18.	Leadroom and noseroom fulfill similar framing (compositional) functions.	○ 188	○ 189

SECTION TOTAL ________

PROBLEM-SOLVING APPLICATIONS

1. Locate the pan and the tilt drag controls and the pan-and-tilt lock controls of the mounting head. Adjust them so that you can pan and tilt the camera as smoothly as necessary. When do you need to use the lock mechanism?

2. Pedestal up and down to see how high and low the camera will go. What can happen if you move the camera too fast to either end of vertical travel?

3. You are to set up a tripod on an uneven, slightly sloping field. How can you make sure that the tripod is level?

4. The director wants you to follow the new mayor up the flight of stairs in city hall without shaking the ENG/EFP camera. What camera mount would you suggest?

5. You are shooting a documentary-style program, and you don't have a tripod at the shoot. What focal length would be most effective if you want relatively steady shots? What other techniques could you consider for getting steady shots?

6. You have been asked to recommend the camera mounting equipment for a four-camera studio production. The director wants one camera to get a moving overhead shot and another camera that can physically move around the floor in a smooth arcing motion. The other two cameras will be shooting a variety of typical long shots and close-ups. Another consideration is that there will be only three camera operators available during the production. List the mounting equipment you would recommend for each camera and why.

7. During the remote coverage of the World Computer Fair, the novice director tells you to zoom in to an ECU of a laptop display and then arc the tripod dolly around the display table to show the other computers. What are the potential problems, if any?

8. In a multicamera studio dance program, the same director tells you, the operator of camera 3, that you should listen only to calls that concern your camera and ignore commands to all the other cameras. Do you agree? If so, why? If not, why not?

9. As you are leaving the studio, the producer states that a spare battery for your ENG/EFP camera is not needed because the story you are to cover will have, at best, a 20-second slot in the newscast. What is your response?

10. The producer tells you to be sure to keep enough headroom when framing an ECU. Do you agree? If so, why? If not, why not?

11. The director tells you, the camera operator, to change from an over-the-shoulder shot to a cross-shot. How can you accomplish such a shot change?

Course No. ______________ Date ______________ Name ______________________________

Audio: Sound Pickup

REVIEW OF KEY TERMS

Match each term with its appropriate definition by filling in the corresponding bubble.

1. **unidirectional**
2. **flat response**
3. **omnidirectional**
4. **pickup pattern**
5. **polar pattern**
6. **condenser microphone**
7. **dynamic microphone**
8. **cardioid**
9. **ribbon microphone**
10. **impedance**
11. **frequency response**
12. **system microphone**

A. A microphone whose sound pickup device consists of a thin band that vibrates with the sound pressures within a magnetic field

A ○1 ○2 ○3 ○4 ○5 ○6 ○7 ○8 ○9 ○10 ○11 ○12

B. A microphone whose diaphragm consists of a plate that vibrates with the sound pressure against another fixed plate (the backplate)

B ○1 ○2 ○3 ○4 ○5 ○6 ○7 ○8 ○9 ○10 ○11 ○12

C. A microphone that can pick up sounds better from one direction—the front—than from the sides or back

C ○1 ○2 ○3 ○4 ○5 ○6 ○7 ○8 ○9 ○10 ○11 ○12

PAGE TOTAL

1. unidirectional	**5. polar pattern**	**9. ribbon microphone**
2. flat response	**6. condenser microphone**	**10. impedance**
3. omnidirectional	**7. dynamic microphone**	**11. frequency response**
4. pickup pattern	**8. cardioid**	**12. system microphone**

D. The territory around the microphone within which the microphone can "hear" well, or has optimal sound pickup

D ○ 1 ○ 2 ○ 3 ○ 4 ○ 5 ○ 6 ○ 7 ○ 8 ○ 9 ○ 10 ○ 11 ○ 12

E. A microphone whose sound pickup device consists of a diaphragm that is attached to a movable coil

E ○ 1 ○ 2 ○ 3 ○ 4 ○ 5 ○ 6 ○ 7 ○ 8 ○ 9 ○ 10 ○ 11 ○ 12

F. A type of resistance to a signal flow: high-Z or low-Z

F ○ 1 ○ 2 ○ 3 ○ 4 ○ 5 ○ 6 ○ 7 ○ 8 ○ 9 ○ 10 ○ 11 ○ 12

G. The range of frequencies a microphone can hear and reproduce

G ○ 1 ○ 2 ○ 3 ○ 4 ○ 5 ○ 6 ○ 7 ○ 8 ○ 9 ○ 10 ○ 11 ○ 12

H. A specific pickup pattern of unidirectional microphones

H ○ 1 ○ 2 ○ 3 ○ 4 ○ 5 ○ 6 ○ 7 ○ 8 ○ 9 ○ 10 ○ 11 ○ 12

PAGE TOTAL

Course No. ____________ Date ____________ Name ____________________

I. The measure of a microphone's ability to hear equally well over its entire frequency range

I ○1 ○2 ○3 ○4 ○5 ○6 ○7 ○8 ○9 ○10 ○11 ○12

J. The two-dimensional representation of a microphone pickup pattern

J ○1 ○2 ○3 ○4 ○5 ○6 ○7 ○8 ○9 ○10 ○11 ○12

K. Uses a base and different heads for various pickup patterns

K ○1 ○2 ○3 ○4 ○5 ○6 ○7 ○8 ○9 ○10 ○11 ○12

L. A microphone that can pick up sounds equally well from all directions

L ○1 ○2 ○3 ○4 ○5 ○6 ○7 ○8 ○9 ○10 ○11 ○12

PAGE TOTAL ________

SECTION TOTAL ________

REVIEW OF HOW MICROPHONES HEAR

1. Fill in the bubbles whose numbers correspond with the polar patterns in the figure below.

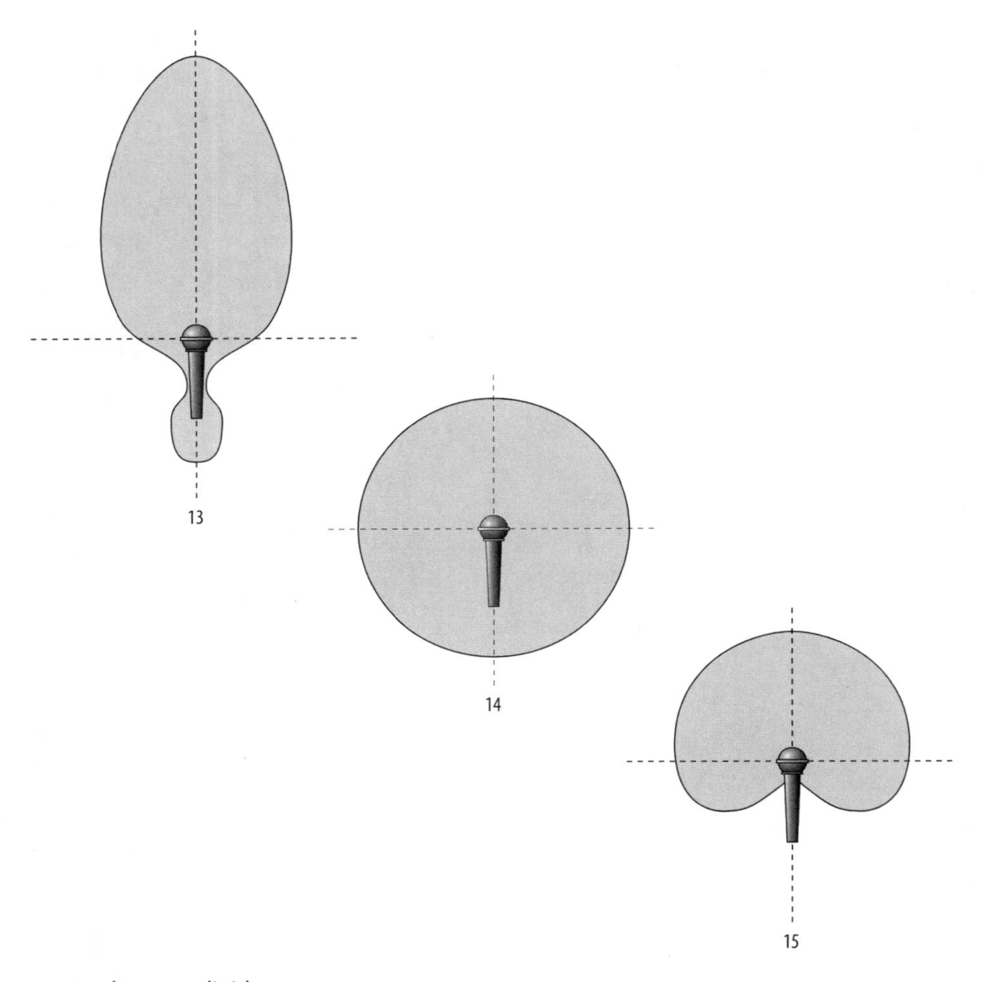

a. hypercardioid

b. cardioid

c. omnidirectional

1a ○ 13 ○ 14 ○ 15

1b ○ 13 ○ 14 ○ 15

1c ○ 13 ○ 14 ○ 15

PAGE TOTAL

Course No. ______________ Date ____________________ Name ____

Select the correct answers and fill in the bubbles with the corresponding numbers.

2. The two-dimensional representation of a microphone's sound pickup area is the (16) *safe area* (17) *polar pattern* (18) *pickup pattern.*

3. Normally, shotgun microphones have (19) *an omnidirectional* (20) *an extremely directional* (21) *a nondirectional* pickup pattern.

4. Select the three types of microphones as classified by their generating element: (22) *dynamic* (23) *unidirectional* (24) *cardioid* (25) *ribbon* (26) *condenser* (27) *hypercardioid.* ***(Fill in three bubbles.)***

5. To eliminate sudden breath pops when speaking close to the microphone, we use a (28) *pop filter* (29) *windscreen* (30) *frequency filter.*

6. Microphones that require a battery or phantom power for their output signal are (31) *dynamic* (32) *ribbon* (33) *condenser.*

7. Faraway speech sounds are picked up best with (34) *a shotgun mic* (35) *an omnidirectional hand mic* (36) *a ribbon mic.*

8. The hand microphones used in ENG normally have (37) *an omnidirectional* (38) *a cardioid* (39) *a hyper- or supercardioid* pickup pattern.

9. The element in a microphone that converts sound waves into an electrical signal is the (40) *sound-generating element* (41) *sound-amplifying chip* (42) *pickup device.*

10. In general, dynamic microphones are (43) *equally sensitive as* (44) *less rugged than* (45) *more rugged than* ribbon microphones.

2	○ 16	○ 17	○ 18
3	○ 19	○ 20	○ 21
4	○ 22	○ 23	○ 24
	○ 25	○ 26	○ 27
5	○ 28	○ 29	○ 30
6	○ 31	○ 32	○ 33
7	○ 34	○ 35	○ 36
8	○ 37	○ 38	○ 39
9	○ 40	○ 41	○ 42
10	○ 43	○ 44	○ 45

PAGE TOTAL ______

SECTION TOTAL ______

REVIEW OF HOW MICROPHONES ARE USED

Select the correct answers and fill in the bubbles with the corresponding numbers.

1. The large shotgun microphone on a perambulator boom has (46) *an omnidirectional* (47) *a hyper- or supercardioid* (48) *a cardioid* pickup pattern.

2. For an optimal pickup of an acoustic guitar, you should use a (49) *dynamic* (50) *condenser* (51) *dynamic cardioid* microphone.

3. The most appropriate mics for the voice pickup of a four-member news team (two anchors, a weathercaster, and a sportscaster) are (52) *lavaliers* (53) *desk mics* (54) *boom mics.*

4. You are to set up microphones for a six-member panel discussion. All participants sit in a row at a table. Normally, you would use (55) *desk mics* (56) *hand mics* (57) *boom mics* for this production.

5. Parabolic microphones are especially effective for picking up (58) *especially soft sounds* (59) *faraway sounds* (60) *extremely close sounds.*

6. To achieve a good and efficient voice pickup during the video recording of four people sitting around a table in a small office, talking about effective sound handling in ENG/EFP, you should use a (61) *parabolic reflector mic* (62) *boundary mic* (63) *large shotgun mic.*

7. When you're setting up the mics for a panel show, the audio engineer advises you to place them in such a way that they will not cause "multiple-microphone interference." This means that the mics must be placed so that they will not (64) *block the faces of the panel members* (65) *cancel some of one another's frequencies* (66) *multiply the ambient noise.*

8. When doing a live report from an accident scene, the most practical mic is a (67) *stand mic* (68) *lavalier mic* (69) *hand mic.*

9. The sound pickup of a dramatic scene of two people sitting at the dinner table is best done with a (70) *fishpole and pencil mic* (71) *fishpole and large shotgun mic* (72) *plant mic.*

10. For miking a kick drum, the microphone that is least subject to input overload has a (73) *ribbon* (74) *condenser* (75) *dynamic* sound-generating element.

1	○ 46	○ 47	○ 48
2	○ 49	○ 50	○ 51
3	○ 52	○ 53	○ 54
4	○ 55	○ 56	○ 57
5	○ 58	○ 59	○ 60
6	○ 61	○ 62	○ 63
7	○ 64	○ 65	○ 66
8	○ 67	○ 68	○ 69
9	○ 70	○ 71	○ 72
10	○ 73	○ 74	○ 75

SECTION TOTAL

Course No. ____________ Date ____________ Name ____________

REVIEW QUIZ

*Mark the following statements as true or false by filling in the bubbles in the **T** (for true) or **F** (for false) column.*

1. The pickup pattern of a system mic can be changed by attaching a different head.
2. If two desk microphones are too close together, their sound pickup may be compromised by multiple-microphone interference.
3. Dual redundancy refers to a backup microphone in case the first mic fails.
4. Because the boundary, or pressure zone, microphone needs a sound-reflecting surface, it should not be used as a hanging mic.
5. Dynamic mics are generally less sensitive to shock and temperature extremes than ribbon mics.
6. Because wireless microphones operate on their own frequency, they are immune to interference from other radio frequencies.
7. A windsock fulfills the identical function as a pop filter.
8. A hand mic clipped to a desk stand can serve as a desk mic.
9. All professional microphones use three-pronged XLR connectors.
10. Blowing into a microphone is a good way to test whether it is turned on.
11. Condenser mics are especially good for the pickup of a bass drum.
12. Using an impedance transformer (a direct box) allows you to play an electric guitar into a mixer.
13. Because lavalier microphones are highly sensitive, they work best when hidden under a shirt or blouse.
14. The parabolic reflector microphone is especially appropriate for intimate, high-quality sound pickup with a high degree of sound presence.
15. Lavaliers can have a dynamic or condenser sound-generating element.
16. Once a microphone is turned off, it is relatively immune to physical shock.

	T	F
1	○ 76	○ 77
2	○ 78	○ 79
3	○ 80	○ 81
4	○ 82	○ 83
5	○ 84	○ 85
6	○ 86	○ 87
7	○ 88	○ 89
8	○ 90	○ 91
9	○ 92	○ 93
10	○ 94	○ 95
11	○ 96	○ 97
12	○ 98	○ 99
13	○ 100	○ 101
14	○ 102	○ 103
15	○ 104	○ 105
16	○ 106	○ 107

SECTION TOTAL

PROBLEM-SOLVING APPLICATIONS

1. You are responsible for the audio pickup of the live remote coverage at the airport during the Thanksgiving rush. Basically, you will have a reporter walking among the people waiting at the ticket counters, briefly interviewing some of the travelers. What type of mic would you use? Why?

2. You are in charge of audio for a show that consists of several intimate numbers by a singer and a small band. After the rehearsal an observer in the control room tells you that the singer holds the mic much too close to her mouth and that she should hold the mic lower and sing *across* rather than *into* it. What is your reaction? Why?

3. You are to provide optimal sound pickup for a preschool children's live-recorded show. The show consists of a host who moves among five to seven children seated on little chairs. The chairs are grouped around a small rug on which the children also play or dance from time to time. The dance music and other recorded audio portions are piped into the studio through the S.A. system. What microphone setup would you suggest for the host and the children? What problems might the S.A. system cause, if any?

4. An official at your former high school asks you to help with the audio for the championship basketball game. Somehow, so the official claims, the visiting spectators seem much louder on television than the home audience, although the latter is actually much larger and noisier than the guests. What can you do to accurately reflect the supportive cheering of the two sides? What specific microphone setups would you use?

5. You are doing a documentary on police patrols in your city. You first want to hear the conversation and the police radio inside the patrol car and then capture the sounds of conversations, yelling, or any other audio when the officers leave the patrol car to confront a suspect. What microphones would you need for optimal sound pickup in these situations?

6. You are to conduct an interview with the university president in her office. What microphones would you use? Why?

7. You are responsible for the pickup of an important live interview with a famous musician. The TD urges you to be sure to have an effective backup in case one of your lavalier mics fails. What is the easiest way to do this?

Course No. ________ Date ________ Name ________

Audio: Sound Control

REVIEW OF KEY TERMS

Match each term with its appropriate definition by filling in the corresponding bubble.

1. **digital audiotape**
2. **sound perspective**
3. **ambience**
4. **automatic gain control**
5. **mix-minus**
6. **equalization**
7. **sweetening**
8. **VU meter**
9. **calibrate**
10. **figure/ground**
11. **MP3**
12. **mixing**

A. A common codec for digital audio

A ○1 ○2 ○3 ○4 ○5 ○6 ○7 ○8 ○9 ○10 ○11 ○12

B. Emphasizing the most important sound source over other sounds

B ○1 ○2 ○3 ○4 ○5 ○6 ○7 ○8 ○9 ○10 ○11 ○12

C. A variety of quality adjustments of recorded sound in postproduction

C ○1 ○2 ○3 ○4 ○5 ○6 ○7 ○8 ○9 ○10 ○11 ○12

PAGE TOTAL

1. digital audiotape	**5. mix-minus**	**9. calibrate**
2. sound perspective	**6. equalization**	**10. figure/ground**
3. ambience	**7. sweetening**	**11. MP3**
4. automatic gain control	**8. VU meter**	**12. mixing**

D. Regulates the audio or video levels automatically, without using pots

D ○1 ○2 ○3 ○4 ○5 ○6 ○7 ○8 ○9 ○10 ○11 ○12

E. Controlling the audio signal by emphasizing certain frequencies and eliminating others

E ○1 ○2 ○3 ○4 ○5 ○6 ○7 ○8 ○9 ○10 ○11 ○12

F. Combining two or more sounds in specific proportions as determined by the event context

F ○1 ○2 ○3 ○4 ○5 ○6 ○7 ○8 ○9 ○10 ○11 ○12

G. Far sounds go with long shots; close sounds with close-ups

G ○1 ○2 ○3 ○4 ○5 ○6 ○7 ○8 ○9 ○10 ○11 ○12

H. Making all VU meters respond in the same way to a specific audio signal

H ○1 ○2 ○3 ○4 ○5 ○6 ○7 ○8 ○9 ○10 ○11 ○12

PAGE TOTAL

Course No. ______ Date ______ Name ______

I. Encodes and records sound signals in digital form

J. Measures the relative loudness of sound

K. Type of multiple audio feed missing the part that is being recorded

L. Environmental sounds

I	○ 1	○ 2	○ 3	○ 4
	○ 5	○ 6	○ 7	○ 8
	○ 9	○ 10	○ 11	○ 12
J	○ 1	○ 2	○ 3	○ 4
	○ 5	○ 6	○ 7	○ 8
	○ 9	○ 10	○ 11	○ 12
K	○ 1	○ 2	○ 3	○ 4
	○ 5	○ 6	○ 7	○ 8
	○ 9	○ 10	○ 11	○ 12
L	○ 1	○ 2	○ 3	○ 4
	○ 5	○ 6	○ 7	○ 8
	○ 9	○ 10	○ 11	○ 12

PAGE TOTAL ______

SECTION TOTAL ______

REVIEW OF STUDIO AND FIELD AUDIO PRODUCTION EQUIPMENT

Select the correct answers and fill in the bubbles with the corresponding numbers.

1. A television audio console lets you (13) *synchronize audio and video in postproduction* (14) *punch up the audio source with the corresponding video* (15) *adjust the volume of each audio input.*

2. All professional video-editing systems let you control and mix (16) *only one audio track* (17) *two or more audio tracks* (18) *no audio tracks unless you interface it with special audio software.*

3. The trim control on the modules of studio consoles regulates the strength of the (19) *outgoing audio signal* (20) *final mix* (21) *incoming audio signal.*

4. The appropriate place that marks the beginning of the "overload zone" is (22) *–5 VU* (23) *–2 VU* (24) *0 VU.*

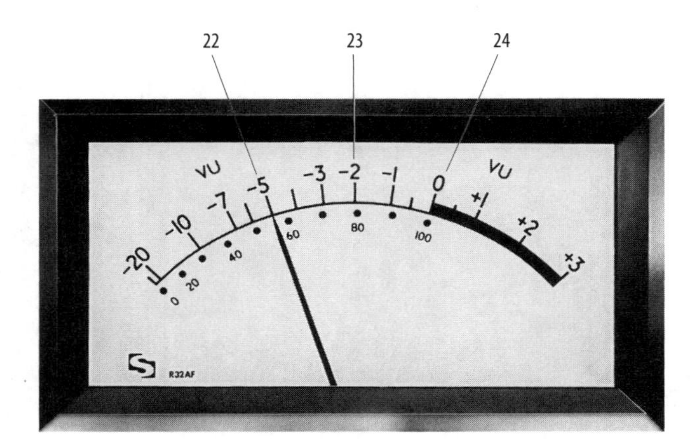

1	○ 13	○ 14	○ 15
2	○ 16	○ 17	○ 18
3	○ 19	○ 20	○ 21
4	○ 22	○ 23	○ 24

PAGE TOTAL

Course No. ______________ Date ______________ Name ______________________________

5. Fill in the bubble whose number identifies the type of head in the head assembly below:

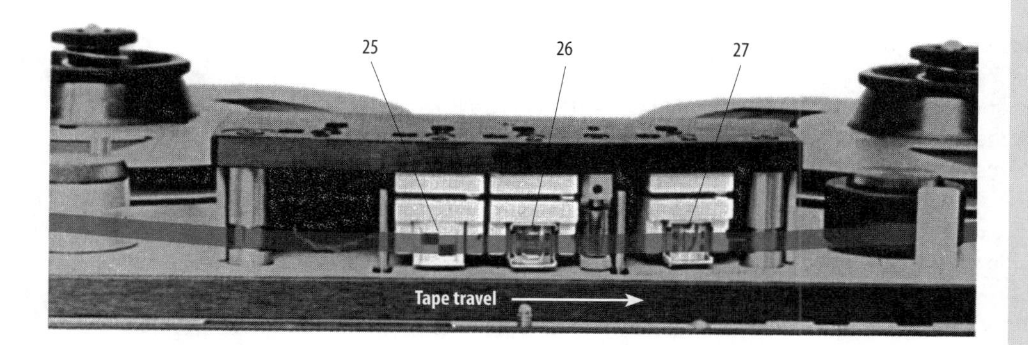

PHOTO: WILLIAM STORM

a. playback head

b. record head

c. erase head

5a	◯ 25	◯ 26	◯ 27
5b	◯ 25	◯ 26	◯ 27
5c	◯ 25	◯ 26	◯ 27

6. Most professional video- and sound-editing software lets you (28) *only see* (29) *only hear* (30) *both see and hear* the audio track.

7. A 16 × 2 audio console has (31) *16 inputs and 2 outputs* (32) *16 slide faders and 2 monitor systems* (33) *16 VU meters and 2 mix buses.*

8. Phantom power means that the power is (34) *virtual but not real* (35) *not necessary* (36) *not supplied by battery but by some other source.*

9. When an incoming mic signal is routed to the mic-level input, it (37) *will be distorted* (38) *must be amplified* (39) *will sound just right.*

10. The equalization controls on console modules (40) *bring all sounds to the same volume level* (41) *emphasize or de-emphasize certain frequencies* (42) *bring all incoming sounds to line-level strength.*

11. The advantage of memory cards over a DAT recorder is that they (43) *have no moving parts* (44) *can store digital and analog audio signals* (45) *can record MP3.*

12. The advantage of hard disks over a DAT recorder is that the audio information can be (46) *digital* (47) *analog or digital* (48) *accessed randomly.*

6	◯ 28	◯ 29	◯ 30
7	◯ 31	◯ 32	◯ 33
8	◯ 34	◯ 35	◯ 36
9	◯ 37	◯ 38	◯ 39
10	◯ 40	◯ 41	◯ 42
11	◯ 43	◯ 44	◯ 45
12	◯ 46	◯ 47	◯ 48

PAGE TOTAL ______

SECTION TOTAL ______

REVIEW OF AUDIO CONTROL

Select the correct answers and fill in the bubbles with the corresponding numbers.

1. Audio-system calibration normally refers to (49) *adjusting the audio input VU meter of the VR to the VU meter of the console output* (50) *adjusting the zoom lens so that it stays in focus* (51) *having the VU meter of the VR peak at a much higher level than the VU meter of the console output.*

2. The proper steps for audio system calibration are:

(52) *1. Activate the control tone on the console or mixer.*
2. Turn up the volume control for the incoming sound on the VR to 0 VU.
3. Bring up the control tone fader on the console or mixer to 0 VU.
4. Bring up the master fader on the console or mixer to 0 VU.

(53) *1. Turn up the volume control for the incoming sound on the VR to 0 VU.*
2. Bring up the control tone fader on the console or mixer to 0 VU.
3. Bring up the master fader on the console or mixer to 0 VU.
4. Activate the control tone on the console or mixer.

(54) *1. Activate the control tone on the console or mixer.*
2. Bring up the master fader on the console or mixer to 0 VU.
3. Bring up the control tone fader on the console or mixer to 0 VU.
4. Turn up the volume control for the incoming sound on the VR to 0 VU.

3. When using a CD player as an additional audio source, it must be connected to the (55) *mic* (56) *line* (57) *neither the mic nor the line* input on the mixer.

4. The audio control tone, which gives a reference level of the recorded material, should be set at (58) *0 VU* (59) *+3 VU* (60) *–3 VU.*

5. The AGC (61) *discriminates automatically between figure and ground* (62) *works especially well in noisy surroundings* (63) *automatically boosts audio levels if they fall below preset levels.*

6. When recording sound during an outdoor EFP, you should (64) *avoid all ambient sounds* (65) *record ambient sounds on a separate track* (66) *re-create the ambient sounds in postproduction.*

1	○ 49	○ 50	○ 51
2	○ 52	○ 53	○ 54
3	○ 55	○ 56	○ 57
4	○ 58	○ 59	○ 60
5	○ 61	○ 62	○ 63
6	○ 64	○ 65	○ 66

PAGE TOTAL

Course No. ______________ Date ______________ Name ______________

7. The correct patches as shown in the figure are: (67) (68) (69) (70) (71) (72).
(Multiple answers are possible.)

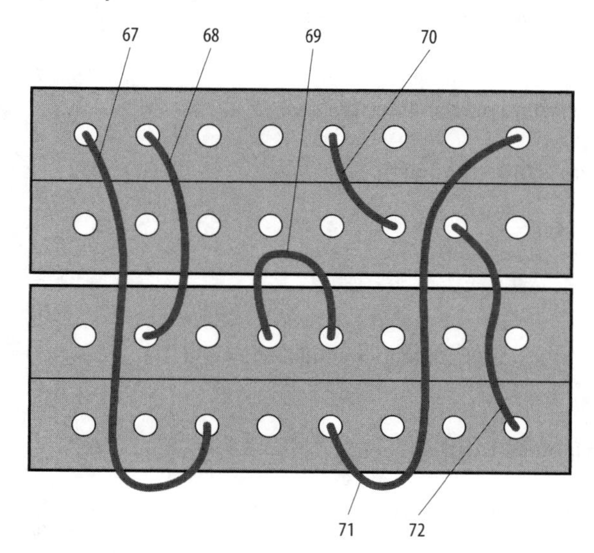

8. A memory card can store (73) *only analog* (74) *only digital* (75) *both analog and digital* audio signals.

9. Indicate the most common place to cut the digital audio track shown below.

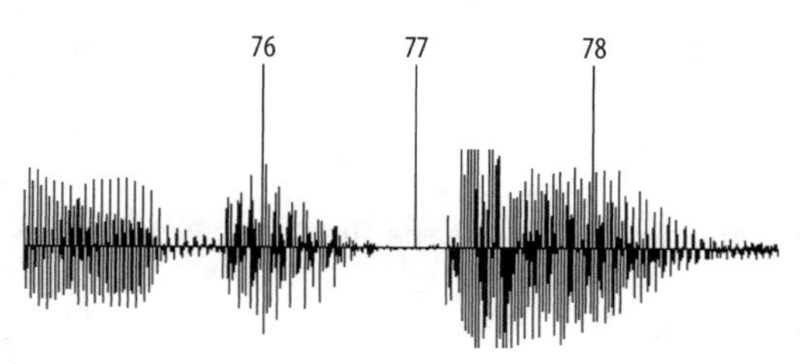

10. Taking a level is (79) *not necessary when the AGC is engaged* (80) *not necessary when recording digital sound* (81) *always necessary.*

11. The least critical speaker placement in a 5.1 surround-sound system is the (82) *front-center speaker* (83) *front-side speakers* (84) *subwoofer.*

12. Recording several minutes of room tone or ambient field sounds during a field production is especially important for (85) *establishing continuity in postproduction* (86) *helping set volume levels in postproduction* (87) *boosting aesthetic energy.*

7	○ 67	○ 68	○ 69
	○ 70	○ 71	○ 72
8	○ 73	○ 74	○ 75
9	○ 76	○ 77	○ 78
10	○ 79	○ 80	○ 81
11	○ 82	○ 83	○ 84
12	○ 85	○ 86	○ 87

PAGE TOTAL

SECTION TOTAL

REVIEW QUIZ

*Mark the following statements as true or false by filling in the bubbles in the **T** (for true) or **F** (for false) column.*

1. The VU meter or the PPM will give an accurate reading of sound perspective.
2. The .1 speaker in the 5.1 surround sound system is the subwoofer.
3. Environmental sounds are always interfering in EFP.
4. All analog recording systems are tape-based.
5. In EFP we should try to mix all sound inputs as much as possible to minimize the need for postproduction mixing.
6. I/O consoles have an output channel for each input channel.
7. On large multichannel consoles, each input channel has its own quality controls.
8. In contrast to large audio consoles, audio mixers have only one input but several outputs.
9. Small digital stereo recorders use an SD memory card for the storage of audio signals.
10. Digital audio signals can be recorded on tape or computer disks.
11. An audio field mixer has quality controls similar to those of a console.
12. One function of the audio console is to route the combined signals to a specific output.
13. When recording digital audio, the master fader level should be kept somewhat below 0 VU.
14. A flat response means that the dynamics are equalized.
15. DAT recorders allow random access.
16. Adding rhythmic sound is one of the primary techniques for establishing visual continuity.
17. A solo switch on the console lets you listen to a single incoming sound while silencing all others.

	T	F
1	○ 88	○ 89
2	○ 90	○ 91
3	○ 92	○ 93
4	○ 94	○ 95
5	○ 96	○ 97
6	○ 98	○ 99
7	○ 100	○ 101
8	○ 102	○ 103
9	○ 104	○ 105
10	○ 106	○ 107
11	○ 108	○ 109
12	○ 110	○ 111
13	○ 112	○ 113
14	○ 114	○ 115
15	○ 116	○ 117
16	○ 118	○ 119
17	○ 120	○ 121

SECTION TOTAL

Course No. ______ Date ______ Name ______

PROBLEM-SOLVING APPLICATIONS

1. When setting up for video-recording a small rock group, you notice that the microphones and other audio sources exceed the number of inputs on the audio console. What can you do?

2. During rehearsal of the same production, you discover that the audio inputs that need the most attention are widely spread apart on the board. How can you get them closer together on the console so that their respective volume controls are adjacent to one another?

3. During a small segment of an EFP in an auto assembly plant, the novice director tells you to be especially careful to mix the ambient sounds and the voices of the reporter and the plant supervisor with the portable mixer so as to facilitate postproduction editing. What is your response?

4. During the digital recording of a concert, the VU meters occasionally peak into the +2 red zone. The director is very concerned about overmodulation. What is your response?

5. Some of the incoming audio signals during a rock concert are so hot (strong) that they bend the needle even at very low fader settings. What can you do to correct this problem without adjusting the source?

6. You have been asked to calibrate the audio system in your studio. Briefly describe the process step-by-step.

7. The director of the evening news would like you, the audio technician, to construct a playlist of all the bumpers for an automated and sequenced playback during the newscast. What piece of widely used audio equipment do you need to accomplish this assignment?

8. You, the audio technician, overhear the floor manager telling the anchorpersons that they do not need to be on the I.F.B. system because it is, after all, his job to relay messages to the talent. What is your response?

9. How can you control the input levels before they reach the camcorder?

10. During postproduction the director insists on laying in a low but highly rhythmical music track, even under the dialogue, to boost the aesthetic energy of the scene. What is your reaction?

Course No. ______________ Date ______________ Name ______________

Lighting

REVIEW OF KEY TERMS

Match each term with its appropriate definition by filling in the corresponding bubble.

1. **ND filter**
2. **baselight**
3. **softlight**
4. **reflected light**
5. **incident light**
6. **cookie**
7. **foot-candle**
8. **lumen**
9. **dimmer**
10. **floodlight**
11. **fluorescent**
12. **spotlight**
13. **incandescent**
14. **lux**
15. **barn doors**

A. A lighting instrument that produces diffused light with a relatively undefined beam edge

A ○1 ○2 ○3 ○4 ○5 ○6 ○7 ○8 ○9 ○10 ○11 ○12 ○13 ○14 ○15

B. Light that is bounced off the illuminated object

B ○1 ○2 ○3 ○4 ○5 ○6 ○7 ○8 ○9 ○10 ○11 ○12 ○13 ○14 ○15

C. A lighting instrument that produces directional, relatively undiffused light

C ○1 ○2 ○3 ○4 ○5 ○6 ○7 ○8 ○9 ○10 ○11 ○12 ○13 ○14 ○15

PAGE TOTAL ______

1. ND filter	**6. cookie**	**11. fluorescent**
2. baselight	**7. foot-candle**	**12. spotlight**
3. softlight	**8. lumen**	**13. incandescent**
4. reflected light	**9. dimmer**	**14. lux**
5. incident light	**10. floodlight**	**15. barn doors**

D. Even, nondirectional (diffused) light necessary for the camera to operate optimally

D ○1 ○2 ○3 ○4 ○5 ○6 ○7 ○8 ○9 ○10 ○11 ○12 ○13 ○14 ○15

E. Metal flaps in front of lighting instruments that control the spread of the light beam

E ○1 ○2 ○3 ○4 ○5 ○6 ○7 ○8 ○9 ○10 ○11 ○12 ○13 ○14 ○15

F. A metal cutout for a pattern projection

F ○1 ○2 ○3 ○4 ○5 ○6 ○7 ○8 ○9 ○10 ○11 ○12 ○13 ○14 ○15

G. Light that strikes the object directly from its source

G ○1 ○2 ○3 ○4 ○5 ○6 ○7 ○8 ○9 ○10 ○11 ○12 ○13 ○14 ○15

H. The American unit of measurement of illumination, or the amount of light that falls on an object

H ○1 ○2 ○3 ○4 ○5 ○6 ○7 ○8 ○9 ○10 ○11 ○12 ○13 ○14 ○15

PAGE TOTAL

Course No. ______ Date ______ Name ______

I. Light produced by a glowing tungsten filament

I				
1	2	3	4	5
6	7	8	9	10
11	12	13	14	15

J. The intensity of one candle (or any other light source radiating isotropically)

J				
1	2	3	4	5
6	7	8	9	10
11	12	13	14	15

K. Lamps that generate light by activating a gas-filled tube

K				
1	2	3	4	5
6	7	8	9	10
11	12	13	14	15

L. European standard for measuring light intensity

L				
1	2	3	4	5
6	7	8	9	10
11	12	13	14	15

M. Floodlight that produces extremely diffused light

M				
1	2	3	4	5
6	7	8	9	10
11	12	13	14	15

N. A device that controls light intensity

N				
1	2	3	4	5
6	7	8	9	10
11	12	13	14	15

PAGE TOTAL ______

1. ND filter	**6. cookie**	**11. fluorescent**
2. baselight	**7. foot-candle**	**12. spotlight**
3. softlight	**8. lumen**	**13. incandescent**
4. reflected light	**9. dimmer**	**14. lux**
5. incident light	**10. floodlight**	**15. barn doors**

O. Reduces incoming light without distorting colors

O ○ 1 ○ 2 ○ 3 ○ 4 ○ 5
○ 6 ○ 7 ○ 8 ○ 9 ○ 10
○ 11 ○ 12 ○ 13 ○ 14 ○ 15

PAGE TOTAL

SECTION TOTAL

Course No. ______________ Date ______________ Name ______________________________

REVIEW OF STUDIO LIGHTING INSTRUMENTS AND CONTROLS

Select the correct answers and fill in the bubbles with the corresponding numbers.

1. Fill in the bubbles whose numbers correspond with the numbers identifying the various parts of the spotlight shown below.

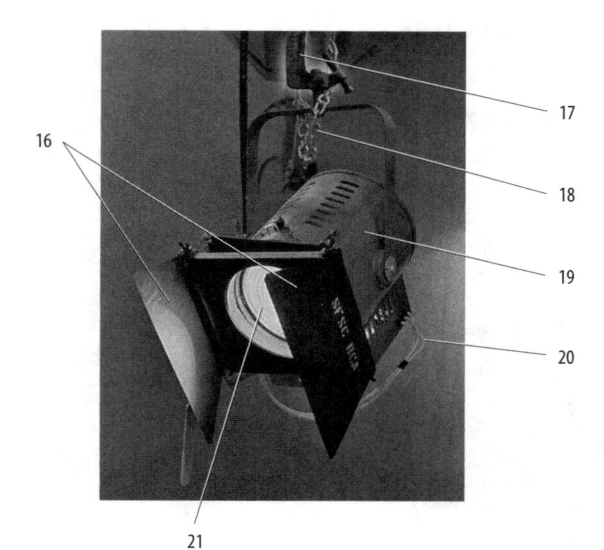

PHOTO: HERBERT ZETTL

a. lamp housing

1a ◯ 16 ◯ 17 ◯ 18 ◯ 19 ◯ 20 ◯ 21

b. power cord

1b ◯ 16 ◯ 17 ◯ 18 ◯ 19 ◯ 20 ◯ 21

c. safety chain

1c ◯ 16 ◯ 17 ◯ 18 ◯ 19 ◯ 20 ◯ 21

d. two-way barn doors

1d ◯ 16 ◯ 17 ◯ 18 ◯ 19 ◯ 20 ◯ 21

e. Fresnel lens

1e ◯ 16 ◯ 17 ◯ 18 ◯ 19 ◯ 20 ◯ 21

f. C-clamp

1f ◯ 16 ◯ 17 ◯ 18 ◯ 19 ◯ 20 ◯ 21

PAGE TOTAL

2. You can make a fluorescent light beam somewhat directional by attaching (22) *an egg crate* (23) *a filter* (24) *barn doors.*

3. Scoops have (25) *a Fresnel lens* (26) *a plain lens* (27) *no lens.*

4. A dimmer controls the (28) *voltage flowing to the lamp* (29) *wattage of the lamp* (30) *amperes flowing to the lamp.*

5. To flood (spread) the light beam of a Fresnel spotlight, you need to move the lamp-reflector unit (31) *toward* (32) *away from* the lens.

6. With the use of the patchboard (or computer patching), you (33) *must link only one instrument* (34) *can link several instruments* (35) *must link all available instruments simultaneously* to a specific dimmer.

7. Fill in the bubbles whose numbers correspond with the appropriate lighting instruments shown below.

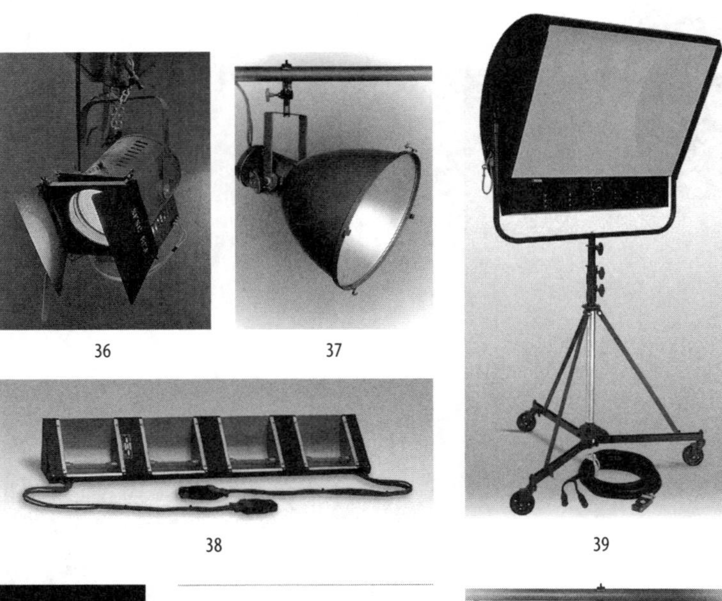

36 37 38 39

40

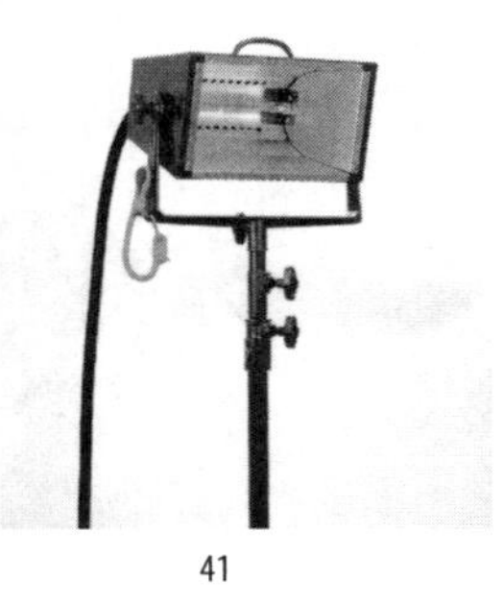

41

42

2	○ 22	○ 23	○ 24
3	○ 25	○ 26	○ 27
4	○ 28	○ 29	○ 30
5	○ 31	○ 32	
6	○ 33	○ 34	○ 35

PAGE TOTAL

Course No. ____________ Date ____________ Name ____________

a. strip, or cyc, light

b. fluorescent floodlight bank

c. Fresnel spotlight

d. scoop

e. ellipsoidal spotlight

f. softlight

g. broad

8. To diffuse the light beam of a scoop even more, you can attach (43) *an egg crate* (44) *a scrim* (45) *color media.*

9. The beam of softlights can be adjusted by (46) *an egg crate* (47) *a focus control* (48) *moving the lamp assembly toward or away from the reflector.*

7a ○ 36 ○ 37 ○ 38 ○ 39 ○ 40 ○ 41 ○ 42

7b ○ 36 ○ 37 ○ 38 ○ 39 ○ 40 ○ 41 ○ 42

7c ○ 36 ○ 37 ○ 38 ○ 39 ○ 40 ○ 41 ○ 42

7d ○ 36 ○ 37 ○ 38 ○ 39 ○ 40 ○ 41 ○ 42

7e ○ 36 ○ 37 ○ 38 ○ 39 ○ 40 ○ 41 ○ 42

7f ○ 36 ○ 37 ○ 38 ○ 39 ○ 40 ○ 41 ○ 42

7g ○ 36 ○ 37 ○ 38 ○ 39 ○ 40 ○ 41 ○ 42

8 ○ 43 ○ 44 ○ 45

9 ○ 46 ○ 47 ○ 48

PAGE TOTAL

SECTION TOTAL

REVIEW OF FIELD LIGHTING INSTRUMENTS AND CONTROLS

Select the correct answers and fill in the bubbles with the corresponding numbers.

1. The diffusion umbrella fulfills a similar function to (49) *a dimmer* (50) *a soft box* (51) *barn doors.*

2. The easiest way to reduce the light intensity of a portable spot is to (52) *use a small dimmer* (53) *use a smaller lamp* (54) *move the light farther away from the object.*

3. One common way to diffuse the light of an open-face spot is to (55) *attach a scrim to the barn doors* (56) *attach color media* (57) *use a dimmer.*

4. Portable light stands must always be secured with (58) *a safety cable* (59) *sandbags* (60) *counterweights.*

5. Fill in the bubbles whose numbers correspond with the appropriate instruments shown below.

61

62

63

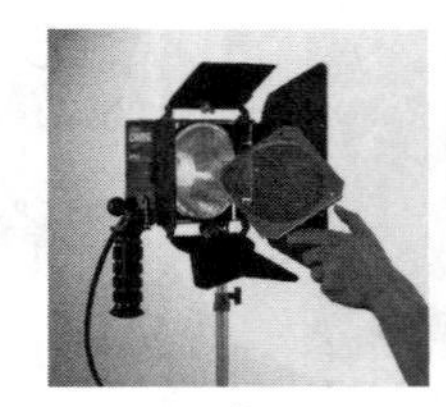
64

65

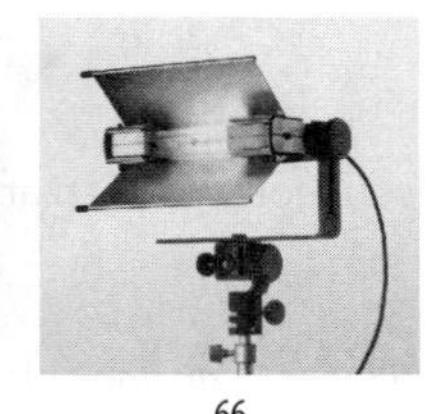
66

67

68

1	○ 49	○ 50	○ 51
2	○ 52	○ 53	○ 54
3	○ 55	○ 56	○ 57
4	○ 58	○ 59	○ 60

PAGE TOTAL

Course No. ______________ Date ________________ Name ______________________________

a. Chinese lantern

b. open-face spot

c. portable fluorescent

d. internal reflector light

e. camera light

f. soft box

g. V-light

h. small Fresnel spot

5a ○ 61 ○ 62 ○ 63 ○ 64 ○ 65 ○ 66 ○ 67 ○ 68

5b ○ 61 ○ 62 ○ 63 ○ 64 ○ 65 ○ 66 ○ 67 ○ 68

5c ○ 61 ○ 62 ○ 63 ○ 64 ○ 65 ○ 66 ○ 67 ○ 68

5d ○ 61 ○ 62 ○ 63 ○ 64 ○ 65 ○ 66 ○ 67 ○ 68

5e ○ 61 ○ 62 ○ 63 ○ 64 ○ 65 ○ 66 ○ 67 ○ 68

5f ○ 61 ○ 62 ○ 63 ○ 64 ○ 65 ○ 66 ○ 67 ○ 68

5g ○ 61 ○ 62 ○ 63 ○ 64 ○ 65 ○ 66 ○ 67 ○ 68

5h ○ 61 ○ 62 ○ 63 ○ 64 ○ 65 ○ 66 ○ 67 ○ 68

PAGE TOTAL

SECTION TOTAL

REVIEW OF LIGHT INTENSITY, LAMPS, AND COLOR MEDIA

Select the correct answers and fill in the bubbles with the corresponding numbers.

1. One foot-candle is approximately (69) *10* (70) *100 lux* (71) *1.*
2. When measuring baselight, you need to read (72) *incident* (73) *reflected* (74) *directional light.*
3. When measuring incident light, you point the foot-candle or lux meter (75) *toward the set* (76) *toward the camera lens* (77) *close to the lighted object.*
4. When reading reflected light, you point the light meter (78) *into the lights* (79) *close to the lighted object* (80) *toward the camera lens.*
5. The beam of softlights (81) *can be adjusted by moving the lamp-reflector unit toward or away from the reflector* (82) *can be adjusted by attaching a Fresnel lens* (83) *cannot be sharply focused.*
6. To flood (spread) the light beam of a Fresnel spotlight, you need to move the lamp-reflector unit (84) *away from* (85) *toward* the lens.
7. Colors are frequently distorted by (86) *inadequate baselight levels* (87) *low-contrast lighting* (88) *lack of shadows.*
8. Quartz lamps fall into the (89) *fluorescent* (90) *incandescent* (91) *HMI* category.
9. The advantage of a quartz lamp is that it (92) *burns at a lower temperature* (93) *does not change color temperature over time* (94) *will not burn out over time.*
10. When aiming a colored light beam on a colored object, they (95) *mix additively* (96) *mix subtractively* (97) *do not mix.*
11. When aiming two different-colored light beams on a neutral-colored background area, they (98) *mix additively* (99) *mix subtractively* (100) *do not mix.*
12. Most fluorescent tubes (101) *burn at exactly 3,200K and 5,600K* (102) *approximate the indoor and outdoor color temperature standards* (103) *do not burn with a color temperature at all.*
13. When putting a red and a green color media in front of the same instrument, you will (104) *generate a yellow light beam* (105) *generate a cool red beam* (106) *block all light coming from the instrument.*

1	○ 69	○ 70	○ 71
2	○ 72	○ 73	○ 74
3	○ 75	○ 76	○ 77
4	○ 78	○ 79	○ 80
5	○ 81	○ 82	○ 83
6	○ 84	○ 85	
7	○ 86	○ 87	○ 88
8	○ 89	○ 90	○ 91
9	○ 92	○ 93	○ 94
10	○ 95	○ 96	○ 97
11	○ 98	○ 99	○ 100
12	○ 101	○ 102	○ 103
13	○ 104	○ 105	○ 106

SECTION TOTAL

Course No. ____________ Date ____________ Name ____________

REVIEW QUIZ

*Mark the following statements as true or false by filling in the bubbles in the **T** (for true) or **F** (for false) column.*

1. A sliding rod and pantograph fulfill similar functions.
2. Focusing a light results in sharper shadows.
3. Barn doors are primarily used for intensity control.
4. To illuminate a large area with even light, we use a variety of Fresnel spots.
5. Portable fluorescent banks are used to illuminate areas with even light.
6. You can use egg crates to further soften the beam of softlights.
7. The shutters on an ellipsoidal spot can shape its beam.
8. One effective method of turning a spotlight into a floodlight is to shine its beam into a diffusion umbrella.
9. Incident light can be measured by pointing the light meter into the lights or toward the camera lens.
10. Regardless of the type of dimmer control, all patching must be done with patch cords for each instrument.
11. An HMI light needs an external ballast to function.
12. When necessary, the beam of softlights can be focused.
13. LED lights can be used to illuminate small areas for close-ups.
14. A flag has a similar function to barn doors.
15. LED lights throw a sharp light beam.
16. The inverse square law is independent of how much the light is collimated.
17. A reflector can substitute for a fill light.

	T	F
1	○ 107	○ 108
2	○ 109	○ 110
3	○ 111	○ 112
4	○ 113	○ 114
5	○ 115	○ 116
6	○ 117	○ 118
7	○ 119	○ 120
8	○ 121	○ 122
9	○ 123	○ 124
10	○ 125	○ 126
11	○ 127	○ 128
12	○ 129	○ 130
13	○ 131	○ 132
14	○ 133	○ 134
15	○ 135	○ 136
16	○ 137	○ 138
17	○ 139	○ 140

PAGE TOTAL ____

SECTION TOTAL ____

PROBLEM-SOLVING APPLICATIONS

1. You are asked to raise the baselight level in a classroom for optimal camera performance. Even though the small portable spotlights are in the maximum flood position, the additional illumination is not even. What other methods do you have available to achieve further diffusion?

2. You are asked to produce extremely sharp beams that reflect as precise pools of light on the studio floor. What type of lighting instruments would you use?

3. When checking the general baselight level and the amount of foot-candles (or lux) falling on the subject, the lighting assistant first stands next to the lighted subject and points the light meter toward the principal camera position. Will the assistant's action produce the desired results? If so, why? If not, why not?

4. You are asked to assemble a lighting kit that will be useful for lighting indoor interviews in small rooms, such as hotel rooms and offices. What instruments and other necessary equipment would you recommend?

5. You are asked to dim all spotlights simultaneously and then do the same thing immediately thereafter with all floodlights. How can you best accomplish this task?

6. The producer asked you to shine a yellow light on a blue commercial display to make one side look "greenish." What is your reaction?

7. The basketball coach of the local high school asks the television production teacher to flood the gym with HMI lights. What are your concerns, if any?

8. When video-recording a commercial for a local jewelry store, you are asked by the owner to try LED lights for illuminating a diamond ring because he wants a minimum of shadows. What is your reaction?

Course No. ______________ Date ______________ Name ______________________________

Techniques of Television Lighting

REVIEW OF KEY TERMS

Match each term with its appropriate definition by filling in the corresponding bubble.

1. **photographic lighting principle**
2. **side light**
3. **key light**
4. **light plot**
5. **fill light**
6. **cameo lighting**
7. **floor plan**
8. **falloff**
9. **background light**
10. **kicker light**
11. **low-key**
12. **silhouette lighting**
13. **high-key**
14. **back light**

A. Illuminates the set, set pieces, and backdrops

A ○1 ○2 ○3 ○4 ○5 ○6 ○7 ○8 ○9 ○10 ○11 ○12 ○13 ○14

B. Dark background, with a few selective light sources on the scene

B ○1 ○2 ○3 ○4 ○5 ○6 ○7 ○8 ○9 ○10 ○11 ○12 ○13 ○14

C. Illumination from behind and above the subject and opposite the camera

C ○1 ○2 ○3 ○4 ○5 ○6 ○7 ○8 ○9 ○10 ○11 ○12 ○13 ○14

PAGE TOTAL

1. **photographic lighting principle**
2. **side light**
3. **key light**
4. **light plot**
5. **fill light**
6. **cameo lighting**
7. **floor plan**
8. **falloff**
9. **background light**
10. **kicker light**
11. **low-key**
12. **silhouette lighting**
13. **high-key**
14. **back light**

D. The speed with which a light picture portion turns into shadow area

D ○ 1 ○ 2 ○ 3 ○ 4 ○ 5 ○ 6 ○ 7 ○ 8 ○ 9 ○ 10 ○ 11 ○ 12 ○ 13 ○ 14

E. Unlighted subject in front of a brightly illuminated background

E ○ 1 ○ 2 ○ 3 ○ 4 ○ 5 ○ 6 ○ 7 ○ 8 ○ 9 ○ 10 ○ 11 ○ 12 ○ 13 ○ 14

F. Light background and ample light on the scene

F ○ 1 ○ 2 ○ 3 ○ 4 ○ 5 ○ 6 ○ 7 ○ 8 ○ 9 ○ 10 ○ 11 ○ 12 ○ 13 ○ 14

G. The triangular arrangement of the three major light sources used to illuminate a subject

G ○ 1 ○ 2 ○ 3 ○ 4 ○ 5 ○ 6 ○ 7 ○ 8 ○ 9 ○ 10 ○ 11 ○ 12 ○ 13 ○ 14

H. Lighted subject in front of a dark background

H ○ 1 ○ 2 ○ 3 ○ 4 ○ 5 ○ 6 ○ 7 ○ 8 ○ 9 ○ 10 ○ 11 ○ 12 ○ 13 ○ 14

PAGE TOTAL

Course No. ____________ Date ______________ Name ______________________________

3. What major light sources were used to illuminate the host of a sports show in the following four pictures? In the diagrams, circle the instrument(s) used; then fill in the bubbles whose numbers correspond with the instruments used to light the subject.

a.

PHOTO: EDWARD AIONA

21

22

23

24

Camera

3a ○ 21 ○ 22 ○ 23 ○ 24

b.

PHOTO: EDWARD AIONA

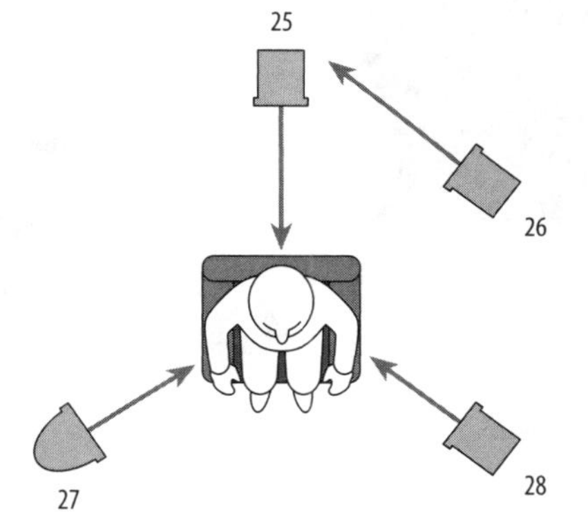

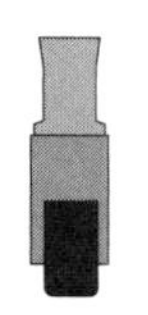

Camera

3b ○ 25 ○ 26 ○ 27 ○ 28

PAGE TOTAL ______

c.

PHOTO: EDWARD AIONA

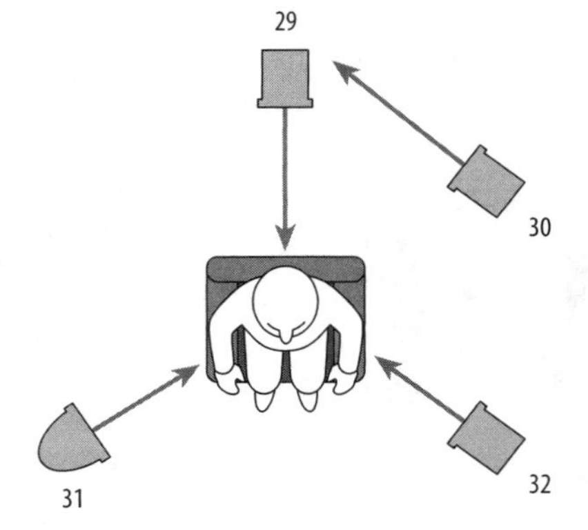

3c ◯ ◯ ◯ ◯
29 30 31 32

d.

PHOTO: EDWARD AIONA

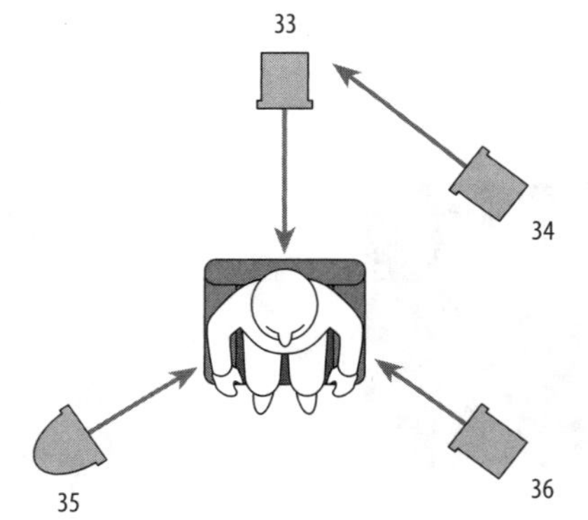

3d ◯ ◯ ◯ ◯
33 34 35 36

PAGE TOTAL

Course No. ______ Date ______ Name ______

I. Additional light that illuminates shadow areas and thereby reduces falloff

J. Principal source of illumination

K. A plan that shows each lighting instrument used relative to the scene to be lighted

L. Directional light from the side of an object

M. Directional light coming from the side and the back of the subject, usually from below

N. A diagram of scenery and major properties drawn onto a grid

I ○1 ○2 ○3 ○4 ○5 ○6 ○7 ○8 ○9 ○10 ○11 ○12 ○13 ○14

J ○1 ○2 ○3 ○4 ○5 ○6 ○7 ○8 ○9 ○10 ○11 ○12 ○13 ○14

K ○1 ○2 ○3 ○4 ○5 ○6 ○7 ○8 ○9 ○10 ○11 ○12 ○13 ○14

L ○1 ○2 ○3 ○4 ○5 ○6 ○7 ○8 ○9 ○10 ○11 ○12 ○13 ○14

M ○1 ○2 ○3 ○4 ○5 ○6 ○7 ○8 ○9 ○10 ○11 ○12 ○13 ○14

N ○1 ○2 ○3 ○4 ○5 ○6 ○7 ○8 ○9 ○10 ○11 ○12 ○13 ○14

PAGE TOTAL

SECTION TOTAL

REVIEW OF LIGHTING TECHNIQUES

Select the correct answers and fill in the bubbles with the corresponding numbers.

1. The arrangement of the lighting instruments shown in the following figure is generally called (15) *photographic lighting principle* (16) *four-point lighting* (17) *field lighting principle.*

2. Fill in the bubbles whose numbers correspond with the functions of the lighting instruments shown in the figure below and whether they are usually (S) *spotlights* or (F) *floodlights.*

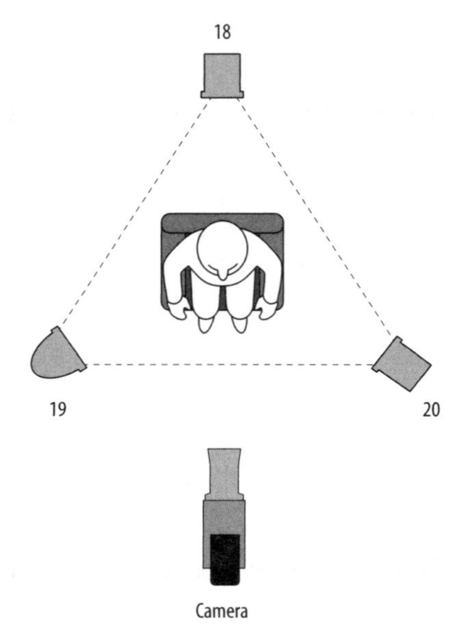

a. fill

b. back

c. key

1 ○ 15 ○ 16 ○ 17

2a ○ 18 ○ 19 ○ 20
○ S ○ F

2b ○ 18 ○ 19 ○ 20
○ S ○ F

2c ○ 18 ○ 19 ○ 20
○ S ○ F

PAGE TOTAL

Course No. ________ Date ________ Name ________________

3. What major light sources were used to illuminate the host of a sports show in the following four pictures? In the diagrams, circle the instrument(s) used; then fill in the bubbles whose numbers correspond with the instruments used to light the subject.

a.

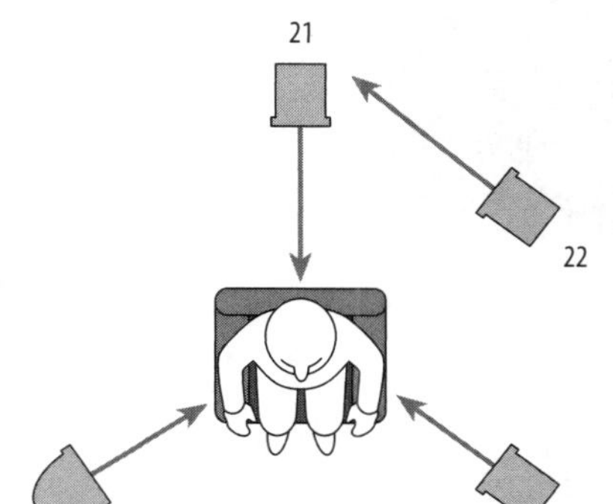

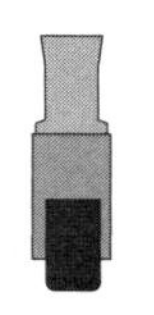

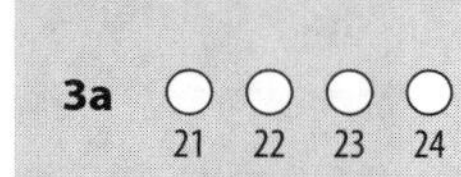

PHOTO: EDWARD AIONA

b.

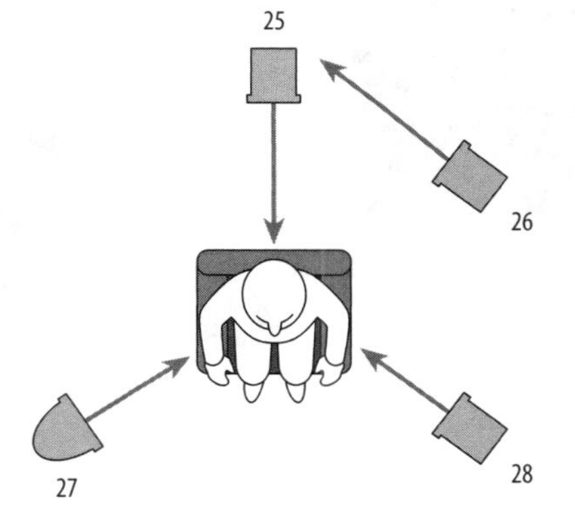

3b ○ 25 ○ 26 ○ 27 ○ 28

PHOTO: EDWARD AIONA

PAGE TOTAL

c.

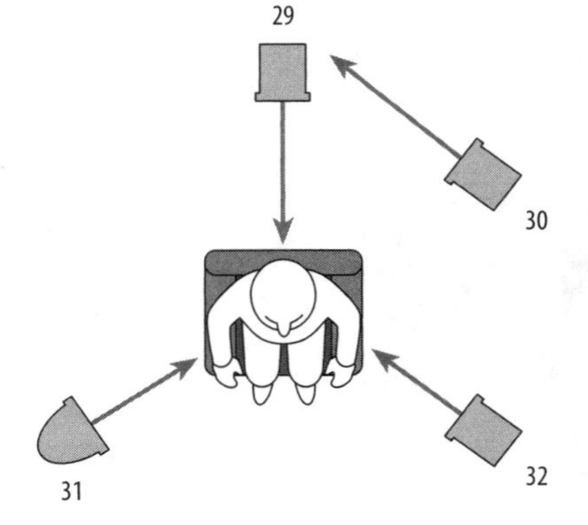

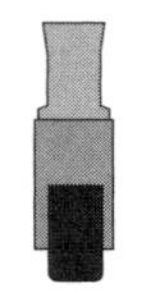

3c ◯ ◯ ◯ ◯
29 30 31 32

PHOTO: EDWARD AIONA

d.

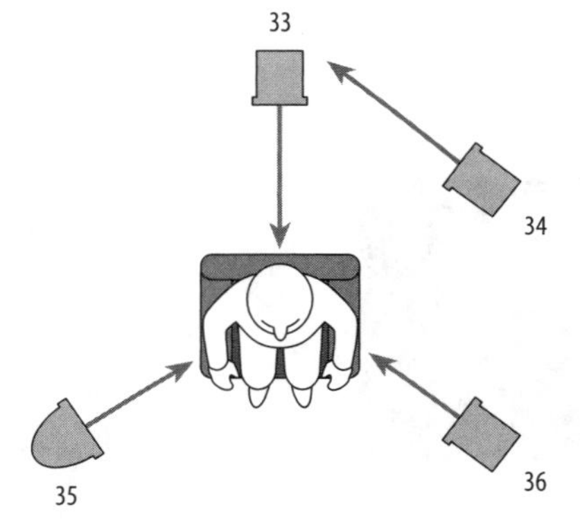

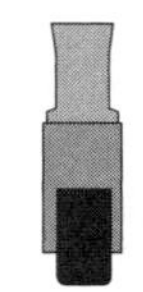

3d ◯ ◯ ◯ ◯
33 34 35 36

PHOTO: EDWARD AIONA

PAGE TOTAL

Course No. ________ Date ________ Name ________

4. You are to evaluate normal lighting setups. In the following six figures, cross out the lighting instruments that are unnecessary or most likely to interfere with the intended lighting effects; then fill in the bubbles whose numbers correspond with the instruments *needed*.

a. cameo lighting

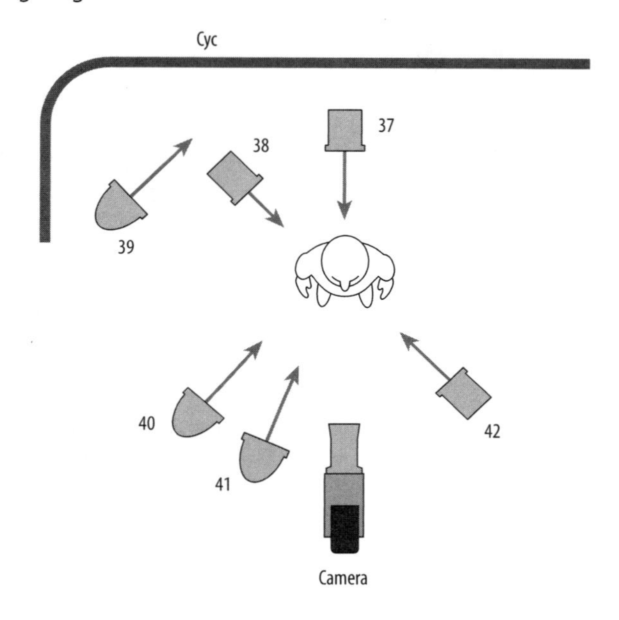

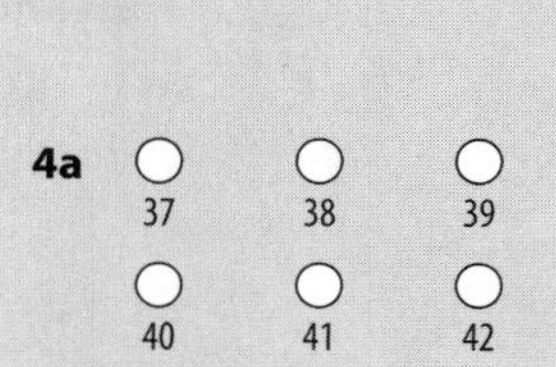

b. small still-life lighting

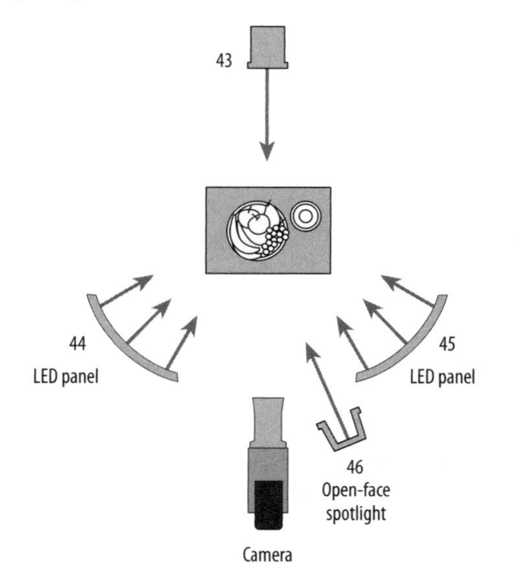

4b ○ 43 ○ 44 ○ 45 ○ 46

PAGE TOTAL

c. chroma-key-area lighting

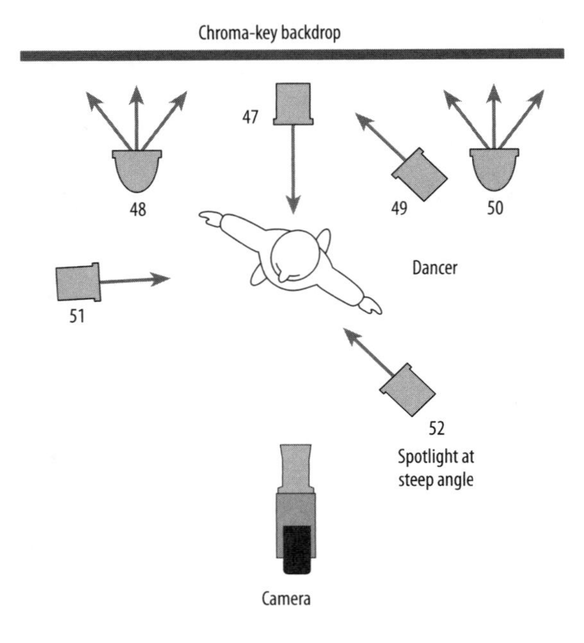

d. newscast

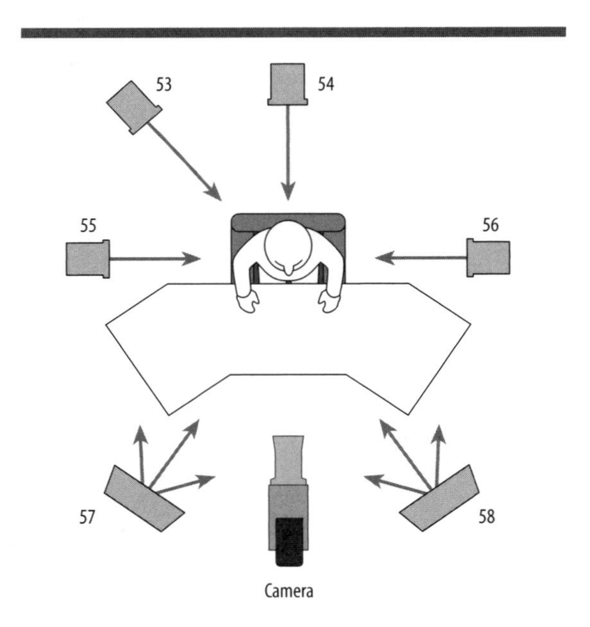

4c ○ 47 ○ 48 ○ 49 ○ 50 ○ 51 ○ 52

4d ○ 53 ○ 54 ○ 55 ○ 56 ○ 57 ○ 58

PAGE TOTAL

Course No. ____________ Date ____________ Name ____________

e. speaker and audience

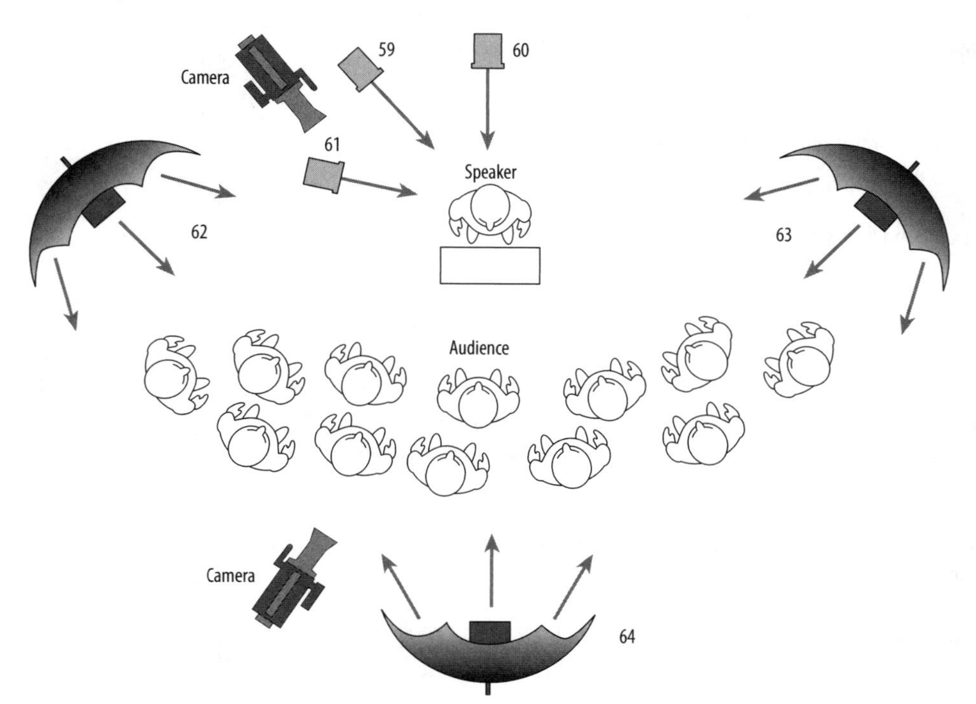

f. dancer in silhouette

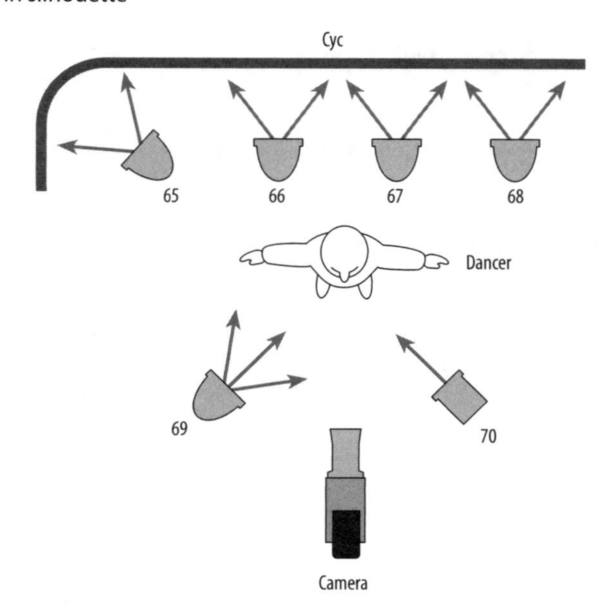

4e ○ 59 ○ 60 ○ 61
○ 62 ○ 63 ○ 64

4f ○ 65 ○ 66 ○ 67
○ 68 ○ 69 ○ 70

PAGE TOTAL

5. Excessive dimming of incandescent lights (more than 10 percent of full power) will (71) *not affect* (72) *decrease* (73) *increase* the color temperature. This means that the white light will (74) *remain basically unchanged* (75) *turn reddish* (76) *turn bluish.* ***(Fill in two bubbles.)***

6. To figure the total wattage that a circuit can safely carry, you should multiply the number of amps by (77) *15* (78) *75* (79) *100.*

7. To light the backdrop for a chroma key, you need (80) *Fresnel spotlights* (81) *ellipsoidal spotlights* (82) *floodlights.*

8. The color temperature of a light can be raised by using (83) *an orange gel* (84) *a light-blue gel* (85) *an amber gel.*

9. Having somebody stand in front of a brightly illuminated building will (86) *provide much needed back light* (87) *cause an undesirable silhouette effect* (88) *help separate the person from the background.*

10. When shooting an ENG interview in bright sunlight, the most convenient fill light is (89) *an HMI spot* (90) *a quartz scoop* (91) *a reflector.*

11. The standard color temperature for outdoor light in video is (92) *5,600K* (93) *3,600K* (94) *3,200K.*

12. These lights fulfill similar functions: (95) *key and fill* (96) *back light and kicker* (97) *background light and kicker.*

13. To make a model's hair look especially glamorous, you need a high-intensity (98) *key light* (99) *back light* (100) *background light.*

14. The usual power rating per circuit of ordinary household wall outlets is (101) *15 amps* (102) *50 amps* (103) *150 amps.*

15. To achieve fast falloff, you need to use primarily (104) *spotlights* (105) *floodlights* (106) *fluorescent lights.*

16. To achieve a sharp shadow of a person cast onto a staircase wall, you need a (107) *focused Fresnel spot* (108) *Chinese lantern* (109) *softlight.*

5	○ 71	○ 72	○ 73
	○ 74	○ 75	○ 76
6	○ 77	○ 78	○ 79
7	○ 80	○ 81	○ 82
8	○ 83	○ 84	○ 85
9	○ 86	○ 87	○ 88
10	○ 89	○ 90	○ 91
11	○ 92	○ 93	○ 94
12	○ 95	○ 96	○ 97
13	○ 98	○ 99	○ 100
14	○ 101	○ 102	○ 103
15	○ 104	○ 105	○ 106
16	○ 107	○ 108	○ 109

PAGE TOTAL

SECTION TOTAL

Course No. ____________ Date ______________ Name ______________________________

REVIEW QUIZ

*Mark the following statements as true or false by filling in the bubbles in the **T** (for true) or **F** (for false) column.*

1. The background light must strike the background from the same side as the key light.
2. Low-key lighting is best achieved with low-hanging softlights.
3. The photographic lighting principle, or triangle lighting, uses a key light, a kicker light, and a back light.
4. In multiple-function lighting, the key light can act as a back light, and the side light as key, depending on the position of the camera relative to the subject.
5. Back lights and background lights fulfill similar functions.
6. The less fill light, the slower the falloff.
7. An LED panel simulates more a spotlight than a floodlight.
8. Color temperature measures the relative reddishness and bluishness of white light.
9. All cameo lighting is highly directional.
10. In most cases, a reflector can substitute for a fill light.
11. Low-key lighting means that the lighting is soft and even, with extremely slow falloff.
12. Cast shadows can suggest a specific locale.
13. Plugging portable lights into different wall outlets means that they are automatically on different power circuits.
14. Two side lights can function similarly to a key and a fill.
15. High-key lighting means that the key light strikes the subject from above eye level.

	T	F
1	○ 110	○ 111
2	○ 112	○ 113
3	○ 114	○ 115
4	○ 116	○ 117
5	○ 118	○ 119
6	○ 120	○ 121
7	○ 122	○ 123
8	○ 124	○ 125
9	○ 126	○ 127
10	○ 128	○ 129
11	○ 130	○ 131
12	○ 132	○ 133
13	○ 134	○ 135
14	○ 136	○ 137
15	○ 138	○ 139

SECTION TOTAL ______

PROBLEM-SOLVING APPLICATIONS

1. You are asked to do the lighting for a shampoo commercial. The director wants you to make the model's blond hair look especially brilliant and glamorous. Which of the three instruments of the lighting triangle needs special attention to achieve the desired result?

2. The show is a brief address by the CEO. Prepare a light plot for the floor plan shown at right. Sketch the type and the locations of the instruments used, as well as the general direction of the light beams.

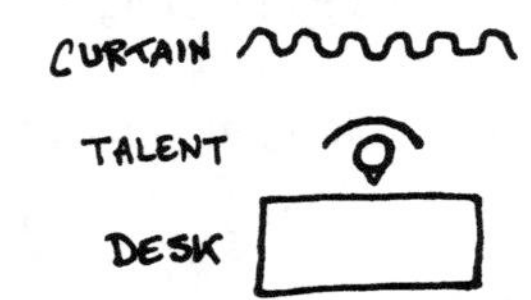

3. You (the camera operator) and a field reporter are sent by the assignment editor to the Plaza Hotel to interview a famous soprano in her room. The field reporter will remain off-camera during the entire interview. The lighting in the hotel room is inadequate, so you need additional lighting. Besides the camera light, you have only one Omni light at your disposal. Where would you place the Omni light? Why?

4. You are the LD for an indoor springtime fashion show. The studio audience is seated along both sides of the runway. The novice director suggests low-key lighting to give the show some extra sparkle. Do you agree with the director's suggestion? If so, why? If not, why not? What are your recommendations?

5. You are the LD for a dance number that plays in front of a light-gray cyc. The dancers wear off-white leotards. The choreographer wants them to appear first in cameo as illuminated figures against a dark blue background and then, in one continuous take, as black silhouettes moving against a bright red background. Can you fulfill the choreographer's request? If so, how? If not, why not?

6. You will be shooting an interview with the CEO of a large software company. Her office has a large window without any curtains. How would you light this interview while taking advantage of the daylight coming through the window? List specific lights, their locations, and all necessary support equipment.

7. You are covering the dedication of a new library. It is a cloudless, sunny day with the sun reflecting off the brilliant white building. The dedication is planned to happen right in front of the building. What are your concerns regarding lighting? What would you suggest to minimize some of the problems?

8. You are to light a two-anchor news set in which the two co-anchors (a dark-haired man and a blond woman) sit side-by-side. The man is worried about his wrinkles, especially because his co-anchor has perfectly smooth skin. What lighting would you suggest? Draw a rough light plot that indicates the type and the approximate locations of the instruments used.

Course No. ____________ Date ____________ Name ____________

Video-recording and Storage Systems

REVIEW OF KEY TERMS

Match each term with its appropriate definition by filling in the corresponding bubble.

1. Y/color difference component system
2. composite system
3. field log
4. control track
5. time base corrector
6. JPEG
7. tapeless video recorder
8. Y/C component system
9. codec
10. ESS system
11. memory card
12. compression
13. MPEG-2
14. framestore synchronizer

A. All digital video recorders that record or store information on a hard drive or read/write optical disc

A
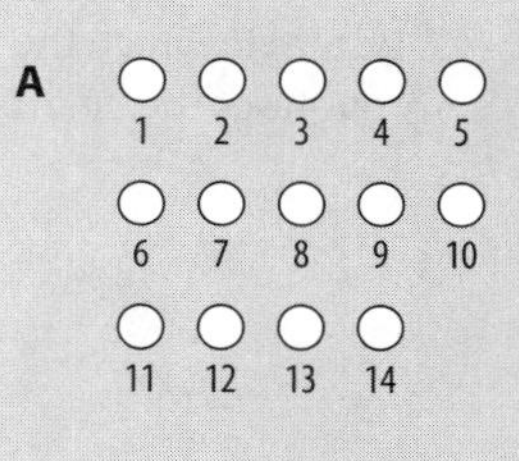

B. A list of shots taken during the recording

B

C. A solid-state read/write digital storage media that has no moving parts

C

PAGE TOTAL

1. Y/color difference component system	**5. time base corrector**	**10. ESS system**
2. composite system	**6. JPEG**	**11. memory card**
3. field log	**7. tapeless video recorder**	**12. compression**
4. control track	**8. Y/C component system**	**13. MPEG-2**
	9. codec	**14. framestore synchronizer**

D. Electronic accessory to a VTR that makes playbacks or transfers stable

D ○ 1 ○ 2 ○ 3 ○ 4 ○ 5 ○ 6 ○ 7 ○ 8 ○ 9 ○ 10 ○ 11 ○ 12 ○ 13 ○ 14

E. A video signal in which luminance (Y), chrominance (C), and sync information are encoded into a single signal

E ○ 1 ○ 2 ○ 3 ○ 4 ○ 5 ○ 6 ○ 7 ○ 8 ○ 9 ○ 10 ○ 11 ○ 12 ○ 13 ○ 14

F. A system in which the luminance (Y) and the R–Y and B–Y signals are kept separate throughout the video-recording process

F ○ 1 ○ 2 ○ 3 ○ 4 ○ 5 ○ 6 ○ 7 ○ 8 ○ 9 ○ 10 ○ 11 ○ 12 ○ 13 ○ 14

G. A compression system generally used for digital television

G ○ 1 ○ 2 ○ 3 ○ 4 ○ 5 ○ 6 ○ 7 ○ 8 ○ 9 ○ 10 ○ 11 ○ 12 ○ 13 ○ 14

H. Image stabilization and synchronization system that stores and reads out one complete video frame

H ○ 1 ○ 2 ○ 3 ○ 4 ○ 5 ○ 6 ○ 7 ○ 8 ○ 9 ○ 10 ○ 11 ○ 12 ○ 13 ○ 14

PAGE TOTAL

Course No. ______ Date ______ Name ______

I. The track on a videotape that contains synchronization information

J. Reducing the amount of digital data for recording or transmission

K. A system in which the luminance (Y) and chrominance (C) signals are kept separate during the encoding and decoding processes but are recorded together

L. A video compression method mostly for still pictures

M. A specific compression standard

N. An electronic device that can grab and digitally store a single frame from any video source

I	1	2	3	4	5
	6	7	8	9	10
	11	12	13	14	
J	1	2	3	4	5
	6	7	8	9	10
	11	12	13	14	
K	1	2	3	4	5
	6	7	8	9	10
	11	12	13	14	
L	1	2	3	4	5
	6	7	8	9	10
	11	12	13	14	
M	1	2	3	4	5
	6	7	8	9	10
	11	12	13	14	
N	1	2	3	4	5
	6	7	8	9	10
	11	12	13	14	

PAGE TOTAL

SECTION TOTAL

REVIEW OF TAPE-BASED AND TAPELESS VIDEO RECORDING

Select the correct answers and fill in the bubbles with the corresponding numbers.

1. Digital recording systems are (15) *nonlinear* (16) *linear* (17) *linear or nonlinear, depending on the recording system.*

2. In contrast to digital recordings, analog recordings (18) *experience quality loss* (19) *remain the same* (20) *lose only color hue* from one generation to the next.

3. Digital video signals can be recorded (21) *only on a computer disk* (22) *only on videotape* (23) *on tape as well as disk.*

4. The NTSC signal is based on a (24) *Y/C component* (25) *composite* (26) *Y/color difference component signal.*

5. Disk-based recording systems are always (27) *linear only* (28) *nonlinear only* (29) *both linear and nonlinear.*

6. All tapeless systems (30) *have no moving parts* (31) *can record digital information only* (32) *can record video only.*

7. An ESS system stores individual frames in (33) *digital only* (34) *analog only* (35) *both digital and analog form.*

8. The type of compression that looks for redundancies from one frame to the next is (36) *interframe* (37) *intraframe* (38) *lossless.*

9. You can use a ¼-inch DV mini-cassette for recording (39) *standard digital signals only* (40) *HDTV only* (41) *standard and HDTV signals.*

10. The control track is essential for (42) *videotape recordings only* (43) *memory card recordings only* (44) *both tape and memory card recordings.*

11. An ESS system can store (45) *individual video frames only* (46) *short video clips only* (47) *both frames and clips.*

12. The Y/C component signal separates the (48) *color and luminance signals* (49) *audio and video signals* (50) *yellow and cyan color signals.*

13. The best color fidelity is achieved through a (51) *4:2:2* (52) *4:1:1* (53) *4:0:0* sampling ratio.

1	○ 15	○ 16	○ 17
2	○ 18	○ 19	○ 20
3	○ 21	○ 22	○ 23
4	○ 24	○ 25	○ 26
5	○ 27	○ 28	○ 29
6	○ 30	○ 31	○ 32
7	○ 33	○ 34	○ 35
8	○ 36	○ 37	○ 38
9	○ 39	○ 40	○ 41
10	○ 42	○ 43	○ 44
11	○ 45	○ 46	○ 47
12	○ 48	○ 49	○ 50
13	○ 51	○ 52	○ 53

PAGE TOTAL

Course No. ______ Date ______ Name ______

14. The illustration below shows a (54) *Y/C component signal* (55) *Y/color difference component signal* (56) *composite signal.*

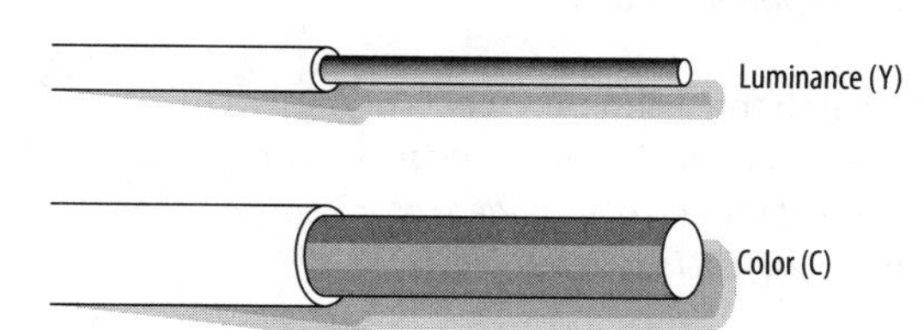

15. Intraframe compression eliminates (57) *various frames* (58) *temporal redundancy in each frame* (59) *spatial redundancy in each frame.*

16. The digital recording system with no moving parts uses (60) *optical discs* (61) *hard drives* (62) *memory cards.*

17. One of the functions that distinguishes a server from a standard large-capacity hard drive is that the server can (63) *record large amounts of digital data* (64) *supply different material to various clients simultaneously* (65) *allow random access of clips.*

14	○ 54	○ 55	○ 56
15	○ 57	○ 58	○ 59
16	○ 60	○ 61	○ 62
17	○ 63	○ 64	○ 65

PAGE TOTAL ______

SECTION TOTAL ______

REVIEW OF HOW VIDEO RECORDING IS DONE

Select the correct answers and fill in the bubbles with the corresponding numbers.

1. Because color bars help the video-record operator match the technical aspects of the playback VR and the playback monitor, you should record them (66) *at the beginning of the video recording* (67) *right after the video leader* (68) *at the end of the video recording* for at least (69) *10 seconds* (70) *30 seconds* (71) *5 minutes.* ***(Fill in two bubbles.)***

2. The field log is normally kept by the (72) *VR operator* (73) *TD* (74) *VO.*

3. To protect a cassette recording from being accidentally erased, the cassette tab (75) *must be in place or in the closed position* (76) *must be removed or in the open position* (77) *cannot prevent erasure.*

4. Nonlinear editing requires the transfer to the computer hard drive (78) *of only analog source media* (79) *of both analog and digital source media* (80) *of only digital source media.*

5. The clapboard aids in (81) *starting the countdown* (82) *synchronizing audio and video* (83) *marking the first video frame.*

6. Two essential items on a slate or clapboard are (84) *producer and date* (85) *title and take number* (86) *title and executive producer.*

7. When using a tapeless recording media, locating certain takes for recording checks is (87) *quicker than* (88) *slower than* (89) *about the same as* using videotape.

8. When beginning a recording, (90) *servers* (91) *memory cards* (92) *videotape recorders* need to reach operating speed to avoid video breakup.

9. Leader numbers are especially helpful for accurate cueing of (93) *servers* (94) *memory cards* (95) *videotape.*

10. You cannot import clips in a nonlinear editing system if it does not recognize the (96) *name* (97) *length* (98) *codec* of the clip.

11. The video leader (99) *should* (100) *should not* contain a 0 VU audio tone and (101) *should* (102) *should not* include the first frame of the recording. ***(Fill in two bubbles.)***

1	○ 66	○ 67	○ 68
	○ 69	○ 70	○ 71
2	○ 72	○ 73	○ 74
3	○ 75	○ 76	○ 77
4	○ 78	○ 79	○ 80
5	○ 81	○ 82	○ 83
6	○ 84	○ 85	○ 86
7	○ 87	○ 88	○ 89
8	○ 90	○ 91	○ 92
9	○ 93	○ 94	○ 95
10	○ 96	○ 97	○ 98
11	○ 99	○ 100	
	○ 101	○ 102	

SECTION TOTAL

Course No. ______ Date ______ Name ______

REVIEW QUIZ

*Mark the following statements as true or false by filling in the bubbles in the **T** (for true) or **F** (for false) column.*

1. The advantage of using a memory card for recording is that it can store more information than videotape.
2. You can use the SMPTE time code to mark frames of analog as well as digital recordings.
3. In a Y/color difference component system, the Y and R–Y/B–Y signals are kept separate throughout the entire recording process.
4. If they are lossy, all codecs are the same.
5. You can use videotape to record both analog and digital signals.
6. You can use a hard disk to record both analog and digital signals.
7. When dubbing videotapes, analog systems produce much more noise in subsequent generations than do digital ones.
8. A control track pulse marks each video frame.
9. The Y/C component system means that the color yellow has been added to the color signals.
10. You should use prerecorded color bars for the video leader.
11. A codec signifies a specific digital compression standard.
12. When a camera feeds a switcher in addition to its own VR, it is no longer an iso camera.
13. All tapeless storage systems allow random access.
14. MPEG-2 is an intraframe compression technique.
15. When digital information is stored on videotape, it allows random access.
16. A framestore synchronizer and a TBC fulfill similar functions.
17. All videotapes provide at least two audio tracks.
18. A 4:1:1 sampling ratio means that the color luminance signal is sampled four times as often as each color signal.

	T	F
1	○ 103	○ 104
2	○ 105	○ 106
3	○ 107	○ 108
4	○ 109	○ 110
5	○ 111	○ 112
6	○ 113	○ 114
7	○ 115	○ 116
8	○ 117	○ 118
9	○ 119	○ 120
10	○ 121	○ 122
11	○ 123	○ 124
12	○ 125	○ 126
13	○ 127	○ 128
14	○ 129	○ 130
15	○ 131	○ 132
16	○ 133	○ 134
17	○ 135	○ 136
18	○ 137	○ 138

SECTION TOTAL ______

PROBLEM-SOLVING APPLICATIONS

1. Before purchasing a new camcorder, why should you inquire about the codec it uses for recording?

2. You are asked to produce a brief instructional video on diagnosing specific skin rashes. The physician in charge insists that you use recording equipment that has a 4:2:2 sampling standard. Why is she so insistent about the sampling?

3. Because the extreme conditions under which the digital movie will be shot necessitate extensive audio postproduction and ADR, the editor insists on equipment that uses intraframe rather than interframe compression. Why do you think the editor specifies the compression system?

4. The department head asks you to defend the switch to a totally tapeless operation. What are your major arguments?

5. The novice digital cinema director tells you not to bother with videotape because it loses quality each time you dub it to another storage media. What is your reply?

6. How do intraframe and interframe compression relate to spatial and temporal redundancy?

Course No. ____________ Date ____________ Name ____________

Switching, or Instantaneous Editing

REVIEW OF KEY TERMS

Match each term with its appropriate definition by filling in the corresponding bubble.

1. **delegation controls**
2. **program bus**
3. **key bus**
4. **fader bar**
5. **preview/preset bus**
6. **downstream keyer**
7. **switching**
8. **genlock**
9. **M/E bus**
10. **chroma key**
11. **auto-transition**
12. **DVE**
13. **super**
14. **wipe**
15. **matte key**

A. Control that allows a title to be keyed over the line-out image as it leaves the switcher

A ○1 ○2 ○3 ○4 ○5 ○6 ○7 ○8 ○9 ○10 ○11 ○12 ○13 ○14 ○15

B. A button that triggers the function of a fader bar

B ○1 ○2 ○3 ○4 ○5 ○6 ○7 ○8 ○9 ○10 ○11 ○12 ○13 ○14 ○15

C. Rows of buttons used to select the upcoming video and route it to the preview monitor

C ○1 ○2 ○3 ○4 ○5 ○6 ○7 ○8 ○9 ○10 ○11 ○12 ○13 ○14 ○15

PAGE TOTAL

1. delegation controls	**6. downstream keyer**	**11. auto-transition**
2. program bus	**7. switching**	**12. DVE**
3. key bus	**8. genlock**	**13. super**
4. fader bar	**9. M/E bus**	**14. wipe**
5. preview/preset bus	**10. chroma key**	**15. matte key**

D. A change from one video source to the next

D ○1 ○2 ○3 ○4 ○5 ○6 ○7 ○8 ○9 ○10 ○11 ○12 ○13 ○14 ○15

E. A row of buttons that can serve a mix or an effects function

E ○1 ○2 ○3 ○4 ○5 ○6 ○7 ○8 ○9 ○10 ○11 ○12 ○13 ○14 ○15

F. Electronically cut-in title whose letters are filled with shades of gray or a specific color

F ○1 ○2 ○3 ○4 ○5 ○6 ○7 ○8 ○9 ○10 ○11 ○12 ○13 ○14 ○15

G. A double exposure of two images

G ○1 ○2 ○3 ○4 ○5 ○6 ○7 ○8 ○9 ○10 ○11 ○12 ○13 ○14 ○15

H. Controls on a switcher that assign specific functions to a bus

H ○1 ○2 ○3 ○4 ○5 ○6 ○7 ○8 ○9 ○10 ○11 ○12 ○13 ○14 ○15

PAGE TOTAL

Course No. ________ Date ________ Name ________

I. A lever on the switcher that activates preset functions such as dissolves, fades, and wipes of varying speeds

I ○1 ○2 ○3 ○4 ○5 ○6 ○7 ○8 ○9 ○10 ○11 ○12 ○13 ○14 ○15

J. Effect that uses color (usually blue or green) for the backdrop, which is replaced by the background image

J ○1 ○2 ○3 ○4 ○5 ○6 ○7 ○8 ○9 ○10 ○11 ○12 ○13 ○14 ○15

K. A bus used to select the video source to be inserted into a background image

K ○1 ○2 ○3 ○4 ○5 ○6 ○7 ○8 ○9 ○10 ○11 ○12 ○13 ○14 ○15

L. Transition in which the new image is revealed in a pattern or shape

L ○1 ○2 ○3 ○4 ○5 ○6 ○7 ○8 ○9 ○10 ○11 ○12 ○13 ○14 ○15

M. The bus on a switcher whose inputs are directly switched to the line-out

M ○1 ○2 ○3 ○4 ○5 ○6 ○7 ○8 ○9 ○10 ○11 ○12 ○13 ○14 ○15

N. Visual effects generated by computer or the software in the switcher

N ○1 ○2 ○3 ○4 ○5 ○6 ○7 ○8 ○9 ○10 ○11 ○12 ○13 ○14 ○15

PAGE TOTAL ________

1. delegation controls	**6. downstream keyer**	**11. auto-transition**
2. program bus	**7. switching**	**12. DVE**
3. key bus	**8. genlock**	**13. super**
4. fader bar	**9. M/E bus**	**14. wipe**
5. preview/preset bus	**10. chroma key**	**15. matte key**

O. Synchronization of video sources or origination sources to prevent picture breakup

O ○ 1 ○ 2 ○ 3 ○ 4 ○ 5 ○ 6 ○ 7 ○ 8 ○ 9 ○ 10 ○ 11 ○ 12 ○ 13 ○ 14 ○ 15

PAGE TOTAL

SECTION TOTAL

Course No. ____________ Date ________________ Name ______________________________

REVIEW OF BASIC SWITCHER LAYOUT AND OPERATION

1. Fill in the bubbles whose numbers correspond with the appropriate parts of the switcher shown in the following figure.

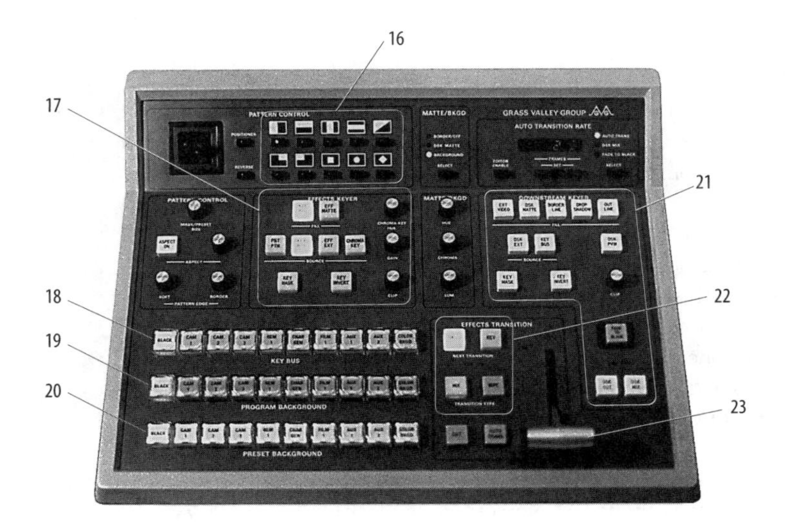

PHOTO: EDWARD AIONA

a. program bus

b. preview/preset bus

c. fader bar

d. key bus

e. delegation controls (mix/effects transition)

f. wipe pattern selector

g. downstream keyer controls

h. key/matte controls

1a ○ 16 ○ 17 ○ 18 ○ 19 ○ 20 ○ 21 ○ 22 ○ 23

1b ○ 16 ○ 17 ○ 18 ○ 19 ○ 20 ○ 21 ○ 22 ○ 23

1c ○ 16 ○ 17 ○ 18 ○ 19 ○ 20 ○ 21 ○ 22 ○ 23

1d ○ 16 ○ 17 ○ 18 ○ 19 ○ 20 ○ 21 ○ 22 ○ 23

1e ○ 16 ○ 17 ○ 18 ○ 19 ○ 20 ○ 21 ○ 22 ○ 23

1f ○ 16 ○ 17 ○ 18 ○ 19 ○ 20 ○ 21 ○ 22 ○ 23

1g ○ 16 ○ 17 ○ 18 ○ 19 ○ 20 ○ 21 ○ 22 ○ 23

1h ○ 16 ○ 17 ○ 18 ○ 19 ○ 20 ○ 21 ○ 22 ○ 23

PAGE TOTAL ______

Select the correct answers and fill in the bubbles with the corresponding numbers.

2. To select the functions of a specific bus or buses, you need to activate the (24) *joystick* (25) *wipe mode selectors* (26) *delegation controls.*

3. The program bus will direct the selected video source to the (27) *preview monitor* (28) *mix bus* (29) *line-out.*

4. To dissolve from C3 (camera 3) to VR (with the auto-transition in the *off* position), you (30) *press the VR button on the preview bus; then press the cut button* (31) *press the VR button on the preview bus; then move the fader bar to the opposite position* (32) *press the VR button on the key bus; then move the fader bar to the opposite position.*

5. To have C3 appear on the preview monitor before switching to it from C1, you need to (33) *press C3 on the preview bus* (34) *press C3 on the preview bus; then press the key button* (35) *press C3 on the program bus; then move the fader bar to the opposite position.*

6. To switch from C1 to C3 by pressing only one button, you need to press the (36) *C3 button on the preview bus* (37) *C3 button on the key bus* (38) *C3 button on the program bus.* ***(This assumes that the appropriate buses have already been delegated a mix/effects function.)***

7. Assuming that the final C.G. credits are keyed with the DSK, you can go to black by (39) *pressing the black button on the program bus* (40) *pressing the black button on the key bus* (41) *pressing the black button in the downstream keyer section.*

2	○ 24	○ 25	○ 26
3	○ 27	○ 28	○ 29
4	○ 30	○ 31	○ 32
5	○ 33	○ 34	○ 35
6	○ 36	○ 37	○ 38
7	○ 39	○ 40	○ 41

PAGE TOTAL

Course No. ______________ Date ____________________ Name ____

8. Identify the proper preview and line monitor images you would expect to see from the switcher output by filling in the corresponding bubble. The highlighted buttons on the following switcher have already been pressed. C1 is focused on the host, C2 on the dancers (see the monitor images below).

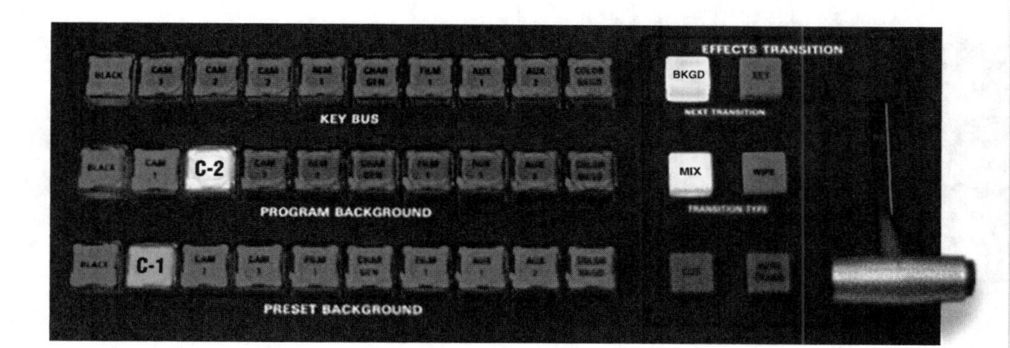

PHOTOS: EDWARD AIONA

42

Preview

Line

43

Preview

Line

44

Preview

Line

8 ○ 42 ○ 43 ○ 44

PAGE TOTAL ______

9. Identify the proper preview and line monitor images you would expect to see from the switcher output and fill in the corresponding bubble.

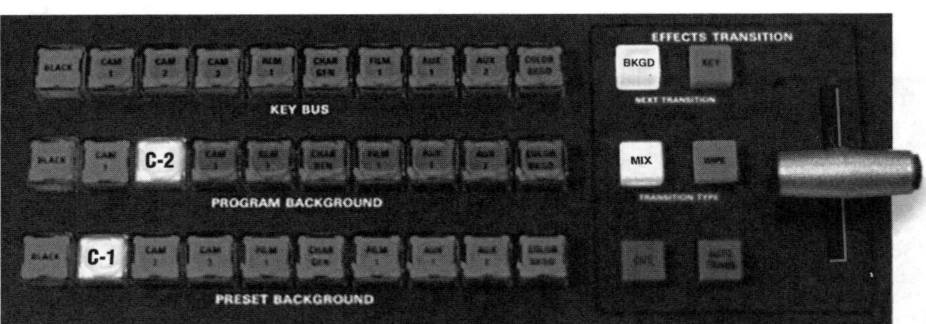

45

Preview

Line

46

Preview

Line

47

Preview

Line

9 ○ 45 ○ 46 ○ 47

PAGE TOTAL

PHOTOS: EDWARD AIONA

Course No. ________ Date ________ Name ________

10. Identify the proper preview and line monitor images you would expect to see from the switcher output at the end of the previous dissolve and fill in the corresponding bubble.

10 ○ 48 ○ 49 ○ 50

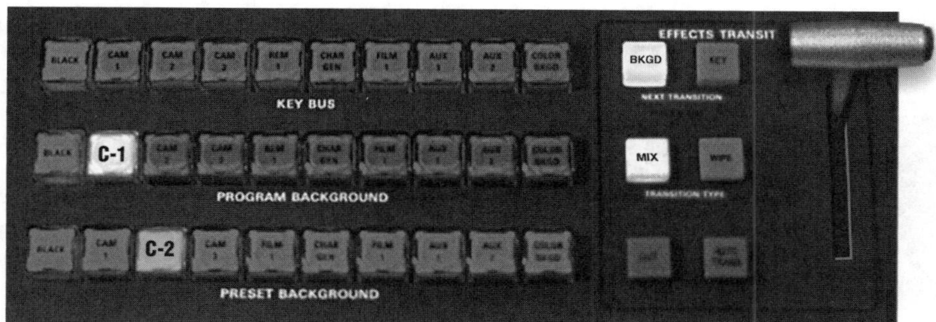

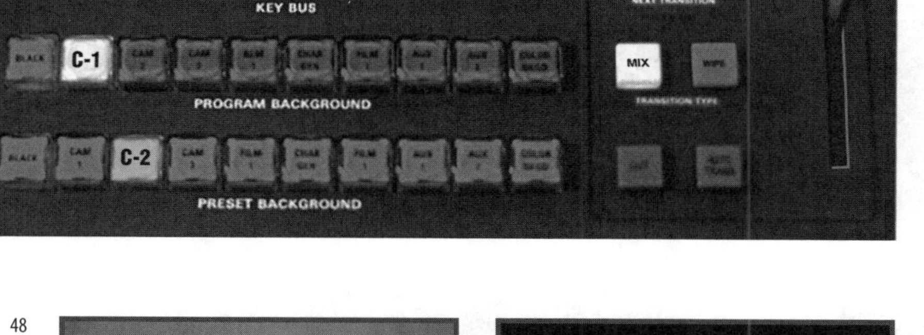

48

Preview

Line

49

Preview

Line

50

Preview

Line

PAGE TOTAL

SECTION TOTAL

PHOTOS: EDWARD AIONA

REVIEW OF ELECTRONIC EFFECTS AND SWITCHER FUNCTIONS

1. Fill in the bubbles whose numbers correspond with the appropriate electronic effects illustrated in the following figures.

51

52

53

54

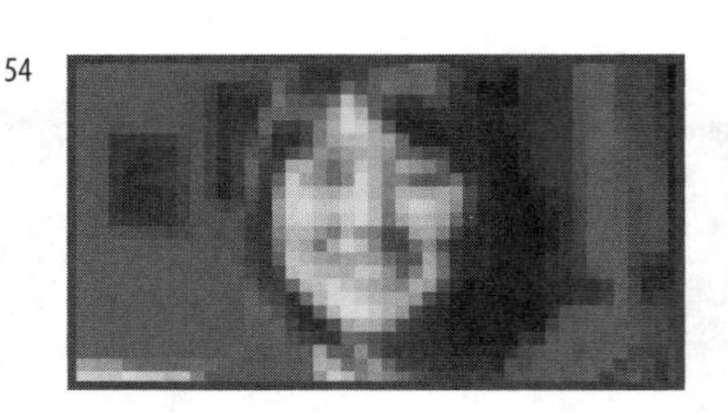

55

56

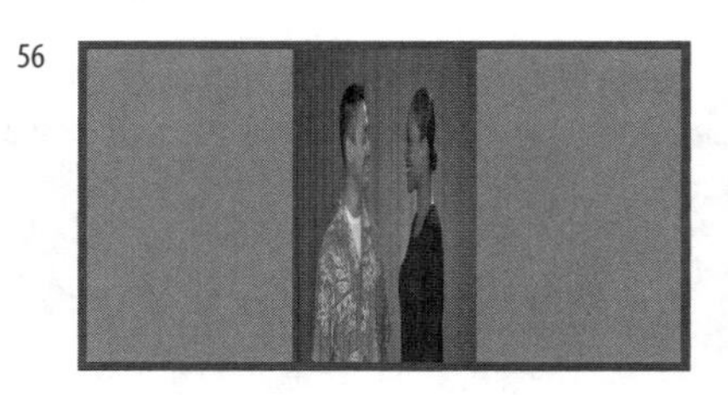

57

58

59

PHOTOS: EDWARD AIONA

Course No. ______________ Date ______________ Name ______________

a. vertical stretching

b. echo effect

c. horizontal wipe

d. peel effect

e. posterization

f. shrinking

g. vertical wipe

h. multiple frames

i. mosaic

1a ○ 51 ○ 52 ○ 53 ○ 54 ○ 55 ○ 56 ○ 57 ○ 58 ○ 59

1b ○ 51 ○ 52 ○ 53 ○ 54 ○ 55 ○ 56 ○ 57 ○ 58 ○ 59

1c ○ 51 ○ 52 ○ 53 ○ 54 ○ 55 ○ 56 ○ 57 ○ 58 ○ 59

1d ○ 51 ○ 52 ○ 53 ○ 54 ○ 55 ○ 56 ○ 57 ○ 58 ○ 59

1e ○ 51 ○ 52 ○ 53 ○ 54 ○ 55 ○ 56 ○ 57 ○ 58 ○ 59

1f ○ 51 ○ 52 ○ 53 ○ 54 ○ 55 ○ 56 ○ 57 ○ 58 ○ 59

1g ○ 51 ○ 52 ○ 53 ○ 54 ○ 55 ○ 56 ○ 57 ○ 58 ○ 59

1h ○ 51 ○ 52 ○ 53 ○ 54 ○ 55 ○ 56 ○ 57 ○ 58 ○ 59

1i ○ 51 ○ 52 ○ 53 ○ 54 ○ 55 ○ 56 ○ 57 ○ 58 ○ 59

PAGE TOTAL

2. Fill in the bubbles whose numbers correspond with the button you would have to press on the pattern selector to create the various effects illustrated in the following figures.

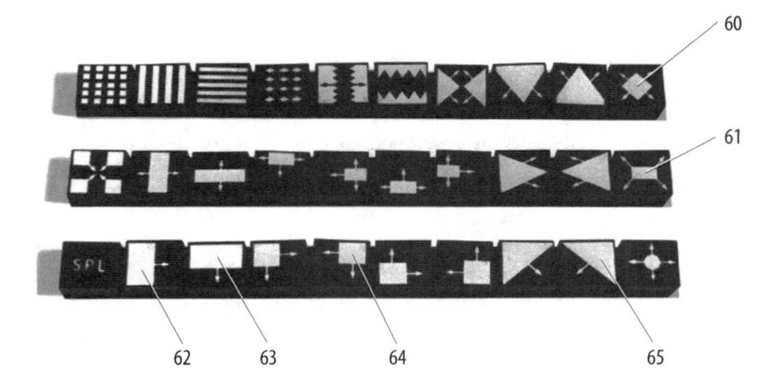

a.

b.

2a ◯ 60 ◯ 61 ◯ 62 ◯ 63 ◯ 64 ◯ 65

2b ◯ 60 ◯ 61 ◯ 62 ◯ 63 ◯ 64 ◯ 65

PAGE TOTAL

PHOTOS: EDWARD AIONA

Course No. ____________ Date ____________ Name ____________________

c.

2c ◯ 60 ◯ 61 ◯ 62 ◯ 63 ◯ 64 ◯ 65

d.

2d ◯ 60 ◯ 61 ◯ 62 ◯ 63 ◯ 64 ◯ 65

e.

2e ◯ 60 ◯ 61 ◯ 62 ◯ 63 ◯ 64 ◯ 65

f.

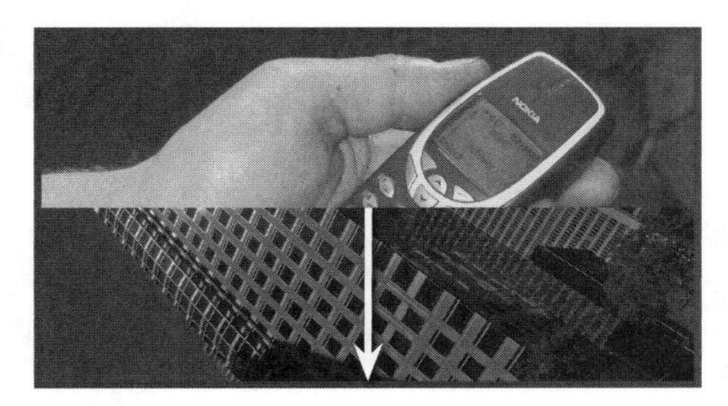

2f ◯ 60 ◯ 61 ◯ 62 ◯ 63 ◯ 64 ◯ 65

PAGE TOTAL ____

PHOTOS: EDWARD AIONA

3. Fill in the bubbles whose numbers correspond with the appropriate key effects shown in the following figures.

66

67

68

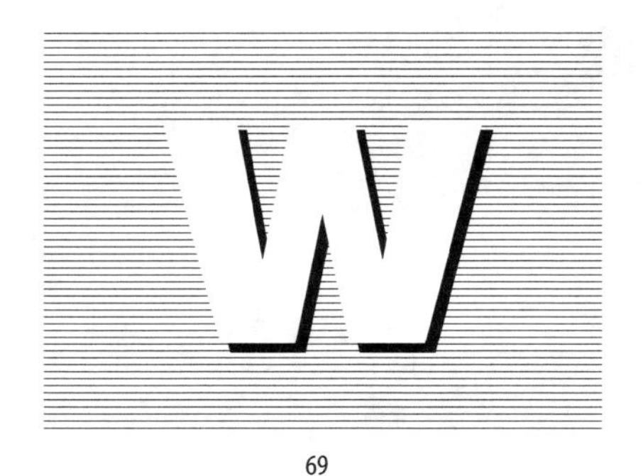

69

a. edge mode

b. outline mode

c. drop-shadow mode

d. matte key

3a ○ 66 ○ 67 ○ 68 ○ 69

3b ○ 66 ○ 67 ○ 68 ○ 69

3c ○ 66 ○ 67 ○ 68 ○ 69

3d ○ 66 ○ 67 ○ 68 ○ 69

PAGE TOTAL

Course No. ______________ Date ______________ Name ______________

Select the correct answers and fill in the bubbles with the corresponding numbers.

4. The two most frequently used backdrop colors for studio chroma-key effects are (70) *green and blue* (71) *blue and red* (72) *yellow and blue.*

5. A peel effect (73) *reveals the picture underneath* (74) *strips the picture of detail* (75) *covers the current picture.*

6. To create echo, stretching, and compression effects, you need (76) *DVE* (77) *a CCU* (78) *a TBC.*

7. A secondary frame (screen within a screen) (79) *must have a 4 × 3 horizontal aspect ratio* (80) *must have a 16 × 9 HDTV aspect ratio* (81) *can have a vertical aspect ratio.*

8. Shrinking effects differ from box wipes because (82) *they maintain the total picture and aspect ratio during the reduction* (83) *the reduction can get smaller than a box wipe* (84) *they need a blue background.*

9. When the talent insists on wearing blue for a standard chroma-key effect, the backdrop color must be (85) *blue* (86) *green* (87) *yellow.*

10. During posterization (88) *the brightness values are reduced* (89) *the saturation is reversed* (90) *all brightness values are reversed.*

4	○ 70	○ 71	○ 72
5	○ 73	○ 74	○ 75
6	○ 76	○ 77	○ 78
7	○ 79	○ 80	○ 81
8	○ 82	○ 83	○ 84
9	○ 85	○ 86	○ 87
10	○ 88	○ 89	○ 90

PAGE TOTAL ______

SECTION TOTAL ______

REVIEW QUIZ

Mark the following statements as true or false by filling in the bubbles in the ***T*** *(for true) or* ***F*** *(for false) column.*

	Statement	T	F
1.	You can achieve a split-screen effect simply by stopping a horizontal wipe midway.	○ 91	○ 92
2.	In a mosaic effect, you can change the size of the tiles (image squares) electronically.	○ 93	○ 94
3.	Through delegation controls, you can assign a preview function to the program bus.	○ 95	○ 96
4.	A drop shadow makes lettering look three-dimensional.	○ 97	○ 98
5.	With DVE you can change the size and the aspect ratio of a picture insert without losing any portion of the picture.	○ 99	○ 100
6.	The downstream keyer can add a title key only if there are no other keys present in the line-out picture.	○ 101	○ 102
7.	Electronically cutting out portions of a background image and filling them with color is called a superimposition.	○ 103	○ 104
8.	The auto-transition fulfills the same function as the fader bar.	○ 105	○ 106
9.	The switcher has a separate button for each video input.	○ 107	○ 108
10.	Assuming that the DSK is inactive, the black button from the program bus will put the line into black.	○ 109	○ 110
11.	The preview bus can also be used as a mix bus, if so delegated.	○ 111	○ 112
12.	Assuming that you are not using auto-transition, the speed of a dissolve depends on how fast the fader bar is moved up or down.	○ 113	○ 114
13.	The DSK black button will put the switcher output to black regardless of what source feeds the line-out.	○ 115	○ 116
14.	It is impossible to have both preset and program buses activated at the same time.	○ 117	○ 118
15.	The clip control allows you to adjust the hue and the saturation of the image selected on the program bus.	○ 119	○ 120

SECTION TOTAL

Course No. ____________ Date ____________ Name ____________

PROBLEM-SOLVING APPLICATIONS

1. The director asks you, the TD, to superimpose a long shot of a dancer over a close-up of her face. The director wants to first have the long shot be the more prominent image and then slowly shift the emphasis to the close-up. How, if at all, can you accomplish such an effect?

2. The director wants you to perform four dissolves that are identical in speed from one dancer to the next. Are such identical dissolves possible? If so, how can they be accomplished? If not, why not?

3. You have keyed the credits over the base picture with the downstream keyer. When the director calls for a fade to black, you press the black button on the program bus. What will you see on the line monitor? Why?

4. The director wants you to do fast cuts between the CUs of the interviewer and the guest during a lively discussion. How can you best accomplish such fast cuts between the two?

5. When checking the chroma key of a weathercaster standing in front of a weather map, you, the TD, discover that the key is not "clean." The talent's dark hair seems to have a blue and purple halo, and the outline of her head and shoulders is not sharp against the map. Assuming that the problem does not lie with the chroma-key equipment, what is the problem? What, if anything, can you do to minimize or eliminate it?

6. The preview monitor shows that the outline-mode title key is hard to read over the busy background. How could you, the TD, make the title more readable without changing the font or the background?

7. When you, the TD, preview the key of a C.G. title, the white letters tear at the edges. How can you correct this problem?

8. The AD informs you, the director, that a dancer, who is supposed to be chroma-keyed over a videotaped landscape scene, wears a saturated medium-blue leotard. The AD is very concerned about this, but the TD assures you that he has already taken care of the problem. What was the potential problem? What did the TD do to solve it?

9. The new news producer would like you, the director, to use a circle wipe to close each story. How would you respond? Why?

10. The same director suggests peel effects between the stories of a headline news teaser. How would you respond? Why?

Course No. ______ Date ______ Name ______

Design

REVIEW OF KEY TERMS

Match each term with its appropriate definition by filling in the corresponding bubble.

1. **color compatibility**
2. **windowbox**
3. **props**
4. **floor plan pattern**
5. **scanning area**
6. **essential area**
7. **C.G.**
8. **floor plan**
9. **pillarbox**
10. **flat**
11. **letterbox**
12. **graphics generator**

A. The picture area usually seen on the camera viewfinder and the preview monitor

A ○1 ○2 ○3 ○4 ○5 ○6 ○7 ○8 ○9 ○10 ○11 ○12

B. Furniture and other objects used for set decorations or by actors or performers

B ○1 ○2 ○3 ○4 ○5 ○6 ○7 ○8 ○9 ○10 ○11 ○12

C. Colors with enough brightness contrast for good monochrome reproduction

C ○1 ○2 ○3 ○4 ○5 ○6 ○7 ○8 ○9 ○10 ○11 ○12

PAGE TOTAL ______

1. color compatibility	**5. scanning area**	**9. pillarbox**
2. windowbox	**6. essential area**	**10. flat**
3. props	**7. C.G.**	**11. letterbox**
4. floor plan pattern	**8. floor plan**	**12. graphics generator**

D. Computer software that allows a designer to draw, color, animate, store, and retrieve images electronically

D ○ 1 ○ 2 ○ 3 ○ 4 ○ 5 ○ 6 ○ 7 ○ 8 ○ 9 ○ 10 ○ 11 ○ 12

E. The section of the television picture, centered within the scanning area, that the home viewer sees

E ○ 1 ○ 2 ○ 3 ○ 4 ○ 5 ○ 6 ○ 7 ○ 8 ○ 9 ○ 10 ○ 11 ○ 12

F. Fitting a 16 × 9 aspect ratio into a 4 × 3 screen without cropping or distortion

F ○ 1 ○ 2 ○ 3 ○ 4 ○ 5 ○ 6 ○ 7 ○ 8 ○ 9 ○ 10 ○ 11 ○ 12

G. A plan of the studio floor with the grid but without a set design

G ○ 1 ○ 2 ○ 3 ○ 4 ○ 5 ○ 6 ○ 7 ○ 8 ○ 9 ○ 10 ○ 11 ○ 12

H. A smaller frame positioned in the center of the TV screen.

H ○ 1 ○ 2 ○ 3 ○ 4 ○ 5 ○ 6 ○ 7 ○ 8 ○ 9 ○ 10 ○ 11 ○ 12

PAGE TOTAL

Course No. ______ Date ______ Name ______

I. A piece of standing scenery used as a background or to simulate a wall

I ○1 ○2 ○3 ○4 ○5 ○6 ○7 ○8 ○9 ○10 ○11 ○12

J. A diagram of scenery and major set properties drawn on a grid

J ○1 ○2 ○3 ○4 ○5 ○6 ○7 ○8 ○9 ○10 ○11 ○12

K. Fitting a 4×3 aspect ratio into a 16×9 screen without cropping or distortion

K ○1 ○2 ○3 ○4 ○5 ○6 ○7 ○8 ○9 ○10 ○11 ○12

L. A dedicated computer that electronically produces letters, numbers, and simple graphic images for video display

L ○1 ○2 ○3 ○4 ○5 ○6 ○7 ○8 ○9 ○10 ○11 ○12

PAGE TOTAL ______

SECTION TOTAL ______

REVIEW OF TELEVISION GRAPHICS

Select the correct answers and fill in the bubbles with the corresponding numbers.

1. On a grayscale *1* represents (13) *TV white* (14) *TV black* (15) *100 percent reflectance.*

2. Color compatibility refers to using colors that differ distinctly as to (16) *hue* (17) *saturation* (18) *brightness.*

3. Dead zones are (19) *uninteresting picture areas* (20) *the empty vertical bars when showing standard TV on HDTV* (21) *a sound problem in studio areas.*

4. The whiteboard writing shown in the photo below is (22) *appropriate* (23) *inappropriate* because it (24) *is within the scanning area* (25) *does not permit good CUs.* ***(Fill in two bubbles.)***

5. All lettering must be contained within the (26) *scanning area* (27) *screen area* (28) *essential area.*

6. The standard television aspect ratio is (29) *4 × 3* (30) *8 × 12* (31) *16 × 9*. For HDTV it is (32) *4 × 3* (33) *8 × 12* (34) *16 × 9*. ***(Fill in two bubbles.)***

7. Normally, low-energy colors are used more for the (35) *foreground* (36) *middleground* (37) *background* in a scene.

8. To store a great many video frames for instant access, you need (38) *an ESS system* (39) *DVE* (40) *an SEG.*

9. The aesthetic energy of a color is principally determined by (41) *hue and brightness* (42) *the color itself* (43) *saturation and brightness.*

1	○ 13	○ 14	○ 15
2	○ 16	○ 17	○ 18
3	○ 19	○ 20	○ 21
4	○ 22	○ 23	
	○ 24	○ 25	
5	○ 26	○ 27	○ 28
6	○ 29	○ 30	○ 31
	○ 32	○ 33	○ 34
7	○ 35	○ 36	○ 37
8	○ 38	○ 39	○ 40
9	○ 41	○ 42	○ 43

PAGE TOTAL

PHOTO: EDWARD AIONA

Course No. ________ Date ________ Name ________

10. Fill in the bubbles whose numbers correspond with the appropriate aspect ratio or frame adjustments in the figure.

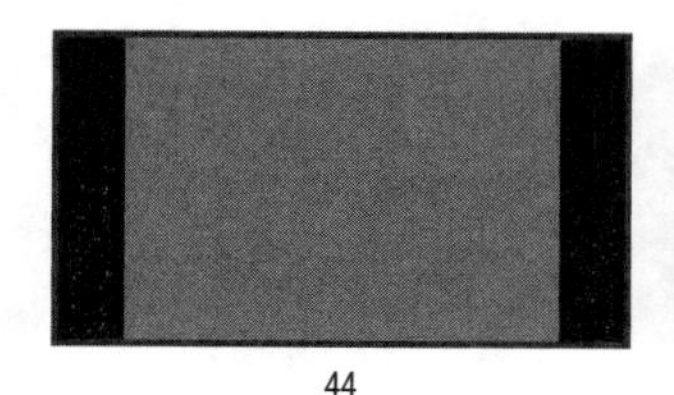
44

45

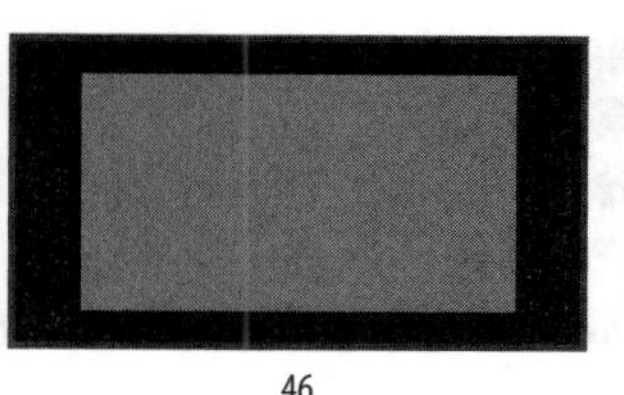
46

a. windowbox

b. letterbox

c. pillarbox

10a ○ 44 ○ 45 ○ 46

10b ○ 44 ○ 45 ○ 46

10c ○ 44 ○ 45 ○ 46

11. The vertically oriented diagram below is (47) *acceptable* (48) *not acceptable* for shooting with a studio camera because (49) *it is not in proper aspect ratio* (50) *the camera can tilt in a close-up.* ***(Fill in two bubbles.)***

11 ○ 47 ○ 48 ○ 49 ○ 50

PAGE TOTAL

12. The edge distortion as shown in the figure below is called (51) *aliasing* (52) *anti-aliasing* (53) *pixel distortion.*

12 ○ 51 ○ 52 ○ 53

13. Fill in the bubbles whose numbers correspond with the appropriate title areas in the following figure.

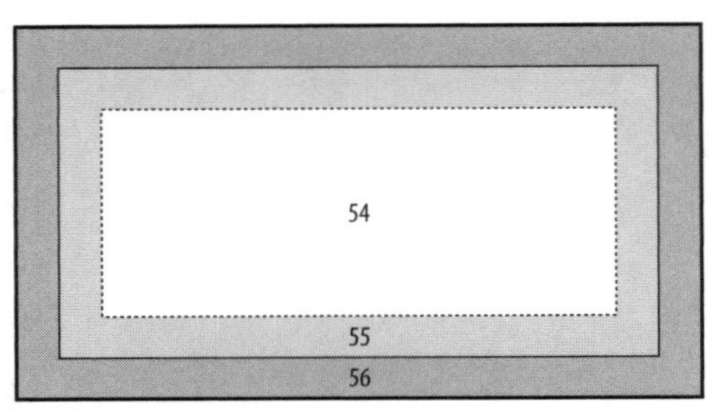

a. total graphic screen area

13a ○ 54 ○ 55 ○ 56

b. essential area

13b ○ 54 ○ 55 ○ 56

c. scanning area

13c ○ 54 ○ 55 ○ 56

14. Fill in the bubbles whose numbers correspond with the type of digital distortion that results from adjusting one aspect ratio to fit another.

57

58

a. a 16 × 9 shot viewed full-screen on a 4 × 3 monitor

14a ○ 57 ○ 58

b. a 4 × 3 shot viewed full-screen on a 16 × 9 monitor

14b ○ 57 ○ 58

PAGE TOTAL

PHOTOS: EDWARD AIONA

Course No. ______________ Date ______________ Name ______________

15. The following six figures show various television graphics displayed on well-adjusted preview monitors. These monitors show the entire scanning area. For each figure state whether you would (59) *accept* (60) *not accept* the television graphic because it has (61) *inappropriate style* (62) *enough contrast between figure and ground* (63) *scattered information* (64) *good grouping of words* (65) *letters that are too small* (66) *information that lies outside the essential area* (67) *letters that get lost in the busy background.* ***(Fill in the two bubbles that seem most appropriate for each graphic.)***

Nuclear
Crisis

a.

Design by
Gary Palmatier

b.

c.

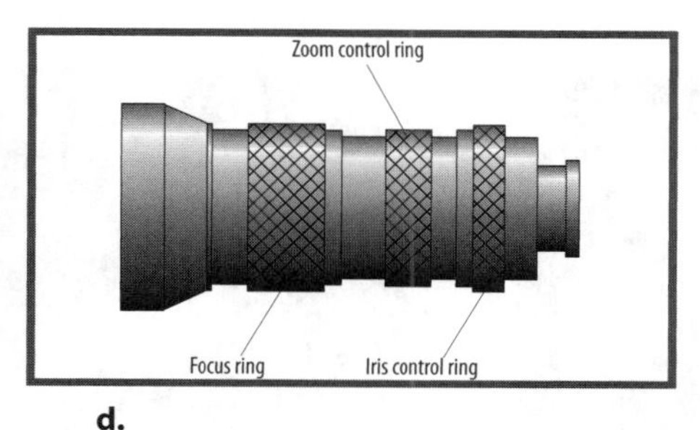

d.

e.

Crew
Susan Walters
Robaire Ream
Cathy Linberg
Karen Austin
Deirdre Cavanaugh
Elizabeth von Radics
Ryan E. Vesely
Mike Mollett
Ken Baird
Dory Schaeffer
Stacey Purviance

f.

15a ○ 59 ○ 60
○ 61 ○ 62 ○ 63 ○ 64
○ 65 ○ 66 ○ 67

15b ○ 59 ○ 60
○ 61 ○ 62 ○ 63 ○ 64
○ 65 ○ 66 ○ 67

15c ○ 59 ○ 60
○ 61 ○ 62 ○ 63 ○ 64
○ 65 ○ 66 ○ 67

15d ○ 59 ○ 60
○ 61 ○ 62 ○ 63 ○ 64
○ 65 ○ 66 ○ 67

15e ○ 59 ○ 60
○ 61 ○ 62 ○ 63 ○ 64
○ 65 ○ 66 ○ 67

15f ○ 59 ○ 60
○ 61 ○ 62 ○ 63 ○ 64
○ 65 ○ 66 ○ 67

PAGE TOTAL ______

SECTION TOTAL ______

PHOTO: HERBERT ZETTL

REVIEW OF SCENERY AND SCENIC DESIGN

1. For the simple sets shown below, select the floor plan shown on the facing page that most closely corresponds and fill in the appropriate bubbles. ***(Assume that the camera shoots straight-on. Note that there are floor plans that do not match any of the set photos. The floor plans are not to scale.)***

a.

b.

c.

d.

e.

f.

1a ○ 68 ○ 69 ○ 70 ○ 71 ○ 72 ○ 73 ○ 74 ○ 75 ○ 76

1b ○ 68 ○ 69 ○ 70 ○ 71 ○ 72 ○ 73 ○ 74 ○ 75 ○ 76

1c ○ 68 ○ 69 ○ 70 ○ 71 ○ 72 ○ 73 ○ 74 ○ 75 ○ 76

1d ○ 68 ○ 69 ○ 70 ○ 71 ○ 72 ○ 73 ○ 74 ○ 75 ○ 76

1e ○ 68 ○ 69 ○ 70 ○ 71 ○ 72 ○ 73 ○ 74 ○ 75 ○ 76

1f ○ 68 ○ 69 ○ 70 ○ 71 ○ 72 ○ 73 ○ 74 ○ 75 ○ 76

PAGE TOTAL

PHOTOS: HERBERT ZETTL

Course No. ______ Date ______ Name ______

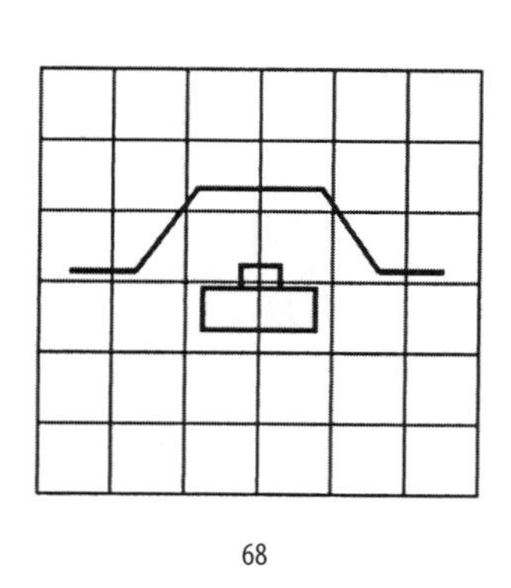

68

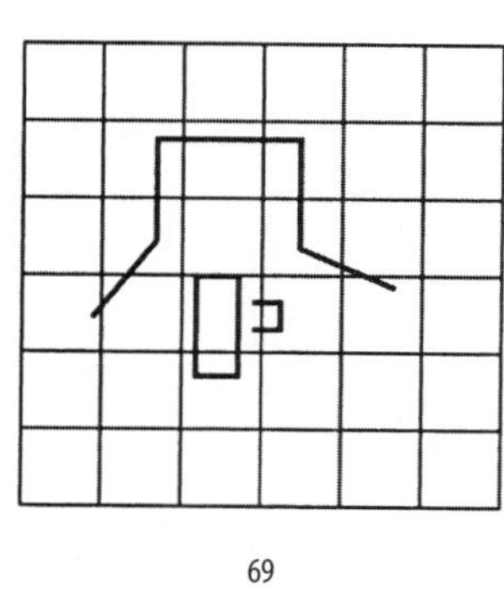

69

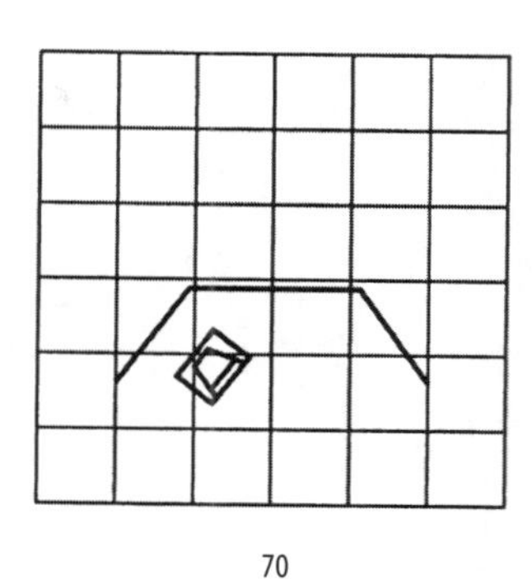

70

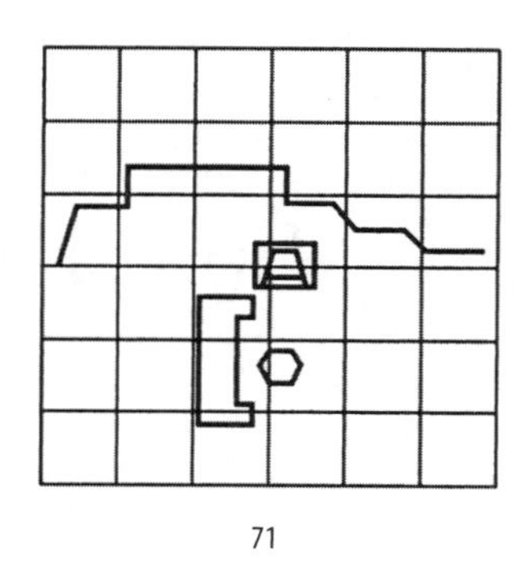

71

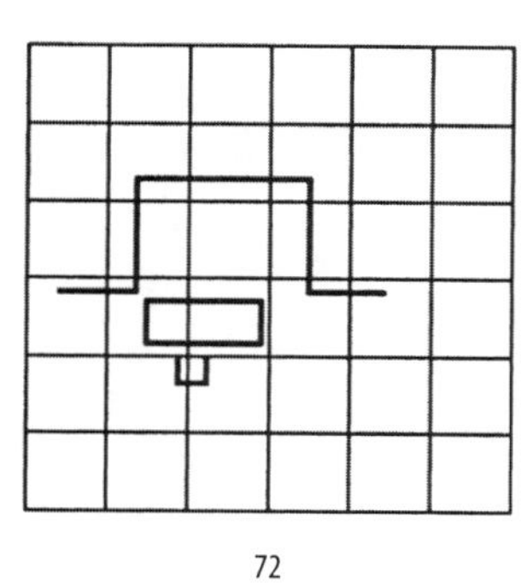

72

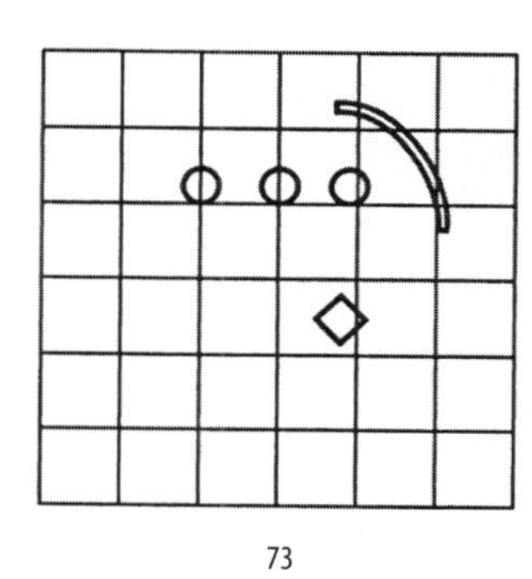

73

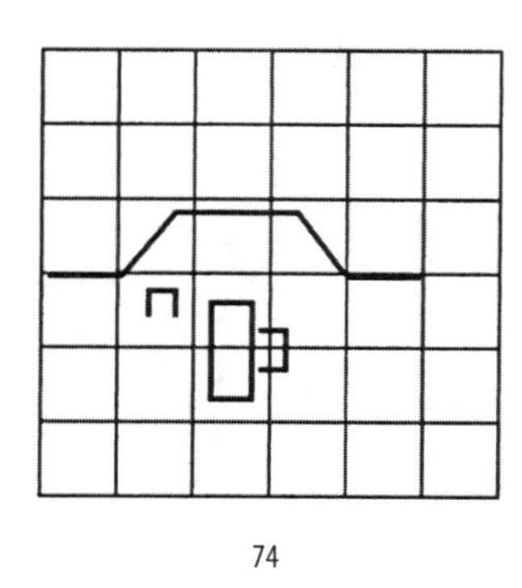

74

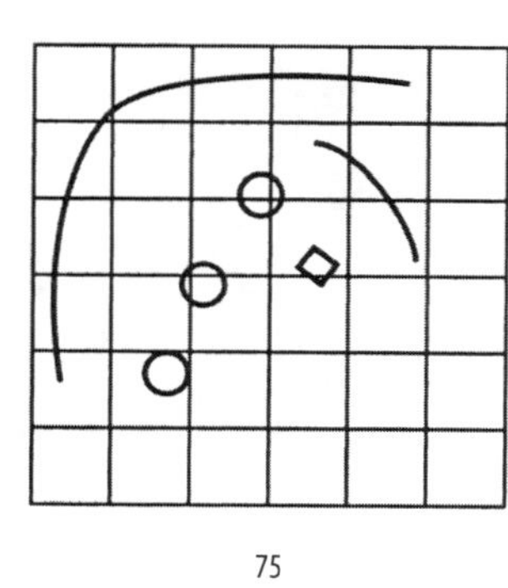

75

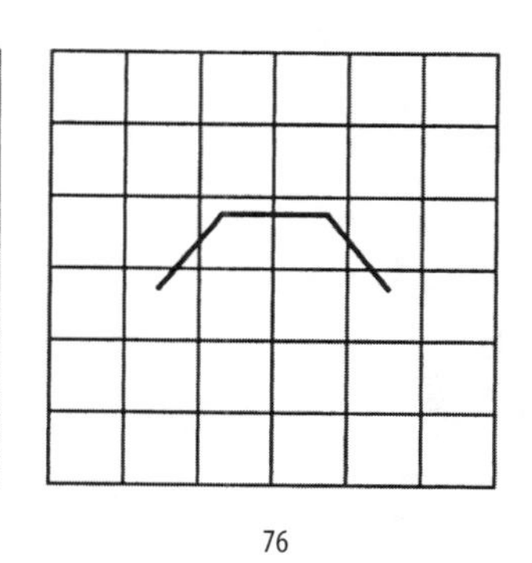

76

2. Fill in the bubbles whose numbers correspond with the numbers identifying the various set pieces shown below.

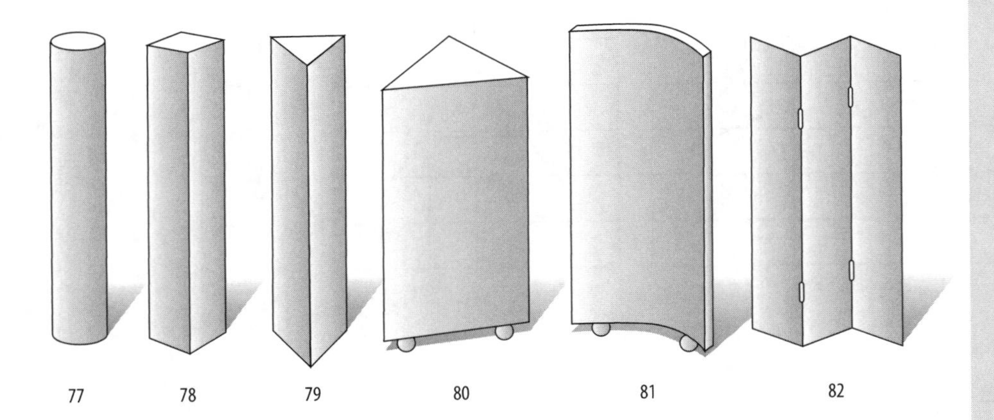

a. sweep

b. round pillar

c. periaktos

d. screen

e. square pillar

f. pylon

2a	○ 77	○ 78	○ 79
	○ 80	○ 81	○ 82
2b	○ 77	○ 78	○ 79
	○ 80	○ 81	○ 82
2c	○ 77	○ 78	○ 79
	○ 80	○ 81	○ 82
2d	○ 77	○ 78	○ 79
	○ 80	○ 81	○ 82
2e	○ 77	○ 78	○ 79
	○ 80	○ 81	○ 82
2f	○ 77	○ 78	○ 79
	○ 80	○ 81	○ 82

PAGE TOTAL

Course No. ______ Date ______ Name ______

Select the correct answers and fill in the bubbles with the corresponding numbers.

3. The usual height for standard set units is (83) *7 feet* (84) *10 feet* (85) *14 feet*. For studios with low ceilings, it is (86) *6 feet* (87) *8 feet* (88) *12 feet*. ***(Fill in two bubbles.)***

4. To elevate scenery, properties, or action areas, we use (89) *periaktoi* (90) *platforms* (91) *pylons.*

5. Pictures and draperies are (92) *set dressings* (93) *set decorations* (94) *hand props.*

6. The continuous piece of canvas or muslin along two, three, or even all four studio walls to form a uniform background is referred to as (95) *a drop* (96) *canvas backing* (97) *a cyclorama.*

7. The standard backgrounds to simulate interior and exterior walls are called (98) *cycs* (99) *flats* (100) *drops.*

3	○ 83	○ 84	○ 85
	○ 86	○ 87	○ 88
4	○ 89	○ 90	○ 91
5	○ 92	○ 93	○ 94
6	○ 95	○ 96	○ 97
7	○ 98	○ 99	○ 100

PAGE TOTAL

SECTION TOTAL

REVIEW QUIZ

Mark the following statements as true or false by filling in the bubbles in the ***T*** *(for true) or* ***F*** *(for false) column.*

1. A good floor plan will aid the LD in the lighting design.

2. The energy of a color is determined primarily by hue.

3. A periaktos looks like a large pylon.

4. A black drape makes an ideal chroma-key backdrop.

5. There is an inevitable picture loss when wide-screen movies are shown in their true aspect ratio on a traditional (4 x 3) television screen.

6. Distinctly different hues (such as red and green) guarantee good brightness contrast.

7. For normal screen titles, all written information must extend beyond the scanning area.

8. A floor plan must show the location of flats but can omit the set properties.

9. The scanning area is contained within the essential area.

10. Hardwall scenery is preferred for permanent sets.

11. Pillarboxing is used to fit a 4 × 3 aspect ratio into a 16 × 9 screen without distortion.

12. Screen clutter can be avoided by grouping related information in specific screen areas.

13. So long as there are distinct colors, brightness differences are relatively unimportant in digital cinema.

14. Writing across a whiteboard from edge to edge facilitates CUs of the entire text.

15. Bold lettering is especially important for mobile media displays.

16. Digitally stretching a 4 × 3 scene to fill a 16 × 9 screen will make the people in it appear fat.

	T	F
1	○ 101	○ 102
2	○ 103	○ 104
3	○ 105	○ 106
4	○ 107	○ 108
5	○ 109	○ 110
6	○ 111	○ 112
7	○ 113	○ 114
8	○ 115	○ 116
9	○ 117	○ 118
10	○ 119	○ 120
11	○ 121	○ 122
12	○ 123	○ 124
13	○ 125	○ 126
14	○ 127	○ 128
15	○ 129	○ 130
16	○ 131	○ 132

SECTION TOTAL

Course No. ______________ Date ______________ Name ______________

PROBLEM-SOLVING APPLICATIONS

1. You are asked to direct a variety of shows and evaluate the location sketch or floor plans (see **a** through **c**). Please be specific as to the potential problems in scale (sets and props), camera accessibility and acceptable shots, lighting, and talent traffic.

   a. Here is a floor plan for a two-camera live-recorded production of a panel discussion by six prominent businesspeople and a moderator.

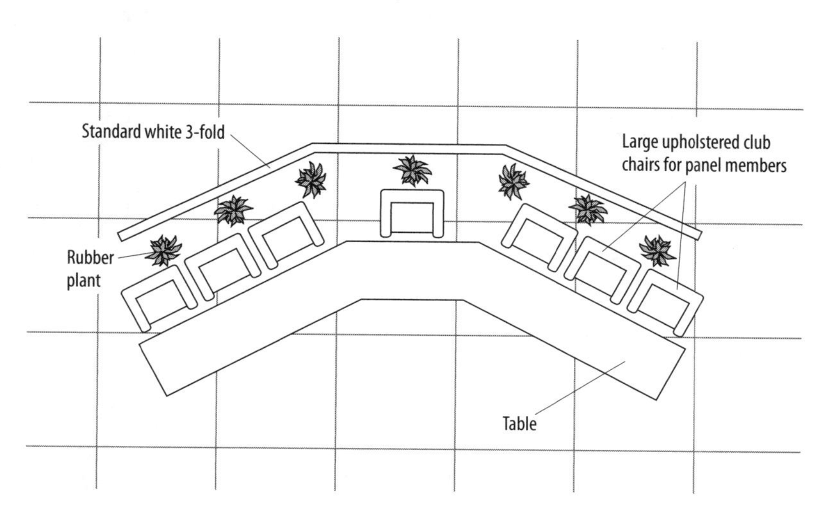

   b. This floor plan is for a two-camera live interview set for a morning show.

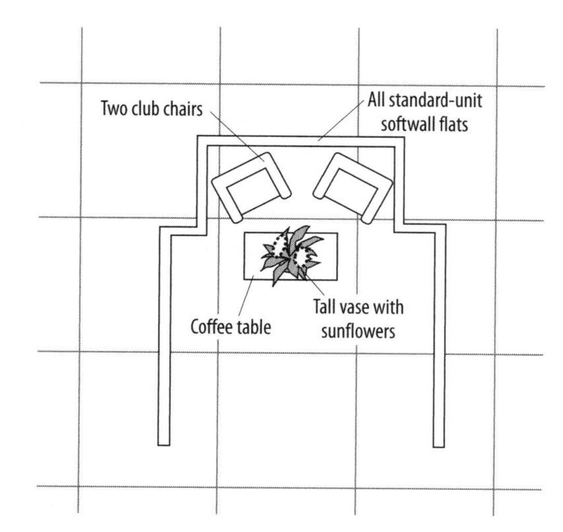

c. This location sketch shows the office of the CEO, who would like to make her monthly 4 p.m. live-satellite TV report from behind her desk.

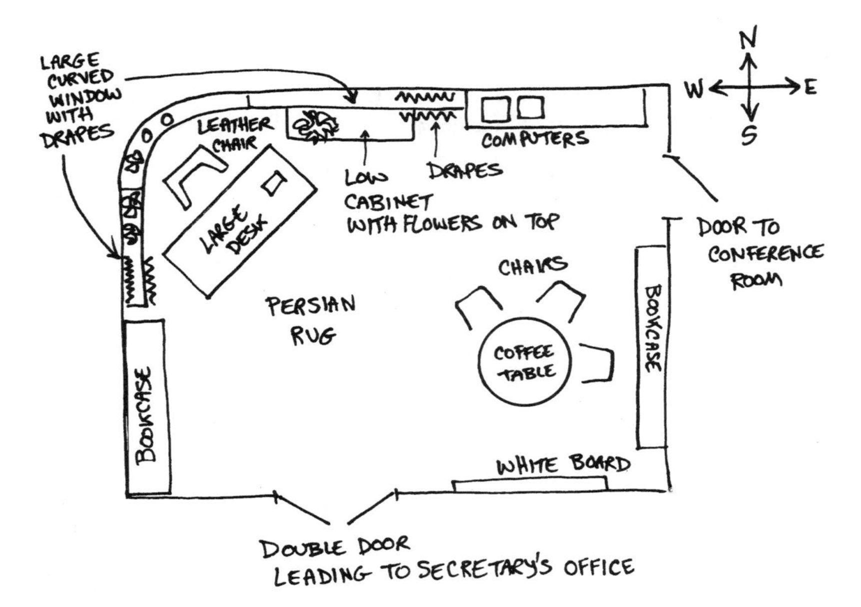

2. Draw a floor plan for a weekly interview show dealing with the art and media scene in your city. The host will interview guests from stage, screen, and radio. Include a detailed prop list. (Use one of the floor plan patterns provided at the back of this book.)

3. Draw a floor plan for a morning news set. The anchors are a woman and a man, and the news content is geared more toward local gossip than international politics. (Use one of the floor plan patterns provided at the back of this book.)

4. The general manager of your corporation would like you to use a highly detailed photo of the latest computer design as the background for the opening and closing titles. Can you accommodate the request and still make the titles optimally readable?

5. The art director proudly shows you the dancing Chinese-like lettering he has created with his titling software for the name identification of the new Chinese consul. Would you use such a title key? If so, why? If not, why not?

Course No. ____________ Date ____________ Name ____________

Television Talent

REVIEW OF KEY TERMS

Match each term with its appropriate definition by filling in the corresponding bubble.

1. **performer**
2. **talent**
3. **actor**
4. **makeup**
5. **blocking**
6. **cue card**
7. **pan-stick**
8. **teleprompter**

A. Carefully worked-out position, movement, and actions by the talent

B. A person who appears on-camera in a nondramatic role

C. A person who appears on-camera in a dramatic role

D. A large, hand-lettered card that contains on-air copy

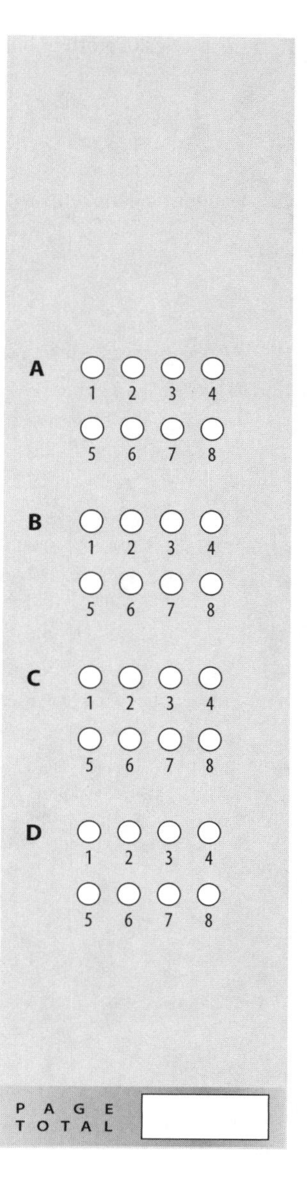

1. performer	**4. makeup**	**7. pan-stick**
2. talent	**5. blocking**	**8. teleprompter**
3. actor	**6. cue card**	

E. All people who regularly appear on television

E ○1 ○2 ○3 ○4 ○5 ○6 ○7 ○8

F. Cosmetics used to enhance, correct, or change appearance.

F ○1 ○2 ○3 ○4 ○5 ○6 ○7 ○8

G. Foundation makeup with a grease base

G ○1 ○2 ○3 ○4 ○5 ○6 ○7 ○8

H. Also known as auto-cue

H ○1 ○2 ○3 ○4 ○5 ○6 ○7 ○8

PAGE TOTAL

SECTION TOTAL

Course No. ______________ Date ______________ Name ______________________________

REVIEW OF PERFORMING TECHNIQUES

1. The following pictures show various *time* cues given to the talent by the floor manager. From the list below, select the specific cue illustrated and fill in the bubble with the corresponding number.

(9) 5 minutes left
(10) 30 seconds left
(11) 15 seconds left
(12) on time
(13) stretch
(14) cut
(15) wind up
(16) standby
(17) cue
(18) speed up

a.

b.

c.

d.

1a ○ 9 ○ 10 ○ 11 ○ 12 ○ 13 ○ 14 ○ 15 ○ 16 ○ 17 ○ 18

1b ○ 9 ○ 10 ○ 11 ○ 12 ○ 13 ○ 14 ○ 15 ○ 16 ○ 17 ○ 18

1c ○ 9 ○ 10 ○ 11 ○ 12 ○ 13 ○ 14 ○ 15 ○ 16 ○ 17 ○ 18

1d ○ 9 ○ 10 ○ 11 ○ 12 ○ 13 ○ 14 ○ 15 ○ 16 ○ 17 ○ 18

PAGE TOTAL

PHOTOS: EDWARD AIONA

(9) 5 minutes left	(13) stretch	(16) standby
(10) 30 seconds left	(14) cut	(17) cue
(11) 15 seconds left	(15) wind up	(18) speed up
(12) on time		

e.

f.

g.

h.

i.

j.

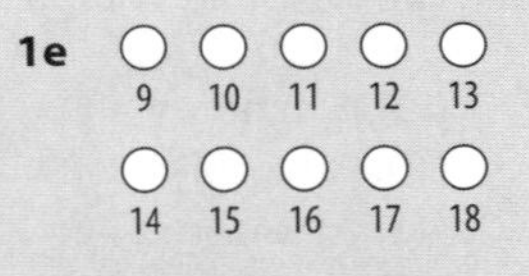

1e ○ 9 ○ 10 ○ 11 ○ 12 ○ 13 ○ 14 ○ 15 ○ 16 ○ 17 ○ 18

1f ○ 9 ○ 10 ○ 11 ○ 12 ○ 13 ○ 14 ○ 15 ○ 16 ○ 17 ○ 18

1g ○ 9 ○ 10 ○ 11 ○ 12 ○ 13 ○ 14 ○ 15 ○ 16 ○ 17 ○ 18

1h ○ 9 ○ 10 ○ 11 ○ 12 ○ 13 ○ 14 ○ 15 ○ 16 ○ 17 ○ 18

1i ○ 9 ○ 10 ○ 11 ○ 12 ○ 13 ○ 14 ○ 15 ○ 16 ○ 17 ○ 18

1j ○ 9 ○ 10 ○ 11 ○ 12 ○ 13 ○ 14 ○ 15 ○ 16 ○ 17 ○ 18

PAGE TOTAL

Course No. ______________ Date ______________ Name ______________

2. The following pictures show various *directional* and *audio* cues given to the talent by the floor manager. From the list below, select the specific cue illustrated and fill in the bubble with the corresponding number.

(19) keep talking
(20) step back
(21) OK
(22) walk
(23) speak up
(24) closer to mic
(25) tone down
(26) closer

a.

b.

c.

d.

PHOTOS: EDWARD AIONA

2a ◯ 19 ◯ 20 ◯ 21 ◯ 22 ◯ 23 ◯ 24 ◯ 25 ◯ 26

2b ◯ 19 ◯ 20 ◯ 21 ◯ 22 ◯ 23 ◯ 24 ◯ 25 ◯ 26

2c ◯ 19 ◯ 20 ◯ 21 ◯ 22 ◯ 23 ◯ 24 ◯ 25 ◯ 26

2d ◯ 19 ◯ 20 ◯ 21 ◯ 22 ◯ 23 ◯ 24 ◯ 25 ◯ 26

PAGE TOTAL

(19) keep talking
(20) step back
(21) OK
(22) walk
(23) speak up
(24) closer to mic
(25) tone down
(26) closer

e.

f.

2e ○ 19 ○ 20 ○ 21 ○ 22 ○ 23 ○ 24 ○ 25 ○ 26

2f ○ 19 ○ 20 ○ 21 ○ 22 ○ 23 ○ 24 ○ 25 ○ 26

g.

h.

2g ○ 19 ○ 20 ○ 21 ○ 22 ○ 23 ○ 24 ○ 25 ○ 26

2h ○ 19 ○ 20 ○ 21 ○ 22 ○ 23 ○ 24 ○ 25 ○ 26

PAGE TOTAL

PHOTOS: EDWARD AIONA

Course No. ______ Date ______ Name ______

Select the correct answers and fill in the bubbles with the corresponding numbers.

3. From the list below, select the microphone most appropriate for the various performance and acting tasks and fill in the bubbles with the corresponding numbers.

(27) lavalier
(28) boom mic
(29) hand mic
(30) fishpole mic
(31) stand mic
(32) desk mic
(33) wireless hand mic
(34) wireless lavalier

a. lead guitarist with a rock band, who also sings and talks to the audience from the stage

3a ○ 27 ○ 28 ○ 29 ○ 30 ○ 31 ○ 32 ○ 33 ○ 34

b. two actors doing an outdoor scene

3b ○ 27 ○ 28 ○ 29 ○ 30 ○ 31 ○ 32 ○ 33 ○ 34

c. singer who is also doing brief dance steps, accompanied by a large band

3c ○ 27 ○ 28 ○ 29 ○ 30 ○ 31 ○ 32 ○ 33 ○ 34

d. interview with a celebrity at a busy airport gate

3d ○ 27 ○ 28 ○ 29 ○ 30 ○ 31 ○ 32 ○ 33 ○ 34

e. moderating a panel discussion with six people

3e ○ 27 ○ 28 ○ 29 ○ 30 ○ 31 ○ 32 ○ 33 ○ 34

f. multiple-camera O/S shots involving two actors in a soap opera

3f ○ 27 ○ 28 ○ 29 ○ 30 ○ 31 ○ 32 ○ 33 ○ 34

g. sounds of breathing and skis on snow during a downhill race

3g ○ 27 ○ 28 ○ 29 ○ 30 ○ 31 ○ 32 ○ 33 ○ 34

h. news anchors who remain seated throughout a studio newscast

3h ○ 27 ○ 28 ○ 29 ○ 30 ○ 31 ○ 32 ○ 33 ○ 34

PAGE TOTAL ______

4. When demonstrating a product during a two-camera live show, you should orient the product toward the (35) *close-up camera* (36) *medium-shot camera* and keep looking at the (37) *close-up camera* (38) *medium-shot camera.* ***(Fill in two bubbles.)***

5. When demonstrating a small object, you should (39) *hold it as close to the lens as possible* (40) *keep it as steady as possible on the display table* (41) *lift it up for optimal camera pickup.*

6. When you notice that you're looking into the wrong (not switched on-the-air) camera, you should (42) *look down and then up again into the on-the-air camera* (43) *glance immediately over to the on-the-air camera* (44) *keep looking into the wrong camera until it is punched up on the air.*

7. When asked for an audio level, you should (45) *quickly count to 10* (46) *say one sentence with a slightly lower voice than when on the air* (47) *speak with your on-the-air voice until told that the level has been taken.*

8. For the talent the most accurate indicator of the camera's field of view is the (48) *relative distance between talent and camera* (49) *floor manager's cues* (50) *studio monitor.*

9. When wearing a lavalier mic, you should (51) *maintain your voice level regardless of how far the camera is away from you* (52) *increase your volume when the camera gets farther away from you* (53) *speak more softly when the camera is relatively close to you.*

10. When you receive cues during the actual video recording that differ from the rehearsed ones, you should (54) *execute the action as rehearsed* (55) *promptly follow the floor manager's cues* (56) *check with the director.*

11. When making an on-camera announcement in the studio, you should normally take your opening cue from the (57) *camera tally light* (58) *camera operator* (59) *floor manager.*

4	○ 35	○ 36	
	○ 37	○ 38	
5	○ 39	○ 40	○ 41
6	○ 42	○ 43	○ 44
7	○ 45	○ 46	○ 47
8	○ 48	○ 49	○ 50
9	○ 51	○ 52	○ 53
10	○ 54	○ 55	○ 56
11	○ 57	○ 58	○ 59

PAGE TOTAL

SECTION TOTAL

Course No. ____________ Date ____________ Name ____________

REVIEW OF ACTING TECHNIQUES

Select the correct answers and fill in the bubbles with the corresponding numbers.

1. When you notice that the boom mic has not quite caught up with you, you should (60) *wait for the mic* (61) *continue with the dialogue* (62) *speak louder.*

2. When repeating action for close-ups (such as drinking a glass of milk), you (63) *use the opportunity to improve on what you have done in the long or medium shots* (64) *get a new glass and have it refilled for each close-up* (65) *use the same props and have your glass filled to the level just before the close-up.*

3. When acting for television, you should project your motions and emotions as you would on the stage (66) *when there is a prolonged dialogue pause* (67) *never* (68) *every time the camera is relatively far away.*

4. When blocked in the camera-far position in an O/S shot, you must make sure that you see the (69) *tally light* (70) *floor manager* (71) *camera lens.*

5. A "blocking map" is (72) *a rough map drawn by the floor manager* (73) *a mental map to remember prominent positions* (74) *the lines drawn on the floor by the AD.*

6. Television plays are video-recorded (75) *in the order of scenes from the beginning to the end of the script* (76) *in brief scenes, grouped by location, characters involved, and so forth* (77) *according to the mood of the director.*

7. When auditioning for a television drama, you should (78) *apply your theatre technique to show that you have stage training* (79) *wear something unusual so the director will remember you* (80) *internalize the role as much as possible.*

8. After the blocking rehearsal with the director, you (81) *can make minor changes if the camera operator concurs* (82) *must keep the exact blocking as rehearsed* (83) *can suggest a different blocking during the video recording.*

9. In a standard studio-recorded daytime serial, the television camera looks at you primarily in (84) *long shots* (85) *close-ups* (86) *low-level shots.*

10. When on a close-up, you should (87) *slow down* (88) *accelerate* (89) *not change* the speed of your on-camera actions.

11. When the camera is relatively far away from you, you should (90) *exaggerate your facial expressions* (91) *exaggerate your gestures* (92) *maintain your close-up acting style.*

1	○ 60	○ 61	○ 62
2	○ 63	○ 64	○ 65
3	○ 66	○ 67	○ 68
4	○ 69	○ 70	○ 71
5	○ 72	○ 73	○ 74
6	○ 75	○ 76	○ 77
7	○ 78	○ 79	○ 80
8	○ 81	○ 82	○ 83
9	○ 84	○ 85	○ 86
10	○ 87	○ 88	○ 89
11	○ 90	○ 91	○ 92

SECTION TOTAL ____

REVIEW OF MAKEUP AND CLOTHING

Select the correct answers and fill in the bubbles with the corresponding numbers.

1. When applying makeup, the ideal lighting conditions are the same as or close to those of (93) *your customary dressing room* (94) *the actual production environment* (95) *normal 3,200K studio lights.*

2. Under high-color-temperature lighting, use (96) *cool* (97) *warm* (98) *neutral* makeup colors.

3. As a weathercaster you can wear green so long as the chroma-key backdrop is (99) *blue* (100) *green* (101) *white.*

4. One of the most widely used makeup foundations is (102) *cake* (103) *grease base* (104) *pan-stick.*

5. When working with a small single-chip camcorder under low-light conditions, you should avoid wearing (105) *red* (106) *green* (107) *blue.*

6. You can counteract a heavy five-o'clock shadow by applying a light layer of (108) *yellow* (109) *skin-colored* (110) *bluish* pan-stick.

7. Clothing with thin, highly contrasting stripes or checkered patterns is (111) *acceptable* (112) *not acceptable* because (113) *the digital camera CCD can handle such a contrast* (114) *it provides exciting patterns* (115) *it causes moiré color vibrations* (116) *it is too detailed for the camera to see.* ***(Fill in two bubbles.)***

8. The dress of a pop singer has many rhinestones that sparkle under the colored stage lights. This dress is (117) *acceptable* (118) *not acceptable* because (119) *the color camera can handle small areas of bright light* (120) *there is too much brightness contrast* (121) *it will cause moiré patterns* (122) *it will help raise the baselight level.* ***(Fill in two bubbles.)***

1	○ 93	○ 94	○ 95	
2	○ 96	○ 97	○ 98	
3	○ 99	○ 100	○ 101	
4	○ 102	○ 103	○ 104	
5	○ 105	○ 106	○ 107	
6	○ 108	○ 109	○ 110	
7	○ 111	○ 112		
	○ 113	○ 114	○ 115	○ 116
8	○ 117	○ 118		
	○ 119	○ 120	○ 121	○ 122

SECTION TOTAL

Course No. ____________ Date ____________ Name ____________

REVIEW QUIZ

*Mark the following statements as true or false by filling in the bubbles in the **T** (for true) or **F** (for false) column.*

1. When asked to give an audio level, you should count quickly to 10.
2. When you are on the air in a dramatic role, you must follow the rehearsed blocking precisely.
3. What you wear when auditioning for a role is unimportant.
4. When you work with a teleprompter, it is best to move the camera as close to the talent as possible.
5. Talent includes both performers and actors.
6. When the camera is relatively far from you, you should walk toward it for good close-ups.
7. In an unrehearsed show, you can give the crew and the director a verbal warning cue of what you are going to do next.
8. When on a close-up, you must pick up the item you are demonstrating and hold it close to the camera lens.
9. When on a panel, you should reposition the desk mic so that it points directly at you.
10. Television actors always portray someone else.
11. To maintain sound perspective, you should talk louder when the camera is farther away from you and more softly when the camera is close to you.
12. When demonstrating a small object, you can help the director by asking that the camera move a little closer.

	T	F
1	○ 123	○ 124
2	○ 125	○ 126
3	○ 127	○ 128
4	○ 129	○ 130
5	○ 131	○ 132
6	○ 133	○ 134
7	○ 135	○ 136
8	○ 137	○ 138
9	○ 139	○ 140
10	○ 141	○ 142
11	○ 143	○ 144
12	○ 145	○ 146

SECTION TOTAL ____

PROBLEM-SOLVING APPLICATIONS

1. To practice blocking, write down a series of moves that carry you around your kitchen. For example, you can start at the stove, then get the teakettle out of the cupboard, fill it with water, and put it on the stove, go back to pick up the telephone, put down the telephone to answer the door, and so forth. Try to hit the same marks each time you go through the routine. If possible, have a friend video-record your blocking maneuvers from the same camera position. You can then compare the recordings and check how accurate your blocking was. As part of the same exercise, you can use various props (kitchen utensils) to see how the camera's field of view (LS to ECU) will influence your handling of them.

2. Use a product of your choice and video-record your pitch. What do you like about your performance? What don't you like? How could you improve your performance?

3. Pretend that you, person A, are receiving a telephone call from person B. In this scene we see and hear only person A (you). Using exactly the same dialogue (see the script on the following page), adapt your delivery and acting style to at least two of the following circumstances:

   a. B calls to tell you that he/she has just got an exciting new job.

   b. B calls to tell you that he/she has just lost his/her job.

   c. B has just had an accident with your new car.

   d. B has broken the engagement.

   e. B has won first prize in a video competition.

   Locate the scene anywhere you like. You may do well to write the other part of the phone conversation so that you can "listen" to the virtual B part of the dialogue and respond more convincingly verbally and nonverbally.

4. Have a friend take close-ups of you when you do the phone exercise and compare your expressions when the phone call brings happy news and unhappy news.

5. An experienced stage director, who is directing an adaptation of a stage play for TV, tells the lead actress that her facial expressions are not big enough to "reach the last row." What is your reaction? Be specific.

Course No. ______ Date ______ Name ______

PHONE CONVERSATION

PERSON A

Hello?

Hi.

Fine, and you?

Good.

No.

No, really. It's always good to hear from you.

I beg your pardon?

You must be kidding.

Yes.

No.

What does Alex say to all this?

No. Should I?

I don't know.

Perhaps.

You want me to come over now?

Yes. Really.

Well, this changes things somewhat.

I think so.

I'm not so sure.

Yes. No. I...

All right. But not...

OK.

If you think this is...

Definitely.

Good-bye... When?

No. Really.

Good-bye.

Course No. ________ Date ________ Name ________

The Director in Production

REVIEW OF KEY TERMS

Match each term with its appropriate definition by filling in the corresponding bubble.

1. **schedule time**
2. **intercom**
3. **camera rehearsal**
4. **dress rehearsal**
5. **single-camera directing**
6. **dry run**
7. **running time**
8. **walk-through**
9. **multicamera directing**

A. A communication system widely used by all production and technical personnel so that they can communicate with one another during a show

A ○ 1 ○ 2 ○ 3 ○ 4 ○ 5 ○ 6 ○ 7 ○ 8 ○ 9

B. Duration of a program or program segment

B ○ 1 ○ 2 ○ 3 ○ 4 ○ 5 ○ 6 ○ 7 ○ 8 ○ 9

C. The times when a program starts and stops

C ○ 1 ○ 2 ○ 3 ○ 4 ○ 5 ○ 6 ○ 7 ○ 8 ○ 9

D. A full rehearsal with cameras and other pieces of production equipment

D ○ 1 ○ 2 ○ 3 ○ 4 ○ 5 ○ 6 ○ 7 ○ 8 ○ 9

PAGE TOTAL ________

1. schedule time	**4. dress rehearsal**	**7. running time**
2. intercom	**5. single-camera directing**	**8. walk-through**
3. camera rehearsal	**6. dry run**	**9. multicamera directing**

E. An orientation session on the set with the production crew and talent

E ○1 ○2 ○3 ○4 ○5 ○6 ○7 ○8 ○9

F. Full rehearsal with talent made-up and dressed

F ○1 ○2 ○3 ○4 ○5 ○6 ○7 ○8 ○9

G. The coordination of one camera for takes that are separately recorded for postproduction

G ○1 ○2 ○3 ○4 ○5 ○6 ○7 ○8 ○9

H. A rehearsal without equipment

H ○1 ○2 ○3 ○4 ○5 ○6 ○7 ○8 ○9

I. The simultaneous coordination of two or more cameras for instantaneous editing

I ○1 ○2 ○3 ○4 ○5 ○6 ○7 ○8 ○9

PAGE TOTAL

SECTION TOTAL

Course No. ____________ Date ____________ Name ____________

REVIEW OF DIRECTOR'S TERMINOLOGY

1. **Director's visualization cues.** From the list below, select the cue necessary to adjust the picture on the left screen to the picture on the right screen (in pairs from **a** through **l**) and fill in the bubbles with the corresponding numbers.

(10) pan right	(15) dolly out	(19) pan left
(11) tilt down	(16) truck right	(20) pedestal down or crane down
(12) zoom in	(17) arc left	(21) tilt up
(13) zoom out	(18) pedestal up or crane up	
(14) dolly in		

a.

1a ○ 10 ○ 11 ○ 12 ○ 13 ○ 14 ○ 15 ○ 16 ○ 17 ○ 18 ○ 19 ○ 20 ○ 21

b.

1b ○ 10 ○ 11 ○ 12 ○ 13 ○ 14 ○ 15 ○ 16 ○ 17 ○ 18 ○ 19 ○ 20 ○ 21

c.

1c ○ 10 ○ 11 ○ 12 ○ 13 ○ 14 ○ 15 ○ 16 ○ 17 ○ 18 ○ 19 ○ 20 ○ 21

PAGE TOTAL

PHOTOS: EDWARD AIONA

PHOTOS: EDWARD AIONA

PHOTOS: HERBERT ZETTL

(10) pan right	(15) dolly out	(19) pan left
(11) tilt down	(16) truck right	(20) pedestal down or crane down
(12) zoom in	(17) arc left	(21) tilt up
(13) zoom out	(18) pedestal up or crane up	
(14) dolly in		

d.

1d ○ 10 ○ 11 ○ 12 ○ 13 ○ 14 ○ 15 ○ 16 ○ 17 ○ 18 ○ 19 ○ 20 ○ 21

e.

1e ○ 10 ○ 11 ○ 12 ○ 13 ○ 14 ○ 15 ○ 16 ○ 17 ○ 18 ○ 19 ○ 20 ○ 21

f.

1f ○ 10 ○ 11 ○ 12 ○ 13 ○ 14 ○ 15 ○ 16 ○ 17 ○ 18 ○ 19 ○ 20 ○ 21

PHOTOS: EDWARD AIONA

PAGE TOTAL

Course No. ______________ Date ______________________ Name ____

PHOTOS: EDWARD AIONA

(10) pan right
(11°tilt down
(12) zoom in
(13) zoom out
(14) dolly in
(15) dolly out
(16) truck right
(17) arc left
(18) pedestal up or crane up
(19) pan left
(20) pedestal down or crane down
(21) tilt up

l.

1l ○ 10 ○ 11 ○ 12 ○ 13 ○ 14 ○ 15 ○ 16 ○ 17 ○ 18 ○ 19 ○ 20 ○ 21

2. Director's sequencing cues.

a. From the list below, select the director who uses the correct sequence of cues for the opening of a two-camera (C1 and C2) interview and fill in the corresponding bubble. ***(There is a title key for the guest. Assume that the crew has received a general standby cue and that bars and tone have already been recorded on the tape by the AD.)***

(22) *Director A:* "Ready to take C.G. Slate. Take slate. Ready black. Black. Beeper. Ready to come up on one CU of host—take one. Cue host. Ready two [on guest]. Take two. Cue guest. Key title. Take one."

(23) *Director B:* "Ready to roll video recorder. Roll video recorder. Ready C.G. Slate. Take C.G. Read slate. Ready black. Ready beeper. Beeper. To black. Change page [C.G.]. One, CU of host. Ready to come up on one. Open mic, cue host, up on one. Two, CU of guest. Ready two. Ready key C.G. Take two, key. Lose key. Ready one, two-shot. Take one."

(24) *Director C:* "Ready to roll video recorder. Roll video recorder. Ready C.G. Slate. Read slate. Ready black. Ready beeper. To black. Beeper. Ready to come up on one. Up on one. Ready two. Take two. Key. Lose key. Ready one. Take one."

2a ○ 22 ○ 23 ○ 24

PAGE TOTAL

b. From the list below, select the correct director's cues for transitions by filling in the corresponding bubbles. ***(Multiple answers are possible.)***

(25) Ready to take camera two. Take camera two.

(26) Ready three. Take three.

(27) Ready one. Dissolve to one.

(28) Ready to go to black. Go to black.

(29) Ready wipe. Dissolve to two.

(30) Ready to change C.G. page. Change page.

2b ○ 25 ○ 26 ○ 27 ○ 28 ○ 29 ○ 30

c. From the list below, select the correct director's cues to the floor manager by filling in the corresponding bubbles. The talent are Mary, Lisa, John, and Larry. ***(Multiple answers are possible.)***

(31) Ready to cue Mary. Cue Mary.

(32) Ready to cue him. Cue him.

(33) Make him talk faster.

(34) Move her stage-right.

(35) Turn the sculpture counterclockwise.

(36) Give Lisa the wind-up.

2c ○ 31 ○ 32 ○ 33 ○ 34 ○ 35 ○ 36

3. Director's cues to floor manager concerning the positioning of props. Select the appropriate cue to the floor manager to adjust the position of the prop shown on the left screen to that of the right screen and fill in the bubbles with the corresponding numbers.

(37) turn the sculpture counterclockwise
(38) turn the sculpture clockwise

3 ○ 37 ○ 38

PAGE TOTAL

PHOTOS: HERBERT ZETTL

4. **Director's cues to the floor manager concerning the positioning of talent.** From the list below, select the appropriate cue to the floor manager to adjust the position of the talent shown on the left screen to that of the right screen (in pairs **a** through **c**) and fill in the bubbles with the corresponding numbers.

(39) have woman turn to her left
(40) have camera arc right

a.

4a ○ 39 ○ 40

(41) have talent turn in [toward the camera]
(42) have talent move left

b.

4b ○ 41 ○ 42

(43) pan right
(44) move talent to camera right

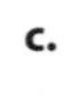

c.

4c ○ 43 ○ 44

PAGE TOTAL

SECTION TOTAL

Course No. ______________ Date ______________ Name ______________________________

REVIEW OF REHEARSAL TECHNIQUES

Select the correct answers and fill in the bubbles with the corresponding numbers.

1. When doing a walk-through/camera rehearsal combination from the studio floor, you should give (45) *only the camera cues* (46) *only the talent cues* (47) *all cues as though you were directing from the control room.*

2. If pressed for time, you should call for (48) *an uninterrupted camera rehearsal* (49) *a walk-through/camera rehearsal combination* (50) *a blocking rehearsal.*

3. Blocking rehearsals are most efficiently conducted from (51) *the control room* (52) *the studio floor or rehearsal hall* (53) *the actual studio set.*

4. When scheduling "notes" segments in your time line, you need to also schedule (54) *additional talent rehearsal time* (55) *reset time* (56) *additional technical rehearsal time.*

5. When engaged in a standard EFP, you need not worry about (57) *talent and technical walk-throughs* (58) *cross-overs from one location to the next* (59) *an extensive intercom setup.*

6. Camera rehearsal is conducted (60) *similar to a dress rehearsal* (61) *for cameras only* (62) *for all technical operations but without talent.*

7. When doing a single-camera ENG or EFP, you should (63) *always get a fair amount of cutaways* (64) *get cutaways only if you think your shots will not cut together well* (65) *not bother with cutaways if you have plenty of time for postproduction.*

8. When calling for a "take," you should pause between the "ready" and the "take" cues (66) *as little as possible* (67) *until you see the TD put his finger on the correct switcher button* (68) *for at least five seconds.*

9. Rehearsals that combine walk-throughs and camera rehearsal are most efficiently conducted from the (69) *studio floor* (70) *rehearsal hall* (71) *control room.*

10. When breaking down an EFP script for a single-camera production, you should (72) *try to maintain the narrative order of the scenes* (73) *combine all scenes that play at the same location and/or with the same talent* (74) *start with the most interesting parts to take advantage of the talent's creative energy.*

1	○ 45	○ 46	○ 47
2	○ 48	○ 49	○ 50
3	○ 51	○ 52	○ 53
4	○ 54	○ 55	○ 56
5	○ 57	○ 58	○ 59
6	○ 60	○ 61	○ 62
7	○ 63	○ 64	○ 65
8	○ 66	○ 67	○ 68
9	○ 69	○ 70	○ 71
10	○ 72	○ 73	○ 74

SECTION TOTAL ______

REVIEW OF MULTICAMERA DIRECTING

1. Assume that the following six shots represent a sequence of video inputs on the preview monitor in the order you will switch them to the line-out. Using the floor plan below, specify the cameras and the other video inputs used for the shots. Note that one video input does not originate in the studio and another uses two video sources simultaneously. ***(Multiple answers are possible.)***

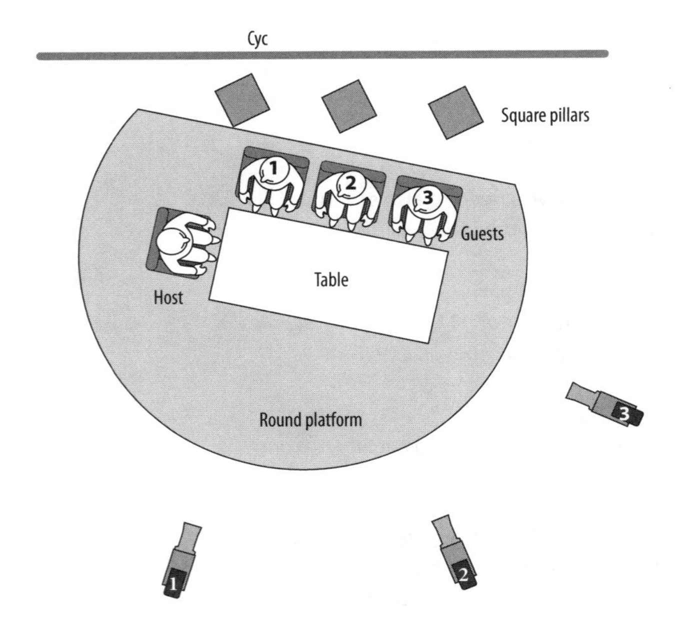

Sources: (75) camera 1

(76) camera 2

(77) camera 3

(78) video recording 1

(79) C.G.

a. 

Host

1a ○ 75 ○ 76 ○ 77 ○ 78 ○ 79

PAGE TOTAL

Course No. ______________ Date ______________ Name ______________

b.

Guest 1

1b ◯ 75 ◯ 76 ◯ 77 ◯ 78 ◯ 79

c.

1c ◯ 75 ◯ 76 ◯ 77 ◯ 78 ◯ 79

d.

Guest 2

1d ◯ 75 ◯ 76 ◯ 77 ◯ 78 ◯ 79

e.

Guest 3

1e ◯ 75 ◯ 76 ◯ 77 ◯ 78 ◯ 79

f.

1f ◯ 75 ◯ 76 ◯ 77 ◯ 78 ◯ 79

PAGE TOTAL ______

SECTION TOTAL ______

PHOTOS: BROADCAST AND ELECTRONIC COMMUNICATION ARTS DEPARTMENT AT SAN FRANCISCO STATE UNIVERSITY

REVIEW OF TIMING

Select the correct answers and fill in the bubbles with the corresponding numbers.

1. Subjective time refers to (80) *the running time of the show segment* (81) *how fast or slow it seems to move* (82) *how the parts of the segment relate to one another.*

2. In the noon newscast, you must switch to the first 3-minute satellite feed at exactly 3:45 minutes into the show and the second feed at 15:15 minutes after the end of the first one. You need to switch to the remote feeds at:

   (83) 12:03:45 and 12:19:00

   (84) 12:03:45 and 12:22:00

   (85) 12:03:45 and 12:15:15

3. The log indicates that the *Women: Face-to-face* program ends at 11:26:30. A large part of the program is taken up by a fashion show. The fashion coordinator would like a 1-minute, a 30-second, and a 15-second cue, as well as a cut at the end of her segment. The regular program host, who follows the fashion show with a 1½-minute closing, would like a 30-second and a 15-second cue and a cut at the end of the program. From the list below, select the correct clock times for the cues.

   (86) 11:24:00, 11:24:30, 11:24:45, 11:25:00 and 11:26:00, 11:26:15, 11:26:30

   (87) 11:24:30, 11:25:00, 11:25:45, 11:26:00 and 11:26:15, 11:26:30

   (88) 11:22:00, 11:23:00, 11:24:00, 11:24:30 and 11:25:00, 11:26:00, 11:26:15

1 ○ 80 ○ 81 ○ 82

2 ○ 83 ○ 84 ○ 85

3 ○ 86 ○ 87 ○ 88

SECTION TOTAL

Course No. ________ Date ________ Name ________

REVIEW OF SINGLE-CAMERA DIRECTING

Select the correct answers and fill in the bubbles with the corresponding numbers.

1. When directing single-camera style, you (89) *can visualize each shot independently of all others* (90) *can do away with cutaways* (91) *need to be concerned about continuity of widely dispersed shots.*

2. In single-camera directing, camera placement is (92) *not important because you are shooting from various angles anyway* (93) *very important to ensure continuity* (94) *determined primarily by where the actors are.*

3. When directing single-camera style, you should (95) *rehearse each shot before the take* (96) *rehearse all shots before ever activating the camera* (97) *rehearse the shots in the order of the storyboard narrative.*

4. When directing single-camera style in the studio, you need to rehearse and direct each shot (98) *always from the control room* (99) *always from the studio floor* (100) *from either the control room or the studio floor.*

5. When directing single-camera style in the field, you need (101) *the camera hooked up to a field monitor* (102) *to use the camera viewfinder to correct shots* (103) *neither a field monitor nor a viewfinder.*

6. In single-camera directing, you should slate (104) *each take* (105) *each shot* (106) *each scene.*

7. When recording out of sequence, you will find (107) *a large external viewfinder* (108) *a storyboard* (109) *the PA's notes* especially helpful for your visualization.

1	◯ 89	◯ 90	◯ 91
2	◯ 92	◯ 93	◯ 94
3	◯ 95	◯ 96	◯ 97
4	◯ 98	◯ 99	◯ 100
5	◯ 101	◯ 102	◯ 103
6	◯ 104	◯ 105	◯ 106
7	◯ 107	◯ 108	◯ 109

SECTION TOTAL ________

REVIEW QUIZ

*Mark the following statements as true or false by filling in the bubbles in the **T** (for true) or **F** (for false) column.*

1. When directing a studio show, the S.A. system is more appropriate than the P.L. system.
2. Keeping accurate running time is more important in directing a live multicamera show than a single-camera EFP.
3. When directing from the control room, you should address the name of the camera operator rather than the camera number to get efficient camera action.
4. During a walk-through/camera rehearsal combination, the director rehearses primarily from the studio floor.
5. When directing a fully scripted show, you should pay more attention to the script than the preview or line monitors.
6. Even with an efficient intercom system, the switcher should always be located right next to the director's position.
7. When directing a daily newscast, you do not need a floor plan to preplan the camera shots.
8. You should tell the floor manager whenever there is a technical problem that you need to solve from the control room.
9. Even when doing an EFP, you should check the video recording to see whether the preceding scene was properly recorded before moving to the next location.
10. To save time in a studio rehearsal, you should use the S.A. system as often as possible.
11. When directing a single-camera EFP, a properly working intercom system is one of the most essential setup items.
12. When doing an EFP, the talent and technical walk-throughs are less important than when doing a studio show.
13. Cutaways are especially important in film-style shooting.
14. An external monitor that carries the camera's video greatly facilitates single-camera directing.
15. The floor manager's cues are especially important when engaged in single-camera EFP.

	T	F
1	○ 110	○ 111
2	○ 112	○ 113
3	○ 114	○ 115
4	○ 116	○ 117
5	○ 118	○ 119
6	○ 120	○ 121
7	○ 122	○ 123
8	○ 124	○ 125
9	○ 126	○ 127
10	○ 128	○ 129
11	○ 130	○ 131
12	○ 132	○ 133
13	○ 134	○ 135
14	○ 136	○ 137
15	○ 138	○ 139

SECTION TOTAL

Course No. ________ Date ________ Name ________

PROBLEM-SOLVING APPLICATIONS

1. Mark a scene from a fully scripted TV play and practice calling the shots.

2. During the video recording of the first scene of a demanding outdoor EFP for a car commercial, the audio person suggests doing a retake because she picked up a brief, distant jet sound. Would you recommend a retake? If so, why? If not, why not?

3. During the evening news, the wrong video-recorded story comes up. What can you do?

4. During an O/S sequence in a multicamera dramatic production, one of the actors has trouble hitting the blocking marks and is frequently obscured by the camera-near person. What advice would you give the actor?

5. Get a published script of an episode of a drama or soap opera and mark it for three-camera and single-camera directing.

6. The director uses one set of commands during rehearsal but switches to another when doing the on-the-air show. Which potential problems do you foresee, if any? Be specific.

7. The producer suggests that you should not waste valuable time by doing a walk-through/camera rehearsal from the studio floor but skip right to the camera rehearsal from the control room. What is your reaction? Why?

8. When checking all the intercom systems before a remote live multicamera telecast of a large political gathering at city hall, the talent's I.F.B. interrupts itself from time to time. What backup cueing device would you recommend that close to airtime?

9. When pressed for time, the director decides to conduct the rehearsal from the studio floor. Would you agree or disagree with such a move? Be specific. How, if at all, would the director's request affect the studio equipment and the control room activities?

10. The floor manager expresses her concern to you, the director, about the lack of adequate intercom facilities for an EFP of the local garden show. How would you respond?

11. The line producer of a complex commercial considers the director's explaining the process message to talent and crew a waste of time. What is your reaction? Be specific.

Course No. ________ Date ________ Name ________

Field Production and Big Remotes

REVIEW OF KEY TERMS

Match each term with its appropriate definition by filling in the corresponding bubble.

1. **broadband**
2. **direct broadcast satellite**
3. **downlink**
4. **microwave relay**
5. **remote survey**
6. **iso camera**
7. **mini-link**
8. **field production**
9. **big remote**
10. **instant replay**
11. **Ku-band**
12. **uplink**

A. A variety of information sent simultaneously over a fiber-optic cable

A ○1 ○2 ○3 ○4 ○5 ○6 ○7 ○8 ○9 ○10 ○11 ○12

B. A signal transport from the remote location to the station or transmitter in various transmission steps

B ○1 ○2 ○3 ○4 ○5 ○6 ○7 ○8 ○9 ○10 ○11 ○12

C. Repeating a key play or an important event for the viewer, through playing back by videotape or disk-stored video, immediately after its live occurrence

C ○1 ○2 ○3 ○4 ○5 ○6 ○7 ○8 ○9 ○10 ○11 ○12

PAGE TOTAL

1. broadband	**5. remote survey**	**9. big remote**
2. direct broadcast satellite	**6. iso camera**	**10. instant replay**
3. downlink	**7. mini-link**	**11. Ku-band**
4. microwave relay	**8. field production**	**12. uplink**

D. A high-frequency signal used by satellites

D ○1 ○2 ○3 ○4 ○5 ○6 ○7 ○8 ○9 ○10 ○11 ○12

E. A production outside the studio to televise live and/or live-record a large scheduled event

E ○1 ○2 ○3 ○4 ○5 ○6 ○7 ○8 ○9 ○10 ○11 ○12

F. A satellite with a high-powered transponder

F ○1 ○2 ○3 ○4 ○5 ○6 ○7 ○8 ○9 ○10 ○11 ○12

G. Any production that occurs outside of the studio

G ○1 ○2 ○3 ○4 ○5 ○6 ○7 ○8 ○9 ○10 ○11 ○12

H. A preproduction on-location investigation of the existing facilities of a scheduled telecast away from the studio

H ○1 ○2 ○3 ○4 ○5 ○6 ○7 ○8 ○9 ○10 ○11 ○12

PAGE TOTAL

Course No. ______________ Date ______________ Name ______________________________

I. An earth station transmitter used to send video and audio signals to a satellite

J. An antenna and equipment that receives the signals coming from a satellite

K. Often used in sports remotes; feeds into the switcher and its own video recorder

L. A setup of several small microwave transmitters and receivers to transport the television signal around obstacles

I	○ 1	○ 2	○ 3	○ 4
	○ 5	○ 6	○ 7	○ 8
	○ 9	○ 10	○ 11	○ 12
J	○ 1	○ 2	○ 3	○ 4
	○ 5	○ 6	○ 7	○ 8
	○ 9	○ 10	○ 11	○ 12
K	○ 1	○ 2	○ 3	○ 4
	○ 5	○ 6	○ 7	○ 8
	○ 9	○ 10	○ 11	○ 12
L	○ 1	○ 2	○ 3	○ 4
	○ 5	○ 6	○ 7	○ 8
	○ 9	○ 10	○ 11	○ 12

PAGE TOTAL

SECTION TOTAL

REVIEW OF FIELD PRODUCTION

Select the correct answers and fill in the bubbles with the corresponding numbers.

1. Instant replays are most common in (13) *big remotes* (14) *EFP* (15) *ENG.*
2. Using multiple cameras or camcorders in EFP means that they (16) *shoot a scene simultaneously* (17) *must feed a switcher* (18) *run in sync.*
3. The most flexible type of field production that needs little or no preproduction is (19) *ENG* (20) *EFP* (21) *big remotes.*
4. The walk-through rehearsal is least important for (22) *ENG* (23) *EFP* (24) *big remotes.*
5. A floor manager is most important in (25) *ENG* (26) *EFP* (27) *big remotes.*
6. The directing procedure that most closely resembles multicamera studio production is (28) *ENG* (29) *EFP* (30) *big remotes.*
7. The field production that affords the most control over the event is (31) *ENG* (32) *EFP* (33) *big remotes.*
8. The remote system least likely to use signal transmission equipment is (34) *ENG* (35) *EFP* (36) *big remotes.*
9. A complex intercommunication system is most important for (37) *ENG* (38) *EFP* (39) *big remotes.*
10. Because the camera setup is done by technical personnel, the director is (40) *not needed* (41) *very important* (42) *consulted only in emergencies* for the specific locations of the key cameras.
11. A big-remote survey requires (43) *only a production survey* (44) *only a technical survey* (45) *both a production and a technical survey.*
12. The normal transmission equipment in ENG vans is (46) *a microwave transmitter* (47) *a satellite uplink* (48) *fiber-optic cable.*
13. To ensure access to the event location, you need (49) *a contact person* (50) *a written statement from the producer* (51) *an OK from the chief of police.*
14. The field production that almost always requires careful postproduction is (52) *ENG* (53) *big remotes* (54) *EFP.*

1	○ 13	○ 14	○ 15
2	○ 16	○ 17	○ 18
3	○ 19	○ 20	○ 21
4	○ 22	○ 23	○ 24
5	○ 25	○ 26	○ 27
6	○ 28	○ 29	○ 30
7	○ 31	○ 32	○ 33
8	○ 34	○ 35	○ 36
9	○ 37	○ 38	○ 39
10	○ 40	○ 41	○ 42
11	○ 43	○ 44	○ 45
12	○ 46	○ 47	○ 48
13	○ 49	○ 50	○ 51
14	○ 52	○ 53	○ 54

SECTION TOTAL

Course No. ____________ Date ____________ Name ____________

REVIEW OF BIG REMOTES

1. Analyze the following five location sketches for field productions and big remotes. Evaluate the type and the position of each camera by the criteria listed below and fill in the bubbles with the corresponding numbers. ***(Multiple answers are possible.)***

(55) camera position OK
(56) wrong or unnecessary camera position
(57) inappropriate camera type
(58) cable hazard
(59) lighting problems (shooting into the sun or against another strong light source)

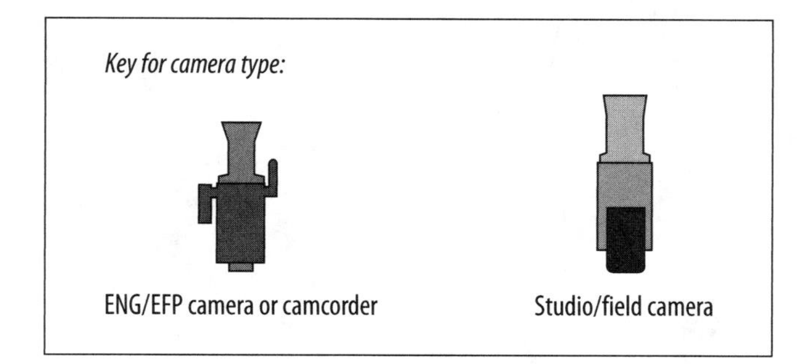

(55) camera position OK
(56) wrong or unnecessary camera position
(57) inappropriate camera type
(58) cable hazard
(59) lighting problems (shooting into the sun or against another strong light source)

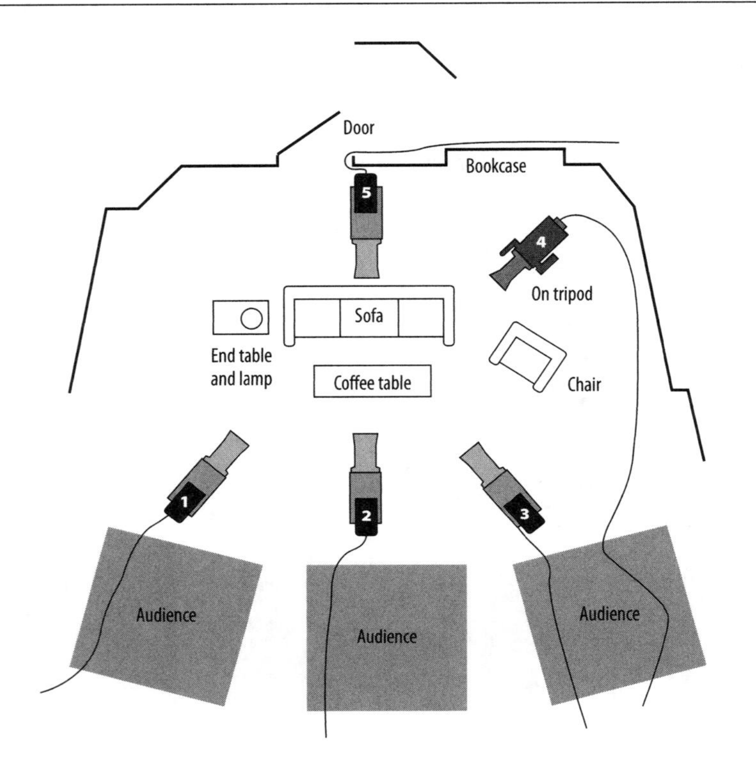

Play

Video recording of two performances of a high-school play (situation comedy) with a live audience, minimal postproduction, and the use of a large remote truck

a. comments on C1

b. comments on C2

c. comments on C3

d. comments on C4

e. comments on C5

1a C1 ○ 55 ○ 56 ○ 57 ○ 58 ○ 59

1b C2 ○ 55 ○ 56 ○ 57 ○ 58 ○ 59

1c C3 ○ 55 ○ 56 ○ 57 ○ 58 ○ 59

1d C4 ○ 55 ○ 56 ○ 57 ○ 58 ○ 59

1e C5 ○ 55 ○ 56 ○ 57 ○ 58 ○ 59

PAGE TOTAL

Course No. ______ Date ______ Name ______

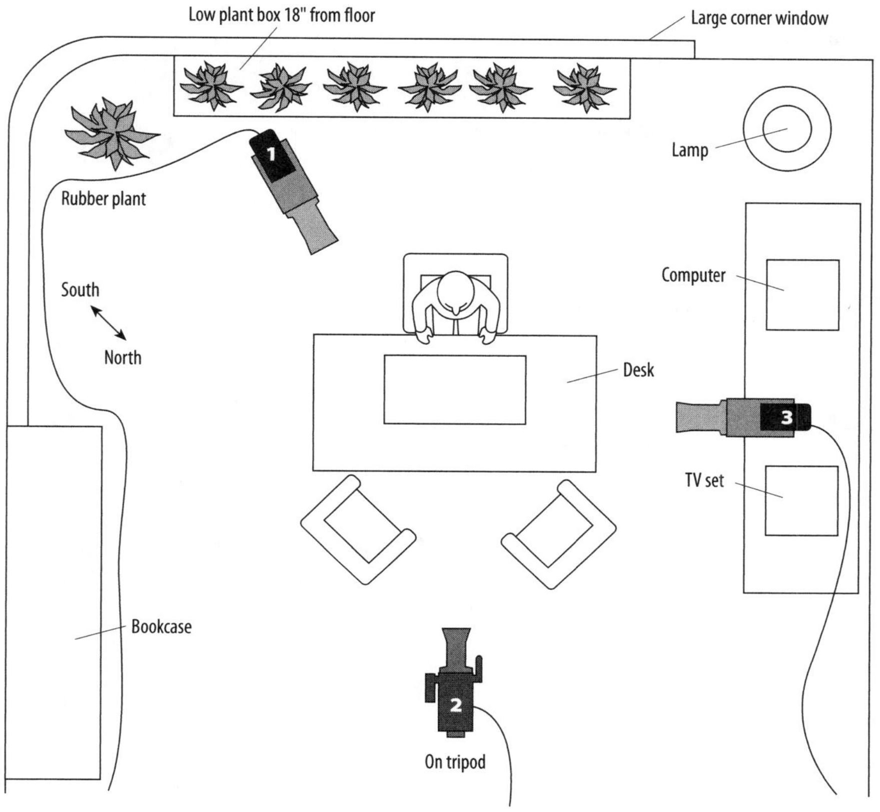

Low plant box 18" from floor
Large corner window
Lamp
Rubber plant
1
Computer
South
North
Desk
3
TV set
Bookcase
2
On tripod

EFP of Company President's Address to Employees
Live-recorded for minimal postproduction
Recording date: July 15
Recording time: 2:30 p.m. to 4:30 p.m.
Place: President's office, Tower Building, 34th floor

f. comments on C1

g. comments on C2

h. comments on C3

1f **C1** ○ 55 ○ 56 ○ 57 ○ 58 ○ 59

1g **C2** ○ 55 ○ 56 ○ 57 ○ 58 ○ 59

1h **C3** ○ 55 ○ 56 ○ 57 ○ 58 ○ 59

PAGE TOTAL

(55) camera position OK
(56) wrong or unnecessary camera position
(57) inappropriate camera type
(58) cable hazard
(59) lighting problems (shooting into the sun or against another strong light source)

In stands on tripod
1
Audience bleachers
Bench
Referee
Bench
Sun
4
High
3
Low
2
On field (mobile)

Tennis Match
Live coverage of tennis match

i. comments on C1

j. comments on C2

k. comments on C3

l. comments on C4

1i C1 ○ 55 ○ 56 ○ 57 ○ 58 ○ 59

1j C2 ○ 55 ○ 56 ○ 57 ○ 58 ○ 59

1k C3 ○ 55 ○ 56 ○ 57 ○ 58 ○ 59

1l C4 ○ 55 ○ 56 ○ 57 ○ 58 ○ 59

PAGE TOTAL

Course No. ________ Date ________ Name ________

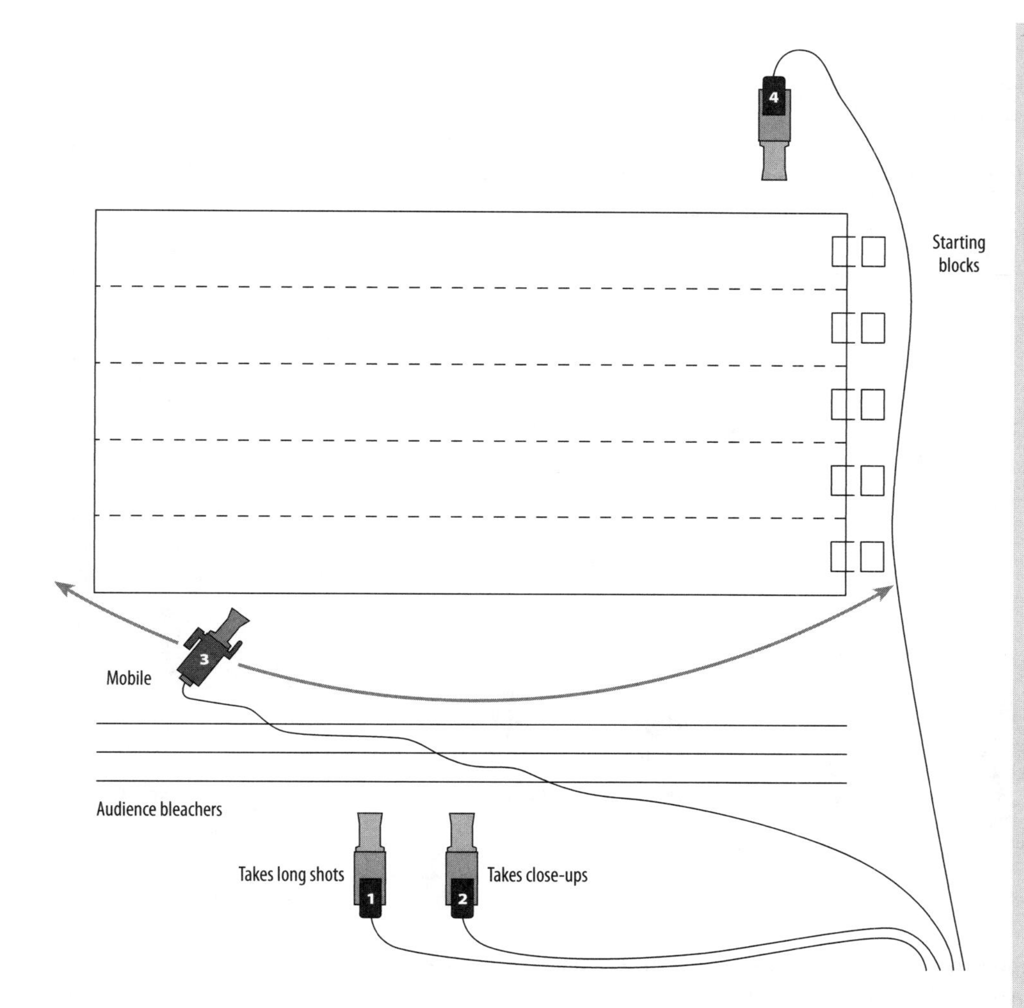

Swim Meet
Live telecast of state swim meet; large indoor pool

m. comments on C1

n. comments on C2

o. comments on C3

p. comments on C4

1m C1 ○ 55 ○ 56 ○ 57 ○ 58 ○ 59

1n C2 ○ 55 ○ 56 ○ 57 ○ 58 ○ 59

1o C3 ○ 55 ○ 56 ○ 57 ○ 58 ○ 59

1p C4 ○ 55 ○ 56 ○ 57 ○ 58 ○ 59

PAGE TOTAL

(55) camera position OK
(56) wrong or unnecessary camera position
(57) inappropriate camera type
(58) cable hazard
(59) lighting problems (shooting into the sun or against another strong light source)

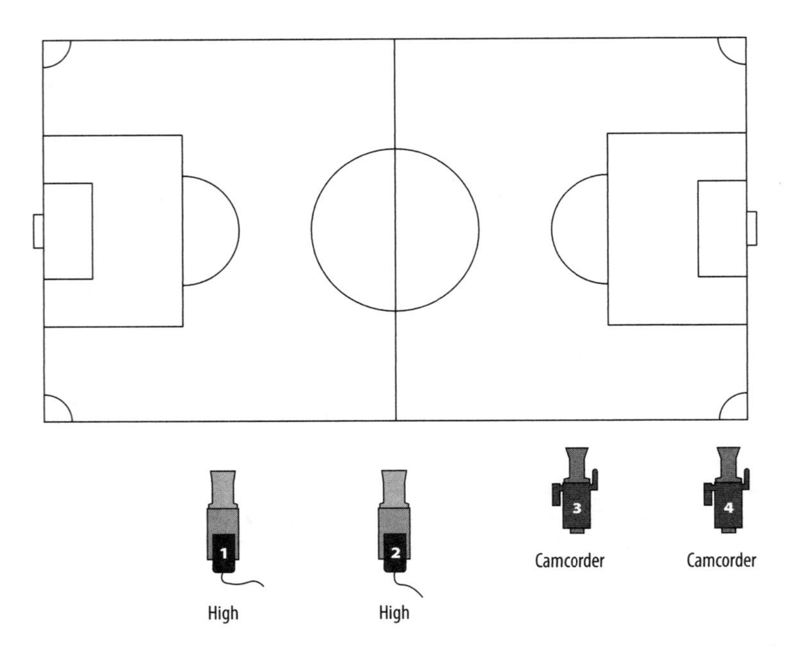

Soccer Practice
EFP of soccer practice for a show that demonstrates the beauty and the grace of a soccer game; heavy postproduction with effects and sound track

q. comments on C1

r. comments on C2

s. comments on C3

t. comments on C4

1q C1 ○ 55 ○ 56 ○ 57 ○ 58 ○ 59

1r C2 ○ 55 ○ 56 ○ 57 ○ 58 ○ 59

1s C3 ○ 55 ○ 56 ○ 57 ○ 58 ○ 59

1t C4 ○ 55 ○ 56 ○ 57 ○ 58 ○ 59

PAGE TOTAL

SECTION TOTAL

Course No. ______________ Date ____________________ Name ____

REVIEW OF FACILITIES REQUESTS

Evaluate the equipment facilities requests for the three EFPs described below. Identify the ***wrong*** *equipment and the items* ***not*** *needed and fill in the bubbles with the corresponding numbers.* ***Multiple answers are possible.***

1. Live stand-up traffic report from downtown during the afternoon rush hour

Facilities Request 1
(60) camcorder
(61) 2 video recorders
(62) shotgun mic (camera mic)
(63) hand mic
(64) 3 portable lighting kits
(65) audiotape recorder
(66) microwave transmission equipment
(67) I.F.B. intercom
(68) C.G.

1 ○ 60 ○ 61 ○ 62 ○ 63 ○ 64 ○ 65 ○ 66 ○ 67 ○ 68

2. Midmorning recording of a brief dance number in front of city hall for a music video using ENG/EFP cameras, *not* camcorders

Facilities Request 2
(69) 3 video recorders
(70) 6 shotgun mics
(71) ESS
(72) 3 ENG/EFP cameras
(73) 3 RCUs, connecting cables, and portable monitors
(74) large audio mixer
(75) portable lighting kit
(76) P.A. audiotape playback system
(77) C.G.

2 ○ 69 ○ 70 ○ 71 ○ 72 ○ 73 ○ 74 ○ 75 ○ 76 ○ 77

3. Taped interview of a media scholar in his hotel room for news item

Facilities Request 3
(78) camcorder
(79) iso video recorder
(80) portable lighting kit
(81) RCU
(82) 2 lavalier mics
(83) portable audio mixer
(84) recording media
(85) batteries
(86) preview monitors

3 ○ 78 ○ 79 ○ 80 ○ 81 ○ 82 ○ 83 ○ 84 ○ 85 ○ 86

SECTION TOTAL ______

REVIEW OF SIGNAL TRANSPORT SYSTEMS

Select the correct answers and fill in the bubbles with the corresponding numbers.

1. When stringing cables is prohibited, an EFP camera can use (87) *its own small transmitter* (88) *a C-band uplink* (89) *a Ku-band satellite to transmit its signal over short distances.*

2. The C-band uplink and downlink dishes are (90) *the same as* (91) *smaller than* (92) *larger than* the ones for the Ku-band.

3. A mini-link refers to (93) *a small uplink* (94) *a small downlink* (95) *several microwave links to transport the signal around an obstacle.*

4. A microwave signal (96) *can* (97) *cannot* be blocked by big buildings or mountains.

5. EFP makes (98) *more-frequent use of satellite transmission than* (99) *less-frequent use of satellite transmission than* (100) *about the same amount of satellite transmission as* big remotes.

6. Small uplink trucks use (101) *the Ku-band* (102) *the C-band* (103) *their own satellite frequency* for signal transmission.

7. The Ku-band operates on a frequency that is (104) *higher than* (105) *lower than* (106) *the same as* the frequency for the C-band and is (107) *more stable in bad weather* (108) *less stable in bad weather* (109) *immune to weather conditions.* ***(Fill in two bubbles.)***

1	○ 87	○ 88	○ 89
2	○ 90	○ 91	○ 92
3	○ 93	○ 94	○ 95
4	○ 96	○ 97	
5	○ 98	○ 99	○ 100
6	○ 101	○ 102	○ 103
7	○ 104	○ 105	○ 106
	○ 107	○ 108	○ 109

SECTION TOTAL

Course No. ________________ Date ____________________ Name ______________________________________

REVIEW QUIZ

*Mark the following statements as true or false by filling in the bubbles in the **T** (for true) or **F** (for false) column.*

1. Cameras used for the regular coverage of a remote telecast cannot be used for instant replay.
2. A careful audio setup is as important as the camera setup in big remotes.
3. Big remote trucks usually contain an audio control, a video-recording control, a program control, and a technical control with signal transmission equipment.
4. You need a switcher when using three EFP cameras as multiple isos.
5. When shooting single-camera EFP for postproduction, you do not need extensive intercom systems.
6. So long as you use an I.F.B. system, the floor manager is unnecessary for big remotes.
7. The contact person is important only in preproduction.
8. So long as you have a good transmission system, you do not need video recorders in the remote truck.
9. Remote surveys are relatively unimportant for ENG.
10. The C.G. operation is especially important during a live broadcast of a football game.
11. To make the remote telecast as exciting as possible, you should use as many cameras as are available.
12. So long as you have good headsets, you do not need other intercom systems on big remotes.
13. If possible, you should do the survey for an outdoor remote during the time the actual production will take place.
14. Remote surveys are relatively unimportant for EFP.

	T	F
1	○ 110	○ 111
2	○ 112	○ 113
3	○ 114	○ 115
4	○ 116	○ 117
5	○ 118	○ 119
6	○ 120	○ 121
7	○ 122	○ 123
8	○ 124	○ 125
9	○ 126	○ 127
10	○ 128	○ 129
11	○ 130	○ 131
12	○ 132	○ 133
13	○ 134	○ 135
14	○ 136	○ 137

SECTION TOTAL ______

PROBLEM-SOLVING APPLICATIONS

1. To get a good overhead shot of a parade, you, the director, would like to place one of the cameras on the balcony of a twentieth-floor window of a nearby hotel. The TD informs you that the hotel manager has nothing against your renting the room for the day and setting up the camera, but she will not allow any cable runs either inside or outside the hotel. What would you suggest?

2. The producer learns at the last minute that the president of the European Union will arrive at the international airport and wants you to cover her arrival live. According to the producer, you should have no problem with the transmission because the station's transmitter is in line-of-sight of the airport. What field production method would you recommend? Specifically, what equipment and personnel would you need to accomplish this assignment?

3. You are the director for the live multicamera coverage of a large computer convention. While you're giving instructions to the talent to wind up her interview with one of the computer experts, her I.F.B. fails. How else can you communicate with her while she is on the air?

4. Conduct a detailed remote survey for the live coverage of one of the following: (1) a football game, (2) a track meet, (3) a basketball game, (4) a baseball game, (5) a concert of a symphony orchestra, (6) an outdoor rock concert, or (7) a modern dance performance in a city park. Be sure to include all major production items, such as camera placement, audio and lighting requirements, intercom and transmission systems, power source, and so forth.

5. Prepare location sketches and facilities requests for the remote or EFP selected in the previous question.

Postproduction

Course No. ______ Date ______ Name ______

Postproduction Editing: How It Works

REVIEW OF KEY TERMS

Match each term with its appropriate definition by filling in the corresponding bubble.

1. **clip**
2. **source media**
3. **vector**
4. **linear editing**
5. **off-line editing**
6. **on-line editing**
7. **capture**
8. **edit master recording**
9. **split edit**
10. **NLE**
11. **window dub**
12. **ADR**
13. **slate**
14. **shot**
15. **time code**
16. **take**
17. **EDL**
18. **VR log**

A. Consists of edit-in and edit-out points expressed in time code numbers

B. Editing that uses tape-based systems

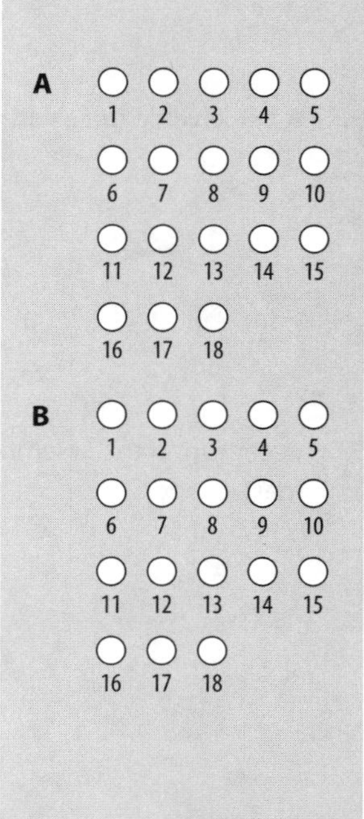

PAGE TOTAL

1. clip	7. capture	13. slate
2. source media	8. edit master recording	14. shot
3. vector	9. split edit	15. time code
4. linear editing	10. NLE	16. take
5. off-line editing	11. window dub	17. EDL
6. on-line editing	12. ADR	18. VR log

C. The interval between two transitions

C ○1 ○2 ○3 ○4 ○5 ○6 ○7 ○8 ○9 ○10 ○11 ○12 ○13 ○14 ○15 ○16 ○17 ○18

D. A "bumped-down" copy of all source recordings with the time code keyed over each frame

D ○1 ○2 ○3 ○4 ○5 ○6 ○7 ○8 ○9 ○10 ○11 ○12 ○13 ○14 ○15 ○16 ○17 ○18

E. A perceivable force with a direction and a magnitude

E ○1 ○2 ○3 ○4 ○5 ○6 ○7 ○8 ○9 ○10 ○11 ○12 ○13 ○14 ○15 ○16 ○17 ○18

F. Recapturing the assembled shots at a higher resolution or creating a high-quality edit master tape

F ○1 ○2 ○3 ○4 ○5 ○6 ○7 ○8 ○9 ○10 ○11 ○12 ○13 ○14 ○15 ○16 ○17 ○18

PAGE TOTAL

Course No. ____________ Date ________________ Name ______________________________

G. Process that produces the EDL, a videotape not used for broadcast, or low-resolution capture

G ○1 ○2 ○3 ○4 ○5 ○6 ○7 ○8 ○9 ○10 ○11 ○12 ○13 ○14 ○15 ○16 ○17 ○18

H. Transferring of video and audio information to a computer hard drive

H ○1 ○2 ○3 ○4 ○5 ○6 ○7 ○8 ○9 ○10 ○11 ○12 ○13 ○14 ○15 ○16 ○17 ○18

I. Similar repeated shots taken during video recording or filming

I ○1 ○2 ○3 ○4 ○5 ○6 ○7 ○8 ○9 ○10 ○11 ○12 ○13 ○14 ○15 ○16 ○17 ○18

J. Allows instant random access to and easy rearrangement of shots

J ○1 ○2 ○3 ○4 ○5 ○6 ○7 ○8 ○9 ○10 ○11 ○12 ○13 ○14 ○15 ○16 ○17 ○18

K. A shot or brief sequence of shots captured on the hard drive

K ○1 ○2 ○3 ○4 ○5 ○6 ○7 ○8 ○9 ○10 ○11 ○12 ○13 ○14 ○15 ○16 ○17 ○18

PAGE TOTAL ______

1. clip	**7. capture**	**13. slate**
2. source media	**8. edit master recording**	**14. shot**
3. vector	**9. split edit**	**15. time code**
4. linear editing	**10. NLE**	**16. take**
5. off-line editing	**11. window dub**	**17. EDL**
6. on-line editing	**12. ADR**	**18. VR log**

L. Audio precedes the shot or bleeds into the next one

L ○ 1 ○ 2 ○ 3 ○ 4 ○ 5 ○ 6 ○ 7 ○ 8 ○ 9 ○ 10 ○ 11 ○ 12 ○ 13 ○ 14 ○ 15 ○ 16 ○ 17 ○ 18

M. The media that contains the final on-line edit

M ○ 1 ○ 2 ○ 3 ○ 4 ○ 5 ○ 6 ○ 7 ○ 8 ○ 9 ○ 10 ○ 11 ○ 12 ○ 13 ○ 14 ○ 15 ○ 16 ○ 17 ○ 18

N. The recording devices (videotape, hard disk, optical disc, or memory card) that hold the recorded material

N ○ 1 ○ 2 ○ 3 ○ 4 ○ 5 ○ 6 ○ 7 ○ 8 ○ 9 ○ 10 ○ 11 ○ 12 ○ 13 ○ 14 ○ 15 ○ 16 ○ 17 ○ 18

O. A list of all consecutive shots recorded during a production with in and out time code numbers and other information

O ○ 1 ○ 2 ○ 3 ○ 4 ○ 5 ○ 6 ○ 7 ○ 8 ○ 9 ○ 10 ○ 11 ○ 12 ○ 13 ○ 14 ○ 15 ○ 16 ○ 17 ○ 18

PAGE TOTAL

Course No. ______________ Date ______________ Name ______________________________

P. A device that provides essential production information recorded at the beginning of each take

Q. The synchronization of speech with the lip movements of the speaker in postproduction

R. Gives each video frame a specific address

P				
○ 1	○ 2	○ 3	○ 4	○ 5
○ 6	○ 7	○ 8	○ 9	○ 10
○ 11	○ 12	○ 13	○ 14	○ 15
○ 16	○ 17	○ 18		

Q				
○ 1	○ 2	○ 3	○ 4	○ 5
○ 6	○ 7	○ 8	○ 9	○ 10
○ 11	○ 12	○ 13	○ 14	○ 15
○ 16	○ 17	○ 18		

R				
○ 1	○ 2	○ 3	○ 4	○ 5
○ 6	○ 7	○ 8	○ 9	○ 10
○ 11	○ 12	○ 13	○ 14	○ 15
○ 16	○ 17	○ 18		

PAGE TOTAL ______

SECTION TOTAL ______

REVIEW OF NONLINEAR EDITING

Select the correct answers and fill in the bubbles with the corresponding numbers.

1. Nonlinear systems (19) *allow* (20) *do not allow* random access to the source material and use (21) *videotape* (22) *disk-based storage systems.* ***(Fill in two bubbles.)***

2. The operational principle of nonlinear editing is (23) *copying images from a source to a record device* (24) *rearranging audio and video data files* (25) *transferring digital data from a VTR to a hard drive.*

3. Audio/video capture from a camcorder to a hard drive is normally done via (26) *FireWire or HDMI cable* (27) *S-video cable* (28) *coaxial cable.*

4. The three major components of a nonlinear editing system are (29) *source media, editing software, digitizer* (30) *source VTR, software, edit master VTR* (31) *source media, computer, editing software.*

5. A VR log must include the (32) *exact shot designation* (33) *in and out time code numbers* (34) *approximate length of each clip.*

6. When the audio precedes a shot or blends into another, it is commonly called (35) *a split edit* (36) *delayed audio* (37) *out-of-phase video.*

7. The time code frames and seconds roll over at (38) *59 frames, 59 seconds* (39) *29 frames, 59 seconds* (40) *29 frames, 29 seconds.*

8. When you want the time code to indicate the correct elapsed clock time even for a long running time, you should record the shots in (41) *drop frame mode* (42) *non–drop frame mode* (43) *PAL time code.*

9. The VR log helps (44) *organize the source material* (45) *eliminate unimportant cutaways* (46) *locate specific shots during the editing process.*

10. Audio transcriptions are important especially when editing (47) *a fully scripted drama* (48) *an interview* (49) *a fully scripted 30-second commercial.*

11. When editing video to audio, you use (50) *video as the A-roll and audio as the B-roll* (51) *audio as the A-roll and video as the B-roll* (52) *no A and B rolls.*

1	○ 19	○ 20	
	○ 21	○ 22	
2	○ 23	○ 24	○ 25
3	○ 26	○ 27	○ 28
4	○ 29	○ 30	○ 31
5	○ 32	○ 33	○ 34
6	○ 35	○ 36	○ 37
7	○ 38	○ 39	○ 40
8	○ 41	○ 42	○ 43
9	○ 44	○ 45	○ 46
10	○ 47	○ 48	○ 49
11	○ 50	○ 51	○ 52

PAGE TOTAL

Course No. ______ Date ______ Name ______

12. When during editing you are looking for a movement with a prominent screen-left direction, you should consult the (53) *vector column in the VR log* (54) *vector notation on the time line* (55) *storyboard.*

13. In nonlinear editing, the time line refers to the (56) *length of the clip* (57) *production schedule* (58) *video and audio tracks with their clips.*

14. When importing source footage into the NLE system for off-line editing, you need to determine the (59) *nature of the source material* (60) *codec* (61) *drop frame or non–drop frame designation.*

15. When importing uncompressed high-definition video, your files will (62) *take longer and require more storage space* (63) *require more storage space but take less time* (64) *require about the same time and storage space as compressed video.*

12	○ 53	○ 54	○ 55
13	○ 56	○ 57	○ 58
14	○ 59	○ 60	○ 61
15	○ 62	○ 63	○ 64

PAGE TOTAL ______

16. Fill in the bubbles whose numbers correspond with the appropriate features of the generic nonlinear editing interface as shown in the following figure:

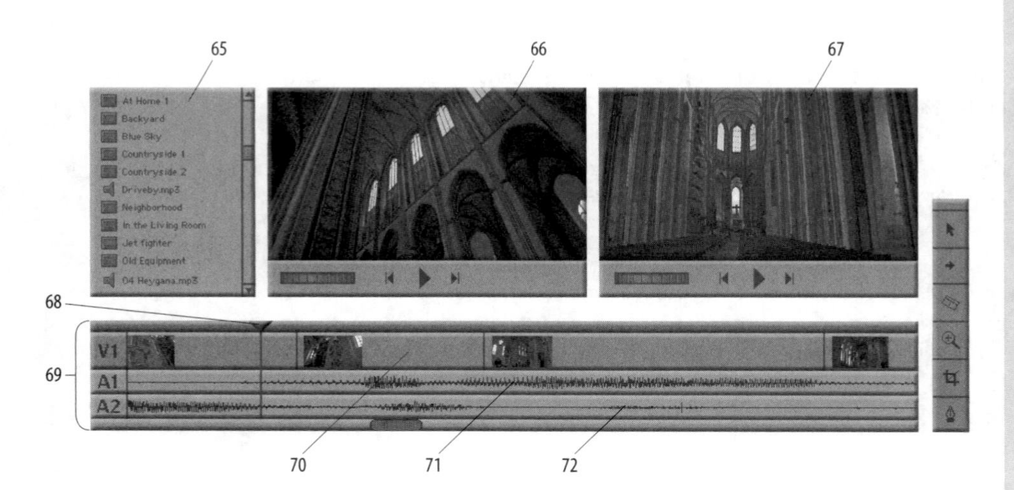

PHOTOS: HERBERT ZETTL

a. time line

b. playhead and scrubber bar

c. video track

d. audio track 1

e. audio track 2

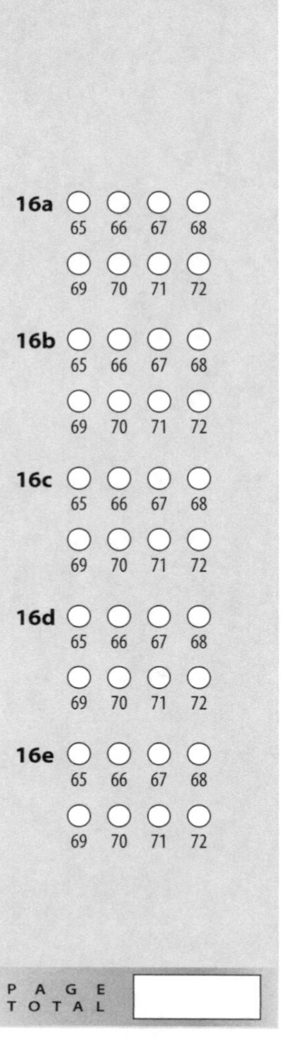

Course No. ____________ Date ____________ Name ____________

f. project panel

g. record monitor

h. source monitor

16f ○ 65 ○ 66 ○ 67 ○ 68
○ 69 ○ 70 ○ 71 ○ 72

16g ○ 65 ○ 66 ○ 67 ○ 68
○ 69 ○ 70 ○ 71 ○ 72

16h ○ 65 ○ 66 ○ 67 ○ 68
○ 69 ○ 70 ○ 71 ○ 72

PAGE TOTAL

SECTION TOTAL

REVIEW OF LINEAR EDITING

Select the correct answers and fill in the bubbles with the corresponding numbers.

1. When using SMPTE time code, you (73) *can* (74) *cannot* add it later over existing source tapes; it (75) *does* (76) *does not* necessarily show the actual time during which the production took place. (***Fill in two bubbles.***)

2. Identify mistakes in the pulse-count display in this figure and fill in the bubbles with the corresponding numbers. (***Multiple answers are possible.***)

3. Select the correct pulse-count display that exhibits the actual edit-in point of the edit master tape shown in the following figure and fill in the bubble with the corresponding number.

Tape is 15:25 minutes in from start

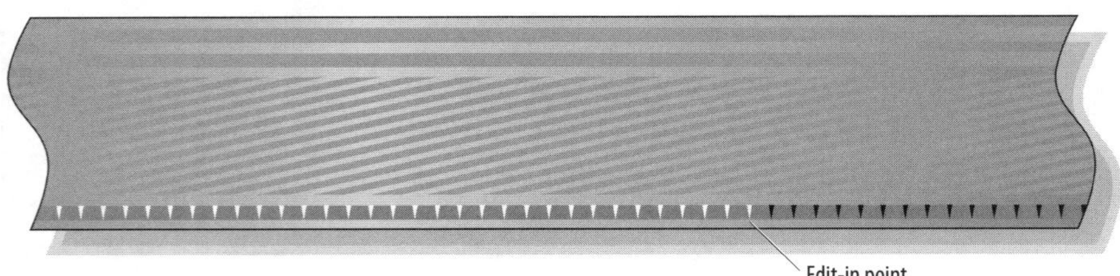

4. The operational principle of linear editing is (85) *file management* (86) *transferring data from a VTR to a hard drive* (87) *copying selected portions of the source tapes.*

5. In the assemble editing mode, the record VTR (88) *will* (89) *will not* copy the control track of the source tape, so you (90) *need* (91) *do not need* to prerecord a continuous control track on the edit master tape. (***Fill in two bubbles.***)

1 ○ 73 ○ 74 ○ 75 ○ 76

2 ○ 77 ○ 78 ○ 79 ○ 80

3 ○ 81 ○ 82 ○ 83 ○ 84

4 ○ 85 ○ 86 ○ 87

5 ○ 88 ○ 89 ○ 90 ○ 91

SECTION TOTAL

Course No. ____________ Date ____________ Name ____________

REVIEW QUIZ

Mark the following statements as true or false by filling in the bubbles in the ***T*** *(for true) or* ***F*** *(for false) column.*

	Statement		T	F
1.	If your source tapes are digital, they don't have to be dubbed to the hard drive of an NLE.	1	◯ 92	◯ 93
2.	Offline editing in NLE means to initially import the source material into the NLE system at a relatively high compression ratio.	2	◯ 94	◯ 95
3.	You can use the AB-roll concept of linear editing in nonlinear editing as well.	3	◯ 96	◯ 97
4.	A split edit in nonlinear editing refers to editing audio separately from video.	4	◯ 98	◯ 99
5.	In nonlinear editing, the video and audio frames are not actually sequenced but told by the NLE in what order to play back.	5	◯ 100	◯ 101
6.	The vector column in a good VR log can show the screen direction of somebody's gaze or movement.	6	◯ 102	◯ 103
7.	Linear editing can be done only with VTRs.	7	◯ 104	◯ 105
8.	The less compression of video and audio files, the more storage space they require.	8	◯ 106	◯ 107
9.	Linear editing works on the same basic principle as nonlinear editing.	9	◯ 108	◯ 109
10.	Changing the shot sequence in the middle of a videotape is difficult regardless of whether the information is analog or digital.	10	◯ 110	◯ 111
11.	Linear editing can be done only when the source tapes hold analog material.	11	◯ 112	◯ 113
12.	Off-line linear editing means that lower-quality equipment is used.	12	◯ 114	◯ 115
13.	All NLE systems offer at least two audio tracks.	13	◯ 116	◯ 117
14.	Linear editing equipment allows random access to the source material.	14	◯ 118	◯ 119
15.	You can add time code to videotape as late as postproduction.	15	◯ 120	◯ 121
16.	Contrary to linear editing, an EDL is not practical for nonlinear editing.	16	◯ 122	◯ 123
17.	The principle of nonlinear editing is file management.	17	◯ 124	◯ 125
18.	Replacing a brief clip sequence in the middle of an edited project is quite difficult and time consuming in nonlinear editing.	18	◯ 126	◯ 127

SECTION TOTAL ____

PROBLEM-SOLVING APPLICATIONS

1. The news producer tells you, the editor, not to bother with an audio transcription of the recent two-hour interview with the mayor because he needs only about 20 seconds of a few memorable sound bites. What is your reaction? Why?

2. The novice director warns you, the editor, that the new client is known to change her mind frequently and may require substantive editing changes right in the middle of the show. The director is worried that such major changes may cause serious time delays. Assuming that you are working with a nonlinear editing system, what would you tell the director? Be specific.

3. The same director tells you to be sure to capture all source tapes at the highest resolution even for an off-line rough-cut because "once in the computer, you are stuck with what you imported." What is your reaction? Why?

4. Even with your new nonlinear editing system, it is cumbersome to find shots that show the new car model traveling in specific screen directions. The producer suggests that you note the various vectors when logging the source footage. What does she mean? How can doing this help you locate the desired shots?

5. The director is a big fan of nonlinear editing because fixing mistakes in postproduction is "now a snap." What is your reaction? Give specific examples.

6. The producer hands you a number of source media from a nature videographer. The program idea is to match the video of the movement of various wild animals to the tempo and the feel of some classical music pieces. He wants you to especially emphasize and juxtapose shots in which the animals move in opposite directions—much like the music. Which logging element would facilitate your editing job?

Course No. ____________ Date ____________ Name ____________

Editing Functions and Principles

REVIEW OF KEY TERMS

Match each term with its appropriate definition by filling in the corresponding bubble.

1. **continuity editing**
2. **vector**
3. **vector line**
4. **graphic vector**
5. **jump cut**
6. **motion vector**
7. **complexity editing**
8. **mental map**
9. **cutaway**
10. **montage**
11. **index vector**

A. A shot that is inserted to facilitate continuity

A ○ 1 ○ 2 ○ 3 ○ 4 ○ 5 ○ 6 ○ 7 ○ 8 ○ 9 ○ 10 ○ 11

B. A vector created by someone looking or something pointing unquestionably in a specific direction

B ○ 1 ○ 2 ○ 3 ○ 4 ○ 5 ○ 6 ○ 7 ○ 8 ○ 9 ○ 10 ○ 11

C. The preservation of visual continuity from shot to shot

C ○ 1 ○ 2 ○ 3 ○ 4 ○ 5 ○ 6 ○ 7 ○ 8 ○ 9 ○ 10 ○ 11

PAGE TOTAL ____

1. continuity editing	**5. jump cut**	**9. cutaway**
2. vector	**6. motion vector**	**10. montage**
3. vector line	**7. complexity editing**	**11. index vector**
4. graphic vector	**8. mental map**	

D. The juxtaposition of two or more shots to generate a third overall idea, which may not be contained in any one

D ○ 1 ○ 2 ○ 3 ○ 4 ○ 5 ○ 6 ○ 7 ○ 8 ○ 9 ○ 10 ○ 11

E. A perceivable force with a direction and a magnitude

E ○ 1 ○ 2 ○ 3 ○ 4 ○ 5 ○ 6 ○ 7 ○ 8 ○ 9 ○ 10 ○ 11

F. The juxtaposition of shots that helps intensify the screen event

F ○ 1 ○ 2 ○ 3 ○ 4 ○ 5 ○ 6 ○ 7 ○ 8 ○ 9 ○ 10 ○ 11

G. Juxtaposing shots that violate the established continuity

G ○ 1 ○ 2 ○ 3 ○ 4 ○ 5 ○ 6 ○ 7 ○ 8 ○ 9 ○ 10 ○ 11

H. Established by two people facing each other or through a prominent movement in a specific direction

H ○ 1 ○ 2 ○ 3 ○ 4 ○ 5 ○ 6 ○ 7 ○ 8 ○ 9 ○ 10 ○ 11

PAGE TOTAL

Course No. ____________ Date ________________ Name ______________________________

I. Virtual image of where things are or are supposed to be in on- and off-screen space

J. Created by an object actually moving or perceived as moving on-screen

K. Created by lines or by stationary elements in such a way as to suggest a line

I	◯ 1	◯ 2	◯ 3	◯ 4
	◯ 5	◯ 6	◯ 7	◯ 8
	◯ 9	◯ 10	◯ 11	
J	◯ 1	◯ 2	◯ 3	◯ 4
	◯ 5	◯ 6	◯ 7	◯ 8
	◯ 9	◯ 10	◯ 11	
K	◯ 1	◯ 2	◯ 3	◯ 4
	◯ 5	◯ 6	◯ 7	◯ 8
	◯ 9	◯ 10	◯ 11	

PAGE TOTAL ______

SECTION TOTAL ______

REVIEW OF CONTINUITY EDITING PRINCIPLES

Select the correct answers and fill in the bubbles with the corresponding numbers.

1. You are given a storyboard to assist you in your single-camera EFP of a conversation between a man and a woman (see the following figure). For each storyboard pair, indicate whether the shots (12) *can* (13) *cannot* be edited together, assuming normal continuity-editing principles.

a.

b.

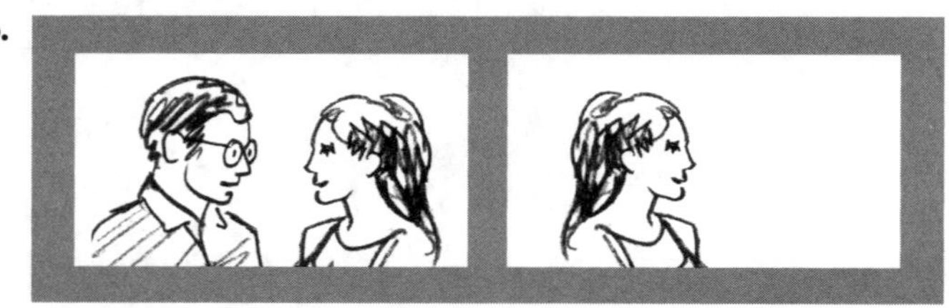

c.

d.

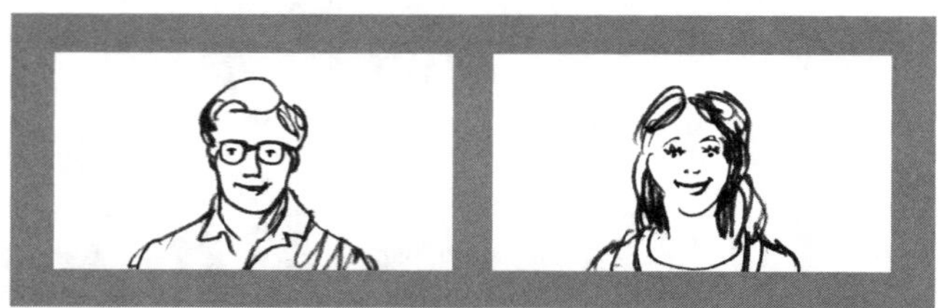

e.

1a ○ 12 ○ 13

1b ○ 12 ○ 13

1c ○ 12 ○ 13

1d ○ 12 ○ 13

1e ○ 12 ○ 13

PAGE TOTAL

2. From the screen images below (repeated on the following page), select the sequence pair you would get when cutting from camera 1 to camera 2 as shown in the diagrams of the camera positions.

14

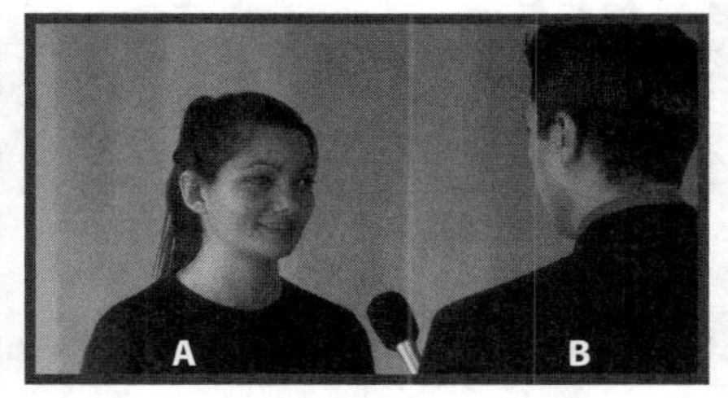

15

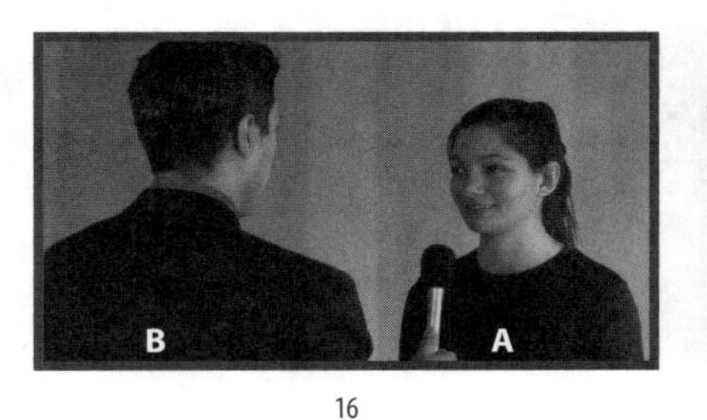

16

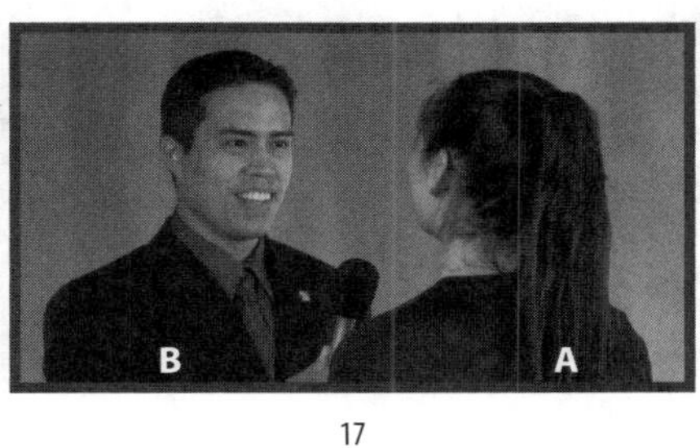

17

Also indicate whether continuity is (18) *good* or (19) *bad*.

a. camera setup A

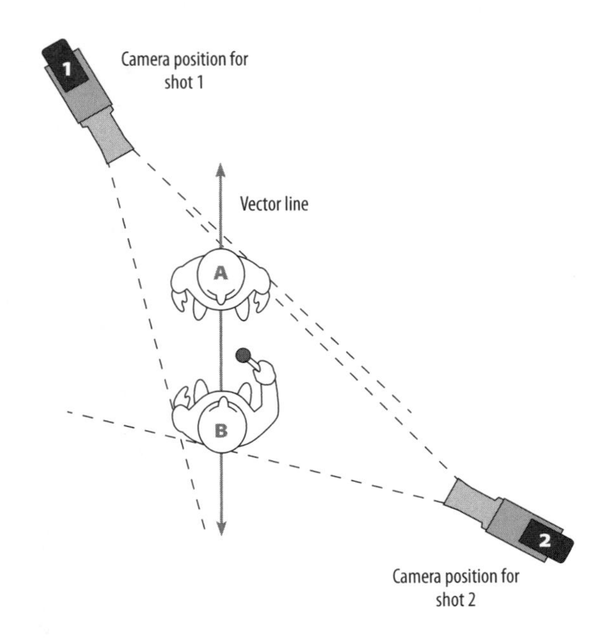

2a ○ 14 ○ 15 ○ 16 ○ 17
○ 18 ○ 19

PAGE TOTAL

14

15

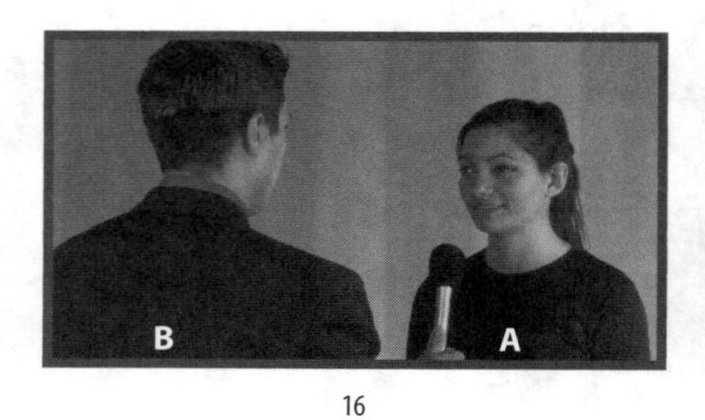

16

17

PHOTOS: EDWARD AIONA

Indicate whether continuity is (18) *good* or (19) *bad.*

b. camera setup B

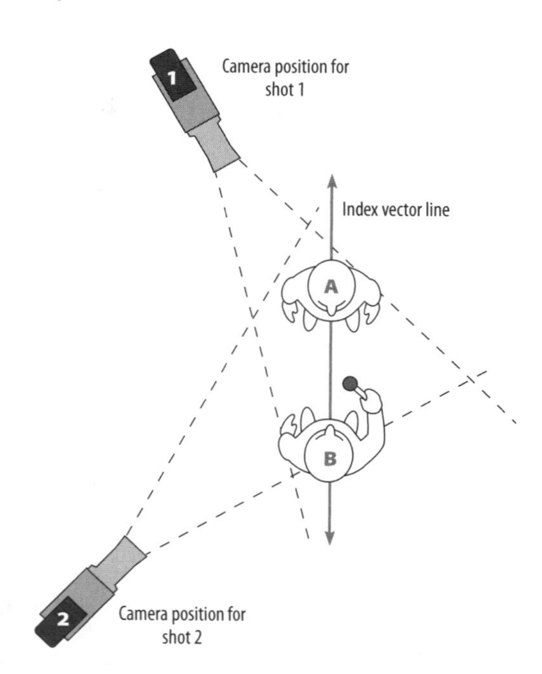

2b ○ 14 ○ 15 ○ 16 ○ 17
○ 18 ○ 19

PAGE TOTAL

Course No. ______________ Date ______________ Name ______________

3. In the following four diagrams, select the camera that is in the ***wrong*** place for proper continuity editing and fill in the corresponding bubble.

a. cutting between person A and person B during a conversation

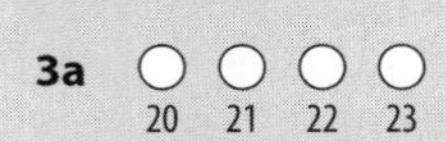

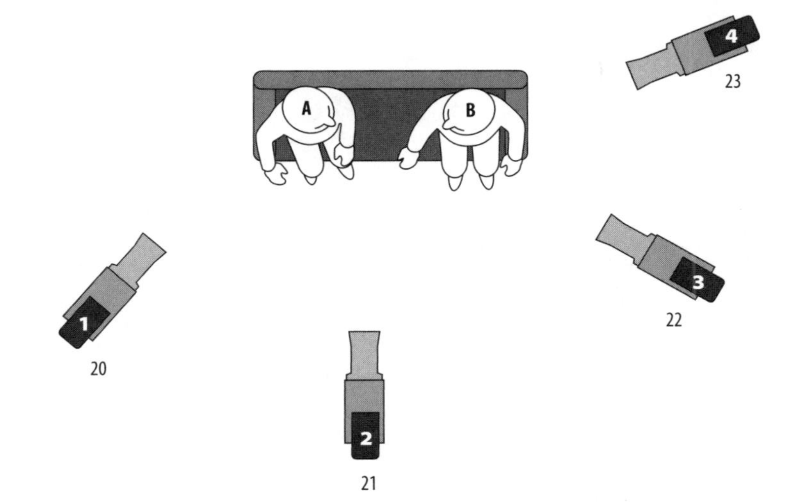

b. editing a dramatic car chase

3b ○ 24 ○ 25 ○ 26

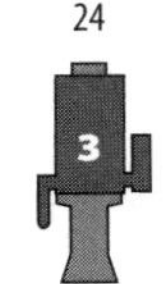

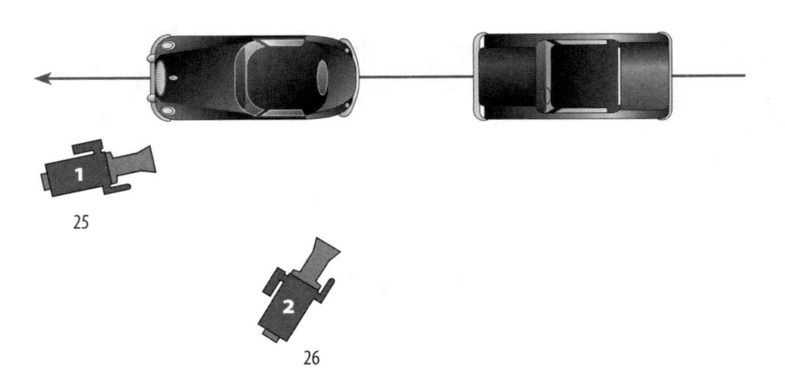

PAGE TOTAL

c. cutting from camera 2 to a different point of view of the university president and her husband during a reception

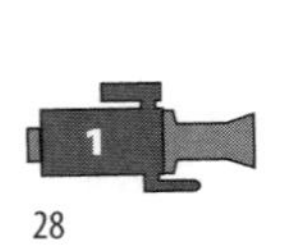

d. cutting from two-shots of piano player and singer to CUs

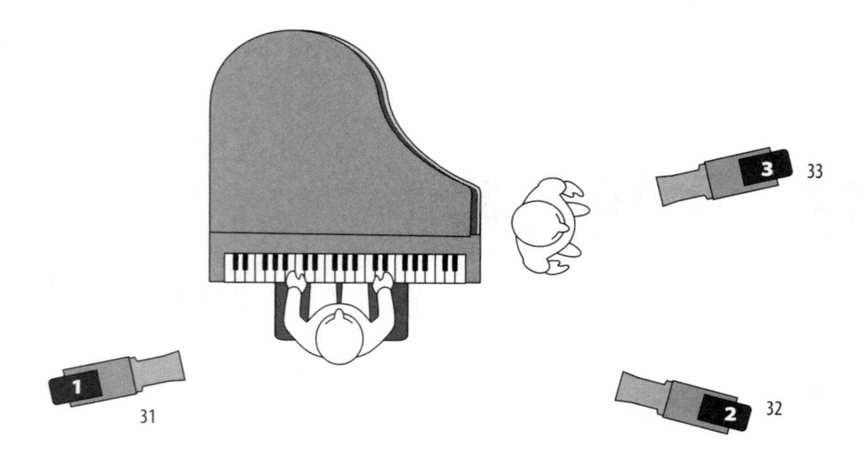

Course No. ______________ Date ______________________ Name ____

4. In the following diagram of a simple interview, select the two cameras that will facilitate optimal cross-shooting and fill in the bubbles with the corresponding numbers.

34

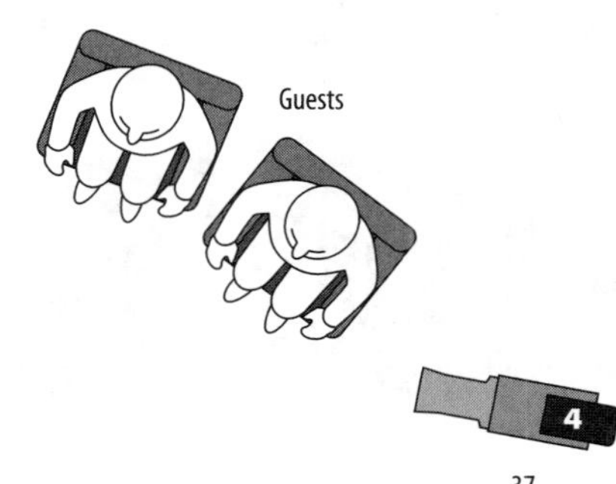

37

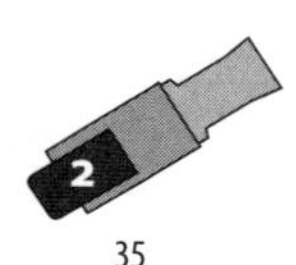

35

36

4 ○ 34 ○ 35 ○ 36 ○ 37

PAGE TOTAL

5. From the nine frame grabs of source clips below, select four shots to tell the story of a woman getting into her car and driving off. Fill in the bubbles with the numbers of the shots you selected in the order you would edit them together.

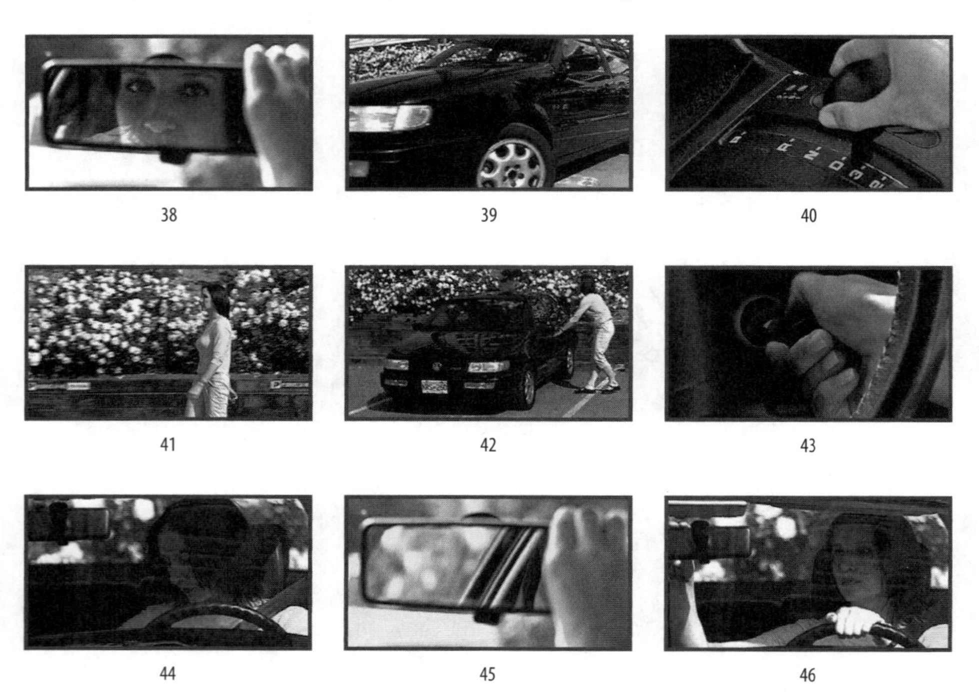

38 39 40

41 42 43

44 45 46

a. shot 1

b. shot 2

c. shot 3

d. shot 4

5a ○ 38 ○ 39 ○ 40 ○ 41 ○ 42 ○ 43 ○ 44 ○ 45 ○ 46

5b ○ 38 ○ 39 ○ 40 ○ 41 ○ 42 ○ 43 ○ 44 ○ 45 ○ 46

5c ○ 38 ○ 39 ○ 40 ○ 41 ○ 42 ○ 43 ○ 44 ○ 45 ○ 46

5d ○ 38 ○ 39 ○ 40 ○ 41 ○ 42 ○ 43 ○ 44 ○ 45 ○ 46

PAGE TOTAL

Course No. ______________ Date ____________________ Name ______________________________________

6. For each of the following shot sequences, fill in the appropriate bubbles to indicate whether the sequence (47) *maintains* or (48) *disturbs* the mental map. If the mental map is disturbed, also indicate whether the major reason is a (49) *position switch* or a (50) *vector problem*. ***(Multiple answers are possible.)***

a.

Shot 1

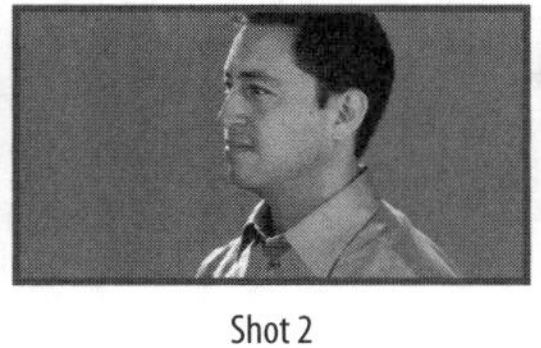
Shot 2

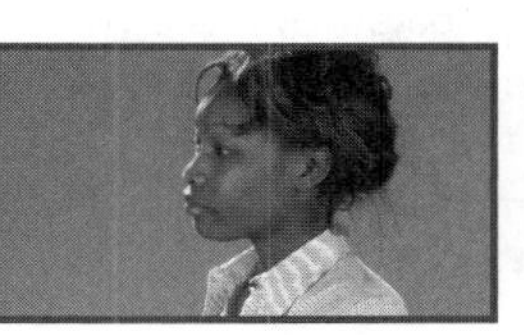
Shot 3

PHOTOS: EDWARD AIONA

6a

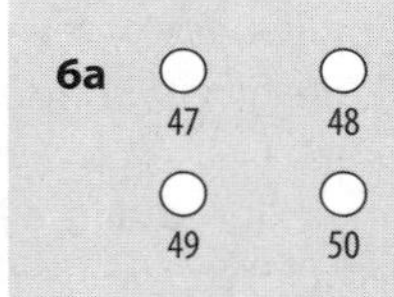

b.

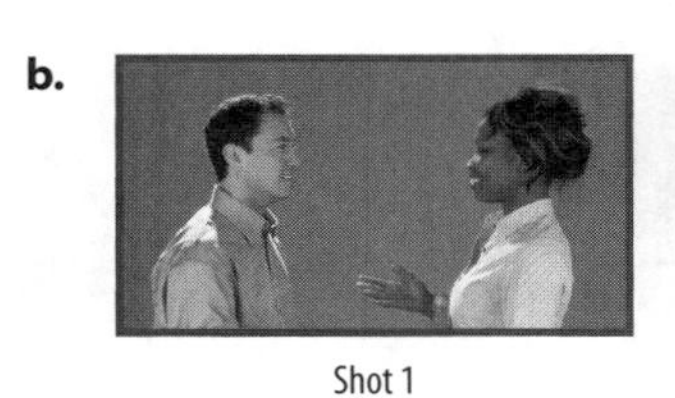
Shot 1

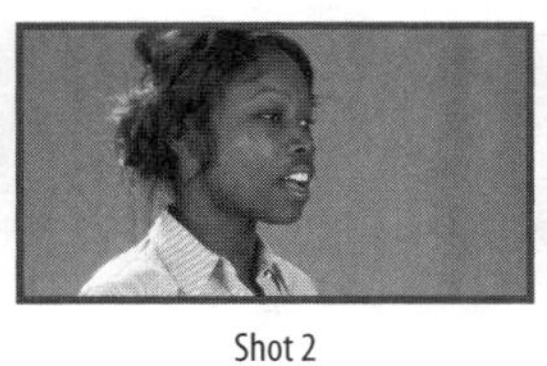
Shot 2

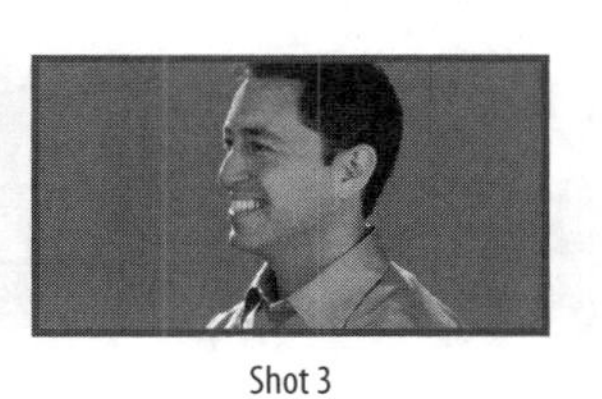
Shot 3

PHOTOS: EDWARD AIONA

6b

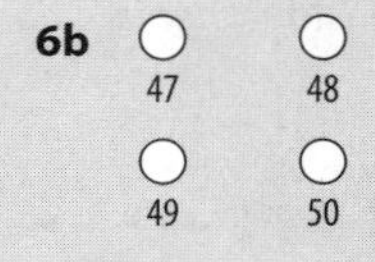

c.

Shot 1

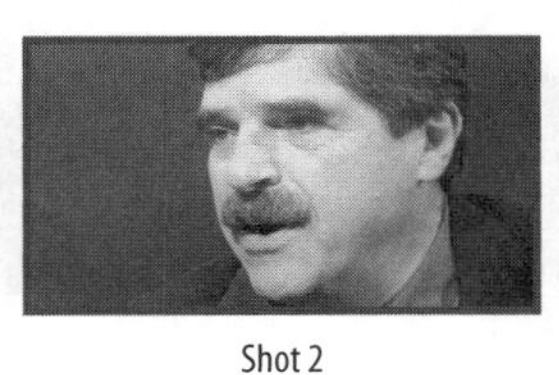
Shot 2

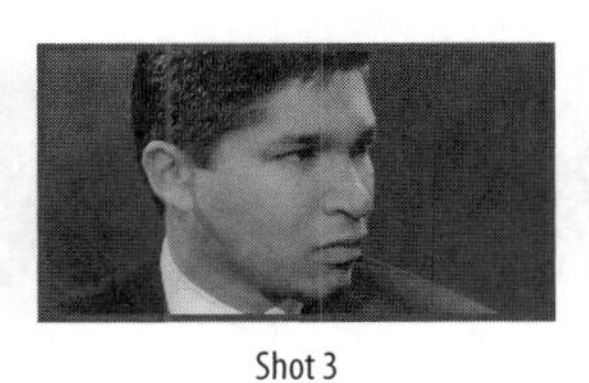
Shot 3

PHOTOS: JOHN VELTRI

6c

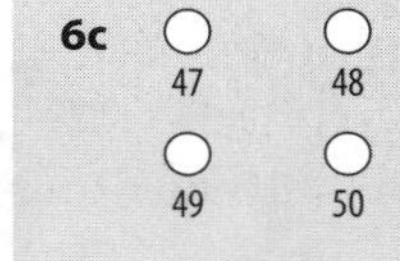

d.

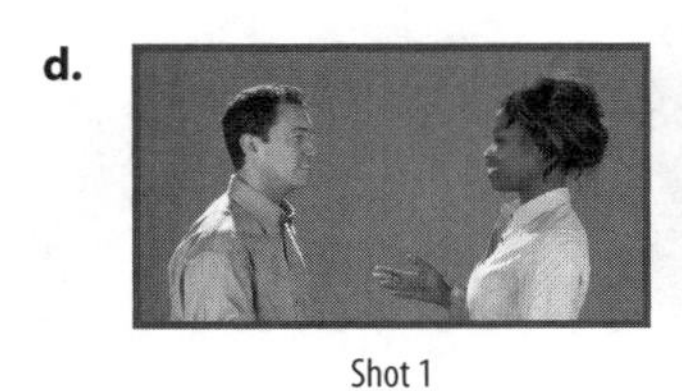
Shot 1

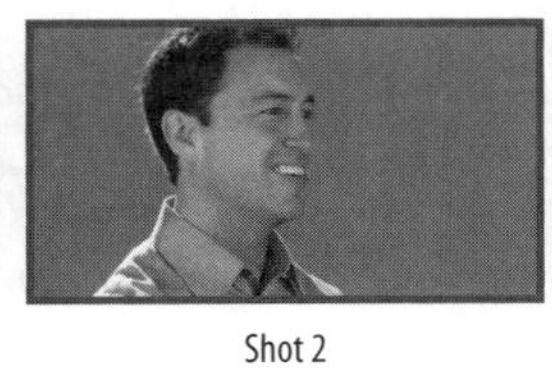
Shot 2

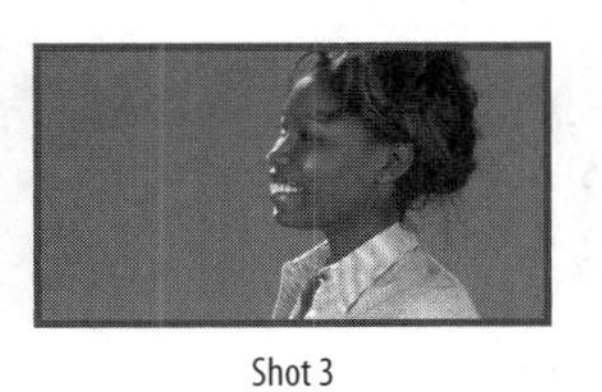
Shot 3

PHOTOS: EDWARD AIONA

6d

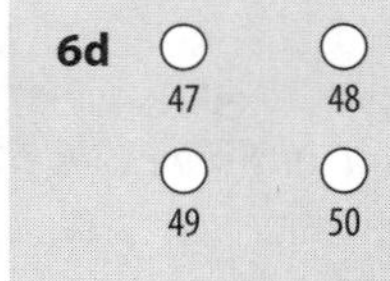

PAGE TOTAL

SECTION TOTAL

REVIEW OF COMPLEXITY EDITING

Select the correct answers and fill in the bubbles with the corresponding numbers.

1. The simultaneity of several separate events can best be shown with (51) *multiple screens* (52) *flashbacks* (53) *flashforwards.*

2. In complexity editing, a jump cut (54) *clearly signals an editing mistake* (55) *should never be used* (56) *can be used as an intensifier.*

3. A series of quick cuts between camera 1 and camera 2 in the figure below would be appropriate in (57) *continuity editing only* (58) *both continuity and complexity editing* (59) *instantaneous editing.*

4. A filmic shorthand in which a rhythmic series of seemingly unrelated shots generates new meaning is called a (60) *montage* (61) *clip* (62) *sequence.*

5. Complexity editing (63) *can occasionally break with continuity principles* (64) *must adhere to continuity principles* (65) *does not consider continuity principles.*

6. In complexity editing, DVE (66) *should be avoided* (67) *can be used to intensify a scene* (68) *can be used to clarify a scene.*

1	○ 51	○ 52	○ 53
2	○ 54	○ 55	○ 56
3	○ 57	○ 58	○ 59
4	○ 60	○ 61	○ 62
5	○ 63	○ 64	○ 65
6	○ 66	○ 67	○ 68

PAGE TOTAL

7. Assuming that you intend to construct a montage of a car fleeing the police, quick cuts among all four cameras shown in the figure below would be appropriate (69) *in continuity editing* (70) *in complexity editing* (71) *under no circumstances.*

7 ○ 69 ○ 70 ○ 71

8. In the context of continuity editing, cutting from shot 1 to shot 2 as shown below is (72) *acceptable* (73) *unacceptable* because (74) *the motion vectors are continuing* (75) *the edit would cause a jump cut.*

8 ○ 72 ○ 73 ○ 74 ○ 75

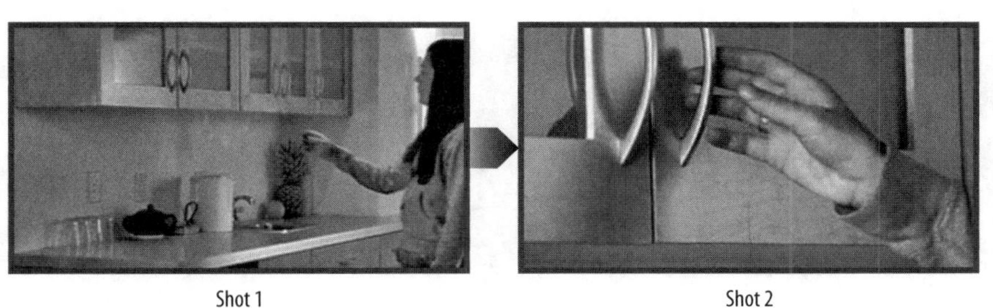

Shot 1 Shot 2

PAGE TOTAL

SECTION TOTAL

REVIEW QUIZ

*Mark the following statements as true or false by filling in the bubbles in the **T** (for true) or **F** (for false) column.*

	Statement	T	F
1.	The vector line extends from the camera to the horizon.	○ 76	○ 77
2.	When cutting from an MS to a CU of somebody sitting down, continuity is best preserved by cutting after the person is seated.	○ 78	○ 79
3.	A blurred still shot of a car represents a motion vector.	○ 80	○ 81
4.	Somebody pointing at an object constitutes an index vector.	○ 82	○ 83
5.	Editing must always be done in the context of ethics—the principles of right conduct.	○ 84	○ 85
6.	If the move is properly motivated, the vector line can be crossed in continuity editing.	○ 86	○ 87
7.	Two of the major editing functions are to shorten and to combine.	○ 88	○ 89
8.	Subject continuity means that we can recognize a person from one shot to the next.	○ 90	○ 91
9.	A cutaway can be any shot so long as it does not project a vector.	○ 92	○ 93
10.	Index and motion vectors play an important role in continuity editing.	○ 94	○ 95
11.	So long as we can recognize a person, it does not matter even in continuity editing that she appears on screen-left in one shot and on screen-right in the next.	○ 96	○ 97
12.	Ethical considerations are the purview of the director and have no place in the busy news editing room.	○ 98	○ 99
13.	A shrinking circle wipe is an especially effective way to close a documentary on a flood disaster.	○ 100	○ 101
14.	A jump cut occurs when the subject has moved his head even slightly from one shot to the next.	○ 102	○ 103
15.	A jump cut may be used effectively in complexity editing.	○ 104	○ 105
16.	Sound is an important factor in maintaining continuity.	○ 106	○ 107
17.	A mental map helps viewers to organize on- and off-screen space.	○ 108	○ 109

SECTION TOTAL

Course No. ______ Date ______ Name ______

PROBLEM-SOLVING APPLICATIONS

1. Select a scene from any type of television show or film that demonstrates complexity editing. Be specific.

2. Use a camcorder and ad-lib a scene in which your crossing the line contributes to an intensified experience.

3. The news director encourages you to use a peel effect for transitions in a headline news teaser. What is your reaction? Be specific.

4. The staging for a presidential debate shows two candidates side by side, facing the audience; a moderator is in the middle, facing the candidates, with his back to the audience. The primary cameras are located in the audience, pointing at the stage. One camera is backstage, exactly opposite the moderator. It is to get three-shots in which we see the backs of the candidates and the moderator as he addresses the candidates. Assuming that the objective is seamless continuity, do you have any concerns about this setup? Be specific.

5. The producer tells you not to worry about using a stock shot of videographers for a necessary cutaway in the editing of a news conference. What is your reaction? Be specific.

Scale: ¼" = 1'

Property List

Scale: ¼" = 1'

Property List

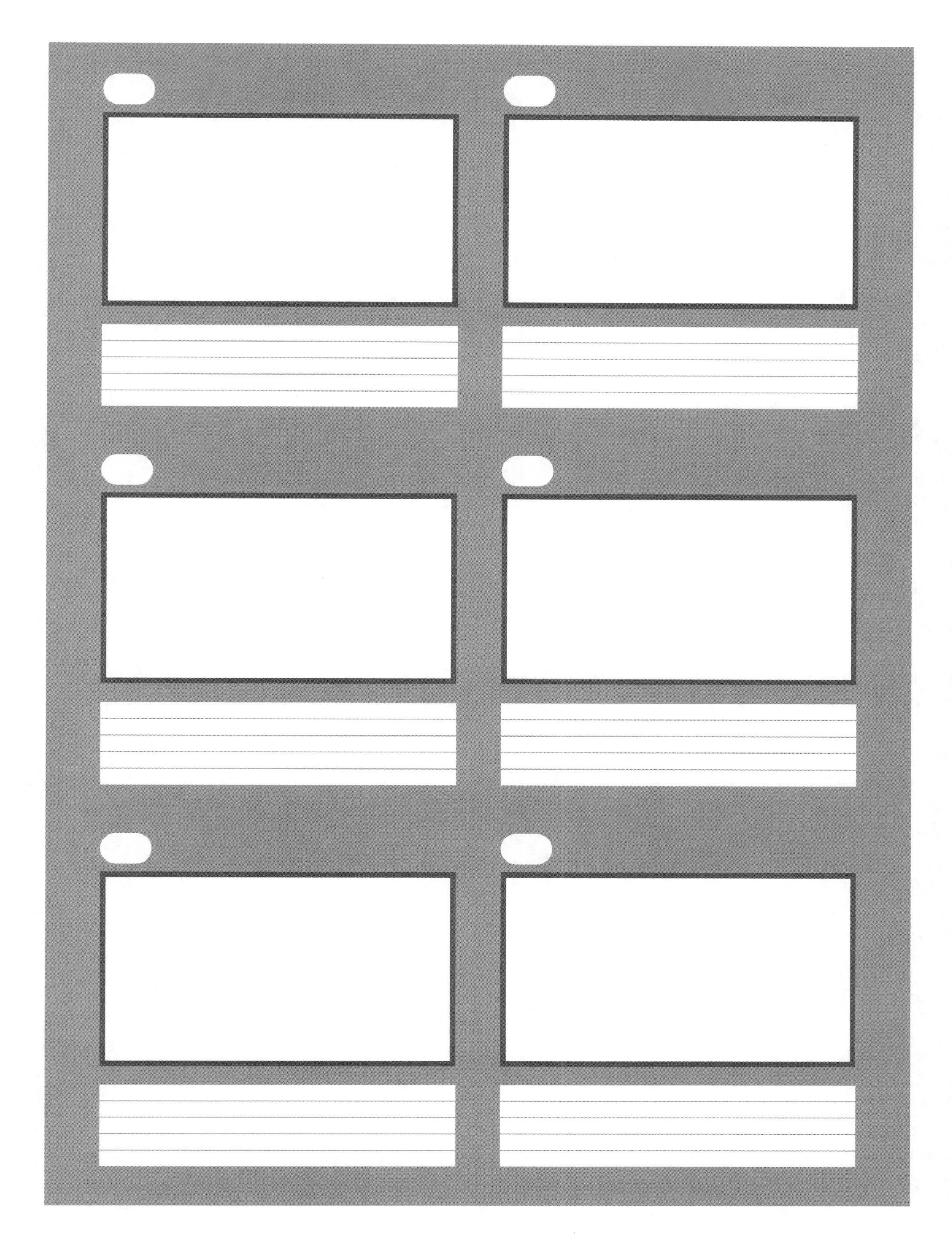

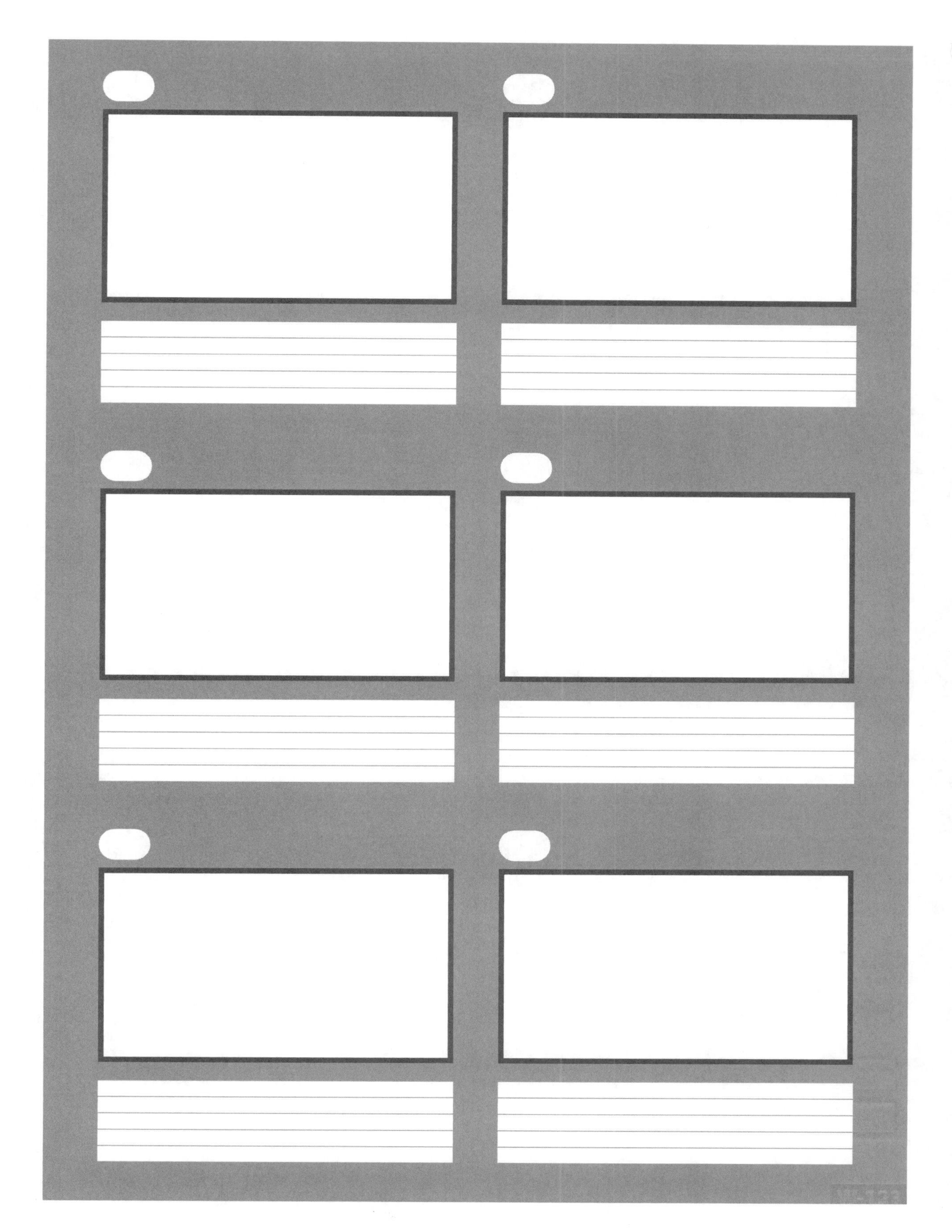